Lecture Notes in Computer Science 4639

Commenced Publication in 1973
Founding and Former Series Editors:
Gerhard Goos, Juris Hartmanis, and Jan van Leeuwen

T0189913

Erzsébet Csuhaj-Varjú
Zoltán Ésik (Eds.)

Fundamentals of Computation Theory

16th International Symposium, FCT 2007
Budapest, Hungary, August 27-30, 2007
Proceedings

 Springer

Volume Editors

Erzsébet Csuhaj-Varjú
Hungarian Academy of Sciences
Computer and Automation Research Institute
Budapest, Hungary
E-mail: csuhaj@sztaki.hu

Zoltán Ésik
University of Szeged
Department of Computer Science
Szeged, Hungary
E-mail: ze@inf.u-szeged.hu

Library of Congress Control Number: 2007932400

CR Subject Classification (1998): F.1, F.2, F.4.1, I.3.5, G.2

LNCS Sublibrary: SL 1 – Theoretical Computer Science and General Issues

ISSN	0302-9743
ISBN-10	3-540-74239-5 Springer Berlin Heidelberg New York
ISBN-13	978-3-540-74239-5 Springer Berlin Heidelberg New York

Springer is a part of Springer Science+Business Media

springer.com

© Springer-Verlag Berlin Heidelberg 2007
Printed in Germany

Typesetting: Camera-ready by author, data conversion by Scientific Publishing Services, Chennai, India
Printed on acid-free paper   SPIN: 12108152   06/3180   5 4 3 2 1 0

Preface

The Symposium on Fundamentals of Computation Theory was established in 1977 for researchers interested in all aspects of theoretical computer science, in particular in algorithms, complexity, and formal and logical methods. It is a biennial conference, which has previously been held in Poznan (1977), Wendisch-Rietz (1979), Szeged (1981), Borgholm (1983), Cottbus (1985), Kazan (1987), Szeged (1989), Gosen-Berlin (1991), Szeged (1993), Dresden (1995), Kraków (1997), Iasi (1999), Riga (2001), Malmö (2003), and Lübeck (2005).

The 16th International Symposium on Fundamentals of Computation Theory (FCT 2007) was held in Budapest, August 27–30, 2007, and was jointly organized by the Computer and Automation Research Institute of the Hungarian Academy of Sciences and the Institute of Computer Science, University of Szeged.

The suggested topics of FCT 2007 included, but were not limited to, automata and formal languages, design and analysis of algorithms, computational and structural complexity, semantics, logic, algebra and categories in computer science, circuits and networks, learning theory, specification and verification, parallel and distributed systems, concurrency theory, cryptography and cryptographic protocols, approximation and randomized algorithms, computational geometry, quantum computation and information, and bio-inspired computation.

The Programme Committee invited lectures from Ahmed Bouajjani (Paris), Oscar H. Ibarra (Santa Barbara), László Lovász (Budapest), and Philip J. Scott (Ottawa) and, from the 147 submissions, selected 39 papers for presentation at the conference and inclusion in the proceedings. This volume contains the texts or the abstracts of the invited lectures and the texts of the accepted papers.

We would like to thank the members of the Programme Committee for the evaluation of the submissions and their subreferees for their excellent cooperation in this work. We are grateful to the contributors to the conference, in particular to the invited speakers for their willingness to present interesting new developments.

Finally, we thank Zsolt Gazdag and Szabolcs Iván for their technical assistance during the preparation of these proceedings, and all those whose work behind the scenes contributed to this volume.

June 2007

Erzsébet Csuhaj-Varjú
Zoltán Ésik

Organization

Programme Committee

Jiri Adámek (Braunschweig, Germany)
Giorgio Ausiello (Rome, Italy)
Jean Berstel (Marne-la-Vallée, France)
Flavio Corradini (Camerino, Italy)
Erzsébet Csuhaj-Varjú (Budapest, Hungary), co-chair
Zoltán Ésik (Szeged, Hungary), co-chair
Jozef Gruska (Brno, Czech Republic)
Masahito Hasegawa (Kyoto, Japan)
Juraj Hromkovic (Zurich, Switzerland)
Anna Ingólfsdóttir (Reykjavik, Iceland)
Masami Ito (Kyoto, Japan)
Frédéric Magniez (Paris, France)
Catuscia Palamidessi (Palaiseau, France)
Gheorghe Păun (Bucharest, Romania and Seville, Spain)
Jean-Éric Pin (Paris, France)
Alexander Rabinovich (Tel-Aviv, Israel)
R. Ramanujam (Chennai, India)
Wojciech Rytter (Warsaw, Poland)
Arto Salomaa (Turku, Finland)
David A. Schmidt (Manhattan, KS, USA)
Alex Simpson (Edinburgh, UK)
Michael Sipser (Cambridge, MA, USA)
Colin Stirling (Edinburgh, UK)
Howard Straubing (Chestnut Hill, MA, USA)
György Turán (Chicago, IL, USA)
Thomas Wilke (Kiel, Germany)

Steering Committee

Bogdan S. Chlebus (Warsaw, Poland and Denver, USA)
Zoltán Ésik (Szeged, Hungary)
Marek Karpinski (Bonn, Germany), chair
Andrzej Lingas (Lund, Sweden)
Miklos Santha (Paris, France)
Eli Upfal (Providence, RI, USA)
Ingo Wegener (Dortmund, Germany)

Additional Referees

Scott Aaronson
Farid Ablayev
Luca Aceto
Dimitris Achlioptas
Ruben Agadzanyan
Jean-Paul Allouche
Roberto Amadio
Amihood Amir
Ricardo Baeza-Yates
Nikhil Bansal
Franco Barbanera
David A. Mix Barrington
Ezio Bartocci
Ingo Battenfeld
Paweł Baturo
Maurice H. ter Beek
Radim Bělohlávek
Dietmar Berwanger
Amitava Bhattacharya
Yngvi Björnsson
Zoltán Blázsik
Luc Boasson
Hans-Joachim
    Böckenhauer
Andreas Brandstädt
Diletta Cacciagrano
Nicola Cannata
Silvio Capobianco
Venanzio Capretta
Olivier Carton
Matteo Cavaliere
Sourav Chakraborty
Jean-Marc
    Champarnaud
Arkadev Chattopadhyay
Parimal P. Chaudhuri
Vincent Conitzer
José R. Correa
Bruno Courcelle
Pierluigi Crescenzi
Maxime Crochemore
Stefan Dantchev
Bhaskar DasGupta

Jürgen Dassow
Clelia De Felice
Camil Demetrescu
Mariangiola
    Dezani-Ciancaglini
Michael Domaratzki
Alberto Peinado
    Domínguez
Manfred Droste
Stefan Dziembowski
Bruno Escoffier
Olivier Finkel
Maurizio Gabbrielli
Peter Gacs
Ricard Gavaldà
Blaise Genest
Robert Gilman
Amy Glen
Judy Goldsmith
Éva Gombás
Martin Grohe
Jiong Guo
Dan Gutfreund
Michel Habib
Vesa Halava
Bjarni V. Halldórsson
Magnús M. Halldórsson
Joseph Y. Halpern
Laszlo Hars
Reinhold Heckmann
Yoram Hirshfeld
Petr Hliněný
Štěpán Holub
Hendrik Jan Hoogeboom
Tamás Horváth
Brian Howard
Mihály Hujter
Mathilde Hurand
Atsushi Igarashi
Lucian Ilie
Csanád Imreh
Chuzo Iwamoto
Klaus Jansen

Emmanuel Jeandel
Yoshihiko Kakutani
Haim Kaplan
Marc Kaplan
Christos Kapoutsis
Juhani Karhumäki
Marek Karpinski
Shin-ya Katsumata
Zurab Khasidashvili
Jürgen Koslowski
Łukasz Kowalik
Mirek Kowaluk
Richard Kralovic
Manfred Kudlek
Werner Kuich
Arvind Kumar
Michal Kunc
Sonia L'Innocente
Anna Labella
Marie Lalire
Sophie Laplante
Peeter Laud
Troy Lee
Stefano Leonardi
Tal Lev-Ami
Nutan Limaye
Kamal Lodaya
Christof Löding
Sylvain Lombardy
Elena Losievskaja
Jack Lutz
Meena Mahajan
Sebastian Maneth
Pierre McKenzie
Paul-André Melliès
Massimo Merro
Antoine Meyer
Mehdi Mhalla
Sounaka Mishra
Tobias Mömke
Raúl Montes-de-Oca
Fabien de Montgolfier
Kenichi Morita

Mohammad Reza
Mousavi
František Mráz
Marcin Mucha
Loránd Muzamel
Kedar Namjoshi
Sebastian Nanz
Zoltán L. Németh
Jean Néraud
Frank Neven
Brian Nielsen
Vivek Nigam
Hidenosuke Nishio
Mitsunori Ogihara
Yoshio Okamoto
Alexander Okhotin
Carlos Olarte
Bernhard Ömer
Marion Oswald
Sang-il Oum
Prakash Panangaden
Madhusudan
Parthasarathy
Pawel Parys
Ion Petre
Wojciech Plandowski
Martin Plesch
Andrea Polini
Sanjiva Prasad

Pavel Pudlák
Tomasz Radzik
Stanisław P.
Radziszowski
George Rahonis
Klaus Reinhardt
Eric Rémila
Vladimir Rogojin
Andrea Roli
Martin Rötteler
Michel de Rougemont
Michaël Rusinowitch
Jan Rutten
Marie-France Sagot
Miklos Santha
Jayalal Sarma
Benoît Saussol
Marcus Schaefer
Manfred Schimmler
Sebastian Seibert
Olivier Serre
Ehud Shapiro
Jeong Seop Sim
Sunil Simon
Jerzy Skurczyński
Michiel Smid
Jiří Srba
George Steiner
S. P. Suresh

Mario Szegedy
Luca Tesei
Carlo Toffalori
Masafumi Toyama
Vladimir Trifonov
Angelo Troina
Tomasz Truderung
Zsolt Tuza
Sándor Vágvölgyi
Frank D. Valencia
Leslie Valiant
Moshe Vardi
György Vaszil
Laurent Viennot
S. Vijayakumar
Leonardo Vito
Heiko Vogler
Laurent Vuillon
Bill Wadge
Magnus Wahlstrom
Thomas Worsch
Peng Wu
Yasushi Yamashita
Mihalis Yannakakis
Shoji Yuen
Eugen Zălinescu
Mario Ziman

Table of Contents

Rewriting Systems with Data

A Framework for Reasoning About Systems with Unbounded Structures over Infinite Data Domains[*]

Ahmed Bouajjani[1], Peter Habermehl[1,2], Yan Jurski[1], and Mihaela Sighireanu[1]

[1] LIAFA, CNRS & U. Paris 7, Case 7014, 2 place Jussieu, 75251 Paris 05, France
[2] LSV, CNRS & ENS Cachan, 61 av Président Wilson, 94235 Cachan, France

Abstract. We introduce a uniform framework for reasoning about infinite-state systems with unbounded control structures and unbounded data domains. Our framework is based on constrained rewriting systems on words over an infinite alphabet. We consider several rewriting semantics: factor, prefix, and multiset rewriting. Constraints are expressed in a logic on such words which is parametrized by a first-order theory on the considered data domain. We show that our framework is suitable for reasoning about various classes of systems such as recursive sequential programs, multithreaded programs, parametrized and dynamic networks of processes, etc. Then, we provide generic results (1) for the decidability of the satisfiability problem of the fragment $\exists^*\forall^*$ of this logic provided that the underlying logic on data is decidable, and (2) for proving inductive invariance and for carrying out Hoare style reasoning within this fragment. We also show that the reachability problem is decidable for a class of prefix rewriting systems with integer data.

1 Introduction

Software verification requires in general reasoning about infinite-state models. The sources of infinity in software models are multiple. They can be related for instance to the complex control these system may have due, e.g., to recursive procedure calls, communication through fifo channels, dynamic creation of concurrent processes, or the consideration of a parametric number of parallel processes. Other important sources of infinity are related to the manipulation of variables and (dynamic) data structures ranging over infinite data domains such as integers, reals, arrays, heap structures like lists and trees, etc.

In the last few years, a lot of effort has been devoted to the development of theoretical frameworks for the formal modeling and the automatic analysis of several classes of software systems. Rewriting systems (on words or terms), as well as related automata-based frameworks, have been shown to be adequate for reasoning about various classes of systems such as recursive programs, multithreaded programs, parametrized or dynamic networks of identical processes, communicating systems through fifo-channels, etc. (see, e.g., [11,4,13] for survey

---

[*] Partially supported by the French ANR project ACI-06-SETI-001.

E. Csuhaj-Varjú and Z. Ésik (Eds.): FCT 2007, LNCS 4639, pp. 1–22, 2007.
© Springer-Verlag Berlin Heidelberg 2007

papers). These works address in general the problem of handling systems with complex control structures, but where the manipulated data range of finite domains, basically booleans. Other existing works address the problem of handling models with finite control structures, but which manipulate variables over infinite data domains such as counters, clocks, etc., or unbounded data structures (over finite alphabets) such as stacks, queues, limited forms of heap memory (e.g., lists, trees), etc. [2,14,29,8,6,26,25,27,12,15]. Notice that the boundary between systems with infinite control and systems with infinite data is not sharp. For instance, recursive programs can be modeled as prefix rewrite systems which are equivalent to pushdown systems, and (classes of) multithreaded programs can be modeled using multiset rewrite systems which are equivalent to Petri nets and to vector addition systems (a particular class of counter machines).

As already said, in all the works mentionned above, only one source of infinity is taken into account (while the others are either ignored or abstracted away). Few works dealing with different sources of infinity have been carried out nevertheless, but the research on this topic is still in its emerging phase [5,1,21,19,17,3,16]. In this paper, we propose a uniform framework for reasoning about infinite-state systems with both unbounded control structures and unbounded data domains. Our framework is based on word rewriting systems over infinite alphabets where each element is composed from a label over a finite set of symbols and a vector of data in a potentially infinite domain. Words over such an alphabet are called data words and rewriting systems on such words are called data word rewriting systems (DWRS for short). A DWRS is a set of rewriting rules with constraints on the data carried by the elements of the words.

The framework we propose allows to consider different rewriting semantics and different theories on data, and allows also to apply in a generic way decision procedures and analysis techniques. The rewriting semantics we consider are either the factor rewriting semantics (which consists in replacing any factor in the word corresponding to the left hand side or a rule by the right hand side), as well as the prefix and the multiset rewriting semantics. The constraints in the rewriting systems are expressed in a logic called DWL which is an extension of the monadic first-order theory of the natural ordering on positive integers (corresponding to positions on the word) with a theory on data allowing to express the constraints on the data values at each position of the word. The theory on data, which is a parameter of the logic DWL, can be any fist-order theory such as Presburger arithmetics, or the first-order theory on reals.

We show that this framework is expressive enough to model various classes of infinite-state systems. Prefix rewriting systems are used to model recursive programs with global and local variables over infinite data domains. Factor rewriting systems are used for modeling parametrized networks of processes with a linear topology (i.e., there is a total ordering between the identities of the processes). This is for instance the case of various parallel and/or distributed algorithms. (We give as an example a model for the Lamport's Bakery algorithm for mutual exclusion.) Multiset rewriting systems can be used for modeling multithreaded programs or dynamic/parametrized networks where the information about identities of processes is not relevant. This is the case for various systems such as cache coherence protocols (see, e.g., [23]).

We address the decidability of the satisfiability problem of the logic DWL. We show that this problem is undecidable for very weak theories on data already for the fragment of $\forall^*\exists^*$ formulas. On the other hand, we prove the generic result that whenever the underlying theory on data has a decidable satisfiability problem, the fragment of $\exists^*\forall^*$ formulas of DWL has also a decidable satisfiability problem.

Then, we address the issue of automatic analysis of DWRS models. We provide two kinds of results. First, we consider the problem of carrying out post and pre condition reasoning based on computing immediate successors and immediate predecessors of sets of configurations. We prove, again in a generic way, that the fragment of $\exists^*\forall^*$ formulas in DWL is effectively closed under the computation of post and pre images by rewriting systems with constraints in $\exists^*\forall^*$. We show how this result, together with the decidability result of the satisfiability problem in $\exists^*\forall^*$, can be used for deciding whether a given assertion is an inductive invariant of a system, or whether the specification of an action is coherent, that is, the execution of an action starting from the pre condition leads to configurations satisfying the post condition. The framework we present here generalizes the one we introduced recently in [16] based on constrained multiset rewriting systems. Our generalization to word factor and prefix rewriting systems allows to deal in a uniform and natural way with a wider class of systems where reasoning about linearly ordered structures is needed.

Finally, we consider the problem of solving the reachability problem for a subclass of DWRS. We provide a new decidability result of this problem for the class of context-free prefix rewriting systems (i.e., where the left hand side of each rule is of size 1) over the data domain of integers with difference constraints. (Extensions of this class lead to undecidabilty.) This results generalizes a previous result we have established few years ago in [17] for a more restricted class of systems where not all difference constraints were allowed.

Related work: Regular model checking has been defined as a uniform framework for reasoning about infinite-state systems [29,28,18,4]. However, this framework is based on finite-state automata and transducers over finite alphabets which does not allow to deal in a simple and natural way with systems with both unbounded control and data domains. The same holds for similar frameworks based on word/tree rewriting systems over a finite alphabet (e.g., [11,13]).

Works on the analysis of models for systems with two sources of infinity such as networks of infinite-stat processes are not very numerous in the literature. In [5], the authors consider the case of networks of 1-clock timed systems and show that the verification problem for a class of safety properties is decidable under some restrictions on the used constraints. Their approach has been extended in [21,19] to a particular class of multiset rewrite systems with constraints (see also [3] for recent developments of this approach). In [17], we have considered the case of prefix rewrite systems with integer data which can be seen as models of recursive programs with one single integer parameter. Again, under some restrictions on the used arithmetical constraints, we have shown the decidability of the reachability problem. The result we prove in section 7 generalization our previous result of [17].

Recently, we have defined a generic framework for reasoning about parametrized and dynamic networks of infinite-state processes based on constrained multiset rewrite systems [16]. The work we present generalizes that work to other classes of rewriting systems.

In a series of papers, Pnueli et al. developed an approach for the verification of parameterized systems combining abstraction and proof techniques (see, e.g., [7]). In [7], the authors consider a logic on (parametric-bound) *arrays* of integers, and they identify a fragment of this logic for which the satisfiability problem is decidable. In this fragment, they restrict the shape of the formula (quantification over indices) to formulas in the fragment $\exists^*\forall^*$ similarly to what we do, and also the class of used arithmetical constraints on indices and on the associated values. In a recent work by Bradley and al. [20], the satisfiability problem of the logic of unbounded arrays with integers is investigated and the authors provide a new decidable fragment, which is incomparable to the one defined in [7], but again which imposes similar restrictions on the quantification alternation in the formulas, and on the kind of constraints that can be used. In contrast with these works, our decidable logical fragment has a weaker ability of expressing ordering constraints on positions (used, e.g., to represent identities of processes in parametrized/dynamic networks), but allows *any* kind of data, provided that the used theory on the considered data domain is decidable. For instance, we can use in our logic general Presburger constraints whereas [7] and [20] allow limited classes of constraints.

Let us finally mention that there are recent works on logics (first-order logics, or temporal logics) over finite/infinite structures (words or trees) over infinite alphabets (which can be considered as abstract infinite data domains) [10,9,24]. The obtained positive results so far concern logics with limited data domain (basically infinite sets with only equality, or sometimes with an ordering relation), and are based on reduction to complex problems such as reachability in Petri nets. Contrary to these works, our approach is to prefer weakening the first-order/model language for describing the structures while preserving the capacity of expressing constraints on data. We believe that this approach could be more useful in practice since it allows to cover a large class of applications as this paper tries to show.

2   A Logic for Reasoning About Words over Data Domains

2.1   Preliminaries

Let Σ be a finite alphabet, and let $\mathbb{D}$ be a potentially infinite *data domain*. For a given $N \in \mathbb{N}$ such that $N \geq 1$, words over $\Sigma \times \mathbb{D}^N$ are called N-dim *data words*. Let $(\Sigma \times \mathbb{D}^N)^*$ (resp. $(\Sigma \times \mathbb{D}^N)^\omega$) be the set of finite (resp. infinite) data words, and let $(\Sigma \times \mathbb{D}^N)^\infty$ be the union of these two sets. Given a data word σ, we denote by $|\sigma|$ the (finite or infinite) length of σ. A word $\sigma \in (\Sigma \times \mathbb{D}^N)^\infty$ can be considered as a mapping from $[0, |\sigma|)$ to $\Sigma \times \mathbb{D}^N$, i.e., $\sigma = \sigma(0)\sigma(1)\ldots$. Given $e = (A, d_1, \ldots, d_N) \in \Sigma \times \mathbb{D}^N$, let $label(e)$ denote the element A and let $data(e)$ denote the vector $(d_1, \ldots, d_N)$. For $k \in \{1, \ldots, N\}$, $data_k(e)$ denotes the elements d_k of e. These notations are generalized in the obvious manner to words over $\Sigma \times \mathbb{D}^N$.

2.2   A First-Order Logic over Data Words

We introduce herefater the *data word logic* (DWL for short) which is a first order logic allowing to reason about data words by considering the labels as well as the data values at each of their positions. The logic DWL is parameterized by a (first-order) logic on the considered data domain $\mathbb{D}$, i.e., by the set of operations and the set of basic predicates (relations) allowed on elements of $\mathbb{D}$.

Let Ω be a finite set of functions over $\mathbb{D}$, and let $\varXi$ be a finite set of relations over $\mathbb{D}$. Consider also a set of *position variables* $\mathcal{I}$ ranging over positive integers and a set of *data variables* $\mathcal{D}$ ranging over data values in $\mathbb{D}$, and assume that $\mathcal{I} \cap \mathcal{D} = \emptyset$. Then, the set of *terms* of $\mathsf{DWL}(\mathbb{D}, \Omega, \varXi)$ is given by the grammar:

$$t ::= u \mid \delta_k[x] \mid o(t_1, \ldots, t_n)$$

where $k \in \{1, \ldots, N\}$, $x \in \mathcal{I}$, $u \in \mathcal{D}$, and $o \in \Omega$. The set of *formulas* of $\mathsf{DWL}(\mathbb{D}, \Omega, \varXi)$ is given by:

$$\varphi ::= 0 < x \mid x < y \mid A[x] \mid r(t_1, \ldots, t_n) \mid \neg\varphi \mid \varphi \vee \varphi \mid \exists u.\ \varphi \mid \exists x.\ \varphi$$

where $x, y \in \mathcal{I}$, $u \in \mathcal{D}$, $A \in \Sigma$, and $r \in \varXi$.

As usual, boolean connectives such as conjunction $\wedge$ and implication $\Rightarrow$ are defined in terms of disjunction $\vee$ and negation $\neg$, and universal quantification $\forall$ is defined as the dual of existential quantification $\exists$. We also define equality $=$ and disequality $\neq$ in terms of $<$ and boolean connectives. Let $x = 0$ be an abbreviation of $\neg(0 < x)$ and let $x = y$ be an abbreviation of $\neg(x < y) \wedge \neg(y < x)$. We also write as usual $t \leq t'$ for $t < t' \vee t = t'$, where t and t' represent either position variables or 0. Then, let $t \neq t'$ be an abbreviation of $\neg(t = t')$, for $t, t' \in \mathcal{I} \cup \{0\}$. We denote by $\mathsf{DWL}_=$ the set of DWL formulas where the only comparisons constraints between position variables, and between position variables and 0 are equality or disequality constraints.

The notions of bound and free variables are defined as usual in first-order logic. Given a formula φ, the set of free variables in φ is denoted $FV(\varphi)$.

Formulas are interpreted on finite or infinite words over the alphabet $\Sigma \times \mathbb{D}^N$. Intuitively, position variables correspond to positions in the considered word. The formula $A[x]$ is true if A is the label of the element at the position corresponding to the position variable x. The term $\delta_k[x]$ represents the kth data value attached to the element at the position corresponding to x. Terms are built from such data values and from data variables by applying operations in Ω. Formulas of the form $r(t_1, \ldots, t_n)$ allow to express constraints on data values at different positions of the word.

Formally, we define a satisfaction relation between such models and formulas. Let $\sigma \in (\Sigma \times \mathbb{D}^N)^\infty$. In order to interpret open formulas, we need valuations of position and data variables. Given $\mu : \mathcal{I} \to \mathbb{N}$ and $\nu : \mathcal{D} \to \mathbb{D}$, the satisfaction relation is inductively defined as follows:

$$\sigma \models_{\mu,\nu} 0 < x \text{ iff } 0 < \mu(x)$$
$$\sigma \models_{\mu,\nu} x < y \text{ iff } \mu(x) < \mu(y)$$
$$\sigma \models_{\mu,\nu} A[x] \text{ iff } label(\sigma(\mu(x))) = A$$
$$\sigma \models_{\mu,\nu} r(t_1, \ldots, t_m) \text{ iff } r(\langle t_1 \rangle_{\sigma,\mu,\nu}, \ldots, \langle t_m \rangle_{\sigma,\mu,\nu})$$
$$\sigma \models_{\mu,\nu} \neg\varphi \text{ iff } \sigma \not\models_{\mu,\nu} \varphi$$
$$\sigma \models_{\mu,\nu} \varphi_1 \vee \varphi_2 \text{ iff } \sigma \models_{\mu,\nu} \varphi_1 \text{ or } \sigma \models_{\mu,\nu} \varphi_2$$
$$\sigma \models_{\mu,\nu} \exists u.\ \varphi \text{ iff } \exists d \in \mathbb{D}.\ \sigma \models_{\mu,\nu[u \leftarrow d]} \varphi$$
$$\sigma \models_{\mu,\nu} \exists x.\ \varphi \text{ iff } \exists i \in \mathbb{N}.\ i < |\sigma| \text{ and } \sigma \models_{\mu[x \leftarrow i],\nu} \varphi$$

where the mapping $\langle \cdot \rangle_{\sigma,\mu,\nu}$, associating to each term a data value, is inductively defined as follows:

$$\langle u \rangle_{\sigma,\mu,\nu} = \nu(u)$$
$$\langle \delta_k[x] \rangle_{\sigma,\mu,\nu} = data_k(\sigma(\mu(x)))$$
$$\langle o(t_1, \ldots, t_n) \rangle_{\sigma,\mu,\nu} = o(\langle t_1 \rangle_{\sigma,\mu,\nu}, \ldots, \langle t_n \rangle_{\sigma,\mu,\nu})$$

Given a formula φ, let $[\![\varphi]\!]_{\mu,\nu} = \{\sigma \in (\Sigma \times \mathbb{D}^N)^\infty : \sigma \models_{\mu,\nu} \varphi\}$. A formula φ is *satisfiable* if and only if there exist valuations μ and ν such that $[\![\varphi]\!]_{\mu,\nu} \neq \emptyset$. The subscripts of $\models$ and $[\![\cdot]\!]$ are omitted in the case of a closed formula.

2.3   Quantifier Alternation Hierarchy

A formula is in *prenex form* if it is written $Q_1 z_1 Q_2 z_2 \ldots Q_m z_m.\ \varphi$ where (1) $Q_1, \ldots, Q_m \in \{\exists, \forall\}$, (2) $z_1, \ldots, z_m \in \mathcal{I} \cup \mathcal{D}$, and φ is a quantifier-free formula. It can be proved that for every formula φ, there exists an equivalent formula φ' in prenex form.

We consider two families $\{\Sigma_n\}_{n \geq 0}$ and $\{\Pi_n\}_{n \geq 0}$ of sets of formulas defined according to the alternation depth of existential and universal quantifiers in their prenex form:

- $\Sigma_0 = \Pi_0$ is the set of formulas where all quantified variables are in $\mathcal{D}$,
- For $n \geq 0$, Σ_{n+1} (resp. Π_{n+1}) is the set of formulas $Q z_1 \ldots z_m.\ \varphi$ where $z_1, \ldots, z_m \in \mathcal{I} \cup \mathcal{D}$, Q is the existential (resp. universal) quantifier $\exists$ (resp. $\forall$), and φ is a formula in Π_n (resp. Σ_n).

It can be seen that, for every $n \geq 0$, Σ_n and Π_n are closed under conjunction and disjunction, and that the negation of a Σ_n formula is a Π_n formula and vice versa. For every $n \geq 0$, let $B(\Sigma_n)$ denote the set of all boolean combinations of Σ_n formulas. Clearly, $B(\Sigma_n)$ subsumes both Σ_n and Π_n, and is included in both Σ_{n+1} and Π_{n+1}.

2.4 Data Independent Formulas

A DWL formula is *data independent* if it does not contain occurrences of data predicates of the form $r(t_1, \ldots, t_n)$ and of quantification over data variables. Syntactically, the set of data independent formulas is the same as the set of formulas of the monadic first-order logic over integers with the usual ordering relation. (Projections on the alphabet Σ of their models define star-free regular languages.) Interpreted over data words, these formulas satisfy the following closure properties: for every data words σ and σ', and for every data independent formula φ, if $label(\sigma) = label(\sigma')$, then $\sigma \models_{\mu,\nu} \varphi$ if and only if $\sigma' \models_{\mu,\nu} \varphi$.

3 The Satisfiability Problem

We investigate in this section the decidability of the satisfiability problem of DWL. First, we can prove that the logic is undecidable for very simple data theories starting from the fragment Π_2. The proof is by a reduction of the halting problem of Turing machines. The idea is to encode a computation of a machine, seen as a sequence of tape configurations, as a data word. Each position corresponds to a cell in the tape of the machine at some configuration in the computation. We associate to each position (1) a positive integer value corresponding to its rank in a configuration, and (2) a label encoding informations (ranging over a finite domain) such as the contents of the cell, the fact that a cell corresponds to the location of the head, and the control state of the machine. Then, using DWL formulas in the Π_2 (i.e., $\forall^* \exists^*$) fragment, it is possible to express that two consecutive configurations correspond indeed to a valid transition of the machine. Intuitively, this is possible because these formulas allow to relate each cell at some configuration to the corresponding cell at the next configuration. We need for that to use the ordering on positions to talk about successive configurations, and the equality on the values attached to positions to relate cells with the same rank in these successive configurations. For the logic $\mathsf{DWL}_=$, since we do not have an ordering on positions, we need to attach another value to position representing their rank in the configurations.

Theorem 1. *The satisfiability problem of the fragment Π_2 is undecidable for* $\mathsf{DWL}(\mathbb{N}, =)$ *and* $\mathsf{DWL}_=(\mathbb{N}, 0, <)$.

Then, the main result of this section is that whenever the underlying theory on data has a decidable satisfiability problem, the fragment Σ_2 has also a decidable satisfiability problem.

Theorem 2. *If the satisfiability problem for* $\mathsf{FO}(\mathbb{D}, \Omega, \Xi)$ *is decidable, then the satisfiability problem of the fragment Σ_2 of* $\mathsf{DWL}(\mathbb{D}, \Omega, \Xi)$ *is also decidable.*

The rest of the section is devoted to the proof of theorem above. We show that the satisfiability problem in the fragment Σ_2 of $\mathsf{DWL}(\mathbb{D}, \Omega, \Xi)$ can be reduced to the satisfiability problem in logic on data $\mathsf{FO}(\mathbb{D}, \Omega, \Xi)$.

First of all, we need to introduce a slight modification in the definition of data words: So far, we have considered that a data word σ is total mappings from the interval $[0, |\sigma|)$ to the alphabet $\Sigma \times \mathbb{D}^N$. Let us consider now that a word σ is a total mapping from a set of natural numbers S_σ to the alphabet $\Sigma \times \mathbb{D}^N$, where S_σ is not necessarily an interval. Clearly, there is an isomorphism π_σ from $[0, |\sigma|)$ to S_σ, and this isomorphism is monotonic. Let us denote $[\sigma]$, for every word σ, the (unique) mapping from $[0, |\sigma|)$ to $\Sigma \times \mathbb{D}^N$ such that, for every $i \in [0, |\sigma|)$, $[\sigma](i) = \sigma(\pi_\sigma(i))$.

Furthermore, assume that in the definition of the satisfaction relation $\models$ between data words and DWL formulas, the last line (the case of existential quantification over position variables) is substituted by: $\sigma \models_{\mu,\nu} \varphi$ iff $\exists i \in S_\sigma. \sigma \models_{\mu[x \leftarrow i],\nu} \varphi$. Then, it can be checked that the following holds.

Lemma 1. *For every data word σ, for every DWL formula φ, and for every position/data variable valuations μ and ν, we have $\sigma \models_{\mu,\nu} \varphi$ iff $[\sigma] \models_{\mu,\nu} \varphi$.*

The lemma above implies that, for every two data words σ and σ' such that $[\sigma] = [\sigma']$, we have $\sigma \models_{\mu,\nu} \varphi$ iff $\sigma' \models_{\mu,\nu} \varphi$, for every φ, μ, and ν.

Before starting the proof, we need to introduce a syntactical form of Σ_n formulas, for any $n \geq 1$. We say that a formula in such a fragment is in *special form* if it is a finite disjunction of formulas of the form

$$\exists x_1, \ldots, x_n \exists \mathbf{u} \forall \mathbf{y}. \left(\left(\bigwedge_{1 \leq i < j \leq n} x_i < x_j \right) \wedge \varphi \right)$$

where $\mathbf{x} = (x_1, \ldots, x_n)$ and $\mathbf{y}$ are position variables, and $\mathbf{u}$ is a vector of data variables. It is easy to show that every formula in the fragment Σ_n has an equivalent Σ_n formula in special form.

We are now ready to stat the proof of Theorem 2. Let φ be a DWL formula, and assume w.l.o.g. that φ is closed, in special form, and given by:

$$\varphi = \exists \mathbf{x}. \exists \mathbf{u}. \forall \mathbf{y}. \psi$$

where $\mathbf{x}$ and $\mathbf{y}$ are vectors of position variables, $\mathbf{u}$ is a vector of data variables. Assume also that φ is satisfiable, which means that there is a data word σ such that $\sigma \models \varphi$.

Then, let Θ be the set of all possible (partial or total) mappings between the variables in $\mathbf{y}$ and the variables in $\mathbf{x}$. Then, we have $\sigma \models \exists \mathbf{x}. \exists \mathbf{u}. \varphi^{(1)}$ where

$$\varphi^{(1)} = \bigwedge_{\theta \in \Theta} \forall \mathbf{y}. \left(\left(\left(\bigwedge_{y \in dom(\theta)} y = \theta(y) \right) \wedge \left(\bigwedge_{y \notin dom(\theta)} \bigwedge_{x \in \mathbf{x}} y \neq x \right) \right) \Rightarrow \psi \right) \quad (1)$$

This means that there are positions $\mathbf{i}$ in the domain of σ, and there are data values $\mathbf{d}$, such that

$$\sigma \models_{\mu,\nu} \varphi^{(1)} \quad (2)$$

where μ and ν are valuations associating $\mathbf{i}$ with $\mathbf{x}$ and $\mathbf{d}$ with $\mathbf{u}$, respectively.

Consider now the data word $\sigma' = \sigma|_{\mathbf{i}}$, i.e., the subword of σ corresponding to the positions in $\mathbf{i}$. Then, it can be seen that (2) implies that:

$$\sigma' \models_{\mu,\nu} \bigwedge_{\substack{\theta \in \Theta \\ dom(\theta)=\mathbf{y}}} \forall \mathbf{y}. \left(\bigwedge_{y \in \mathbf{y}} y = \theta(y) \Rightarrow \psi \right) \tag{3}$$

which is equivalent to $\sigma' \models \varphi^{(2)}$ where

$$\varphi^{(2)} = \exists \mathbf{x}. \, \exists \mathbf{u}. \bigwedge_{\substack{\theta \in \Theta \\ dom(\theta)=\mathbf{y}}} \bigwedge_{y \in \mathbf{y}} \psi[\theta(y)/y] \tag{4}$$

Conversely, every minimal model (according to the size of its domain) of the formula $\varphi^{(2)}$ above is necessarily a model of the formula $\exists \mathbf{x}. \, \exists \mathbf{u}. \, \varphi^{(1)}$, which is equivalent to the formula φ. Therefore, we have reduced the satisfiability problem of Σ_2 to the satisfiability problem in Σ_1.

The last step of the proof is to reduce the satisfiability problem of the Σ_1 formula $\varphi^{(2)}$ to the satisfiability problem of a pure data formula in $\mathsf{FO}(\mathbb{D}, \Omega, \Xi)$. For that, we must get rid of the comparisons between position variables, and of constraints on position labels.

Since the formula φ in special form, the values associated with the position variables $\mathbf{x}$ are in the same order as their indices. Then, let $\varphi^{(3)} = \exists \mathbf{x}. \, \exists \mathbf{u}. \, \psi'$ be the formula obtained from $\varphi^{(2)}$ by replacing each constraint $x_i < x_j$ by $true$ if $i < j$, or by $false$ otherwise. The formula $\varphi^{(3)}$ is equivalent to the formula $\varphi^{(2)}$ but has no comparison constraints between position variables. Now, since the alphabet Σ is finite, we can build an equivalent formula to $\varphi^{(3)}$ which has no label constraints: we consider a disjunction on all possible mappings λ from $\mathbf{x}$ to Σ. For each of these mapping λ, we replace in $\varphi^{(3)}$ each occurrence of a formula $A[x]$ by $true$ if $\lambda(x) = A$, or by $false$ otherwise. Let $\varphi^{(4)}$ be the so obtained formula.

Finally, we define a $\mathsf{FO}(\mathbb{D}, \Omega, \Xi)$ formula which is satisfiable if and only if $\varphi^{(4)}$ is satisfiable. This formula is obtained by replacing in $\varphi^{(4)}$ terms involving positions variables by data variables: for each variable $x \in \mathbf{x}$ and for each rank $k \in \{1, \ldots, N\}$, we associate a fresh data variable $v_{x,k}$. Then, we remove in $\varphi^{(4)}$ the quantification of $\mathbf{x}$ and we substitute each occurrence of a term $\delta_k(x)$ by the variable $v_{x,k}$.

4 Rewriting Systems over Data Words

4.1 Rewriting Rules

A *data word rewriting rule* over the logic DWL has the form:

$$A_0 \cdots A_n \mapsto B_0 \cdots B_m \; : \; \varphi$$

where $A_i, B_j \in \Sigma$ for all $i \in \{0, \ldots, n\}$ and $j \in \{0, \ldots, m\}$, and φ is a DWL formula such that (1) $FV(\varphi) = \{x_0, \ldots, x_n\} \cup \{y_0, \ldots, y_m\}$, and (2) all the occurrences in φ of the variables y_j are in terms of the form $\delta_k(y_j)$ for $0 \leq k \leq N$. We assume in this definition that the left hand side of a rule has at least one symbol ($n \geq 0$), and that its right hand side can be empty. When $B_0 \cdots B_m$ is empty, the formula φ has only free variables $\{x_0, \ldots, x_n\}$ related to the left hand side of the rule.

Intuitively, the application of a rewriting rule to a data word σ (leading to a new word σ') consists in replacing in σ a subword γ such that $label(\gamma)$ is equal to $A_0 \cdots A_n$ by another word γ' such that $label(\gamma')$ is equal to $B_0 \cdots B_m$, provided that the formula φ, relating the data values in σ with data values in γ', is satisfied. Each variable x_k (resp. y_k) represents the position in σ (resp. σ') of the kth elements of γ (resp. γ'). The formula φ can constrain the positions corresponding to elements of γ as well as their attached data values w.r.t. data values at other position in σ. Moreover, the formula φ can constrain the data values in the new word by relating these data values with data values attached to positions in σ.

4.2   Rewriting Semantics

A *rewriting system* is given by a set of rewrite rules and a rewriting semantics. Several rewriting relations between words can be considered depending on the adopted semantics of rewriting. Given a set Δ of data word rewriting rules, we consider here four relations $\Rightarrow_{\Delta,f}$, $\Rightarrow_{\Delta,p}$, and $\Rightarrow_{\Delta,m}$ corresponding respectively to factor, prefix, and multiset rewriting. Subscripts are omitted whenever the considered rewriting system and/or rewriting semantics are know from the context.

Let us start by defining the semantics of factor and prefix rewriting. For that, let us fix a rewrite system Δ. Then, for every $\sigma, \sigma' \in (\Sigma \times \mathbb{D}^N)^*$, we have $\sigma \Rightarrow_f \sigma'$ (resp. $\sigma \Rightarrow_p \sigma'$) if and only if there exists a rewrite rule "$A_0 \cdots A_n \mapsto B_0 \cdots B_m : \varphi$" and there exist data words $\alpha, \beta, \gamma, \gamma' \in (\Sigma \times \mathbb{D}^N)^*$ such that

- *factor rewriting:* $\sigma = \alpha\gamma\beta$ and $\sigma' = \alpha\gamma'\beta$,
- *prefix rewriting:* $\sigma = \gamma\beta$, $\sigma' = \gamma'\beta$, and $|\alpha| = 0$,

with $label(\gamma) = A_0 \cdots A_n$, $label(\gamma') = B_0 \cdots B_m$ and

$$\sigma \models \varphi[(|\alpha| + i)/x_i]_{0 \leq i \leq n}[data_k(\gamma'(j))/\delta_k[y_j]]_{0 \leq k \leq N, 0 \leq j \leq m}$$

Now, in order to define the multiset rewriting relation, we consider the equivalence relation between words which abstracts away the ordering between symbols: Given $\sigma, \sigma' \in (\Sigma \times \mathbb{D}^N)^*$, we have $\sigma \simeq \sigma'$ if and only if there exists a permutation π of $\{0, \ldots, |\sigma| - 1\}$ such that $\sigma(\pi(0)) \cdots \sigma(\pi(|\sigma| - 1)) = \sigma'$. Then, for every $\sigma, \sigma' \in (\Sigma \times \mathbb{D}^N)^*$, we have $\sigma \Rightarrow_m \sigma'$ if and only if $\exists \theta, \theta' \in (\Sigma \times \mathbb{D}^N)^*$ such that $\sigma \simeq \theta$, $\theta \Rightarrow_f \theta'$, and $\theta' \simeq \sigma'$.

It can be seen that for every σ, σ' such that $\sigma \simeq \sigma'$, and for every formula φ in DWL$_=$, we have $\sigma \models \varphi$ if and only if $\sigma' \models \varphi$. This fact is not true in general for DWL formulas. Therefore, in the case of multiset rewriting, we assume naturally that all the constraints in the rewriting rules are in DWL$_=$.

Given a set of rewriting rules Δ, the corresponding factor, prefix, and multiset rewriting system are denoted Δ_f, Δ_p, and Δ_m, respectively. Let $\mathsf{DWRS}_\sharp$ be the class of all $\sharp$-rewriting systems, for $\sharp \in \{f, p, m\}$.

5 Models of Infinite-State Systems

5.1 Recursive Programs with Data

We show hereafter that sequential programs with recursive procedure calls can be translated into prefix rewriting systems. We consider that a program has several procedures, and we assume that it uses a set of global variables $\mathbf{g} = (g_1, \ldots, g_N)$ and that each procedure has a set of local variables $\mathbf{l} = (l_1, \ldots, l_M)$. (We assume w.l.o.g. that the local variables are the same for all procedures, all of them ranging over some data domain $\mathbb{D}$.)

A program is given by its inter-procedural control flow graphs (ICFG for short) which is a collection of control flow graphs (CFG), one for each of its procedures. Nodes in the CFG of a procedure represent control points in its source code, and edges represent transitions from a control point to another one. We assume that each procedure Π has an initial node n_{in}^Π. Edges in CFGs are labeled by statements which can be either (1) tests over the values of the global/local variables, (2) assignments of the global/local variables, (3) procedure calls, or (4) procedure returns leading to a termination control point. Variables are assigned values of expressions built from global and local variables using a set of operations Ω. Tests over variables are first-order assertions based on a set of predicates Ξ.

Consider an ICFG, and let $\mathcal{N}$ be the set of its nodes. We associate with the considered ICFG a prefix rewriting systems over the alphabet $(\mathcal{N} \cup \{G\}) \times \mathbb{D}^{N+M}$ where G is a special symbol, N is the number of global variables, and M is the number of local variables of each procedure. Indeed, we consider that a configuration of the recursive program defined by the ICFG is represented by a finite word of the form $(G, \mathbf{d_0})(n_1, \mathbf{d_1})(n_2, \mathbf{d_2}) \cdots (n_\ell, \mathbf{d_\ell})$ where $n_i \in \mathcal{N}$ for all $i \geq 1$ and $\mathbf{d_i} \in \mathbb{D}^{N+M}$ for all $i \geq 0$. The element $(G, \mathbf{d_0})$ at position 0 of the word is used to store the value of the global variables: we assume that for every $k \in \{1, \ldots, N\}$, the value of the variable g_k is equal to the kth element of the vector $\mathbf{d_0}$. Moreover, the rest of the word $(n_1, \mathbf{d_1})(n_2, \mathbf{d_2}) \cdots (n_\ell, \mathbf{d_\ell})$ represents the call stack of the program. In the element $(n_i, \mathbf{d_i})$ of this stack, n_i represents the point at which the control of the program will return after all the calls higher in stack (i.e., of index less than i in our word representation) will be done, and $\mathbf{d_i}$ represents the values of the local variables which must be restored when the control will reach the point n_i: we assume that for every $k \in \{N+1, \ldots, N+M\}$, the value of the variable l_k is equal to the kth element of the vector $\mathbf{d_i}$. Then, the set of rewriting rules of the system associated with the considered ICFG is defined as follows:

Test: $n \xrightarrow{\varphi(\mathbf{g},\mathbf{l})} n'$ where φ is a $\mathsf{FO}(\mathbb{D}, \Omega, \Xi)$ formula, is modeled by:

$$Gn \mapsto Gn' \ : \ \varphi\zeta \wedge \varphi_{id}$$

where ζ is the substitution $[\delta_k[x_0]/g_k]_{1 \leq k \leq N}[\delta_k[x_1]/l_k]_{N+1 \leq k \leq N+M}$, and

$$\varphi_{id} = \bigwedge_{i=1}^{N} \bigwedge_{j=N+1}^{N+M} \delta_i[y_0] = \delta_i[x_0] \wedge \delta_j[y_1] = \delta_j[x_1]$$

Assignment: $n \xrightarrow{(\mathbf{g},\mathbf{l}):=\mathbf{t}(\mathbf{g},\mathbf{l})} n'$ where $\mathbf{t}$ is a vector of Ω-terms, is modeled by:

$$Gn \mapsto Gn' \ : \ \bigwedge_{i=1}^{N} \bigwedge_{j=N+1}^{N+M} \delta_i[y_0] = t_i\zeta \wedge \delta_j[y_1] = t_j\zeta$$

where ζ is the substitution defined in the previous case.

Procedure call: $n \xrightarrow{\mathsf{call}(\Pi)} n'$ is modeled by:

$$Gn \mapsto Gn_{in}^{\Pi} n' \ : \ \varphi_{id}'$$

where

$$\varphi_{id}' = \bigwedge_{i=1}^{N} \bigwedge_{j=N+1}^{N+M} \delta_i[y_0] = \delta_i[x_0] \wedge \delta_j[y_2] = \delta_j[x_1]$$

Procedure return: $n \xrightarrow{\mathsf{return}} n'$ is modeled by:

$$Gn \mapsto G \ : \ \bigwedge_{i=1}^{N} \delta_i[y_0] = \delta_i[x_0]$$

More general prefix rewriting systems can be used in order to handle applications where stack inspection is needed. Indeed, the side constraints we allow in the rewriting rules can be used for the expression of global conditions on the stack content that must be satisfied before the execution of certain actions. This is important for modeling various control access and resource-usage scenarios. For instance, operations on security-critical objects can be executed only if certain conditions are satisfied, e.g., (1) all procedures in the call stack have a certain permission, or (2) a "privileged" procedure is present in the call stack and all procedures higher in the stack have a permission. These constraints can be expressed as DWL (data independent) formulas in the fragment Σ_2:

$$\forall x. \ (x_1 \leq x \Rightarrow \mathtt{perm}[x])$$

$$\exists x. \ (x_1 \leq x \wedge \mathtt{privilege}[x] \wedge \forall y. \ ((x_1 \leq y \wedge y < x) \Rightarrow \mathtt{perm}[y]))$$

5.2   Dynamic/Parametrized Networks of Processes

Unbounded networks of identical processes can be modeled using rewriting systems. We assume that each process is defined by an extended automaton, i.e., a finite-control machine manipulating a set of variables $\mathbf{v} = (v_1, \ldots, v_N)$ ranging over some given data domain $\mathbb{D}$. More precisely, an extended automaton is defined by a finite set of control locations $\mathcal{Q}$, and a set of transitions between these locations. Each transition is labeled by a statement which can be either a test over the values of the variables, or an assignments of the variables. As in section 5.1, assigned values to variables are defined using expressions built from variables and a set of operations Ω, and tests are first-order assertions based on a set of predicates Ξ.

Consider a network of n processes, where n is an arbitrary positive integer (greater than 1). We represent a configuration of such a network by a word of length n over the alphabet $\mathcal{Q} \times \mathbb{D}^N$. Then, to reason uniformly about networks with an arbitrary number of processes, (1) we consider the set of all finite words over $\mathcal{Q} \times \mathbb{D}^N$ as possible configurations, and (2) we model the dynamics of the whole family of networks with an arbitrary size by means of a rewriting system. We use different rewriting semantics depending on the topology of the network. In general, using factor rewriting systems allows to reason about networks with a linear topology, i.e., where processes are arranged sequentially (or sometimes as a ring). This corresponds to the case where an ordering is assumed between the process identities (inducing a notion of neighborhood). Multiset rewriting systems are used when the ordering between process is not relevant. This is the case of many systems such as cache coherence protocols [23] and some classes of multithreaded programs [22,16].

Data rewriting systems we consider allow to model various communication (and synchronization) schemas between processes (e.g., shared variables, rendezvous), tests on local and global configurations, as well as dynamic creation and deletion of processes.

As an example, we give hereafter the model corresponding to (a simplified version of) the Lamport's Bakery protocol for mutual exclusion. As usual in such protocols, the algorithms handle a set of processes which compete for entering into a critical section. The model of each process is a machine with tree control locations: nocs, req, and cs. The location nocs correspond to activities of the processes outside the critical section. When the process needs to enter the critical section, it takes a ticket with a number (a positive integer) which is bigger that the number of all existing tickets, and moves to the control location req. Then, the process waits at this location for his turn to enter the critical section, that is, until the number on its tickets become the smallest of all numbers on existing tickets. In case of a conflict (since it may happen actually that two processes obtain the same ticket number), the process with the smallest rank (identity) enters the critical section. Then, the process can exit the critical section and return to the control location nocs.

The Bakery protocol can be modeled by the following factor rewriting system Δ_{bakery} defined over the alphabet $\{\texttt{nocs}, \texttt{req}, \texttt{cs}\} \times \mathbb{N}$.

$$\texttt{nocs} \mapsto \texttt{req} : \forall i.\ \delta[y_0] > \delta[i]$$
$$\texttt{req} \mapsto \texttt{cs} : \forall i.\ \big(\delta[i] > 0 \Rightarrow (\delta[x_0] < \delta[i]\ \vee\ \delta[x_0] = \delta[i] \wedge x_0 < i)\big) \wedge$$
$$\delta[y_0] = \delta[x_0]$$
$$\texttt{cs} \mapsto \texttt{nocs} : \delta[y_0] = 0$$

Notice that Δ_{bakery} is a system of the class $\mathsf{DWRS}_f[\Pi_1]$ since all side constraints in the rule are universally quantified formulas.

6   Post and Pre Condition Reasoning

We address in this section the problem of checking the validity of assertions on the configurations of systems modeled by data word rewriting systems. We show that the fragment Σ_2 of DWL is effectively closed under the computation of one (forward or backward) rewriting step of rewriting systems in $\mathsf{DWRS}[\Sigma_2]$ (for the three considered ewriting semantics). We show how to use this result in checking inductive invariance of given assertions, and for carrying out Hoare-style reasoning about our models.

6.1   post and pre Operators

We define hereafter the operators of immediate successors and immediate predecessors. Let Δ be a set of data word rewriting rules over the alphabet $\Sigma \times \mathbb{D}^N$. Then, for every finite data word $\sigma \in (\Sigma \times \mathbb{D}^N)^*$, we define, for any $\sharp \in \{f, p, m\}$:

$$\mathsf{post}_{\Delta,\sharp}(\sigma) = \{\sigma' \in (\Sigma \times \mathbb{D}^N)^*\ :\ \sigma \Rightarrow_{\Delta,\sharp} \sigma'\}$$
$$\mathsf{pre}_{\Delta,\sharp}(\sigma) = \{\sigma' \in (\Sigma \times \mathbb{D}^N)^*\ :\ \sigma' \Rightarrow_{\Delta,\sharp} \sigma\}$$

representing, respectively, the set of immediate successors and predecessors of σ in the rewrite system $\Delta_\sharp$. Then, let $\mathsf{post}^*_{\Delta,\sharp}$ and $\mathsf{pre}^*_{\Delta,\sharp}$ be the reflexive-transitive closure of $\mathsf{post}_{\Delta,\sharp}$ and $\mathsf{pre}_{\Delta,\sharp}$ respectively, i.e., $\mathsf{post}^*_{\Delta,\sharp}(\sigma)$ (resp. $\mathsf{pre}^*_{\Delta,\sharp}(\sigma)$) is the set of all successors (resp. predecessors) of σ in $\Delta_\sharp$. These definitions can be generalized straightforwardly to sets of words.

6.2   Computing post and pre Images

The main result of this section is the following:

Theorem 3. *Let $\Delta_\sharp$ be a rewriting system in $\mathsf{DWRS}_\sharp[\Sigma_n]$, for $\sharp \in \{f, p, m\}$ and $n \geq 2$. Then, for every DWL closed formula φ in the fragment Σ_n, the sets $\mathsf{post}_{\Delta,\sharp}(\llbracket \varphi \rrbracket)$ and $\mathsf{pre}_{\Delta,\sharp}(\llbracket \varphi \rrbracket)$ are effectively definable by DWL formulas in the same fragment Σ_n.*

The rest of the section is devoted to the proof of the theorem above. Let us consider first the problem of computing **post** images in the case of a factor rewriting system.

Let $\exists \mathbf{z}.\ \phi$ be a formula in $\Sigma_{\geq 2}$, and let $\tau = A_0 \ldots A_n \mapsto B_0 \ldots B_m\ :\ \varphi(\mathbf{x}, \mathbf{y})$ be a data rewriting rule, with $\mathbf{x} = \{x_0, \ldots, x_n\}$ and $\mathbf{y} = \{y_0, \ldots, y_m\}$. We suppose w.l.o.g. that the sets of variables $\mathbf{x}, \mathbf{y}$, and $\mathbf{z}$ are disjoint.

By definition of the factor rewriting semantics, the positions associated with the variables $\mathbf{x}$ are consecutive and correspond to a factor $A_0 \ldots A_n$ in the rewritten word. We strengthen the constraint φ of the rule τ in order to make this fact explicit. Then, we define the formula

$$\varphi^{(1)} = \varphi \wedge \Big(\bigwedge_{i \in [0, n-1]} \neg (\exists t.\ x_i < t < x_{i+1}) \Big) \wedge \bigwedge_{i \in [0,n]} A_i[x_i]$$

By definition of data word rewriting systems, all the occurrences of positions variables $\mathbf{y}$ in the constraint φ are used in terms of the form $\delta_k[y]$. Then, we can eliminate all occurrences of all variables in $\mathbf{y}$ by replacing each $\delta_k[y]$ in $\varphi^{(1)}$ by a fresh data variable in a vector $\mathbf{v}$. Let $\xi : \mathbf{y} \times [1, N] \to \mathbf{v}$ be the bijective mapping such that $\delta_k[y]$ is replaced by $\xi(y, k)$. We define:

$$\varphi^{(2)} = \varphi^{(1)}[\delta_k[y] \leftarrow \xi(y, k)]_{y \in \mathbf{y}, k \in [1, N]}$$

Then, the rule τ can be applied only on words satisfying

$$\exists \mathbf{z}.\ \phi \wedge \exists \mathbf{x}.\ \exists \mathbf{v}.\ \varphi^{(2)} \tag{5}$$

This formula could be written in special form: For every vector $\mathbf{t}$ of (fresh) position variables such that $|\mathbf{x}| \leq |\mathbf{t}| \leq |\mathbf{x}| + |\mathbf{z}|$, consider the formula

$$\bigvee_{\theta \in \Theta} \exists \mathbf{t}.\ \exists \mathbf{v}.\ \Big(\bigwedge_{t_i, t_j \in \mathbf{t}, i < j} t_i < t_j \Big) \wedge (\phi \wedge \varphi^{(2)})[\mathbf{x} \leftarrow \theta(\mathbf{x}), \mathbf{z} \leftarrow \theta(\mathbf{z})] \tag{6}$$

where Θ is the set of all total mappings from $\mathbf{x} \cup \mathbf{z}$ to $\mathbf{t}$. Then, the formula (5) is equivalent to the disjunction of all the formulas (6) for all the possible vectors $\mathbf{t}$ defined as above. Let us focus in the sequel on one disjunct of the resulting formula. Then, consider that such a disjunct is the formula:

$$\psi = \exists t_1 \ldots t_{p-1} \exists t_p \ldots t_{p+n} \exists t_{p+n+1} \ldots t_q \exists \mathbf{v}.\ \phi^{(1)}$$

with $\forall i \in [0, n], \theta(x_i) = t_{p+i}, p \geq 1, p + n \leq q$.

Let σ be a model of ψ. By definition of factor rewriting, the rule τ eliminates from σ the factor corresponding to the position associated with the variables $t_p..t_{p+n}$, and insert at position t_p a new word of length m (labeled $B_0 \cdots B_m$). By Lemma 1, we can assume that the distance between the positions corresponding to t_p and t_{p+n+1} in σ is at least $m + 1$. Therefore, there is enough room for inserting new positions in σ corresponding to the right hand the rule. These positions will be associated with $\mathbf{y}$.

The formula $\phi^{(2)}$ below gives the constraints on positions and labels resulting from the insertion of right hand side of the rule τ:

$$\phi^{(2)} = \left(\bigwedge_{i \in [0,m]} B_i(y_i) \right) \wedge t_{p-1} < y_0 \wedge y_m < t_{p+n+1}$$

$$\wedge \left(\bigwedge_{\substack{i,j \in [0,m] \\ i<j}} y_i < y_j \wedge \neg(\exists x.\, y_i < x < y_j) \right) \wedge \left(\bigwedge_{\substack{k \in [1,N] \\ y \in \mathbf{y}}} \delta_k(y) = \xi(y,k) \right)$$

Let $\mathbf{w}$ be a new data variable vector of length $n \cdot N$, and let η be a bijective mapping from $\{t_p, \ldots, t_{p+n}\} \times [1,N]$ to $\mathbf{w}$. (We use the mapping η for substituting occurrences of terms $\delta_k[x]$ in $\phi^{(1)}$ by fresh data variables.)

Then, the formula corresponding to $\mathsf{post}_{\tau,f}(\llbracket \psi \rrbracket)$ is given by:

$$\exists y_1 \ldots y_m \exists \mathbf{w} \exists t_1 \ldots t_{p-1} \exists t_{p+n+1} \ldots t_q \exists \mathbf{v}.\ \phi^{(3)} \wedge \phi^{(2)}$$

where the formula $\phi^{(3)}$ is the result of the application to $\phi^{(1)}$ of a transformation $\ominus$ defined inductively in Table 1:

$$\phi^{(3)} = \phi^{(1)} \ominus (t_p \ldots t_{p+n}, y_0 \ldots y_m, \mathtt{lab}, \eta, \{t_1, \ldots, t_{p-1}\}, \{t_{p+n+1}, \ldots, t_q\})$$

where for all $i \in [0,n]$, $\mathtt{lab}(t_{p+i}) = A_i$.

The first parameter of the operator $\ominus$, called $\mathbf{x}$, is a set of position variables that are deleted. The second parameter, called $\mathbf{y}$, is a set of position variables that are not concerned by the constraint. The third parameter of $\ominus$, the mapping $\mathtt{lab}$, associates with position variables in $\mathbf{x}$ their label in $A_0, \ldots, A_n$. The fourth parameter, η, associates with each position variable $x \in \mathbf{x}$ and each integer $k \in [1,N]$ a variable $\eta(x,k)$ in $\mathbf{v}$. The last parameters, Inf and Sup, are sets of position variables which are ordered, by the context, before resp. after the variables in $\mathbf{x}$. Intuitively $\ominus$ deletes from a formula all occurences of the variables in $\mathbf{x}$ and all constraints concerning them and preserves all constraints concerning the rest of the configuration.

Notice that the obtained formula remains in the same fragment as the original formula since only a prefix of existential quantification is added.

It is easy to adapt the construction above in order to deal with prefix rewriting or multiset rewriting semantics. Indeed, prefix rewriting is particular case where the rewriting position is always the position 0. For multiset rewriting, the construction is simplified since ordering constraints are not used (see [16]).

Finally, let us mention that it is possible to define a symmetrical (and very similar) construction for $\mathsf{pre}_{\tau,\varphi}$ images.

6.3 Application in Verification

Invariance checking consists in deciding whether a given property (1) is satisfied by the set of initial configurations, and (2) is stable under the transition relation of a system. Formally, given a rewriting system Δ and a closed formula φ_{init} defining the set of initial configurations, we say that a closed formula φ

Table 1. The operation $\ominus$

$$(0 < z) \ominus (\mathbf{x}, \mathbf{y}, \mathtt{lab}, \eta, Inf, Sup) = \begin{cases} 0 < z & \text{if } z \in Inf \\ \text{false} & \text{if } z \in Sup \text{ or } z = x_i \in \mathbf{x} \text{ with } i > 0 \\ 0 < y_0 & \text{if } z \text{ is } x_0 \end{cases}$$

$$(z < z') \ominus (\mathbf{x}, \mathbf{y}, \mathtt{lab}, \eta, Inf, Sup) = \begin{cases} z < z' & \text{if } z, z' \in Inf \text{ or } z, z' \in Sup \\ \text{true} & \text{if } z \in Inf \text{ and } (z' \in Sup \text{ or } z' \in \mathbf{x}) \\ & \text{or } z \in \mathbf{x} \; z' \in Sup \\ & \text{or } z, z' = x_i, x_j \in \mathbf{x} \text{ with } i < j \\ \text{false} & \text{otherwise} \end{cases}$$

$$A[z] \ominus (\mathbf{x}, \mathbf{y}, \mathtt{lab}, \eta, Inf, Sup) = \begin{cases} \text{true} & \text{if } z \in \mathbf{x} \text{ and } \mathtt{lab}(z) = A \\ \text{false} & \text{if } z \in \mathbf{x} \text{ and } \mathtt{lab}(z) \neq A \\ A[z] & \text{otherwise} \end{cases}$$

$$r(\ldots, t_i, \ldots) \ominus (\mathbf{x}, \mathbf{y}, \mathtt{lab}, \eta, Inf, Sup) = r(\ldots, t_i[\delta_k(x) \leftarrow \eta(x, k)]_{x \in \mathbf{x}}, \ldots)$$

$$(\neg \varphi) \ominus (\mathbf{x}, \mathbf{y}, \mathtt{lab}, \eta, Inf, Sup) = \neg(\varphi \ominus (\mathbf{x}, \mathbf{y}, \mathtt{lab}, \eta, Inf, Sup))$$

$$(\varphi_1 \vee \varphi_2) \ominus (\mathbf{x}, \mathbf{y}, \mathtt{lab}, \eta, Inf, Sup) = \varphi_1 \ominus (\mathbf{x}, \mathbf{y}, \mathtt{lab}, \eta, Inf, Sup) \vee \\ \varphi_2 \ominus (\mathbf{x}, \mathbf{y}, \mathtt{lab}, \eta, Inf, Sup)$$

$$(\exists u. \; \varphi) \ominus (\mathbf{x}, \mathbf{y}, \mathtt{lab}, \eta, Inf, Sup) = \exists u. \; (\varphi \ominus (\mathbf{x}, \mathbf{y}, \mathtt{lab}, \eta, Inf, Sup))$$

$$(\exists z. \; \varphi) \ominus (\mathbf{x}, \mathbf{y}, \mathtt{lab}, \eta, Inf, Sup) = \exists z. \bigwedge_{y \in \mathbf{y}} (z \neq y) \wedge (\varphi \ominus (\mathbf{x}, \mathbf{y}, \mathtt{lab}, \eta, Inf \cup \{z\}, Sup)) \vee \\ \exists z. \bigwedge_{y \in \mathbf{y}} (z \neq y) \wedge (\varphi \ominus (\mathbf{x}, \mathbf{y}, \mathtt{lab}, \eta, Inf, Sup \cup \{z\})) \vee \\ \bigvee_{x \in \mathbf{x}} \varphi[z \leftarrow x]) \ominus (\mathbf{x}, \mathbf{y}, \mathtt{lab}, \eta, Inf, Sup)$$

is an *inductive invariant* of (Δ, φ_{in}) if and only if (1) $[\![\varphi_{init}]\!] \subseteq [\![\varphi]\!]$, and (2) $\mathsf{post}_\Delta([\![\varphi]\!]) \subseteq [\![\varphi]\!]$. Clearly, (1) is equivalent to $[\![\varphi_{init}]\!] \cap [\![\neg\varphi]\!] = \emptyset$, and (2) is equivalent to $\mathsf{post}_\Delta([\![\varphi]\!]) \cap [\![\neg\varphi]\!] = \emptyset$. (Notice that this fact is also equivalent to $[\![\varphi]\!] \cap \mathsf{pre}_\Delta([\![\neg\varphi]\!]) = \emptyset$.)

Corollary 1. *The problem whether a formula $\varphi \in B(\Sigma_1)$ is an inductive invariant of (Δ, φ_{init}), where $\Delta \in \mathsf{DWRS}[\Sigma_2]$ and $\varphi_{init} \in \Sigma_2$, is decidable.*

For example, consider the system $\Delta_{bakery} \in \mathsf{DWRS}_f[\Pi_1]$ introduced in section 5.2. To prove that mutual exclusion is ensured, we check that the formula $\varphi_{mutex} = \forall x, y. \; x \neq y \Rightarrow \neg(\mathtt{cs}[x] \wedge \mathtt{cs}[y])$ (i.e., it is impossible to have two different processes in the critical section simultaneously) is implied by an inductive invariant φ_{inv} of $(\Delta_{bakery}, \varphi_{init})$ where $\varphi_{init} = \forall x. \; \mathtt{nocs}[x]$ (i.e., all processes are idle). For that, we consider the formula

$$\varphi_{inv} = \forall x. \; \mathtt{cs}[x] \Rightarrow \\ \delta[x] \neq 0 \wedge \forall y. \; x \neq y \Rightarrow (\delta[y] = 0 \vee \delta[x] < \delta[y] \vee \delta[x] = \delta[y] \wedge x < y)$$

Notice that all formulas φ_{mutex}, φ_{init}, and φ_{inv} are in the fragment Π_1. Then, the validity of $\varphi_{inv} \Rightarrow \varphi_{mutex}$ can be checked automatically by Theorem 2 since it is a $B(\Sigma_1)$ formula, and the inductive invariance of φ_{inv} for $(\Delta_{bakery}, \varphi_{init})$ can be decided by Corollary 1.

Hoare-style reasoning consists in, given two properties expressed by formulas φ_1 and φ_2, and a given set of rules Δ, deciding whether starting from configurations satisfying φ_1, the property φ_2 necessarily hold after the application of the rules in Δ. Formally, this consists in checking that $\mathsf{post}_\Delta(\llbracket\varphi_1\rrbracket) \subseteq \llbracket\varphi_2\rrbracket$. In that case, we say that $(\varphi_1, \Delta, \varphi_2)$ constitutes a Hoare triple.

Corollary 2. *The problem whether $(\varphi_1, \Delta, \varphi_2)$ is a Hoare triple, where $\varphi_1 \in \Sigma_2$, $\Delta \in \mathsf{DWRS}[\Sigma_2]$ and $\varphi_2 \in \Pi_2$, is decidable.*

7   Reachability Analysis for Integer Context-Free Systems

In this section, we show that for restricted word rewriting systems (called $\mathsf{CFS_{DL}}$), the reachability problem of sets described by data-independent formulas is decidable. We consider a class of context-free prefix rewriting rules with integer data and constraints in the difference logic. To show decidability, we use a slight generalization of $\mathbb{Z}$-input 1-counter machines introduced in [17] to represent set of finite data words (subsets of $(\Sigma \times \mathbb{Z})^*$). Then, we show that given $\mathsf{CFS_{DL}}$ Δ, and given a set of data words described by a $\mathbb{Z}$-input 1-counter machine M, it is possible to compute a machine M' representing the set of all reachable words (by the iterative application of rules in Δ). This allows then to prove decidability of the reachability problem for $\mathsf{CFS_{DL}}$.

In the sequel, we consider the logic DWL based on *difference logic* (DL) given as $\mathsf{DWL}(\mathbb{Z}, \{0\}, \{\leq_k : k \in \mathbb{Z}\})$ where for every $u, v, k \in \mathbb{Z}$, $(u, v) \in \leq_k$ iff $u - v \leq k$. Then, *context-free systems with difference constraints* ($\mathsf{CFS_{DL}}$) are sets Δ of data word rewriting rules with one symbol on the left-hand side and zero, one or two symbols on the right-hand side. The formulas φ appearing in the rules are from $\mathsf{DWL}(\mathbb{Z}, \{0\}, \leq_k : k \in \mathbb{Z})$.

A $\mathbb{Z}$-input 1-counter machine[1] M is described by a finite set of states Q, an initial state $q_0 \in Q$, a final state $q_f \in Q$, a non-accepting state *fail* $\in Q$, and a counter c that contains initially 0. The initial configuration is given by the tuple $(q_0, 0)$. It reads pieces of input of the form $S(i)$ where S is a symbol out of Σ and $i \in \mathbb{Z}$ is an integer number. The instructions have the following form (q is different from q_f and *fail*):

1. $(q : c := c + 1; \mathsf{goto}\ q')$
2. $(q : c := c - 1; \mathsf{goto}\ q')$
3. $(q : \mathsf{If}\ c \geq 0\ \mathsf{then\ goto}\ q'\ \mathsf{else\ goto}\ q'')$.
4. $(q : \mathsf{If}\ c = 0\ \mathsf{then\ goto}\ q'\ \mathsf{else\ goto}\ q'')$.
5. $(q : \mathsf{Read\ input}\ S(i).\ \mathsf{If}\ S = X\ \mathsf{and}\ i = K\ \mathsf{then\ goto}\ q'\ \mathsf{else\ goto}\ q'')$.
6. $(q : \mathsf{Read\ input}\ S(i).\ \mathsf{If}\ S = X\ \mathsf{and}\ i \# c + K\ \mathsf{then\ goto}\ q'\ \mathsf{else\ goto}\ q'')$,
7. $(q : \mathsf{If}\ P(c)\ \mathsf{then\ goto}\ q'\ \mathsf{else\ goto}\ q'')$, where P is a unary Presburger predicate.

where $\# \in \{\leq, \geq, =\}$, $X \in \Sigma$ and $K \in \mathbb{Z}$ is an integer constant.

---

[1] This definition generalizes the one in [17] by allowing difference constraints in the read instructions.

The language $L(M) \subseteq (\Sigma \times \mathbb{Z})^*$ is defined in a straightforward manner. It is easy to see that for a data independent formula φ one can construct a machine M_φ whose language is $[\![\varphi]\!]$.

For any M we have the following theorem.

Theorem 4. *Let Δ be a $\mathsf{CFS_{DL}}$ and M a $\mathbb{Z}$-input 1-counter machine. Then a $\mathbb{Z}$-input 1-counter machine M' with $L(M') = \mathsf{post}^*_{\Delta,p}(L(M))$ can be effectively constructed.*

The proof is done in several steps and follows the line of the proof given in in [17] for less general classes of rewriting systems and of counter machines M.

- The set $\{d \mid X(d) \Rightarrow^*_p \epsilon\}$ can be characterized by a Presburger formula with one free variable (difference + modulo constraints). This is done by using a translation to alternating one-counter automata.
- The decreasing rules (with ϵ on the right-hand side) of Δ can be eliminated from Δ. To do this, modulo constraints have to be added to the difference logic. Modulo constraints can be eliminated by coding the information in the control states.
- The set $post^*_{\Delta,p}(L(M))$ is then computed by (1) putting M into a special form and (2) applying saturation rules adding a finite number of new transitions to it.

Now we can state the main result of this section.

Theorem 5. *The problem $\mathsf{post}^*_{\Delta,p}([\![\varphi_1]\!]) \cap [\![\varphi_2]\!] = \emptyset$ is decidable for a $\mathsf{CFS_{DL}}$ Δ and two data independent formulas φ_1 and φ_2.*

We give a sketch of the proof. We (1) construct a machine M for $[\![\varphi_1]\!]$, (2) obtain the machine M' for $\mathsf{post}^*_{\Delta,p}([\![\varphi_1]\!])$ using theorem 4, (3) observe that intersection with $[\![\varphi_2]\!]$ can be done by computing the regular set over Σ corresponding to $[\![\varphi_2]\!]$ and restricting M' to words in this set and (4) observe that emptiness of a $\mathbb{Z}$-input 1-counter machine is decidable since emptiness of 1-counter machines is decidable.

$\mathsf{CFS_{DL}}$ allow to model recursive programs with one integer parameter. However, having only one symbol in the left-hand side of rules does not allow to model return values. Let us therefore consider extensions of $\mathsf{CFS_{DL}}$ with more than one symbol in the left-hand side of rules. If we allow rewrite rules with two symbols in the left-hand side where only the data attached to the first appears in constraints, the reachability problem is already undecidable[2]. This model corresponds to having integer return values. On the other hand, we can model return values from a finite domain by adding to the rules of $\mathsf{CFS_{DL}}$ a symbol to the beginning of the left and the right hand sides, provided the constraints do not use the data attached to these symbols. For this extension the reachability problem can be shown to be still decidable.

---

[2] A two-counter machine can be simulated : one counter is coded as the data value, the other one is coded by the number of symbols.

8   Conclusion

We have presented a generic framework for reasoning about infinite-state systems with unbounded control structures manipulating data over infinite domains. This framework extend and unify several of our previous works [11,17,16].

The framework we propose is based on constrained rewriting systems on words over infinite alphabets. The constraints are expressed in a logic which is parametrized by a theory on the considered data domain. We provide generic results for the decidability of the satisfiability problem of the fragment Σ_2 of this logic, and for proving inductive invariance and for carrying out Hoare style reasoning within this fragment.

We have shown that our framework can be used for handling a wide class of systems: recursive sequential programs, multithreaded programs, distributed algorithms, etc. Actually, it is not difficult to consider other rewriting semantics than those considered in the paper. For instance, all our results extend quite straightforwardly to cyclic rewriting allowing to deal with fifo queues. Therefore, our framework can also be used to reason about communicating systems through fifo channels which may contain data over infinite domains. This is particularly useful for handling in a parametric way communication protocols where message have sequence numbers (such as the sliding window protocol). Another potential application of our framework concern programs manipulating dynamic linked lists.

Ongoing and future work include the extension of our framework to rewriting systems on more general structures like trees and some classes of graphs.

References

1. Abdulla, P., Nylen, A.: Timed Petri Nets and BQOs. In: Colom, J.-M., Koutny, M. (eds.) ICATPN 2001. LNCS, vol. 2075. Springer, Heidelberg (2001)
2. Abdulla, P.A., Cerans, K., Jonsson, B., Tsay, Y.-K.: General decidability theorems for infinite-state systems. In: Proc. of LICS'96, pp. 313–321 (1996)
3. Abdulla, P.A., Delzanno, G.: On the Coverability Problem for Constrained Multiset Rewriting. In: Proc. of AVIS'06, Satellite workshop of ETAPS'06, Vienna, Austria (2006)
4. Abdulla, P.A., Jonsson, B., Nilsson, M., Saksena, M.: A Survey of Regular Model Checking. In: Gardner, P., Yoshida, N. (eds.) CONCUR 2004. LNCS, vol. 3170. Springer, Heidelberg (2004)
5. Abdulla, P.A., Jonsson, B.: Verifying networks of timed processes (extended abstract). In: Steffen, B. (ed.) ETAPS 1998 and TACAS 1998. LNCS, vol. 1384. Springer, Heidelberg (1998)
6. Annichini, A., Asarin, E., Bouajjani, A.: Symbolic techniques for parametric reasoning about counter and clock systems. In: Emerson, E.A., Sistla, A.P. (eds.) CAV 2000. LNCS, vol. 1855. Springer, Heidelberg (2000)
7. Arons, T., Pnueli, A., Ruah, S., Xu, J., Zuck, L.D.: Parameterized Verification with Automatically Computed Inductive Assertions. In: Berry, G., Comon, H., Finkel, A. (eds.) CAV 2001. LNCS, vol. 2102. Springer, Heidelberg (2001)

8. Boigelot, B.: Symbolic Methods for Exploring Infinite State Space. PhD thesis, Faculté des Sciences, Université de Liège, vol. 189 (1999)
9. Bojanczyk, M., David, C., Muscholl, A., Schwentick, Th., Segoufin, L.: Two-variable logic on data trees and XML reasoning. In: Proc. of PODS'06. ACM Press, New York (2006)
10. Bojanczyk, M., Muscholl, A., Schwentick, Th., Segoufin, L., David, C.: Two-variable logic on words with data. In: Proc. of LICS'06. IEEE, New York (2006)
11. Bouajjani, A.: Languages, Rewriting systems, and Verification of Infinte-State Systems. In: Orejas, F., Spirakis, P.G., van Leeuwen, J. (eds.) ICALP 2001. LNCS, vol. 2076. Springer, Heidelberg (2001)
12. Bouajjani, A., Bozga, M., Habermehl, P., Iosif, R., Moro, P., Vojnar, T.: Programs with Lists Are Counter Automata. In: Ball, T., Jones, R.B. (eds.) CAV 2006. LNCS, vol. 4144. Springer, Heidelberg (2006)
13. Bouajjani, A., Esparza, J.: Rewriting Models for Boolean Programs. In: Pfenning, F. (ed.) RTA 2006. LNCS, vol. 4098. Springer, Heidelberg (2006)
14. Bouajjani, A., Esparza, J., Maler, O.: Reachability analysis of pushdown automata: Application to model-checking. In: Mazurkiewicz, A., Winkowski, J. (eds.) CONCUR 1997. LNCS, vol. 1243. Springer, Heidelberg (1997)
15. Bouajjani, A., Habermehl, P., Rogalewicz, A., Vojnar, T.: Abstract Tree Regular Model Checking of Complex Dynamic Data Structures. In: Graf, S., Zhang, W. (eds.) ATVA 2006. LNCS, vol. 4218. Springer, Heidelberg (2006)
16. Bouajjani, A., Jurski, Y., Sighireanu, M.: A generic framework for reasoning about dynamic networks of infinite-state processes. In: TACAS'07. LNCS (2007)
17. Bouajjani, A., Habermehl, P., Mayr, R.: Automatic Verification of Recursive Procedures with one Integer Parameter. Theoretical Computer Science 295 (2003)
18. Bouajjani, A., Jonsson, B., Nilsson, M., Touili, T.: Regular Model Checking. In: Emerson, E.A., Sistla, A.P. (eds.) CAV 2000. LNCS, vol. 1855. Springer, Heidelberg (2000)
19. Bozzano, M., Delzanno, G.: Beyond Parameterized Verification. In: Katoen, J.-P., Stevens, P. (eds.) ETAPS 2002 and TACAS 2002. LNCS, vol. 2280. Springer, Heidelberg (2002)
20. Bradley, A.R., Manna, Z., Sipma, H.B.: What's decidable about arrays? In: Emerson, E.A., Namjoshi, K.S. (eds.) VMCAI 2006. LNCS, vol. 3855. Springer, Heidelberg (2005)
21. Delzanno, G.: An assertional language for the verification of systems parametric in several dimensions. Electr. Notes Theor. Comput. Sci. 50(4) (2001)
22. Delzanno, G., Raskin, J.-F., Van Begin, L.: Towards the automated verification of multithreaded java programs. In: Katoen, J.-P., Stevens, P. (eds.) ETAPS 2002 and TACAS 2002. LNCS, vol. 2280, pp. 173–187. Springer, Heidelberg (2002)
23. Delzanno, G.: Constraint-based Verification of Parameterized Cache Coherence Protocols. Formal Methods in System Design 23(3) (2003)
24. Demri, S., Lazic, R.: LTL with the freeze quantifier and register automata. In: Proc. of LICS'06. IEEE, New York (2006)
25. Finkel, A., Leroux, J.: How to compose presburger-accelerations: Applications to broadcast protocols. In: Agrawal, M., Seth, A.K. (eds.) FST TCS 2002: Foundations of Software Technology and Theoretical Computer Science. LNCS, vol. 2556. Springer, Heidelberg (2002)

26. Finkel, A., Schnoebelen, Ph.: Well-structured transition systems everywhere! Theor. Comput. Sci. 256(1-2), 63–92 (2001)
27. Habermehl, P., Iosif, R., Vojnar, T.: Automata-Based Verification of Programs with Tree Updates. In: Hermanns, H., Palsberg, J. (eds.) TACAS 2006 and ETAPS 2006. LNCS, vol. 3920. Springer, Heidelberg (2006)
28. Kesten, Y., Maler, O., Marcus, M., Pnueli, A., Shahar, E.: Symbolic Model Checking with Rich Assertional Languages. In: Grumberg, O. (ed.) CAV 1997. LNCS, vol. 1254. Springer, Heidelberg (1997)
29. Wolper, P., Boigelot, B.: Verifying systems with infinite but regular state spaces. In: Vardi, M.Y. (ed.) CAV 1998. LNCS, vol. 1427. Springer, Heidelberg (1998)

Spiking Neural P Systems: Some Characterizations*

Oscar H. Ibarra and Sara Woodworth

Department of Computer Science
University of California, Santa Barbara, CA 93106, USA
{ibarra,swood}@cs.ucsb.edu

Abstract. We look at the recently introduced neural-like systems, called SN P systems. These systems incorporate the ideas of spiking neurons into membrane computing. We study various classes and characterize their computing power and complexity. In particular, we analyze asynchronous and sequential SN P systems and present some conditions under which they become (non-)universal. The non-universal variants are characterized by monotonic counter machines and partially blind counter machines and, hence, have many decidable properties. We also investigate the language-generating capability of SN P systems.

Keywords: Spiking neural P system, asynchronous mode, sequential mode, partially blind counter machine, semilinear set, language generator.

1 Introduction

The area of membrane computing [15,17] is a recent field of research looking to biological cells as a motivation for computation. Many membrane computing models have been defined with a plethora of interesting results. Recently a new model called spiking neural P system (SN P system) was introduced in [11] and this model has already been investigated in a series of papers (see the P Systems Web Page [17]). SN P systems were inspired by the natural processes of spiking neurons in our brain. These neurons process information that our brain receives from the environment [5,12,13].

An SN P system consists of a set of neurons (membranes) connected by synapses. The structure is represented as a directed graph where the directed edges represent the synapses and the nodes represent the neurons. The system has only a single unit of information referred to as the spike and represented by symbol a. The spikes are stored in the neurons. When a neuron fires, it sends a spike along each outgoing synapse which is then received and processed by the neighboring neurons. A neuron can also send a spike to the environment (thus, the environment can be considered as a neighbor). Such a neuron will be called an output neuron. A neuron fires by means of firing rules that are associated with each neuron. Firing rules are of the form $E/a^j \rightarrow a; d$ where E

---

* This research was supported in part by NSF Grants CCF-0430945 and CCF-0524136.

E. Csuhaj-Varjú and Z. Ésik (Eds.): FCT 2007, LNCS 4639, pp. 23–37, 2007.

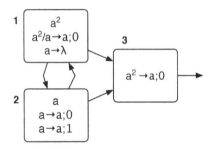

Fig. 1. SN P system generating the set $Q = \{n|n \geq 1\}$

is a regular expression over the symbol a, $j \geq 1$ is the number of spikes consumed by processing the rule, and $d \geq 0$ is the delay between the time the rule
is applied and the time the neuron emits its spike. A rule is applicable when the
neuron contains a^n spikes, and $a^n \in L(E)$ = language denoted by E. If a rule is
used and the neuron fires with a delay, the neuron is said to be 'closed' during
the delay (otherwise the neuron is 'open'). A closed neuron is unable to receive
spikes (all spikes received by the neuron during this time are lost) and unable
to fire any rule. For notational convenience, firing rules of the form $a^j/a^j \rightarrow a; d$
are written as $a^j \rightarrow a; d$. A second rule type of the form $a^j \rightarrow \lambda$ (referred to
as forgetting rules) are also allowed in the neurons and are used to remove a^j
spikes from the neuron without emitting a spike. These rules are applicable if
the neuron currently contains exactly a^j spikes. Forgetting rules are restricted
such that $a^j \notin L(E)$ for any E in the same neuron. Hence forgetting rules are
disjoint from firing rules in each neuron. (In this paper, for convenience we refer
to forgetting rules as firing rules which emit no spike.)

SN P systems operate in a nondeterministic, maximally parallel manner (like
other P system models) using a global clock. However, each neuron is only able
to fire at most one rule per step since the rule must cover all the spikes currently
in the neuron. It is possible that two (or more) rules are applicable in a given
step. In this case, the applied rule is selected nondeterministically. The system
operates in a maximally parallel manner in that at each step, all neurons that
are fireable, must fire (applying some rule).

SN P systems are able to generate output in many different manners. Here, we
designate an output neuron which has a synapse to the environment. A number n
is said to be generated if n spikes (represented by a^n) are sent to the environment
during the computation and the computation eventually halts, i.e., it reaches a
configuration where all neurons are open but no neuron is fireable. Many other
methods of generating output have been defined and analyzed, but these other
methods do not make sense in terms of asynchronous SN P systems.

Figure 1 shows an example of an SN P system. This system generates the set
$Q = \{n|n \geq 1\}$. During the first step, neuron 1 is fireable using rule $a^2/a \rightarrow a; 0$.
Neuron 2 is fireable using either of its rules. If the first rule $(a \rightarrow a; 0)$ is used,
two spikes are sent to neuron 3 and the configuration of neuron 1 and neuron 2
remain the same after the first step. This continues (with neuron 3 emitting a

spike to the environment) in each additional step as long as neuron 2 picks the first of its two fireable rules. Once neuron 2 chooses to fire the second rule (rather than the first), neuron 2 becomes closed for one time step. This causes the spike sent to it to be lost. Hence, in the next step, neuron 1 contains only a single spike which is forgotten and neuron 3 does not spike. In this step, neuron 2 emits its spikes. In the following step, neuron 1 again forgets its spike and neuron 3 can again spike. Now, no spikes exist in the system and all neurons are open, which is a halting configuration. The output of the computation is the number of times neuron 2 chose to fire its first rule plus 1.

Variations to the standard model of SN P system have been studied. In particular, an extended form of spiking rules was introduced and studied in [3,14]. An *extended rule* has the form $E/a^j \rightarrow a^p; d$. This rule operates in the same manner as before except that firing sends p spikes along each outgoing synapse (and these p spikes are received simultaneously by each neighboring neuron). Clearly, when $p = 1$, the extended rules reduce to the standard (or non-extended) rules in the original definition. Note also that forgetting rules are just a special case of firing rules, i.e., when $p = 0$.

For the most part, we will deal with SN P systems with extended rules in this paper. We will consider systems with three types of neurons:

1. A neuron is *bounded* if every rule in the neuron is of the form $a^i/a^j \rightarrow a^p; d$, where $1 \le j \le i$, $p \ge 0$, and $d \ge 0$. There can be several such rules in the neuron. These rules are called *bounded rules*.
2. A neuron is *unbounded* if every rule in the neuron is of the form $a^i(a^k)^*/a^j \rightarrow a^p; d$, where $i \ge 0, k \ge 1, j \ge 1, p \ge 0, d \ge 0$. Again, there can be several such rules in the neuron. These rules are called *unbounded rules*.
3. A neuron is *general* if it can have *general rules*, i.e., bounded as well as unbounded rules.

One can allow rules like $\alpha_1 + ... + \alpha_n \rightarrow a^p; d$ in the neuron, where all α_i's have bounded (resp., unbounded) regular expressions as defined above. But such a rule is equivalent to putting n rules $\alpha_i \rightarrow a^p : d$ $(1 \le i \le n)$ in the neuron. It is known that any regular set over a 1-letter symbol a can be expressed as a finite union of regular sets of the form $\{a^i(a^j)^k \mid k \ge 0\}$ for some $i, j \ge 0$. Note such a set is finite if $j = 0$. We can define three types of SN P systems:

1. *Bounded SN P system* – a system in which every neuron is bounded.
2. *Unbounded SN P system* – a system in which every neuron is either bounded or unbounded.
3. *General SN P system* – a system with general neurons (i.e., each neuron can contain both bounded and unbounded rules).

Let $k \ge 1$. A *k-output SN P system* has k output neurons, $O_1, ..., O_k$. We say that the system generates a k-tuple $(n_1, ..., n_k) \in N^k$ if, starting from the initial configuration, there is a sequence of steps such that each output neuron O_i generates (sends out to the environment) exactly n_i spikes and the system eventually halts.

We will consider systems with delays and systems without delays (i.e., $d = 0$ in all rules) in this paper.

2 Asynchronous General SN P Systems

In the standard (i.e., synchronized) model of an SN P system, all neurons fire at each step of the computation whenever they are fireable. This synchronization is quite powerful: It known that a set $Q \subseteq N^1$ is recursively enumerable if and only if it can be generated by a 1-output general SN P system (with or without delays) [11,8]. This result holds for systems with standard rules or extended rules, and it generalizes to systems with multiple outputs. Thus, such systems are universal.

In [1] the computational power of SN P systems that operate in an asynchronous mode was introduced and studied. In an *asynchronous SN P system*, we do not require the neurons to fire at each step. During each step, any number of fireable neurons are fired (including the possibility of firing no neurons). When a neuron is fireable it may (or may not) choose to fire during the current step. If the neuron chooses not to fire, it may fire in any later step as long as the rule is still applicable. (The neuron may still receive spikes while it is waiting which may cause the neuron to no longer be fireable.) Hence there is no restriction on the time interval for firing a neuron. Once a neuron chooses to fire, the appropriate number of spikes are sent out after a delay of exactly d time steps and are received by the neighboring neurons during the step when they are sent.

Before proceeding further, we recall the definition of a counter machine. A nondeterministic multicounter machine (CM) M is a nondeterministic finite automaton with a finite number of counters (it has no input tape). Each counter can only hold a nonnegative integer. The machine starts in a fixed initial state with all counters zero. During computation, each counter can be incremented by 1, decremented by 1, or tested for zero. A distinguished set of k counters (for some $k \geq 1$) is designated as the output counters. The output counters are non-decreasing (i.e., cannot be decremented). A k-tuple $(n_1, ..., n_k) \in N^k$ is generated if M eventually halts in an accepting state, all non-output counters zero, and the contents of the output counters are $n_1, ..., n_k$, respectively. We will refer to a CM with k output counters (the other counters are auxiliary counters) as a k-output CM.

It is well-known that a set $Q \subseteq N^k$ is generated by a k-output CM if and only if Q is recursively enumerable. Hence, k-output CMs are universal.

The following result was recently shown in [1]. It says that SN P systems which operate in an asynchronous mode of computation are still universal provided the neurons are allowed to use extended rules.

Theorem 1. *A set $Q \subseteq N^k$ is recursively enumerable if and only if it can be generated by an asynchronous k-output general SN P system with extended rules. The result holds for systems with or without delays.*

It remains an open question whether the above result holds for the case when the system uses only standard (i.e., non-extended) rules.

3 Asynchnronous Unbounded SN P Systems with Extended Rules

In this section, we will examine unbounded SN P systems (again assuming the use of extended rules). Recall that these systems can only use bounded and unbounded neurons (i.e., no general neurons are allowed). In contrast to Theorem 1, these systems can be characterized by partially blind multicounter machines (PBCMs).

A *partially blind k-output multicounter machine* (k-output PBCM) [7] is a k-output CM, where the counters cannot be tested for zero. (The output counters are non-decreasing.) The counters can be incremented by 1 or decremented by 1, but if there is an attempt to decrement a zero counter, the computation aborts (i.e., the computation becomes invalid). Again, by definition, a successful generation of a k-tuple requires that the machine enters an accepting state with all non-output counters zero.

It is known that k-output PBCMs can be simulated by k-dimensional vector addition systems, and vice-versa [7]. (Hence, such counter machines are not universal.) In particular, a k-output PBCM can generate the reachability set of a vector addition system.

3.1 Systems Without Delays

In [1], asynchronous unbounded SN P systems without delays were investigated. The systems considered in [1] are restricted to halt in a pre-defined configuration. Specifically, a computation is valid if, at the time of halting, the numbers of spikes that remain in the neurons are equal to pre-defined values; if the system halts but the neurons do not have the pre-defined values, the computation is considered invalid and the output is ignored. These systems were shown to be equivalent to PBCMs in [1]. However, it was left as an open question whether the 'pre-defined halting' requirement was necessary to prove this result. Here we show that the result still holds even if we do not have this condition. Note that for these systems, firing zero or more neurons at each step is equivalent to firing one or more neurons at each step (otherwise, since there are no delays, the configuration stays the same when no neuron is fired).

Theorem 2. *A set $Q \subseteq N^k$ is generated by a k-output PBCM if and only if it can be generated by an asynchronous k-output unbounded SN P system without delays. Hence, such SN P systems are not universal.*

Note that by Theorem 1, if we allow both bounded rules and unbounded rules to be present in the neurons, SN P systems become universal.

Again, it remains an open question whether the above theorem holds for the case when the system uses only standard rules.

It is known that PBCMs with only one output counter can only generate semilinear sets of numbers. Hence:

Corollary 1. *Asynchronous 1-output unbounded SN P systems without delays can only generate semilinear sets of numbers.*

The results in the following corollary can be obtained using Theorem 2 and the fact that they hold for k-output PBCMs.

Corollary 2. *1. The family of k-tuples generated by asynchronous k-output unbounded SN P systems without delays is closed under union and intersection, but not under complementation.*
2. The membership, emptiness, infiniteness, disjointness, and reachability problems are decidable for asynchronous k-output unbounded SN P systems without delays; but containment and equivalence are undecidable.

3.2   Systems with Delays

In Theorem 2, we showed that restricting an asynchronous SN P system without delays to contain only bounded and unbounded neurons gives us a model equivalent to a PBCM. However, it is possible that allowing delays would give additional power. For asynchronous unbounded SN P systems with delays, we can no longer assume that firing zero or more neurons at each step is equivalent to firing one or more neurons at each step.

Note that not every step in a computation has at least one neuron with a fireable rule. In a given configuration, if no neuron is fireable but at least one neuron is closed, we say that the system is in a *dormant* step. If there is at least one fireable neuron in a given configuration, we say the system is in a *non-dormant* step. (Of course, if a given configuration has no fireable neuron, and all neurons are open, we are in a halting configuration.) Thus, an SN P system with delays might be dormant at some point in the computation until a rule becomes fireable. However, the clock will keep on ticking. Interestingly, the addition of delays does not increase the power of the system.

Theorem 3. *A set $Q \subseteq N^k$ is generated by a k-output PBCM if and only if it can be generated by an asynchronous k-output unbounded SN P system with delays.*

This result contrasts the result in [8] which shows that synchronous unbounded SN P systems with delays and standard rules (but also standard output) are universal.

1-Asynchronous Unbounded SN P Systems. Define a 1-*asynchronous unbounded SN P system* with delays as an asynchronous unbounded SN P system with delays where we require that *in every non-dormant step at least one rule is applied in the system*. Thus, 1-asynchronous unbounded SN P systems differ from the previously defined asynchronous unbounded SN P system with delays because here *idle steps (where no neuron fires) only occur when the system is dormant*.

For a PBCM to be able to simulate a 1-asynchronous unbounded SN P system, it must be able to distinguish a non-dormant step from a dormant step. Thus when the PBCM nondeterministically guesses to apply no rules in a given step, it must guarantee that it is a dormant step. The method of simulating unbounded neurons with a PBCM in Theorem 2 does not have this ability. In fact, no simulation of a 1-asynchronous unbounded SN P system (with delays) by a PBCM is possible since it can be shown that this system is universal.

Theorem 4. *1-asynchronous unbounded k-output SN P systems (with delays) are universal.*

Note that if a 1-asynchronous unbounded SN P system has no delay, the system is equivalent to the asynchronous unbounded SN P system studied in Theorem 2.

Strongly Asynchronous Unbounded SN P Systems. Define a *strongly asynchronous unbounded SN P system* as an asynchronous unbounded SN P system which has the property that in a valid computation, *every step of the computation has at least one fireable neuron*, unless the SN P system is in a halting configuration. Otherwise (i.e., there is a step in which there is no fireable rule), the computation is viewed as invalid and no output from such a computation is included in the generated set.

Since we are guaranteed to have at least one fireable neuron at each step (meaning all steps are non-dormant), it is natural to also require at least one neuron fires at each step. We call this model a *strongly 1-asynchronous unbounded SN P system*. Interestingly, with these restrictions we again find equivalence with PBCMs.

Theorem 5. *Strongly 1-asynchronous unbounded k-output SN P systems with delays and k-output PBCMs are equivalent.*

Again note a strongly 1-asynchronous unbounded SN P systems without delays reverts to the asynchronous unbounded SN P system model studied in Theorem 2.

4   Asynchronous Bounded SN P Systems

We consider in this section, asynchronous SN P systems, where the neurons can only use bounded rules. We show that these bounded SN P systems with extended rules generate precisely the semilinear sets.

A *k-output monotonic* CM is a nondeterministic machine with k counters, all of which are output counters. The counters are initially zero and can only be incremented by 1 or 0 (they cannot be decremented). When the machine halts in an accepting state, the k-tuple of values in the k-counter is said to be generated by the machine. Clearly, a k-output monotonic CM is a special case of a PBCM, where all the counters are output counters and all the instructions are addition instructions.

It is known that a set $Q \subseteq N^k$ is semilinear if and only if it can be generated by a k-output monotonic CM [6].

We can show the following:

Theorem 6. $Q \subseteq N^k$ *can be generated by a k-output monotonic CM if and only if it can be generated by a k-output asynchronous bounded SN P system with extended rules. The result holds for systems with or without delays.*

Proof. To show that a k-output asynchronous bounded SN P system with extended rules Π can be simulated by a k-output monotonic CM $\mathcal{M}$ is straightforward. All of the rules and the configuration of the system can be simulated by the finite control. (This is because each neuron has a bounded number of 'useful' spikes.) The counters in $\mathcal{M}$ are used to store the number of spikes sent to the environment by the output neurons. For each emitted spike by output neuron O_i, the corresponding counter c_i is incremented by one.

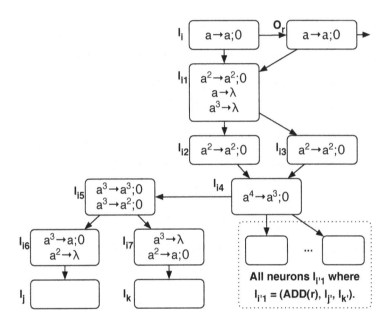

Fig. 2. Asynchronous bounded SN P module simulating an instruction of the form $l_i = (\text{ADD}(r), l_j, l_k)$

To show that a k-output monotonic CM $\mathcal{M}$ can be simulated by a k-output asynchronous bounded SN P system with extended rules Π, we give a construction from $\mathcal{M}$ to Π. This construction works by creating an asynchronous bounded SN P system module to simulate each instruction in $\mathcal{M}$. We assume without loss of generality that $\mathcal{M}$ contains only a single HALT instruction.

Simulating an instruction of the form $l_i = (\text{ADD}(r), l_j, l_k)$ is done by creating the module shown in Figure 2. This module is initiated when a single spike is sent to neuron l_i. Now neuron l_i will eventually fire sending a spike to neuron r and neuron l_{i1}. Neuron O_r will then eventually fire sending a spike to the environment along with a spike to neuron l_{i1} and each neuron $l_{i'1}$ where $l_{i'} = (\text{ADD}(r), l_{j'}, l_{k'})$. Now, if l_{i1} has previously forgotten its spike from neuron l_i, the computation will not continue. In this case the computation will not halt do to the infinite looping of the HALT module (described below). If, l_{i1} has not forgotten its previous spike, it is now fireable. (Also each neuron $l_{i'1}$ is able to forget its spike sent by neuron O_r.) Once neuron l_{i1} eventually fires, it sends spikes to neurons l_{i2} and l_{i3} which are used to multiply the spikes so that neuron l_{i4} eventually contains a^4 spikes.

When neuron l_{i4} spikes, three spikes are sent to neuron l_{i5} along with sending three spikes to each neuron $l_{i'1}$ where $l_{i'} = (\text{ADD}(r), l_{j'}, l_{k'})$. Neuron l_{i5} is used to initiate the nondeterministic choice of which instruction will be executed next. The three spikes sent to each $l_{i'1}$ guarantee that if the previous spike sent by neuron O_r was not forgotten, additional spikes will not cause these neurons to execute their associated rule. These three spikes can again be forgotten by the neurons, but if they do not do this before receiving additional spikes, the neuron will become unusable (since they will have surpassed their bound).

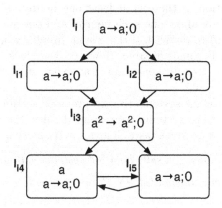

Fig. 3. Asynchronous bounded SN P module simulating an instruction of the form $l_i = (\text{HALT})$

Neuron l_{i5} nondeterministically picks the next rule by either emitting two or three spikes when it is fired. If three spikes are emitted, instruction l_j will be simulated next. If two spikes are emitted, instruction l_k will be simulated next. Again, if neurons l_{i6} and l_{i7} do not properly forget their spikes (when they should forget rather than fire) before more spikes are sent to them (if the rule is executed again), the neuron becomes unusable which could stop the simulation. If the simulation stops, the HALT instruction will never execute and the computation will never halt.

Simulating an instruction of the form $l_i = (\text{HALT})$ is done by creating the module shown in Figure 3. This module is initiated when a single spike is sent to neuron l_i. This neuron eventually fires sending a spike to neuron l_{i1} and neuron l_{i2} which both spike sending a spike to neuron l_{i3}. (This is done to increase the number of spikes.) Neuron l_{i3} will then eventually spike emitting two spikes to neuron l_{i4} and neuron l_{i5} causing them to no longer be fireable and halting the computation. If this instruction is never executed, neuron l_{i4} and neuron l_{i5} will continuously pass a single spike between them causing the system to never halt. Hence, if any of the previous simulation of the ADD instructions does not operate 'correctly', the halt instruction will never be executed and the computation will never halt.   □

At present, we do not know whether Theorem 6 holds when the the system is restricted to use only standard (non-extended) rules. However, we can show the result holds for synchronous bounded SN P systems using only standard rules.

5   Sequential SN P Systems

Sequential SN P systems are another closely related model introduced in [10]. These are systems that operate in a sequential mode. This means that at every step of the computation, if there is at least one neuron with at least one rule that is fireable, we only allow one such neuron and one such rule (both nondeterministically chosen) to be fired. If there is no fireable rule, then the system is dormant until a rule becomes fireable. However, the clock will keep on ticking. The system is called *strongly sequential* if at every step, there is at least one neuron with a fireable rule.

Unlike for asynchronous systems (in the previous section), where the results relied on the fact that the systems use extended rules, the results here hold for systems that use standard rules (as well as for systems that use extended rules).

Theorem 7. *The following results hold for systems with delays.*

1. *Sequential k-output unbounded SN P systems with standard rules are universal.*
2. *Strongly sequential k-output general SN P systems with standard rules are universal.*
3. *Strongly sequential k-output unbounded SN P systems with standard rules and k-output PBCMs are equivalent.*

The above results also hold for systems with extended rules.

Item 3 in the above results improves the result found in [10] which required a special halting configuration similar to the halting configuration in [1]. Here, we find this halting requirement is not necessary.

6   SN P Systems as Language Generators

In this section, we use the SN P system as a language generator as in two recent papers [2,4]. Consider an SN P system Π with output neuron, O, which is bounded. Interpret the output from O as follows. At times when O spikes, a is interpreted to be 1, and at times when it does not spike, interpret the output to be 0. We say a binary string $x = a_1...a_n$, where $n \geq 1$, is generated by Π if starting in its initial configuration, it outputs x and halts. We assume that the SN P systems use standard rules.

6.1   Regular Languages

It was recently shown in [2] that for any finite binary language F, the language $F1$ (i.e., with a supplementary suffix of 1) can be generated by a bounded SN P system. The following shows that, in general, this result does not hold when F is an infinite regular language.

Observation 1. *Let $F = 0^*$. Then $F1$ cannot be generated by a bounded SN P system.*

To generate $F1$, the SN P system must be able to generate the string 1, meaning there must initially be at least one spike in the output membrane (say a^n, for some n) and there must be a rule of the form $E/a^j \rightarrow a; 0$ where $a^n \in E$. Therefore, there cannot be a forgetting rule of the form $a^n \rightarrow \lambda$ in the same neuron. To generate the strings in 0^+1 (i.e., there is at least one 0), the neuron must contain a rule of the form $E/a^j \rightarrow a; d$ where $a^n \in E$ and $d \geq 1$. (This is necessary so the rule of the form $E/a^j \rightarrow a; 0$ in the output neuron is not used.) Let d_0 be the maximum d in all such rules. Clearly the output neuron must spike within $d_0 + 1$ steps from the beginning of the computation. Hence, there is no way to produce the string $0^{d_0+1}1$.

It is interesting to note that by just modifying the previous language 0^* to always begin with at least one zero (so $F = 0^+$) we can generate $F1$.

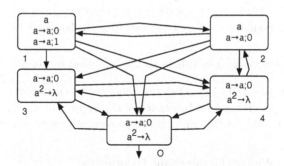

Fig. 4. Bounded SN P system generating the language $F1 = 0^+1$

Observation 2. *Let $F = 0^+$. Then $F1$ can be generated by a bounded SN P system. Thus, it is possible to generate some languages where F is an infinite language.*

We give a bounded SN P system which generates $F1$ (shown in Figure 4). Here the output neuron initially contains no spikes guaranteeing it will not spike during the first step. Both neurons 1 and 2 will fire during the first step. If neuron 1 chooses to fire the rule $a \rightarrow a; 0$ then two spikes are received by neurons 3, 4, and O. These spikes are forgotten during the next step causing the output to be 0. This is repeated until neuron 1 chooses to fire rule $a \rightarrow a; 1$. This will cause neurons 3, 4, and O to receive one spike at this time step. This will cause all three neurons to fire during the next time step (when neuron 1 also fires). This causes the output to be 1 and leaves neurons 2, 3, 4, and O with three spikes. No further neuron is fireable causing the system to halt after producing 0^+1.

In contrast to Observation 1, we have:

Observation 3. *Let $F = 0^*$. Then $1F$ can be generated by a bounded SN P system.*

We give a bounded SN P system which generates $1F$ (shown in Figure 5). Here, the output neuron will spike during the first time step outputting a 1. The additional two neurons are used to determine how many 0s will follow. Neurons 1 and 2 will fire until both choose to nondeterministically use rule $a^2 \rightarrow a; 0$ which will cause both to contain only a single spike in the next time step causing the system to halt. Since neurons 1 and 2 are not connected to neuron O, after the first step, neuron O will never fire causing the remainder of the output to be 0's.

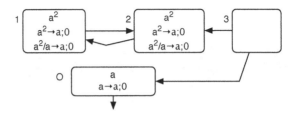

Fig. 5. Bounded SN P system generating the language $1F = 10^*$

Observation 3 actually generalizes to the following rather surprising result:

Theorem 8. *Let $L \subseteq (0+1)^*$. Then the language $1L$ (i.e., with a supplementary prefix 1) can be generated by a bounded SN P system if and only if L is regular. (The result holds also for $0L$, i.e., the supplementary prefix is 0 instead of 1.)*

6.2   Another Way of Generating Languages

Now we define a new way of "generating" a string. Under this new definition, various classes of languages can be obtained. We say a binary string $x = a_1...a_n$, where $n \geq 0$, is generated by Π if it outputs $1x10^d$, for some d which may depend on x, and halts. Thus, in the generation, Π outputs a 1 before generating x, followed by 10^d for some d. (**Note the prefix 1 and the suffix 10^d are not considered as part of the string.**) The set $L(\Pi)$ of binary strings generated by Π is called the language generated by Π. We can show the following:

1. When there is no restriction, Π can generate *any unary recursively enumerable* language. Generalizing, for $k \geq 1$, any recursively enumerable language

$L \subseteq 0^*10^*...10^*$ (k occurrences of 0^*s) can be generated by an unrestricted SN P system.

There are variants of the above result. For example, for $k \geq 1$, let $w_1, ..., w_k$ be fixed (not necessarily distinct) non-null binary strings. Let $L \subseteq w_1^*...w_k^*$ be a bounded language. Then L can be generated by an unrestricted SN P system if and only it is recursively enumerable.

2. There are non-bounded binary languages, e.g., $L = \{xx^r \mid x \in \{0,1\}^+\}$ (the set of even-length palindromes), that cannot be generated by unrestricted SN P systems. However, interestingly, as stated in item 5 below, the complement of L can be generated by a very simple SN P system.

3. Call the SN P system *linear spike-bounded* if the number of spikes in the unbounded neurons at any time during the computation is at most $O(n)$, where n is the length of the string generated. Define *polynomial spike-bounded* SN P system in the same way. Then linear spike-bounded SN P systems and polynomial spike-bounded SN P systems are weaker than unrestricted SN P systems. In fact, $S(n)$ spike-bounded SN P systems are weaker than unrestricted SN P systems, for any recursive bound $S(n)$.

4. Call the SN P system *1-reversal* if every unbounded neuron satisfies the property that once it starts "spiking" it will no longer receive future spikes (but may continue computing). Generalizing, an SN P system is *r-reversal* for some fixed integer $r \geq 1$, if every unbounded neuron has the property that the number of times its spike size changes values from nonincreasing to nondecreasing and vice-versa during any computation is at most r. There are languages generated by linear spike-bounded SN P systems that cannot be generated by r-reversal SN P systems. For example, $L = \{1^{k^2} \mid k \geq 1\}$ can be generated by the former but not by the latter.

5. Let $w_1, ..., w_k$ be fixed (not necessarily distinct) non-null binary strings. For every string x in $w_1^*...w_k^*$, let $\psi(x) = (\#(x)_1, ..., \#(x)_k)$ in N^k, where $\#(x)_i =$ number of occurrences of w_i in x. Thus, $\psi(x)$ is the Parikh map of x with respect to $(w_1, ..., w_k)$. For $L \subseteq w_1^*...w_k^*$, let $\psi(L) = \{\psi(x) \mid x \in L\}$. A 1-reversal (resp. r-reversal) SN P system can generate L if and only if $\psi(L)$ is semilinear.

Similarly, a 1-reversal (resp., r-reversal) SN P system can generate a language $L \subseteq 0^*10^*...10^*$ (k occurrences of 0^*s) if and only if $\{(n_1, ..., n_k) \mid 0^{n_1} 1...10^{n_k} \in L\}$ is semilinear.

While the language $L = \{xx^r \mid x \in \{0,1\}^+\}$ cannot be be generated by an unrestricted SN P system (by item 2), its complement can be generated by a 1-reversal SN P system.

6. 1-reversal (resp., r-reversal) SN P system languages are closed under some language operations (e.g., union) but not under some operations (e.g., complementation). Many standard decision problems (e.g., membership, emptiness, disjointness) for 1-reversal (resp., r-reversal) SN P systems are decidable, but not for some questions (e.g., containment and equivalence).

7   Conclusion

Theorem 2 solved an open question in [1]. We showed that the special halting configuration with regard to the number of spikes within each neuron is not necessary to prove equivalence to PBCMs.

We also looked at conditions under which an asynchronous unbounded SN P system becomes universal. It is interesting to note that *all* of the following are needed to gain universality:

1. Delays in the system are needed.
2. Dormant steps in the system are needed.
3. At least one neuron must fire at every step if the system is not in a dormant state.

Clearly we must have 1 in order to have 2. If we just have 1 and 2, Theorem 3 shows the system is not universal. If we just have 1 and 3, Theorem 5 shows the system is not universal.

An interesting open question is whether these results hold if we restrict the SN P system to use only standard (i.e., non-extended) rules.

For sequential SN P systems, we were able to obtain similar results with the use of only standard rules. If the system is strongly sequential with only bounded and unbounded neurons, the model is only as powerful as PBCMs. However, if we allow either general neurons or dormant steps, the system becomes universal.

We also showed that asynchronous bounded SN P systems with extended rules are equivalent to monotonic counter machines which are known to generate precisely the semilinear sets.

Finally, we presented some results concerning the language generating capability of SN P systems.

References

1. Cavaliere, M., Egecioglu, O., Ibarra, O.H., Ionescu, M., Păun, Gh., Woodworth, S.: Asynchronous spiking neural P systems; Decidability and Undecidability 2006 (submitted)
2. Chen, H., Freund, R., Ionescu, M., Păun, Gh., Pérez-Jiménez, M.J.: On string languages generated by spiking neural P systems. In: Proc. 4th Brainstorming Week on Membrane Computing, pp. 169–194 (2006)
3. Chen, H., Ionescu, M., Ishdorj, T.-O., Păun, A., Păun, Gh., Pérez-Jiménez, M.J.: Spiking neural P systems with extended rules: Universality and languages, Natural Computing (special issue devoted to DNA12 Conf.) (to appear)
4. Chen, H., Ionescu, M., Păun, A., Păun, Gh., Popa, B.: On trace languages generated by (small) spiking neural P systems. In: Pre-proc. 8th Workshop on Descriptional Complexity of Formal Systems (June 2006)
5. Gerstner, W., Kistler, W.: Spiking Neuron Models. Single Neurons, Populations, Plasticity. Cambridge Univ. Press, Cambridge (2002)
6. Harju, T., Ibarra, O., Karhumaki, J., Salomaa, A.: Some decision problems concerning semilinearity and commutation. Journal of Computer and System Sciences 65, 278–294 (2002)

7. Greibach, S.: Remarks on blind and partially blind one-way multicounter machines. Theoretical Computer Science 7(3), 311–324 (1978)
8. Ibarra, O.H., Păun, A., Păun, Gh., Rodríguez-Patón, A., Sosik, P., Woodworth, S.: Normal forms for spiking neural P systems. Theoretical Computer Science (to appear)
9. Ibarra, O.H., Woodworth, S.: Characterizations of some restricted spiking neural P systems. In: Hoogeboom, H.J., Păun, Gh., Rozenberg, G., Salomaa, A. (eds.) WMC 2006. LNCS, vol. 4361, pp. 424–442. Springer, Heidelberg (2006)
10. Ibarra, O.H., Woodworth, S., Yu, F., Păun, A.: On spiking neural P systems and partially blind counter machines. In: Calude, C.S., Dinneen, M.J., Păun, Gh., Rozenberg, G., Stepney, S. (eds.) UC 2006. LNCS, vol. 4135, pp. 113–129. Springer, Heidelberg (2006)
11. Ionescu, M., Păun, Gh., Yokomori, T.: Spiking neural P systems. Fundamenta Informaticae 71(2-3), 279–308 (2006)
12. Maass, W.: Computing with spikes. Special Issue on Foundations of Information Processing of TELEMATIK 8(1), 32–36 (2002)
13. Maass, W., Bishop, C. (eds.): Pulsed Neural Networks. MIT Press, Cambridge (1999)
14. Păun, A., Păun, Gh.: Small universal spiking neural P systems. In: BWMC2006. BioSystems, vol. II, pp. 213–234, 2006 (in press)
15. Păun, Gh.: Membrane Computing – An Introduction. Springer, Heidelberg (2002)
16. Păun, Gh., Pérez-Jiménez, M.J., Rozenberg, G.: Spike trains in spiking neural P systems. Intern. J. Found. Computer Sci. 17(4), 975–1002 (2006)
17. The P Systems Web Page: http://psystems.disco.unimib.it

Approximating Graphs by Graphs and Functions
(Abstract)

László Lovász

Department of Computer Science
Eötvös Loránd University
Pázmány Péter sétány 1/C, H-1117 Budapest, Hungary
lovasz@cs.elte.hu

In many areas of science huge networks (graphs) are central objects of study: the internet, the brain, various social networks, VLSI, statistical physics. To study these graphs, new paradigms are needed: What are meaningful questions to ask? When are two huge graphs "similar"? How to "scale down" these graphs without changing their fundamental structure and algorithmic properties? How to generate random examples with the desired properties? A reasonably complete answer can be given in the case when the huge graphs are dense (in the more difficult case of sparse graphs there are only partial results).

E. Csuhaj-Varjú and Z. Ésik (Eds.): FCT 2007, LNCS 4639, p. 38, 2007.
© Springer-Verlag Berlin Heidelberg 2007

Traces, Feedback, and the Geometry of Computation
(Abstract)

Philip J. Scott

Department of Mathematics & Statistics, University of Ottawa,
Ottawa, Ontario K1N 6N5, Canada
phil@site.uottawa.ca

The notion of feedback and information flow is a fundamental concept arising in many classical areas of computing. In the late 1980s and early 1990s, an algebraic structure dealing with cyclic operations emerged from various fields, including flowchart schemes, dataflow with feedback, iteration theories, action calculi in concurrency theory, proof theory (Linear Logic and Geometry of Interaction), and network algebra, as well as in pure mathematics. This structure, now known as a "traced monoidal category", was formally introduced in an influential paper of Joyal, Street and Verity (1996) from current work in topology and knot theory. However these authors were also keenly aware of its potential applicability. Since then, such structures – with variations – have been pursued in several areas of mathematics, logic and theoretical computer science, including game semantics, quantum programming languages and quantum protocols, and computational biology. We shall survey applications in logic and theoretical computer science and discuss progress towards an abstract geometry of algorithms.

E. Csuhaj-Varjú and Z. Ésik (Eds.): FCT 2007, LNCS 4639, p. 39, 2007.

A Largest Common d-Dimensional Subsequence of Two d-Dimensional Strings

Abdullah N. Arslan

Department of Computer Science
University of Vermont
Burlington, VT 05405, USA
aarslan@cs.uvm.edu

Abstract. We introduce a definition for a *largest common d-dimensional subsequence of two d-dimensional strings* for $d \geq 1$. Our purpose is to generalize the well-known definition of a longest common subsequence of linear strings for dimensions higher than one. We prove that computing a largest common two-dimensional subsequence of two given two-dimensional strings is NP-complete. We present an algorithm for the case of the problem when the definition is weakened.

Keywords: common subsequence, multi-dimensional string, image, NP, dynamic programming.

1 Introduction

A subsequence of a given string S is a string that is obtained from S by selecting a sequence of symbols from S in their order of appearance in S. The longest common subsequence of given two linear strings is a common subsequence of these strings with the biggest length.

The approximate string matching and edit distance techniques have been applied to two dimensional shape recognition [8,11,2]. The longest common subsequence problem for strings in high dimensions can be applied to many problems such as finding common subimages in given two images, finding a given image in video where time could be regarded as the third dimension, image recognition and classification in medical imaging, and finding common structures in given three-dimensional protein structures.

We introduce a definition for a *largest common d-dimensional subsequence of two d-dimensional strings*, which generalizes the definition of a longest common subsequence of linear strings for higher dimensions. Our definition requires that common symbols selected from each sequence preserve their relative position ordering along each dimension simultaneously. Figure 1 shows a simple example for two-dimensional strings. The two-dimensional string in Part (c) is a common subsequence of both two-dimensional strings S_1 (shown in Part (a)), and S_2 (shown in Part (b)). We label the matching symbols with the same numbers. The position ordering in the common subsequence is $5 < 2 < 1 = 3 < 4 < 6$ on dimension one (horizontal dimension), and $1 < 2 = 3 = 4 < 5 = 6$ on dimension

E. Csuhaj-Varjú and Z. Ésik (Eds.): FCT 2007, LNCS 4639, pp. 40–51, 2007.

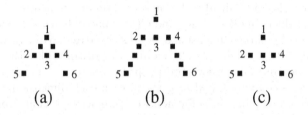

Fig. 1. A simple example for the definition of a largest common subsequence in two dimensions

two (vertical dimension). These orderings are correct for the matching symbol positions in both S_1, and S_2 on all dimensions (dimensions one and two in this case).

Figure 2 contains some example cases that illustrate the capability of our definition on identifying largest common subimages of two images. In Part (a), if the position orders for matching symbols are not preserved then the true common subimage that we show in Part $(a.2)$ may be overshadowed by another subimage of the same size as we depict such an example image in Part $(a.1)$. Our definition is powerful enough to capture the similarity between an image and its enlarged version as we show in Part (b). In this case, position order preservation in matching symbols makes it possible to realize that the original image is contained in the enlarged version as a subimage.

Under our definition, computing a largest common d-dimensional subsequence of two d-dimensional strings in NP-hard when $d > 1$. We show this by proving that the decision version of the problem for $d = 2$ is NP-complete. We also show that if we weaken this definition such that the relative ordering of selected

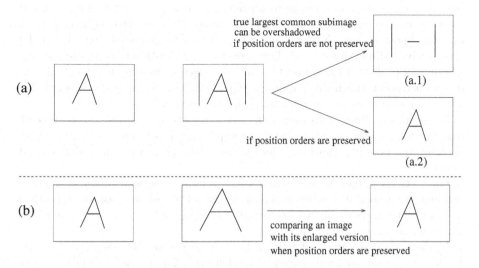

Fig. 2. Sample input images, and their common subimages

symbol positions is constrained to be preserved on one dimension only then we can solve the problem in $O(n^3)$ time, and $O(n^2)$ space when $d = 2$, where n is the common size of the two d-dimensional strings compared. We also show how to modify this algorithm for $d > 2$. The modified algorithm takes the same $O(n^3)$ asymptotic time, and $O(n^2)$ asymptotic space, where $n = m^d$ is the common size of the images compared, and m is the size on each dimension.

The outline of this paper is as follows: we summarize previous work on multi-dimensional string editing in Section 2. We give our definition of a largest common subsequence for multi-dimensional strings in Section 3. We show, in Section 4, that computing a largest common subsequence under our definition is NP-hard. In Section 5, we present our algorithm for a weaker definition of a longest common subsequence. We include our remarks in Section 6. We summarize our results in Section 7.

2   Previous Related Work

String editing for linear strings is a very well-studied, and well-understood process because the edit operations insert, delete, and substitute form a natural basis for changing one string to another [10]. This frame-work of string editing gives rise to the definition of a longest common subsequence which can be computed efficiently [3].

Several studies in the literature attempted to define edit distance between strings in high dimensions. Krithivasan and Sitalakshmi [7] defined edit distance between two images of the same size as the sum of the edit distances between all pairs of rows in the same positions in both strings. Baeza-Yates [1] proposed an edit distance model for two dimensional strings. This model allows deletion of prefixes of rows or columns in each string (image), changing a prefix of a row (or a column) in one image to a prefix of a row (or a prefix of a column if columns are compared) in the other image using character level edit operations between linear strings. Navarro and Baeza-Yates [9] extended this edit model recursively to dimensions $d \geq 3$. They develop an algorithm to compute the edit distance in $O(d!m^{2d})$ time, and $O(d!m^{2d-1})$ space if each input string's size in each dimension is m. In this paper, we express the time complexity on input size n (e.g. for images of size $m \times m$, the input size $n = m^2$).

For strings of dimensions higher than one, the string edit distance is difficult to define because there is no commonly accepted single set of edit operations that explains well possible transformations between any given two multi-dimensional strings.

Baeza-Yates [1] had the following remark: a *largest common subimage* of given two images can possibly be defined as a largest set of disjoint position-pairs that match exactly in both images such that a suitable order given by the position values is the same for both images.

In this paper, we give a definition for a *largest common subsequence of two given strings* that naturally extends the definition for the case of linear strings to higher dimensions. In our definition, we require that the order of matching positions is preserved on every dimension simultaneously (as we show it in an example in Figure 1). This way the integral composition (or geometry) of the subsequence is similar to those in the two strings from which this common subsequence is obtained as we illustrate on some examples in Figure 2. This is an essential property in our definition which has not been considered to this extent in existing models in [7,1,9].

3   Definitions

We give our definitions for dimension two for simplicity. They can be extended to higher dimensions easily.

An $n \times m$ string S is a *two dimensional string* (an image) that we represent as a matrix

$$\begin{pmatrix} s_{1,1} & s_{1,2} & \cdots & s_{1,m} \\ s_{2,1} & s_{2,2} & \cdots & s_{2,m} \\ \cdots & \cdots & \cdots & \cdots \\ s_{n,1} & s_{n,2} & \cdots & s_{n,m} \end{pmatrix}$$

where each $s_{i,j} \in \Sigma \cup \{\epsilon\}$ where ϵ denotes the null string. We also use $\{s_{i,j}\}$ to denote S. The null-symbol ϵ may be used to represent background of the image.

The size of S is $\sum_{s_{i,j} \neq \epsilon} 1$. That is, the size of S is the number of non-null symbols in S.

Let $(x_1, x_2, \ldots, x_d).k$ denote the kth component x_k of a given point (i.e. position) $(x_1, x_2, \ldots, x_d)$ in d-dimensional space.

Let $i_1 \leq_1 i_2 \leq_2 \ldots \leq_{k-1} i_k$ be a sequence s where each i_j is an integer for $1 \leq j \leq k$, and each $\leq_r$ is in $\{'<', '='\}$. We say that s is a *correct ordering* if for every r, $1 \leq r \leq k-1$, $i_r < i_{r+1}$ when $\leq_r = '<'$, and $i_r = i_{r+1}$ when $\leq_r = '='$.

An $n' \times m'$ string $S' = \{s'_{i,j}\}$ is a *subsequence* of an $n \times m$ string $S = \{s_{i,j}\}$ if the following are true:

- for every $p' = (i', j')$ there exists a unique index $f(p') = (i, j)$ such that $s_{i',j'} = s_{i,j}$,
- let I' be the set of non-null symbol positions in S', and let I be the set of matching positions in S, i.e. $I = \{f(p') \mid p' \in I'\}$. Let z denote the size of I' (or I since $|I'| = |I|$). For every dimension k, $1 \leq k \leq 2$, if the ordering in I' is $p'_1.k \leq_1 p'_2.k \leq_2 \ldots \leq_{z-1} p'_z$ where each p'_ℓ is a unique (i'_ℓ, j'_ℓ), and $\leq_\ell$ is $'='$ if $p'_\ell.k = p'_{\ell+1}.k$, and $'<'$ if $p'_\ell.k < p'_{\ell+1}.k$ for all ℓ, $1 \leq \ell \leq z-1$. This ordering is correct for $f(p')_1.k \leq_1 f(p')_2.k \leq_2 \ldots \leq_{z-1} f(p'_z)$. That is, on every dimension, the position ordering of indices in I' is identical with the ordering of the corresponding indices in S. For the example, in Figure 1, the string in Part (c) is a common subsequence of both strings in parts (a) and (b). Matching positions are labelled with the same numbers for simplicity. The orderings $5' <' 2' <' 1' =' 3' <' 4' <' 6$ on dimension one, and $1' <' 2' =' 3' =' 4' <' 5' =' 6$ on dimension two obtained from the common subsequence in Part (c) are all correct for the strings in parts (a) and (b).

We note that a subsequence S' of S can be obtained by first deleting a set of rows, and columns from S, and selecting the symbols that appear in the remaining string such that the unselected symbols are all ϵ's. It is easy to verify that when S' is obtained from S this way, the ordering of the positions of matching symbols are identical in both S' and S.

For given two-dimensional strings S_1 and S_2, where S_1 is an $n_1 \times m_1$, and S_2 is an $n_2 \times m_2$ string, S' is a largest two-dimensional common subsequence of S_1 and S_2 if S' is a two-dimensional common subsequence of both S_1 and S_2 with the largest possible size.

Our definition makes it possible to capture the true subimages for the examples shown in Figure 2. Part $(a.2)$ in the figure, since the position ordering distinguishes between $<$, and $=$, the two skew edges of letter A do not match straight vertical lines, and a false match shown in Part $(a.1)$ is avoided. The position order preservation on every dimension also functions as an invariant when one image is an enlarged version of the other. Part (b) has an example where the largest common subimage is the original image when the image and its enlarged version are compared.

We can verify that our definition of largest common subsequence can easily be generalized for strings of dimensions higher than two.

4   Complexity of Computing A Largest Common Two-Dimensional Subsequence of Given Two Two-Dimensional Strings

We consider the following decision problem, $LCS\text{-}IMAGE$: given two two-dimensional strings X and Y, is the largest size for a two-dimensional common subsequence of these strings $\geq J$? We show that this problem is NP-complete.

This problem is clearly in NP since for a given two-dimensional sequence S, and for each non-null symbol position in S, matching positions in both X and Y, we can verify that the size of S is $\geq J$, the ordering of positions in S for each dimension is correct for orderings of matching positions in both X and Y. This verification can be done in polynomial time.

We show next that an NP-complete problem, $CLIQUE$, is polynomial time reducible to $LCS\text{-}IMAGE$, where $CLIQUE$ is the following well-known problem [5]: given an undirected graph G, and integer K, does G contain a clique of size $\geq K$ (is there a complete subgraph of G with K or more vertices?).

We construct from a given graph $G = (V, E)$, where V is the set of vertices, and E is the set of edges, two-dimensional strings $X = \{x_{i,j}\}$ and $Y = \{y_{i,j}\}$ as follows:

– for all (i, j), $1 \leq i, j \leq |V|$,

$$x_{i,j} = \begin{cases} 1, \text{ if } i = j \text{ or } (i, j) \in E; \\ 0, \text{ otherwise.} \end{cases}$$

That is, X is the string corresponding to the adjacency matrix for G, where the diagonal is entirely composed of 1's,
- and Y is a $K \times K$ string where every symbol is a 1, i.e. for all (i, j), $y_{i,j} = 1$.

We can create X and Y in time polynomial in the size of graph G.

G has a clique of size K iff Y is a common subsequence of X and Y, i.e. the size of the common subsequence of X and Y is $\geq K^2$.

If G has a clique of size K then let $C = \{i_1, i_2, \ldots, i_K\}$ be the set of vertices in this clique. By keeping row i_j, and column i_j for all $i_j \in C$, and deleting all other rows, and columns, we can obtain Y as a subsequence of X (note that we set all $x_{i,i} = 1$ in X, and when Y is obtained from X this way the orders of matching positions are correct for all dimensions). Therefore, in this case the largest common subsequence of X and Y is Y whose size is K^2.

If the size of a largest common subsequence of X and Y is $\geq K^2$, then this common subsequence can only be the string Y. In this case, since the order of matching positions in X has to be the same as that in Y, which is a $K \times K$ string entirely composed of 1s, on every dimension, and since the only way to get a 1 in position (i, i) for all i's is to choose the same set of numbers for the rows, and columns, we can see that the subsequence Y in X corresponds to a set of identical row and column positions $C = \{i_1, i_2, \ldots, i_K\}$. From the construction, we can verify that the vertices in C form a K-clique in G.

Clearly, since the problem is NP-complete for two-dimensional strings, it is NP-hard for dimensions higher than two.

5   Computing A Largest Weak Common Subsequence

In this section, we weaken the definition of a common subsequence of two d-dimensional (for $d > 1$) strings such that the order of matching positions is required to be preserved on only one given dimension. More formally, for a given dimension k, $1 \leq k \leq d$, a d-dimensional string S' is a subsequence of a d-dimensional string S when the ordering of matching positions in S' on dimension k is a correct ordering with respect to $<$, and $=$ operators for the ordering of matching positions in S on dimension k. We use the adjective "weak" to distinguish a largest common subsequence, and the problem of finding a largest common subsequence in this definition from those that use the original definition.

We refer to a largest weak common subsequence (subimage when $d = 2$) by $lwcs$, and we call the problem of finding it as the $lwcs$ problem. This case of the problem can still have many practical applications such as comparing two images of objects moving in one dimension. For the examples in Figure 2, this weak definition is still powerful enough to find letter A as the largest common image in Part (a), and it can realize if one image is a widened or a lengthened version of another image.

We present a dynamic programming algorithm for computing a largest weak common subsequence of given two two-dimensional strings. The algorithm can easily be modified for strings in dimensions higher than two.

Without loss of generality, we choose the horizontal direction to represent the dimension on which the ordering of matching positions is required to be preserved.

We use $S[i][j]$ to represent the symbol $s_{i,j}$ in S. We denote by $S[i][j..k]$ the substring of row i in S starting at column j and ending at column k. $S[(i,j)..(k,l)]$ is the rectangular subimage of S whose top-left corner position is (i,j) and bottom-right position is (k,l).

Let S_1, and S_2 be two given images of size $n_1 \times m_1$, and $n_2 \times m_2$, respectively. We compute a largest weak common subimage between all subimages of S_1 and S_2. When we compare two images, the left-most columns are not considered for convenience in a future step. For this reason, we also imagine an additional column, column 0, on the dimension that the ordering is preserved (the dimension shown horizontally). For example, in Figure 3, two subimages $S_1[1, i - \Delta_i, n_1, i]$ and $S_2[1, j - \Delta_j, n_2, j]$ are shown. In the *lwcs* computation, the left-most columns $S_1[1..n_1][i - \Delta_i]$ and $S_2[1..n_2][j - \Delta_j]$ do not contribute to the total *lwcs* size. We use empty circles on these positions to highlight this fact.

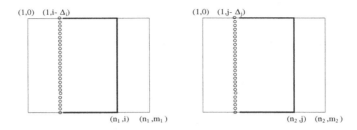

Fig. 3. Regions involved when we compare two subimages from each image

Let $R_{i-\Delta_i,i,j-\Delta_j,j}$ be the *lwcs* size between two images $S_1[(1, i - \Delta_i)..(n_1, i)]$ and $S_2[(1, j - \Delta_j)..(n_2, j)]$.

The boundary conditions are $R_{i,i,j-\Delta_j,j} = 0$ for all i, Δ_j, j, $0 \le i \le m_1$, $1 \le \Delta_j \le m_2$, $\Delta_j \le j \le m_2$. Symmetrically, $R_{i-\Delta_i,i,j,j} = 0$ for all i, Δ_i, j, $1 \le \Delta_i \le m_1$, $\Delta_i \le i \le m_1$, $0 \le j \le m_2$.

We compute $R_{i-\Delta_i,i,j-\Delta_j,j}$ for all Δ_i, Δ_j, i, j. $R_{0,m_1,0,m_2}$ is the *lwcs* size for S_1 and S_2.

We first perform a step what we call the *base step*. Our definition of largest weak common subsequence dictates that there is only one dimension on which the ordering is required to be preserved. There is no requirement involving the other dimensions. This freedom on other dimensions implies that a *maximal matching* between two strings is sought. This gives rise to a well-known graph problem if we represent each symbol in the strings by a vertex. A *matching M* of an undirected graph $G = (V, E)$, where V and E are the sets of vertices, and edges respectively, is a subset of edges with the property that no two edges in M share the same vertex. In this base step, we consider all columns from each image pairwise as we depict in Figure 4. For all i, $1 \le i \le m_1$, for all j, $1 \le j \le m_2$, we create a *bipartite graph* $G_{i,j} = (V, E)$ and compute the cardinality of a *maximal bipartite matching* on $G_{i,j}$ and store it in $R_{i-1,i,j-1,j}$. A graph $G = (V, E)$ is bipartite if the set of vertices V can be partitioned into two sets A, and B such that each edge in E has one vertex in A, and one vertex in B. The maximum cardinality bipartite matching problem on a given bipartite graph seeks a matching in the graph with the maximum cardinality. In the base step, we create $G_{i,j}$ as follows:

- we set $V = A \cup B$ where $A = \{S_1[k][i] \mid 1 \le k \le n_1\}$, and $B = \{S_2[k][j] \mid 1 \le k \le n_2\}$: we create a vertex for each symbol in column i of S_1 to obtain set A, and create a vertex for each symbol in column j of S_2 to obtain set B. $A \cup B$ is the set of vertices in $G_{i,j}$;
- we set $E = \{(S_1[k_1][i], S_2[k_2][j]) \mid S_1[k_1][i] = S_2[k_2][j], S_1[k_1][i] \in A, S_2[k_2][j] \in B\}$: the edges in E have one end-point in A and the other end-point in B, where the symbols corresponding to these vertices match.

We can use any maximum cardinality bipartite matching algorithm in this step (for example, the algorithm proposed by Hopcroft and Karp [6] which was shown to be a special case of the maximum-flow algorithm applied to a simple network by Even and Tarjan [4]).

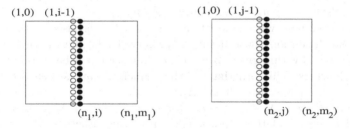

Fig. 4. All columns of both images are considered pairwise, and for every-pair the maximum cardinality bipartite matching problem is solved on the graph created

For all $\Delta_i, \Delta_j, i, j, 1 \le \Delta_i \le m_1, 1 \le \Delta_j \le m_2, \Delta_i \le i \le m_1, \Delta_j \le j \le m_2$ we compute $R_{i-\Delta_i,i,j-\Delta_j,j}$ as follows:

For all $i', j', 1 \le i' \le i, 1 \le j' \le j$, and $i' \ne i$ or $j' \ne j$, but not both $i' = i$ and $j' = j$, we consider two parts of the subimage $S_1[1, i - \Delta_i, n_1, i]$ separated by column i', and similarly, two parts of the subimage $S_2[1, j - \Delta_j, n_2, j]$ separated by column j' as we show in Figure 5. We compute $R_{i-\Delta_i,i,j-\Delta_j,j}$ as the maximum of the following:

- $R_{i',i,j',j} + R_{i-\Delta_i,i',j-\Delta_j,j'}$: We can combine the order-preserved *lwcs* obtained between regions 1 of S_1 and 1 of S_2 as shown in Figure 5 Part a. Note that because the left-most columns are not counted in the *lwcs* computation between two subimages, the intersecting lines in these regions belong to only one region, and they are not counted more than once. Figure 5 depicts this by using empty circles on the leftmost columns.

- $R_{i',i,j-\Delta_j,j}$: If region 1 of S_1 and the entire subimage of S_2 give an *lwcs* then region 2 of S_1 is avoided because otherwise the ordering will be violated. This case is shown in Part b of the figure;
- $R_{i-\Delta_i,i',j-\Delta_j,j}$: Similarly, an *lwcs* can possibly be obtained between region 2 of S_1 and the entire subimage of S_2, and region 1 of S_1 is avoided. This case is shown in Part c of the figure;
- $R_{i-\Delta_i,i,j',j}$: This case is shown in Part d of the figure, and it is symmetric to Part b;
- $R_{i-\Delta_i,i,j-\Delta_j,j}$: This case is shown in Part e of the figure, and it is symmetric to Part c.

Note that we do not consider all possible cases in Figure 5. Cases we do not consider either violate the ordering, or they cannot have an optimal *lwcs*. For example, pairing region 1 of S_1 with 2 of S_2, and region 2 of S_1 with region 1 of S_1 violates the ordering whenever both have more than zero matches. For another example, pairing region 1 of S_1, and region 2 of S_2 alone cannot yield a weak common subimage larger than that can be obtained in Part (b) in Figure 5 where region 1 of S_1 is paired with the entire subimage of S_2 if we think inductively in increasing subimage sizes for pairings.

We claim that the $R_{i-\Delta_i,i,j-\Delta_j,j}$ we compute is the *lwcs* size between the subimages $S_1[0, i - \Delta_i, n_1, i]$ and $S_2[0, j - \Delta_j, n_2, j]$.

For the correctness, we use induction. The induction is based on the subimage sizes on the dimension the ordering is preserved (the dimension represented horizontally), and within given sizes on this dimension it is based on all possible subimage pairings. This ordering for the induction can be generated by the following nested loops: For all $\Delta_i, \Delta_j, i, j, i', j'$, $1 \le \Delta_i \le m_1$, $1 \le \Delta_j \le m_2$, $\Delta_i \le i \le m_1$, $\Delta_j \le j \le m_2$, $1 \le i' \le i$, $1 \le j' \le j$, and not both $i' = i$ and $j' = j$ (the latter conditions are necessary otherwise this would be a self-referencing problem).

For the base case, we consider all pairwise comparisons between columns of both images, and store the maximum matching cardinalities. Clearly, our claim is true in the base case.

When one of the subimages is a column, and the other one includes more than one columns, for the subimage with a single column, region 1 is the single column of this image, and region 2 when paired with any region in the other subimage will produce a 0, because of the boundary conditions we set. In this case, the single column region of one image, and one of the regions in the other subimage will produce an *lwcs* for this pairing of subimages. Obviously, the ordering is preserved: essentially all matches are in one column in this case.

When each subimage paired for comparison is wider than a single column, then there exist a column a^* in S_1, and b^* in S_2 such that when we consider regions on both sides of these separator columns they pair in a way to maximize the size of a common subsequence. In our algorithm, we consider all possible cases which can yield a largest common subsequence, and we only allow pairings with which order is preserved. Optimality, and preservation of the ordering are implied

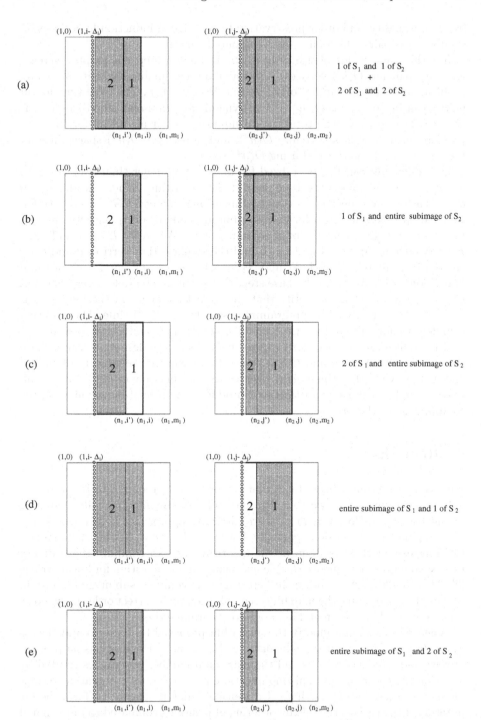

Fig. 5. Computing $lwcs$ of images $S_1[1, i - \Delta_i, n_1, i]$ and $S_2[1, j - \Delta_j, n_2, j]$: All cases for pairing parts of subimages in both images that can yield an optimal $lwcs$

by the optimality, and order preservation by our strong induction hypothesis on smaller sized pairs which have been computed earlier.

If both S_1 and S_2 are of size $n = m \times m$, in the base step, the graph we create for every pair has $O(m)$ vertices. The maximum cardinality bipartite matching problem for each pair can be solved in time $O(m^{2.5})$ [6,4]. The total time for all pairs spent in this step is $O(m^{4.5})$. The dynamic programming step takes longer, $\Theta(n^3) = \Theta(m^6)$ time. The total time complexity of our algorithm for the $lwcs$ problem when $d = 2$ is therefore $O(n^3)$, and it requires $O(n^2)$ space since all these steps can be performed using $O(n^2)$ space.

Our algorithm can be generalized for dimensions higher than two. For this, we only need to change the base step of the algorithm. When the dimension $d > 2$, each column becomes a $(d-1)$-dimensional string. In $G_{i,j} = (A \cup B, E)$, we create a vertex in set A for each symbol in column i of S_1, and we create a vertex in set B for each symbol in column j of S_2. We create E by including in E every edge (a, b), $a \in A$ and $b \in B$ and the symbols that vertices a and b are created from match. If both S_1 and S_2 are of size $n = m^d$, and each of the d dimensions is of size m, in the base step, the graph we create for every pair has $O(m^{d-1})$ vertices. We can verify that the total time spent in the base step is $O(m^{2.5d-0.5})$. The dynamic programming step takes $O(n^3)$ time. The resulting complexities for this case are the same, i.e. $O(n^3)$ time, and $O(n^2)$ space.

If in addition to the $lwcs$ size, an optimal $lwcs$ is also desired we can achieve this within $O(n^3)$ time, and $O(n^2)$ space, by storing the separator columns i' and j' that yielded optimal $lwcs$ sizes in the pairings involved. Then we can have a simple recursive algorithm that produces an optimal $lwcs$, following the computation of the $lwcs$ size.

6   Remarks

Since our definition of $lwcs$ states that the ordering is required to be preserved only on one dimension, we solved a maximum cardinality bipartite matching in the base step. However, in practice alternate approaches help preserve the ordering in other dimensions partially. For example, if $d = 2$ (when we compare two images), in the base step of the algorithm we can compute the ordinary longest common subsequence (lcs) length using an lcs algorithm for linear strings [3]. This should perform better in detecting true common subimage because in the lcs of two-columns, the matching symbols keep their correct ordering whereas the maximum bipartite matching does not guarantee this.

When $d > 2$, we can modify the algorithm proposed by Navarro and Baeza-Yates [9] to compute not the edit distance based on edit operations on prefixes of rows and columns, but instead the maximum number of matches created by applying these operations. This modification, although does not guarantee complete order preservation on all $d-1$ dimensions simultaneously, performs better in terms of the satisfaction of the ordering of positions compared to our original solution for the base step that uses maximum cardinality bipartite matching. We can use this modified algorithm in the base step of our algorithm.

We can easily verify that these replacements in the base step do not increase the time or space complexity of our *lwcs* algorithm.

7 Conclusion

We introduce a definition for a *largest common d-dimensional subsequence of two d-dimensional strings* for $d \geq 1$. In our definition we require that orders of matching positions are identical on all d dimensions simultaneously in the strings compared. We show that under this definition, the problem of finding a largest common subsequence of two given multi-dimensional strings is NP-hard when $d \geq 2$. However, if we weaken the definition so that the ordering of matching positions is required to be preserved on only one given dimension, then the problem has a polynomial time solution. We present a dynamic programming algorithm for this case.

References

1. Baeza-Yates, R.A.: Similarity in two dimensional strings. In: Hsu, W.-L., Kao, M.-Y. (eds.) COCOON 1998. LNCS, vol. 1449, pp. 319–328. Springer, Heidelberg (1998)
2. Bunke, H., Buhler, U.: Applications of approximate string matching to 2d shape recognition. Pattern Recognition 26(12), 1797–1812 (1993)
3. Bengroth, L., Hakkonen, H., Raita, T.: A survey of longest common subsequence algorithms. In: Proc. SPIRE, pp. 39–48 (2000)
4. Even, S., Tarjan, R.E.: Network flow and testing graph connectivity. J. SIAM Comp. 4(4), 507–512 (1975)
5. Garey, M.R., Johnson, D.S.: Computers and intractability: a guide to the theory of NP-completeness. W. H. Freeman and Compnay, New York (1979)
6. Hopcroft, J.E., Karp, R.M.: An $n^{5/2}$ algorithm for maximum matching in bipartite graphs. J. SIAM Comp. 2, 225–231 (1973)
7. Krithivasan, K., Sitalakshmi, R.: Efficient two-dimensional pattern matching in the presence of erriors. Information Sciences 43, 169–184 (1987)
8. Klein, P.N., Sebastian, T.B., Kimia, B.B.: Shape matching using edi-distance: an implementation. In: Proc. SODA, pp. 781–790 (2001)
9. Navarro, G., Baeza-Yates, R.A.: Fast multi-dimensional approximate matching. In: Crochemore, M., Paterson, M.S. (eds.) Combinatorial Pattern Matching. LNCS, vol. 1645, pp. 243–257. Springer, Heidelberg (1999)
10. Wagner, R.A., Fisher, M.J.: The string-to-string correction problem. J. of ACM 21(1), 168–173 (1974)
11. Wu, W.-Y., Wang, M.-J.: Two-dimensional object recognition through two-stage string matching. IEEE Transactions on Image Processing 8(7), 978–981 (1999)

Analysis of Approximation Algorithms for k-Set Cover Using Factor-Revealing Linear Programs*

Stavros Athanassopoulos, Ioannis Caragiannis, and Christos Kaklamanis

Research Academic Computer Technology Institute &
Department of Computer Engineering and Informatics
University of Patras, 26500 Rio, Greece

Abstract. We present new combinatorial approximation algorithms for k-set cover. Previous approaches are based on extending the greedy algorithm by efficiently handling small sets. The new algorithms further extend them by utilizing the natural idea of computing large packings of elements into sets of large size. Our results improve the previously best approximation bounds for the k-set cover problem for all values of $k \geq 6$. The analysis technique could be of independent interest; the upper bound on the approximation factor is obtained by bounding the objective value of a *factor-revealing* linear program.

1 Introduction

Set cover is a fundamental combinatorial optimization problem with many applications. Instances of the problem consist of a set of elements V and a collection S of subsets of V and the objective is to select a subset of S of minimum cardinality so that every element is covered, i.e., it is contained in at least one of the selected sets.

The natural greedy algorithm starts with an empty solution and augments it until all elements are covered by selecting a set that contains the maximum number of elements that are not contained in any of the previously selected sets. Denoting by n the number of elements, it is known by the seminal papers of Johnson [10], Lovasz [13], and Chvátal [1] that the greedy algorithm has approximation ratio H_n. The tighter analysis of Slavík [14] improves the upper bound on the approximation ratio to $\ln n - \ln \ln n + O(1)$. Asymptotically, these bounds are tight due to a famous inapproximability result of Feige [3] which states that there is no $(1 - \epsilon) \ln n$-approximation algorithm for set cover unless all problems in NP have deterministic algorithms running in subexponential time $O\left(n^{\text{polylog}(n)}\right)$.

An interesting variant is k-set cover where every set of S has size at most k. Without loss of generality, we may assume that S is closed under subsets. In this case, the greedy algorithm can be equivalently expressed as follows:

---

* This work was partially supported by the European Union under IST FET Integrated Project 015964 AEOLUS.

E. Csuhaj-Varjú and Z. Ésik (Eds.): FCT 2007, LNCS 4639, pp. 52–63, 2007.

Greedy phases: For $i = k$ down to 1 do:
Choose a maximal collection of disjoint i-sets.

An i-set is a set that contains exactly i previously uncovered elements and a collection T of disjoint i-sets is called maximal if any other i-set intersects some of the sets in T.

A tight bound of H_k on the approximation ratio of the greedy algorithm is well known in this case. Since the problem has many applications for particular values of k and due to its interest from a complexity-theoretic viewpoint, designing algorithms with improved second order terms in their approximation ratio has received much attention. Currently, there are algorithms with approximation ratio $H_k - c$ where c is a constant. Goldschmidt et al. [4] were the first to present a modified greedy algorithm with $c = 1/6$. This value was improved to $1/3$ by Halldórsson [6] and to $1/2$ by Duh and Fürer [2]. Recently, Levin [12] further improved the constant to $98/195 \approx 0.5026$ for $k \geq 4$. On the negative side, Trevisan [15] has shown that, unless subexponential-time deterministic algorithms for NP exist, no polynomial-time algorithm has an approximation ratio of $\ln k - \Omega(\ln \ln k)$.

The main idea that has been used in order to improve the performance of the greedy algorithm is to efficiently handle small sets. The algorithm of Goldschmidt et al. [4] uses a matching computation to accommodate as many elements as possible in sets of size 2 when no set of size at least 3 contains new elements. The algorithms of Halldórsson [5,6] and Duh and Fürer [2] handle efficiently sets of size 3. The algorithm of [2] is based on a semi-local optimization technique. Levin's improvement [12] extends the algorithm of [2] by efficiently handling of sets of size 4.

A natural but completely different idea is to replace the phases of the greedy algorithm associated with large sets with set-packing phases that also aim to maximize the number of new elements covered by large maximal sets. The approximation factor is not getting worse (due to the maximality condition) while it has been left as an open problem in [12] whether it leads to any improvement in the approximation bound. This is the aim of the current paper: we show that by substituting the greedy phases in the algorithms of Duh and Fürer with packing phases, we obtain improved approximation bounds for every $k \geq 6$ which approaches $H_k - 0.5902$ for large values of k.

In particular, we will use algorithms for the k-set packing problem which is defined as follows. An instance of k-set packing consists of a set of elements V and a collection S of subsets of V each containing exactly k elements. The objective is to select as many as possible disjoint sets from S. When $k = 2$, the problem is equivalent to maximum matching in graphs and, hence, it is solvable in polynomial time. For $k \geq 3$, the problem is APX-hard [11]; the best known inapproximability bound for large k is asymptotically $O\left(\frac{\log k}{k}\right)$ [7]. Note that any maximal collection of disjoint subsets yields a $1/k$-approximate solution. The best known algorithms have approximation ratio $\frac{2-\epsilon}{k}$ for any $\epsilon > 0$ [8] and are based on local search; these are the algorithms used by the packing phases of our algorithms.

Our analysis is based on the concept of factor-revealing LPs which has been introduced in a different context in [9] for the analysis of approximation algorithms for facility location. We show that the approximation factor is upper-bounded by the maximum objective value of a factor-revealing linear program. Hence, no explicit reasoning about the structure of the solution computed by the algorithms is required. Instead, the performance guarantees of the several phases are used as black boxes in the definition of the constraints of the LP.

The rest of the paper is structured as follows. We present the algorithms of Duh and Fürer [2] as well as our modifications in Section 2. The analysis technique is discussed in Section 3 where we present the factor-revealing LP lemmas. Then, in Section 4, we present a part of the proofs of our main results; proofs that have been omitted from this extended abstract will appear in the final version of the paper. We conclude in Section 5.

2 Algorithm Description

In this section we present the algorithms considered in this paper. We start by giving an overview of the related results in [2]; then, we present our algorithms and main statements.

Besides the greedy algorithm, local search algorithms have been used for the k-cover problem, in particular for small values of k. Pure local search starts with any cover and works in steps. In each step, the current solution is improved by replacing a constant number of sets with a (hopefully smaller) number of other sets in order to obtain a new cover. Duh and Fürer introduced the technique of semi-local optimization which extends pure local search. In terms of 3-set cover, the main idea behind semi-local optimization is that once the sets of size 3 have been selected, computing the minimum number of sets of size 2 and 1 in order to complete the covering can be done in polynomial time by a matching computation. Hence, a semi-local (s, t)-improvement step for 3-set cover consists of the deletion of up to t 3-sets from the current cover and the insertion of up to s 3-sets and the minimum necessary 2-sets and 1-sets that complete the cover. The quality of an improvement is defined by the total number of sets in the cover while in case of two covers of the same size, the one with the smallest number of 1-sets is preferable. Semi-local optimization for $k \geq 4$ is much similar; local improvements are now defined on sets of size at least 3 while 2-sets and 1-sets are globally changed. The analysis of [2] shows that the best choice of the parameters (s, t) is $(2, 1)$.

Theorem 1 (Duh and Fürer [2]). *Consider an instance of k-set cover whose optimal solution has a_i i-sets. Then, the semi-local $(2, 1)$-optimization algorithm has cost at most $a_1 + a_2 + \sum_{i=3}^{k} \frac{i+1}{3} a_i$.*

Proof (outline). The proof of [2] proceeds as follows. Let b_i be the number of i-sets in the solution. First observe that $\sum_{i=1}^{k} i b_i = \sum_{i=1}^{k} i a_i$. Then, the following two properties of semi-local $(2, 1)$-optimization are proved: $b_1 \leq a_1$ and $b_1 + b_2 \leq \sum_{i=1}^{k} a_i$. The theorem follows by summing the three inequalities.   □

This algorithm has been used as a basis for the following algorithms that approximate k-set cover. We will call them $\mathsf{GSLI}_{k,\ell}$ and $\mathsf{GRSLI}_{k,\ell}$, respectively.

Algorithm $\mathsf{GSLI}_{k,\ell}$.
Greedy phases: For $i = k$ down to $\ell + 1$ do:
    Choose a maximal collection of i-sets.
Semi-local optimization phase: Run the semi-local $(2, 1)$-optimization
    algorithm on the remaining instance.

Theorem 2 (Duh and Fürer [2]). *Algorithm* $\mathsf{GSLI}_{k,4}$ *has approximation ratio* $H_k - 5/12$.

Algorithm $\mathsf{GRSLI}_{k,\ell}$
Greedy phases: For $i = k$ down to $\ell + 1$ do:
    Choose a maximal collection of i-sets.
Restricted phases: For $i = \ell$ down to 4 do:
    Choose a maximal collection of disjoint i-sets so that the choice of
    these i-sets does not increase the number of 1-sets in the final solu-
    tion.
Semi-local optimization phase: Run the semi-local optimization al-
    gorithm on the remaining instance.

Theorem 3 (Duh and Fürer [2]). *Algorithm* $\mathsf{GRSLI}_{k,5}$ *has approximation ratio* $H_k - 1/2$.

We will modify the algorithms above by replacing each greedy phase with a packing phase for handling sets of not very small size. We use the local search algorithms of Hurkens and Schrijver [8] in each packing phase. The modified algorithms are called $\mathsf{PSLI}_{k,\ell}$ and $\mathsf{PRSLI}_{k,\ell}$, respectively.

A local search algorithm for set packing uses a constant parameter p (informally, this is an upper bound on the number of local improvements performed at each step) and, starting with an empty packing Π, repeatedly updates Π by replacing any set of $s < p$ sets of Π with $s + 1$ sets so that feasibility is maintained and until no replacement is possible. Clearly, the algorithm runs in polynomial time. It has been analyzed in [8] (see also [5] for related investigations).

Theorem 4 (Hurkens and Schrijver [8]). *The local search t-set packing algorithm that performs at most p local improvements at each step has approximation ratio* $\rho_t \geq \frac{2(t-1)^r - t}{t(t-1)^r - t}$, *if* $p = 2r - 1$ *and* $\rho_t \geq \frac{2(t-1)^r - 2}{t(t-1)^r - 2}$, *if* $p = 2r$.

As a corollary, for any constant $\epsilon > 0$, we obtain a $\frac{2-\epsilon}{t}$-approximation algorithm for t-set packing by using $p = O(\log_t 1/\epsilon)$ local improvements. Our algorithms $\mathsf{PSLI}_{k,\ell}$ and $\mathsf{PRSLI}_{k,\ell}$ simply replace each of the greedy phases of the algorithms $\mathsf{GSLI}_{k,\ell}$ and $\mathsf{GRSLI}_{k,\ell}$, respectively, with the following packing phase:

Packing phases: For $i = k$ down to $\ell + 1$ do:
    Select a maximal collection of disjoint i-sets using a $\frac{2-\epsilon}{i}$-approximation
    local search i-set packing algorithm.

Our first main result (Theorem 5) is a statement on the performance of algorithm $\mathsf{PSLI}_{k,\ell}$ (for $\ell = 4$ which is the best choice).

Theorem 5. *For any constant $\epsilon > 0$, algorithm $\mathsf{PSLI}_{k,4}$ has approximation ratio at most $H_{k/2} + \frac{1}{6} + \epsilon$ for even $k \geq 6$, at most $2H_k - H_{\frac{k-1}{2}} - \frac{11}{9} + \epsilon$ for $k \in \{5, 7, 9, 11, 13\}$, and at most $H_{\frac{k-1}{2}} + \frac{1}{6} + \frac{2}{k} - \frac{1}{k-1} + \epsilon$ for odd $k \geq 15$.*

Algorithm $\mathsf{PSLI}_{k,4}$ outperforms algorithm $\mathsf{GSLI}_{k,4}$ for $k \geq 5$, algorithm $\mathsf{GRSLI}_{k,5}$ for $k \geq 19$, as well as the improvement of Levin [12] for $k \geq 21$. For large values of k, the approximation bound approaches $H_k - c$ with $c = \ln 2 - 1/6 \approx 0.5264$. Algorithm $\mathsf{PSLI}_{k,\ell}$ has been mainly included here in order to introduce the analysis technique. As we will see in the next section, the factor-revealing LP is simpler in this case. Algorithm $\mathsf{PRSLI}_{k,\ell}$ is even better; its performance (for $\ell = 5$) is stated in the following.

Theorem 6. *For any constant $\epsilon > 0$, algorithm $\mathsf{PRSLI}_{k,5}$ has approximation ratio at most $2H_k - H_{\frac{k-1}{2}} - \frac{77}{60} + \epsilon$ for odd $k \geq 7$, at most $\frac{461}{240} + \epsilon$ for $k = 6$, and at most $2H_k - H_{k/2} - \frac{77}{60} + \frac{2}{k} - \frac{1}{k-1} + \epsilon$ for even $k \geq 8$.*

Algorithm $\mathsf{PRSLI}_{k,5}$ achieves better approximation ratio than the algorithm of Levin [12] for every $k \geq 6$. For example, the approximation ratio of $\frac{461}{240} \approx 1.9208$ for 6-set cover improves the previous bound of $H_6 - \frac{98}{195} \approx 1.9474$. For large values of k, the approximation bound approaches $H_k - c$ with $c = 77/60 - \ln 2 \approx 0.5902$. See Table 1 for a comparison between the algorithms discussed in this section.

Table 1. Comparison of the approximation ratio of the algorithms $\mathsf{GSLI}_{k,4}$, $\mathsf{PSLI}_{k,4}$, $\mathsf{GRSLI}_{k,5}$, the algorithm in [12] and algorithm $\mathsf{PRSLI}_{k,5}$ for several values of k

k	$\mathsf{GSLI}_{k,4}$ [2]	$\mathsf{PSLI}_{k,4}$	$\mathsf{GRSLI}_{k,5}$ [2]	[12]	$\mathsf{PRSLI}_{k,5}$
3	1.3333	1.3333	1.3333	1.3333	1.3333
4	1.6667	1.6667	1.5833	1.5808	1.5833
5	1.8667	1.8444	1.7833	1.7801	1.7833
6	2.0333	2	1.95	1.9474	1.9208
7	2.1762	2.1429	2.0929	2.0903	2.0690
8	2.3012	2.25	2.2179	2.2153	2.1762
9	2.4123	2.3524	2.3290	2.3264	2.2917
10	2.5123	2.45	2.4290	2.4264	2.3802
19	3.1311	3.0453	3.0477	3.0452	2.9832
20	3.1811	3.0956	3.0977	3.0952	3.0305
21	3.2287	3.1409	3.1454	3.1428	3.0784
22	3.2741	3.1865	3.1908	3.1882	3.1217
50	4.0825	3.9826	3.9992	3.9966	3.9187
75	4.4847	4.3814	4.4014	4.3988	4.3178
100	4.7707	4.6659	4.6874	4.6848	4.6021
large k	$H_k - 0.4167$	$H_k - 0.5264$	$H_k - 0.5$	$H_k - 0.5026$	$H_k - 0.5902$

3   Analysis Through Factor-Revealing LPs

Our proofs on the approximation guarantee of our algorithms essentially follow by computing upper bounds on the objective value of factor-revealing linear programs whose constraints capture simple invariants maintained in the phases of the algorithms.

Consider an instance (V, S) of k-set cover. For any phase of the algorithms associated with i ($i = \ell, ..., k$ for algorithm $\mathsf{PSLI}_{k,\ell}$ and $i = 3, ..., k$ for algorithm $\mathsf{PRSLI}_{k,\ell}$), consider the instance (V_i, S_i) where V_i contains the elements in V that have not been covered in previous phases and S_i contains the sets of S which contain only elements in V_i. Denote by $\mathcal{OPT}^i$ an optimal solution of instance (V_i, S_i); we also denote the optimal solution $\mathcal{OPT}^k$ of $(V_k, S_k) = (V, S)$ by $\mathcal{OPT}$. Since S is closed under subsets, without loss of generality, we may assume that $\mathcal{OPT}^i$ contains disjoint sets. Furthermore, it is clear that $|\mathcal{OPT}^{i-1}| \leq |\mathcal{OPT}^i|$ for $i \leq k$, i.e., $|\mathcal{OPT}^i| \leq |\mathcal{OPT}|$.

For a phase of algorithm $\mathsf{PSLI}_{k,\ell}$ or $\mathsf{PRSLI}_{k,\ell}$ associated with i, denote by $a_{i,j}$ the ratio of the number of j-sets in $\mathcal{OPT}^i$ over $|\mathcal{OPT}|$. Since $|\mathcal{OPT}^i| \leq |\mathcal{OPT}|$, we obtain that

$$\sum_{j=1}^{i} a_{i,j} \leq 1. \tag{1}$$

The i-set packing algorithm executed on packing phase associated with i includes in i-sets the elements in $V_i \backslash V_{i-1}$. Since $V_{i-1} \subseteq V_i$, their number is

$$|V_i \backslash V_{i-1}| = |V_i| - |V_{i-1}| = \left(\sum_{j=1}^{i} j a_{i,j} - \sum_{j=1}^{i-1} j a_{i-1,j} \right) |\mathcal{OPT}|. \tag{2}$$

Denote by ρ_i the approximation ratio of the i-set packing algorithm executed on the phase associated with i. Since at the beginning of the packing phase associated with i, there exist at least $a_{i,i}|\mathcal{OPT}|$ i-sets, the i-set packing algorithm computes at least $\rho_i a_{i,i}|\mathcal{OPT}|$ i-sets, i.e., covering at least $i\rho_i a_{i,i}|\mathcal{OPT}|$ elements from sets in $\mathcal{OPT}^i$. Hence, $|V_i \backslash V_{i-1}| \geq i\rho_i a_{i,i}|\mathcal{OPT}|$, and (2) yields

$$\sum_{j=1}^{i-1} j a_{i-1,j} - \sum_{j=1}^{i-1} j a_{i,j} - i(1 - \rho_i)a_{i,i} \leq 0. \tag{3}$$

So far, we have defined all constraints for the factor-revealing LP of algorithm $\mathsf{PSLI}_{k,\ell}$. Next, we bound from above the number of sets computed by algorithm $\mathsf{PSLI}_{k,\ell}$ as follows. Let t_i be the number of i-sets computed by the i-set packing algorithm executed at the packing phase associated with $i \geq \ell + 1$. Clearly,

$$t_i = \frac{1}{i}|V_i \backslash V_{i-1}| = \left(\frac{1}{i} \sum_{j=1}^{i} j a_{i,j} - \frac{1}{i} \sum_{j=1}^{i-1} j a_{i-1,j} \right) |\mathcal{OPT}|. \tag{4}$$

By Theorem 1, we have that

$$t_\ell \leq \left(a_{\ell,1} + a_{\ell,2} + \sum_{j=3}^{\ell} \frac{j+1}{3} a_{\ell,j} \right) |\mathcal{OPT}|. \tag{5}$$

Hence, by (4) and (5), it follows that the approximation guarantee of algorithm $\mathsf{PSLI}_{k,\ell}$ is

$$\frac{\sum_{i=4}^{k} t_i}{|\mathcal{OPT}|} \leq \sum_{i=\ell+1}^{k} \frac{1}{i} \left(\sum_{j=1}^{i} j a_{i,j} - \sum_{j=1}^{i-1} j a_{i-1,j} \right) + a_{\ell,1} + a_{\ell,2} + \sum_{j=3}^{\ell} \frac{j+1}{3} a_{\ell,j}$$

$$= \frac{1}{k} \sum_{j=1}^{k} j a_{k,j} + \sum_{i=\ell+1}^{k-1} \frac{1}{i(i+1)} \sum_{j=1}^{i} j a_{i,j} + \frac{\ell}{\ell+1} a_{\ell,1}$$

$$+ \frac{\ell-1}{\ell+1} a_{\ell,2} + \sum_{j=3}^{\ell} \left(\frac{j+1}{3} - \frac{j}{\ell+1} \right) a_{\ell,j} \tag{6}$$

Hence, an upper bound on the approximation ratio of algorithm $\mathsf{PSLI}_{k,\ell}$ follows by maximizing the right part of (6) subject to the constraints (1) for $i = \ell, ..., k$ and (3) for $i = \ell + 1, ..., k$ with variables $a_{i,j} \geq 0$ for $i = \ell, ..., k$ and $j = 1, ..., i$. Formally, we have proved the following statement.

Lemma 1. *The approximation ratio of algorithm* $\mathsf{PSLI}_{k,\ell}$ *when a* ρ_i-*approximation* i-*set packing algorithm is used at phase* i *for* $i = \ell + 1, ..., k$ *is upperbounded by the maximum objective value of the following linear program:*

$$\text{maximize } \frac{1}{k} \sum_{j=1}^{k} j a_{k,j} + \sum_{i=\ell+1}^{k-1} \frac{1}{i(i+1)} \sum_{j=1}^{i} j a_{i,j} + \frac{\ell}{\ell+1} a_{\ell,1}$$

$$+ \frac{\ell-1}{\ell+1} a_{\ell,2} + \sum_{j=3}^{\ell} \left(\frac{j+1}{3} - \frac{j}{\ell+1} \right) a_{\ell,j}$$

$$\text{subject to } \sum_{j=1}^{i} a_{i,j} \leq 1, i = \ell, ..., k$$

$$\sum_{j=1}^{i-1} j a_{i-1,j} - \sum_{j=1}^{i-1} j a_{i,j} - i(1 - \rho_i) a_{i,i} \leq 0, i = \ell + 1, ..., k$$

$$a_{i,j} \geq 0, i = \ell, ..., k, j = 1, ..., i$$

Each packing or restricted phase of algorithm $\mathsf{PRSLI}_{k,\ell}$ satisfies (3); a restricted phase associated with $i = 4, ..., \ell$ computes a maximal i-set packing and, hence, $\rho_i = 1/i$ in this case.

In addition, the restricted phases impose extra constraints. Denote by b_1, b_2, and b_3 the ratio of the number of 1-sets, 2-sets, and 3-sets computed by the semi-local optimization phase over $|\mathcal{OPT}|$, respectively. The restricted phases guarantee that the number of the 1-sets in the final solution does not increase, and, hence,

$$b_1 \leq a_{i,1}, \text{ for } i = 3, ..., \ell. \tag{7}$$

Following the proof of Theorem 1, we obtain $b_1 + b_2 \leq a_{3,1} + a_{3,2} + a_{3,3}$ while it is clear that $b_1 + 2b_2 + 3b_3 = a_{3,1} + 2a_{3,2} + 3a_{3,3}$. We obtain that the number t_3 of sets computed during the semi-local optimization phase of algorithm PRSLI$_{k,\ell}$ is

$$
\begin{aligned}
t_3 &= (b_1 + b_2 + b_3)|\mathcal{OPT}| \\
&\leq \left(\frac{b_1}{3} + \frac{b_1 + b_2}{3} + \frac{b_1 + 2b_2 + 3b_3}{3} \right) |\mathcal{OPT}| \\
&\leq \left(\frac{1}{3}b_1 + \frac{2}{3}a_{3,1} + a_{3,2} + \frac{4}{3}a_{3,3} \right) |\mathcal{OPT}|.
\end{aligned}
\tag{8}
$$

Reasoning as before, we obtain that (4) gives the number t_i of i-sets computed during the packing or restricted phase associated with $i = 4, ..., k$. By (4) and (8), we obtain that the performance guarantee of algorithm PRSLI$_{k,\ell}$ is

$$
\begin{aligned}
\frac{\sum_{i=3}^{k} t_i}{|\mathcal{OPT}|} &\leq \sum_{i=4}^{k} \frac{1}{i} \left(\sum_{j=1}^{i} ja_{i,j} - \sum_{j=1}^{i-1} ja_{i-1,j} \right) + \frac{2}{3}a_{3,1} + a_{3,2} + \frac{4}{3}a_{3,3} + \frac{1}{3}b_1 \\
&= \frac{1}{k} \sum_{j=1}^{k} ja_{k,j} + \sum_{i=4}^{k-1} \frac{1}{i(i+1)} \sum_{j=1}^{i} ja_{i,j} + \frac{5}{12}a_{3,1} + \frac{1}{2}a_{3,2} \\
&\quad + \frac{7}{12}a_{3,3} + \frac{1}{3}b_1
\end{aligned}
\tag{9}
$$

Hence, an upper bound on the approximation ratio of algorithm PRSLI$_{k,\ell}$ follows by maximizing the right part of (9) subject to the constraints (1) for $i = 3, ..., k$, (3) for $i = 4, ..., k$, and (7), with variables $a_{i,j} \geq 0$ for $i = 3, ..., k$ and $j = 1, ..., i$, and $b_1 \geq 0$. Formally, we have proved the following statement.

Lemma 2. *The approximation ratio of algorithm* PRSLI$_{k,\ell}$ *when a ρ_i-approximation i-set packing algorithm is used at phase i for $i = \ell + 1, ..., k$ is upperbounded by the maximum objective value of the following linear program:*

$$\text{maximize } \frac{1}{k} \sum_{j=1}^{k} ja_{k,j} + \sum_{i=4}^{k-1} \frac{1}{i(i+1)} \sum_{j=1}^{i} ja_{i,j} + \frac{5}{12}a_{3,1} + \frac{1}{2}a_{3,2} + \frac{7}{12}a_{3,3} + \frac{1}{3}b_1$$

$$\text{subject to } \sum_{j=1}^{i} a_{i,j} \leq 1, i = 3, ..., k$$

$$\sum_{j=1}^{i-1} ja_{i-1,j} - \sum_{j=1}^{i-1} ja_{i,j} - i(1 - \rho_i)a_{i,i} \leq 0, i = \ell + 1, ..., k$$

$$\sum_{j=1}^{i-1} ja_{i-1,j} - \sum_{j=1}^{i-1} ja_{i,j} - (i-1)a_{i,i} \leq 0, i = 4, ..., \ell$$

$$b_1 - a_{i,1} \leq 0, i = 3, ..., \ell$$

$$a_{i,j} \geq 0, i = 3, ..., k, j = 1, ..., i$$

$$b_1 \geq 0$$

4   Proofs of Main Theorems

We are now ready to prove our main results. We can show that the maximum objective values of the factor-revealing LPs for algorithms $PSLI_{k,4}$ and $PRSLI_{k,5}$ are upper-bounded by the values stated in Theorems 5 and 6. In order to prove this, it suffices to find feasible solutions to the dual LPs that have these values as objective values. In the following, we prove Theorem 5 by considering the case when k is even. The proof for odd k as well as the proof of Theorem 6 (which is slightly more complicated) will appear in the final version of the paper.

Proof of Theorem 5. The dual of the factor-revealing LP of algorithm $PSLI_{k,4}$ is:

$$\text{minimize} \sum_{i=4}^{k} \beta_i$$

$$\text{subject to } \beta_4 + \gamma_5 \geq \frac{4}{5}$$

$$\beta_4 + 2\gamma_5 \geq \frac{3}{5}$$

$$\beta_4 + 3\gamma_5 \geq \frac{11}{15}$$

$$\beta_4 + 4\gamma_5 \geq \frac{13}{15}$$

$$\beta_i + j\gamma_{i+1} - j\gamma_i \geq \frac{j}{i(i+1)}, i = 5, ..., k-1, j = 1, ..., i-1$$

$$\beta_i + i\gamma_{i+1} - (i-2+\epsilon)\gamma_i \geq \frac{1}{i+1}, i = 5, ..., k-1$$

$$\beta_k - j\gamma_k \geq \frac{j}{k}, j = 1, ..., k-1$$

$$\beta_k - (k-2+\epsilon)\gamma_k \geq 1$$

$$\beta_i \geq 0, i = 4, ..., k$$

$$\gamma_i \geq 0, i = 5, ..., k$$

We consider only the case of even k. We set $\gamma_k = \frac{1}{k(k-1)}$ and $\gamma_{k-1} = 0$. If $k \geq 8$, we set $\gamma_i = \gamma_{i+2} + \frac{2}{i(i+1)(i+2)}$ for $i = 5, ..., k-2$. We also set $\beta_k = 1 + (k-2+\epsilon)\gamma_k$, $\beta_4 = \frac{13}{15} - 4\gamma_5$ and

$$\beta_i = \frac{1}{i+1} - i\gamma_{i+1} + (i-2+\epsilon)\gamma_i$$

for $i = 5, ..., k-1$.

We will show that all the constraints of the dual LP are satisfied. Clearly, $\gamma_i \geq 0$ for $i = 5, ..., k$. Observe that $\gamma_{k-1} + \gamma_k = \frac{1}{k(k-1)}$. If $k \geq 8$, by the definition of γ_i for $i = 5, ..., k-2$, we have that

$$\gamma_i + \gamma_{i+1} - \frac{1}{i(i+1)} = \gamma_{i+1} + \gamma_{i+2} + \frac{2}{i(i+1)(i+2)} - \frac{1}{i(i+1)}$$

$$= \gamma_{i+1} + \gamma_{i+2} - \frac{1}{(i+1)(i+2)}$$

and, hence,

$$\gamma_i + \gamma_{i+1} - \frac{1}{i(i+1)} = \gamma_{k-1} + \gamma_k - \frac{1}{k(k-1)} = 0,$$

i.e., $\gamma_i + \gamma_{i+1} = \frac{1}{i(i+1)}$ for $i = 5, ..., k-1$. Now, the definition of β_i's yields

$$\begin{aligned}
\beta_i &= \frac{1}{i+1} - i\gamma_{i+1} + (i - 2 + \epsilon)\gamma_i \\
&= \frac{i-1}{i(i+1)} - (i-1)\gamma_{i+1} + (i-1)\gamma_i + \epsilon\gamma_i \qquad (10) \\
&\geq \frac{j}{i(i+1)} - j\gamma_{i+1} + j\gamma_i \\
&\geq 0
\end{aligned}$$

for $i = 5, ..., k-1$ and $j = 1, ..., i-1$. Hence, all the constraints on β_i for $i = 5, ..., k-1$ are satisfied.

The constraints on β_k are also maintainted. Since $\gamma_k = \frac{1}{k(k-1)} \leq \frac{1}{k}$ we have that $\beta_k = 1 + (k - 2 + \epsilon)\gamma_k \geq \frac{k-1}{k} + (k-1)\gamma_k + \epsilon\gamma_k \geq \frac{j}{k} + j\gamma_k \geq 0$ for $j = 1, ..., k-1$. It remains to show that the constraints on β_4 are also satisfied. It suffices to show that $\gamma_5 \leq 1/45$. This is clear when $k = 6$. If $k \geq 8$, consider the equalities $\gamma_i + \gamma_{i+1} = \frac{1}{i(i+1)}$ for odd $i = 5, ..., k-3$ and $-\gamma_i - \gamma_{i+1} = -\frac{1}{i(i+1)}$ for even $i = 6, ..., k-2$. Summing them, and since $\gamma_{k-1} = 0$, we obtain that

$$\begin{aligned}
\gamma_5 &= \sum_{i=3}^{k/2-1} \left(\frac{1}{2i(2i-1)} - \frac{1}{2i(2i+1)} \right) \\
&= \sum_{i=3}^{k/2-1} \left(\frac{1}{2i-1} - \frac{1}{2i} - \frac{1}{2i} + \frac{1}{2i+1} \right) \\
&= \sum_{i=3}^{k/2-1} \left(\frac{1}{2i-1} + \frac{1}{2i+1} \right) - \sum_{i=3}^{k/2-1} \frac{1}{i} \\
&= -\frac{1}{5} + \sum_{i=3}^{k/2-1} \frac{2}{2i-1} + \frac{1}{k-1} - H_{k/2-1} + \frac{3}{2} \\
&= -\frac{1}{5} + 2H_{k-2} - H_{k/2-1} - \frac{8}{3} + \frac{1}{k-1} - H_{k/2-1} + \frac{3}{2}
\end{aligned}$$

$$= 2H_{k-2} - 2H_{k/2-1} + \frac{1}{k-1} - \frac{41}{30}$$

$$\leq 2\ln 2 - \frac{41}{30}$$

$$\leq 1/45$$

The first inequality follows since $2H_{k-2} - 2H_{k/2-1} + \frac{1}{k-1}$ is increasing on k (this can be easily seen by examining the original definition of γ_5) and since $\lim_{t\to\infty} H_t/\ln t = 1$.

We have shown that all the constraints of the dual LP are satisfied, i.e., the solution is feasible. In order to compute the objective value, we use the definition of β_i's and equality (10). We obtain

$$\sum_{i=4}^{k} \beta_i = \beta_4 + \beta_5 + \sum_{i=3}^{k/2-1} (\beta_{2i} + \beta_{2i+1}) + \beta_k$$

$$= \frac{13}{15} - 4\gamma_5 + \frac{2}{15} + 4\gamma_5 - 4\gamma_6 + \sum_{i=3}^{k/2-1} \left(\frac{1}{2i+1} - 2i\gamma_{2i+1} + (2i-2)\gamma_{2i} \right.$$

$$\left. + \frac{2i}{(2i+1)(2i+2)} - 2i\gamma_{2i+2} + 2i\gamma_{2i+1} \right) + 1 + (k-2)\gamma_k + \epsilon \sum_{i=5}^{k} \gamma_i$$

$$= 2 + \sum_{i=3}^{k/2-1} \frac{1}{i+1} - 4\gamma_6 + \sum_{i=3}^{k/2-1} ((2i-2)\gamma_{2i} - 2i\gamma_{2i+2}) + (k-2)\gamma_k$$

$$+ \epsilon \sum_{i=5}^{k} \gamma_i$$

$$\leq H_{k/2} + 1/6 + \epsilon$$

where the last inequality follows since $\sum_{i=5}^{k} \gamma_i \leq \sum_{i=5}^{k} \frac{1}{i(i-1)} = \sum_{i=5}^{k} \left(\frac{1}{i-1} - \frac{1}{i} \right) = 1/4 - 1/k$.

By duality, $\sum_{i=4}^{k} \beta_i$ is an upper bound on the maximum objective value of the factor-revealing LP. The theorem follows by Lemma 1.   □

5   Extensions

We have experimentally verified using Matlab that our upper bounds are tight in the sense that they are the maximum objective values of the factor-revealing LPs (ignoring the ϵ term in the approximation bound). Our analysis technique can also be used to provide simpler proofs of the results in [2] (i.e., Theorems 2 and 3); this is left as an exercise to the reader. The several cases that are considered in the proofs of [2] are actually included as constraints in the factor-revealing LPs which are much simpler than the ones for algorithms $\mathsf{PSLI}_{k,\ell}$ and $\mathsf{PRSLI}_{k,\ell}$. Furthermore, note that we have not combined our techniques with the recent algorithm of Levin [12] that handles sets of size 4 using a restricted local search phase. It is tempting to conjecture that further improvements are possible.

References

1. Chvátal, V.: A greedy hueristic for the set-covering problem. Mathematics of Operations Research 4, 233–235 (1979)
2. Duh, R., Fürer, M.: Approximation of k-set cover by semi local optimization. In: Proceedings of the 29th Annual ACM Symposium on Theory of Computing (STOC '97), pp. 256–264. ACM Press, New York (1997)
3. Feige, U.: A threshold of $\ln n$ for approximating set cover. Journal of the ACM 45(4), 634–652 (1998)
4. Goldschmidt, O., Hochbaum, D., Yu, G.: A modified greedy heuristic for the set covering problem with improved worst case bound. Information Processing Letters 48, 305–310 (1993)
5. Halldórsson, M.M.: Approximating discrete collections via local improvements. In: Proceedings of the 6th Annual ACM/SIAM Symposium on Discrete Algorithms (SODA '95), pp. 160–169 (1995)
6. Halldórsson, M.M.: Approximating k-set cover and complementary graph coloring. In: Cunningham, W.H., Queyranne, M., McCormick, S.T. (eds.) Integer Programming and Combinatorial Optimization. LNCS, vol. 1084, pp. 118–131. Springer, Heidelberg (1996)
7. Hazan, E., Safra, S., Schwartz, O.: On the complexity of approximating k-set packing. Computational Complexity 15(1), 20–39 (2006)
8. Hurkens, C.A.J., Schrijver, A.: On the size of systems of sets every t of which have an SDR, with an application to the worst-case ratio of heuristics for packing problems. SIAM Journal on Discrete Mathematics 2(1), 68–72 (1989)
9. Jain, K., Mahdian, M., Markakis, E., Saberi, A., Vazirani, V.V.: Greedy facility location algorithms analyzed using dual fitting with factor-revealing LP. Journal of the ACM 50(6), 795–824 (2003)
10. Johnson, D.S.: Approximation algorithms for combinatorial problems. Journal of Computer and System Sciences 9, 256–278 (1974)
11. Kann, V.: Maximum bounded 3-dimensional matching is MAX SNP-complete. Information Processing Letters 37, 27–35 (1991)
12. Levin, A.: Approximating the unweighted k-set cover problem: greedy meets local search. In: Erlebach, T., Kaklamanis, C. (eds.) WAOA 2006. LNCS, vol. 4368, pp. 290–310. Springer, Heidelberg (2007)
13. Lovász, L.: On the ratio of optimal integral and fractional covers. Discrete Mathematics 13, 383–390 (1975)
14. Slavík, P.: A tight analysis of the greedy algorithm for set cover. Journal of Algorithms 25, 237–254 (1997)
15. Trevisan, L.: Non-approximability results for optimization problems on bounded degree instances. In: Proceedings of the 33rd Annual ACM Symposium on Theory of Computing (STOC '01), pp. 453–461. ACM Press, New York (2001)

A Novel Information Transmission Problem and Its Optimal Solution

Eric Bach[1,*] and Jin-Yi Cai[2,**]

[1] Computer Sciences Department, University of Wisconsin
Madison, WI 53706, USA
bach@cs.wisc.edu
[2] Computer Sciences Department, University of Wisconsin
Madison, WI 53706, USA
jyc@cs.wisc.edu

Abstract. We propose and study a new information transmission problem motivated by today's internet. Suppose a real number needs to be transmitted in a network. This real number may represent data or control and pricing information of the network. We propose a new transmission model in which the real number is encoded using Bernoulli trials. This differs from the traditional framework of Shannon's information theory. We propose a natural criterion for the quality of an encoding scheme. Choosing the best encoding reduces to a problem in the calculus of variations, which we solve rigorously. In particular, we show there is a unique optimal encoding, and give an explicit formula for it.

We also solve the problem in a more general setting in which there is prior information about the real number, or a desire to weight errors for different values non-uniformly.

Our tools come mainly from real analysis and measure-theoretic probability, but there is also a connection to classical mechanics. Generalizations to higher dimensional cases are open.

1 Introduction

In Shannon's information theory and the theory of error correcting codes, the following communication model is basic. Two parties A and B share a line of transmission, on which one can send an *ordered* sequence of bits. The receiver gets another *ordered* sequence of bits, possibly corrupted. While this corruption can change, omit, or locally transpose bits, by and large the order of the bits is kept intact.[1] Of course this model was very much motivated by the teletype networks of Shannon's day.

---

* Supported by NSF CCF-0523680, CCF-0635355, and a Vilas Research Associate Award.
** Supported by NSF CCR-0511679.

[1] Most work has focused on the so-called discrete memoryless channel, in which only bit changes are allowed. The model of [13] allows arbitrary changes, but only on fixed-length blocks.

With today's internet, one might revisit this model. When a message is sent from one node to another, it has no fixed path. Abstractly, one might imagine a model in which symbols are being sent in a highly parallel and non-deterministic fashion with no particular fixed route. The receiver receives these symbols in some probabilistic sense but in no particular order.

Suppose we still consider sending bit sequences. Then if arbitrary re-orderings are allowed, then only the cardinality, or what amounts to the same thing, the fraction of 1's observed, will matter. Furthermore, if some omissions occur probabilistically then even this fraction is only meaningful approximately. Thus, with arbitrary re-ordering of the bits, it severely restricts the ways by which information may be meaningfully conveyed.

Instead of sending bit sequences, what about sending a real number holistically? Let's consider the following new model of information transmission. Two parties A and B have access to a one-way communication medium, and A wishes to transmit a real number x to B. The medium may transmit signals, with some probabilistic error, in large multiplicity but in no particular order. By normalizing we assume $0 \leq x \leq 1$, and think of x as a probability. Communication is done by the following process. Party A can send a large number of i.i.d. samples from a Bernoulli distribution to B. The receiver observes these bits and estimates x. (The Bernoulli distribution, on the samples generated a priori, accounts for the probabilistic nature of errors and losses of signals due to the communication medium.)

The new information transmission problem is the following. We may think of the Bernoulli random variable as an "encoding" of x, through its mean value. Then what does it mean to be a good encoding scheme? How do we evaluate encoding strategies, and is there an optimal one? We note that x is only transmitted completely in the limit, so the answers must be asymptotic.

Although abstract, this problem is motivated by concrete current research in computer networking. As is familiar, messages are broken up into small packets which are then sent more or less independently along different routes. These routes can vary with time of day, system load, etc., so the network must maintain and transmit information about their quality.

We can think of a particular route as consisting of ℓ links, say $v_{i-1} \rightarrow v_i$ for $i = 1, \ldots, \ell$. Each link has an associated number p_i, $0 \leq p_i \leq 1$. For example, p_i could be a normalized cost or a measure of congestion for using the i-th link. The network can determine through observation the average $x = (\sum_i p_i)/\ell$ for a particular route, allowing the routing protocol to take this into account so as to avoid congestion.

To allow efficient estimation of this average, researchers have investigated the possibility of using current packet designs, which already specify a bit called the Explicit Congestion Notification (ECN) bit. Each link on a route may set this bit to 0 or 1 as it sees fit, for every packet it handles. This bit then gets transmitted to the next link, which may be reset again. Recently, networking researchers have focused on a class of protocols using ECN (so-called *one-bit protocols*), which can be defined mathematically as follows. The link $v_{i-1} \rightarrow v_i$

receives a bit $X_{i-1} \in \{0,1\}$ from the previous link; based on X_{i-1} and p_i it uses randomization to produce X_i. The last node can observe X_ℓ many times and combine these observations to produce an estimate for x.

Several protocols of this type appear in the literature [2,10]. What they have in common is that the expected value of X_ℓ is some function f of the average x. The observer then tries to infer x from the observed approximate value of $f(x)$. This is an example of our new model of information transmission, in that, one produces a collection of 0-1 random variables all with the expected value equal to some function f of some number x. The receiver observes these 0-1 random variables, in no particular order and with probabilistic losses and delays. From an observational record the receiver tries to infer x.

Since one can imagine more general schemes using this idea, there is no reason to expect developments to stop with ECN. For inspiration, we look to Shannon, who did not waste time optimizing teletype codes, but rather went on to study general methods of symbolic encoding. It is compelling, therefore, to develop a theory applicable to all of the more general schemes, and ask if there is any choice of f that is in some sense optimal. In this paper, we answer this question affirmatively, under conditions on f that are as general as could be desired.

2   The Formalized Problem and a Guide to Its Solution

Initially, A and B agree on a transformation function f. To send $x \in [0,1]$, to B, the transmitter A generates random bits, which are i.i.d. 0-1 random variables with expected value $y = f(x)$. The receiver B gets n of these, say $Y_1, \ldots, Y_n$, and uses $f^{-1}\left(\frac{1}{n}\sum_{i=1}^{n} Y_i\right)$ to estimate x. For this to work, f must be strictly monotonic, say increasing. Also, f should map 0 to 0 and 1 to 1, to avoid loss of bandwidth.

We now outline our criterion for evaluating f, and justify its choice. Let $g = f^{-1}$ and $\bar{Y} = n^{-1}\sum_{i=1}^{n} Y_i$. If g is smooth, then by the strong law of large numbers, $g(\bar{Y}) \to x$, a.e. We expect $g(\bar{Y}) - x$ to be $\Theta(n^{-1/2})$, so the natural measure for the error is $\mathbf{E}[n(g(\bar{Y}) - x)^2]$. By the mean value theorem, we should have $n(g(\bar{Y}) - x)^2 \approx g'(y)^2[n(\bar{Y} - y)^2]$, and $\mathbf{E}\left[n(\bar{Y} - y)^2\right] = y(1 - y)$. Thus, we expect

$$\mathbf{E}\left[n(g(\bar{Y}) - x)^2\right] \to g'(y)^2 y(1 - y). \tag{1}$$

Written in terms of f, this is $\frac{f(x)(1-f(x))}{f'(x)^2}$. Thus, we should try to minimize

$$\int_0^1 \frac{f(x)(1 - f(x))}{f'(x)^2}\,dx, \tag{2}$$

over a suitable class of functions f. The optimal choice turns out to be

$$f = \frac{1 - \cos(\pi x)}{2}. \tag{3}$$

In particular, the optimal choice is *not* the identity function, as one might naively suppose. Nor is the naive choice even close: its value of (2) exceeds the optimum by about 64%.

In the rest of this paper, we carry out this argument in a rigorous way. The interchange of limits and integration is not trivial, because we want it to hold for the optimum curve, for which the integrand is unbounded. Also, we derive the optimal curve using the calculus of variations. But as is typical with the calculus of variations, this derivation only suggests optimality. (Euler's mathematics may have been brilliant, but it lacked a certain rigor.) As with the Dirichlet problem [7, p. 119], the hard part is to prove optimality. We will do this by an independent argument, under very general conditions on the curve. Our tools come mainly from real analysis and measure-theoretic probability, in particular Lebesgue's convergence theorems, Fatou's lemma, and uniform integrability.

The rest of the paper is organized as follows. In Section 3 we prove (1), and then show that the limit of its average (over possible values of x) is given by (2), for the particular choice (3). In Section 4 we prove that (3) actually minimizes (2). Section 5 treats these problems in a more general setting in which the receiver has prior information about x, or wishes to weight errors for different x differently. The full version of this paper includes two appendices. The more difficult proof of a general limit theorem is given in Appendix 1, where we show that the average of (1) has a limit, for a wide class of transformations. In Appendix 2 we connect our variational problems to classical mechanics.

3 Two Convergence Theorems

3.1 Notation

We call f *admissible* if $f \in C[0,1]$ (continuous), and is strictly increasing, with $f(0) = 0$ and $f(1) = 1$. Let $g = f^{-1}$ be its inverse function (also admissible). Since f and g are increasing, f' and g' exist a.e. [12]. Whenever $f'(x) \neq 0$, then at $y = f(x)$, $g'(y)$ exists and $g'(y) = 1/f'(x)$. If $f'(x) = 0$, we say g' has a *singularity* at y.

Our class of functions is the natural one to consider, for f can only be computable if it is continuous, as is well known [6].

Let $Y_1, Y_2, \ldots, Y_n$ be i.i.d. 0-1 random variables with $\Pr[Y_i = 1] = y$, and let $\overline{Y} = \frac{\sum_{i=1}^{n} Y_i}{n}$ be their sample mean. We also let $\hat{Y} = \left(\frac{\sum_{i=1}^{n}(Y_i - y)}{\sqrt{n}} \right)^2$, so that $n(\overline{Y} - y)^2 = \hat{Y}$. Note that $0 \leq \hat{Y} \leq n$.

We will find it convenient to use measure theory notation. Accordingly, let $\Omega = \{0,1\}^n$, with the measure μ induced by n Bernoulli trials with success probability y. Then, for example, $\mathbf{E}[\hat{Y}] = \int_{\Omega} \hat{Y} \, d\mu = \mathbf{Var}(Y_1) = y(1 - y)$.

For a choice of f as above, it will be convenient to let

$$F_n(y) = \int_{\Omega} n(g(\overline{Y}) - g(y))^2 \, d\mu, \tag{4}$$

and $\alpha = n(g(\overline{Y}) - g(y))^2$. Since $\alpha \leq n$, we have $0 \leq F_n(y) \leq n$.

3.2   A Pointwise Convergence Theorem

Theorem 1. *Let f be admissible and $0 < y < 1$. If $g'(y)$ exists, we have $\lim_{n\to\infty} F_n(y) = (g'(y))^2 y(1 - y)$, where F_n is given by (4). Therefore, the convergence is almost everywhere (a.e.).*

Proof. The proof is easiest when g' is continuous in an interval around y, so we assume this first. Then, for any $\epsilon > 0$, there exists a $\delta > 0$ such that if $|y' - y| \le \delta$ then $|(g'(y'))^2 - (g'(y))^2| \le \frac{\epsilon}{2y(1-y)}$. For this δ, let $B_\delta = \{\omega \in \Omega \mid |\overline{Y} - y| > \delta\}$.

Since $\int_\Omega \hat{Y} \, d\mu = y(1 - y)$, we have $F_n(y) - (g'(y))^2 y(1 - y) = I_1 + I_2 + I_3$, where

$$I_1 = \int_{\Omega - B_\delta} [\alpha - (g'(y))^2 \hat{Y}] \, d\mu; \quad I_2 = \int_{B_\delta} \alpha \, d\mu; \quad I_3 = -\int_{B_\delta} (g'(y))^2 \hat{Y} \, d\mu.$$

We will estimate these three integrals separately.

For I_1, by the mean value theorem (MVT), there exists some $\xi = \xi(y, \overline{Y})$ which lies between y and $\overline{Y}$, such that $g(\overline{Y}) - g(y) = g'(\xi)(\overline{Y} - y)$. Thus, $\alpha = (g'(\xi))^2 \hat{Y}$. Note that $n(\overline{Y} - y)^2 = \hat{Y}$, and on $\Omega - B_\delta$, $|\xi - y| \le \delta$, we have $|(g'(\xi))^2 - (g'(y))^2| \le \frac{\epsilon}{2y(1-y)}$. It follows that $|I_1|$ is at most

$$\int_{B_\delta^c} |(g'(\xi))^2 - (g'(y))^2| \hat{Y} \, d\mu \le \frac{\epsilon}{2y(1 - y)} \int_{B_\delta^c} \hat{Y} \, d\mu \le \frac{\epsilon}{2y(1 - y)} \int_\Omega \hat{Y} \, d\mu = \frac{\epsilon}{2}.$$

By the Chernoff bound [3], $\mu(B_\delta) < 2e^{-2\delta^2 n}$, so $|I_2| \le n \int_{B_\delta} d\mu = n\mu(B_\delta) < 2ne^{-2\delta^2 n}$, and since $\hat{Y} \le n$, we have $|I_3| \le (g'(y))^2 n \int_{B_\delta} d\mu \le 2n(g'(y))^2 e^{-2\delta^2 n}$. Combining these three estimates, we get

$$|F_n(y) - (g'(y))^2 y(1 - y)| = |I_1 + I_2 + I_3| \le \frac{\epsilon}{2} + 2ne^{-2\delta^2 n}(1 + (g'(y))^2) < \epsilon,$$

for sufficiently large n. Since ϵ was arbitrary, we get Theorem 1.

We indicate briefly how to modify this proof to work at any $y \ne 0, 1$ where $g'(y)$ exists. Only I_1 needs to be reconsidered. Suppose first that $g'(y) > 0$. Then there is a $\delta > 0$ such that for $\overline{Y}$ within δ of y (but not equal to y), we have

$$\frac{g(\overline{Y}) - g(y)}{\overline{Y} - y} = g'(y)(1 + \eta),$$

with $|\eta| \le \epsilon/(6g'(y)^2 y(1 - y))$, and $|\eta| \le 1$. Then (even allowing $\overline{Y} = y$),

$$\alpha = n(g(\overline{Y}) - g(y))^2 = ng'(y)^2 (\overline{Y} - y)^2 (1 + \eta)^2.$$

Plug this into I_1, and expand $(1 + \eta)^2$. The main terms will cancel, and we can estimate η^2 by $|\eta|$, we find

$$|I_1| \le \int_\Omega \frac{\epsilon}{2y(1 - y)} \hat{Y} = \frac{\epsilon}{2}.$$

We handle $g' = 0$ similarly, but with

$$\frac{g(\overline{Y}) - g(y)}{\overline{Y} - y} = \eta,$$

where $\eta^2 \leq \epsilon/(2y(1-y))$. The case $g' < 0$ is forbidden by monotonicity. □

3.3 Convergence for the Optimal Transformation

Our information transmission problem is concerned with minimizing the limit of

$$\int_0^1 \int_\Omega n(g(\overline{Y}) - x)^2 \, d\mu \, dx,$$

for an unknown function $y = f(x)$, where $g = f^{-1}$. Assuming the relevant integrals exist, we can write this entirely in terms of its inverse function g,

$$\int_0^1 g'(y) \int_\Omega n(g(\overline{Y}) - g(y))^2 \, d\mu \, dy.$$

In this section, we evaluate the limit of this for the optimal f. A corresponding theorem for general f was stated in [2], and proved in [1]. This result, however, assumed $g'(y)$ to be continuous on $[0, 1]$, and in particular bounded on this interval. While adequate for the class of functions realizable in the on-line setting for the ECN bit in a network, this assumption is not satisfied by our optimal function f. In particular, our particular $g'(y)$ is unbounded near 0 and 1, making the resulting proof much more difficult. A proof for the general case is provided in Appendix 1 of the full paper.

In the remainder of this section, we let $f(x) = (1 - \cos \pi x)/2$. We note that f is smooth and strictly increasing. Its inverse function $g(y)$ is continuously differentiable except at 0 and 1. Explicitly,

$$(g'(y))^2 = \frac{1}{\pi^2 y(1-y)}; \tag{5}$$

this has a pole of order 1 at $y = 0$ and $y = 1$. Let $\tilde{F}_n(y) = g'(y)F_n(y)$.

Theorem 2. *For $f(x) = (1 - \cos \pi x)/2$, we have*

$$\lim_{n \to \infty} \int_0^1 \tilde{F}_n(y) \, dy = \int_0^1 \lim_{n \to \infty} \tilde{F}_n(y) \, dy = \int_0^1 (g'(y))^3 y(1-y) \, dy.$$

Proof. Observe that there is a symmetry between the first and the second half of the interval, by the map $y \mapsto 1 - y$, and therefore we will only need to evaluate $\lim_{n \to \infty} \int_0^{1/2} \tilde{F}_n(y) \, dy$.

Let $\delta_n = \frac{8 \log n}{n}$. Then

$$\int_0^{1/2} \tilde{F}_n(y) \, dy = \int_0^{\delta_n} \tilde{F}_n(y) \, dy + \int_0^{1/2} F_n^*(y) \, dy, \tag{6}$$

where $F_n^*(y) = \tilde{F}_n(y) \mathbf{1}_{[\delta_n, 1/2]}$, and $\mathbf{1}$ denotes the indicator function. Our strategy will be to prove that the first term has the limit 0, and use Lebesgue's dominated convergence theorem to evaluate the limit of the second.

Let $y < \delta_n$. As $F_n(y)$ is itself an integral, we may (as with Gaul) divide it into three parts:

$$F_n(y) = \int_{\overline{Y} \leq y} \alpha \, d\mu + \int_{y < \overline{Y} \leq 1/2} \alpha \, d\mu + \int_{\overline{Y} > 1/2} \alpha \, d\mu. \tag{7}$$

We will show that the contributions of each part in the integral $\int_0^{\delta_n} \tilde{F}_n(y) \, dy$ goes to 0.

If $\overline{Y} \leq y$, by the monotonicity of g we get $(g(\overline{Y}) - g(y))^2 \leq (g(y))^2 = x^2$. It is easy to check by elementary calculus that $1 - \cos t \geq t^2/4$ for $0 \leq t \leq \pi/3$. then $y = f(x) = (1 - \cos \pi x)/2 \geq \frac{\pi^2}{8} x^2$, for $0 \leq x \leq 1/3$. It follows that, for $0 \leq y \leq 1/4$,

$$\int_{\overline{Y} \leq y} \alpha \, d\mu \leq nx^2 \int_\Omega d\mu = nx^2 \leq \frac{8ny}{\pi^2}.$$

So, there is a $c > 0$ such that for $\overline{Y} \leq y$ and sufficiently large n,

$$\int_0^{\delta_n} g'(y) \int_{\overline{Y} \leq y} \alpha \, d\mu \, dy \leq cn \int_0^{\delta_n} \sqrt{y} \, dy = \frac{2c}{3} n \delta_n^{3/2} \longrightarrow 0. \tag{8}$$

For $y < \overline{Y} \leq 1/2$, by MVT, there exists some $\xi = \xi(y, \overline{Y})$ such that $g(\overline{Y}) - g(y) = g'(\xi)(\overline{Y} - y)$, satisfying $y \leq \xi \leq \overline{Y} \leq 1/2$. By the explicit formula for g' we have $(g'(\xi))^2 \leq \frac{2}{\pi^2 y}$. Thus

$$\int_{y < \overline{Y} \leq 1/2} \alpha \, d\mu \leq \frac{2}{\pi^2 y} \int_\Omega \hat{Y} \, d\mu \leq \frac{2}{\pi^2}.$$

Then

$$\int_0^{\delta_n} g'(y) \int_{y < \overline{Y} \leq 1/2} \alpha \, d\mu \, dy \leq \frac{2g(\delta_n)}{\pi^2} \longrightarrow 0. \tag{9}$$

Finally we treat $\overline{Y} > 1/2$. From the Chernoff bound, we have

$$\int_{\overline{Y} > 1/2} \alpha \, d\mu \leq n\mu(\overline{Y} > 1/2) < ne^{-n/8}.$$

Therefore

$$\int_0^{\delta_n} g'(y) \int_{\overline{Y} > 1/2} \alpha \, d\mu \, dy < ne^{-n/8} \int_0^{\delta_n} g'(y) \, dy = ne^{-n/8} g(\delta_n) \longrightarrow 0. \tag{10}$$

Combining (8)–(10) with (7), we get $\lim_{n \to \infty} \int_0^{\delta_n} \tilde{F}_n(y) \, dy = 0$.

We now consider the second integral in (6). Our first goal is to bound $F_n(y)$ independently of n on $\delta_n \leq y \leq 1/2$.

Let B denote the event that $[\overline{Y} < y/2 \text{ or } \overline{Y} > 3/4]$. Inspired by King Solomon, we now divide F_n into two:

$$F_n(y) = \int_B \alpha \, d\mu + \int_{B^c} \alpha \, d\mu.$$

By the Chernoff bound [3], and $y \geq \delta_n$,

$$\mu(B) < e^{-yn/8} + e^{-n/8} < 2/n.$$

It follows that

$$\int_B \alpha \, d\mu \leq n\mu(B) < 2. \tag{11}$$

On B^c, by the mean value theorem (MVT), there exists some $\xi = \xi(y, \overline{Y})$ which lies between y and $\overline{Y}$, such that $g(\overline{Y}) - g(y) = g'(\xi)(\overline{Y} - y)$. Therefore $\alpha = (g'(\xi))^2 \hat{Y}$. Since $\overline{Y} \in B^c$, we have $y/2 \leq \overline{Y} \leq 3/4$. Combining this with $y \leq 1/2$, we get $y/2 \leq \xi \leq 3/4$. Using this in (5), we see that $(g'(\xi))^2 \leq \frac{8}{\pi^2 y}$. Then

$$\int_{B^c} \alpha \, d\mu \leq \frac{8}{\pi^2 y} \int_{B^c} \hat{Y} \, d\mu \leq \frac{8}{\pi^2 y} \int_\Omega \hat{Y} \, d\mu = \frac{8(1-y)}{\pi^2} \leq \frac{8}{\pi^2}. \tag{12}$$

From (11) and (12) we see that for $y \geq \delta_n$, $F_n(y) \leq \frac{8}{\pi^2} + 2 < 3$. This implies that

$$|F_n^*| \leq 3g'(y),$$

and since g' is integrable on $[0, 1/2]$ (near 0, g' is of order $1/\sqrt{y}$) we can apply dominated convergence to get

$$\lim_{n \to \infty} \int_0^{1/2} F_n^*(y) \, dy = \int_0^{1/2} \lim_{n \to \infty} \tilde{F}_n(y) \, dy = \int_0^{1/2} (g'(y))^3 y(1-y) \, dy. \qquad \square$$

4 Deriving the Optimal Transformation

We consider the following optimization problem. Let

$$I_y = \int_0^1 \frac{y(1-y)}{(y')^2} \, dx.$$

We seek a smooth increasing function y, satisfying the boundary conditions $y(0) = 0$ and $y(1) = 1$, that minimizes I_y. (Note that we are now letting y stand for a function, instead of a value.)

We use the calculus of variations to get a guess for y. Form the Euler-Lagrange equation

$$\frac{\partial L}{\partial y} - \frac{d}{dx} \frac{\partial L}{\partial y'} = 0,$$

with $L(y, y') = y(1-y)/(y')^2$. Then, $y(x) = (1 - \cos \pi x)/2$ is a solution matching the boundary conditions, for which $I_y = 1/\pi^2$. (Integrability is to be expected here, since L did not involve x explicitly. See [4].)

More work is needed to prove this is optimal. Recall that y is *admissible* if it is in $C[0,1]$, strictly increasing, with $y(0) = 0$, and $y(1) = 1$.

Theorem 3. *For any admissible function y, we have*

$$\int_0^1 \frac{y(1-y)}{(y')^2} dx \geq \frac{1}{\pi^2},$$

with equality iff $y = (1 - \cos \pi x)/2$. The case where the integral is infinite is not excluded.

Proof. Define a new admissible function θ by $y(x) = (1 - \cos \pi\theta(x))/2$. Since θ increases, θ' exists a.e., and at any point x of differentiability, $y'(x) = \frac{\pi}{2} \sin \pi\theta(x) \cdot \theta'(x)$, by the mean value theorem. Also $y(x) \neq 0, 1$ except for $x = 0, 1$, so we have (using $\sin^2 + \cos^2 = 1$) $\frac{(y')^2}{y(1-y)} = \pi^2 \theta'$, a.e.

We may assume that θ' is positive a.e. and $I_y < +\infty$, as otherwise the theorem is true. Then, by Jensen's inequality,

$$\int_0^1 \frac{y(1-y)}{(y')^2} dx = \frac{1}{\pi^2} \int_0^1 \frac{1}{(\theta')^2} dx \geq \frac{1}{\pi^2 \left(\int_0^1 \theta'(x) dx \right)^2}.$$

(To apply this, we need $\theta' \in L^1[0,1]$, which is true. See [12, Ex. 13, p. 157].)

We have $\int_0^1 \theta' \leq \theta(1) - \theta(0) = 1$, with equality iff θ is absolutely continuous (AC). (Combine [11, Thm. 2, p. 96] and [12, Thm. 7.18].) This gives the inequality of the theorem.

We may assume that θ is AC (otherwise, the inequality is strict). If θ' is not constant a.e., then the Jensen inequality is strict and we are done. On the other hand, if the inequality becomes equality, we have to have $\theta'(x) = c$ a.e. Then, $\theta(x) = \int_0^x c = cx$, so $c = 1$ and $\theta = x$ (everywhere!), giving the theorem. □

Remarks. It is possible that the integral is infinite; this happens, for example, if $y = x^3$. Also, without the monotonicity condition, the minimum need not exist. Consider, for example, $y_n = \sin^2((n+1)\pi x)$. Then we have $0 \leq y_n \leq 1$, with $y_n(0) = 0$ and $y_n(1) = 1$. However, $\int_0^1 y_n(1 - y_n)(y_n')^{-2} dx = 1/(4\pi^2(n+1)^2) \to 0$.

Theorem 4. *Let f be any admissible function. If $f \neq (1 - \cos \pi x)/2$, there is a constant $\delta_f > 0$ with the following property. For sufficiently large n,*

$$\int_0^1 \mathbf{E}[n(g(\bar{Y}) - y)^2] dx \geq \frac{1}{\pi^2} + \delta_f.$$

Proof. By Fatou's lemma [12] and Theorem 1,

$$\liminf_{n \to \infty} \int_0^1 \mathbf{E}[n(g(\bar{Y}) - y)^2] dx \geq \int_0^1 \lim_{n \to \infty} \mathbf{E}[n(g(\bar{Y}) - y)^2] dx = \int_0^1 g'(y)^2 y(1-y) dx.$$

But this is strictly greater than the corresponding integral for $f = (1 - \cos \pi x)/2$, which is $1/\pi^2$.   □

5   Modeling Prior Information and Non-uniform Penalties

In this section we generalize our model to let the the receiver have prior information about the transmitter's value x. To convey this information, we use a weight function φ ("prior density" in Bayesian jargon) that we assume differentiable and positive on $(0, 1)$.

Such a weight function also allows us to weight errors differently, depending on the value of x. For example, to send 0 and receive 0.1 might be much worse than to send 0.5 and receive 0.6, and the weight function can reflect this.

We are thus led to the more general problem of choosing an admissible y to minimize

$$\int_0^1 \frac{w(y)\varphi(x)}{(y')^2} dx.$$

For simplicity and clarity of exposition we will assume y smooth, i.e. $y \in C^1[0, 1]$.

Again, we begin with a variational approach. If L is the integrand, then

$$\frac{\partial L}{\partial y} - \frac{d}{dx}\frac{\partial L}{\partial y'} = 3w'\varphi(y')^{-2} + 2w\varphi'(y')^{-3} - 6w\varphi(y')^{-4}y''. \qquad (13)$$

On the other hand,

$$\frac{d}{dx}\left(w\varphi^\alpha(y')^\beta\right) = w'\varphi^\alpha(y')^{\beta+1} + \alpha w\varphi^{\alpha-1}\varphi'(y')^\beta + \beta w\varphi^\alpha(y')^{\beta-1}y''. \qquad (14)$$

The coefficients of (13) and (14) are proportional provided that $(3 : 2 : -6) = (1 : \alpha : \beta)$. Therefore, for $\alpha = 2/3$ and $\beta = -2$, we can put the Euler-Lagrange equation in the form

$$\varphi^{1/3}(y')^{-1}\frac{d}{dx}\left(w\varphi^{2/3}(y')^{-2}\right) = 0.$$

This implies that $w(y)\varphi(x)^{2/3} = c(y')^2$, for some constant c. If we take the square root of both sides and then separate variables, we see that

$$\int \varphi^{1/3}dx = c_1 \int \frac{dy}{\sqrt{w(y)}} + c_2. \qquad (15)$$

This relation plus the boundary conditions $y(0) = 0$, $y(1) = 1$ will determine y.

When $w(y) = y(1 - y)$ we can integrate the right hand side and solve for y to obtain $y = \frac{1 - \cos(A\Phi(x)+B)}{2}$, where $\Phi(x) = \int_0^x \varphi(t)^{1/3}dt$. The optimal function will not change if we multiply φ by a constant, so let us normalize φ so that $\Phi(1) = 1$. Clearly Φ is monotonic, and $\Phi(0) = 0$. From the boundary conditions, we get $A = \pi$ and $B = 0$, so $y = \frac{1 - \cos(\pi\Phi(x))}{2}$.

Optimality now can be proved as before. First, for our choice of y we have

$$\int_0^1 \frac{y(1-y)\varphi(x)}{(y')^2}dx = \int_0^1 \frac{\varphi(x)}{\pi^2\Phi'(x)^2}dx = \frac{1}{\pi^2}\int_0^1 \varphi(x)^{1/3}dx = \frac{1}{\pi^2}.$$

Now, suppose y is any other function. Then there is a function θ, increasing from 0 to 1 on [0,1], for which

$$y = \frac{1 - \cos(\pi\theta(\Phi(x)))}{2}.$$

Then

$$\int_0^1 \frac{y(1-y)\varphi(x)}{(y')^2}dx = \frac{1}{\pi^2}\int_0^1 \frac{\varphi(x)^{1/3}}{[\theta'(\Phi(x))]^2}dx.$$

Since $\int_0^1 \varphi^{1/3} = 1$, we can apply Jensen's inequality to get

$$\int_0^1 \frac{\varphi(x)^{1/3}}{[\theta'(\Phi(x))]^2}dx \geq \left[\int_0^1 \theta'(\Phi(x))\varphi(x)^{1/3}dx\right]^{-2} = [\theta(1) - \theta(0)]^{-2} = 1.$$

It follows from the considerations above that any admissible C^1 function is optimal with respect to some weight. Indeed, let the equation of the path be $y = (1 - \cos(\pi\theta(x)))/2$, where θ increases from 0 to 1. Then we may take $\varphi = (\theta')^3$.

6   Open Problems

One way to generalize our information transmission problem is to consider a higher dimensional analog of it.

In the problem we have just addressed, there is one real number $x \in [0,1]$ that A wishes to transmit to B. A natural 2-dimensional version of it is this: We have a point x on the convex hull Δ of $\{(1,0,0),(0,1,0),(0,0,1)\}$. That is, $x = p_1e_1 + p_2e_2 + p_3e_3$, where $p_1, p_2, p_3 \geq 0$ and $p_1 + p_2 + p_3 = 1$. The transmitter A can generate i.i.d. random variables with three outcomes, perhaps Red, White, and Blue with probabilities q_1, q_2 and q_3. Of course, $(q_1, q_2, q_3) \in \Delta$ as well. Now the transmitter A and the receiver B must choose beforehand a transformation f which maps Δ to itself, with an inverse g. Then, in the same formulation of this paper, what would be the optimal transformation function f, if one exists?

This problem is open, as is the analogous problem for any higher dimension. We don't have any significant results to report, but we can make two remarks.

First, the Euler-Lagrange equation is a nonlinear PDE with 95 terms, leading to some pessimism about the possibility of a closed form solution. (Recall that with all problems in the calculus of variations, even if the Euler-Lagrange equation is solved, we still do not have a guarantee of optimality.) It might be amenable to numerical approximations.

Second, some of the naive functions from Δ to Δ are not optimal.

Acknowledgements

We thank John Gubner and Jack Lutz for useful comments on this work.

References

1. Adler, J., Cai, J.-Y., Shapiro, J.K., Towsley, D.: Estimate of congestion price using probabilistic packet marking. Technical Report UM-TR-200223, UMASS-Amherst, (2002), See `http://www-net.cs.umass.edu/~jshapiro/um-tr-2002-23.pdf`
2. Adler, J., Cai, J.-Y., Shapiro, J.K., Towsley, D.: Estimate of congestion price using probabilistic packet marking. In: Proc. INFOCOMM 2003, pp. 2068–2078 (2003)
3. Alon, N., Spencer, J.: The Probabilistic Method. Wiley-Interscience, Chichester (1992)
4. Gelfand, I.M., Fomin, S.V.: Calculus of Variations. Prentice-Hall, Englewood Cliffs (1963)
5. Grimmett, G.R., Stirzaker, D.R.: Probability and Random Processes, 2nd edn. Oxford Univ. Press, Oxford (1992)
6. Ko, K.: Computational Complexity of Real Functions. Birkhauser (1991)
7. Körner, T.W.: Fourier Analysis. Cambridge Univ. Press, Cambridge (1990)
8. Lanczos, C.: The Variational Principles of Mechanics, 3rd edn. Univ. Toronto Press (1966)
9. Landau, L.D., Lifshitz, E.M.: Mechanics, 3rd edn. Pergamon Press, London (1976)
10. Low, S.H., Lapsley, D.E.: Optimization flow control, I: Basic algorithm and convergence. IEEE/ACM Transactions on Networking 7, 861–875 (1999)
11. Royden, H.L.: Real Analysis. Macmillan, NYC (1968)
12. Rudin, W.: Real and Complex Analysis. McGraw-Hill, New York (1974)
13. Verdú, S., Han, T.S.: A general formula for channel capacity. IEEE Transactions on Information Theory 4, 1147–1157 (1994)

Summaries of Appendices

The full version of this paper contains two appendices, which we summarize here.

In Appendix 1, we investigate the convergence properties of $\int_0^1 \tilde{F}_n(y) \, dy$, as $n \to \infty$. We show that under very general conditions on an admissible f, $\lim_{n\to\infty} \int_0^1 \tilde{F}_n(y) \, dy$ exists, and

$$\lim_{n\to\infty} \int_0^1 \tilde{F}_n(y) \, dy = \int_0^1 \lim_{n\to\infty} \tilde{F}_n(y) \, dy = \int_0^1 (g'(y))^3 y(1-y) \, dy.$$

In Appendix 2, we discuss our variational problems using the language and methods of classical mechanics [8,9]. In particular, it follows from Hamilton-Jacobi theory that our variational problems can always be reduced to quadrature [4].

Local Testing of Message Sequence Charts Is Difficult*

Puneet Bhateja[1], Paul Gastin[2],
Madhavan Mukund[1], and K. Narayan Kumar[1]

[1] Chennai Mathematical Institute, Chennai, India
{puneet,madhavan,kumar}@cmi.ac.in
[2] LSV, ENS Cachan & CNRS, France
Paul.Gastin@lsv.ens-cachan.fr

Abstract. Message sequence charts are an attractive formalism for specifying communicating systems. One way to test such a system is to substitute a component by a test process and observe its interaction with the rest of the system. Unfortunately, local observations can combine in unexpected ways to define implied scenarios not present in the original specification. Checking whether a scenario specification is closed with respect to implied scenarios is known to be undecidable when observations are made one process at a time. We show that even if we strengthen the observer to be able to observe multiple processes simultaneously, the problem remains undecidable. In fact, undecidability continues to hold even without message labels, provided we observe two or more processes simultaneously. On the other hand, without message labels, if we observe one process at a time, checking for implied scenarios is decidable.

1 Introduction

Message Sequence Charts (MSCs) [7] are an appealing visual formalism that are used in a number of software engineering notational frameworks such as SDL [15] and UML [4]. A collection of MSCs is used to capture the scenarios that a designer might want the system to exhibit (or avoid).

A standard way to generate a set of MSCs is via Hierarchical (or High-level) Message Sequence Charts (HMSCs) [10]. Without losing expressiveness, we consider only a subclass of HMSCs called Message Sequence Graphs (MSGs). An MSG is a finite directed graph in which each node is labeled by an MSC. An MSG defines a collection of MSCs by concatenating the MSCs labeling each path from an initial vertex to a terminal vertex.

A natural way to test a distributed implementation against an MSG specification is to substitute test processes for one or more components and record the interactions between the test process(es) and the rest of the system. We refer to this form of testing of distributed message-passing systems as *local testing*.

---

* Partially supported by *Timed-DISCOVERI*, a project under the Indo-French Networking Programme.

E. Csuhaj-Varjú and Z. Ésik (Eds.): FCT 2007, LNCS 4639, pp. 76–87, 2007.

The implementation is said to pass a local test if the observations at the test process(es) are consistent with the MSG specification.

An important impediment to local testing is the possibility of implied scenarios. Let $T = \{P_1, P_2, \ldots, P_k\}$ be a collection of subsets of processes. We say that an MSC M is T-implied by an MSC language $\mathcal{L}$ if the projections of M onto each subset $P_i \in T$ agree with the projections onto P_i of some good MSC $M_{P_i} \in \mathcal{L}$. Implied scenarios have been studied in [2,3], where the observations are restricted to individual processes rather than arbitrary subsets.

Let T_k denote the set of all subsets of processes of size k. We say that an MSC language $\mathcal{L}$ is k-testable if every T_k-implied scenario is already present in $\mathcal{L}$. In other words, if a specification is k-testable, it is possible to accurately test an implementation by performing a collection of local tests with respect to T_k. On the other hand, if $\mathcal{L}$ is not k-testable, even an exhaustive set of local tests with respect to T_k cannot rule out an undesirable implied scenario.

It has been shown in [3] that 1-testability is undecidable, even for regular MSG specifications. (The results of [3] are formulated in the context of distributed synthesis, but they can also be interpreted in terms of local testing.) We extend the results of [3] to show that for any n, k-testability of an MSG specification with n processes is undecidable, for all $k \in \{1, 2, \ldots, n-1\}$.

We also consider MSG specifications over n processes without message labels. Somewhat surprisingly, k-testability remains undecidable for $k \in \{2, \ldots, n-1\}$. However, for unlabelled MSG specifications, 1-testability is decidable.

The paper is organized as follows. We begin with preliminaries about MSCs, before we formally define k-testability in Section 3. The next section establishes various undecidability results. In Section 5, we show that 1-testability is decidable for unlabelled MSG specifications. We conclude with a brief discussion.

2   Preliminaries

2.1   Message Sequence Charts

Let $\mathcal{P} = \{p, q, r, \ldots\}$ be a finite set of processes (agents) that communicate with each other through messages via reliable FIFO channels using a finite set of message types $\mathcal{M}$. For $p \in \mathcal{P}$, let $\Sigma_p = \{p!q(m), p?q(m) \mid p \neq q \in \mathcal{P}, m \in \mathcal{M}\}$ be the set of communication actions in which p participates. The action $p!q(m)$ is read as p *sends the message m to q* and the action $p?q(m)$ is read as p *receives the message m from q*. We set $\Sigma = \bigcup_{p \in \mathcal{P}} \Sigma_p$. We also denote the set of *channels* by $Ch = \{(p, q) \in \mathcal{P}^2 \mid p \neq q\}$.

Labelled posets. A Σ-labelled poset is a structure $M = (E, \leq, \lambda)$ where $(E, \leq)$ is a partially ordered set and $\lambda : E \to \Sigma$ is a labelling function. For $e \in E$, let $\downarrow e = \{e' \mid e' \leq e\}$. For $p \in \mathcal{P}$ and $a \in \Sigma$, we set $E_p = \{e \mid \lambda(e) \in \Sigma_p\}$ and $E_a = \{e \mid \lambda(e) = a\}$, respectively. For $(p, q) \in Ch$, we define the relation $<_{pq}$:

$$e <_{pq} e' \stackrel{\text{def}}{=} \exists m \in \mathcal{M} \text{ such that } \lambda(e) = p!q(m),\ \lambda(e') = q?p(m) \text{ and}$$
$$|\downarrow e \cap E_{p!q(m)}| = |\downarrow e' \cap E_{q?p(m)}|$$

The relation $e <_{pq} e'$ says that channels are FIFO with respect to *each message*—if $e <_{pq} e'$, the message m read by q at e' is the one sent by p at e.

Finally, for each $p \in \mathcal{P}$, we define the relation $\leq_{pp} = (E_p \times E_p) \cap \leq$, with $<_{pp}$ standing for the largest irreflexive subset of $\leq_{pp}$.

Definition 1. *An MSC over $\mathcal{P}$ is a finite Σ-labelled poset $M = (E, \leq, \lambda)$ where:*

1. *Each relation $\leq_{pp}$ is a linear (total) order.*
2. *If $p \neq q$ then for each $m \in \mathcal{M}$, $|E_{p!q(m)}| = |E_{q?p(m)}|$.*
3. *If $e <_{pq} e'$, then $|{\downarrow}e \cap \left(\bigcup_{m \in \mathcal{M}} E_{p!q(m)} \right)| = |{\downarrow}e' \cap \left(\bigcup_{m \in \mathcal{M}} E_{q?p(m)} \right)|$.*
4. *The partial order $\leq$ is the reflexive, transitive closure of $\bigcup_{p,q \in \mathcal{P}} <_{pq}$.*

The second condition ensures that every message sent along a channel is received. The third condition says that every channel is FIFO across all messages.

In diagrams, the events of an MSC are presented in *visual order*. The events of each process are arranged in a vertical line and messages are displayed as horizontal or downward-sloping directed edges. Fig. 1 shows an example with three processes $\{p, q, r\}$ and six events $\{e_1, e_1', e_2, e_2', e_3, e_3'\}$ corresponding to three messages—m_1 from p to q, m_2 from q to r and m_3 from p to r.

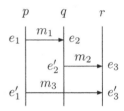

Fig. 1. An MSC

For an MSC $M = (E, \leq, \lambda)$, we let $\mathrm{lin}(M) = \{\lambda(\pi) \mid \pi$ is a linearization of $(E, \leq)\}$. For instance, $p!q(m_1)\ q?p(m_1)\ q!r(m_2)\ p!r(m_3)\ r?q(m_2)\ r?p(m_3)$ is one linearization of the MSC in Fig. 1.

MSC languages. An *MSC language* is a set of MSCs. We can also regard an MSC language $\mathcal{L}$ as a word language over Σ given by $\mathrm{lin}(\mathcal{L}) = \bigcup\{\mathrm{lin}(M) \mid M \in \mathcal{L}\}$.

Definition 2. *An MSC language $\mathcal{L}$ is said to be a* regular MSC language *if the word language $\mathrm{lin}(\mathcal{L})$ is a regular language over Σ.*

Let M be an MSC and $B \in \mathbb{N}$. We say that $w \in \mathrm{lin}(M)$ is B-bounded if for every prefix v of w and for every channel $(p, q) \in Ch$, $\sum_{m \in \mathcal{M}} |\pi_{p!q(m)}(v)| - \sum_{m \in \mathcal{M}} |\pi_{q?p(m)}(v)| \leq B$, where $\pi_\Gamma(v)$ denotes the projection of v on $\Gamma \subseteq \Sigma$. This means that along the execution of M described by w, no channel ever contains more than B-messages. We say that M is (universally) B-bounded if every $w \in \mathrm{lin}(M)$ is B-bounded. An MSC language $\mathcal{L}$ is B-bounded if every $M \in \mathcal{L}$ is B-bounded. Finally, $\mathcal{L}$ is bounded if it is B-bounded for some B.

We then have the following result [5].

Theorem 3. *If an MSC language $\mathcal{L}$ is regular then it is bounded.*

2.2   Message Sequence Graphs

Message sequence graphs (MSGs) are finite directed graphs with designated initial and terminal vertices. Each vertex in an MSG is labelled by an MSC. The edges represent (asynchronous) MSC concatenation, defined as follows.

Let $M_1 = (E^1, \leq^1, \lambda_1)$ and $M_2 = (E^2, \leq^2, \lambda_2)$ be a pair of MSCs such that E^1 and E^2 are disjoint. The *(asynchronous) concatenation* of M_1 and M_2 yields the MSC $M_1 \circ M_2 = (E, \leq, \lambda)$ where $E = E^1 \cup E^2$, $\lambda(e) = \lambda_i(e)$ if $e \in E^i$, $i \in \{1, 2\}$, and $\leq \, = (\leq^1 \cup \leq^2 \cup \bigcup_{p \in \mathcal{P}} E_p^1 \times E_p^2)^*$.

A *Message Sequence Graph* is a structure $\mathcal{G} = (Q, \rightarrow, Q_{in}, F, \Phi)$, where Q is a finite and nonempty set of states, $\rightarrow \, \subseteq Q \times Q$, $Q_{in} \subseteq Q$ is a set of initial states, $F \subseteq Q$ is a set of final states and Φ labels each state with an MSC.

A *path* π through an MSG $\mathcal{G}$ is a sequence $q_0 \rightarrow q_1 \rightarrow \cdots \rightarrow q_n$ such that $(q_{i-1}, q_i) \in \, \rightarrow$ for $i \in \{1, 2, \ldots, n\}$. The MSC generated by π is $M(\pi) = M_0 \circ M_1 \circ M_2 \circ \cdots \circ M_n$, where $M_i = \Phi(q_i)$. A path $\pi = q_0 \rightarrow q_1 \rightarrow \cdots \rightarrow q_n$ is a *run* if $q_0 \in Q_{in}$ and $q_n \in F$. The language of MSCs accepted by $\mathcal{G}$ is $L(\mathcal{G}) = \{M(\pi) \mid \pi \text{ is a run through } \mathcal{G}\}$. We say that an MSC language $\mathcal{L}$ is *MSG-definable* if there exists and MSG $\mathcal{G}$ such that $\mathcal{L} = L(\mathcal{G})$.

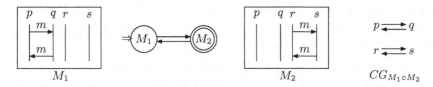

$$M_1 \qquad\qquad\qquad\qquad\qquad\qquad M_2 \qquad\qquad CG_{M_1 \circ M_2}$$

Fig. 2. A message sequence graph

An example of an MSG is depicted in Fig. 2. The initial state is marked $\Rightarrow$ and the final state has a double line. The language $\mathcal{L}$ defined by this MSG is *not* regular: $\mathcal{L}$ projected to $\{p!q(m), r!s(m)\}^*$ consists of $\sigma \in \{p!q(m), r!s(m)\}^*$ such that $|\sigma\!\upharpoonright_{p!q(m)}| = |\sigma\!\upharpoonright_{r!s(m)}| \geq 1$, which is not a regular string language.

In general, it is undecidable whether an MSG describes a regular MSC language [5]. However, a sufficient condition for the MSC language of an MSG to be regular is that the MSG be *locally synchronized*.

Communication graph. For an MSC $M = (E, \leq, \lambda)$, let CG_M, *the communication graph of* M, be the directed graph $(\mathcal{P}, \mapsto)$ where:

– $\mathcal{P}$ is the set of processes of the system.
– $(p, q) \in \, \mapsto$ iff there exists an $e \in E$ with $\lambda(e) = p!q(m)$.

M is said to be *com-connected* if CG_M consists of one nontrivial strongly connected component and isolated vertices.

Locally synchronized MSGs. The MSG $\mathcal{G}$ is *locally synchronized* [12] (or *bounded* [1]) if for every loop $\pi = q \rightarrow q_1 \rightarrow \cdots \rightarrow q_n \rightarrow q$, the MSC $M(\pi)$ is com-connected. In Fig. 2, $CG_{M_1 \circ M_2}$ is not com-connected, so the MSG is not locally synchronized. We have the following result for MSGs [1].

Theorem 4. *If $\mathcal{G}$ is locally synchronized, $L(\mathcal{G})$ is a regular MSC language.*

3   Locally Testable MSC Languages

In local testing, we substitute test process(es) for one or more components and
record the interactions between the test process(es) and the rest of the system.
The implementation is said to pass a local test if the observations at the test
process(es) are consistent with the MSG specification. An important impediment
to local testing is the possibility of implied scenarios.

Definition 5. *Let $M = (E, \leq, \lambda)$ be an MSC and $P \subseteq \mathcal{P}$ a set of processes. The
P-observation of M, $M{\upharpoonright}_P$, is the collection of local observasions $\{(E_p, \leq_{pp})\}_{p \in P}$,
where $\leq_{pp} = \leq \cap (E_p \times E_p)$. The collection $\{(E_p, \leq_{pp})\}_{p \in P}$ can also be viewed as a
labelled partial order $(E_P, \leq_P)$ where $E_P = \bigcup_{p \in P} E_p$ and $\leq_P = \left(\bigcup_{p,q \in P} <_{pq} \right)^*$.*

*Let $T \subseteq 2^{\mathcal{P}}$ be a family of subsets of processes. An MSC M is said to be
T-implied by an MSC-language $\mathcal{L}$ if for every subset $P \in T$ there is an MSC
$M_P \in \mathcal{L}$ such that $M_P{\upharpoonright}_P = M{\upharpoonright}_P$.*

*We denote by T_k the set $\{P \subseteq \mathcal{P} \mid |P| = k\}$ of all subsets of $\mathcal{P}$ of size k and
we say that an MSC is k-implied if it is T_k-implied.*

Fig. 3 illustrates the idea of implied scenarios. The MSC M' is 1-implied by
$\{M_1, M_2\}$. However, M' is not 2-implied by $\{M_1, M_2\}$ because the $\{p, s\}$-
observation of M' does not match either M_1 or M_2.

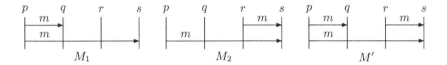

Fig. 3. An example of implied scenarios

We are interested in checking the global behaviour of a distributed implementation by testing it locally against an MSG specification. For this to be meaningful, the MSG should be closed with respect to implied scenarios generated by
the test observations. This leads to the following definition.

Definition 6. *Let $|\mathcal{P}| = n$. An MSG $\mathcal{G}$ is said to be k-testable if every scenario
M that is k-implied by $L(\mathcal{G})$ is already a member of $L(\mathcal{G})$.*

We have the following negative result from [3] (adapted to our context).

Theorem 7. *Let $\mathcal{G}$ be a locally-synchronized MSG, so that $L(\mathcal{G})$ is a regular
MSC language. It is undecidable whether $L(\mathcal{G})$ is 1-testable.*

This result is somewhat surprising, since the analogous problem for synchronous
systems is decidable [16]. The root cause of this undecidability is the fact that
even when a MSC language $\mathcal{L}$ is regular, and hence B-bounded for some B,
the set of scenarios implied by $\mathcal{L}$ may not be bounded. An example is shown in
Fig. 4—all messages are labelled m and labels are omitted.

Since M_1 and M_2 are both com-connected, the language $(M_1 + M_2)^*$ is a regular MSG-definable language.

On the other hand, for each $k \in \mathbb{N}$, the MSC in which the p-observation matches $M_1^{2k} M_2^k$ and the q-observation matches $M_2^k M_1^{2k}$ has a global cut where the channel (p, q) has capacity $k + 1$. The figure shows the case $k = 2$. The dotted line marks the global cut where the channel (p, q) has maximum capacity.

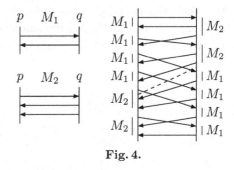

Fig. 4.

4   Undecidability

We know from [3] that 1-testability is undecidable for regular MSG-definable languages. The example in Fig. 3 suggests that it might be possible to determine the smallest $k < n$ such that an MSG specification with n processes is k-testable. (Observe that every MSC language over n processes is trivially n-testable.) Unfortunately, this is not the case. For all $k < n$, the problem of determining whether a regular MSG specification is k-testable is undecidable.

The undecidability proofs in this section use reductions from the *Modified Post's Correspondence Problem* (MPCP) [6]. An instance of MPCP is a collection $\{(v_1, w_1), (v_2, w_2) \ldots, (v_r, w_r)\}$ of pairs of words over an alphabet Σ. A solution is a sequence $i_2 i_3 \ldots i_m$ of indices from $\{1, 2, \ldots, r\}$ such that $v_1 v_{i_2} v_{i_3} \cdots v_{i_m} = w_1 w_{i_2} w_{i_3} \cdots w_{i_m}$. It is proved in [6] that checking whether an instance of MPCP admits a solution is undecidable. A careful examination of the proof in [6] shows that MPCP is undecidable even under following assumptions:

1. For each word u in the list $\{(v_1, w_1), (v_2, w_2) \ldots, (v_r, w_r)\}$, $1 \leq |u| \leq 4$.
2. w_1 is a strict prefix of v_1 and is shorter by at least 2 letters.
3. If the instance has a solution then it has a solution of the form $i_2 i_3 \ldots i_m$ such that $w_1 w_{i_2} \ldots w_{i_k}$ is a strict prefix of $v_1 v_{i_2} \ldots v_{i_k}$ for each $k < m$.

Theorem 8. *For $3 \leq k \leq n$, $(k - 1)$-testability is undecidable for regular 1-bounded MSG-definable languages over n processes.*

Proof. Let $\Delta = \{(v_1, w_1), (v_2, w_2), \ldots, (v_r, w_r)\}$ be an instance of MPCP satisfying the assumptions described above. For each pair (v_ℓ, w_ℓ), we construct k MSCs M_{v_ℓ}, M_{w_ℓ} and $\{M_{v_\ell, w_\ell}^j \mid 1 < j < k\}$ over processes $\{1, 2, \ldots, n\}$, such that only processes $\{1, 2, \ldots, k\}$ are active in these k MSCs. The message alphabet for these MSCs is the alphabet of the MPCP instance along with the integers $\{1, 2, \ldots, r\}$. In the definition below, v_ℓ^j and w_ℓ^j are the jth symbols in the strings v_ℓ and w_ℓ, respectively. Also, $i \xrightarrow{m} j$ denotes the MSC generated by the sequence $i!j(m) \, j?i(m)$ where i sends message m to j. For $m \in \mathcal{M}$ and $i < j$ we define

$$N_m^{i,j} = (i \xrightarrow{m} i+1) \cdots (j-1 \xrightarrow{m} j)(j \xrightarrow{m} j-1) \cdots (i+1 \xrightarrow{m} i).$$

In this MSC, the message m is sent from i to j through the intermediate processes $i+1, \ldots, j-1$ and an acknowledgment is sent back from j to i through the same route. We also let $N_\ell = (k \xrightarrow{\ell} 1)$ and define for $1 < j < k$ the MSCs

$$M_{v_\ell} = N_\ell N^{1,k}_{v^1_\ell} \cdots N^{1,k}_{v^{|v_\ell|}_\ell}$$

$$M_{w_\ell} = N_\ell N^{1,k}_{w^1_\ell} \cdots N^{1,k}_{w^{|w_\ell|}_\ell}$$

$$M^j_{v_\ell, w_\ell} = N_\ell N^{1,j}_{v^1_\ell} N^{j,k}_{w^1_\ell} \cdots N^{1,j}_{v^{|w_\ell|}_\ell} N^{j,k}_{w^{|w_\ell|}_\ell} N^{1,j}_{v^{|w_\ell|+1}_\ell} \cdots N^{1,j}_{v^{|v_\ell|}_\ell} \qquad \text{if } |w_\ell| \le |v_\ell|$$

$$M^j_{v_\ell, w_\ell} = N_\ell N^{1,j}_{v^1_\ell} N^{j,k}_{w^1_\ell} \cdots N^{1,j}_{v^{|v_\ell|}_\ell} N^{j,k}_{w^{|v_\ell|}_\ell} N^{j,k}_{w^{|v_\ell|+1}_\ell} \cdots N^{j,k}_{w^{|w_\ell|}_\ell} \qquad \text{otherwise.}$$

Since each word in the MPCP instance is nonempty, each of these MSCs is com-connected, so any MSG whose node labels are drawn from this set of MSCs is guaranteed to be locally-synchronized. For $1 < j < k$, we define

$$L_v = M_{v_1}\{M_{v_\ell} \mid 1 \le \ell \le r\}^*$$
$$L_w = M_{w_1}\{M_{w_\ell} \mid 1 \le \ell \le r\}^*$$
$$L^j_{v,w} = M^j_{v_1, w_1}\{M^j_{v_\ell, w_\ell} \mid 1 \le \ell \le r\}^*$$
$$L_\Delta = L_v \cup L_w \cup \bigcup_{1<j<n} L^j_{v,w}.$$

Claim. Δ has a solution iff L_Δ is not $(k-1)$-testable.

Let $i_2, i_3, \ldots, i_m$ be a solution of Δ that satisfies Condition 3 listed above. Let $v_1 v_{i_2} \ldots v_{i_m} = w_1 w_{i_2} \ldots w_{i_m} = a_1 a_2 \ldots a_\ell$. Then, we first construct the MSC $M' = N^{1,k}_{a_1} N^{1,k}_{a_2} \cdots N^{1,k}_{a_\ell}$. In M', we insert events labelled $k!1(1)$, $k!1(i_2), \ldots, k!1(i_m)$ into k so as to partition its communications with $k-1$ as $w_1, w_{i_2}, \ldots, w_{i_m}$. Finally, we insert events labelled $1?k(1)$, $1?k(i_2)$, $\ldots$, $1?k(i_m)$ into 1 to partition its communications with 2 as $v_1, v_{i_2}, \ldots, v_{i_m}$. Call this MSC M. To observe that M is indeed a valid MSC, we note that for each $j < m$, $w_1 w_{i_2} \ldots w_{i_j}$ is a prefix of $v_1 v_{i_2} \ldots v_{i_j}$, so the receive event $1?k(i_j)$ inserted into 1 can occur later than the corresponding send event $k!1(i_j)$ inserted into n.

It is easy to verify that $M\restriction_{\{1,2,\ldots,k-1\}} = (M_{v_1} M_{v_{i_2}} \cdots M_{v_{i_m}})\restriction_{\{1,2,\ldots,k-1\}}$. Similarly, $M\restriction_{\{2,3,\ldots,k\}} = (M_{w_1} M_{w_{i_2}} \cdots M_{w_{i_m}})\restriction_{\{2,3,\ldots,k\}}$. Finally, for $1 < j < k$ we have $M\restriction_{\{1,\ldots,j-1,j+1,\ldots,k\}} = (M^j_{v_1 w_1} M^j_{v_{i_2} w_{i_2}} \cdots M^j_{v_{i_m} w_{i_m}})\restriction_{\{1,\ldots,j-1,j+1,\ldots,k\}}$. Thus M is $(k-1)$-implied by L_Δ.

To see that M is not already in L_Δ, simply observe that there is at least one event in M between the second $k!1$ event and the second $1?k$ event and this is not the case for any MSC in L.

Conversely, suppose there is an MSC $M \notin L_\Delta$ that is $(k-1)$-implied by L_Δ. The MSC M must be of one of the following two types:

Type 1. $M\restriction_{\{j\}} \notin (\{N^{1,k}_m \mid m \in \mathcal{M}\}^*)\restriction_{\{j\}}$ for some $1 < j < k$.

Type 2. $M\restriction_{\{j\}} \in (\{N^{1,k}_m \mid m \in \mathcal{M}\}^*)\restriction_{\{j\}}$ for all $1 < j < k$.

If M is of type 1 as witnessed by j, it must be the case that $M{\upharpoonright}_{\{1,2,\ldots,k-1\}} = (M^j_{v_{i_1} w_{i_1}} M^j_{v_{i_2} w_{i_2}} \cdots M^j_{v_{i_m} w_{i_m}}){\upharpoonright}_{\{1,2,\ldots,k-1\}}$. Similarly, we also have $M{\upharpoonright}_{\{2,3,\ldots,k\}} = (M^j_{v_1 w_1} M^j_{v_{i_2} w_{i_2}} \cdots M^j_{v_{i_m} w_{i_m}}){\upharpoonright}_{\{2,3,\ldots,k\}}$. Hence, $M = M^j_{v_1 w_1} M^j_{v_{i_2} w_{i_2}} \cdots M^j_{v_{i_m} w_{i_m}}$, which in turn implies that $M \in L_\Delta$ thus contradicting our initial assumption. Therefore M cannot be of type 1.

On the other hand, if M is of type 2, we show that if M is 1-implied by L_Δ then either $M \in L_v \cup L_w$ or Δ has a solution. Note that this is a stronger result since we only assume that M is 1-implied instead of $(k-1)$-implied.

We have $(L^j_{v,w}){\upharpoonright}_1 = (L_v){\upharpoonright}_1$ and $(L^j_{v,w}){\upharpoonright}_k = (L_w){\upharpoonright}_k$. Hence, $M{\upharpoonright}_1 \in (L_v \cup L_w){\upharpoonright}_1$ and $M{\upharpoonright}_k \in (L_v \cup L_w){\upharpoonright}_k$. Using in addition the fact that M is of type 2, we deduce that if we remove from M the messages from k to 1 we obtain an MSC $M' = N^{1,k}_{a_1} N^{1,k}_{a_2} \cdots N^{1,k}_{a_\ell}$ for some word $a_1 a_2 \cdots a_\ell$.

Now, if $M \notin L_v \cup L_w$ then we must have $M{\upharpoonright}_1 \in (L_v){\upharpoonright}_1$ and $M{\upharpoonright}_k \in (L_w){\upharpoonright}_k$ (otherwise the second message from k to 1 would induce a cycle in the MSC). Therefore, the sequence of messages from k to 1 parses on the left the sequence $a_1 a_2 \cdots a_\ell$ into some $v_1 v_{i_2} \cdots v_{i_m}$ and on the right the same sequence into $w_1 w_{i_2} \cdots w_{i_m}$ and Δ has a solution. $\qquad \square$

Remark 9. We can modify the proof to obtain the undecidability of 1-testability even for regular 1-bounded MSG-definable languages over $n \geq 3$ processes. Below, we get down to 2 processes but the regular language is only 4-bounded.

4.1   Undecidability of 1-Testability for 2 Processes

The argument in [3] shows that 1-testability is undecidable for regular MSG-definable languages with four processes. We tighten this result to show that 1-testability is undecidable for regular MSG-definable languages over 2 processes.

Theorem 10. *For $n \geq 2$, 1-testability is undecidable for regular 4-bounded MSG-definable languages over n processes.*

Proof. As before, let $\Delta = \{(v_1, w_1), (v_2, w_2), \ldots, (v_r, w_r)\}$ be an instance of MPCP satisfying the assumptions (1–3) stated earlier. With each word $v_i = a_1 a_2 \ldots a_k$ we associate an MSC M_{v_i} as indicated in Fig. 5.

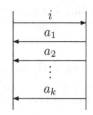

Fig. 5.

Similarly we construct the MSCs M_{w_i}. First, observe that each of these MSCs is com-connected, so any MSG that uses these MSCs as node labels is locally synchronized. Also, from assumption 1 of the MPCP instance, the MSCs are 4-bounded and therefore, any language generated by these MSCs is 4-bounded.

Let $L_v = \{M_{v_i} \mid 1 \leq i \leq r\}$ and $L_w = \{M_{w_i} \mid 1 \leq i \leq r\}$. Consider the MSG-definable regular language $L_\Delta = M_{v_1}.(L_v)^* + M_{w_1}.(L_w)^*$.

If M is any MSC in L then $M{\upharpoonright}_1$ is a word of the form $1!2(1)\ 1?2(x_1)\ 1!2(i_2)\ 1?2(x_{i_2})\ \cdots\ 1!2(i_k)\ 1?2(x_{i_k})$ where either each x_{i_j} is v_{i_j} or each x_{i_j} is w_{i_j}. A similar property holds for $M{\upharpoonright}_2$.

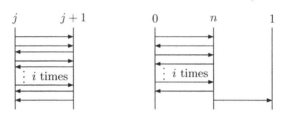

Fig. 6.

Suppose there is a 1-implied MSC M that is not in L_Δ. Then, $M\lceil_1 = (M_{w_1}M_{w_{i_2}}\cdots M_{w_{i_k}})\lceil_1$ and $M\lceil_2 = (M_{v_1}M_{v_{i_2}}\cdots M_{v_{i_k}})\lceil_2$. It follows that the MPCP instance Δ has a solution.

Conversely, from any solution $i_2i_3\ldots i_k$ to the MPCP instance Δ, it is quite easy to construct a 1-implied scenario where p_1 witnesses the $w_1w_{i_2}\cdots w_{i_k}$ and p_2 witnesses the $v_1v_{i_2}\cdots v_{i_k}$. □

Finally, we turn our attention to MSGs over a singleton message alphabet. As we shall see in the next section, 1-testability is decidable for locally-synchronized MSG languages over singleton alphabets. However, k-testability is undecidable for any $k > 1$.

Theorem 11. *Let n and k be any two integers with $n > 2$ and $1 < k < n-1$. There is a constant B such that the problem of deciding whether a B-bounded HMSC language over a singleton alphabet is k-testable is undecidable.*

Proof. (Sketch) Following the proof of Theorem 10, it suffices to prove the result for $k = n-2$. We modify the reduction used in the proof of Theorem 8 to use a singleton message alphabet. Let us assume that the message alphabet is $\{a_1, a_2, \ldots a_k\}$. A communication a_i between process j and $j + 1$ is replaced by the communication pattern at the left of Fig. 6. Since $k > 1$, any subset containing j and $j + 1$ would witness that the communication between j and $j + 1$ is uniquely and correctly parsed. We still have to deal with the message from process n to 1. We add an additional process 0 and simulate the act of sending i from n to 1 by the MSC at the right of Fig. 6. Since $k > 1$, the pair $\{n, 0\}$ will jointly witness that i is sent from n to 1. □

5   Decidability

In this section we consider the 1-testability problem for regular MSC languages where the message alphabet for each channel is a singleton. In this case, we may omit the message content in any event. Throughout this section, we write $p!q$ and $q?p$ rather than $p!q(m)$ and $q?p(m)$.

Proper and complete words. For a word w and a letter a, $\#_a(w)$ denotes the number of times a appears in w. We say that $\sigma \in \Sigma^*$ is *proper* if for every prefix τ of σ and every pair p, q of processes, $\#_{p!q}(\tau) \geq \#_{q?p}(\tau)$. We say that σ is *complete* if σ is proper and $\#_{p!q}(\sigma) = \#_{q?p}(\sigma)$ for every pair p, q of processes.

Every linearization of any MSC is a complete word and every complete word is the linearization of a unique MSC.

Suppose L is the set of linearizations of a MSC language. Let $L_p = \{w\!\upharpoonright_p \mid w \in L\}$. Let, $\textit{1-closure}(L) = \{w \mid w \text{ is complete and } \forall p.\ w\!\upharpoonright_p \in L_p\}$. Observe that $\textit{1-closure}(L)$ is the set of 1-implied words of L.

Let L be the set of linearizations of some regular MSC language over a singleton message alphabet. From any finite automaton $A = (Q, \Sigma, \delta, i, F)$ accepting L we can easily construct for each $p \in \mathcal{P}$ an automaton $A_p = (Q_p, \Sigma\!\upharpoonright_p, \delta_p, i_p, F_p)$ that accepts L_p. Note that $\textit{1-closure}(L)$ is exactly the set of complete words accepted by the (free) product $\prod_p A_p$ of these automata. The product automaton accepts a regular language. The difficulty is in ensuring that a word that is accepted is complete. However, since the message alphabet is a singleton, it suffices to keep track of the number of sent and as yet unreceived messages along any channel. This leads us naturally to the following idea.

From these automata $(A_p)_{p \in \mathcal{P}}$, we construct a labelled Petri net N whose firing sequences are related to words in $\textit{1-closure}(L)$.[1]

1. The set of places is $\bigcup_{p \in \mathcal{P}} Q_p \cup \{c_{pq} \mid p, q \in \mathcal{P}\}$.
2. The set of transtions is $\bigcup_{p \in \mathcal{P}} \delta_p$.
3. The transition $(s, p!q, t) \in \delta_p$ removes a token from the place s and deposits a token each at the places t and c_{pq}.
4. The transition $(s, q?p, t) \in \delta_p$ removes a token each from the places s and c_{qp} and deposits a token at t.
5. The initial marking has one token in each place i_p, $p \in \mathcal{P}$, corresponding to the initial states of the automata A_p.
6. The label on the transition (s, x, t) is $x \in \Sigma$.

From the definition of N it follows that in any reachable marking, for any $p \in \mathcal{P}$, exactly one place in Q_p has a token. We say that a marking of this net is *final* if every place of the form c_{pq} is empty and for each $p \in \mathcal{P}$ there is $f_p \in F_p$ such that f_p is marked. There are only finitely many final markings.

It is quite easy to observe that a word $w \in \textit{1-closure}(L)$ if and only if there is a firing sequence labelled w from the initial marking to some final marking. This leads us naturally to the following proposition:

Proposition 12. *Let B be any integer. We can decide if 1-closure(L) contains a word that is not B-bounded.*

Proof. The set of markings where exactly one of the places of the form c_{pq} has $B + 1$ tokens (and all other places have at most B tokens) is finite. Since reachability is decidable for Petri nets [8,9], we can check for each such marking χ whether χ is reachable from the initial marking and if some final marking is reachable from χ. □

Now, if the given MSC language $\mathcal{L}$ is regular we can compute a bound B from its presentation such that $\mathcal{L}$ is B-bounded. Using the proposition above, we can

---

[1] Due to lack of space, we are constrained to omit basic definitions concerning Petri nets. See [14] for a detailed introduction.

check if *1-closure*(L) contains words that are not B-bounded. If the answer is yes, then L is not 1-testable. On the other hand, if there are no words in *1-closure*(L) that violate the B bound on any channel, we can look for 1-implied scenarios using the following proposition.

Proposition 13. *Let L be a B-bounded MSC regular language. We can decide if* 1-closure(L) *contains any B-bounded words not in L.*

Proof (Sketch). Construct the net N corresponding to the product automaton $\prod_p A_p$ as described earlier. Explore all reachable configurations in which each place in $\{c_{pq} \mid p, q \in \mathcal{P}\}$ has no more than B tokens. This results in a finite automaton that accepts all the B-bounded words in *1-closure*(L). □

From the two propositions described above, we conclude that:

Theorem 14. *The 1-testability problem for regular MSC languages over a singleton message alphabet is decidable.*[2]

In fact, in this case we can even decide if *1-closure*(L) is regular.

Theorem 15. *Let L be a regular MSC language over a singleton message alphabet. Then, it is decidable whether* 1-closure(L) *is regular.*

Proof. We reduce this to the Intermediate Marking Problem (IMP) for Petri nets, which is known to be decidable [17].

Consider the Petri net constructed above. Define an *intermediate marking* to be one that can be reached from the initial configuration and from which some final marking is reachable. If the number of intermediate markings is finite, there is a bound B such that along any firing sequence from the initial marking to a final marking, no place ever contains more than B tokens. In other words, if w is the word generated by some firing sequence from the initial to a final configuration then the number of unreceived messages at any prefix of w is bounded by B. Thus, the language *1-closure*(L) is the language of a bounded Petri net and hence regular.

On the other hand, if the number of intermediate markings is infinite, we may conclude that for any B there is a word $w \in$ *1-closure*(L) which has a prefix with B sent and as yet unreceived messages. Thus *1-closure*(L) is not B-bounded for any B and hence not regular. □

6   Discussion

We have seen in this paper that developing a framework for locally testing MSC based specifications is hard. This is because MSG-based specifications permit unintended implied scenarios that cannot, in general, be detected algorithmically.

There are two approaches to attack the problem of local testing in light of this bottleneck. One is to characterize structural conditions for k-testability. This

---

[2] This theorem can also be viewed as a special case of the result proved in [11] that 1-testability is decidable for MSCs without fifo channels, but our proof for this special case is simpler than the general proof in [11].

is analogous to identifying locally synchronized MSGs as those that generate regular MSC specifications, even though the general problem of checking whether an MSG specification describes a regular MSC language is undecidable [5].

Another tactic would be to recognize that practical implementations always work with bounded buffers and impose an upper bound B on the buffer size. The set of B-bounded MSCs in the k-closure of a regular MSC language is again regular, so the B-bounded k-testability problem is decidable for all regular MSG-definable languages. The focus could now be on efficiently identifying the smallest k for which an MSG specification is k-testable. Another interesting problem is to identify a minimal set of tests to validate a k-testable specification.

References

1. Alur, R., Yannakakis, M.: Model checking of message sequence charts. In: Baeten, J.C.M., Mauw, S. (eds.) CONCUR 1999. LNCS, vol. 1664, pp. 114–129. Springer, Heidelberg (1999)
2. Alur, R., Etessami, K., Yannakakis, M.: Inference of message sequence graphs. IEEE Trans. Software Engg. 29(7), 623–633 (2003)
3. Alur, R., Etessami, K., Yannakakis, M.: Realizability and Verification of MSC Graphs. Theor. Comput. Sci. 331(1), 97–114 (2005)
4. Booch, G., Jacobson, I., Rumbaugh, J.: Unified Modeling Language User Guide. Addison-Wesley, London, UK (1997)
5. Henriksen, J.G., Mukund, M., Narayan Kumar, K., Sohoni, M., Thiagarajan, P.S.: A Theory of Regular MSC Languages. Inf. Comp. 202(1), 1–38 (2005)
6. Hopcroft, J.E., Ullman, J.D.: Introduction to Automata Theory, Languages and Computation. Addison-Wesley, London, UK (1979)
7. ITU-TS Recommendation Z.120: Message Sequence Chart (MSC). ITU-TS, Geneva (1997)
8. Kosaraju, S.R.: Decidability of Reachability in Vector Addition Systems. Proc 14th ACM STOC, 267–281 (1982)
9. Mayr, E.W.: An Algorithm for the General Petri Net Reachability Problem. SIAM J. Comput 13(3), 441–460 (1984)
10. Mauw, S., Reniers, M.A. (eds.): High-level message sequence charts. In: Proc SDL'97, pp. 291–306. Elsevier, Amsterdam (1997)
11. Morin, R.: Recognizable Sets of Message Sequence Charts. In: Alt, H., Ferreira, A. (eds.) STACS 2002. LNCS, vol. 2285, pp. 523–534. Springer, Heidelberg (2002)
12. Muscholl, A., Peled, D.: Message sequence graphs and decision problems on Mazurkiewicz traces. In: Kutyłowski, M., Wierzbicki, T., Pacholski, L. (eds.) MFCS 1999. LNCS, vol. 1672, pp. 81–91. Springer, Heidelberg (1999)
13. Muscholl, A., Peterson, H.: A note on the commutative closure of star-free languages. Information Processing Letters 57(2), 71–74 (1996)
14. Reisig, W., Rozenberg, G. (eds.): Lectures on Petri Nets I: Basic Models, Advances in Petri Nets. LNCS, vol. 1491. Springer, Heidelberg (1998)
15. Rudolph, E., Graubmann, P., Grabowski, J.: Tutorial on message sequence charts. Computer Networks and ISDN Systems — SDL and MSC 28 (1996)
16. Thiagarajan, P.S.: A Trace Consistent Subset of PTL. In: Lee, I., Smolka, S.A. (eds.) CONCUR 1995. LNCS, vol. 962, pp. 438–452. Springer, Heidelberg (1995)
17. Wimmel, H.: Infinity of Intermediate States Is Decidable for Petri Nets. In: Cortadella, J., Reisig, W. (eds.) ICATPN 2004. LNCS, vol. 3099, pp. 426–434. Springer, Heidelberg (2004)

On Notions of Regularity for Data Languages*

Henrik Björklund and Thomas Schwentick

University of Dortmund

Abstract. Motivated by considerations in XML theory and model checking, data strings have been introduced as an extension of finite alphabet strings which carry, at each position, a symbol *and* a data value from an infinite domain. Previous work has shown that it is not easy to come up with an expressive yet decidable automata model for data languages. Recently, such an automata model, *data automata*, was introduced. This paper introduces a simpler but equivalent model and investigates its expressive power, algorithmic and closure properties and some extensions.

1 Introduction

Regular string languages are clearly one of the most fundamental concepts in (Theoretical) Computer Science. They have applications in basically all branches of Computer Science. It can be argued that the following properties are the basis of their success: (1) Expressiveness, (2) Decidability, (3) Efficiency, (4) Closure properties, and (5) Robustness. The notion of regularity has been generalized to other structures, such as infinite strings and finite or infinite trees. Recent applications of regular languages are in Model Checking and XML processing.

- In model checking, whether a formula holds in a system is checked on the product of the system automaton and an automaton obtained from the formula. The step from the "real" system to its finite state representation usually involves many abstractions, especially with respect to data values. Even though this approach has been successful and found its way into large scale industrial applications, it has some inherent shortcomings. As an example, n identical processes with m states each give rise to an overall model size of m^n. If the number of processes is unbounded and/or unknown in advance the finite state approach fails. Decidability can sometimes still be obtained by restricting the problem in various ways [9,1].
- In XML document processing, regular concepts occur in various contexts. First, most applications restrict the structure of the allowed documents to conform to a certain specification (DTD or XML Schema), which can be modeled as a regular tree language. Second, navigation (XPath) and transformation (XSLT) languages have tight connections to various tree automata models and other regular description mechanisms (see, e.g., [13]).

  These approaches abstract away from attributes of XML documents, and concentrate on their structure. This is not always enough: a *schema* should

---

* This work was supported by the DFG Grant SCHW678/3-1.

E. Csuhaj-Varjú and Z. Ésik (Eds.): FCT 2007, LNCS 4639, pp. 88–99, 2007.

also allow definitions of restrictions on the data values through integrity constraints. This problem has been addressed (see, e.g., [3]), but as in the case of model checking, the methods largely rely on a case-to-case analysis.

Thus, in both settings, the finite state abstraction leads to interesting results but does not address all problems arising in applications. It would already be a big step if each position could carry a *data value* in addition to its label. This paper is part of a broader research program which aims at studying such extensions in a systematic way. We concentrate on the setting, where data values can only be tested for equality. Furthermore, we only consider finite *data strings*.

Quite a number of specification mechanisms for data languages have been suggested, such as *register automata* [11], *pebble automata* [14], *quasi-regular expressions* [12], *data automata* [4], and *LTL with freeze quantifier* [6]. For an overview of these models, see [16]. There are also investigations which assume more knowledge of the data [5,17]. Two observations are immediate from this work: the classes of data languages are very heterogeneous, i.e., pairs of defined classes are often incomparable, and one quickly obtains undecidable mechanisms.

Thus, the question remains whether there is a decent notion of regularity for data languages. The results so far are not very promising. Maybe there is no single class of data languages sharing all required properties. Rather there could be several classes fulfilling the requirements only to a certain extent. The research dedicated to this question therefore has to study a broad variety of models in order to identify important concepts.

In this paper, our requirements on expressive power are guided by the goal of model checking in the presence of an unbounded number of processes. Each computation naturally gives rise to a data string, where the data values represent process numberss. We aim at describing *global properties* of the computation, taking the whole string into consideration, as well as *local properties* which concern the actions of individual processes. As an example, consider processes sharing a printer. Consider three kinds of events: a print job can be requested (r), start (s), and terminate (t). A global property could be that a started job must terminate before the next job can start, giving rise to a regular constraint of the form $(r^* sr^* tr^*)^*$. A natural local property is stated by the regular expression $(rst)^*$, i.e., each process goes arbitrarily often through a request-start-terminate cycle. Thus, we are interested in mechanisms which are at least able to specify global properties as well as local properties through regular (finite alphabet) languages R_{glob} and R_{loc}, respectively. Formalisms will differ in their ability to coordinate the local and global properties.

Register automata [11] are a quite natural decidable model for data languages.[1] They are able to deal with any regular global properties, but are weak when it comes to local properties.

Another natural approach to specification is through logics. Data strings can be modeled as finite structures. Due to the limited access to data values, they can be represented by an equivalence relation. Although first-order logic on data strings is undecidable, the two-variable fragment has decidable satisfiability [4].

---

[1] The authors called the model *finite memory automata*, but we adopt the name *register automata*, used in the later literature.

The latter paper introduced a new automaton model for data strings, *data automata (DAs)*. As they have the expressive power described above and decidable emptiness, they fulfill, to some extent, the requirements (1) expressiveness and (2) decidability. Requirements (3)-(5) were not studied in depth in [4], but a characterization of the class $\mathcal{R}$ of data languages accepted by DAs in terms of an existential MSO logic was given. Thus a certain robustness was established. We study $\mathcal{R}$ and some extensions and restrictions more thoroughly.

Contributions. First, we address the robustness of $\mathcal{R}$, and show that there are a couple of simplifications of DAs which do not affect their expressive power (cf. 3.4 and 3.6). We arrive at the new, equivalent model of *class memory automata (CMAs)*. Next, in one of our main results, we confirm the expressiveness of $\mathcal{R}$ by showing that it (strictly) captures all data languages accepted by register automata (4.1). We then turn to the complexity of model checking, and first consider register automata. Even though their data complexity is polynomial, the combined complexity is NP-complete [15]. The number k of registers turns out to be crucial.With respect to k, the problem is $W[1]$-complete (5.3). For CMAs, the data complexity is already NP-complete (5.1).

The high data complexity of CMAs suggests the consideration of *deterministic* CMAs. The data and combined complexity of model checking become polynomial (5.1). Even though deterministic CMAs can express regular global and local properties, they are considerably weaker than CMAs, as they neither capture register automata (4.2) nor two-variable logics (4.3). Allowing deterministic CMAs to operate in a two-way fashion results in undecidability (5.4).

$\mathcal{R}$ is closed under union, intersection, product, and concatenation, but neither under complement nor under Kleene star. The former follows from the undecidability of universality for register automata [14], the latter is Proposition 6.1. Deterministic CMAs are closed under intersection but not under union, concatenation, or complement. To obtain a deterministic model closed under Boolean operations, we add certain Presburger conditions to the acceptance conditions (6.3). Despite its closure under negation this model is still decidable.

Since $\mathcal{R}$ is still unable to handle a number of natural properties, we investigate how much it can be extended, while preserving decidability. More precisely, we consider two extensions of CMAs which allow more interaction between global and local properties: one model with a synchronization mechanism and one with the ability to "reset" information seen for a data value. Returning to our printer example, these automata can handle, e.g., restarts of the system.

Most proofs have been left out, and will appear in the full version of the paper. An overview of the classes under consideration is given by Figure 1.

2   Preliminaries

Data words. Let Σ be a finite alphabet and Δ an infinite set. A **data word** is a finite sequence over $\Sigma \times \Delta$. A **data language** is a set of such words. If $w = (a_1, d_1) \ldots (a_n, d_n)$, then $\text{str}(w) = a_1 \ldots a_n$ is the **string projection** of w. The *marked string projection* $\text{mstr}(w)$ is the string $(a_1, b_1) \cdots (a_n, b_n)$ over $\Sigma \times \{0, 1\}$ for which $b_i = 1$ iff $d_i = d_{i-1}$ ($b_1 = 0$) For each data value d, the set

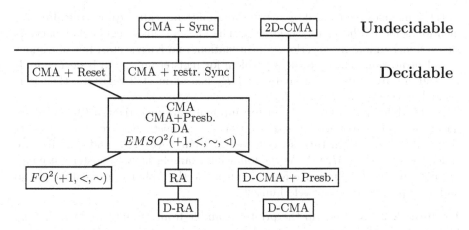

Fig. 1. A schematic picture of inclusions among classes of data languages. The lowermost three (branches of) classes are pairwise incomparable.

of all positions with value d is called a **class** of w, the string induced by these positions is called a **class string**. A position j is called the **class successor** of a position i (denoted $i \lhd j$), if both have the same data value, and there is no other position with the same value between i and j. Unless otherwise stated data values can only be compared with respect to equality. In the sequel, we will assume w.l.o.g. that all data languages and automata we investigate are defined over the same data set Δ, which contains all data values used in examples and proofs. In particular $\mathbb{N} \subseteq \Delta$. We will also talk about data languages over Σ, where Σ is an finite alphabet, implicitly assuming that the data set is Δ.

Register automata. Register automata (RAs) were introduced in [11] and have later been studied in, e.g., [14,6]. They were defined for sequences of data values only, but the generalization to data words is straightforward. RAs have registers in which they can store data values which can later be compared with the data value of the current position. We extend the notion of [11].

Definition 2.1 ([11,14]). A **register automaton** over finite alphabet Σ is a tuple $R = (Q, q_0, F, k, \tau_0, P)$, where Q is a finite set of states, q_0 is the initial state, F are the accepting states, k is the number of registers, $\tau_0 : \{1, \ldots, k\} \to \Delta \cup \{\bot\}$ is the initial register assignment ($\bot \notin \delta$ is a special value, indicating an empty assignment), and P is a finite set of transitions. There are two kinds of transitions, *write* and *read* transitions. A **read transition** $(i, p, a) \to q$ can be applied if the current state is p, the next input symbol is a and the next input data value is already stored in register i. It preservs register contents and goes to state q. A **write transition** $(p, a) \to (q, i)$ can be applied if the current state is p, the next input symbol is a and the next input data value d is currently not stored in any register. It writes d into register i and goes to state q. R is **deterministic** if for each state p and letter a there is exactly one transition $(p, a) \to (q, i)$, and for each register i at most one transition $(i, p, a) \to q$. A **run** on a data string w is a sequence $(q_0, \tau_0), \ldots, (q_n, \tau_n)$ of configurations, defined in the obvious way.

In [11] languages recognized by register automata are called **quasi-regular**. We mention some of the results from [11]: The class of quasi-regular languages is closed under union, intersection, concatenation, and Kleene star, but not under complementation. The emptiness problem for register automata is decidable. If R_1 and R_2 are register automata, and R_2 has at most 2 registers, then it is decidable whether $L(R_1) \subseteq L(R_2)$.

In [14] different versions of register automata are investigated (2-way, alternating, etc.). In particular, it is shown that deterministic RA are strictly weaker than RA, which are in turn strictly weaker than 2-way RA, and that RA are strictly weaker than MSO^*. As the following example shows, register automata are not sufficiently expressive for our purposes: their ability to combine global and local properties is severely limited.

Example 2.2. We take up the printer example from the introduction. Let L_0 be the set of **valid traces**, i.e., the data words whose string projection matches the expression $(r^*sr^*tr^*)^*$ and for which each class string satisfies $(rst)^*$. There is no register automaton for L_0.

For the sake of a contradiction, assume that register automaton R accepts L_0. Let k be the number of registers of R. Let w be the data string of the form $(r, 1) \cdots (r, k+1)(s, 1)(t, 1) \cdots (s, k+1)(t, k+1)$. As $w \in L_0$, R has an accepting run ρ on w. After reading the first $k + 1$ positions of w, there is at least one data value $d \in \{1, \ldots, k+1\}$ which does not occur in any register of R. We can conclude that R also accepts the string $w' \notin L_0$ resulting from w by replacing $(s, d), (t, d)$ with $(s, k+2), (t, k+2)$.

3 Data and Class-Memory Automata

Definition 3.1 ([4]). A **data automaton** D is a pair (A, B), where A is a nondeterministic letter-to-letter transducer (the **base automaton**) with a finite **output alphabet** Γ and B an NFA (the **class automaton**). A data word $w = w_1 \ldots w_n$ over Σ is accepted by D if there is an accepting run of A on the marked string projection mstr(w), yielding an output string $g_1 \ldots g_n$, such that B accepts each string $g_{i_1} \ldots g_{i_k}$ induced by a class of w.   □

Example 3.2. We construct a data automaton $D = (A, B)$ for the language L_0 from Example 2.2. The transducer A makes sure that the string projection matches $(r^*sr^*tr^*)^*$, and copies its input to the class strings. The class automaton B verifies that each class string matches $(rst)^*$. The resulting automata are shown in Figure 2.

Data automata were used as a tool in the decidability proof of [4]. We show here that their definition can be simplified. First, we show that it is not necessary that A reads the marked string projection mstr(w) (indicating where the data value changes): the expressive power is the same if it only reads str(w). Second, we exhibit an equivalent model, class memory automata, which combines the two automata into a single one. We illustrate the coloring technique used in the proof of the first result with an example.

Fig. 2. A data automaton for the language L_0. A has states i (idle) and a (active). B has states c (computing), w (waiting) and p (printing). States i and c are the initial and final states of A and B.

Example 3.3. We consider the language L_1 of traces in which the pattern $(t,d)(r,d)$ does not occur, i.e., after a print job of a process terminates it can request the next one only after some other event occurred. We note that a DA whose base automaton reads $\mathrm{mstr}(w)$ can easily accept this language by simply avoiding the pattern $(t,b)(r,0)$, for $b \in \{0,1\}$.

However, it is less obvious how a data automaton can proceed if its base automaton only sees $\mathrm{str}(w)$. Intuitively, the class automaton has no clue whether between a t and the subsequent r of some class some other event occurred. On the other hand, the base automaton does not know about data values whatsoever.

Nevertheless, by working together, the base and class automaton can accept L_1 by using the color technique explained next. The idea is that the base automaton guesses a color (black or yellow), for each t-position and each r-position, such that the following two conditions hold.

(1) Every r-position shares the color of the previous t-position in the same class (if it exists).
(2) If an r-position immediately follows a t-position they have different colors.

Obviously, if colors can be assigned such that (1) and (2) hold, then $w \in L_1$. Furthermore, condition (1) can be checked by the class automaton, condition (2) by the base automaton.

It remains to show that for each $w \in L_1$ a coloring fulfilling (1) and (2) can be found. To this end, we associate with w a directed graph $G(w)$ whose vertices are the positions of w carrying t or r and which has an edge from i to j if

(i) $j < i$, j carries t and i is the next position of w in the class of j (carrying r), or
(ii) $j = i + 1$, i carries t and j carries r.

The intuitive meaning of an edge (i,j) is that the color of i determines the color of j. Observe that each node in $G(w)$ has in-degree at most one. Furthermore, there are no cycles. Thus, we can assign colors as follows: (1) Each node of in-degree 0 gets the color black. (2) Whenever there is an edge (i,j) and i is a t-position which is already colored then j gets a different color of i. (3) Whenever there is an edge (i,j) and i is an r-position which is already colored then j gets the same color as i. Clearly, this leads to a coloring respecting conditions (1) and (2). Figure 3 gives an illustration.

Proposition 3.4. *For every data automaton, there is an equivalent data automaton for which the transducer A only reads $\mathrm{str}(w)$ instead of $\mathrm{mstr}(w)$.*

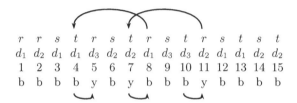

r	r	s	t	r	s	t	r	s	t	r	s	t	s	t
d_1	d_2	d_1	d_1	d_3	d_2	d_2	d_1	d_3	d_3	d_2	d_1	d_1	d_2	d_2
1	2	3	4	5	6	7	8	9	10	11	12	13	14	15
b	b	b	b	y	b	y	b	b	b	y	b	b	b	b

Fig. 3. A data string w with its graph $G(w)$ and the induced coloring

We now introduce class memory automata, which are conceptually simpler than data automata but equally expressive. Moreover, they have a meaningful notion of determinism but we will see that the deterministic variant is less expressive.

Definition 3.5. A **class-memory automaton** C is a tuple $(Q, \Sigma, \delta, q_I, F_L, F_G)$, where Q is a finite set of states, Σ is a finite alphabet, q_I is the initial state, and

- $\delta : (Q \times \Sigma \times (Q \cup \{\bot\})) \to P(Q)$ is a **transition function**;
- F_G and F_L are the sets of globally and locally accepting states, respectively;

The semantics of class memory automata (CMA) is defined through the notion of class-memory functions. Such a function simply assigns to every data value d the state of the automaton that was assumed after reading the last (previous) position with value d. More formally, a **class-memory function** is a function $f : \Delta \to Q \cup \{\bot\}$ such that $f(d) \neq \bot$ for only finitely many d. A **configuration** of C is a pair (q, f) where $q \in Q$ and f is a class-memory function. We call q the **global state** of C and $f(d)$ the **local state** of d. The initial configuration of A is (q_I, f_I), where $f_I(d) = \bot$ for all $d \in \Delta$. When reading a pair $(a, d) \in \Sigma \times \Delta$, the automaton can go from configuration (q, f) to (q', f') if (1) $q' \in \delta(q, a, f(d))$, (2) $f'(d) = q'$, and (3) for all $d' \neq d$, $f'(d') = f(d')$. The automaton accepts if, for the final configuration (q, f), $q \in F_G$ and $f(d) \in F_L \cup \{\bot\}$, for all $d \in \Delta$. A CMA is **deterministic** if each $\delta(p, a, q)$ is a singleton.

CMAs are similar to the automata studied in [2] for *nested words*. However, stated in the terms of this paper, in [2] each data value can only appear twice and the edges between positions with equal value must be nested.

Proposition 3.6. *Data automata and CMAs are expressively equivalent.*

It should be stressed that it is due to Proposition 3.4 that CMAs do not need to read extended symbols indicating data value changes.

4   Expressiveness

In this section, we compare the expressive power of CMAs with that of RAs. The main result is that CMAs are strictly stronger than RAs. Remarkably, his result does not carry over to the deterministic counterparts.

Theorem 4.1. *CMAs are strictly more expressive than register automata.*

Proof. The set L_0 of valid traces is not recognized by any RA (Example 2.2) but by a CMA. It remains to show that for every RA, we can construct an equivalent CMA. Let $R = (Q, q_0, F, k, \tau_0, P)$ be a fixed RA with k registers. Without loss of generality, we can assume that each state q determines whether it is reached by a read or a write transition. Let $\rho = (q_0, \tau_0), \ldots, (q_n, \tau_n)$ be a run of R on some input $w = (a_1, d_1) \cdots (a_n, d_n)$. Note that, after each step i, d_i is stored in some register of R, i.e., $\tau_i(j) = d_i$, for some j. We say that a transition *closes* a register j if either it is the last transition for this register in the run or there is no transition reading from j before the next write to j.

Intuitively, the CMA C guesses, for each transition, whether it closes the register. To ensure that the guesses are correct, C makes use of the coloring technique that was already used in Example 3.3 and Proposition 3.6.

More precisely, the states of C are of the form (q, l, S, p), where $q \in Q$, $l \in \{1, \ldots, k\}$, p stores some information to be specified below, and S is a subset of $\{\text{open(black)}, \text{open(yellow)}, \text{close(black)}, \text{close(yellow)}\}$. Intuitively, it corresponds to a configuration of R with state q, in which the last transition affected register l, and in which this transition was a write iff open(b) $\in S$, for some b and it closed the register iff close(c) $\in S$, for some c. We show that C can be constructed such that the following holds.

Claim. C has a run $\rho = (q_0, l_0, S_0, p_0), \ldots, (q_n, l_n, S_n, p_n)$ on w, fulfilling condition (1)-(3) below if and only if R has an accepting run on w.

A position is *opening* if open(b) $\in S_i$, for some b and *closing* if close(c) $\in S_i$, for some c.

(1) The transitions of C are consistent with the transition relation of R, i.e., for each i, $0 < i \leq n$, R has a read transition $(l_i, q_{i-1}, a_i) \to q_i$ or a write transition $(q_{i-1}, a_i) \to (q_i, l_i)$. Furthermore, i is opening if and only if the latter applies.
(2) For each position i, there is an opening position $j \leq i$ and a closing position $j' \geq i$ with (a) $l_i = l_j = l_{j'}$, (b) $d_i = d_j = d_{j'}$, and (c) for all positions m, $j < m < j'$ it holds either $l_m \neq l_i$ and $d_m \neq d_i$ or $l_m = l_i$, $d_m = d_i$ and $S_m = \emptyset$.
(3) If open(b) $\in S_i$, for some b, then either there is no $j < i$ with $d_j = d_i$, or the following two conditions hold.
    (a) For the largest position $j < i$ with $d_j = d_i$, close(b) $\in S_j$.
    (b) If the largest position $m < i$ with $l_m = l_j$ is closing then close(c) $\in S_m$, for $c \neq b$.

The proof of the claim is left out. We note how conditions (1)-(3) can be checked by C: (1) is straightforward. (2) can be checked by the local states. For each data value, the sequence of opening, closing and other positions must be ok. Condition (3a) can be also checked by using local states whereas (3b) uses the global state. The necessary information is stored in the p-component of the states of C. $\square$

In the next section, we show that Model Checking for CMAs is computationally expensive, due to the possible non-determinism. Thus, it is natural to consider deterministic CMAs. Although they can check useful properties, they are not as powerful as unrestricted CMAs.

Proposition 4.2. *Deterministic RAs and deterministic CMAs are expressively incomparable.*

Data languages can also be described in terms of logic. It is shown in [4] that emptiness for data automata is decidable, and that they capture $FO^2(+1, <, \sim)$, that is, the two-variable fragment of first order logic with the usual string predicates $+1$ and $<$, and the $\sim$-predicate, which is true for two positions in the same class. Actually, marked data automata are shown to be expressively equivalent to $EMSO^2(+1, <, \sim, \lhd)$, i.e., existential monadic second-order logic with the class successor $\lhd$ as additional predicate. By Propositions 3.4 and 3.6 and their constructive proofs, all these results carry over to unmarked data automata and CMAs. Consider the $FO^2(+1, \sim)$-formula $\Psi \equiv \forall x \, \forall y \, (x + 1 = y \wedge t(x) \wedge r(y)) \rightarrow x \not\sim y$. As Ψ defines L_1 we can conclude the following.

Proposition 4.3. *Deterministic CMAs cannot express all $FO^2(+1, \sim)$-definable properties.*

5   Algorithmic Properties

The *model checking problem* for automata asks whether a data word w is in the language $L(A)$, for an automaton A. If A is fixed, we refer to the complexity of the problem as *data complexity*. If A is considered as part of the input we speak about *combined complexity*.

Proposition 5.1. *(a) For deterministic CMAs and deterministic RAs, data and combined complexity are polynomial.*
(b) The data complexity (and thus also the combined complexity) of model checking for CMAs is NP-complete.

For RAs, the data and combined complexity are (probably) different.

Proposition 5.2. *The data complexity of model checking for RAs is polynomial.*

Proof. Consider an RA R with k registers. The number of possible configuration of R on input w is polynomial, more precisely bounded by $|Q| \cdot \binom{|w|}{k} \cdot k!$. Thus, one can check $w \in L(R)$ in polynomial time by inductively computing the set of reachable configurations, for each position of w.   □

Intuitively, RAs have a lower complexity as they can only store information about a bounded number of classes. The combined complexity of model checking for RAs is, however, NP-complete [15]. Clearly, the number of registers is crucial for the complexity of model checking. The following result shows that the problem is indeed hard w.r.t. this parameter.

Proposition 5.3. *The combined complexity of model checking for RAs, parameterized by the number of registers, is $W[1]$-hard.[2]*

---

[2] For an introduction to fixed-parameter complexity and $W[1]$, see, e.g., [8,7,10].

No parameterized upper bound for this problem is yet known, except that it belongs to XP (as can be seen from the proof of Proposition 5.2).

2-Way Deterministic CMAs. Since deterministic CMAs are clearly weaker than general CMAs, it is natural to ask whether we can allow them to move both ways. (Transitions depend on the current state and the target state of the previous transition at a position with the same data value.) Unfortunately, this extension does not preserve decidability.

Theorem 5.4. *Emptiness for 2-way deterministic CMAs is undecidable.*

6   Closure Properties

For automata-theoretic approaches to static analysis tasks and verification, closure properties are of great importance, since they facilitate modular reasoning.

Proposition 6.1. *(a) The class $\mathcal{R}$ of languages accepted by class-memory automata is effectively closed under intersection, union and concatenation.*
(b) It is not closed under complementation and Kleene star.

For (a), closure under union and intersection for data automata was shown in [4]. Closure under concatenation is shown by a construction that lets two CMAs run after each other. The proof of (b) shows that if $\mathcal{R}$ were closed under Kleene star, the halting problem for 2-counter machines would reduce to CMA emptiness.

Proposition 6.2. *The class of languages recognized by deterministic CMAs is effectively closed under intersection. It is not closed under union, concatenation, or Kleene star.*

Presburger conditions. Instead of just requiring that the memory states for all data values are locally accepting, we can generalize the acceptance condition in the following way. Suppose that CMA C has states $Q = \{q_1, \ldots, q_m\}$. Each computation ρ of C with final configuration (p, f) induces a function $g : Q \to \mathbb{N}$, where $g(q)$ is the number of data values d with $f(d) = q$.

We consider atomic formulas of two kinds: (1) q, where $q \in Q$ and (2) $(q_1 + \cdots + q_k \mod c) = c'$, where the q_i are from Q and c, c' are constant numbers. A configuration (p, f) fulfills q iff $p = q$. It fulfills $(q_1 + \cdots + q_k \mod c) = c'$ iff $(g(q_1) + \cdots + g(q_k) \mod c) = c'$.

A **Presburger CMA** C is a CMA with a Boolean combination Φ of such formulas. A run of C is accepting if its final configuration satisfies Φ.

Proposition 6.3. *(a) For each Presburger CMA there is an equivalent CMA.*
(b) The class of languages accepted by deterministic Presburger CMAs is closed under Boolean operations (intersection, union, and complementation).
(c) For each deterministic Presburger CMA there is an equivalent 2-way deterministic CMA.

7   CMA with Synchronization and Reset

The expressive power of CMAs is sufficient to handle many parameterized verification tasks, but there are also properties which they cannot express. We next
investigate ways of strengthening the expressive power, while maintaining decidability of the emptiness problem. CMAs can combine global regular properties
with local regular properties (of class strings). The "communication" between
the global and the local properties is limited: the global automaton can send
information to a class only when the class occurs in the input. Our next model
allows *synchronous* communication with all processes.

Definition 7.1. A *CMA with synchronization* is a CMA $C = (Q, \Sigma, \delta, q_I, g,$
$F_L, F_G)$ equipped with a *synchronization function* $g : Q \times \mathcal{P}(Q)$. Some of the
transitions *apply* g. When such a transition is taken from a configuration the
automaton first changes state and updates the memory function for the current
data value as usual, assuming a configuration (q, f). Then, it updates the class
memory function by setting $f(d)$ to some state in $g(f(d))$, unless $f(d) = \bot$.

Theorem 7.2. *Emptiness for CMAs with synchronization is undecidable.*

The proof is by reduction from the 2-counter machine halting problem. We can,
however, allow synchronization, if we restrict the synchronization function.

Definition 7.3. Let C be a CMA with synchronization, Q the states of C, and g
its synchronization function. Consider the graph $G_g = (Q, E)$ such that there is
an edge from p to q if and only if $q \in g(p)$. A subset E' of E defines a permutation
on Q if it is functional and bijective (each state in Q has exactly one incoming
and one outgoing edge in E'). We say that C has **restricted synchronization**
if there is a subset of E that induces a permutation on Q.

Example 7.4. Consider a variation L_s of the language L_0 from Examples 2.2
and 3.2 where we use an additional symbol n (network failure). When a network
failure occurs, some printer jobs that have been requested may disappear and
thus these requests have to be repeated. The network failure notifications are
sent by a special network process. Thus, L_s is the set of data words w such that

1. there is exactly one class string matching n^*,
2. each other class string of w matches $(rst + r)^*$,
3. if $i \lhd j$ and both i and j carry label r, there must be a position k with
   $i < k < j$ that has label n, and
4. $str(w)$ matches $((r + n)^* s(r + n)^* t(r + n)^*)^*$.

It is easy to construct a CMA with restricted synchronization that accepts L_s.

Proposition 7.5. *(a) CMAs with restricted synchronization are strictly*
   stronger than CMAs.
(b) Emptiness for CMAs with restricted synchronization is decidable.

Another way of restricting the synchronization, which is very natural when considering verification problems, is to allow only synchronization transitions that, for each class, drop all information computed so far.

Definition 7.6. A class-memory automaton *with reset* is a CMA with synchronization function g such that $g(q) = \{\bot\}$ for all states q.

Proposition 7.7. *(a) CMA with reset are strictly stronger than ordinary CMA. (b) Emptiness for CMA with reset is decidable.*

Acknowledgements. We thank Mikolaj Bojańczyk, Anca Muscholl and Luc Segoufin for many valuable discussions.

References

1. Abdulla, P., Jonsson, B., Nilsson, M., Saksena, M.: A survey of regular model checking. In: Gardner, P., Yoshida, N. (eds.) CONCUR 2004. LNCS, vol. 3170, pp. 35–48. Springer, Heidelberg (2004)
2. Alur, R., Madhusudan, P.: Adding nesting structure to words. In: Ibarra, O.H., Dang, Z. (eds.) DLT 2006. LNCS, vol. 4036, pp. 1–13. Springer, Heidelberg (2006)
3. Arenas, M., Fan, W., Libkin, L.: Consistency of XML specifications. In: Bertossi, L., Hunter, A., Schaub, T. (eds.) Inconsistency Tolerance. LNCS, vol. 3300, pp. 15–41. Springer, Heidelberg (2005)
4. Bojańczyk, M., Muscholl, A., Schwentick, T., Segoufin, L., David, C.: Two-variable logic on words with data. In: LICS'06, pp. 7–16 (2006)
5. Bouyer, P., Petit, A., Thérien, D.: An algebraic approach to data languages and timed languages. Information and Computation 182(2), 137–162 (2003)
6. Demri, S., Lazić, R.: LTL with the freeze quantifier and register automata. In: LICS'06, pp. 17–26 (2006)
7. Downey, R.G.: Parameterized complexity for the skeptic. In: CCC'03, pp. 147–169 (2003)
8. Downey, R.G., Fellows, M.R.: Parameterized Complexity. Springer, Heidelberg (1999)
9. Emerson, E., Namjoshi, K.: Reasoning about rings. In: POPL'95, pp. 85–94 (1995)
10. Flum, J., Grohe, M.: Parameterized Complexity Theory. Springer, Heidelberg (2006)
11. Kaminski, M., Francez, N.: Finite-memory automata. TCS 132(2), 329–363 (1994)
12. Kaminski, M., Tan, T.: Regular expressions for languages over infinite alphabets. In: Chwa, K.-Y., Munro, J.I.J. (eds.) COCOON 2004. LNCS, vol. 3106, pp. 171–178. Springer, Heidelberg (2004)
13. Neven, F.: Automata, logic, and XML. In: Bradfield, J.C. (ed.) CSL 2002 and EACSL 2002. LNCS, vol. 2471, pp. 2–26. Springer, Heidelberg (2002)
14. Neven, F., Schwentick, T., Vianu, V.: Finite state machines for strings over infinite alphabets. ACM transactions on computational logic 15(3), 403–435 (2004)
15. Sakamoto, H., Ikeda, D.: Intractability of decision problems for finite-memory automata. TCS 231(2), 297–308 (2000)
16. Segoufin, L.: Automata and logics for words and trees over an infinite alphabet. In: Ésik, Z. (ed.) CSL 2006. LNCS, vol. 4207, pp. 41–57. Springer, Heidelberg (2006)
17. Wilke, T.: Automaten und Logiken zur Beschreibung zeitabhängiger Systeme. PhD thesis, University of Kiel (1994)

FJMIP: A Calculus for a Modular Object Initialization

Viviana Bono[1],* and Jarosław D.M. Kuśmierek[2],**

[1] Torino University, Department of Computer Science
[2] Warsaw University, Institute of Informatics

Abstract. In most mainstream object-oriented languages, the object initialization protocol is based on constructors, where different constructors of the same class are, in fact, overloaded variants of the same method. This approach has some disadvantages: it forces an exponential growth of the code with respect to the number of properties, it may cause duplication of code, and it may create unnecessary code dependencies.

To the best of our knowledge, the literature lacks formal proposals that model non-trivial object initialization protocols.

In this paper we present a calculus (called FJMIP), which is an extension of the Igarashi-Pierce-Wadler Featherweight Java and models a novel object initialization protocol. Our calculus is reasonably simple, but it offers two benefits: (*i*) it formalizes a modular way of initializing objects that does not suffer from the previous mentioned flaws, while still being an expressive object initialization protocol; (*ii*) as a by-product, it allowed us to introduce a novel technique to prove that our object initialization process actually initializes all the fields of an object.

1   Introduction

Most object-oriented class-based languages are equipped with some form of object initialization protocol. This protocol describes two aspects:

- what kind of information must be supplied to a class, to create and initialize an object. A class may support more than one variant of object initialization, which means that there may be more than one accepted set of such information;
- what code is executed during this initialization. The sequence of instructions which should be executed depends on the kind of information supplied, therefore if the class supports distinct sets of information to be supplied during initialization, then, for any such a set, a different sequence of instructions must be executed.

Usually, initialization protocols are specified by a list of *constructors*. Each constructor corresponds to one accepted set of information and every constructor consists of: (*i*) a list of parameters (names and types), specifying a set of

---

* Partly supported by MIUR Cofin '06 EOS DUE project.
** Partly supported by the Polish government grant 3 T11C 002 27 and by SOFTLAB - Poland, Warsaw, Jana Olbrachta 94.

information required to initialize an object; (*ii*) a body, containing a list of instructions which should be executed in order to initialize an object.

This traditional constructor approach has many different disadvantages: (*i*) it makes the number of constructors grow exponentially with respect to the number of properties, and this is noticeable especially when a class contains many different properties, each of them with multiple variants of initialization; (*ii*) very often, it enforces the duplication of the initialization code; (*iii*) it causes unnecessary code dependencies.

Our novel initialization protocol is based on the idea of splitting big constructor declarations into smaller and composable parts, called *initialization modules* (or *ini modules*). Ini modules still allow a static verification of the declarations of the initialization parameters and of the object creation expressions, but: (*i*) they need less coding than constructors; (*ii*) they do not enforce copying of the code (as the constructors do).

In order to know more about ini modules, we direct the reader to the papers [9,3] that contain: (*i*) a detailed description of the motivations for introducing ini modules; (*ii*) a description of the full version of our initialization protocol, extending the one presented here by allowing more flexibility in the declaration of the ini modules, expressions evaluating into default values for ini modules parameters, exceptions declarations to be thrown by the ini modules; (*iii*) a detailed comparison of our proposal with related proposals of initialization protocols present in the literature; (*iv*) the presentation of JavaMIP, which is an extension of Java with ini modules; (*v*) a seamless integration of the new approach with mainstream languages (such as Java); (*vi*) and the description of a JavaMIP working implementation equipped with some evaluation benchmarks. However, the papers [9,3] do not deal with the formalization of this approach.

In the literature, there exist a few proposals of formal calculi modelling well the concepts of class and object, and the operations on them, such as the method call. Those calculi are usually equipped with type-checking systems, and proven sound via some well-known properties, such as the subject reduction. However, most of those calculi do not deal with the formalization of any non-trivial object initialization protocol. We believe that this depends mainly on the fact that classical constructors are almost always modelled as overloaded variants of the same method.

In this paper we present FJMIP, which is an extension of FJ [8], that models our novel object initialization protocol based on ini modules. FJMIP is reasonably simple, but it has some unusual properties: the formalization of the step-by-step initialization process (which concerns the change of state of the object being initialized) is done without the typical but complicated usage of heap and references. Additionally, we use *null* values during the the initialization process to represent fields not yet initialized, and we prove that, at the end of the process, no field will have a *null* value, i.e., no field remains uninitialized. This is a strong property that our modular approach enjoys, which cannot be verified for most object-oriented languages.

In order to read this paper, some knowledge on FJ would be preferable, but: (i) our initialization protocol is orthogonal to the other features of FJ (in particular, to the semantics of method invocation); (ii) FJ is a functional subset of Java. Therefore, the reader needs only some familiarity with Java itself.

2   FJMIP Syntax

In the definition of the FJMIP calculus, we will use the following notations:

- $\bar{a}$ to denote a *set* of elements $\{a_1, ..., a_n\}$.
- $\vec{a}$ to denote a *sequence* of elements $(a_1, ..., a_n)$. We assume that every sequence can be implicitly converted to a corresponding set.
- $A \cdot B$ to denote a *concatenation* of two sequences. In contexts in which a is an element, we will use $A \cdot a$ to denote $A \cdot (a)$, and $a \cdot A$ instead of $(a) \cdot A$.
- $t[e/x]$ denotes the term t where every occurrence of x is substituted by e.
- $t[\vec{e}/\vec{x}]$ (defined if the length of $\vec{e}$ is equal to the length of $\vec{x}$) denotes the parallel substitution of a sequence of variables by a sequence of expressions.
- $s[i \leftarrow e]$ denotes a sequence, which in every position has the same element as in s, except for position i in which it has e. For example, $(a, b)[2 \leftarrow h] = (a, h)$.
- $\uplus$ denotes the union of disjoint (tagged) sets. A *tagged set* is a set of pairs (name,type).

We present now the FJMIP context-free grammar (which was derived from the FJ one, and it is a subset of the full JavaMIP grammar, [9]). The shaded productions are used only during the evaluation, and they are not part of the "programmer syntax":

$$
\begin{aligned}
L \quad &::= \texttt{class } C \texttt{ extends } C \texttt{ \{ } \overrightarrow{C\ f};\ RIM;\ \overrightarrow{OIM};\ \overrightarrow{M}\texttt{ \} } \\
M \quad &::= C\ m\ (\overrightarrow{C\ x}) \texttt{ \{ return } e \texttt{ \} } \\
RIM &::= \texttt{required } C\ (\overrightarrow{C\ p}) \texttt{ initializes () \{ } \overrightarrow{\texttt{this}.f = e;}\texttt{ \} } \\
OIM &::= \texttt{optional } C\ (\overrightarrow{C\ p}) \texttt{ initializes } (\vec{p}) \texttt{ \{ new}[\overrightarrow{p := e}]\texttt{ \} } \\
IM \quad &::= RIM \mid OIM \\
e \quad &::= x \mid e.f \mid e.m(\vec{e}) \mid (C)\ e \mid \texttt{new } C[\overrightarrow{p := e}] \mid \\
& \texttt{new } C(\vec{e}) \mid \texttt{new } C(\overrightarrow{e \mid \texttt{null}})[\overrightarrow{p := e}][\overrightarrow{IM}]
\end{aligned}
$$

The productions define the following syntactic domains:

L: declarations of classes. A declaration consists of the name of the class, the name of the parent class, and the members of the class, which are fields, initialization modules (one *required* and some *optional*) and methods;

M: methods;

$IM = RIM \cup OIM$: declarations of initialization modules, required and optional[1];

e: terms (also called expressions).

---

[1] In the production for OIM, the 2nd and 3rd occurrences of p will be always instantiated with the same sequence of variables (see Sections 4 and 5).

Symbols C, D, E, T range over class names, symbols f, g range over field names, symbol m indicates a method identifier, p, q, r range over names of initialization parameters (which are a special case of variable), and e, t range over expressions. Moreover, we use the meta-keyword **mod** to denote one of the keywords **required** or **optional**. We identify terms up to an equivalence relation, including permutation over sets of assignments.

Methods, fields, method invocation, field lookup, and casts are the same as for FJ. However, an FJMIP class does not have the FJ constructor, but contains, besides field and method declarations, the declaration of *ini modules*, whose execution will induce a modular object initialization process. Each class must contain exactly one required ini module and can contain some optional ini modules.

A required ini module takes some *input* parameters $\overrightarrow{C\,p}$ that may be used in the expressions $\bar{e}$ to initialize the class fields $\bar{f}$. An optional ini module takes some *input* parameters (listed before the keyword **initializes**) that may be used to initialize the *output* parameters (listed after) via the $\bar{e}$ expressions. Those *output* parameters are *input* parameters of other modules declared above in the same class or up in the hierarchy, thus their initialization causes the "following" ini module to be executed, according to a semantical order which will be explained later in the paper.

The execution of the ini modules is triggered by an expression of the form **new** $C\,[\overrightarrow{p := e}]$, that will invoke the first ini module, i.e., the one that will *consume* (some of) the $\bar{p}$ parameters, explicitly initialized in the **new**. Note that here we use square brackets instead of parentheses, as it is instead in **new** $C(\bar{e})$, to obtain a non-ambiguous syntax in the case of empty sets of parameters.

An expression representing an object during the initialization process, called the *intermediate form*, has the syntax **new** $C(\overrightarrow{e \mid \texttt{null}})[\overrightarrow{p := e}][\overrightarrow{IM}]$. It (i) describes the state of the object (that is, for every field, there is a corresponding expression, or a **null** if the field is not yet initialized); (ii) lists the parameters not yet consumed; (iii) and keeps track of those ini modules which are to be executed in order to consume the parameters. The execution proceeds from the last ini module in the sequence.

An initialized object is represented by the expression **new** $C(\overrightarrow{e})$. In particular, the $\overrightarrow{e}$ are the initialization expressions for the fields. The expression **new** $C(\overrightarrow{e})$ is part of the original FJ object syntax, therefore the original FJ reduction rules apply to it.

As it is for FJ, a *class table* CT is a mapping from class names C to class declarations, and an FJMIP (FJ) program is a pair (CT, e) of a class table and an expression. Likewise in the presentation of FJ, we assume a fixed class table CT (on which we assume that the same sanity checks of FJ are done).

Below we present an example of FJMIP program (using the type **float** only for the example's sake, even though there are no primitive types in FJMIP), to which we will refer in the sequel:

```
/* Class of points definable by two different coordinate systems: */
class Pt { float x,y;
  required Pt (float x, float y) initializes ()
    { this.x = x; this.y=y; }                    //RPt - main ini module for Pt
  optional Pt (float angle, float rad) initializes (x,y)
    { new[x:=cos(angle)*rad, y:=sin(angle)*rad];}                    //OPt
}
```

```
/* Class of colored points whose color is definable by two different color palettes:*/
class ClPt extends Pt { float r, g, b;
  required ClPt (float r, float g, float b) initializes ()
    { this.r = r; this.g = g; this.b = b; } //RClPt - main ini module for ClPt
  optional ClPt (float c, float m, float yc, float k) initializes (r,g,b)
    { new[r:=..., g:=..., b:=...]; }    //OClPt - calls the main ini module of ClPt
}
```

```
/* An expression creating a ClPt object with polar coordinates and CMYK palette: */
new ClPt [angle:=1.2, rad:=4, c:=0, m:=1, yc:=1, k:=0]
```

```
/* A class extending the set of options of the inherited property (color): */
class HSBClPt extends ClPt {
  optional HSBClPt(float h, float s, float b) initializes(r,g,b) {...}
}
```

```
/* Another initialization of fields via the (colored) offset of ColorVector: */
class VectorClPt extends ClPt {
  optional VectorClPt(ColorVector c) initializes(x,y,r,g,b) {...}
}
```

3   Auxiliary Functions

We inherit from FJ the auxiliary functions for field lookup (*fields*), for method type lookup (*mtype*), and for method body lookup (*mbody*). We define our own ones to deal with ini modules.

Initialization modules lookup. The function *IModules* takes a class name as a parameter, and returns the sequence of all ini modules declared in this class and in all of the class' ancestors.

$$IModules(\texttt{Object}) = \epsilon \quad \frac{CT(C) = \texttt{class } C \texttt{ extends } D\{\overrightarrow{E\,f};\ RIM;\ \overrightarrow{OIM};\ \overrightarrow{M}\}}{IModules(C) = RIM \cdot IModules(D) \cdot \overrightarrow{OIM}}$$

In the example from Section 2, $IModules(\texttt{ClPt}) = (\text{R}_{\texttt{ClPt}}, \text{R}_{\texttt{Pt}}, \text{O}_{\texttt{Pt}}, \text{O}_{\texttt{ClPt}})$.

The order in the resulting sequence of ini modules determines the order of execution of the ini modules themselves, as it will be shown in the operational semantics. The required ini modules of the hierarchy are placed in the *IModules* result in the reverse order with respect to the order of declaration. This is because required ini modules are executed after all the superclass ini modules. Instead, optional ini modules are executed starting from the given class.

Notice that we fixed a precise execution order of ini modules for simplicity, but the full version of our approach, [9], allows to define any possible combination of execution among them. Moreover, note that our present choice is as restrictive as the one for Java, where the *super* call must be the first instruction.

Activated module lookup. The function *activated* takes as arguments a class name and a set of parameters. It returns a sequence of ini module declarations from that class and its ancestors which are activated by the given set of parameters. This function uses the above defined function $IModules$ and another one, *activated'*. The function *activated'* takes a sequence of ini modules and a set of parameters and returns a sequence of ini modules, which is a subsequence of the one passed as the first argument. In the example from Section 2, $activated(\texttt{C1Pt}, \{\texttt{angle}, \texttt{rad}, \texttt{r}, \texttt{g}, \texttt{b}\}) = (\texttt{R}_{\texttt{C1Pt}}, \texttt{R}_{\texttt{Pt}}, \texttt{O}_{\texttt{Pt}})$.

$$activated(C, \overline{p}) = activated'(IModules(C), \overline{p})$$

$$activated'(\epsilon, \emptyset) = \epsilon \qquad \frac{aIM = \texttt{mod}\,C\,(\overrightarrow{D\,q})\,\texttt{initializes}(\overrightarrow{r})\{...\} \quad \overrightarrow{q} \cap \overline{p} = \emptyset \neq \overrightarrow{q}}{activated'(\overrightarrow{IM} \cdot aIM,\ \overline{p}) = activated'(\overrightarrow{IM},\ \overline{p})}$$

$$\frac{aIM = \texttt{mod}\,C\,(\overrightarrow{D\,q})\,\texttt{initializes}(\overrightarrow{r})\{...\} \quad \overrightarrow{q} \cap \overline{p} = \overrightarrow{q}}{activated'(\overrightarrow{IM} \cdot aIM,\ \overline{p}) = activated'(\overrightarrow{IM},\ (\overline{p} - \overrightarrow{q}) \uplus \overrightarrow{r}) \cdot aIM}$$

The first set of parameters $\overline{p}$ triggers the lookup, that looks for an ini module aIM whose input parameters are all included within $\overline{p}$, then the lookup proceeds recursively by looking for the ini modules that are activated by what remains of the $\overline{p}$ plus the output parameters of aIM. Notice that this function performs also a correctness check: it checks that all parameters, starting from $\overline{p}$, are consumed by some ini modules belonging to the given class and its ancestors. The condition $\overrightarrow{q} \cap \overline{p} = \emptyset \neq \overrightarrow{q}$ ensures that *activated* is indeed a function.

Notice that the order of activation of the ini modules depends strictly on the order in which those are declared in the classes. This may look like a limitation, but the declaration order inducing the correct activation of the ini modules is enforced by the typing rules (see Section 5), therefore the programmer is guided by the compiler.

Class initialization parameter declaration lookup. The function *params* looks up the set of all input parameters of the ini modules of the given class and its ancestors. Every parameter is equipped with its type.

$$ip(\texttt{mod}\,C\,(\overrightarrow{D\,p})\,\texttt{initializes}\,(\overrightarrow{q})\,\{...\}) = \overline{D\,p}$$

$$params(\texttt{Object}) = \emptyset \qquad \frac{CT(C) = \texttt{class}\,C\,\texttt{extends}\,D\{\,...\,\overline{IM}; \overrightarrow{M};\,\}}{params(C) = params(D) \uplus ip(IM_1) \uplus ... \uplus ip(IM_n)}$$

Note that the union of disjoin sets $\uplus$ works as a consistency check: there cannot exist two declarations of input parameters with the same name in the ini modules belonging to the hierarchy of a given class. In the example from Section 2, $params(\mathtt{ClPt}) = \{\mathtt{float\ x}, \mathtt{float\ y}, \mathtt{float\ angle}, \mathtt{float\ rad}, \mathtt{float\ r}, \mathtt{float}$ $\mathtt{g}, \mathtt{float\ b}, \mathtt{float\ c}, \mathtt{float\ m}, \mathtt{float\ yc}, \mathtt{float\ k}\}$

Required initialization module lookup. The function $Rmodules$, when applied to a class name, returns the set of declarations of required ini modules found in this class and all its ancestors. In the example from Section 2, $RModules(\mathtt{ClPt})$ $= \{\mathtt{R_{ClPt}}, \mathtt{R_{Pt}}\}$.

$$Rmodules(\mathtt{Object}) = \emptyset \qquad \frac{CT(C) = \mathtt{class}\ C\ \mathtt{extends}\ D\ \{...\ RIM...\}}{Rmodules(C) = Rmodules(D) \cup \{RIM\}}$$

Expected input parameter lookup. The function $input$ takes a sequence of ini modules, and returns the list of all input parameters from such modules that are not matched with the output parameters of modules appearing *after* in the sequence itself. This function is defined only if all output parameters of a module match the input parameters of modules appearing *before* it in the sequence itself. In the example from Section 2, $input((\mathtt{R_{ClPt}}, \mathtt{R_{Pt}}, \mathtt{O_{Pt}})) = \{\mathtt{angle}, \mathtt{rad}, \mathtt{r}, \mathtt{g}, \mathtt{b}\}$.

$$input(\epsilon) = \emptyset \qquad \frac{aIM = \mathtt{mod}\ C\ (\overrightarrow{D\,q})\ \mathtt{initializes}\ (\overrightarrow{r})\{...\}\quad input(\overrightarrow{IM}) = \overrightarrow{A} \uplus \overrightarrow{r}}{input(\overrightarrow{IM} \cdot aIM) = \overrightarrow{A} \uplus \overrightarrow{q}}$$

Property 1. For every class C and set of parameters $\overline{p}$ such that $activated(C, \overline{p})$ is defined, $input(activated(C, \overline{p})) = \overline{p}$ holds.

Initialized field lookup. The function $RIMFields$ takes a list of ini modules and returns the sequence of fields initialized by the required ini modules. We recall that in FJMIP only required ini modules initialize the fields. In the example from Section 2, $RIMFields((\mathtt{R_{ClPt}}, \mathtt{R_{Pt}}, \mathtt{O_{Pt}})) = (\mathtt{x}, \mathtt{y}, \mathtt{r}, \mathtt{g}, \mathtt{b})$.

$$RIMfields(\epsilon) = \epsilon \qquad \frac{aIM = \mathtt{optional}\ C\ ...}{RIMFields(\overrightarrow{IM} \cdot aIM) = RIMFields(\overrightarrow{IM})}$$

$$\frac{aIM = \mathtt{required}\ C\ (...)\ \mathtt{initializes}\ ()\ \{\ \overrightarrow{\mathtt{this}.f = e;}\ \}}{RIMFields(\overrightarrow{IM} \cdot aIM) = RIMFields(\overrightarrow{IM}) \cdot \overrightarrow{f}}$$

4   Operational Semantics

We inherit all of the reduction rules of FJ and add rules responsible for the execution of the initialization process. We start with the rule triggering the object initialization process:

$$\mathtt{new}\ C[\overline{p := e}] \rightarrow \mathtt{new}\ C(\mathtt{null}^{|fields(C)|})\ [\overline{p := e}]\ [activated(C, \overline{p})]\ (StartRed)$$

This rule evaluates an FJMIP **new** expression into an intermediate form, in order to start the actual field initialization process. This intermediate form is built from: (i) a sequence of **null** values, one for each of the fields of C; (ii) the set of parameters $\overline{p}$ that trigger the initialization process, with their initialization expressions; (iii) the set of ini modules taken from class C and its ancestors activated by $\overline{p}$; this set of ini modules must be executed in order to initialize the fields of the created object.

The next rule is responsible for the execution of optional ini modules:

$$\frac{\begin{array}{c} OIM = \texttt{optional}\ C(\overrightarrow{q})\ \texttt{initializes}(\overrightarrow{r})\ \{\texttt{new}(\overline{r := e})\} \\ \overline{p^1} = \{q_1 \texttt{:=} t_1,\ ...,\ q_k := t_k\} \uplus \overline{p} \\ \overline{p^2} = \{r_1 \texttt{:=} e_1[\overrightarrow{t}/\overrightarrow{q}],\ ...,\ r_l \texttt{:=} e_l[\overrightarrow{t}/\overrightarrow{q}]\} \uplus \overline{p} \end{array}}{\texttt{new}\ C(\overrightarrow{e_{st}})\,[\overline{p^1}]\,[\overline{IM} \cdot OIM]\ \rightarrow\ \texttt{new}\ C(\overrightarrow{e_{st}})\,[\overline{p^2}]\,[\overline{IM}]}\ (OptionalRed)$$

The last-in-the-sequence optional ini module of the intermediate form is executed. The intermediate form reduces into another intermediate form which: (i) has the same sequence of field initialization expressions $\overrightarrow{e_{st}}$ (we recall that optional ini modules do not initialize fields); (ii) has one less ini module to execute; (iii) has an updated set of initialization parameters, containing those parameters among the $\overline{p^1}$ which are not yet consumed (that is, not corresponding to any input parameter of the OIM), plus the output parameters of OIM, whose initialization expressions now "contain" the initialization terms supplied to the OIM's input parameters (see $\overline{p^2}$). Note that the union of disjoint sets here is important: it prevents the execution of an ini module whose output parameters are already present in the set of the calculated parameters.

The next rule is responsible for the execution of required ini modules:

$$\frac{\begin{array}{c} fields(C) = \overline{C\ f} \\ RIM = \texttt{required}\ D(\overrightarrow{q})\ \texttt{initializes}()\ \{\texttt{this}.f_{j_1} \texttt{=} e_1\,;...;\ \texttt{this}.f_{j_n} \texttt{=} e_n\,;\} \\ \overline{p^1} = \{q_1 := t_1, ..., q_k := t_k\} \uplus \overline{p} \\ \overrightarrow{e_{st}''} = \overrightarrow{e_{st}}[j_1 \leftarrow e_1[\overrightarrow{t}/\overrightarrow{q}],\ ...,\ j_n \leftarrow e_n[\overrightarrow{t}/\overrightarrow{q}]] \end{array}}{\texttt{new}\ C(\overrightarrow{e_{st}})\,[\overline{p^1}]\,[\overline{IM} \cdot RIM]\ \rightarrow\ \texttt{new}\ C(\overrightarrow{e_{st}''})\,[\overline{p}]\,[\overline{IM}]}\ (RequiredRed)$$

The last-in-the-sequence required ini module of the intermediate form is executed. The intermediate form reduces into another intermediate form which: (i) has more fields initialized: a subset $f_{j_1}, ..., f_{j_n}$ of the fields $\overrightarrow{f}$ has been initialized (see $\overrightarrow{e_{st}''}$) by the expressions in RIM's body (substituted for the corresponding **null** place holder); (ii) has one less ini module to execute; (iii) has a reduced set of initialization parameters, containing all parameters from the set $\overline{p^1}$, minus the input parameter of RIM.

Finally, a rule concluding the object initialization process turns the last intermediate form into a FJ object expressions (to which the FJ reduction rules apply):

$$\texttt{new}\ C(\overrightarrow{e})\,[\,]\,[\,]\ \rightarrow\ \texttt{new}\ C(\overrightarrow{e})\quad (EndRed)$$

FJMIP *values* are the ones of FJ, i.e., `new` $C(\overrightarrow{v})$.
Note that the execution of the object initialization could get stuck in three cases:

- not enough initialization parameters have been supplied in order to execute all the required ini modules (we recall that all required ini modules of the hierarchy must be executed to have a fully initialized object);
- too many initialization parameters have been supplied, which would prevent the set of expected input parameters not to be empty after the execution of the ini modules (see the auxiliary function *input*);
- a parameter is supplied in two ways at the same time, for example, it is an output parameter of two activated optional modules.

The reduction sequence for the example from Section 2 is as follows:

```
new ClPt[angle:=1.2, rad:=4, r:=1, g:=1, b:=0] →
new ClPt(null, null, null, null, null)[angle:=1.2, rad:=4, r:=1, g:=1, b:=0]
        [R_ClPt,R_Pt,O_Pt] →
new ClPt(null, null, null, null, null)[x:=1.45, y:=3.73, r:=1, g:=1, b:=0]
        [R_ClPt,R_Pt] →
new ClPt(1.45, 3.73, null, null, null)[r:=1, g:=1, b:=0][R_ClPt] →
new ClPt(1.45, 3.73, 1, 1, 0)[][] →
new ClPt(1.45, 3.73, 1, 1, 0)
```

5   Type Checking

Types are all induced by classes, therefore they are named by classes, as in FJ. The subtyping relation <: of FJ is inherited by FJMIP, as well as all the type checking rules with the following modification and additions:

- a rule for class declaration is extended by a check responsible for ini modules;
- a rule for the FJMIP object creation expression is added. The original typing rule for the FJ `new(...)` expression is kept, because it belongs also to FJMIP;
- a rule for the intermediate form is added.

In the typing rules, we use the following judgement forms:

- $\overline{M}$ `OK IN` $\overline{C}$: methods $\overline{M}$ type check correctly in the environment endowed by class C.
- $\overline{IM}$ `OK_with` $\overline{C\,p}$: (*i*) ini modules $\overline{IM}$ type check correctly in the environment endowed by the classes of a hierarchy containing ini modules whose input parameters include the $\overline{C\,p}$; (*ii*) every output parameter of modules $\overline{IM}$ either is contained in $\overline{p}$ or corresponds to an input parameter of one of the $\overline{IM}$ (if this is the case, this one was declared before in the code of the same class, or in a superclass). The typing rules introduced below will detail the meaning of this judgment.

- $\overline{p:D} \vdash e : C$: expression e has type C in the context where $\overline{p}$ are assumed to be of types $\overline{D}$. A context can be abbreviated by the symbol Γ.
- class C extends D { ... } OK: the declaration of class C is correct.
- CT OK: for every class C in the domain of CT, we have $CT(C)$ OK.

In the sequel, when we will say that a certain judgement holds, we mean that this judgement is derivable in the system we are defining.

The following rule checks class declarations:

$$
\begin{array}{cl}
(1.a) & \overline{M} \text{ OK IN } C \\
(1.b) & \overline{IM} \text{ OK\_with } params(D) \\
(1.c) & params(C) = \_ \\
\hline
& \text{class } C \text{ extends } D \text{ \{ } \overline{C\,f}; \ \overline{IM}; \ \overline{M} \text{ \} OK}
\end{array} \quad (ClassDecl)
$$

This rule is analogous to the corresponding FJ rule but with one extra judgement, responsible for checking the ini modules. Premise $(1.b)$ type checks the expressions inside the modules and triggers the recursive check on the output parameters of C's ini modules $\overline{IM}$, that are to be consumed by the input parameters of some ini modules declared before in the same class, or by some ini modules of the superclass or of any other ancestor up in the hierarchy. The fact that we start this check from $params(D)$ ensures that the first declared ini module in C will be checked against the ini modules of the ancestors only. Note that the initialization parameters declared in the hierarchy must be distinct and this is implied by the fact that otherwise $params(C)$ would be undefined (premise $(1.c)$).

The following rule checks optional ini module declarations:

$$
\begin{array}{cl}
(2.a) & OIM = \text{optional } C \ (\overrightarrow{D\,p}) \text{ initializes}(\overrightarrow{g})\{\text{new}(\overrightarrow{g := e})\} \\
(2.b) & CT(C) = \text{class } C \text{ extends } E \ \{...\} \\
(2.c) & \overline{T\,g} \subseteq \overline{s} \\
(2.d) & \overline{p:D} \vdash \overline{e:C'} \qquad \overrightarrow{C'} <: \overrightarrow{T} \\
(2.e) & \overline{IM} \text{ OK\_with } \overline{s} \cup \overline{D\,p} \\
\hline
& OIM \cdot \overrightarrow{IM} \quad \text{OK\_with } \overline{s}
\end{array} \quad (OptionalOK)
$$

Premise $(2.c)$ says that the output parameters of the module OIM are contained in the set $\overline{s}$ and are tagged with some types $\overline{T}$, premise $(2.d)$ says that the expressions assigned to those output parameters must be type-compatible with $\overline{T}$, and premise $(2.e)$ says that remaining modules must be checked against the set of parameters extended with the set of input parameters of OIM.

The following rule checks required ini module declarations:

$$
\begin{array}{cl}
(2.f) & RIM = \text{required } C \ (\overrightarrow{D\,p}) \text{ initializes () \{ } \overrightarrow{\text{this.}f = e;} \text{ \} } \\
(2.g) & CT(C) = \text{class } C \text{ extends } E \ \{ \ \overline{C\,f}; \ ... \ \} \\
(2.h) & \overline{p:D} \vdash \overline{e:C'} \qquad \overrightarrow{C'} <: \overrightarrow{C} \\
(2.i) & \overline{IM} \text{ OK\_with } \overline{s} \cup \overline{D\,p} \\
\hline
& RIM \cdot \overrightarrow{IM} \quad \text{OK\_with } \overline{s}
\end{array} \quad (RequiredOK)
$$

Premise (2.h) says that expressions assigned to fields in the required module must have types matching the declarations of the fields in that class. Analogously to the condition on the standard FJ constructor, the list of fields initialized in the FJMIP required module must be equal to the list of fields declared in the class. Premises (2.e) of rule (*OptionalOK*) and (2.i) of rule (*RequiredOK*) are the recursive calls for the $\mathtt{OK\_with}$ check. In fact, we recall that the activation order for ini modules is dependant on the actual declaration order in the code, therefore the first ini module declared in a class is type-checked against the ini modules of the ancestors only (see rule (*ClassDecl*)), while the others are checked against the ini modules of the ancestors *and* the ini modules declared above in the code of the same class (by rules (*OptionalOK*) and (*RequiredOK*)).

The following axiom makes the type checking definition for ini modules sound:

$$\epsilon \; \mathtt{OK\_with} \; \overline{s} \; (AxiomOK)$$

We state two properties, whose proofs can be found in [2].

Property 2. We assume CT $\mathtt{OK}$. Then, for every class C and set of parameters $\overline{p}$ such that $activated(C, \overline{p})$ is defined, $activated(C, \overline{p})$ $\mathtt{OK\_with}$ $\emptyset$ holds.

Property 3. We assume that the judgement CT $\mathtt{OK}$ holds. If the judgement $\overrightarrow{IM} \cdot M$ $\mathtt{OK\_with}$ $\overline{s}$ holds, then the judgement $\overrightarrow{IM}$ $\mathtt{OK\_with}$ $\overline{s}$ holds as well.

The following rule checks object creation expressions:

$$
\begin{array}{c}
(3.a) \quad Rmodules(C) \subseteq activated(C, \{p_{j_1}, ..., p_{j_k}\}) \\
(3.b) \quad params(C) = \{T_1 \; p_1, ..., T_n \; p_n\} \\
(3.c) \quad \dfrac{\Gamma \vdash e_1 : U_1 \quad U_1 < T_{j_1} \quad ... \quad \Gamma \vdash e_k : U_k \quad U_k <: T_{j_k}}{\Gamma \vdash \mathbf{new} \; C \; [p_{j_1} \mathtt{:=} e_1, \; ..., \; p_{j_k} \; \mathtt{:=} \; e_k] \; : \; C} \; (New)
\end{array}
$$

This rule says that the list of activated modules must contain all the required ini modules of C's hierarchy (premise (3.a)). Moreover, the parameters initialized by the $\mathbf{new}$ must be contained in the set of input parameters of the ini modules from C's hierarchy, and the types of the initialization expressions must match their declaration types (premises (3.b) and (3.c)). Additionally, the fact that $activated(C, \{p_{j_1}, ..., p_{j_k}\})$ is defined guarantees that the passed parameters form a proper set (so there are no duplicates in it), and that no unnecessary parameters are passed.

The following rule checks intermediate forms:

$$
\begin{array}{c}
(4.a) \quad params(C) = \{T_1 \; p_1, ..., T_n \; p_n\} \\
(4.b) \quad \{p_{j_1}, ..., p_{j_k}\} = input(\overrightarrow{IM}) \\
(4.c) \quad \Gamma \vdash e_1 : W_1 \quad W_1 < T_{j_1} \quad ... \quad \Gamma \vdash e_k : W_k \quad W_k <: T_{j_k} \\
(4.d) \quad RIMfields(\overrightarrow{IM}) = (f_j, ..., f_l) \\
(4.e) \quad fields(C) = (U_1 \; f_1, ..., U_l \; f_l) \\
(4.f) \quad \forall_{i<j}. \; g_i \neq \mathtt{null} \; \wedge \; g_i : V_i \; \wedge \; V_i <: U_i \\
(4.g) \quad \forall_{i \geq j}. \; g_i = \mathtt{null} \\
(4.h) \quad \forall_{im \in \overrightarrow{IM}} .im = mod \; D(...) \; \mathtt{initializes} \; ... \; \wedge \; C <: D \\
(4.i) \quad \overrightarrow{IM} \; \mathtt{OK\_with} \; \emptyset \\
\hline
\Gamma \vdash \mathbf{new} \; C \; (\overrightarrow{g}) \; [p_{j_1} := e_1, ..., p_{j_k} := e_k][\overrightarrow{IM}] \; : \; C
\end{array} \; (Intermediate)
$$

This rule ensures that the intermediate forms, which result from FJMIP programs via the execution process, are well typed. Premise (4.b) checks if the set of initialization parameters not yet consumed is a subset of all input parameters of C's hierarchy (with respect to premise (4.a)), and if they correspond exactly to the input parameters of the not-yet-executed ini modules $\overrightarrow{IM}$, so that they will be all consumed. Premise (4.c) ensures that the types of the initialization expressions e_i are compatible with the corresponding parameters' types. Premise (4.f) guarantees that: (i) the fields already initialized are a subset of the fields that must be initialized in the required modules (with respect to premise (4.d)); and that each corresponding initialization expression g_i is typed correctly (with respect to premise (4.e)). Premise (4.g) ensures that the remaining fields are not yet initialized. Premise (4.h) checks if the sequence $\overrightarrow{IM}$ of ini modules is part of the hierarchy of C. Premise (4.i) states that all output parameters of the ini $\overrightarrow{IM}$ modules must type check and be consumed within the $\overrightarrow{IM}$ themselves.

Our calculus enjoys the subject reduction property and a form of type soundness. This ensure that each `null` present in an intermediate form is replaced by the appropriate expression for each field, yielding the property that no field remains uninitialized after the initialization process is finished. The proofs of these properties, and other results, can be found in [2].

Theorem 1 (SR). *If $\Gamma \vdash e : C$ and $e \rightarrow e'$, then $\Gamma \vdash e' : C'$, for some $C <: C'$.*

Theorem 2 (TS). *If $\emptyset \vdash e : C$ and $e \rightarrow e'$ with e' a normal form, then e' is either a value v with $\emptyset \vdash e : D$ and $D <: C$, or an expression containing (D)new $C\,(\overrightarrow{g})$, where $C <: D$ does not hold.*

6   Conclusion and Related Work

In the literature there exist studies of different variants of typed and untyped object-oriented calculi. Some of them are the cornerstones of the theory of object-oriented languages, such as the ones presented in [1,4,5,6]. However, none of those calculi allow any specification of non-trivial object initialization protocols. If we move into the Java-like realm, we find the elegant work by Flatt et al. [7], but also their proposal does not model constructors at all. FJ [8] itself is a functional subset of Java that allow only one trivial constructor per class.

We believe that our FJMIP calculus is the first attempt to model a non-trivial object initialization protocol, one that also offers some practical advantages with respect to the traditional constructor-based one, such as: (i) it reacts better when a superclass is extended: if a parent class is extended with a new optional parameter, then the subclass gains automatically the corresponding set of options of initialization; (ii) it reduces the number of initialization modules from exponential to linear: different points of view of the instantiation can be defined separately; (iii) as a further effect, it discards code duplication: if a subclass must add something to the object initialization protocol, it does not have to reference all the parent constructors; (iv) it discards ambiguities introduced by constructor overloading. Moreover, FJMIP is also sound from the object initialization point of view: we proved that at the end of the initialization process no field will remain uninitialized.

FJMIP has a "bigger brother", JavaMIP [9], which is an extension of Java. The JavaMIP language allows more flexibility in the design of the object initialization protocol. In the paper [9] there is also a description of related work concerning different solutions to the problem of constructors. An implementation of JavaMIP, by Giovanni Monteferrante, together with some working examples and benchmarks, is available at [10].

References

1. Abadi, M., Cardelli, L.: A Theory of Objects. Springer, Heidelberg (1996)
2. Bono, V., Kuśmierek, J.: Featherweight JavaMIP: a calculus for a modular object initialization protocol. Manuscript (2007), available at
   http://www.di.unito.it/~bono/papers/FWJavaMIP.pdf
3. Bono, V., Kuśmierek, J.: Modularizing constructors. In: Proc. TOOLS '07, 2007 (to appear)
4. Bruce, K.B.: Foundations of Object-Oriented Languages: Types and Semantics. MIT Press, Cambridge (2002)
5. Fisher, K., Honsell, F., Mitchell, J.C.: A lambda-calculus of objects and method specialization. Nordic J. of Computing, 1(1), 3–37 (1994). Preliminary version appeared In: Proc. LICS '93, pp. 26–38.
6. Fisher, K., Mitchell, J.C.: A delegation-based object calculus with subtyping. In: Proc. FCT '95, vol. 965, pp. 42–61. Springer, Heidelberg (1995)
7. Flatt, M., Krishnamurthi, S., Felleisen, M.: Classes and Mixins. In: Proc. POPL '98, pp. 171–183. ACM Press, New York (1998)
8. Igarashi, A., Pierce, B., Wadler, P.: Featherweight Java: A minimal core calculus for Java and GJ. ACM SIGPLAN Transactions on Programming Languages and Systems (TOPLAS), 23(3) (May 2001)
9. Kuśmierek, J., Bono, V.: A modular object initialization protocol. Manuscript (2006), available at http://www.di.unito.it/~bono/papers/JavaMIP02.pdf
10. Monteferrante, G.: javamip2java: A preprocessor for the JavaMIP language(2006), Available at http://www.di.unito.it/~bono/papers/javamip/

Top-Down Deterministic Parsing of Languages Generated by CD Grammar Systems

Henning Bordihn[1,*] and György Vaszil[2,**]

[1] Institut für Informatik, Universität Gießen, Arndstraße 2
D-35392 Gießen, Germany
bordihn@informatik.uni-giessen.de
[2] Computer and Automation Research Institute, Hungarian Academy of Sciences
Kende utca 13-17, H-1111 Budapest, Hungary
vaszil@sztaki.hu

Abstract. The paper extends the notion of context-free LL(k) grammars to CD grammar systems using two different derivation modes, examines some of the properties of the resulting language families, and studies the possibility of parsing these languages deterministically, without backtracking.

1 Introduction

Most of the efficient parsers which have been developed are applicable to the class of context-free grammars or subclasses thereof. On the other hand, also non-context-free aspects are encountered in several applications of formal languages [7]. Since the late sixties, a series of grammars has been proposed which use—in the core—context-free productions but posses some additional control on the application of the productions, adding to the generative capacity of context-free grammars [7]. In 1990, the concept of cooperating distributed grammar systems (CDGS, in short) has been introduced in [4] as a model of distributed problem solving. Context-free CDGS consist of several context-free grammars, the components of the system, which jointly derive a formal language, rewriting the sentential form in turns, according to some cooperation protocol (the so-called mode of derivation). In the present paper, two different cooperation protocols are taken into consideration, namely the $=m$-mode and the t-mode, in which a component, once started, has to perform exactly m and as many as possible derivation steps, respectively. This approach enhances the descriptive power of (single) context-free grammars, as well. For a survey on CDGS, see the monograph [5] or the handbook chapter [8].

Our aim is to restrict context-free CDGS so that deterministic one pass no backtrack parsing in a top-down manner becomes possible. The initial steps in

---

* Most of the work was done while the first author was at the University of Potsdam, Institut für Informatik, August-Bebel-Straße 89, D-14482 Potsdam, Germany.
** The second author was supported by a research scholarship of the Alexander von Humboldt Foundation.

E. Csuhaj-Varjú and Z. Ésik (Eds.): FCT 2007, LNCS 4639, pp. 113–124, 2007.

this direction have already been taken by the authors of the present paper in [2]. (For a different approach towards the deterministic parsing of languages generated by CDGS, see also [9].) In order to make these systems deterministic, we first restrict context-free CDGS to leftmost derivations. Two kinds of leftmostness are taken into consideration which, in contrast to context-free grammars, can alter the generative capacity of CDGS; for an extensive investigation of leftmost context-free CDGS see [3]. At second, some sort of $LL(k)$ condition is imposed on the systems which guarantees that for any sentential form of any leftmost derivation, the first k symbols of the resulting string which have not yet been derived to the left of the leftmost nonterminal in the sentential form determine the next step to be performed by the CDGS. That is, according to the derivation mode, a unique sequence of productions from a unique component.

In the present paper, we first define the concept of $LL(k)$ context-free CDGS working in the $=m$- and t-modes of derivation using two different types of leftmost restrictions, and prove some first hierarchical properties for these systems. Then we focus on the $=m$-mode of derivation. We show that $LL(k)$ CDGS working in the $=m$-mode, for any $m \geq 2$, can be simulated by $LL(k)$ CDGS working in the $=2$-mode if the so-called sw-type of leftmostness is imposed, and that $LL(k)$ systems of this type can generate non-semilinear languages. Then we define the notion of a lookup table which, based on pairs of nonterminals and lookahead strings, identifies the component and the sequence of rules needed for the continuation of the derivation according to the $LL(k)$ condition. Opposed to the case of $LL(k)$ context-free grammars, the existence and the effective constructibility of the lookup table is not obvious, but we show that in most cases, if we have the lookup table, then a lookup string of length one is sufficient, and a parsing algorithm of strictly sub-quadratic time complexity (see also [2]) can be given. Finally, we present a decidable condition which implies the effective constructibility of the lookup table.

2   Definitions

We assume the reader to be familiar with the basic notions of formal languages, as contained for example in [11]. The set of positive integers is denoted by $\mathbb{N}$ and the cardinality of a set M is denoted by $\#M$. By V^+ we denote the set of nonempty words over an alphabet V; if the empty word ε is included, then we use the notation V^*. The length, the number of symbols contained by a word $w \in V^*$ is denoted by $|w|$, the set of elements of V^* with length at most k, for some $k \in \mathbb{N}$, is denoted by $V^{\leq k}$. By $\mathrm{pref}_k(w)$, $w \in V^*$, we denote the prefix of length k of w if $|w| \geq k$ or, otherwise, the string w itself. Set union and subtraction are denoted by $\cup$ and $-$, respectively, $\subseteq$ denotes inclusion, while $\subset$ denotes strict inclusion.

A context-free grammar is a quadruple $G = (N, T, P, S)$, where N and T are disjoint alphabets of nonterminals and terminals, respectively, $S \in N$ is the axiom, and P is a finite set of productions of the form $A \to z$, where $A \in N$ and $z \in (N \cup T)^*$.

Now we repeat the definition of a cooperating distributed grammar system, where we restrict ourselves (without further mentioning) to the case of context-free components. A cooperating distributed grammar system of degree n is an $(n + 3)$-tuple $G = (N, T, S, P_1, P_2, \ldots, P_n)$ where N and T are two disjoint alphabets, the alphabet of nonterminals and the alphabet of terminals, respectively, $S \in N$ is the axiom, and for $1 \leq i \leq n$, P_i is a finite set of context-free productions, i.e., productions of the form $A \to z$, $A \in N$, $z \in (N \cup T)^*$. The sets P_i are called components of the system G. Let $\mathrm{dom}(P_i) = \{A \in N \mid A \to z \in P_i\}$, the *domain* of P_i, and let $T_i = (N \cup T) - \mathrm{dom}(P_i)$, $1 \leq i \leq n$. We call a CD grammar system *deterministic*, if for all $1 \leq i \leq n$, P_i contains at most one rule for each nonterminal, that is, for each symbol $A \in N$, the property of $\#\{w \mid A \to w \in P_i\} \leq 1$ holds.

Now we recall the definition of two different types of leftmost rewriting steps from [3]. A *strong leftmost rewriting step* is a direct derivation step that rewrites the leftmost nonterminal of a sentential form. Formally, $x \Longrightarrow_i^s y$ where $x = x_1 A x_2$, $y = x_1 z x_2$, $A \to z \in P_i$, and $x_1 \in T^*$, $x_2, z \in (N \cup T)^*$.

A *weak leftmost rewriting step* is a direct derivation step that rewrites the leftmost nonterminal from the domain of the active component. Formally, $x \Longrightarrow_i^w y$ where $x = x_1 A x_2$, $y = x_1 z x_2$, $A \to z \in P_i$, and $x_1 \in T_i^*$, $x_2, z \in (N \cup T)^*$.

Let $\Longrightarrow_s^{l}{}_i$ and $\Longrightarrow_w^{l}{}_i$ denote l consecutive strong and weak derivation steps of component P_i, $l \geq 1$, respectively. Further, let $\Longrightarrow_\alpha^*{}_i$ denote the reflexive and transitive closure of $\Longrightarrow_\alpha{}_i$, $\alpha \in \{s, w\}$.

Let m be a positive integer. An *m-step derivation* by the ith component in the $\alpha\beta$-type of leftmostness, where $\alpha, \beta \in \{s, w\}$, denoted by $\Longrightarrow_{\alpha\beta}^{=m}{}_i$, is defined as

$$x \Longrightarrow_{\alpha\beta}^{=m}{}_i y, \text{ iff } x \Longrightarrow_\alpha{}_i x_1 \text{ and } x_1 \Longrightarrow_\beta{}_i^{m-1} y,$$

for some $x_1 \in (N \cup T)^*$.

A *t-derivation* by the ith component in the $\alpha\beta$ manner, where $\alpha, \beta \in \{s, w\}$, denoted by $\Longrightarrow_{\alpha\beta}^{t}{}_i$, is defined as

$$x \Longrightarrow_{\alpha\beta}^{t}{}_i y, \text{ iff } x \Longrightarrow_\alpha{}_i x_1, \ x_1 \Longrightarrow_\beta^*{}_i y, \text{ and there is no } z \text{ with } y \Longrightarrow_\beta{}_i z,$$

for some $x_1 \in (N \cup T)^*$.

Let $\mathrm{prod}(x \Longrightarrow_{\alpha\beta}^{\mu}{}_i y)$ denote the set of production sequences which can be used in the derivation step $x \Longrightarrow_{\alpha\beta}^{\mu}{}_i y$, $\mu \in \{t, =m \mid m \geq 1\}$, $\alpha, \beta \in \{s, w\}$. More precisely,

$$(A_0 \to w_0, A_1 \to w_1, \ldots, A_{l-1} \to w_{l-1}) \in \mathrm{prod}(x \Longrightarrow_{\alpha\beta}^{\mu}{}_i y)$$

if and only if there are strings $x_0, x_1, \ldots, x_l$ with $x_{j-1} = z_{j-1} A_{j-1} z'_{j-1}$ and $x_j = z_{j-1} w_{j-1} z'_{j-1}$, and $A_{j-1} \to w_{j-1} \in P_i$, for $1 \leq j \leq l$, $l \geq 1$.

In the following we will consider derivations in the ss- and sw-types of leftmostness. Note that these types of derivations were also considered in [6] where they were called "strong-leftmost" and "weak-leftmost", respectively.

The language generated by a CD grammar system G of degree n in the μ-mode of derivation, $\mu \in \{t, =m \mid m \geq 1\}$, with the γ-type of leftmostness,

$\gamma \in \{ss, sw\}$, is defined by $L_\gamma(G, \mu) = \{ w \in T^* \mid S \overset{\mu}{\underset{\gamma}{\Longrightarrow}}^* w \}$ where $x \overset{\mu}{\underset{\gamma}{\Longrightarrow}}^* y$, for $x, y \in (N \cup T)^*$, denotes the fact that there is a γ-leftmost derivation by G consisting of an arbitrary number of μ-mode derivation steps yielding y from x, that is, either $x = y$, or $x \overset{\mu}{\underset{\gamma}{\Longrightarrow}}_{i_1} x_1 \overset{\mu}{\underset{\gamma}{\Longrightarrow}}_{i_2} x_2 \ldots \overset{\mu}{\underset{\gamma}{\Longrightarrow}}_{i_r} x_r = y$, $1 \le i_j \le n$, $1 \le j \le r$.

Next, we present the definition of an LL(k) condition appropriate for (deterministic) CD grammar systems. It is adopted from the context-free case which can be found, e.g., in [1,10,11].

Definition 1. A CD grammar system $G = (N, T, S, P_1, P_2, \ldots P_n)$, $n \ge 1$, satisfies the LL(k) condition for some $k \ge 1$, if for any two leftmost derivations of type $\alpha \in \{ss, sw\}$ and of mode $\mu \in \{t, =m \mid m \ge 1\}$,

$$S \overset{\mu}{\underset{\alpha}{\Longrightarrow}}^* uXy \overset{\mu}{\underset{\alpha}{\Longrightarrow}}_i uz \overset{\mu}{\underset{\alpha}{\Longrightarrow}}^* uv \text{ and } S \overset{\mu}{\underset{\alpha}{\Longrightarrow}}^* uXy' \overset{\mu}{\underset{\alpha}{\Longrightarrow}}_{i'} uz' \overset{\mu}{\underset{\alpha}{\Longrightarrow}}^* uv'$$

with $u, v, v' \in T^*$, $X \in N$, $y, y', z, z' \in (N \cup T)^*$, if $\text{pref}_k(v) = \text{pref}_k(v')$, then $i = i'$ and $\text{prod}(uXy \overset{\mu}{\underset{\alpha}{\Longrightarrow}}_i uz) = \text{prod}(u'Xy' \overset{\mu}{\underset{\alpha}{\Longrightarrow}}_{i'} u'z')$ is a singleton set.

The idea behind this concept is the following: Given a terminal word uv and a sentential form uXy, $X \in N$ and $y \in (N \cup T)^*$, which has been obtained from S, then the first k letters of v allow to determine the component and the sequence of rules of that component which is to be applied to uXy in order to derive uv.

Note that, also according to the sw-type of leftmostness, X must be the very leftmost occurrence of a nonterminal in the sentential form since in those situations a new component has to start over working on the sentential forms.

By $CD_n(\mu, \alpha)$ with $n \ge 1$, $\mu \in \{t, =m \mid m \ge 1\}$, $\alpha \in \{ss, sw\}$ the family of languages which can be generated by a CD grammar system of degree n working in the α-type of leftmostness and the μ-mode of derivation is denoted. When deterministic components are considered, we replace CD with dCD. If we restrict to grammar systems of degree n which satisfy the LL(k)-condition for some $k \ge 1$, then the families of languages obtained are denoted by $X_n LL(k)(\mu, \alpha)$, $X \in \{CD, dCD\}$. If the number of components is not restricted, we write $CD_*(\mu, \alpha)$, $dCD_*(\mu, \alpha)$, $CD_* LL(k)(\mu, \alpha)$, and $dCD_* LL(k)(\mu, \alpha)$, respectively. Finally, let RE, CF, LL(k), and ET0L denote the families of all recursively enumerable, context-free, context-free LL(k) languages, and the family of languages generated by so called extended interactionless tabled Lindenmayer systems, respectively [11].

Some of the proofs of the results from [3,6] can also be adapted to the deterministic case, thus, we can formulate them as follows.

1. $CD_*(\mu, ss) = dCD_*(\mu, ss) = CF$, for $\mu \in \{t, =m \mid m \ge 1\}$,
2. ET0L $\subset CD_3(t, sw) = CD_*(t, sw)$, and
3. $CD_*(=m, sw) = dCD_*(=m, sw) = RE$, for any $m \ge 2$.

Before continuing, we illustrate the notion of LL(k) CD grammar systems with two examples.

Example 1. Consider $G = (N, \{a, b, c\}, S, P_1, P_2, P_3)$, the deterministic CDGS with $N = \{S, S', S'', A, B, C, A', B', C'\}$, and

$$P_1 = \{S \to S', S' \to S'', S'' \to ABC, A' \to A, B' \to B, C' \to C\},$$
$$P_2 = \{A \to aA', B \to bB', C \to cC'\}, \quad P_3 = \{A \to a, B \to b, C \to c\}.$$

This system generates by leftmost derivations of type sw, in the =3-mode the language $L_1 = \{a^n b^n c^n \mid n \geq 1\}$, and as can easily be checked, satisfies the LL(2) condition, thus, $\{a^n b^n c^n \mid n \geq 1\} \in dCD_3LL(2)(=3, sw)$.

The languages $L_2 = \{wcw \mid w \in \{a, b\}^*\}$ and $L_3 = \{a^n b^m c^n d^m \mid n, m \geq 1\}$ are shown to be in $dCD_4LL(2)(=2, sw)$ and in $dCD_4LL(2)(=2, sw)$ in a similar way, respectively. Furthermore, the respective CDGS generate the same languages in the t-mode of derivation and the sw-type of leftmostness, still satisfying the LL(2) condition.

The CD grammar system of the next example is working in the ss-type of leftmostness and the t-mode of derivation. Although these systems generate only context-free languages, the LL(k) variants are also able to deterministically describe languages (and thus allow their deterministic top-down parsing) which cannot be generated by context-free LL(k) grammars.

Example 2. Consider $G = (\{S, S', A\}, \{a, b, c\}, S, P_1, P_2, P_3, P_4, P_5)$ with components

$$P_1 = \{S \to aS'A\}, \quad P_3 = \{S' \to \varepsilon\}, \quad P_5 = \{A \to c\}.$$
$$P_2 = \{S' \to S\}, \quad\quad P_4 = \{A \to b\},$$

This deterministic system generates by leftmost derivations of type ss, in the t-mode the non-LL(k) language $L_{ss}(G, t) = \{a^n b^n \mid n \geq 1\} \cup \{a^n c^n \mid n \geq 1\}$, and as can easily be checked, satisfies the LL(1) condition.

3 Properties of Languages Generated by LL(k) CD Grammar Systems

First we present the following trivial hierarchies.

Lemma 1. *For any integers $k \geq 1$, $n \geq 1$, $\mu \in \{t, =m \mid m \geq 1\}$, $\alpha \in \{ss, sw\}$, and $X \in \{CD, dCD\}$, we have*

1. $dCD_nLL(k)(\mu, \alpha) \subseteq CD_nLL(k)(\mu, \alpha)$,
2. $X_nLL(k)(\mu, \alpha) \subseteq X_nLL(k+1)(\mu, \alpha)$,
3. $X_nLL(k)(\mu, \alpha) \subseteq X_{n+1}LL(k)(\mu, \alpha)$.

Now we summarize the relationship of context-free LL(k) languages and the languages generated by LL(k) CD grammar systems.

Theorem 2. *For any $k \geq 1$,*

1. $LL(k) = CD_*LL(k)(=1, \alpha) = dCD_*LL(k)(=1, \alpha)$, *for $\alpha \in \{ss, sw\}$,*
2. $LL(k) \subseteq dCD_*LL(k)(=m, ss)$, *for $m \geq 2$,*
3. $LL(k) \subset dCD_*LL(k)(=m, sw)$, *for $m \geq 2$,*
4. $dCD_*LL(k)(t, \alpha) - LL(k) \neq \emptyset$, *for $\alpha \in \{ss, sw\}$.*

Proof. Let $L \in \mathrm{LL}(k)$ for some $k \geq 1$, and let $G = (N, T, P, S)$ be a context-free $\mathrm{LL}(k)$ grammar with $L = L(G)$. Let the rules $r \in P$ be labeled by $1 \leq \mathrm{lab}(r) \leq \#P$. First we construct a deterministic CD grammar system $G' = (N', T, S, P_1, P_2, \ldots, P_n)$ satisfying the $\mathrm{LL}(k)$ condition and generating L in the $=m$-mode of derivation, for any $m \geq 2$, using any of the ss or sw-types of leftmostness, as follows. The number of components of G' is going to be $n = \#P$. Let $N' = N \cup \{ X^{(i)} \mid 1 \leq i \leq m-1, \ X \in N \}$, and for $1 \leq i \leq \#P$, if $i = \mathrm{lab}(X \to \alpha)$, then let

$$P_i = \{ X \to X^{(1)}, X^{(1)} \to X^{(2)}, \ldots, X^{(m-1)} \to \alpha \}.$$

It is easy to see that for any $\mu \in \{=m \mid m \geq 2\}$, $\alpha \in \{\mathrm{ss}, \mathrm{sw}\}$, a rewriting step $x \xrightarrow[\alpha]{\mu}_i y$ in G' is possible if and only if $x \Longrightarrow_i y$ is possible in G, where $\Longrightarrow_i$ denotes a rewriting step on the leftmost nonterminal using rule r with $\mathrm{lab}(r) = i$.

To prove the equality in point 1, note that any CD grammar system $G = (N, T, S, P_1, P_2, \ldots, P_n)$ working in the $=1$-mode is equivalent to a context free grammar (N, T, S, P) where $P = \bigcup_{1 \leq i \leq n} P_i$ using any of the two types of leftmostness in the derivations. Moreover, both of them are equivalent to a deterministic CDGS $G' = (N, T, S, P_1', P_2', \ldots, P_r')$ working in the $=1$-mode, where $r = \#P$, $\#P_i = 1$ for $1 \leq i \leq r$, and $\bigcup_{1 \leq i \leq r} P_i = P$. Then the equality follows from point 1 of Lemma 1.

The strictness of the inclusions in point 3 and the statement of 4 follows from Example 1 and Example 2 above.

Now we study further properties of the families of languages generated by $\mathrm{LL}(k)$ CDGS in the $=m$-mode of derivation, $m \geq 2$, and the sw-type of leftmostness. We show that any language in this family can also be generated by deterministic systems in the $=2$-mode.

Theorem 3. *For any $k \geq 1$, $m \geq 2$,*

$$\mathrm{CD}_* \mathrm{LL}(k)(=m, \mathrm{sw}) = \mathrm{dCD}_* \mathrm{LL}(k)(=m, \mathrm{sw}) = X_* \mathrm{LL}(k)(= 2, \mathrm{sw}),$$

where $X \in \{\mathrm{dCD}, \mathrm{CD}\}$.

Proof. We show that the inclusion $\mathrm{CD}_* \mathrm{LL}(k)(=m_1, \mathrm{sw}) \subseteq \mathrm{dCD}_* \mathrm{LL}(k)(=m_2, \mathrm{sw})$ holds for any $m_2 \geq 2$. Let $G = (N, T, S, P_1, P_2, \ldots, P_n)$ be a CDGS satisfying the $\mathrm{LL}(k)$ condition, $k \geq 1$, in derivation mode $=m_1$ for some $m_1 \geq 2$, with the sw-type of leftmostness. For any $m_2 \geq 2$, we construct a deterministic CDGS G' satisfying the $\mathrm{LL}(k)$ condition, such that $L_{\mathrm{sw}}(G, =m_1) = L_{\mathrm{sw}}(G', =m_2)$. Let the set of terminals and the start symbol of G' be the same as that of G, let the set of nonterminals be defined as $N' = N \cup \{X_{i,j}^{(l)} \mid 1 \leq i \leq n, 1 \leq j \leq m_1, 1 \leq l \leq m_2 - 1\}$, the union being disjoint, and let us define for all P_i, $1 \leq i \leq n$, the rule sets $P_{i,j}$, $1 \leq j \leq r_i$ for some $r_i \geq 1$, in such a way that for all $1 \leq i \leq n$,

- $\bigcup_{1 \le j \le r_i} P_{i,j} = P_i$,
- $\mathrm{dom}(P_{i,j}) = \mathrm{dom}(P_i)$ for all $1 \le j \le r_i$, and
- $P_{i,j}$ is deterministic, that is, for all $A \in \mathrm{dom}(P_i), \#\{w \mid A \to w \in P_{i,j}\} = 1$.

Now for all $i, j, l_i, 1 \le i \le n, 2 \le j \le m_1 - 1, 1 \le l_i \le r_i$, let the components of G' be defined as follows.

$$P_{i,1,l_i} = \{A \to X_{i,1}^{(1)} w \mid A \to w \in P_{i,l_i}\} \cup$$
$$\{X_{i,1}^{(s)} \to X_{i,1}^{(s+1)}, X_{i,1}^{(m_2-1)} \to X_{i,2}^{(1)} \mid 1 \le s \le m_2 - 2\},$$

$$P_{i,j,l_i} = P_{i,l_i} \cup \{X_{i,j}^{(s)} \to X_{i,j}^{(s+1)}, X_{i,j}^{(m_2-1)} \to X_{i,j+1}^{(1)} \mid 1 \le s \le m_2 - 2\},$$

$$P_{i,m_1,l_i} = P_{i,l_i} \cup \{X_{i,m_1}^{(s)} \to X_{i,m_1}^{(s+1)}, X_{i,m_1}^{(m_2-1)} \to \varepsilon \mid 1 \le s \le m_2 - 2\}.$$

To see that the $L_{\mathrm{sw}}(G, =m_1) = L_{\mathrm{sw}}(G', =m_2)$ holds, consider that for $u \in T^*$, G can execute a derivation step $uAy \overset{=m_1}{\underset{sw}{\Longrightarrow}}_i uy'$ if and only if

$$uAy \overset{=m_2}{\underset{sw}{\Longrightarrow}}_{i,1,l_{i,1}} uX_{i,2}^{(1)} y_1 \overset{=m_2}{\underset{sw}{\Longrightarrow}}_{i,2,l_{i,2}} uX_{i,3}^{(1)} y_2 \ldots uX_{i,m_1}^{(1)} y_{m_1-1} \overset{=m_2}{\underset{sw}{\Longrightarrow}}_{i,m_1,l_{i,m_1}} uy'$$

for some $1 \le l_{i,j} \le r_i$ for all $1 \le j \le m_1$, can be executed by G'. Note that the nonterminals $X_{i,j}^{(l)}$ must always be replaced as they appear leftmost in the sentential forms. The leftmostness of these nonterminals also implies that until they have been erased, no new terminal symbols are added to the already derived terminal prefix of the generated string appearing left of these nonterminals, so the rule sequence determined by the LL(k) property at the beginning of an m_1-step derivation of G and these nonterminals of the form $X_{i,j}^{(s)}$ also determine the unique rule sequence for each m_2-step derivation of G' which means that it also satisfies the LL(k) property.

The statements of the theorem are consequences of the inclusion we have proved above, and the results of Lemma 1.

Lemma 4. *For any $X \in \{\mathrm{dCD}, \mathrm{CD}\}$, $\mu \in \{t, =m \mid m \ge 2\}$, there are non-semilinear languages in $X_*\mathrm{LL}(1)(\mu, \mathrm{sw})$.*

Proof. In [2] a deterministic CDGS is presented which satisfies the LL(1) condition and generates a non-semilinear language in the derivation mode $=2$ with the sw-type of leftmostness. By analyzing this system, we can see that it generates the same non-semilinear language also in the t-mode of derivation. Thus, our statement follows from the proof found in [2] and from Theorem 3.

4 Using Lookup Tables

We are going to define the notion of a lookup table for CD grammar systems satisfying the LL(k) condition in some mode μ of derivation using type α of leftmostness, $\mu \in \{t, = m\}$, $\alpha \in \{\mathrm{ss}, \mathrm{sw}\}$. The lookup table determines the component and the sequence of rules which are needed for the continuation of the derivation, according to the definition of the $LL(k)$ condition. The selection of the rules is based on pairs of nonterminals (the leftmost nonterminal in the leftmost derivation) and lookahead strings (the first k terminal letters of the suffix of the resulting terminal word which is derived by the remaining part of the leftmost derivation).

A lookup table can be very useful. First of all, it provides us with the possibility of constructing efficient parsers, that is, algorithms which, when given a grammar G and a terminal string w as input, decide whether w can be generated by G, and if so, reconstruct the derivation.

For $CD_nLL(k)(\mu, ss)$, $\mu \in \{t, = m \mid m \geq 1\}$ slight modifications of the usual top down methods for context-free $LL(k)$ parsing can be used to provide parsing algorithms also for grammar systems of these types. For context-free $LL(k)$ parsing, see, for example, [1].

For CD grammar systems working in the sw-type of leftmostness we need more sophisticated methods while derivations of this type also rewrite nonterminals which are not the leftmost ones in the sentential form. In Figure 1 we present a parsing algorithm for languages in $CD_nLL(k)(= m, sw)$. This algorithm already appeared in [2] where also an initial version of the following theorem was given.

Theorem 5. *If a CD grammar system G satisfying the $LL(k)$ condition in the $=m$ derivation mode using the sw-type of leftmostness is given together with its lookup table, then for $L_{sw}(G, =m)$ a parser can be constructed as presented in Figure 1, which halts on every input word w in $O(n \cdot \log^2 n)$ steps, where n is the length of w.*

In case of deterministic systems, the lookup table is more simple, it only needs to give a component for the pairs of nonterminals and lookahead strings, the exact order of the application of the rules is automatically determined due to the restriction to sw-type of leftmost derivations.

Definition 2. Let $G = (N, T, S, P_1, P_2, \ldots P_n) \in dCD_nLL(k)(\mu, \alpha)$, for some $\mu \in \{t, =m \mid m \geq 1\}$, $\alpha \in \{ss, sw\}$, $n \geq 1$, and $k \geq 1$. The *lookup table* M_G for G is a subset of $N \times T^{\leq k} \times \{P_1, \ldots, P_n\}$ such that for all uXy with

$$S \underset{\alpha}{\overset{\mu}{\Longrightarrow}}{}^* uXy \underset{\alpha}{\overset{\mu}{\Longrightarrow}} uz \underset{\alpha}{\overset{\mu}{\Longrightarrow}}{}^* uv$$

$u, v \in T^*$, $X \in N$, $y, z \in (N \cup T)^*$, the entry $(X, \mathrm{pref}_k(v))$ contains the component which is to be applied to the sentential form uXy.

Clearly, to have a lookup table, it is necessary for a grammar system to satisfy the $LL(k)$ condition. The implication in the other direction, opposed to the context-free case, is not as obvious. Just as unclear is the existence of a general algorithm for the construction of a lookup table when a (deterministic) grammar system is given.

Sometimes it is not difficult to construct a lookup table, as in the case of the grammar system from Example 1, which is as follows:

$$\{(S, aa, P_1), (S, ab, P_1), (A, aa, P_2), (A, ab, P_3), (A', aa, P_1), (A', ab, P_1)\}.$$

1 $step \leftarrow 0$

2 $mainStack \leftarrow push(mainSt, S)$

3 $stacksForN(S) \leftarrow push(stacksForN(S), 0)$

4 **while** $mainStack$ is not empty and there is no ERROR **do**

5   **if** $top(mainStack)$ is a terminal symbol **then**

6   **if** $top(mainStack)$ coincides with the first symbol of $input$ **then**

7       $mainStack \leftarrow pop(mainStack)$

8       $input \leftarrow input$ without its first symbol

9   **else** ERROR

10 **else** $topmost \leftarrow top(mainStack)$

11     $stepOfTopmost \leftarrow top(stacksForN(topmost))$

12     $stacksForN(topmost) \leftarrow pop(stacksForN(topmost))$

13   **if** there exist $(i; pQueue) \in pQueuesLeft$ such that
    $i \geq stepOfTopmost$, $left(first(pQueue)) = topmost$,
    and furthermore, if $(i'; pQueue') \in pQueuesLeft$
    with $left(first(pQueue')) = topmost$, then $i < i'$, **then**

14       $pQueuesLeft \leftarrow pQueuesLeft - \{(i; pQueue)\}$

15       $pToUse \leftarrow first(pQueue)$

16       $pQueue \leftarrow butfirst(pQueue)$

17   **if** $pQueue$ is not empty **then**

18         $pQueuesLeft \leftarrow pQueuesLeft \cup \{(i; pQueue)\}$

19       $mainStack \leftarrow pop(mainStack)$

20       $mainStack \leftarrow push(mainStack, right(pToUse))$

21   **for** each symbol X from $right(pToUse)$ **do**

22   **if** $X \in N$ **then**

23           $stacksForN(X) \leftarrow push(stacksForN(X), step)$

24   **else** $step \leftarrow step + 1$

25       $lookahead \leftarrow$ the next k symbols of $input$

26       $pQueue \leftarrow lookupTable(topmost, lookahead)$

27   **if** $pQueue$ is empty **then**

28   ERROR

29   **else** $pToUse \leftarrow first(pQueue)$

30         $pQueue \leftarrow butfirst(pQueue)$

31   **if** $pQueue$ is not empty **then**

32           $pQueuesLeft \leftarrow pQueuesLeft \cup \{(step, pQueue)\}$

33         $mainStack \leftarrow pop(mainStack)$

34         $mainStack \leftarrow push(mainStack, right(pToUse)$

35   **for** each symbol X from $right(pToUse)$ **do**

36   **if** $X \in N$ **then**

37             $stacksForN(X) \leftarrow push(stacksForN(X), step)$

38 **if** there is no ERROR **then** successful termination

Fig. 1. The parsing algorithm. It uses the variables: $step$, $stepOfTopmost \in \mathbb{N}$; $mainStack$, a stack over $N \cup T$, the "main" stack of the parser; $stacksForN$, an l-tuple of stacks where $l = \#N$, it provides a stack over $\mathbb{N}$ for each nonterminal; $input \in T^*$, the string to be analyzed; $topmost \in N$; $pQueue$, a production queue; $pQueuesLeft \subseteq \mathbb{N} \times PQ$, where PQ denotes the set of all production queues of length at most m, that is, $pQueuesLeft$ is a set of pairs of the form $(i; pq)$ where i is an integer and pq is a production queue; $pToUse \in N \times (N \cup T)^*$, a production.

Now we show that if we assume the existence of a lookup table, then in the case of the sw-type of leftmostness, the length of the necessary lookahead can be decreased to $k = 1$. That is, in contrast to the context-free case, the hierarchies of language classes corresponding to deterministic LL(k) CDGS induced by k collapse, namely to the first level.

Theorem 6. *Given a deterministic CD grammar system G satisfying the LL(k) condition for some derivation mode $\mu \in \{t, = m \mid m \geq 2\}$ with the sw-type of leftmostness, and its lookup table M_G, then $L_{sw}(G, \mu) \in \text{dCD}_*\text{LL}(1)(\mu, sw)$.*

Proof. Given $G = (N, T, S, P_1, P_2, \dots P_n)$, a deterministic CDGS as above, satisfying the LL(k) condition for some $k \geq 2$. Let the look-up table of G denoted by M_G. We construct a context-free CD grammar system H which satisfies the LL(1) condition and for which $L_{sw}(H, \mu) = L_{sw}(G, \mu)$ holds, μ is as above. Let the set of non-terminals for H be the set

$$N' = N \cup \{(X, v), (X, v)^{(i)}(A, v), (A, v)', (A, v)^{(i)}, (\bar{a}, v), (\bar{a}, v)^{(i)} \mid A \in N,$$
$$v \in T^{\leq k}, a \in T, 1 \leq i < m\} \cup \{\bar{a} \mid a \in T\},$$

where X is a new symbol, and let the axiom of H be (S, ε). Let us define for any $\alpha = x_1 x_2 \dots x_t$, $x_i \in N \cup T$, $1 \leq i \leq t$, the string $\bar{\alpha} = \text{b}(x_1)\text{b}(x_2)\dots\text{b}(x_t)$ with $\text{b}(x) = x$ for $x \in N$, and $\text{b}(x) = \bar{x} \in \bar{T}$ for $x \in T$.

We construct the following components.

(1) *Scanning components.* For all $A \in N$ and $u \in T^{\leq k-1}$, $v \in T^{\leq k}$, we have:

$$\{(A, u) \to (A, u)^{(1)}, \dots, (A, u)^{(m-2)} \to (A, u)^{(m-1)}, (A, u)^{(m-1)} \to a(A, ua)'\},$$

and also the components:

$$\{(A, v)' \to (A, v)^{(1)}, \dots, (A, v)^{(m-2)} \to (A, v)^{(m-1)}, (A, v)^{(m-1)} \to (A, v)\}.$$

These collect the look-ahead string v of length at most k symbol by symbol. Their correct, deterministic use is guaranteed by the look-ahead of length one. If $|u| = k$ for some (A, u), or the look-ahead symbol is ε, then these components cannot be used any more.

(2) *Direct simulating components.* For $(A, v, P_i) \in M_G$, $A \to \alpha \in P_i$, $\alpha = uB\beta$ with $u \in T^*$, $B \in N$, $\beta \in (N \cup T)^*$, we have:

$$\{(A, v) \to v_1(X, v_2)B\bar{\beta}\} \cup \bar{P}_i$$

where if $u = vv'$, for some $v' \in T^*$, then $v_1 = v'$ and $v_2 = \varepsilon$. Otherwise, if $v = uu'$, for some $u' \in T^*$, then $v_1 = \varepsilon$ and $v_2 = u'$. Furthermore, $\bar{P}_i$ denotes the set $\{A \to \bar{\alpha} \mid A \to \alpha \in P_i\}$.

If $\alpha = u$ with $u \in T^*$, then we have:

$$\{(A, v) \to v_1(X, v_2)\} \cup \bar{P}_i$$

where if $u = vv'$, for some $v' \in T^*$, then $v_1 = v'$ and $v_2 = \varepsilon$, or otherwise, if $v = uu'$, for some $u' \in T^*$, then $v_1 = \varepsilon$ and $v_2 = u'$, and $\bar{P}_i$ is as above. These components will do the same as the component P_i of G under look-ahead v. It is taken into consideration that either the prefix v of u has already been generated by the scanning components and only the corresponding suffix must be produced in the first step, or the suffix of the scanned look-ahead string which is no part of u is stored in the new nonterminal which is now leftmost.

Note that the scanning components have nonempty look-aheads, the simulating components which rewrite some (A, v) with $|v| < k$, on the other hand, are to be used under empty look-ahead string.

(3) *Look-ahead shifting components.* For all $u \in T^{\leq k}$, we have:

$$\{(X, u) \rightarrow (X, u)^{(1)}, \ldots, (X, u)^{(m-1)} \rightarrow (X, u)^{(m-2)}, (X, u)^{(m-2)} \rightarrow \varepsilon\} \cup$$

$$\{B \rightarrow (B, u) \mid B \in N \cup \bar{T}\}, \text{ and}$$

$$\{(\bar{b}, u) \rightarrow (\bar{b}, u)^{(1)}, \ldots, (\bar{b}, u)^{(m-1)} \rightarrow (\bar{b}, u)^{(m-2)}, (\bar{b}, u)^{(m-2)} \rightarrow b(X, \delta) \mid \bar{b} \in \bar{T},$$

$$\delta = \varepsilon \text{ if } u = \varepsilon, \text{ or } \delta = u' \text{ if } u = bu'\}.$$

By these components the stored look-ahead string u is transferred to the next symbol to the right of the leftmost nonterminal symbol. If this next symbol is from $\bar{T}$, then the corresponding terminal symbol is generated and the rest of the look-ahead is shifted further, if it is from N, a look-ahead string of maximal possible length (k, in general) is supplemented to it with the help of the scanning components. Only then, the next simulation can be performed.

The comments given to the components constructed above show that the equalities $L_{sw}(H, \mu) = L_{sw}(G, \mu)$, $\mu \in \{t, =m \mid m \geq 2\}$, hold, and as H is LL(1), the proof of is complete.

From Theorem 3, and Theorem 6 we obtain the following corollary.

Corollary 7. *Given a deterministic CD grammar system G satisfying the LL(k) condition, $k \geq 1$, and its lookup table for derivation mode $=m$, $m \geq 2$ with type of leftmostness sw. Then $L_{sw}(G, =m) \in \mathrm{dCD}_* \mathrm{LL}(1)(=2, sw)$.*

Now we present a decidable condition that the CD grammar system has to satisfy in order that the lookup table be effectively constructable.

First we need the notion of the FIRST and FOLLOW sets, as defined for example in [1]. Let (N, T, S, P) be a context-free grammar, and let $k \geq 1$. Now, for any $x \in (N \cup T)^*$, let $\mathrm{FIRST}_k(x) = pref_k(L(x))$, where $L(x)$ denotes the set of terminal words that can be derived from x, $L(x) = \{w \in T^* \mid x \Rightarrow_P^* w\}$. Let also, for any $A \in N$, $\mathrm{FOLLOW}_k(A) = \{w \in T^* \mid S \Rightarrow_P^* xAy$ and $w \in \mathrm{FIRST}_k(y)\}$. For an effective construction of these sets, see for example [1].

Definition 3. A deterministic CD grammar system $G = (N, T, S, P_1, \ldots, P_n)$, $n \geq 1$, satisfies the *strong-LL(k) condition* for some $k \geq 1$, if for all i, $1 \leq i \leq n$, and productions $A \rightarrow \alpha \in \bigcup_{1 \leq i \leq n} P_i$, the fact that $A \rightarrow \alpha \in P_i$ implies that $A \rightarrow \alpha \notin P_j$ for all $j \neq i$. Moreover, for all productions $A \rightarrow \alpha$, $A \rightarrow \beta \in \bigcup_{1 \leq i \leq n} P_i$ for some $A \in N$, such that $\alpha \neq \beta$, the condition

$$\mathrm{FIRST}_k(\alpha \mathrm{FOLLOW}_k(A)) \cap \mathrm{FIRST}_k(\beta \mathrm{FOLLOW}_k(A)) = \emptyset$$

holds with respect to the context-free grammar $(N, T, S, \bigcup_{1 \leq i \leq n} P_i)$.

Note that, as we have already mentioned, since the FIRST and FOLLOW sets are effectively constructable, the strong-LL(k) condition is algorithmically decidable. Note also, that all examples presented in the paper satisfy the strong-LL(k) property.

For all context-free grammars (N, T, S, P) which satisfy the strong-LL(k) property, a lookup table $M \subset (N \times T^{\leq k} \times P)$ can be effectively constructed as described for example in [1]. Thus, we can effectively construct a lookup table M_G also for any CDGS $G = (N, T, S, P_1, P_2, \ldots, P_n)$ satisfying the strong LL(k) property by constructing M_H for $H = (N, T, S, \bigcup_{1 \leq i \leq n} P_i)$, and then let $(A, w, P_i) \in M_G$ if and only if $(A, w, A \rightarrow \alpha) \in M_H$ for some $A \rightarrow \alpha \in P_i$.

Corollary 8. *For any deterministic CD grammar system G satisfying the strong-LL(k) property, $k \geq 1$, a derivation mode $\mu \in \{t, = m \mid m \geq 2\}$, we have that $L_{sw}(G, \mu) \in \mathrm{dCD}_n\mathrm{LL}(1)(\mu, sw)$, and moreover, the corresponding deterministic CDGS satisfying the LL(1) condition and the lookup table for deterministic parsing can be effectively constructed.*

References

1. Aho, A.V., Ulmann, J.D.: The Theory of Parsing, Translation, and Compiling, vol. I. Prentice-Hall, Englewood Cliffs, N.J. (1973)
2. Bordihn, H., Vaszil, Gy.: CD grammar systems with LL(k) conditions. In: Csuhaj-Varjú, E., Vaszil, Gy. (eds.) Proceedings of Grammar Systems Week, MTA SZTAKI, Budapest, pp. 95–112 (2004)
3. Bordihn, H., Vaszil, Gy.: On leftmost derivations in CD grammar systems. In: Loos, R., Fazekas, Sz. Zs., Martín-Vide, C. (eds.) Pre-Proceedings of the First International Conference on Language and Automata Theory and Applications, LATA 2007, Universitat Rovira i Virgili, Tarragona, pp. 187–198 (2007)
4. Csuhaj-Varjú, E., Dassow, J.: On cooperating/distributed grammar systems. Journal of Information Processing and Cybernetics EIK, 26(1-2), 49–63 (1990)(Presented at the 4th Workshop on Mathematical Aspects of Computer Science, Magdeburg, 1998)
5. Csuhaj-Varjú, E., Dassow, J., Kelemen, J.,Păun, Gh.: Grammar Systems. A Grammatical Approach to Distribution and Cooperation, Gordon and Breach Science Publishers, Yverdon (1994)
6. Dassow, J., Mitrana, V.: On the leftmost derivation in cooperating grammar systems. Rev. Roumaine Math. Pures Appl. 43, 361–374 (1998)
7. Dassow, J., Păun, Gh.: Regulated Rewriting in Formal Language Theory. Springer, Heidelberg (1989)
8. Dassow, J., Păun, Gh., Rozenberg, G.: Grammar Systems. In: Rozenberg, G., Salomaa, A. (eds.) Handbook of Formal Languages, ch. 4, pp. 155–213. Springer, Heidelberg (1997)
9. Mitrana, V.: Parsability approaches in CD grammar systems. In: Freund, R., Kelemenová, A. (eds.) Proceedings of the International Workshop Grammar Systems 2000, Silesian University at Opava, pp. 165–185 (2000)
10. Rosenkrantz, D.J., Stearns, R.E.: Properties of deterministic top-down grammars. Information and Control 17, 226–256 (1970)
11. Salomaa, A.: Formal Languages. Academic Press, New York (1973)

The Complexity of Membership Problems for Circuits over Sets of Positive Numbers

Hans-Georg Breunig

Universität Würzburg
Institut für Informatik, Am Hubland, 97074 Würzburg, Germany
breunig@informatik.uni-wuerzburg.de

Abstract. We investigate the problems of testing membership in the subset of the positive numbers produced at the output of $(\cup, \cap, {}^-, +, \times)$ combinational circuits. These problems are a natural modification of those studied by McKenzie and Wagner (2003), where circuits computed sets of natural numbers. It turns out that the missing 0 has strong implications, not only because 0 can be used to test for emptiness. We show that the membership problem for the general case and for $(\cup, \cap, +, \times)$ is PSPACE-complete, whereas it is NEXPTIME-hard if one allows 0. Furthermore, testing membership for $(\cap, \times)$ is NL-complete (as opposed to $C_=L$-hard), and several other cases are resolved.

Keywords: Computational complexity, Arithmetic circuits, Combinational circuits.

1 Introduction

Consider arithmetic expressions over the power set of $\mathbb{N}$ using pair-wise addition $(+)$ and pair-wise multiplication $(\times)$ as well as the set operators union $(\cup)$, intersection $(\cap)$, and complement $({}^-)$, e.g. $(\{2,5\} \cup \{1\}) + \overline{\{4\}}$. Such formulae are called *Integer Expressions* and they were introduced by Stockmeyer and Meyer as early as 1973. The corresponding membership problem, that asks whether a given number is in the set computed by the formula, is denoted by N-MEMBER [5]. McKenzie and Wagner exhaustively studied membership problems of formulae and circuits from $\{\cup, \cap, {}^-, +, \times\}$ [3]. It turned out that, depending on the set of allowed operations, these problems belong to a wide range of complexity classes ranging from L to (possibly) undecidabilty. Further related problems for integer circuits were studied in [8][9][6][2].

In this paper we study membership problems for circuits (and formulae), but this time over the positive natural numbers $\mathbb{N}^+$ instead of $\mathbb{N}$. Concerning membership problems for formulae we cannot prove that they become easier when omitting 0. Although the general case probably is undecidable for formulae over $\mathbb{N}$ we obtain the same lower bound for formulae over $\mathbb{N}^+$. Things are different for circuit problems. Not only that the membership problem for the general case is decidable, we can prove it to be in PSPACE as opposed to the NEXPTIME-hardness for circuits over $\mathbb{N}$. Additionally, for circuits having all but complement

E. Csuhaj-Varjú and Z. Ésik (Eds.): FCT 2007, LNCS 4639, pp. 125–136, 2007.

gates, we also show that the membership problem becomes easier when omitting 0 (PSPACE-completeness instead of NEXPTIME-completeness). Another difference is that the membership problem for $\{\cap, \times\}$-circuits is NL-complete while McKenzie and Wagner showed it to be $C_=L$-hard when using $\mathbb{N}$.

2   Preliminaries

If A is a set, then $\|A\|$ denotes the cardinality of A. The length of a string x is denoted by $|x|$.

We use the following complexity classes which can be found in textbooks on computational complexity (e.g. [4]): P, NP, L, NL, and PSPACE as well as FP, FL, and #L.

LOGDCFL: A language A is in LOGDCFL if there exists a Turing machine with an unlimited stack, but a logarithmic bound for the space it may use on the working tape as well as a polynomial bound for its runtime which decides A.

$C_=L$:   A language A is in $C_=L$ if there exist functions $f \in \#L$ and $g \in FL$ such that $x \in A \Leftrightarrow f(x) = g(x)$.

We use the following well-known reducibilities: $\leq_m^{\log}$ and $\leq_m^p$ ([4]). The former is used in this paper unless otherweise noted.

AC^0-reduction: The AC^0-reduction is defined using Boolean circuits. A *circuit family* is a set $\{C_n : n \in \mathbb{N}\}$ where each C_n is a circuit with n Boolean inputs $x_1, \ldots, x_n$ and output gates $y_1, \ldots, y_r$. $\{C_n\}$ has *size* $s(n)$ if each circuit C_n has at most $s(n)$ gates; it has *depth* $d(n)$ if the length of the longest path from an input gate to an output gate is at most $d(n)$. A family $\{C_n\}$ is *uniform* if the function $n \mapsto C_n$ is easy to compute in some sense. A function f is in AC^0 if there is an uniform circuit family $\{C_n\}$ of size $n^{O(1)}$ and depth $O(1)$ consisting of AND, OR, and NOT gates with unbounded in-degree such that for each input x of length n, the output of C_n on input x is $f(x)$.

Let GAP (graph accessibility problem) be the set of all tuples (G, s, t) where G is a graph and there is a path from s to t. It holds that GAP is NL-complete (see [4]). Let GAP$'$ be the set of all elements in GAP where the vertices have in-degree 0 or 2 and where the vertices are ordered topologically. We now show that GAP$'$ is NL-hard. This works by reducing an arbitrary NL-problem A (decided by a Turing machine M) to GAP$'$. Without loss of generality we assume that M has a unique accepting configuration t and all other configurations have exactly two successor configurations. We construct a new machine M' that works like M, but that has a counter for every transition of M. The value of this counter is placed leftmost on the tape, resulting in configurations that are ordered chronologically. This implies that a given configuration is lexicographically smaller than the configurations of all direct and indirect successors. Now, consider the transition graph G' of M' on input x. It holds that M halts on input x with the start configuration s if and only if $(G', s, t) \in$ GAP$'$. Thus GAP$'$ is NL-hard and it follows that $\overline{\text{GAP}'}$ is coNL-hard.

3   Definitions

The set of *positive numbers* is given by $\mathbb{N}^+ =_{\text{def}} 1, 2, 3, \ldots$ and the set of *natural numbers* by $\mathbb{N} =_{\text{def}} \mathbb{N}^+ \cup \{0\}$. $\mathbb{Z}$ denotes the set of *integer numbers*. For two sets $U, V \subseteq \mathbb{Z}$ we define $U \oplus V =_{\text{def}} \{u+v \; : \; u \in U \text{ and } v \in V\}$ and $U \otimes V =_{\text{def}} \{u \cdot v \; : \; u \in U \text{ and } v \in V\}$.

3.1 $(\mathcal{O}, U)$-Circuits

A *circuit* $C = (G, A, g_C)$ is a finite acyclic graph (G, A) with a specified node g_C, the *output gate*. The gates with in-degree 0 are called *input gates*.
We consider different types of circuits. Let $\mathcal{O} \subseteq \{\cup, \cap, ^-, +, \times\}$ and let $U \subseteq \mathbb{Z}$ be decidable in logarithmic space and closed under $+$ and $\times$. An $(\mathcal{O}, U)$-*circuit* $C = (G, A, g_C, \alpha)$ is a *circuit* (G, A, g_C) whose gates have in-degree 0, 1, or 2 and are labeled by the function $\alpha : G \to \mathcal{O} \cup U$ in the following way: Every gate of in-degree 0 has label $\alpha(g) \in U$, every gate g of in-degree 1 has label $\alpha(g) = ^-$, and every gate g of in-degree 2 has label $\alpha(g) \in \{\cup, \cap, +, \times\}$.
For an $(\mathcal{O}, U)$-circuit C we define an interpretation function I_U:

- If g is an input gate with label a, then $I_U(g) =_{\text{def}} \{a\}$.
- If g is a $+$-gate with predecessors g_1, g_2, then $I_U(g) =_{\text{def}} I_U(g_1) \oplus I_U(g_2)$.
- If g is a $\times$-gate with predecessors g_1, g_2, then $I_U(g) =_{\text{def}} I_U(g_1) \otimes I_U(g_2)$.
- If g is an $\cup$-gate with predecessors g_1, g_2, then $I_U(g) =_{\text{def}} I_U(g_1) \cup I_U(g_2)$.
- If g is a $\cap$-gate with predecessors g_1, g_2, then $I_U(g) =_{\text{def}} I_U(g_1) \cap I_U(g_2)$.
- If g is a $^-$-gate with predecessor g_1, then $I_U(g) =_{\text{def}} U \setminus I_U(g_1)$.

The set $I_U(C) =_{\text{def}} I_U(g_C)$ is called the *set computed by* C. An $(\mathcal{O}, U)$-*formula* is an $(\mathcal{O}, U)$-circuit with maximum out-degree 1.
For $\mathcal{O} \subseteq \{\cup, \cap, ^-, +, \times\}$ the *membership problems* for $(\mathcal{O}, U)$-circuits and $(\mathcal{O}, U)$-formulae are defined as
$\mathrm{MC}_U(\mathcal{O}) =_{\text{def}} \{(C, b) : C \text{ is an } (\mathcal{O}, U)\text{-circuit}, b \in U, \text{ and } b \in I_U(C)\}$ and
$\mathrm{MF}_U(\mathcal{O}) =_{\text{def}} \{(C, b) : C \text{ is an } (\mathcal{O}, U)\text{-formula}, b \in U, \text{ and } b \in I_U(C)\}$.

4   General Considerations

The following proposition allows us to compare membership problems of circuits over different domains.

Proposition 1. *Let* $\mathcal{O} \subseteq \{\cup, \cap, +, \times\}$ *and let* $U \subseteq U' \subseteq \mathbb{Z}$ *where* U, U' *are logarithmic-space decidable and closed under* $+, \times$.

1. $\mathrm{MC}_U(\mathcal{O}) \leq_m^{\log} \mathrm{MC}_{U'}(\mathcal{O})$.
2. $\mathrm{MF}_U(\mathcal{O}) \leq_m^{\log} \mathrm{MF}_{U'}(\mathcal{O})$.

Proof. The reduction function f works as follows: Check (in logarithmic space) whether C contains an input value from $U' \setminus U$ or whether $b \in U' \setminus U$. If this is not the case, then output (C, b). Otherwise, output (C_0, b_0), which is not in $\mathrm{MF}_{U''}(\mathcal{O})$ for all $U'' \subseteq \mathbb{Z}$ (think of a syntactically invalid instance for example). So it holds that $(C, b) \in \mathrm{MC}_U(\mathcal{O}) \iff f(C, b) \in \mathrm{MC}_{U'}(\mathcal{O})$. $\qquad\square$

The reason why membership problems for circuits over $\mathbb{N}$ are (in some cases) more difficult than for circuits over $\mathbb{N}^+$, can be found in the possibility to use the question of whether 0 is a member to test for emptiness. Let $\text{EMPTY}_U(\mathcal{O}) =_{\text{def}}$ $\{C : C$ is an $(\mathcal{O}, U)$-circuit and $I_U(C) = \emptyset\}$ and for $0 \in U$ let $0\text{-MC}_U(\mathcal{O}) =_{\text{def}}$ $\{C : (C, 0) \in \text{MC}_U(\mathcal{O})\}$.

Proposition 2. *Let* $\{\cap, \times\} \subseteq \mathcal{O} \subseteq \{\cup, \cap, +, \times\}$.

$$\overline{0\text{-MC}_{\mathbb{N}}(\mathcal{O})} \equiv_m^{AC^0} \text{EMPTY}_{\mathbb{N}}(\mathcal{O})$$

Proof. It holds that $C \notin 0\text{-MC}_{\mathbb{N}}(\mathcal{O}) \iff C \cap 0 \in \text{EMPTY}_{\mathbb{N}}(\mathcal{O})$ and $C \in \text{EMPTY}_{\mathbb{N}} \iff C \times 0 \notin 0\text{-MC}_{\mathbb{N}}(\mathcal{O})$. □

Given $(\{\cup, \cap, ^-, \times\}, \mathbb{N})$-circuits, we want to evaluate C with respect to $\mathbb{N}^+$. Therefore, we define for all input gates $g \in G$ where $\alpha(g) = 0$: $I_{\mathbb{N}^+}(g) =_{\text{def}} \emptyset$.

Lemma 1. *Let* $C = (G, A, g_C, \alpha)$ *be a* $(\{\cup, \cap, ^-, \times\}, \mathbb{N})$-*circuit. For all* $g \in G$ *it holds that*

$$I_{\mathbb{N}^+}(g) = I_{\mathbb{N}}(g) \setminus \{0\}.$$

Proof. We prove this by induction on the structure of C.

IB: Let g be an input gate. If $\alpha(g) = 0$, then $I_{\mathbb{N}^+}(g) = I_{\mathbb{N}}(g) \setminus \{0\} = \emptyset$. If $\alpha(g) > 0$, then $I_{\mathbb{N}^+}(g) = I_{\mathbb{N}}(g) \setminus \{0\} = I_{\mathbb{N}}(g) = \alpha(g)$.
IS: Let $\alpha(g) = ^-$ and let g_1 be the predecessor of g.

$$\begin{aligned}
I_{\mathbb{N}^+}(g) &= \mathbb{N}^+ \setminus I_{\mathbb{N}^+}(g_1) \\
&= \mathbb{N}^+ \setminus (I_{\mathbb{N}}(g_1) \setminus \{0\}) \\
&= \mathbb{N}^+ \setminus (I_{\mathbb{N}}(g_1) \cap \overline{\{0\}}) \\
&= \mathbb{N}^+ \cap (\overline{I_{\mathbb{N}}(g_1)} \cup \{0\}) \\
&= (\mathbb{N}^+ \cap \overline{I_{\mathbb{N}}(g_1)}) \cup (\mathbb{N}^+ \cap \{0\}) \\
&= \overline{I_{\mathbb{N}}(g_1)} \setminus \{0\} = I_{\mathbb{N}}(g) \setminus \{0\}
\end{aligned}$$

Let $\alpha(g) \in \{\cup, \cap, \times\}$ and let g_1 and g_2 be the predecessors of g.

$$\begin{aligned}
\alpha(g) = \cap: \quad I_{\mathbb{N}^+}(g) &= I_{\mathbb{N}^+}(g_1) \cap I_{\mathbb{N}^+}(g_2) \\
&= (I_{\mathbb{N}}(g_1) \cap I_{\mathbb{N}}(g_2)) \setminus \{0\} \ = \ I_{\mathbb{N}}(g) \setminus \{0\}
\end{aligned}$$

$$\begin{aligned}
\alpha(g) = \cup: \quad I_{\mathbb{N}^+}(g) &= I_{\mathbb{N}^+}(g_1) \cup I_{\mathbb{N}^+}(g_2) \\
&= (I_{\mathbb{N}}(g_1) \cup I_{\mathbb{N}}(g_2)) \setminus \{0\} \ = \ I_{\mathbb{N}}(g) \setminus \{0\}
\end{aligned}$$

$$\begin{aligned}
\alpha(g) = \times: \quad I_{\mathbb{N}^+}(g) &= I_{\mathbb{N}^+}(g_1) \otimes I_{\mathbb{N}^+}(g_2) \\
&= (I_{\mathbb{N}}(g_1) \setminus \{0\}) \otimes (I_{\mathbb{N}}(g_2) \setminus \{0\}) \\
&= \{k \cdot l \ : \ k \in I_{\mathbb{N}}(g_1) \wedge l \in I_{\mathbb{N}}(g_2) \wedge k \cdot l \neq 0\} \\
&= (I_{\mathbb{N}}(g_1) \otimes I_{\mathbb{N}}(g_2)) \setminus \{0\} \ = \ I_{\mathbb{N}}(g) \setminus \{0\} \qquad □
\end{aligned}$$

5   Upper Bounds

Compared to the results for membership problems for circuits over $\mathbb{N}$ [3], we will give new upper bounds for $MC_{\mathbb{N}+}(\cap, \times)$, $MC_{\mathbb{N}+}(\cap, +, \times)$, $MC_{\mathbb{N}+}(\cup, \cap, +, \times)$, $MC_{\mathbb{N}+}(\cup, \cap, ^-, +, \times)$, and $MF_{\mathbb{N}+}(\cup, \cap, ^-, +, \times)$.

5.1   Using Known Upper Bounds

Many results that have already been shown for circuits over $\mathbb{N}$ can be adopted as the following corollary states.

Corollary 1. *Let* $\mathcal{O} \subseteq \{\cup, \cap, +, \times\}$.

1. $MC_{\mathbb{N}+}(\mathcal{O}) \leq_m^{\log} MC_{\mathbb{N}}(\mathcal{O})$
2. $MF_{\mathbb{N}+}(\mathcal{O}) \leq_m^{\log} MF_{\mathbb{N}}(\mathcal{O})$

Proof. Follows from Proposition 1.   □

Given the upper bounds from McKenzie and Wagner [3] for membership problems where complement is not used within the respective circuits, we obtain immediately:

Corollary 2

1. $MF_{\mathbb{N}+}(\cap)$, $MF_{\mathbb{N}+}(\cap, \times)$, $MF_{\mathbb{N}+}(\cap, +)$, $MF_{\mathbb{N}+}(\cup)$, $MF_{\mathbb{N}+}(\cup, \cap)$, $MF_{\mathbb{N}+}(\times)$, and $MF_{\mathbb{N}+}(+)$ *are in* L.
2. $MC_{\mathbb{N}+}(\cap)$ *and* $MC_{\mathbb{N}+}(\cup)$ *are in* NL.
3. $MC_{\mathbb{N}+}(+)$ *and* $MC_{\mathbb{N}+}(\cap, +)$ *are in* $C_=L$.
4. $MF_{\mathbb{N}+}(+, \times)$ *and* $MF_{\mathbb{N}+}(\cap, +, \times)$ *are in* LOGDCFL.
5. $MC_{\mathbb{N}+}(\cup, \cap)$ *and* $MC_{\mathbb{N}+}(+, \times)$ *are in* P.
6. $MC_{\mathbb{N}+}(\cup, \times)$, $MF_{\mathbb{N}+}(\cup, \times)$, $MF_{\mathbb{N}+}(\cup, +)$, $MF_{\mathbb{N}+}(\cup, \cap, \times)$, $MC_{\mathbb{N}+}(\cup, +)$, $MF_{\mathbb{N}+}(\cup, \cap, +)$, $MF_{\mathbb{N}+}(\cup, +, \times)$, *and* $MF_{\mathbb{N}+}(\cup, \cap, +, \times)$ *are in* NP.
7. $MC_{\mathbb{N}+}(\cup, \cap, \times)$, $MC_{\mathbb{N}+}(\cup, \cap, +)$, *and* $MC_{\mathbb{N}+}(\cup, +, \times)$ *are in* PSPACE.

5.2   An Upper Bound for $MC_{\mathbb{N}+}(\cup, \cap, ^-, +, \times)$

For $(\{\cup, \cap, ^-, +, \times\}, \mathbb{N})$-circuits the complexity of the membership problem is not known. McKenzie and Wagner proved it to be NEXPTIME-hard.

Given $(\{\cup, \cap, ^-, +, \times\}, \mathbb{N}^+)$-circuits, we show that it suffices to consider only finite sets of numbers. Thus the problem is decidable. We prove that it is even in PSPACE.

Theorem 1. $MC_{\mathbb{N}+}(\cup, \cap, ^-, +, \times)$ *is in* PSPACE.

Proof. Let (C, b) be an input instance of $MC_{\mathbb{N}+}(\cup, \cap, ^-, +, \times)$ where $C = (G, A, g_C, \alpha)$ and $b > 0$. Let $S =_{\text{def}} \{1, \ldots, b\}$.
For a $(\cup, \cap, ^-, +, \times, \mathbb{N}^+)$-circuit C define a modified interpretation I' as follows:

If g is an input gate with label a then $I'(g) = \{a\}$ if $a \geq b$,
$$I'(g) = \emptyset \quad \text{otherwise.}$$

If g is a $^-$-gate with predecessor g_1 then $I'(g) = S \setminus I'(g_1)$.
  Let g be a gate of in-degree 2 with predecessors g_1 and g_2.
If g is a +-gate then $I'(g) = \{k + m : k \in I'(g_1) \wedge m \in I'(g_2)\} \cap S$.
If g is a ×-gate then $I'(g) = \{k \cdot m : k \in I'(g_1) \wedge m \in I'(g_2)\} \cap S$.
If g is a $\cap$-gate then $I'(g) = I'(g_1) \cap I'(g_2)$.
If g is a $\cup$-gate then $I'(g) = I'(g_1) \cup I'(g_2)$.

Claim. For all $g \in G$ it holds that $I'(g) = I_{\mathbb{N}^+}(g) \cap S$.

We show this by induction on the structure of C.
Let $g \in G$ be an input gate. Then $I(g) \subseteq S$ by definition.
Let g be a $^-$-gate with predecessor g_1. Then
$I'(g) = S \setminus I'(g_1) = S \setminus (I_{\mathbb{N}^+}(g_1) \cap S) = S \setminus I_{\mathbb{N}^+}(g_1) = (\mathbb{N} \setminus I_{\mathbb{N}^+}(g_1)) \cap S = I_{\mathbb{N}^+}(g) \cap S$.
Let g be a gate of in-degree 2 with predecessors g_1 and g_2.
If $\alpha(g) = +$:

$$
\begin{aligned}
I'(g) &= \{k + l : k \in I'(g_1) \wedge l \in I'(g_2)\} \cap S \\
&= \{k + l : k \in I_{\mathbb{N}^+}(g_1) \cap S \wedge l \in I_{\mathbb{N}^+}(g_2) \cap S\} \cap S \\
&\overset{(*)}{=} \{k + l : k \in I_{\mathbb{N}^+}(g_1) \wedge l \in I_{\mathbb{N}^+}(g_2)\} \cap S \\
&= I(g) \cap S
\end{aligned}
$$

(*): Since $k, l > 0$ it holds that $k + l \leq b$ implies $k < b$ and $l < b$.
If $\alpha(g) = \times$, then we get a chain of equations that is analogous to the one in the
'+'-case. (*): Since $k, l > 0$ it holds that $k \cdot l \leq b$ implies $k \leq b$ and $l \leq b$.
If $\alpha(g) = \cap$: $I'(g) = I'(g_1) \cap I'(g_2) = (I_{\mathbb{N}^+}(g_1) \cap S) \cap (I_{\mathbb{N}^+}(g_2) \cap S) = I_{\mathbb{N}^+}(g) \cap S$.
If $\alpha(g) = \cup$: $I'(g) = I'(g_1) \cup I'(g_2) = (I_{\mathbb{N}^+}(g_1) \cap S) \cup (I_{\mathbb{N}^+}(g_2) \cap S) = (I_{\mathbb{N}^+}(g_1) \cup I_{\mathbb{N}^+}(g_2)) \cap S = I_{\mathbb{N}^+}(g) \cap S$.
It follows that $I'(C) = I_{\mathbb{N}^+}(C) \cap S$.
  The following shows an alternating decision algorithm for "$b \in I'(C)$". For a
gate g with $\alpha(g) \in \{+, \times, \cup, \cap\}$, let g_1 and g_2 be the predecessor gates. For a
gate g with $\alpha(g) = ^-$, let g_1 be the predecessor gate.

```
g := g_C; a := b; σ := 1
while g is not an input gate do:
    if σ = 1 then
        if α(g) = + then choose ex. c_1, c_2 ∈ S s.t. a = c_1 + c_2;
                          choose univ. i ∈ {1,2}; g := g_i; a := c_i
        if α(g) = × then choose ex. c_1, c_2 ∈ S s.t. a = c_1 · c_2;
                          choose univ. i ∈ {1,2}; g := g_i; a := c_i
        if α(g) = ∪ then choose existentially i ∈ {1,2}; g := g_i
        if α(g) = ∩ then choose universally i ∈ {1,2}; g := g_i
        if α(g) = ⁻ then g := g_1; σ := 0
```

```
if σ = 0 then
    if α(g) = + then choose univ. c₁, c₂ ∈ S s.t. a = c₁ + c₂;
                     choose ex. i ∈ {1,2}; g := gᵢ; a := cᵢ
    if α(g) = × then choose univ. c₁, c₂ ∈ S s.t. a = c₁ · c₂;
                     choose ex. i ∈ {1,2}; g := gᵢ; a := cᵢ
    if α(g) = ∪ then choose universally i ∈ {1,2}; g := gᵢ
    if α(g) = ∩ then choose existentially i ∈ {1,2}; g := gᵢ
    if α(g) = ⁻ then g := g₁; σ := 1
if g is an input gate then if(σ = 1 ⇔ a = α(g)) then accept else reject
```

This alternating algorithm runs in polynomial time, since the number of loop traversals is linearly bounded.   □

Corollary 3. $\mathrm{MF}_{\mathbb{N}^+}(\cup, \cap, ^-, +, \times)$, $\mathrm{MF}_{\mathbb{N}^+}(\cup, \cap, ^-, +)$, $\mathrm{MF}_{\mathbb{N}^+}(\cup, \cap, ^-, \times)$, $\mathrm{MC}_{\mathbb{N}^+}(\cup, \cap, +, \times)$, $\mathrm{MC}_{\mathbb{N}^+}(\cup, \cap, ^-, +)$, *and* $\mathrm{MC}_{\mathbb{N}^+}(\cup, \cap, ^-, \times)$ *are in* PSPACE.

5.3 Circuits with Intersection as the Only Set Operation

Lemma 2. $\mathrm{MC}_{\mathbb{N}^+}(\cap, +, \times)$ *is in* P.

Proof. Consider the input (C, b). The decision algorithm works by evaluating C, starting with the input gates and working downwards. Once there are gates whose interpretation would give a value that is larger than b, we replace it with the empty set.   □

Lemma 3. $\mathrm{MC}_{\mathbb{N}^+}(\cap, \times)$ *is in* NL.

Proof. Let $C = (G, A, g_C, \alpha)$ be an $(\{\cap, \times\}, \mathbb{N}^+)$-circuit with m $\cap$-gates. Choose an $\cap$-gate $g_k \in G$ ($1 \le k \le m$) and let g_k^1 and g_k^2 be the left and right predecessor of g_k, respectively. We construct a new circuit $C_k^i = (G_k^i, A_k^i, g_C, \alpha_k^1)$ ($i \in \{1,2\}$) from C as follows: $G_k^i =_{\mathrm{def}} G \cup \{g'\}$ where g' is a new gate, $A_k^i =_{\mathrm{def}} (A \setminus \{g_k^i, g_k)\}) \cup \{(g', g_k)\}$, and $\alpha_k^i(g') =_{\mathrm{def}} 1$, $\alpha_k^i(g_k) =_{\mathrm{def}} \times$, and $\alpha_k^i(g) =_{\mathrm{def}} \alpha(g)$ for all $g \in G \setminus \{g_k\}$. It holds that

$$b \in I_{\mathbb{N}^+}(C) \iff b \in I_{\mathbb{N}^+}(C_k^1) \text{ and } b \in I_{\mathbb{N}^+}(C_k^2). \tag{$\star$}$$

We already know that $\mathrm{MC}_{\mathbb{N}^+}(\times) \in$ NL (Corollary 2). We now give a coNL-algorithm that decides $\mathrm{MC}_{\mathbb{N}^+}(\cap, \times)$ using $\mathrm{MC}_{\mathbb{N}^+}(\times)$ as oracle. It works by testing $(\star)$ for every $\cap$-gate. Consider the input (C, b) where $C = (G_C, A_C, g_C, \alpha_C)$.

```
output a new input gate g' with label 1 on the oracle tape
output all gates from G_C along with their label, but replace ∩-labels with
the ×-label.
for each (g₁, g₂) ∈ A_C do
    if α_C(g₂) ≠ ∩ then output (v₁, v₂)
    if α_C(g₂) = ∩ then {
        search g₃ ≠ g₁ such that (g₃, g₂) ∈ A_C
        choose i ∈ {1, 3}
```

output (g', g_2) and (g_i, g_2)
}
(let C' be the circuit that is given by the previous output operations)
if $(C', b) \in \mathrm{MC}_{\mathbb{N}+}(\times)$ then accept else reject

This is a logarithmic-space algorithm, since we only need to store two pointers: one for the outer loop and one for the search function. Furthermore, there are only linear-sized questions asked to the oracle.   □

6   Lower Bounds

6.1   NP- and PSPACE-hard Problems

Lemma 4

1. $\mathrm{MF}_{\mathbb{N}+}(\cup, \times)$ *is NP-hard.*
2. $\mathrm{MC}_{\mathbb{N}+}(\cup, \cap, \times)$ *is PSPACE-hard.*
3. $\mathrm{MF}_{\mathbb{N}+}(\cup, \cap, ^-, \times)$ *is PSPACE-hard.*

Proof. All three proofs can essentially be found in the full paper by McKenzie and Wagner.   □

Corollary 4. $\mathrm{MC}_{\mathbb{N}+}(\cup, \times)$, $\mathrm{MF}_{\mathbb{N}+}(\cup, \cap, \times)$, $\mathrm{MF}_{\mathbb{N}+}(\cup, +, \times)$, *and* $\mathrm{MF}_{\mathbb{N}+}(\cup, \cap, +, \times)$ *are NP-hard.*
$\mathrm{MF}_{\mathbb{N}+}(\cup, \cap, ^-, +, \times)$, $\mathrm{MC}_{\mathbb{N}+}(\cup, \cap, ^-, \times)$, $\mathrm{MC}_{\mathbb{N}+}(\cup, \cap, +, \times)$, *and* $\mathrm{MC}_{\mathbb{N}+}(\cup, \cap, ^-, +, \times)$ *are PSPACE-hard.*

The following theorem was proven by Yang [9]. Rather than $\mathrm{MC}_{\mathbb{N}+}(\cup, +, \times)$, he called this problem *Integer Circuit Evaluation.*

Theorem 2 ([9]). $\mathrm{MC}_{\mathbb{N}+}(\cup, +, \times)$ *is PSPACE-hard.*

This theorem also gives a lower bound for circuits over natural numbers [3], since $\mathrm{MC}_{\mathbb{N}+}(\cup, +, \times) \leq_{\mathrm{m}}^{\log} \mathrm{MC}_{\mathbb{N}}(\cup, +, \times)$ (see Proposition 1).

Lemma 5. $\mathrm{MF}_{\mathbb{N}+}(\cup, +)$ *is NP-hard.*

Proof. We show that the NP-complete sum of subset (SOS) problem is logarithmic-space many-one reducible to $\mathrm{MF}_{\mathbb{N}+}(\cup, +)$. The reduction function maps $(a_1, \ldots, a_n, b)$ to $(F, b+n)$ where

$$F =_{\mathrm{def}} (\{a_1 + 1\} \cup \{1\}) + \cdots + (\{a_n + 1\} \cup \{1\}).$$

Now we obtain

$$(a_1, \ldots, a_n, b) \in \mathrm{SOS} \iff \exists\, I \subseteq \{1, \ldots, n\} : \sum_{i \in I} a_i = b$$

$$\iff \exists\, I \subseteq \{1, \ldots, n\} : \sum_{i \in I} (a_i + 1) + \sum_{i \notin I} 1 = b + n. \quad □$$

Corollary 5. $MC_{\mathbb{N}+}(\cup,+)$ *and* $MF_{\mathbb{N}+}(\cup,\cap,+)$ *are NP-hard.*

For $\mathcal{O} \subseteq \{\cup,\cap,^-,+\}$ we define the *generalized membership problems* $MC_U^\star(\mathcal{O})$ and $MF_U^\star(\mathcal{O})$. Here, the sets that are processed by the circuits contain values from $U' =_{def} U^n \cup \{\infty\}$. Note that the addition is defined componentwise and $a + \infty = \infty + a = \infty$ for all $a \in U'$.

Lemma 6. *Let* $\mathcal{O} \subseteq \{\cup,\cap\}$.
  $MC_{\mathbb{N}+}(\mathcal{O} \cup \{\times\}) \leq_m^P MC_{\mathbb{N}}^\star(\mathcal{O} \cup \{+\})$.

Proof. The idea is to split values into (relatively prime) factors and represent them by tuples. Multiplication is done by adding these tuples componentwise. This case is a restriction of the proof found in the full paper by McKenzie and Wagner.   □

The following is also known from their full paper:

Lemma 7. *Let* $\mathcal{O} \subseteq \{\cap\}$.
  $MC_{\mathbb{N}}^\star(\mathcal{O} \cup \{\cup,+\}) \equiv_m^{log} MC_{\mathbb{N}+}(\mathcal{O} \cup \{\cup,+\})$.

From lemmas 6, 7 we obtain: $MC_{\mathbb{N}+}(\cup,\cap,\times) \leq_m^P MC_{\mathbb{N}+}(\cup,\cap,+)$, so:

Corollary 6. $MC_{\mathbb{N}+}(\cup,\cap,+)$ *and* $MC_{\mathbb{N}+}(\cup,\cap,^-,+)$ *are PSPACE-hard with respect to polynomial-time reducibility.*

Lemma 8. $MF_{\mathbb{N}+}(^-,+)$ *is PSPACE-hard.*

Proof. We use a technique that was used by Travers for circuits over integers[6]. Define the following variant of the *sum of subset problem.*

$$QSOS =_{def} \Big\{(a_1,\ldots,a_n,b) \,:\, a_1,\ldots,a_n,b \in \mathbb{N},\ n \equiv 1 \ (mod\ 2),\ and$$

$$\exists_{c_1 \in \{0,1\}} \forall_{c_2 \in \{0,1\}} \cdots \exists_{c_n \in \{0,1\}} \Big((\sum_{i=1}^n c_i \cdot a_i) = b \Big) \Big\}$$

QSOS is PSPACE-complete and $QSOS \leq_m^{log} MF_{\mathbb{N}}(^-,+)$ [6].

We now show that $QSOS \leq_m^{log} MF_{\mathbb{N}+}(^-,+)$. For this we modify the proof in [6] as follows. Since we do not allow 0, we define

$$A_i =_{def} \begin{cases} \overline{1 + \overline{(a_i)}} & \text{if } a_i \geq 1 \\ 1 & \text{otherwise} \end{cases}$$

for the construction of F. Since it holds that

$(a_1,\ldots,a_n,b) \in QSOS$

$$\Longleftrightarrow \exists_{c_1 \in \{0,1\}} \forall_{c_2 \in \{0,1\}} \cdots \exists_{c_n \in \{0,1\}} \Big((\sum_{i=1}^n c_i \cdot a_i) = b \Big)$$

$$\Longleftrightarrow \exists_{a_1' \in \{1,a_1+1\}} \forall_{a_2' \in \{1,a_2+1\}} \cdots \exists_{a_n' \in \{1,a_n+1\}} \Big((\sum_{i=1}^n c_i \cdot a_i') = b + n \Big),$$

we obtain that $(a_1,\ldots,a_n,b) \in QSOS \Longleftrightarrow (F, b+n) \in MF_{\mathbb{N}+}(^-,+)$.   □

Corollary 7. $\mathrm{MF}_{\mathbb{N}+}(\cup, \cap, ^-, +)$ *is* PSPACE-*hard.*

6.2   L-, NL-, and P-hard Problems

Lemma 9. *Let* $\mathcal{O} \subseteq \{\cup, \cap, ^-\}$.

1. $\mathrm{MC}_{\mathbb{N}}(\mathcal{O}) \equiv_m^{\mathrm{AC}^0} \mathrm{MC}_{\mathbb{N}+}(\mathcal{O})$.
2. $\mathrm{MF}_{\mathbb{N}}(\mathcal{O}) \equiv_m^{\mathrm{AC}^0} \mathrm{MF}_{\mathbb{N}+}(\mathcal{O})$.

Proof. "$\leq_m^{\mathrm{AC}^0}$": Given (C, b) as input where $C = (G, A, g_C, \alpha)$, the reducing function f outputs $(C', b+1)$ where $C' = (G, A, g_C, \alpha')$ and $\alpha'(g) =_{\mathrm{def}} \alpha(g) + 1$ for all input gates $g \in G$. It holds that $(C, b) \in \mathrm{MC}_{\mathbb{N}}(\mathcal{O})$ if and only if $f((C, b)) \in \mathrm{MC}_{\mathbb{N}+}(\mathcal{O})$ since $\mathbb{N} \to \mathbb{N}^+$, $x \mapsto x + 1$ is an isomorphism from $(\mathfrak{P}(\mathbb{N}), \cup, \cap, ^-)$ to $(\mathfrak{P}(\mathbb{N}^+), \cup, \cap, ^-)$.
"$\geq_m^{\mathrm{AC}^0}$": We can use the identity function with the only exception that we need to map an instance containing a gate labeled with 0 to an invalid instance. $\quad\square$

Given the results from McKenzie and Wagner [3], we obtain:

Corollary 8. *The following holds with respect to* AC^0-*reducibility.*

1. $\mathrm{MC}_{\mathbb{N}+}(\cup, \cap)$ *and* $\mathrm{MC}_{\mathbb{N}+}(\cup, \cap, ^-)$ *are* P-*hard.*
2. $\mathrm{MC}_{\mathbb{N}+}(\cup)$ *and* $\mathrm{MC}_{\mathbb{N}+}(\cap)$ *are* NL-*hard.*
3. $\mathrm{MF}_{\mathbb{N}+}(\cap)$, $\mathrm{MF}_{\mathbb{N}+}(\cap, \times)$, $\mathrm{MF}_{\mathbb{N}+}(\cap, +)$, $\mathrm{MF}_{\mathbb{N}+}(\cap, +, \times)$, $\mathrm{MF}_{\mathbb{N}+}(\cup)$, $\mathrm{MF}_{\mathbb{N}+}(\cup, \cap)$, *and* $\mathrm{MF}_{\mathbb{N}+}(\cup, \cap, ^-)$ *are* L-*hard.*
4. $\mathrm{MC}_{\mathbb{N}+}(\cup, \cap, ^-)$ *is in* P.
5. $\mathrm{MF}_{\mathbb{N}+}(\cup, \cap, ^-)$ *is in* L.

Lemma 10. $\mathrm{MC}_{\mathbb{N}+}(\times)$ *is* NL-*hard.*

Proof. We show that the coNL-complete problem $\overline{\mathrm{GAP}'}$ is logarithmic-space many-one reducible to $\mathrm{MC}_{\mathbb{N}+}(\times)$. Let $G = (V, A)$ be a directed acyclic graph and $s, t \in V$ be the source and target vertices, respectively. Assume that every vertex has in-degree 0 or 2 and s has in-degree 0 (otherwise $C \in \overline{\mathrm{GAP}'}$ anyway). Now convert G into a $(\{\times\}, \mathbb{N}^+)$-circuit C by labeling every vertex with in-degree 2 by $\times$, the source vertex s by 2, and all other vertices with in-degree 0 by 1. Let t be the output vertex g_C. We obtain

$$
\begin{aligned}
(G, s, t) \in \overline{\mathrm{GAP}'} &\iff \text{there is no path in } G \text{ from } s \text{ to } t \\
&\iff 1 \in I(C) \\
&\iff (C, 1) \in \mathrm{MC}_{\mathbb{N}+}(\times). \qquad\square
\end{aligned}
$$

Corollary 9. $\mathrm{MC}_{\mathbb{N}+}(\cap, \times)$ *is* NL-*hard.*

Lemma 11. $\mathrm{MC}_{\mathbb{N}+}(+)$ *is* $\mathrm{C}_=\mathrm{L}$-*hard.*

Table 1. Summary of results. Bold entries show differences to [3] (cf. footnotes).

$\mathcal{O}$	$MC_{\mathbb{N}+}(\mathcal{O})$ lower bound	$MC_{\mathbb{N}+}(\mathcal{O})$ upper bound	cf.	$MF_{\mathbb{N}+}(\mathcal{O})$ lower bound	$MF_{\mathbb{N}+}(\mathcal{O})$ upper bound	cf.
∪ ∩ ⁻ + ×	**PSPACE²**	**PSPACE¹**	C4, Th1	PSPACE	**PSPACE¹**	C4, C3
∪ ∩ + ×	**PSPACE²**	**PSPACE²**	C4, C3	NP	NP	C4, C2
∪ + ×	PSPACE	PSPACE	Th2, C2	NP	NP	C4, C2
∩ + ×	**$C_=L^4$**	**P^5**	C10, L2	L	LOGDCFL	C8, C2
+ ×	**$C_=L^4$**	P	C10, C2	—	LOGDCFL	C2
∪ ∩ ⁻ +	PSPACE	PSPACE	C6, C3	PSPACE	PSPACE	C7, C3
∪ ∩ +	PSPACE	PSPACE	C6, C2	NP	NP	C5, C2
∪ +	NP	NP	C5, C2	NP	NP	L5, C2
∩ +	$C_=L$	$C_=L$	C10, C2	L	L	C8, C2
+	$C_=L$	$C_=L$	L11, C2	—	L	C2
∪ ∩ ⁻ ×	PSPACE	PSPACE	C4, C3	PSPACE	PSPACE	L4, C3
∪ ∩ ×	PSPACE	PSPACE	L4, C2	NP	NP	C4, C2
∪ ×	NP	NP	C4, C2	NP	NP	L4, C2
∩ ×	**NL^5**	**NL^4**	C9, L3	L	L	C8, C2
×	NL	NL	L10, C2	—	L	C2
∪ ∩ ⁻	P	P	C8, C8	L	L	C8, C8
∪ ∩	P	P	C8, C2	L	L	C8, C2
∪	NL	NL	C8, C2	L	L	C8, C2
∩	NL	NL	C8, C2	L	L	C8, C2

Proof. Let $A \in C_=L$. By definition there exist functions $f \in \#L$ and $g \in FL$ such that $x \in A \iff f(x) = g(x)$.

Let M be a non-deterministic logaritmic-space Turing machine such that $f(x)$ is the number of accepting paths of M on input x. Without loss of generality, we can assume that each configuration of M has exactly two successor configurations and that all computation paths have the same length.

Let C_x be the transition graph of M on input x. Since M is a logarithmic-space Turing machine it holds that every configuration of M on input x has length $\log|x|$. Given a configuration, both successor configurations can be computed in constant-time. Hence, C_x can be constructed in logarithmic space. But we additionally change C_x in such a way that inner nodes are replaced by +-gates and nodes at the end of accepting and rejecting paths are labeled 2 and 1, respectively. Thus, C_x becomes a $(\{+\}, \mathbb{N}^+)$-circuit. We define $f'(x) = I(C_x)$.

---

[1] ?

[2] NEXPTIME

[3] co-R

[4] P

[5] $C_=L$

Let m be the number of paths in C_x and let l be the length of a computation path. It holds that $m = 2^l$ which can be computed in logarithmic space.

Since g is a logarithmic space computable function, the same holds for $g' =_{\text{def}} g + m$. So $x \in A \iff f'(x) = g'(x) \iff g'(x) \in I_{\mathbb{N}^+}(C_x) \iff (C_x, g'(x)) \in \text{MC}_{\mathbb{N}^+}(+)$. $\square$

Corollary 10. $\text{MC}_{\mathbb{N}^+}(\cap, +)$, $\text{MC}_{\mathbb{N}^+}(+, \times)$, *and* $\text{MC}_{\mathbb{N}^+}(\cap, +, \times)$ *are* $\text{C}_=\text{L}$-*hard*.

7   Conclusion and Open Problems

In Table 1 we give an overview of our results. Open problems are apparent. $\text{MC}_{\mathbb{N}^+}(\cap, +, \times)$ and $\text{MC}_{\mathbb{N}^+}(+, \times)$ seem to be rather difficult open problems. It seems unlikely that one can reduce a P-complete problem (like a variant of the Boolean Circuit Evaluation problem) to one of these problems due to the missing 0. On the other hand, P was the best upper bound we could obtain. The gap at $\text{MF}_{\mathbb{N}^+}(\cap, +, \times)$ and $\text{MF}_{\mathbb{N}^+}(+, \times)$ has already been present in [3] for the $\mathbb{N}^+$ counterparts.

Acknowledgments. The author is very grateful to Christian Glaßer, Klaus W. Wagner, Daniel Meister, Bernhard Schwarz, and Stephen Travers for very useful discussions and important hints, and to the anonymous referees for their suggestions.

References

[1] Agrawal, M., Allender, E., Impagliazzo, R., Pitassi, T., Rudich, S.: Reducing the Complexity of Reductions. Computational Complexity 10, 117–138 (2001)

[2] Glaßer, C., Herr, K., Reitwießner, C., Travers, S., Waldherr, M.: Equivalence Problems for Circuits over Sets of Natural Numbers. In: International Computer Science Symposium in Russia (CSR) 2007 (to appear)

[3] McKenzie, P., Wagner, K.W.: The Complexity of Membership Problems for Circuits over Sets of Natural Numbers. In: Alt, H., Habib, M. (eds.) STACS 2003. LNCS, vol. 2607, pp. 571–582. Springer, Heidelberg (2003)

[4] Papadimitriou, C.H.: Computational Complexity. Pearson (1994)

[5] Stockmeyer, L.J., Meyer, A.R.: Word Problems Requiring Exponential Time. In: Proceedings 5th ACM Symposium on the Theory of Computation, pp. 1–9 (1973)

[6] Travers, S.: The Complexity of Membership Problems for Circuits over Sets of Integers. Theor. Comput. Sci. 211–229 (2006)

[7] Vollmer, H.: Introduction to Circuit Complexity. Springer, Heidelberg (1999)

[8] Wagner, K.W.: The Complexity of Problems Concerning Graphs with Regularities. In: Chytil, M.P., Koubek, V. (eds.) Mathematical Foundations of Computer Science 1984. LNCS, vol. 176, pp. 544–552. Springer, Heidelberg (1984)

[9] Yang, K.: Integer circuit evaluation is PSPACE-complete. In: Proceedings 15th Conference on Computational Complexity, pp. 204–211 (2000)

Pattern Matching in Protein-Protein Interaction Graphs

Gaëlle Brevier[1], Romeo Rizzi[2], and Stéphane Vialette[3]

[1] G-SCOP, Grenoble, France
Gaelle.Giberti@imag.fr
[2] Dipartimento di Matematica ed Informatica (DIMI),
Università di Udine, Italy
Romeo.Rizzi@dimi.uniud.it
[3] Laboratoire de Recherche en Informatique (LRI), UMR CNRS 8623
Université Paris-Sud 11, France
Stephane.Vialette@lri.fr

Abstract. In the context of comparative analysis of protein-protein interaction graphs, we use a graph-based formalism to detect the preservation of a given protein complex (pattern graph) in the protein-protein interaction graph (target graph) of another species with respect to (w.r.t.) orthologous proteins. We give an efficient exponential-time randomized algorithm in case the occurrence of the pattern graph in the target graph is required to be exact. For approximate occurrences, we prove a tight inapproximability results and give four approximation algorithms that deal with bounded degree graphs, small ortholog numbers, linear forests and very simple yet hard instances, respectively.

1 Introduction

High-throughput analysis makes possible the study of protein-protein interactions at a genome-wise scale [5,6,16], and comparative analysis tries to determine the extent to which protein networks are conserved among species. Indeed, mounting evidence suggests that proteins that function together in a pathway or a structural complex are likely to evolve in a correlated fashion, and, during evolution, all such functionally linked proteins tend to be either preserved or eliminated in a new species [9].

Protein interactions identified on a genome-wide scale are commonly visualized as protein interaction graphs, where proteins are vertices and interactions are edges [14]. Experimentally derived interaction networks can be extremely complex, so that it is a challenging problem to extract biological functions or pathways from them. However, biological systems are hierarchically organized into functional modules. Several methods have been proposed for identifying functional modules in protein-protein interaction graphs. As observed in [10], cluster analysis is an obvious choice of methodology for the extraction of functional modules from protein interaction networks. Comparative analysis of protein-protein interaction graphs aims at finding complexes that are common

E. Csuhaj-Varjú and Z. Ésik (Eds.): FCT 2007, LNCS 4639, pp. 137–148, 2007.
© Springer-Verlag Berlin Heidelberg 2007

to different species. Kelley *et al.* [7] developed the program PathBlast, which aligns two protein-protein interaction graphs combining topology and sequence similarity. Sharan *et al.* [11] studied the conservation of complexes (they focused on dense, clique-like interaction patterns) that are conserved in *Saccharomyces cerevisae* and *Helicobacter pylori*, and found 11 significantly conserved complexes (several of these complexes match very well with prior experimental knowledge on complexes in yeast only). They actually recast the problem of searching for conserved complexes as a problem of searching for heavy subgraphs in an edge- and node-weighted graph, whose vertices are orthologous protein pairs. A promising computational framework for alignment and comparison of more than one protein network together with a three-way alignment of the protein-protein interaction network of *Caenorhabditis elegans, Drosophila melanogaster* and *Saccharomyces cerevisae* is presented in [12]. The related problem of finding a query path or a query graph that is most similar to a target graph is considered in [19].

In [3], this pattern matching problem was stated as the problem of finding an occurrence of a pattern graph G in a target graphs H w.r.t lists constraints (referred hereafter as the EXACT-(ρ, σ)-MATCHING problem): to each vertex u of G is associated a lists $\mathcal{L}(u)$ of vertices in H and the occurrence of G is H is required to be an injective graph homomorphism ϕ from G to H such that $\phi(u) \in \mathcal{L}(u)$ for each vertex u in G. The two parameters ρ and σ denote the maximum size of a list of G and the maximum number of occurrences of a vertex of H among the lists of G, respectively. Roughly speaking, the rationale of this approach is as follows. First, graph homomorphism only preserves adjacency, and hence can deal with interaction datasets that are missing many true protein interactions. Second, injectivity is required in order to establish a bijective relationship between proteins in the complex and proteins in the occurrence. Finally, graph homomorphism with respect to orthologous links can be easily recasted as list homomorphism: a list of putative orthologs is associated to each protein (vertex) of the complex, and each such protein can only be mapped by the homomorphism to a protein occurring in its list. In the context of comparative analysis of protein-protein interaction graphs, drastic restrictions were imposed on the size of the lists. Some (classical and parameterized) hardness results together with several heuristics for the EXACT-(ρ, σ)-MATCHING problem were presented in [3]. These results were improved in [4]. Of particular importance in the context of computational biology, we investigated in [4] the problem of finding approximate occurrences (the MAX-(ρ, σ)-MATCHING problem), *i.e.*, the injective mapping of G to H were no longer required to be a graph homomorphism but to match as many edges as possible.

Aiming at presenting accurate computational models, we combine state-of-the art approaches to identifying orthologs, *i.e*, genes in different species that originate from a single gene in the last common ancestor of these species for transferring functional information between genes in different organisms with a high degree of reliability [13], and the above mentioned line of research by considering additional structural constraints on the lists: for each distinct vertices u and v of G, either $\mathcal{L}(u) = \mathcal{L}(v)$ or $\mathcal{L}(u) \cap \mathcal{L}(v) = \emptyset$. The obtained problem is modeled by replacing lists by colors: to all vertices of G and H is associated a color and a vertex of G can only be mapped to a vertex of H with the same color.

This paper is organized as follows. We briefly discuss in Section 2 basic notations and definitions that we will use throughout the paper. In Section 3 we give a randomized algorithm for finding an injective mapping w.r.t to the colorings that matches all the edges of the pattern graph. We prove in Section 4 that the problem of finding an injective mapping w.r.t to the colorings that matches as many edges of the pattern graph as possible is hard to approximate even if both the pattern graph and the target graph are linear forests or trees. Section 5 is devoted to approximation. with a focus on two restricted but still hard cases: (i) the pattern graph or the target graph has bounded degree, (ii) the number of occurrences of each color in the target graph is considered to be small and (iii) both the pattern graph and the target graphs are linear forests. Section 6 concludes our word and propose future directions of research.

2   Preliminaries

We assume readers have basic knowledge about graph theory [2] and we shall thus use most conventional terms of graph theory without defining them (we only recall basic notations), we only recall basic notations. Let G be a graph. We write $\mathbf{V}(G)$ for the set of vertices and $\mathbf{E}(G)$ for the set of edges. Also, we write $\mathbf{n}(G)$ for $\#\mathbf{V}(G)$ and $\mathbf{m}(G)$ for $\#\mathbf{E}(G)$. The maximum degree $\Delta(G)$ of a graph G is the largest degree over all vertices. A graph is called a *linear forest* if every component is a path. Let G be a graph together with a coloring $\lambda : \mathbf{V}(G) \to \mathcal{C}$ of its vertices. For any color $c_i \in \mathcal{C}$, we denote by $\mathcal{C}_G(c_i)$ the set of vertices of G that are colored with color c_i, i.e., $\mathcal{C}_G(c_i) = \{u \in \mathbf{V}(G) : \lambda(u) = c_i\}$. The *multiplicity* of λ in G, written $\mathtt{mult}(G, \lambda)$, is the maximum number of occurrences of a color in G, i.e., $\mathtt{mult}(G, \lambda) = \max\{\#\mathcal{C}_G(c_i) : c_i \in \mathcal{C}\}$. Let G and H be two graphs and let $\theta : \mathbf{V}(G) \to \mathbf{V}(H)$ be an injective mapping. The set of edges of G that are preserved in H by θ is written $\mathtt{match}(G, H, \theta)$, i.e., $\mathtt{match}(G, H, \theta) = \{\{u, v\} \in \mathbf{E}(G) : \{\theta(u), \theta(v)\} \in \mathbf{E}(H)\}$. If both G and H are equipped with some colorings $\lambda_G : \mathbf{V}(G) \to \mathcal{C}$ and $\lambda_H : \mathbf{V}(H) \to \mathcal{C}$ of their vertices, a mapping $\theta : \mathbf{V}(G) \to \mathbf{V}(H)$ is said to be *with respect to* (w.r.t.) λ_G and λ_H if $\lambda_G(u) = \lambda_H(\theta(u))$ for every $u \in \mathbf{V}(G)$, i.e., θ is a color preserving mapping. For simplicity, we shall usually abbreviate such a mapping as $\theta : \mathbf{V}(G) \xrightarrow{\lambda_G, \lambda_H} \mathbf{V}(H)$.

We are now in position to formally define the MAX-(ρ, σ)-MATCHING-COLOR problem we are interested in.

MAX–(ρ, σ)–MATCHING–COLOR

- **Input** : Two graphs G and H together with the coloring mappings $\lambda_G :$ $\mathbf{V}(G) \to \mathcal{C}$, $\mathtt{mult}(G, \lambda_G) = \rho$, and $\lambda_H : \mathbf{V}(H) \to \mathcal{C}$, $\mathtt{mult}(H, \lambda_H) = \sigma$.
- **Solution** : An injective mapping $\theta : \mathbf{V}(G) \xrightarrow{\lambda_G, \lambda_H} \mathbf{V}(H)$.
- **Measure** : The number of edges of G matched by the injective mapping θ, *i.e.*, #match(G, H, θ).

We designate by EXACT–(ρ, σ)–MATCHING–COLOR the extremal problem of finding an injective mapping $\theta : \mathbf{V}(G) \xrightarrow{\lambda_G, \lambda_H} \mathbf{V}(H)$ that matches all the edges of G, *i.e.*, θ is required to be an injective graph homomorphism [4]. Also, we call an instance of both the MAX–(ρ, σ)–MATCHING–COLOR and EXACT–(ρ, σ)– MATCHING–COLOR problems *colorful* if $\rho = 1$.

Let $(G, H, \mathcal{C}, \lambda_G, \lambda_H)$ be an instance of the MAX–(ρ, σ)–MATCHING–COLOR. First, a necessary and sufficient condition for an injective mapping to exists is #$\mathcal{C}_G(c_i) \leq$ #$\mathcal{C}_H(c_i)$ for each color $c_i \in \mathcal{C}$. Second, an edge $\{u, v\} \in \mathbf{E}(G)$, $\lambda_G(u) = c_u$ and $\lambda_G(v) = c_v$, is called a *bad edge* if there does not exist distinct $u' \in \mathcal{C}_H(c_u)$ and $v' \in \mathcal{C}_H(c_v)$ such that $\{u', v'\} \in \mathbf{E}(H)$. Clearly, if we remove from G its bad edges, this does not affect the optimal solutions for the MAX– (ρ, σ)–MATCHING–COLOR problem, since bad edges can never be matched. Notice that we can tell bad edges apart in $\mathcal{O}(\sigma^2 \mathbf{m}(G)) = \mathcal{O}(\mathbf{m}(G))$ time, provided σ is a constant. Therefore, throughout the paper, we will consider only trim instances as defined in the following.

Definition 1 (Trim instance). *An instance $(G, H, \mathcal{C}, \lambda_G, \lambda_H)$ of the* MAX– (ρ, σ)–MATCHING–COLOR *or the* EXACT–(ρ, σ)–MATCHING–COLOR *problem is said to be* trim *if the following conditions hold true:*

- *for each color $c_i \in \mathcal{C}$, #$\mathcal{C}_G(c_i) \leq$ #$\mathcal{C}_H(c_i)$, and*
- *for each edge $\{u_i, u_j\} \in \mathbf{E}(G)$, there exists an edge $\{v_i, v_j\} \in \mathbf{E}(H)$ such that $\lambda_G(u_i) = \lambda_H(v_i)$ and $\lambda_G(u_j) = \lambda_H(v_j)$.*

3   Exact Colorful Instances

This section is devoted to the EXACT–$(1, \sigma)$–MATCHING–COLOR problem. On the one hand, both the EXACT–$(1, \sigma)$–MATCHING–COLOR problem for $\Delta(G) \leq 2$ and the EXACT–$(\rho, 2)$–MATCHING–COLOR problem are solvable in polynomial-time for any constant ρ and σ [3]. On the other hand, the EXACT–$(1, 3)$–MATCHING– COLOR problem for $\Delta(G) = 3$ and $\Delta(H) = 4$ is **NP**–complete [4]. We first observe that the EXACT–$(1, \sigma)$–MATCHING–COLOR problem is easily solvable in $\tilde{\mathcal{O}}(\sigma^{\mathbf{n}(G)})$ time: the brute-force algorithm tries all possible injective mappings $\theta : \mathbf{V}(G) \xrightarrow{\lambda_G, \lambda_H} \mathbf{V}(H)$ and returns the best one. We give a faster randomized algorithm (referred hereafter as Algorithm Rand-Exact-Matching-Colors) than runs in $\tilde{\mathcal{O}}(f(\sigma)^{\mathbf{n}(G)})$ expected time, where

$$f(\sigma) = \frac{4\sigma(2\sigma - 2)^3}{4(2\sigma - 2)^3 + 27(2\sigma - 3)}.$$

Observe that $f(\sigma) < \sigma$, for $\sigma > 2$. For the sake of illustration, $f(3) < 2.279$, $f(4) < 3.460$ and $f(5) < 4.578$.

We present here a random walk algorithm for the EXACT–$(1, \sigma)$–MATCHING–COLOR problem similar to [8]. For simplicity, we assume the worst case where each color occurs exactly σ times in graph H. The basic idea is to start with a random injective mapping θ, look at an edge e of G that is not matched by θ, select at random one end-vertex u of e and finally change at random the image of u, i.e., $\theta(u)$. Observe however that, oppositely to satisfiability-like algorithms where changing the assignment of a boolean variable in an unsatisfied clause result in a satisfied clause, the edge e might be here still not matched by the new injective mapping θ.

Algorithm 1: Rand-Exact-Matching-Colors

Input: An instance $(G, H, \mathcal{C}, \lambda_G, \lambda_H)$ of the EXACT–$(1, \sigma)$–MATCHING–COLOR problem.

Output: An occurrence of G in H, i.e., an injective homomorphism
$$\theta : \mathbf{V}(G) \xrightarrow{\lambda_G, \lambda_H} \mathbf{V}(H) \text{ (if such a mapping exists)}.$$

begin

    repeat *terminating whether an occurrence of G in H w.r.t λ_G and λ_H is found.*

        Let $\theta : \mathbf{V}(G) \xrightarrow{\lambda_G, \lambda_H} \mathbf{V}(H)$ be a random injective.

        repeat *up to $3\,\mathbf{n}(G)$ times, terminating whether an occurrence of G in H w.r.t λ_G and λ_H is found.*

            (1) *Choose at random an edge $e \in \mathbf{E}(G)$ that is not matched by θ.*

            (2) *Choose at random one vertex $u \in e$.*

            (3) *Change at random the value of $\theta(u)$ w.r.t λ_G and λ_H.*

        end

    end

end

Fig. 1. Algorithm Rand-Exact-Matching-Colors

Let $(G, H, \mathcal{C}, \lambda_G, \lambda_H)$ be an arbitrary instance of the EXACT–$(1, \sigma)$–MATCHING–COLOR problem, and suppose that there exists an injective homomorphism $\theta_{OPT} : \mathbf{V}(G) \xrightarrow{\lambda_G, \lambda_H} \mathbf{V}(H)$, i.e., $(G, H, \mathcal{C}, \lambda_G, \lambda_H)$ is a YES instance. Without loss of generality we may assume that, for each color $c_i \in \mathcal{C}$, exactly σ vertices of H are colored with color c_i (and hence H has $\sigma \, \#\mathcal{C}$ vertices). Fix any injective mapping $\theta : \mathbf{V}(G) \xrightarrow{\lambda_G, \lambda_H} \mathbf{V}(H)$ and let $\theta_i : \mathbf{V}(G) \xrightarrow{\lambda_G, \lambda_H} \mathbf{V}(H)$ be the injective mapping after the i-th step of the inner loop of Algorithm Rand-Exact-Matching-Colors. Let X_i be the number of vertices $u \in \mathbf{V}(G)$ such that

$\theta_i(u) = \theta_{\mathbf{opt}}(u)$. If $X_i = \mathbf{n}(G)$, Algorithm Rand-Exact-Matching-Colors terminates with an injective homomorphism. Clearly, the algorithm could terminate before $X_i = \mathbf{n}(G)$ by finding a different injective homomorphism, but for our analysis the worst case is that the algorithm only stops when $X_i = \mathbf{n}(G)$.

Suppose $1 \leq X_i \leq \mathbf{n}(G) - 1$. At each step, we choose an edge $e = \{u, v\} \in \mathbf{E}(G)$ that is not matched. Since $(G, H, \mathcal{C}, \lambda_G, \lambda_H)$ is a YES instance, θ_i and $\theta_{\mathbf{opt}}$ disagree on at least one of u and v. Suppose first that θ_i and $\theta_{\mathbf{opt}}$ disagree on exactly one of u and v. Then, the probability of increasing the number of agreements between $\theta_{\mathbf{opt}}$ and θ_{i+1} is $\frac{1}{2\sigma-2}$, the probability of decreasing the number of agreements between $\theta_{\mathbf{opt}}$ and θ_{i+1} is $\frac{\sigma-1}{2\sigma-2}$ and the probability of obtaining the same number of agreements between $\theta_{\mathbf{opt}}$ and θ_{i+1} is $\frac{\sigma-2}{2\sigma-2}$. Suppose now that θ_i and $\theta_{\mathbf{opt}}$ disagree on both u and v. Then, the probability of increasing the number of agreements between $\theta_{\mathbf{opt}}$ and θ_{i+1} is $\frac{2}{2\sigma-2}$, the probability of decreasing the number of agreements between $\theta_{\mathbf{opt}}$ and θ_{i+1} is 0 (θ_i and $\theta_{\mathbf{opt}}$ indeed already both disagree on both vertices) and the probability of obtaining the same number of agreements between $\theta_{\mathbf{opt}}$ and θ_{i+1} is $\frac{2\sigma-4}{2\sigma-2}$.

We now consider a pessimistic stochastic process $(Y_1, Y_2, \ldots)$ defined as follows:

$$\Pr[Y_{i+1} = j + 1 | Y_i = j] \geq \frac{1}{2\sigma - 2}$$
$$\Pr[Y_{i+1} = j - 1 | Y_i = j] \leq \frac{2\sigma - 3}{2\sigma - 2}.$$

This stochastic process is best understood by using the same metaphor as in [8]: consider a particle moving on the integer line, with probability $(2\sigma - 1)^{-1}$ of moving up by one and probability $(2\sigma - 3)(2\sigma - 2)^{-1}$ of moving down by one. Observe that in the pessimistic stochastic process $(Y_1, Y_2, \ldots)$ the particle never stays in place whereas the probability of obtaining the same number of agreements is non-zero in Algorithm Rand-Exact-Matching-Colors. Let r_j be the probability of exactly k "moves down", and $j + k$ "moves up" in a sequence of $2k + j$ moves. We have

$$r_j \geq \left(\frac{2\sigma - 3}{2\sigma - 2}\right)^k \left(\frac{1}{2\sigma - 2}\right)^{j+k}.$$

Now, let q_j be the probability that the algorithm finds an injective homomorphism within $j + 2k \leq 3\,\mathbf{n}(G)$ steps, starting from a random injective mapping $\theta : \mathbf{V}(G) \xrightarrow{\lambda_G, \lambda_H} \mathbf{V}(H)$.

Lemma 1. $q_j \geq \dfrac{\sqrt{3}}{8\sqrt{\pi j}} \left(\dfrac{27(2\sigma - 3)}{4(2\sigma - 2)^3}\right)^j.$

Let p_j be the probability that a random injective mapping $\theta : \mathbf{V}(G) \xrightarrow{\lambda_G, \lambda_H} \mathbf{V}(H)$ has j disagreements with $\theta_{\mathbf{opt}}$. We now derive a lower bound for q, the probability that the process finds an occurrence of G in H w.r.t λ_G and λ_H in $3\,\mathbf{n}(G)$ steps stating from a random injective mapping.

Lemma 2. $q \geq \dfrac{\sqrt{3}}{8\sqrt{\pi\,\mathbf{n}(G)}}\, f(\sigma)^{-\mathbf{n}(G)}$.

Therefore, if we assume that there exists an injective mapping $\theta : \mathbf{V}(G) \xrightarrow{\lambda_G, \lambda_H} \mathbf{V}(H)$, the number of random injective mappings the process tries before finding an occurrence of G in H is a geometric random variable with parameter q. Hence, the expected of random injective mappings tried is q^{-1}, and for each injective mapping the algorithm uses at most $3\,\mathbf{n}(G)$ steps. Thus, the expected number of steps until a solution is found is bounded by $\mathcal{O}(\mathbf{n}(G)^{3/2}\, f(\sigma)^{\mathbf{n}(G)})$. We have thus proved the following.

Proposition 1. *Algorithm* Rand-Exact-Matching-Colors *returns an injective homomorphism* $\theta : \mathbf{V}(G) \xrightarrow{\lambda_G, \lambda_H} \mathbf{V}(H)$ *(if such a solution mapping exists) in* $\tilde{\mathcal{O}}(f(\sigma)^{\mathbf{n}(G)})$ *time, where*

$$f(\sigma) = \frac{4\sigma(2\sigma - 2)^3}{4(2\sigma - 2)^3 + 27(2\sigma - 3)}.$$

4   Hardness Results

The MAX–$(1, 2)$–MATCHING–COLOR problem for bipartite graphs G and H with $\Delta(G) = 3$ and $\Delta(H) = 2$ (resp. $\Delta(G) = 6$ and $\Delta(H) = 5$) is **APX**–hard and is not approximable within ratio 1.0005 (resp. 1.0014), unless $\mathbf{P} = \mathbf{NP}$ [4]. Therefore, there is a natural interest to investigate the complexity issues of the MAX–(ρ, σ)–MATCHING–COLOR problem for restricted graph classes. Unfortunately, as we shall prove here, the MAX–$(3, 3)$–MATCHING–COLOR (resp. MAX–$(2, 2)$–MATCHING–COLOR) problem is **APX**–hard even if both G and H are linear forests (resp. trees with maximum degree 3).

Proposition 2. *The* MAX–$(3, 3)$–MATCHING–COLOR *problem is* **APX**–*hard even if both G and H are linear forests.*

It remains open, however, whether the MAX–(ρ, σ)–MATCHING–COLOR problem for linear forests G and H is polynomial-time solvable in case $\rho < 3$. The rationale of this question stems from the following proposition.

Proposition 3. *The* MAX–$(2, 2)$–MATCHING–COLOR *problem is* **APX**–*hard even if both G and H are trees.*

5   Approximation Algorithms

We proved in Section 4 that the MAX–$(3, 3)$–MATCHING–COLOR problem for linear forests is **APX**–hard. In the light of this negative result, we first focus here on approximating the MAX–(ρ, σ)–MATCH–COLORS problem for bounded degree graphs and give a polynomial-time approximation algorithm that achieves a performance ratio of $2(\Delta_{\min} + 1)$, where $\Delta_{\min} = \min\{\Delta(G), \Delta(H)\}$. Next, we propose a randomized algorithm with performance ratio 4σ. Finally, we give an approximation algorithm that achieves a performance ratio of 4 in case both G and H are linear forests.

5.1   Bounded Degree Graphs

We first consider bounded degree graphs. Let $\mathcal{C} = \{c_1, c_2, \ldots, c_m\}$ be a set of colors and G be a graph whose vertices are colored with colors taken from $\mathcal{C}$. Also, let $A = [a_{i,j}]$ be a symmetric matrix of order m whose entries are natural integers. Consider the problem, referred hereafter as the MAX-MATCHING-WITH-COLOR-CONSTRAINTS (MMWCC) problem, of finding in G a maximum cardinality matching $\mathcal{M} \subseteq \mathbf{E}(G)$ subject to the constraint that, for $1 \leq i \leq j \leq m$, the number of edges in $\mathcal{M}$ having one end-vertex colored c_i and one end-vertex colored c_j is at most $a_{i,j}$. It is clear that a straightforward greedy algorithm delivers a 2-approximation algorithm for the MMWCC problem.

Lemma 3. *The* MMWCC *problem is* **NP**–*complete but is approximable within ratio 2.*

Recall that an *edge coloring* of a graph G is *proper* if no two adjacent edges are assigned the same color. A proper edge coloring with k colors is called a *proper k-edge-coloring* and is equivalent to the problem of partitioning the edge set into k matchings. The smallest number of colors needed in a proper edge coloring of a graph G is *the chromatic index* $\chi'(G)$ [2]. Vizing's theorem [18] states that $\chi'(G) \leq \Delta(G) + 1$ and that such an edge coloring can be found in polynomial-time. Combining Lemma 3 and Vizing's theorem we obtain the following result.

Proposition 4. *For any ρ and σ, the* MAX–(ρ, σ)–MATCH–COLORS *problem is approximable within ratio $2(\Delta_{min} + 1)$, where $\Delta_{min} = \min\{\Delta(G), \Delta(H)\}$.*

5.2   A Randomized Algorithm

We give here a randomized approximation algorithm which achieves a performance ratio of $4\,\sigma$ for the MAX–(ρ, σ)–MATCH–COLORS problem, for any ρ and σ.

Let $\mathcal{C}$ be a set of colors and G be a graph whose vertices are colored with colors taken from $\mathcal{C}$. Define a *legal (ℓ_1, ℓ_2)-labeling* of G to be an assignment to labels $\{\ell_1, \ell_2\}$ to the vertices of G such that, for each color $c_i \in \mathcal{C}$, either $\left\lfloor \frac{\#\mathcal{C}_G(c_i)}{2} \right\rfloor$ or $\left\lceil \frac{\#\mathcal{C}_G(c_i)}{2} \right\rceil$ vertices in $\mathcal{C}_G(c_i)$ are labeled ℓ_1. Of particular importance here is the fact that it is easy to choose at random a legal (ℓ_1, ℓ_2)-labeling of G. Define the *cut induced by a legal (ℓ_1, ℓ_2)-labeling* to be the set of edges that have one end-vertex with label ℓ_1 and one end-vertex with label ℓ_2.

Consider now an arbitrary trim instance $(G, H, \mathcal{C}, \lambda_G, \lambda_H)$ of the MAX–(ρ, σ)–MATCH–COLORS problem and let $\theta_{\mathbf{opt}} : \mathbf{V}(G) \xrightarrow{\lambda_G, \lambda_H} \mathbf{V}(H)$ be an optimal solution. Now, let L be a random legal (ℓ_1, ℓ_2)-labeling of G and $C_L \subseteq \mathbf{E}(G)$ be the cut induced by L. Finally, let $E' = C_L \cap \mathsf{match}(G, H, \theta_{\mathbf{opt}})$. The expected size of E' and the size of $\mathsf{match}(G, H, \theta_{\mathbf{opt}})$ are related by the following lemma.

Lemma 4. $\mathsf{Exp}[\#E'] \geq \dfrac{\#match(G, H, \theta_{\mathbf{opt}})}{2}$.

Combining Lemma 4 with a weighted bipartite matching algorithm yields the following result.

Proposition 5. *There exists a randomized algorithm with expected performance ratio* $4\,\sigma$ *for the* MAX–(ρ, σ)–MATCHING–COLOR *problem.*

5.3   Linear Forests

We proved in Section 4 that the MAX–$(3, 3)$–MATCHING–COLOR problem is **APX**-hard even if both G and H are linear forests. Furthermore, according to Proposition 4, the MAX–(ρ, σ)–MATCHING–COLOR problem for linear forests is approximable within ratio $2(\Delta_{\min} + 1) = 6$. We strengthen this result here by giving an algorithm that achieves a performance ratio of 4 for the MAX–(ρ, σ)–MATCHING–COLOR problem for linear forests. The proof make use of weighted 2-intervals sets. More precisely, our approach is based on the 2-INTERVAL–PATTERN problem [17,1]. This problem, initially motivated by RNA secondary structure prediction, asks to find a maximum cardinality subset of a 2-interval set with respect to some prespecified geometric constraints.

We need some additional definitions. A 2-*interval* [15,17] is the union of two disjoint intervals defined over a line. A 2-interval is denoted by $D = (I, J)$, where I and J are two closed intervals defined over a single line such that I is completely to the left of J. Two 2-intervals $D_1 = (I_1, J_1)$ and $D_2 = (I_2, J_2)$ are *disjoint*, if both 2-intervals share no common point, i.e., $(I_1 \cup J_1) \cap (I_2 \cup J_2) = \emptyset$. A 2-interval $D = (I, J)$ is said to be *balanced* if $|I| = |J|$, i.e., both intervals have the same length. By abuse of notation, a set of balanced 2-interval is also said to be balanced. Let $\mathcal{D}$ be a set 2-intervals. If we associate to each 2-interval $D \in \mathcal{D}$ a weight $\omega(D)$, the weight of $\mathcal{D}$, denoted $\omega(\mathcal{D})$, is defined to be the sum of the weights of all the 2-intervals in $\mathcal{D}$.

Let $(G, H, \mathcal{C}, \lambda_G, \lambda_H)$ be a trim instance of the MAX–(ρ, σ)–MATCHING–COLOR problem where both G and H are linear forests. Let $P_1^G, P_2^G, \ldots, P_k^G$ (resp. $P_1^H, P_2^H, \ldots, P_\ell^H$) be the collection of all paths of G (resp. H). First, we arrange the paths $P_1^G, P_2^G, \ldots, P_k^G$ and next the paths $P_1^H, P_2^H, \ldots, P_\ell^H$ along an horizontal line, arbitrarily. According to this arrangement, we define the *label* (resp. *reversal label*) of any subpath of a path to be string obtained by concatenating the colors (view as letters) of the vertices of the path reading from left to right (resp. right to left). Second, we construct a corresponding set of weighted 2-intervals $\mathcal{D}[G, H]$ as follows. For each pair (Q_i^G, Q_j^H), where Q_i^G is a subpath of length at least one of a path in $\{P_1^G, P_2^G, \ldots, P_k^G\}$ and Q_j^H is a subpath of length at least one of a path in $\{P_1^H, P_2^H, \ldots, P_\ell^H\}$ Q_G^i and Q_j^H having the same length, if the label of Q_i^G is identical to the label of Q_j^H or to the reversal label of Q_j^H, we add to $\mathcal{D}[G, H]$ a 2-interval whose left interval covers all the vertices (and only those vertices) of the subpath Q_i^G and whose right interval covers all the vertices (and only those vertices) of the subpath Q_j^H. The weight of this 2-interval is defined to be the length of the subpath Q_i^G (which also the length of the subpath Q_j^H). Without loss of generality, we may assume that each 2-interval in $\mathcal{D}[G, H]$ is balanced and that two 2-intervals that correspond to two vertex-disjoint pairs of subpaths are disjoint. See Figure 2 for an illustration of this construction. The rationale of this construction stems from the following lemma.

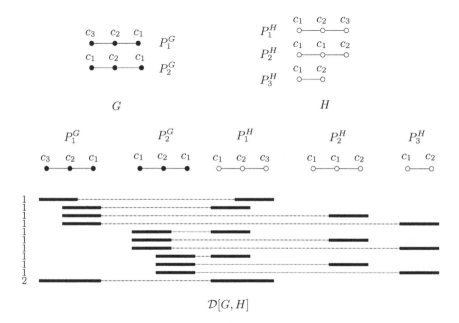

Fig. 2. Constructing a weighted 2-intervals set from an instance $(G, H, \mathcal{C}, \lambda_G, \lambda_H)$ of the MAX–(ρ, σ)–MATCHING–COLOR problem where both G and H are linear forests. The weights of all the 2-intervals in the set $\mathcal{D}[G, H]$ are given in the left part of the figure.

Lemma 5. *There exists a pairwise disjoint subset $\mathcal{D}' \subseteq \mathcal{D}[G, H]$ of weight $\omega(\mathcal{D}')$ if and only if there exists an injective mapping $\theta : \mathbf{V}(G) \xrightarrow{\lambda_G, \lambda_H} \mathbf{V}(H)$ such that $\#match(G, H, \theta) \geq \omega(\mathcal{D}')$.*

According to Lemma 5 it is thus enough to focus on finding a maximum weighted subset of $\mathcal{D}[G, H]$ of disjoint 2-intervals, which is exactly the 2-INTERVAL–PATTERN problem. In [1], an algorithm with performance ratio 4 is proposed for finding a subset of disjoint 2-intervals in a balanced 2-intervals set. We have thus proved the following.

Corollary 1. *For any ρ and σ, the* MAX–(ρ,σ)–MATCHING–COLOR *problem is approximable within ratio* 4 *in case both G and H are linear forests.*

6 Conclusion

In the context of comparative analysis of protein-protein interaction graphs, we considered the problem of finding an occurrence of a given complex in the protein-protein interaction graph of another species. We gave an efficient randomized algorithm in case the mapping is required to be an injective homomorphism. Also, we proved the MAX–(3,3)–MATCHING–COLOR problem for linear forests to be **APX**–hard and we gave an approximation algorithm that achieves a performance ratio of $2(\Delta_{\min} + 1)$, a randomized algorithm with approximation ratio $4\,\sigma$ and a simple approximation algorithm with performance ratio 4 in case both G and H are linear forests.

We mention here some possible directions for future works. First, is it possible to improve the approximation ratio for bounded degree graphs presented in Proposition 4? Second, due to biological constraints, improving Proposition 5 is of particular interest. In particular, does a deterministic or randomized approximation algorithm with performance ratio σ exist for the MAX–(ρ,σ)–MATCHING–COLOR problem?

References

1. Crochemore, M., Hermelin, D., Landau, G., Rawitz, D., Vialette, S.: Approximating the 2-interval pattern problem, Theoretical Computer Science (special issue for Alberto Apostolico) 2006 (to appear)
2. Diestel, R.: Graph theory, 2nd edn. Graduate texts in Mathematics, vol. 173. Springer, Heidelberg (2000)
3. Fagnot, I., Lelandais, G., Vialette, S.: Bounded list injective homomorphism for comparative analysis of protein-protein interaction graphs. In: Proc. 1st Algorithms and Computational Methods for Biochemical and Evolutionary Networks (Comp-BioNets), pp. 45–70. KCL publications (2004)
4. Fertin, G., Rizzi, R., Vialette, S.: Finding exact and maximum occurrences of protein complexes in protein-protein interaction graphs. In: Jedrzejowicz, J., Szepietowski, A. (eds.) MFCS 2005. LNCS, vol. 3618, pp. 328–339. Springer, Heidelberg (2005)
5. Gavin, A.C., Boshe, M., et al.: Functional organization of the yeast proteome by systematic analysis of protein complexes. Nature 414(6868), 141–147 (2002)
6. Ho, Y., Gruhler, A., et al.: Systematic identification of protein complexes in Saccharomyces cerevisae by mass spectrometry. Nature 415(6868), 180–183 (2002)
7. Kelley, B.P., Sharan, R., Karp, R.M., Sittler, T., Root, D.E., Stockwell, B.R., Ideker, T.: Conserved pathways within bacteria and yeast as revealed by global protein network alignment. PNAS 100(20), 11394–11399 (2003)
8. Mitzenmacher, M., Upfal, E.: Probability and computing: Randomized algorithms and probabilistic analysis. Cambridge University Press, Cambridge (2005)
9. Pellegrini, M., Marcotte, E.M., Thompson, M.J., Eisenberg, D., Yeates, T.O.: Assigning protein functions by comparative genome analysis: protein phylogenetic profiles. PNAS 96(8), 4285–4288 (1999)

10. Pereira-Leal, J.B., Enright, A.J., Ouzounis, C.A.: Detection of functional modules from protein interaction networks. Proteins 54(1), 49–57 (2004)
11. Sharan, R., Ideker, T., Kelley, B., Shamir, R., Karp, R.M.: Identification of protein complexes by comparative analysis of yeast and bacterial protein interaction data. In: Proc. 8th annual international conference on Computational molecular biology (RECOMB), pp. 282–289. ACM Press, New York (2004)
12. Sharan, R., Suthram, S., Kelley, R.M., Kuhn, T., McCuin, S., Uetz, P., Sittler, T., Karp, R., Ideker, T.: Conserved patterns of protein interaction in multiple species. PNAS 102(6), 1974–1979 (2004)
13. Tatusov, R.L., Koonin, E.V., Lipman, D.J.: A genomic perspective on protein families. Science 278(5338), 631–637 (1997)
14. Titz, B., Schlesner, M., Uetz, P.: What do we learn from high-throughput protein interaction data? Expert Review of Anticancer Therapy 1(1), 111–121 (2004)
15. Trotter, W.T., Harary, F.: On double and multiple interval graphs. J. Graph Theory 3, 205–211 (1979)
16. Uetz, P., Giot, L., et al.: A comprehensive analysis of protein-protein interactions in Saccharomyces cerevisae. Nature 403(6770), 623–627 (2000)
17. Vialette, S.: On the computational complexity of 2-interval pattern matching. Theoretical Computer Science 312(2-3), 223–249 (2004)
18. Vizing, V.G.: On an estimate of the chromatic class of a p-graph. Diskret. Analiz 3, 23–30 (1964)
19. Yang, Q., Sze, S.-H.: Path matching and graph matching in biological networks. JCB 14(1), 56–67 (2007)

From Micro to Macro: How the Overlap Graph Determines the Reduction Graph in Ciliates*

Robert Brijder, Hendrik Jan Hoogeboom, and Grzegorz Rozenberg

Leiden Institute of Advanced Computer Science, Universiteit Leiden,
Niels Bohrweg 1, 2333 CA Leiden, The Netherlands
rbrijder@liacs.nl

Abstract. The string pointer reduction system (SPRS) and the graph pointer reduction system (GPRS) are important formal models of gene assembly in ciliates. The reduction graph is a useful tool for the analysis of the SPRS, providing valuable information about the way that gene assembly is performed for a given gene. The GPRS is more abstract than the SPRS – not all information present in the SPRS is retained in the GPRS. As a consequence the reduction graph cannot be defined for the GPRS in general, but we show that it can be defined if we restrict ourselves to the so-called realistic overlap graphs (which correspond to genes occurring in nature). Defining the reduction graph within the GPRS allows one to carry over from the SPRS to the GPRS several results that rely on the reduction graph.

1 Introduction

Gene assembly is a process that takes place in unicellular organisms called ciliates, which have two functionally different nuclei: micronucleus (MIC) and macronucleus (MAC). Gene assembly transforms the genome of MIC into the genome of MAC. During gene assembly each gene in its MIC form gets transformed into the same gene in its MAC form. Among the formal models of gene assembly the string pointer reduction system (SPRS) and the graph pointer reduction system (GPRS) [5] are of interest for this paper. The former consist of three types of string rewriting rules operating on so-called legal strings, while the latter consist of three types of graph rewriting rules operating on so-called overlap graphs. The GPRS is an abstraction of the SPRS with some information present in the SPRS lost in the GPRS.

Legal strings represent genes in their micronuclear form. The reduction graph, which is defined for legal strings, is a notion that describes the corresponding gene in its macronuclear form (along with its waste products, the substrings "spliced out" in the process) – it is unique for a given legal string. It has been shown that the reduction graph contains the information needed for the use of string negative rules (one of the three types of string rewriting rules) in the

---

* This research was supported by the Netherlands Organization for Scientific Research (NWO) project 635.100.006 "VIEWS" and NSF grant 0622112.

E. Csuhaj-Varjú and Z. Ésik (Eds.): FCT 2007, LNCS 4639, pp. 149–160, 2007.

transformation of a MIC form of a gene to its MAC form [3,2,1]. Therefore it
would be useful to have a notion of the reduction graph also for the GPRS. However, this is not so straightforward, because as we will show, since the GPRS
loses some information concerning the application of string negative rules, there
is no unique reduction graph for a given overlap graph, cf. Example 6. We will
show that when we restrict ourselves to "realistic" overlap graphs then one gets
a unique reduction graph. These overlap graphs are called realistic since they
correspond to (micronuclear) genes. In this paper, we explicitly define the notion of reduction graph for realistic overlap graphs (within the GPRS) and show
its equivalence with the notion of reduction graph for legal strings (within the
SPRS). Finally, we give a number of direct corollaries of this equivalence, including an answer to an open problem formulated in Chapter 13 in [5]. Due to space
constraints, proofs of our results are omitted – they are given in an extended
version of this paper [4].

2   Gene Assembly in Ciliates

Two models that are used to formalize the process of gene assembly in ciliates
are the string pointer reduction system (SPRS) and the graph pointer reduction
system (GPRS). The SPRS consist of three types of string rewriting rules operating on *legal strings* while the GPRS consist of three types of graph rewriting
rules operating on *overlap graphs*. For the purpose of this paper it is not necessary to recall the string and graph rewriting rules; a complete description of
SPRS and GPRS, as well as a proof of their "weak" equivalence, can be found
in [5]. We do recall the notions of legal string and overlap graph, and we also
recall the notion of realistic string.

    We fix $\kappa \geq 2$, and define the alphabet $\Delta = \{2, 3, \ldots, \kappa\}$. For $D \subseteq \Delta$, we define
$\bar{D} = \{\bar{a} \mid a \in D\}$ and $\Pi_D = D \cup \bar{D}$; also $\Pi = \Pi_\Delta$. The elements of Π will be
called *pointers*. We use the "bar operator" to move from Δ to $\bar{\Delta}$ and back from
$\bar{\Delta}$ to Δ. Hence, for $p \in \Pi$, $\bar{\bar{p}} = p$. For a string $u = x_1 x_2 \cdots x_n$ with $x_i \in \Pi$, the
inverse of u is the string $\bar{u} = \bar{x}_n \bar{x}_{n-1} \cdots \bar{x}_1$. For $p \in \Pi$, we define $\mathbf{p}$ to be p if
$p \in \Delta$, and $\bar{p}$ if $p \in \bar{\Delta}$, i.e., $\mathbf{p}$ is the "unbarred" variant of p. The *domain of u*,
denoted by dom(u), is $\{\mathbf{p} \mid p$ occurs in $v\}$. We say that u is a *legal string* if for
each $p \in$ dom(u), u contains exactly two occurrences (of elements) from $\{p, \bar{p}\}$.
For a pointer p and a legal string u, if both p and $\bar{p}$ occur in u then we say that
both p and $\bar{p}$ are *positive in u*; if on the other hand only p or only $\bar{p}$ occurs in
u, then both p and $\bar{p}$ are *negative in u*. So, every pointer occurring in a legal
string is either positive or negative in it. Therefore, we can define a partition of
dom(u) = pos(u) $\cup$ neg(u), where pos(u) = $\{p \in$ dom(u) $\mid p$ is positive in $u\}$ and
neg(u) = $\{p \in$ dom(u) $\mid p$ is negative in $u\}$.

    Let $u = x_1 x_2 \cdots x_n$ be a legal string with $x_i \in \Pi$ for $1 \leq i \leq n$. For a pointer
$p \in \Pi$, the *p-interval of u* is the substring $x_i x_{i+1} \cdots x_j$ with $\{x_i, x_j\} \subseteq \{p, \bar{p}\}$
and $1 \leq i < j \leq n$. Substrings $x_{i_1} \cdots x_{j_1}$ and $x_{i_2} \cdots x_{j_2}$ *overlap in u* if $i_1 <
i_2 < j_1 < j_2$ or $i_2 < i_1 < j_2 < j_1$. Two distinct pointers $p, q \in \Pi$ *overlap
in u* if the p-interval of u overlaps with the q-interval of u. Thus, two distinct

pointers $p, q \in \Pi$ overlap in u iff there is exactly one occurrence from $\{p, \bar{p}\}$ in the q-interval, or equivalently, there is exactly one occurrence from $\{q, \bar{q}\}$ in the p-interval of u. Also, for $p \in \text{dom}(u)$, we denote the set of all $q \in \text{dom}(u)$ such that p and q overlap in u by $O_u(p)$, and for $0 \le i \le j \le n$, we denote by $O_u(i, j)$ the set of all $p \in \text{dom}(u)$ such that there is exactly one occurrence from $\{p, \bar{p}\}$ in $x_{i+1}x_{i+2} \cdots x_j$. Also, we define $O_u(j, i) = O_u(i, j)$. Intuitively, $O_u(i, j)$ is the set of $p \in \text{dom}(u)$ for which the the substring between "positions" i and j in u contains exactly one representative from $\{p, \bar{p}\}$, where position i for $0 < i < n$ means the "space" between x_i and x_{i+1} in u. For $i = 0$ it is the "space" on the left of x_1, and for $i = n$ it is the "space" on the right of x_n. We have $O_u(i, n) = O_u(0, i)$ for i with $0 \le i \le n$. The symmetric difference of sets X and Y, $(X \backslash Y) \cup (Y \backslash X)$, is denoted by $X \oplus Y$. We denote the symmetric difference of a family of sets $(X_i)_{i \in A}$ by $\bigoplus_{i \in A} X_i$. For $i, j, k \in \{0, \dots, n\}$, we have $O_u(i, j) \oplus O_u(j, k) = O_u(i, k)$.

A *labelled graph* is a 4-tuple $G = (V, E, f, \Gamma)$, where V is a finite set of vertices, $E \subseteq \{\{x, y\} \mid x, y \in V, x \ne y\}$ is a set of edges, Γ a finite set of labels, and $f : V \to \Gamma$ is the labelling function. Labelled graphs $G = (V, E, f, \Gamma)$ and $G' = (V', E', f', \Gamma)$ are *isomorphic*, denoted by $G \approx G'$, if there is a bijection $\alpha : V \to V'$ such that $f(v) = f'(\alpha(v))$ for $v \in V$, and $\{x, y\} \in E$ iff $\{\alpha(x), \alpha(y)\} \in E'$ for $x, y \in V$. Bijection α is then called an *isomorphism from G to G'*.

Definition 1. Let u be a legal string. The *overlap graph of u*, denoted by γ_u, is the labelled graph $(\text{dom}(u), E, \sigma, \{+, -\})$, where for $p, q \in \text{dom}(u)$ with $p \ne q$, $\{p, q\} \in E$ iff p and q overlap in u, and σ is defined by: $\sigma(p) = +$ if $p \in \text{pos}(u)$, and $\sigma(p) = -$ if $p \in \text{neg}(u)$.

Example 1. Let $u = 24535423$ be a legal string. The overlap graph of u is

$$\gamma = (\{2, 3, 4, 5\}, \{\{2, 3\}, \{4, 3\}, \{5, 3\}\}, \sigma, \{+, -\}),$$

where $\sigma(v) = -$ for all vertices v of γ.

Let γ be an overlap graph. Similar to legal strings, we define $\text{dom}(\gamma)$ as the set of vertices of γ, $\text{pos}(\gamma) = \{p \in \text{dom}(\gamma) \mid \sigma(p) = +\}$, $\text{neg}(\gamma) = \{p \in \text{dom}(\gamma) \mid \sigma(p) = -\}$ and for $q \in \text{dom}(u)$, $O_\gamma(q) = \{p \in \text{dom}(\gamma) \mid \{p, q\} \in E\}$.

We define the alphabet $\Theta_\kappa = \{M_i, \bar{M}_i \mid 1 \le i \le \kappa\}$. We say that $\delta \in \Theta_\kappa^*$ is a *micronuclear arrangement* if for each i with $1 \le i \le \kappa$, δ contains exactly one occurrence from $\{M_i, \bar{M}_i\}$. With each string over Θ_κ, we associate a unique string over Π through the homomorphism $\pi_\kappa : \Theta_\kappa^* \to \Pi^*$, thus $\pi_\kappa(uv) = \pi_\kappa(u)\pi_\kappa(v)$ for all $u, v \in \Theta_\kappa^*$, defined by: $\pi_\kappa(M_1) = 2$, $\pi_\kappa(M_\kappa) = \kappa$, $\pi_\kappa(M_i) = i(i+1)$ for $1 < i < \kappa$, and $\pi_\kappa(\bar{M}_j) = \overline{\pi_\kappa(M_j)}$ for $1 \le j \le \kappa$.

We say that string u is a *realistic string* if there is a micronuclear arrangement δ such that $u = \pi_\kappa(\delta)$. Note that every realistic string is a legal string. Realistic strings are most useful for the gene assembly models, since only these legal strings can correspond to genes in ciliates. An overlap graph γ is *realistic* if it is the overlap graph of a realistic string. Not every overlap graph of a legal string is realistic. For example, it can be shown that the overlap graph γ of $u = 24535423$ given in Example 1 is not realistic. In fact, one can show that it is not even *realizable* — there is no isomorphism α such that $\alpha(\gamma)$ is realistic.

3   The Reduction Graph

We now recall the notion of a (full) reduction graph, which was first introduced in [3]. Below we present this graph in a slightly modified form. The reduction graph is a 2-edge coloured graph. A *2-edge coloured graph* is a 5-tuple $G = (V, E_1, E_2, f, \Gamma)$, where both (V, E_1, f, Γ) and (V, E_2, f, Γ) are labelled graphs. The basic notions for labelled graphs carry over to 2-edge coloured graphs.

Definition 2. Let $u = p_1 p_2 \cdots p_n$ with $p_1, \ldots, p_n \in \Pi$ be a legal string. The *reduction graph of* u, denoted by $\mathcal{R}_u$, is a 2-edge coloured graph $(V, E_1, E_2, f, \mathrm{dom}(u))$, where

$$V = \{I_1, I_2, \ldots, I_n\} \cup \{I'_1, I'_2, \ldots, I'_n\},$$

$$E_1 = \{e_1, e_2, \ldots, e_n\} \text{ with } e_i = \{I'_i, I_{i+1}\} \text{ for } 1 \le i \le n-1, e_n = \{I'_n, I_1\},$$

$$E_2 = \{\{I'_i, I_j\}, \{I_i, I'_j\} \mid i, j \in \{1, 2, \ldots, n\} \text{ with } i \ne j \text{ and } p_i = p_j\} \cup \{\{I_i, I_j\}, \{I'_i, I'_j\} \mid i, j \in \{1, 2, \ldots, n\} \text{ and } p_i = \bar{p}_j\}, \text{ and}$$

$$f(I_i) = f(I'_i) = \mathbf{p}_i \text{ for } 1 \le i \le n.$$

The edges of E_1 are called the *reality edges*, and the edges of E_2 are called the *desire edges*. Intuitively, the "space" between p_i and p_{i+1} corresponds to the

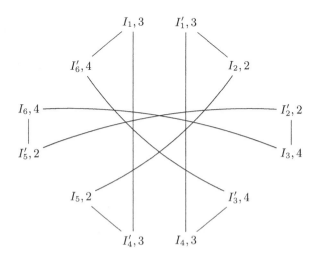

Fig. 1. The reduction graph of u of Example 2

reality edge $e_i = \{I'_i, I_{i+1}\}$. Hence, we say that i is the *position of* e_i, denoted by posn(e_i), for all $i \in \{1, 2, \ldots, n\}$. Note that positions are only defined for reality edges. Since for every vertex v there is a unique reality edge e such that $v \in e$, we also define the *position of* v, denoted by posn(v), as the position of e. Thus, posn(I'_i) = posn(I_{i+1}) = i (while posn(I_1) = n).

Example 2. Let $u = 32\bar{4}3\bar{2}4$ be a legal string. Since $\bar{4}3\bar{2}$ can not be a substring of a realistic string, u is not realistic. The reduction graph $\mathcal{R}_u$ of u is depicted in Figure 1. The labels of the vertices are also shown in this figure. Note the desire edges corresponding to positive pointers (here 2 and 4) cross (in the figure), while those for negative pointers are parallel. Since the exact identity of the vertices in a reduction graph is not essential for the problems considered in this paper, in order to simplify the pictorial representation of reduction graphs we will omit this in the figures. We will also depict reality edges as "double edges" to distinguish them from the desire edges. Figure 2 shows the reduction graph in this simplified representation.

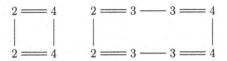

Fig. 2. The reduction graph of u of Example 2 in the simplified representation

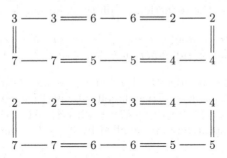

Fig. 3. The reduction graph of u of Example 3

Example 3. Let $u = \pi_7(M_7 M_1 M_6 M_3 M_5 \overline{M_2} M_4) = 726734563\bar{3}245$. Thus, unlike the previous example, u is a realistic string. The reduction graph is given in Figure 3. As usual, the vertices are represented by their labels.

The reduction graph is defined for legal strings. In this paper, we show how to directly construct the reduction graph of realistic string u from only the overlap graph of u. In this way we can define the reduction graph for realistic overlap graphs in a direct way.

Next we consider sets of overlapping pointers corresponding to pairs of vertices in reduction graphs, and start to develop a calculus for these sets that will later enable us to characterize the existence of certain edges in the reduction graph, cf. Theorem 12.

Example 4. We again consider the legal string $u = 32\bar{4}3\bar{2}4$ and its reduction graph $\mathcal{R}_u$ from Example 2. Desire edge $e = \{I_2', I_5'\}$ is connected to reality edges $e_1 = \{I_2', I_3\}$ and $e_2 = \{I_5', I_6\}$ with positions 2 and 5 respectively. We have $O_u(2,5) = \{2,3,4\}$. Also, reality edges $\{I_1', I_2\}$ and $\{I_2', I_3\}$ have positions 1 and 2 respectively. We have $O_u(1,2) = \{2\}$.

Lemma 3. *Let u be a legal string. Let $e = \{v_1, v_2\}$ be a desire edge of $\mathcal{R}_u$ and let p be the label of both v_1 and v_2. Then $O_u(\mathrm{posn}(v_1), \mathrm{posn}(v_2))$ is $O_u(p)$ if p is negative in u, and $O_u(p) \oplus \{p\}$ if p is positive in u.*

Let u be a legal string. We define, for $P \subseteq \mathrm{dom}(u)$, $\Pi_u(P) = (\mathrm{pos}(u) \cap P) \oplus (\bigoplus_{t \in P} O_u(t))$. Similarly, we define $\Pi_\gamma(P)$ for an overlap graph γ.
    Let

$$p_0 = p_1 \quad\text{---}\quad p_1 = p_2 \quad\text{---}\quad p_2 = \cdots = p_n \quad\text{---}\quad p_n = p_{n+1}$$

be a subgraph of $\mathcal{R}_u$, where (as usual) the vertices in the figure are represented by their labels, and let e_1 (e_2, resp.) be the leftmost (rightmost, resp.) edge. Note that e_1 and e_2 are reality edges and therefore $\mathrm{posn}(e_1)$ and $\mathrm{posn}(e_2)$ are defined. Then by Lemma 3 $O_u(\mathrm{posn}(e_1), \mathrm{posn}(e_2)) = \Pi_u(P)$ with $P = \{p_1, \dots, p_n\}$.
    By the definition of the reduction graph the following lemma holds.

Lemma 4. *Let u be a legal string. If I_i and I_i' are vertices of $\mathcal{R}_u$, then $O_u(\mathrm{posn}(I_i), \mathrm{posn}(I_i')) = \{p\}$, where p is the label of I_i and I_i'.*

Example 5. We again consider the legal string u and desire edge e as in the previous example. Since e has vertices labelled by positive pointer 2, by Lemma 3 we have (again) $O_u(2,5) = O_u(2) \oplus \{2\} = \{2,3,4\}$. Also, since I_2 and I_2' with positions 1 and 2 respectively are labelled by 2, by Lemma 4 we have (again) $O_u(1,2) = \{2\}$.

4   The Reduction Graph of Realistic Strings

The next theorem asserts that the overlap graph γ for a realistic string u retains all information of $\mathcal{R}_u$ (up to isomorphism). In fact, the operations on strings that were shown in [7] (and [5]) to keep the overlap graph unchanged, also lead to the same reduction graph. In the sequel of this paper, we will give a method to determine $\mathcal{R}_u$ (up to isomorphism), from γ. Of course, the naive method is to first determine a legal string u corresponding to γ and then to determine the reduction graph of u. However, we present a method that allows one to construct $\mathcal{R}_u$ in a direct way from γ.

Theorem 5. *Let u and v be realistic strings. If $\gamma_u = \gamma_v$, then $\mathcal{R}_u \approx \mathcal{R}_v$.*

This theorem does *not* hold for legal strings in general — the next example illustrates that legal strings having the same overlap graph can have different reduction graphs.

Example 6. Let $u = \pi_\kappa(M_1M_2M_3M_4) = 223344$ be a realistic string and let $v = 234432$ be a legal string. Note that v is not realistic. Legal strings u and v have the same overlap graph γ ($\gamma = (\{2,3,4\}, \varnothing, \sigma, \{+,-\})$, where $\sigma(p) = -$ for $p \in \{2,3,4\}$). Both $\mathcal{R}_u$ and $\mathcal{R}_v$ have four connected components. However, $\mathcal{R}_u$ has three connected components consisting of two vertices and one consisting of six vertices, while $\mathcal{R}_v$ has two connected components consisting of two vertices and two consisting of four vertices. Therefore, $\mathcal{R}_u \not\approx \mathcal{R}_v$.

Definition 6. Let u be a legal string and let $\kappa = |\mathrm{dom}(u)| + 1$. If $\mathcal{R}_u$ contains a subgraph L of the following form:

$$2 \text{---} 2 \Longleftrightarrow 3 \text{---} 3 \Longleftrightarrow \cdots \Longleftrightarrow \kappa \text{---} \kappa$$

where the vertices in the figure are represented by their labels, then we say that u is *rooted* and L is called a *root subgraph of $\mathcal{R}_u$*.

Note that realistic string u with $\mathrm{dom}(u) = \{2, 3, \ldots, 7\}$ from Example 3 is rooted. It turns out that this illustrates a general property.

Theorem 7. *Every realistic string is rooted.*

In the remaining of this paper, we will denote $|\mathrm{dom}(u)| + 1$ by κ for rooted strings, when the rooted string u is understood from the context of considerations. The reduction graph of a realistic string may have more than one root subgraph: it is easy to verify that realistic string $234 \cdots \kappa 234 \cdots \kappa$ for $\kappa \geq 2$ has two root subgraphs.

Example 2 shows that not every rooted string is realistic. The results in the sequel of this paper that consider realistic strings also hold for rooted strings, since we will not be using any properties of realistic string that are not true for rooted strings in general.

The next lemma is essential to prove the main result of this paper.

Lemma 8. *Let u be a rooted string. Let L be a root subgraph of $\mathcal{R}_u$. Let i and j be positions of reality edges in $\mathcal{R}_u$ that are not edges of L. Then $O_u(i,j) = \varnothing$ iff $i = j$.*

The following lemma is easy to verify.

Lemma 9. *Let u be a rooted string. Let L be a root subgraph of $\mathcal{R}_u$. If I_i and I_i' are vertices of $\mathcal{R}_u$, then exactly one of I_i and I_i' belongs to L.*

The following result captures the main idea that allows for the determination of the reduction graph from the overlap graph only. It relies on Lemmas 3, 4, 8, and 9.

Theorem 10. *Let u be a rooted string, let L be a root subgraph of $\mathcal{R}_u$, and let $p, q \in \mathrm{dom}(u)$ with $p < q$. Then there is a reality edge e in $\mathcal{R}_u$ with both vertices not in L, one labelled by p and the other by q iff $\Pi_\gamma(P) = \{p\} \oplus \{q\}$, where $P = \{p+1, \ldots, q-1\} \cup P'$ for some $P' \subseteq \{p, q\}$.*

5   Compressing the Reduction Graph

It is obvious that in reduction graphs one can replace each subgraph $p \relbar\joinrel\relbar p$ (a desire edge with its vertices) by a single vertex labelled by p without losing information. We denote this compressed version of the reduction graph $\mathcal{R}_u$ by $\mathrm{cps}(\mathcal{R}_u)$. Clearly, $\mathcal{R}_u$ can be easily reconstructed from $\mathrm{cps}(\mathcal{R}_u)$ (up to isomorphism).

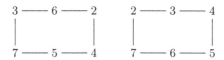

Fig. 4. The labelled graph $\mathrm{cps}(\mathcal{R}_u)$, where $\mathcal{R}_u$ is defined in Example 7. The vertices in the figure are represented by their labels.

Example 7. We are again considering the realistic string u defined in Example 3. The reduction graph of $\mathcal{R}_u$ is depicted in Figure 3. The labelled graph $\mathrm{cps}(\mathcal{R}_u)$ is depicted in Figure 4. Since this graph has just one set of edges, the reality edges are depicted as "single edges" instead of "double edges" as we did for reduction graphs.

In the next section we define the compressed reduction graph directly for overlap graphs and show that it is isomorphic to the compressed reduction graph of the underlying legal string.

6   From Overlap Graph to Reduction Graph

Here we define (compressed) reduction graphs for realistic overlap graphs, inspired by the characterization from Theorem 10. In the remaining part of this section we will show its equivalence with reduction graphs for realistic strings.

Definition 11. Let $\gamma = (Dom_\gamma, E_\gamma, \sigma, \{+, -\})$ be a realistic overlap graph and let $\kappa = |Dom_\gamma| + 1$. The *(compressed) reduction graph* of γ, denoted by $\mathcal{R}_\gamma$, is a labelled graph (V, E, f, Dom_γ), where $V = \{J_p, J_p' \mid 2 \le p \le \kappa\}$, $f(J_p) = f(J_p') = p$, for $2 \le p \le \kappa$, and $e \in E$ iff one of the following conditions hold:

1. $e = \{J'_p, J'_{p+1}\}$ and $2 \le p < \kappa$.
2. $e = \{J_p, J_q\}$, $2 \le p < q \le \kappa$, and $\Pi_\gamma(P) = \{p\} \oplus \{q\}$, where $P = \{p+1, \ldots, q-1\} \cup P'$ for some $P' \subseteq \{p, q\}$.
3. $e = \{J'_2, J_p\}$, $2 \le p \le \kappa$, and $\Pi_\gamma(P) = \{p\}$, where $P = \{2, \ldots, p-1\} \cup P'$ for some $P' \subseteq \{p\}$.
4. $e = \{J'_\kappa, J_p\}$, $2 \le p \le \kappa$, and $\Pi_\gamma(P) = \{p\}$, where $P = \{p+1, \ldots, \kappa\} \cup P'$ for some $P' \subseteq \{p\}$.
5. $e = \{J'_2, J'_\kappa\}$, $\kappa > 3$, and $\Pi_\gamma(P) = \varnothing$, where $P = \{2, \ldots, \kappa\}$.

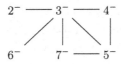

Fig. 5. The overlap graph γ of a realistic string (used in Example 8)

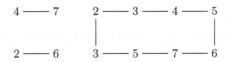

Fig. 6. The reduction graph $\mathcal{R}_\gamma$ of the overlap graph γ of Example 8. The vertices in the figure are represented by their labels.

Example 8. The overlap graph γ in Figure 5 is realistic. Indeed, realistic string $u = \pi_7(M_4 M_3 M_7 M_5 M_2 M_1 M_6) = 453475623267$ has this overlap graph. Clearly, the reduction graph $\mathcal{R}_\gamma$ of γ has the edges $\{J'_p, J'_{p+1}\}$ for $2 \le p < 7$. The following table lists the remaining edges of $\mathcal{R}_\gamma$. The table also states the characterizing conditions for each edge as stated in Definition 11.

Edge	P	Witness
$\{J_2, J_6\}$	$\{3, 4, 5\}$	$\{2, 4, 5, 6, 7\} \oplus \{3, 5\} \oplus \{3, 4, 7\} = \{2, 6\}$
$\{J_2, J_6\}$	$\{2, 3, 4, 5, 6\}$	$\{3\} \oplus \{2, 4, 5, 6, 7\} \oplus \{3, 5\} \oplus \{3, 4, 7\} \oplus \{3\} = \{2, 6\}$
$\{J_4, J_7\}$	$\{5, 6\}$	$\{3, 4, 7\} \oplus \{3\} = \{4, 7\}$
$\{J_4, J_7\}$	$\{4, 5, 6, 7\}$	$\{3, 5\} \oplus \{3, 4, 7\} \oplus \{3\} \oplus \{3, 5\} = \{4, 7\}$
$\{J_3, J_5\}$	$\{4\}$	$\{3, 5\} = \{3, 5\}$
$\{J_5, J'_7\}$	$\{6, 7\}$	$\{3\} \oplus \{3, 5\} = \{5\}$
$\{J'_2, J_3\}$	$\{2\}$	$\{3\} = \{3\}$

We have now completely determined $\mathcal{R}_\gamma$; it is shown in Figure 6. As we have done for reduction graphs of legal strings, in the figures the vertices of reduction graphs of realistic overlap graphs are represented by their labels.

Example 9. In the second example we construct the reduction graph of an overlap graph that contains positive pointers. The overlap graph γ in Figure 7 is realistic. Indeed, realistic string $u = \pi_7(M_7 M_1 M_6 M_3 M_5 \overline{M_2} M_4) = 726734563\overline{3}245$

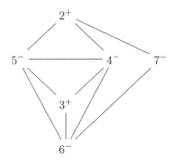

Fig. 7. The overlap graph γ of a realistic string (used in Example 9)

Fig. 8. The reduction graph $\mathcal{R}_\gamma$ of the overlap graph γ of Example 9

introduced in Example 3 has this overlap graph. Again, the reduction graph $\mathcal{R}_\gamma$ of γ has the edges $\{J'_p, J'_{p+1}\}$ for $2 \le p < 7$. The remaining edges are listed in the table below.

Edge	P	Witness
$\{J_3, J_7\}$	$\{4, 5, 6\}$	$\{2, 3, 5, 6\} \oplus \{2, 3, 4, 6\} \oplus \{3, 4, 5, 7\} = \{3, 7\}$
$\{J_3, J_6\}$	$\{3, 4, 5\}$	$\{3\} \oplus \{4, 5, 6\} \oplus \{2, 3, 5, 6\} \oplus \{2, 3, 4, 6\} = \{3, 6\}$
$\{J_2, J_6\}$	$\{2, 3, 4, 5, 6\}$	$\{2\} \oplus \{4, 5, 7\} \oplus \{3\} \oplus \{4, 5, 6\} \oplus \{2, 3, 5, 6\}$
		$\oplus\{2, 3, 4, 6\} \oplus \{3, 4, 5, 7\} = \{2, 6\}$
$\{J_2, J_4\}$	$\{3, 4\}$	$\{3\} \oplus \{4, 5, 6\} \oplus \{2, 3, 5, 6\} = \{2, 4\}$
$\{J_4, J_5\}$	$\{4, 5\}$	$\{2, 3, 5, 6\} \oplus \{2, 3, 4, 6\} = \{4, 5\}$
$\{J_5, J_7\}$	$\{5, 6, 7\}$	$\{2, 3, 4, 6\} \oplus \{3, 4, 5, 7\} \oplus \{2, 6\} = \{5, 7\}$
$\{J'_2, J'_7\}$	$\{2, \ldots, 7\}$	$\{2\} \oplus \{4, 5, 7\} \oplus \ldots \oplus \{2, 6\} = \varnothing$

Again, we have now completely determined the reduction graph; it is shown in Figure 8.

Figures 4 and 8 show, for $u = 726734563\overline{3}\overline{2}45$, that $\mathrm{cps}(\mathcal{R}_u) \approx \mathcal{R}_\gamma$. The next theorem shows that this is a general property for realistic strings u. The proof of the result relies on Theorem 10.

Theorem 12. *Let u be a realistic string. Then, $\mathrm{cps}(\mathcal{R}_u) \approx \mathcal{R}_{\gamma_u}$.*

Example 10. The realistic string $u = 453475623267$ was introduced in Example 8. The reduction graph $\mathcal{R}_\gamma$ of the overlap graph of u is given in Figure 6. The reduction graph $\mathcal{R}_u$ of u is given in Figure 9. It is easy to see that after applying cps to $\mathcal{R}_u$ one obtains a graph that is indeed isomorphic to $\mathcal{R}_\gamma$.

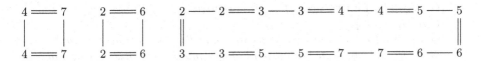

Fig. 9. The reduction graph of u of Example 10. The vertices in the figure are represented by their labels.

This makes clear why there were two proofs for both edges $\{J_2, J_6\}$ and $\{J_4, J_7\}$ in Example 8; each one corresponds to one reality edge in $\mathcal{R}_u$ (outside L).

7   Consequences

Using the previous theorem and [6] (or Chapter 13 in [5]), we can now easily characterize successfulness for realistic overlap graphs in any given $S \subseteq \{Gnr, Gpr, Gdr\}$. The notions of successful reduction, string negative rule and graph negative rule used in this section are defined in [5]. Due to the "weak equivalence" of the string pointer reduction system and the graph pointer reduction system, proved in Chapter 11 of [5], we can, using Theorem 12, restate Theorem 26 in [3] in terms of graph reduction rules.

Theorem 13. *Let u be a realistic string, and N be the number of connected components in $\mathcal{R}_{\gamma_u}$. Then every successful reduction of γ_u has exactly $N - 1$ graph negative rules.*

As an immediate consequence we get the following corollary. It provides a solution to an open problem formulated in Chapter 13 in [5].

Corollary 14. *Let u be a realistic string. Then γ_u is successful in $\{Gpr, Gdr\}$ iff $\mathcal{R}_{\gamma_u}$ is a connected graph.*

Example 11. Every successful reduction of the overlap graph of Example 8 has exactly two graph negative rules, because its reduction graph consist of exactly three connected components. For example $\mathbf{gnr}_4\ \mathbf{gdr}_{5,7}\ \mathbf{gnr}_2\ \mathbf{gdr}_{3,6}$ is a successful reduction of this overlap graph. Similarly, every successful reduction of the overlap graph of Example 9 has exactly one graph negative rule. One example is $\mathbf{gnr}_2\ \mathbf{gpr}_4\ \mathbf{gpr}_5\ \mathbf{gpr}_7\ \mathbf{gpr}_6\ \mathbf{gpr}_3$.

With the help of [6] (or Chapter 13 in [5]) and Corollary 14, we are ready to complete the characterization of successfulness for realistic overlap graphs in any given $S \subseteq \{Gnr, Gpr, Gdr\}$.

Theorem 15. *Let u be a realistic string. Then γ_u is successful in:*

- *$\{Gnr\}$ iff γ_u is a discrete graph (it has no edges) with only negative vertices.*
- *$\{Gnr, Gpr\}$ iff each connected component of γ_u that consists of more than one vertex contains a positive vertex.*

- $\{Gnr, Gdr\}$ iff all vertices of γ_u are negative.
- $\{Gnr, Gpr, Gdr\}$.
- $\{Gdr\}$ iff all vertices of γ_u are negative and $\mathcal{R}_{\gamma_u}$ is a connected graph.
- $\{Gpr\}$ iff each connected component of γ_u contains a positive vertex and $\mathcal{R}_{\gamma_u}$ is a connected graph.
- $\{Gpr, Gdr\}$ iff $\mathcal{R}_{\gamma_u}$ is a connected graph.

8   Discussion

We have shown how to directly construct the reduction graph of a realistic string u (up to isomorphism) from the overlap graph γ of u. This allows one to reconstruct a representation of the macronuclear gene (and its waste products) given only the overlap graph of the micronuclear gene. Moreover, this results allows one to (directly) determine the number n of graph negative rules that are necessary to reduce γ successfully. Along with some results in previous papers, it also allows us to give a complete characterization of the successfulness of γ in any given $S \subseteq \{Gnr, Gpr, Gdr\}$.

It remains an open problem to find a (direct) method to determine this number n for overlap graphs γ in general (not just for realistic overlap graphs).

References

1. Brijder, R., Hoogeboom, H.J., Muskulus, M.: Applicability of loop recombination in ciliates using the breakpoint graph. In: Berthold, M.R., Glen, R.C., Fischer, I. (eds.) CompLife 2006. LNCS (LNBI), vol. 4216, pp. 97–106. Springer, Heidelberg (2006)
2. Brijder, R., Hoogeboom, H.J., Rozenberg, G.: The breakpoint graph in ciliates. In: Berthold, M.R., Glen, R.C., Diederichs, K., Kohlbacher, O., Fischer, I. (eds.) CompLife 2005. LNCS (LNBI), vol. 3695, pp. 128–139. Springer, Heidelberg (2005)
3. Brijder, R., Hoogeboom, H.J., Rozenberg, G.: Reducibility of gene patterns in ciliates using the breakpoint graph. Theor. Comput. Sci. 356, 26–45 (2006)
4. Brijder, R., Hoogeboom, H.J., Rozenberg, G.: How overlap determines the macronuclear genes in ciliates. LIACS Technical Report 2007-02, [arXiv:cs.LO/0702171] (2007)
5. Ehrenfeucht, A., Harju, T., Petre, I., Prescott, D.M., Rozenberg, G.: Computation in Living Cells – Gene Assembly in Ciliates. Springer, Heidelberg (2004)
6. Ehrenfeucht, A., Harju, T., Petre, I., Rozenberg, G.: Characterizing the micronuclear gene patterns in ciliates. Theory of Computing Systems 35, 501–519 (2002)
7. Harju, T., Petre, I., Rozenberg, G.: Formal properties of gene assembly: Equivalence problem for overlap graphs. In: Jonoska, N., Păun, G., Rozenberg, G. (eds.) Aspects of Molecular Computing. LNCS, vol. 2950, pp. 202–212. Springer, Heidelberg (2003)

A String-Based Model for Simple Gene Assembly

Robert Brijder[1], Miika Langille[2], and Ion Petre[2,3]

[1] Leiden Institute of Advanced Computer Science, Universiteit Leiden
Niels Bohrweg 1, 2333 CA Leiden, The Netherlands
rbrijder@liacs.nl
[2] Turku Centre for Computer Science
Department of IT, Åbo Akademi University
Turku 20520, Finland
miika.langille@abo.fi
ion.petre@abo.fi
[3] Academy of Finland

Abstract. The simple intramolecular model for gene assembly in ciliates is particularly interesting because it can predict the correct assembly of all available experimental data, although it is not universal. The simple model also has a confluence property that is not shared by the general model. A previous formalization of the simple model through sorting of signed permutations is unsatisfactory because it effectively ignores one operation of the model and thus, it cannot be used to answer questions about parallelism in the model, or about measures of complexity. We propose in this paper a string-based model in which a gene is represented through its sequence of pointers and markers and its assembly is represented as a string rewriting process. We prove that this string-based model is equivalent to the permutation-based model as far as gene assembly is concerned, while it tracks all operations of the model.

1 Introduction

Gene assembly in ciliates has been subject to extensive combinatorial research in recent years, see [2]. Ciliates are unicellular eukaryotes that organize their genome differently in their two types of nuclei. In micronuclei, genes are split into blocks (called *MDSs*), placed in a shuffled order on the chromosome, separated by non-coding blocks. Moreover, some of the MDSs are even presented in an inverted form. In macronuclei however, genes are contiguous sequences of nucleotides, with all blocks sorted in the orthodox order. The assembly of genes from their micronuclear to their macronuclear form has a definite combinatorial and computational flavor: each MDS M ends with a sequence of nucleotides (called a *pointer*) that is repeated identically in the beginning of the MDS that should follow M in the macronuclear gene.

The exact kinetical mechanisms of gene assembly still remain to be clarified through further laboratory experiments. Two models have been proposed for gene assembly: an intermolecular one, see [7,8] and an intramolecular one,

E. Csuhaj-Varjú and Z. Ésik (Eds.): FCT 2007, LNCS 4639, pp. 161–172, 2007.
© Springer-Verlag Berlin Heidelberg 2007

see [3,10]. The intramolecular model, that we consider in this paper, consists of three operations: ld, hi, and dlad. For a detailed discussion of these operations, including their formalization on various levels of abstraction, we refer to [2]. We consider in this paper the simple variant of the model, in which the sequences manipulated by each operation are minimal, see [5] for a detailed discussion. It turns out that although not universal, the simple model is capable of correctly predicting the assembly of all currently available data on genes in stichotrichous ciliates, see [6]. Also, the model has an interesting confluence property that does not hold in the general model, see [9].

The simple model for gene assembly has been investigated in [9] as a process of sorting signed permutations. A major shortcoming of this formalization is that the ld operations are not explicitly modeled and it is only assumed that they take place eventually at some arbitrary moment of time. Although sufficient for the purpose of [9], this formalization does not allow reasoning about parallelism or complexity of assemblies.

Abstracting the gene to its sequence of pointers has been a solution to this problem in the case of the general model. In this way, the gene assembly process is formalized through a process of string rewriting. Doing the same in the case of simple operations does not work, as shown in Example 9 of this paper: the string-based model is not equivalent with the permutation-based model.

We propose in this paper a simple solution to this problem. Rather than representing a gene only through its sequence of pointers, we preserve also the beginning and ending markers (starting the first MDS and ending the last MDS, respectively). We prove that with this simple addition, the string-based model is equivalent with the permutation-based model.

2   Mathematical Preliminaries

For a finite alphabet $A = \{a_1, \ldots, a_n\}$, we denote by A^* the free monoid generated by A and call any element of A^* a *word*. Let $\overline{A} = \{\overline{a}_1, \ldots, \overline{a}_n\}$, where $A \cap \overline{A} = \emptyset$. For $p, q \in A \cup \overline{A}$, we say that p, q have the same *signature* if either $p, q \in A$, or $p, q \in \overline{A}$ and we say that they have *different signatures* otherwise.

We denote $A^{\maltese} = (A \cup \overline{A})^*$. For any $u \in A^{\maltese}$, $u = x_1 \ldots x_k$, with $x_i \in A \cup \overline{A}$, for all $1 \leq i \leq k$, we denote $\|u\| = \|x_1\| \ldots \|x_k\|$, where $\|a\| = \|\overline{a}\| = a$, for all $a \in A$. We also denote $\overline{u} = \overline{x}_k \ldots \overline{x}_1$, where $\overline{\overline{a}} = a$, for all $a \in A$. For two alphabets A, B, a mapping $f : A^{\maltese} \to B^{\maltese}$ is called a *morphism* if $f(uv) = f(u)f(v)$ and $f(\overline{u}) = \overline{f(u)}$.

A *permutation* π over A is a bijection $\pi : A \to A$. Fixing the order relation $(a_1, a_2, \ldots, a_m)$ over A, we often denote π as the word $\pi(a_1) \ldots \pi(a_m) \in A^*$. A *signed permutation* over A is a string $\psi \in A^{\maltese}$, where $\|\psi\|$ is a permutation over A.

A string $v \in \Sigma^*$ over an (unsigned) alphabet Σ is a *double occurrence string* if every pointer $a \in \mathrm{dom}(v)$ occurs exactly twice in v. A signing of a non-empty double occurrence string is a *legal string*.

Two legal strings, u and v over alphabets A and B, respectively, are said to have the same *structure* if there is a morphism $f : A^\maltese \to B^\maltese$ such that $f(u) = v$, and is denoted by $u \equiv v$.

3 Permutations and Strings

Simple operations have previously been defined in terms of signed permutations, see [9]. This formalism was useful for observing simple operations applied sequentially, with the arrangement of explicit MDSs being clearly visible. However, note that with signed permutations the ld operation is always assumed to occur at some point in the assembly. This is not sufficient for studying some aspects of simple gene assembly, and so we now task ourselves with providing a formal framework for performing simple gene assembly on *legal strings*. Micronuclear and intermediate genes are represented as a sequence of pointers and the operations defined as a *string pointer reduction system*. Moreover, we will show that simple gene assembly on legal strings is equivalent to that on signed permutations.

Legal strings have been used before to formalise the general operations in [2]. This definition is missing one crucial piece of information, necessary to model simple operations in particular. Namely, the absence of markers indicating the beginning and end of the macronuclear gene make it necessary to extend the definition.

Let us now briefly describe the signed permutations, followed then by the extended definition of legal strings and the string pointer reduction system.

3.1 Signed Permutations

Genes may be represented as a sequence of MDSs. Using the alphabet $\Pi_n = \{1, 2, \ldots, n\}$ to denote MDSs, where the numbering is given by the order in which MDSs are assembled in the macronuclear gene. Thus, a micronuclear gene will be a signed permutation over Π_n and a macronuclear gene will be a sorted signed permutation over Π_n.

Definition 1. *We say that a signed permutation π is sorted if either:*

i. $\pi = i(i+1)\ldots n1\ldots(i-1)$, *or*
ii. $\pi = \overline{(i-1)}\ldots\overline{1n}\ldots\overline{(i+1)i}$,

for some $1 \leq i \leq n$. If $i = 1$ we say that π is a linear permutation. We call π circular otherwise. In case i. we say that π is sorted in the orthodox order, while in case ii. we say that π is sorted in the inverted order.

The term circular in the above definition refers to a gene that gets assembled, say in the form $i\ldots n1\ldots(i-1)$, and then gets excised from the chromosome by an ld operation applied on the pointer in the beginning of i and its identical copy at the end of $(i-1)$.

Given this definition for genes, we may now consider gene assembly as a sorting of signed permutations. We will define the simple operations as transformation rules for signed permutations in such a way that a simple operation is applicable on a gene pattern if and only if the corresponding rule is applicable on the associated signed permutation.

As mentioned previously, when using signed permutations to model gene assembly, the ld operation is ignored. Indeed, ld only combines two MDSs i and $(i+1)$ already placed next to each other into a bigger composite MDS. To avoid renaming the alphabet after each ld operation, we will consider that when i and $(i+1)$ are placed next to each other, the operation joining them is already accomplished.

Definition 2. *The molecular model of simple* hi *and simple* dlad *can be formalized as follows.*

i. For each $p \geq 1$, sh_p is defined as follows:

$$\mathsf{sh}_p(xp\ldots(p+i)(\overline{p+k})\ldots(\overline{p+i+1})y) = xp\ldots(p+i)(p+i+1)\ldots(p+k)y,$$
$$\mathsf{sh}_p(x(\overline{p+i})\ldots\overline{p}(p+i+1)\ldots(p+k)y) = xp\ldots(p+i)(p+i+1)\ldots(p+k)y,$$
$$\mathsf{sh}_p(x(p+i+1)\ldots(p+k)(\overline{p+i})\ldots\overline{p}y) = x(\overline{p+k})\ldots(\overline{p+i+1})(\overline{p+i})\ldots\overline{p}y,$$
$$\mathsf{sh}_p(x(\overline{p+k})\ldots(\overline{p+i+1})p\ldots(p+i)y) = x(\overline{p+k})\ldots(\overline{p+i+1})(\overline{p+i})\ldots\overline{p}y,$$

where $k > i \geq 0$ and x, y, z are signed strings over Π_n. We denote $\mathsf{Sh} = \{\mathsf{sh}_i \mid 1 \leq i \leq n\}$.

ii. For each p, $2 \leq p \leq n-1$, sd_p is defined as follows:

$$\mathsf{sd}_p(x\,p\ldots(p+i)\,y\,(p-1)\,(p+i+1)\,z) = xy(p-1)p\ldots(p+i)(p+i+1)z,$$
$$\mathsf{sd}_p(x\,(p-1)(p+i+1)yp\ldots(p+i)z) = x(p-1)p\ldots(p+i)(p+i+1)yz,$$

where $i \geq 0$ and x, y, z are signed strings over Π_n. We also define $\mathsf{sd}_{\overline{p}}$ as follows:

$$\mathsf{sd}_{\overline{p}}(x(\overline{p+i+1})(\overline{p-1})y(\overline{p+i})\ldots\overline{p}z) = x\,(\overline{p+i+1})(\overline{p+i})\ldots\overline{p}(\overline{p-1})yz,$$
$$\mathsf{sd}_{\overline{p}}(x(\overline{p+i})\ldots\overline{p}y(\overline{p+i+1})(\overline{p-1})z) = xy(\overline{p+i+1})(\overline{p+i})\ldots\overline{p}(\overline{p-1})z,$$

where $i \geq 0$ and x, y, z are signed strings over Π_n. We denote $\mathsf{Sd} = \{\mathsf{sd}_i, \mathsf{sd}_{\overline{i}} \mid 1 \leq i \leq n\}$.

We say that a signed permutation π over the set of integers $\{i, i+1, \ldots, i+l\}$ is sortable if there are operations $\phi_1, \ldots, \phi_k \in \mathsf{Sh} \cup \mathsf{Sd}$ such that $(\phi_k \circ \ldots \circ \phi_1)(\pi)$ is a sorted permutation. We say that π is blocked if neither an sh operation, nor an sd one is applicable to π and π is not sorted.

Let $\phi = \phi_k \circ \ldots \circ \phi_1$, $\phi_i \in \mathsf{Sh} \cup \mathsf{Sd}$, for all $1 \leq i \leq k$. We say that ϕ is a strategy for π if $\phi(\pi)$ is either sorted, or blocked. In the former case we say that ϕ is a sorting strategy, while in the latter case we say that ϕ is an unsuccessful strategy for π.

If $\phi = \phi_k \circ \ldots \circ \phi_1$ is a sorting strategy for π, we say that π is Sh-sortable if $\phi_1, \ldots, \phi_k \in \mathsf{Sh}$ and we say that π is Sd-sortable if $\phi_1, \ldots, \phi_k \in \mathsf{Sd}$.

Example 1.   *i. The permutation $\pi_1 = \bar{2}1435$ is sortable. It has the following sorting strategies:* $\mathsf{sh}_1 \circ \mathsf{sd}_4 \circ \mathsf{sh}_1(\pi_1) = 12345$ *and* $\mathsf{sh}_1 \circ \mathsf{sh}_1 \circ \mathsf{sd}_4(\pi_1) = 12345$.

ii. The permutation $\pi_2 = 25\bar{3}14$ is blocked as no operations are applicable and it is not sorted.

iii. The permutation $\pi_3 = 35124$ has two assembly strategies which lead to different results: $\mathsf{sd}_3(\pi_3) = 51234$ *and* $\mathsf{sd}_4(\pi_3) = 34512$.

3.2   Legal Strings

Using signed permutations, we hold the information of explicit MDSs. It is sufficient, however, to consider the gene as a sequence of pointers. Moving to legal strings, we remove all notation pertaining to explicit MDSs, and represent the gene as a string of pointers, with each pointer occurring twice in the gene. Though some information is lost about the gene, the notation is elegant and the characteristics and rules of the model are maintained.

To represent a micronuclear or intermediate gene as a legal string, we need a set of pointers, $\Delta_k = \{2, 3, \ldots, k\}$, and the set of markers $M = \{b, e\}$. Note that each marker will occur only once in the string, and each pointer will occur twice. As with signed permutations, k represents the length of the gene, and we shall assume that it will be clear from the context. Let this be our alphabet $\Sigma = \Delta_k \cup M$.

The definition for legal strings given in [2] did not include the markers, but as it turns out, they are essential to being able to formalise simple operations on legal strings. Additionally, not including the markers meant that it was no longer possible to uniquely obtain a signed permutation from a legal string, as too much information was lost.

Let $a \in \Sigma \cup \overline{\Sigma}$ and let $u \in \Sigma^{\circledast}$ be a legal string. If u contains both substrings a and $\bar{a}$ then a is *positive* in u; otherwise, a is *negative* in u.

Example 2.   *i. Consider the signed string $u_1 = b23423\overline{e4}$ over Σ_4. Clearly, u_1 is legal. Pointer 4 is positive in u_1, while 2 and 3 are negative in u_1.*

ii. The string $u_2 = \overline{43}b234$ is not legal, since 2 has only one occurrence in u_2 and the end marker does not occur at all.

Let $u = a_1 a_2 \ldots a_n \in \Sigma^{\circledast}$ be a legal string over Σ, where $a_i \in \Sigma \cup \overline{\Sigma}$ for each i. For each letter $a \in \operatorname{dom}(u)$, there are indices i and j with $1 \le i < j \le n$ such that $\|a_i\| = a = \|a_j\|$, excepting of course the two markers present. The substring $u_{(a)} = a_i a_{i+1} \ldots a_j$ is the *a–interval* of u. Two different letters $a, b \in \Sigma$ are said to *overlap in u* if the a–interval and the b–interval of u overlap: if $u_{(a)} = a_{i_1} \ldots a_{j_1}$ and $u_{(b)} = a_{i_2} \ldots a_{j_2}$, then either $i_1 < i_2 < j_1 < j_2$ or $i_2 < i_1 < j_2 < j_1$. Moreover, for each letter a, we denote by

$$O_u(a) = \{b \in \Sigma \mid b \text{ overlaps with } a \text{ in } u\} \cup \{a\}.$$

Example 3.   *Let $u = b2342\overline{354}\overline{e}656$ be a legal string. The 5–interval of u is the substring $u_{(5)} = \overline{5}465$, which contains only one occurrence of the pointers 4 and 6,* *but either two or no occurrences of 2 and 3 and so $O_u(5) = \{4,5,6\}$. Note that the marker is not included in the interval. Similarly,*

$$u_{(2)} = 2342, \qquad \text{and 2 overlaps with 3 and 4,}$$
$$u_{(3)} = 3423, \qquad \text{and 3 overlaps with 2 and 4,}$$
$$u_{(4)} = 423\overline{54}, \qquad \text{and 4 overlaps with 2, 3 and 5,}$$
$$u_{(6)} = \overline{6}56, \qquad \text{and 6 overlaps with 5.}$$

We will now consider the transformation from signed permutations to legal strings, carried out by replacing each MDS with the pointers at either end of it. When transforming from signed permutations to legal strings, there is a problem with how to deal with sorted blocks of pointers. Therefore, when doing the transformation, the signed permutation must be assumed to be micronuclear, i.e., no MDSs have been joined together. This allows for the following morphism from a signed permutation π to a legal string u. $\zeta : \Pi_n^{\maltese} \to \Sigma_n^{\maltese}$ is applied to each letter in π, and is defined as follows

i. $i \to i(i+1)$,
ii. $1 \to b2$,
iii. $n \to ne$,

where $1 \le i \le n$.

Example 4. *i. The signed permutation $\pi = 1\overline{2}$, is equivalent to the following legal string $\zeta(\pi) = b2\overline{e2}$.*
ii. The signed permutation $\pi = 2314\overline{7}6\overline{5}$, is equivalent to the following legal string $\zeta(\pi) = 2334b245\overline{e7}6\overline{765}$.

We say that a legal string u is *realistic* if there exists a signed permutation π such that $u = \zeta(\pi)$. It is important to note, however, that not all legal strings are realistic. The following example illustrates this point.

Example 5. *The string $u = be22$ is legal but it is not realistic, since it has no "realistic parsing", i.e., there is no signed permutation which would transform to this legal string.*

Let us first give our full definition of legal strings and the simple operations, and then we will show why the markers are necessary to the abstraction.

3.3   Simple Operations on Signed Strings

Recall that in signed permutations gene assembly was simulated by joining together MDSs to obtain a (circularly) sorted sequence, the macronuclear gene. With the move to legal strings we have now removed all details of individual MDSs, leaving us with a string of pointers. The goal remains the same though, to remove all pointers by matching them together, consequently building the macronuclear gene. We shall now introduce the *string pointer reduction system* for simple operations to formalise the three molecular operations ld, hi and dlad.

A realistic legal string is considered sorted, once all pointers are removed, leaving us with one of the following possible sorted permutations: be, $\overline{eb}$, eb and $\overline{be}$. The first two cases represent linear sortings, and the latter two are considered circular.

In the following we shall consider only realistic legal strings, that is, the strings are from $\Sigma_k^{\bigstar}$ for some $k \geq 2$. Each of the rules below is a function that maps legal strings to legal strings.

i. The *string negative rule* snr for a pointer $p \in \Delta_k$ $(k \geq 2)$ is the equivalent of the ld operation. As with ld, the string negative rule is always simple, and as such, it is the same for the general and simple model. It can only be applied when the pointers are adjacent, i.e., not separated by any other pointers. It can be formalised in the following way

$$\mathsf{snr}_p(u_1ppu_2) = u_1u_2,$$

$$\mathsf{snr}_p(pu_3p) = u_3,$$

where $u_1, u_2 \in \Sigma^{\bigstar}$ and u_3 contains only markers (boundary case). Let

$$\mathsf{Snr} = \{\mathsf{snr}_p \mid p \in \Delta_k, k \geq 2\}$$

be the set of all simple string negative rules.

ii. The *simple string positive rule* sspr for a pointer $p \in \Delta_k(k \geq 2)$ is the equivalent of the sh operation. Recall that in the simple hi operation the pointers may only be separated by one MDS, and thus only a single pointer or marker. This gives us the following formalisation

$$\mathsf{sspr}_p(u_1pu_2\overline{p}u_3) = u_1\overline{u}_2u_3,$$

where $|u_2| = 1$ and $u_1, u_2, u_3 \in \Sigma^{\bigstar}$. Let

$$\mathsf{Sspr} = \{\mathsf{sspr}_p \mid p \in \Delta_k, k \geq 2\}$$

be the set of all simple string positive rules.

iii. The *simple string double rule* ssdr for pointers $p, q \in \Delta_k(k \geq 2)$ is the equivalent of the sd operation. Recall that in the simple dlad operation the first occurrences of p and q must be adjacent, with the same condition applied to the second occurrences. This gives us the following formalisation

$$\mathsf{ssdr}_{p,q}(u_1pqu_2pqu_3) = u_1u_2u_3,$$

where $u_1, u_2, u_3 \in \Sigma^{\bigstar}$. Let

$$\mathsf{Ssdr} = \{\mathsf{ssdr}_{p,q} \mid p, q \in \Delta_k, k \geq 2\}$$

be the set of all simple string double rules.

Note that the operations have now become very simple indeed, with only pointers being removed. The sspr operation is the only one that affects the remaining permutation, inverting the pointer or marker separating the two occurrences.

Example 6. *Consider the following signed permutation* $\pi = \overline{1}2354$. *It has the following realistic legal string* $u = \zeta(\pi) = \overline{2b}23345e45$, *and these simple operations applicable to it.*

i. *The* snr *operation removes two adjacent pointers with the same signature. E.g.,* $\mathsf{snr}_3(u) = \overline{2b}245e45$.
ii. *The* sspr *operation removes two pointers with different signatures, separated by exactly one pointer or marker, inverting the sequence separating them. E.g.,* $\mathsf{sspr}_2(u) = b3345e45$.
iii. *The* ssdr *operation removes two overlapping pointers, where the first occurrences are adjacent, as are the second occurrences. E.g.,* $\mathsf{ssdr}_{4,5}(u) = \overline{2b}233e$.

Example 7. *Let us consider the actin I gene from Sterkiella nova, which has the following signed permutation* $\pi = 346579\overline{2}18$. *The corresponding legal string would be*

$$u = \zeta(\pi) = 34456756789e\overline{32}b289.$$

It has the following assembly strategy using the simple string pointer reduction system.

$$\mathsf{sspr}_2(u) = 34456756789e\overline{3b}89$$

$$\mathsf{ssdr}_{5,6} \circ \mathsf{sspr}_2(u) = 3447789e\overline{3b}89$$

$$\mathsf{snr}_4 \circ \mathsf{ssdr}_{5,6} \circ \mathsf{sspr}_2(u) = 37789e\overline{3b}89$$

$$\mathsf{snr}_7 \circ \mathsf{snr}_4 \circ \mathsf{ssdr}_{5,6} \circ \mathsf{sspr}_2(u) = 389e\overline{3b}89$$

$$\mathsf{ssdr}_{8,9} \circ \mathsf{snr}_7 \circ \mathsf{snr}_4 \circ \mathsf{ssdr}_{5,6} \circ \mathsf{sspr}_2(u) = 3e\overline{3b}$$

$$\mathsf{sspr}_3 \circ \mathsf{ssdr}_{8,9} \circ \mathsf{snr}_7 \circ \mathsf{snr}_4 \circ \mathsf{ssdr}_{5,6} \circ \mathsf{sspr}_2(u) = \overline{e}b$$

It is important to note that the naming of simple operations on legal strings does not follow the same standard as for signed permutations. On signed permutations an operation was always named by the smallest MDS in the composite involved. However, on legal strings we have lost the information of MDSs, and only remaining pointers are shown. Thus an operation must be named according to the pointer on which the fold is made. This is illustrated in the following example.

Example 8. *Let* $u = b2452334e\overline{5}$ *be a realistic legal string corresponding to the signed permutation* $\pi = 1423\overline{5}$. *They have the following equivalent assembly strategies,*

$$\mathsf{snr}_3(u) = b24524e\overline{5}, \qquad (\text{Id operation is assumed here})$$

$$\mathsf{ssdr}_{2,4} \circ \mathsf{snr}_3(u) = b5e\overline{5}, \qquad\qquad \mathsf{sd}_2(\pi) = 1234\overline{5},$$

$$\mathsf{sspr}_5 \circ \mathsf{ssdr}_{2,4} \circ \mathsf{snr}_3(u) = be, \qquad\qquad \mathsf{sh}_1 \circ \mathsf{sd}_2(\pi) = 12345.$$

Let us now show that the operations on legal strings are equivalent to those on signed permutations.

Theorem 1. *For any signed permutations π and π', with $\pi' = \phi_m \circ \ldots \phi_1(\pi)$, $\phi_i \in \mathsf{Sh} \cup \mathsf{Sd}$, let $u = \zeta(\pi)$ and $u' = \zeta(\pi')$. Then $\tau_0(u') = \psi_m \circ \tau_m \circ \ldots \psi_1 \circ \tau_1(u)$, where ψ_i is the equivalent of ϕ_i, and τ_i is a composition of snr operations.*

Proof. We will only prove the claim for $m = 1$. In its full generality, the claim can be proved by iterating the same argument.

Case 1: $\phi_1 \in \mathsf{Sh}$. Then π is of one of the four forms in Definition 2(i). Assume that $\pi = xp\ldots(p+i)\overline{(p+k)}\ldots\overline{(p+i+1)}y$, as the other cases are completely similar. Then $u = \zeta(x)p(p+1)(p+1)\ldots(p+i)(p+i)(p+i+1)\overline{(p+k+1)}\overline{(p+k)}\overline{(p+k)}\ldots\overline{(p+i+2)}\overline{(p+i+2)}\overline{(p+i+1)}\zeta(y)$ and

$$\mathsf{sh}_p(\pi) = xp\ldots(p+k)y,$$
$$\zeta(\mathsf{sh}_p(u)) = \zeta(x)p(p+1)(p+1)\ldots(p+k)(p+k)(p+k+1)$$
$$(\mathsf{sspr}_{p+i+1} \circ \mathsf{snr}_{p+1} \circ \mathsf{snr}_{p+i}\, \mathsf{snr}_{p+i+2} \circ \mathsf{snr}\, p+k)(u) = \zeta(x)p(p+k+1)\zeta(y) =$$
$$= (\mathsf{snr}_{p+1} \circ \ldots \circ \mathsf{snr}_{p+k})(\zeta(\mathsf{sh}_p(u))).$$

Case 2: $\phi_1 \in \mathsf{Sd}$. Then π is of one of the forms in Definition 2(ii). Assume that $\pi = x\,p\ldots(p+i)\,y\,(p-1)\,(p+i+1)\,z$, as the other cases are completely similar. Then $u = \zeta(x)p(p+1)(p+1)\ldots(p+i)(p+i)(p+i+1)\zeta(y)(p-1)p(p+i+1)(p+i+2)\zeta(z)$ and

$$\mathsf{sd}_p(\pi) = xy(p-1)p\ldots(p+i)(p+i+1)z,$$
$$\zeta(\mathsf{sd}_p(\pi)) = \zeta(xy)(p-1)pp\ldots(p+i+1)(p+i+1)(p+i+2)\zeta(z),$$
$$(\mathsf{ssdr}_{p,p+i+1} \circ \mathsf{snr}_{p+1} \circ \ldots \circ \mathsf{snr}_{p+i})(u) = \zeta(x)\zeta(y)(p-1)(p+i+2)\zeta(z) =$$
$$= (\mathsf{snr}_p \circ \ldots \circ \mathsf{snr}_{p+i+1})(\zeta(\mathsf{sd}_p(\pi))).$$

Thus we have shown that the operations on legal strings are equivalent to those for signed permutations. □

Now that we have defined legal strings and formalised simple operations on legal strings we will show why it was necessary to extend the definition of legal strings without markers given in [2]. Consider the definition of the sspr operation $\mathsf{sspr}_p(u_1pu_2\overline{p}u_3) = u_1\overline{u}_2u_3$. The operation sspr_p may only be applied if $|u_2| = 1$, i.e., it contains a single pointer or a single marker. As no markers are recorded in the definition given in [2], u_2 could in fact contain a pointer and a marker, thus making sspr_p inapplicable. The problem is illustrated in the following example.

Example 9. *Let $u = \overline{3}24234$ be a legal string with the markers removed. It would seem that sspr_2 should be applicable to u, as they are separated by only a single pointer. However, u can be obtained from the following signed permutation $\pi = \overline{2}413$. It is clear that sh_1 is not applicable to π because of MDS 4. Since the permutation-based model and the string-based one should be equivalent, this in turn implies that sspr_2 should not be applicable to u.*

3.4   Confluent Strategies on Legal Strings

It was shown in [9] that assembly strategies for a given signed permutation using simple operations are confluent: they are either all successful, or all unsuccessful and moreover, they lead to final results having the same structure. We will now show that the same applies on legal strings. Note that we have no need to define structure for legal strings, as they are in fact isomorphic and thus share the same characteristics and the same operations may be applied to both. Note also that the results of this section are simpler to prove and more general than the similar results in [9].

We will now show that assembly strategies on legal strings are confluent. Lemma 1 will show the case where one of the operations is an snr operation, Lemma 2 considers two sspr operations, and Lemma 3 considers two ssdr operations.

Lemma 1. *Let u be a legal string over Σ_n and $\phi, \psi \in \mathsf{Snr} \cup \mathsf{Sspr} \cup \mathsf{Ssdr}$ be two operations applicable to u. If $\phi \in \mathsf{Snr}$ or $\psi \in \mathsf{Snr}$, then $\phi \circ \psi(u) = \psi \circ \phi(u)$.*

Proof. The proof for this is straightforward as the pointers involved in an snr operation must be adjacent, and thus cannot overlap or affect any other pointers. □

Lemma 2. *Let u be a legal string over Σ_n and $\psi, \phi \in \mathsf{Sspr}$ be two operations applicable to u. Then either $\phi \circ \psi(u) = \psi \circ \phi(u)$, or $\psi(u) \equiv \phi(u)$.*

Proof. Let $\psi = \mathsf{sspr}_p$ and $\phi = \mathsf{sspr}_q$, for some $p \neq q$. If $pq\overline{pq} \not\leq u$ and $qp\overline{qp} \not\leq u$, then clearly $\mathsf{sspr}_q \circ \mathsf{sspr}_p(u) = \mathsf{sspr}_p \circ \mathsf{sspr}_q(u)$.

Now, if $pq\overline{pq} \leq u$ or $qp\overline{qp} \leq u$, then clearly applying one operation makes the other inapplicable, but if $\psi(u) = qq$ and $\phi(u) = pp$ then $\psi(u) \equiv \phi(u)$. □

Note, however, that after applying one of the sspr operations, an snr operation on the remaining pointer becomes available, giving us the following:

$$\mathsf{sspr}_p(u_1 pq\overline{pq} u_2) = u_1 \overline{qq} u_2,$$

$$\mathsf{sspr}_q(u_1 pq\overline{pq} u_2) = u_1 pp u_2,$$

$$\mathsf{snr}_{\overline{q}} \circ \mathsf{sspr}_p(u_1 pq\overline{pq} u_2) = u_1 u_2 = \mathsf{snr}_p \circ \mathsf{sspr}_q(u_1 pq\overline{pq} u_2),$$

for some legal strings u_1, u_2 over Σ_n.

Lemma 3. *Let u be a legal string over Σ_n and $\psi, \phi \in \mathsf{Ssdr}$ be two operations applicable to u. Then either $\phi \circ \psi(u) = \psi \circ \phi(u)$, or $\psi(u) \equiv \phi(u)$.*

Proof. Let $\psi = \mathsf{ssdr}_{p,q}$ and $\phi = \mathsf{ssdr}_{r,s}$, for some $p, q \neq r, s$. If they have a different signature, then clearly $\phi \circ \psi(u) = \psi \circ \phi(u)$. Assume then that $p, q, r, s \in \Sigma_n$. The cases when one or more of p, q, r, s are in $\overline{\Sigma}_n$ are completely similar.

Also, if $p, q \neq r, s$ then $\phi \circ \psi(u) = \psi \circ \phi(u)$, so let $q = r$ (the case when $p = s$ is similar). Thus either $pqsu_1 pqs \leq u$ or $sqpu_1 sqp \leq u$. Both operations cannot be applied, but $\mathsf{ssdr}_{p,q}(u)$ and $\mathsf{ssdr}_{q,s}(u)$ arrive at equivalent results. Indeed, $\mathsf{ssdr}_{p,q}(u) = u_2 su_1 su_3$, $\mathsf{ssdr}_{q,s}(u) = u_2 pu_1 pu_3$ and so, $\mathsf{ssdr}_{p,q}(u) \equiv \mathsf{ssdr}_{q,s}(u)$. □

We can now extend the results above for strategies using all three operations.

Theorem 2. *Let u be a legal string over Σ_n and $\phi, \psi \in \mathsf{Snr} \cup \mathsf{Sspr} \cup \mathsf{Ssdr}$ be two operations applicable to u. Then either $\phi \circ \psi(u) = \psi \circ \phi(u)$, or $\psi(u) \equiv \phi(u)$.*

Proof. Based on the previous three lemmata, we only need to prove the claim in the case $\phi \in \mathsf{Sspr}$, $\psi \in \mathsf{Ssdr}$. Now, if $\phi = \mathsf{sspr}_p$, $\psi = \mathsf{ssdr}_{q,r}$, and $p \neq q, r$, then $\phi \circ \psi(u) = \psi \circ \phi(u)$.

Assume then that $p = q$. Then in sspr_p, the occurrences of p must have different signatures, and in $\mathsf{ssdr}_{p,r}$, all occurrences of p and r must have the same signature, a contradiction. $\qquad\square$

Example 10. *Let $u = b2345623456e$.*

i. *Both $\mathsf{ssdr}_{2,3}$ and $\mathsf{ssdr}_{5,6}$ are applicable to u. Moreover, $\mathsf{ssdr}_{2,3} \circ \mathsf{ssdr}_{5,6}$ and $\mathsf{ssdr}_{5,6} \circ \mathsf{ssdr}_{2,3}$ are also applicable to u and*

$$\mathsf{ssdr}_{2,3} \circ \mathsf{ssdr}_{5,6}(u) = \mathsf{ssdr}_{5,6} \circ \mathsf{ssdr}_{2,3}(u) = b44e.$$

ii. *Both $\mathsf{ssdr}_{2,3}$ and $\mathsf{ssdr}_{3,4}$ are applicable to u. Moreover, applying either operation gives an equivalent result:*

$$\sigma(\mathsf{ssdr}_{2,3}(u)) = b456456e \equiv b256256e = \sigma(\mathsf{ssdr}_{3,4}(u)).$$

We now have the necessary information to show that the result for signed permutations described in [9] also applies on legal strings, namely that simple operations on legal strings are confluent.

Theorem 3. *Let u be a legal string over Σ_n and ϕ, ψ be two strategies for u. Then either ϕ and ψ are both sorting strategies for u, or they are both unsuccessful strategies. Moreover, $\phi(u) \equiv \psi(u)$.*

Proof. Assume that the claim of the theorem is not true and consider a legal string u of minimal length such that $\phi = \phi_1 \ldots \phi_k$ is a successful strategy for u, while $\psi = \psi_1 \ldots \psi_l$ is an unsuccessful one, $\phi_i, \psi_j \in \mathsf{Snr} \cup \mathsf{Sspr} \cup \mathsf{Ssdr}$, $1 \leq i \leq k$, $1 \leq j \leq l$.

It follows from Theorem 2 that either $\phi_k(u) \equiv \psi_l(u)$, or $\phi_k \circ \psi_l(u) = \psi_l \circ \phi_k(u)$. If they are equivalent, then $\phi_k(u)$ or $\psi_l(u)$ would be a smaller counterexample than u contradicting the minimality of u. In the latter case note that due to the minimality of u, it follows that $\phi_k(u)$ has only successful strategies and $\psi_l(\pi)$ has only unsuccessful strategies. Consequently, $\psi_l \circ \phi_k(\pi)$ has both successful and unsuccessful strategies, contradicting the minimality of u. $\qquad\square$

Example 11. *The legal string $u = 45\overline{767}eb2\overline{323}456$ has several sorting strategies. some of them are shown below.*

$$\phi_1(u) = \mathsf{snr}_6 \circ \mathsf{snr}_3 \circ \mathsf{sspr}_2 \circ \mathsf{sspr}_7 \circ \mathsf{ssdr}_{4,5}(u) = eb,$$

$$\phi_2(u) = \mathsf{snr}_4 \circ \mathsf{ssdr}_{5,6} \circ \mathsf{sspr}_7 \circ \mathsf{snr}_2 \circ \mathsf{sspr}_3(u) = eb,$$

$$\phi_3(u) = \mathsf{snr}_6 \circ \mathsf{ssdr}_{4,5} \circ \mathsf{sspr}_7 \circ \mathsf{snr}_3 \circ \mathsf{sspr}_2(u) = eb,$$

$$\phi_4(u) = \mathsf{snr}_4 \circ \mathsf{ssdr}_{5,6} \circ \mathsf{sspr}_7 \circ \mathsf{snr}_3 \circ \mathsf{sspr}_2(u) = eb.$$

Example 12. *The legal string $v = b234678e56782345$ has several unsuccessful strategies. Some of them are shown below.*

$$\psi_1(v) = \mathsf{ssdr}_{6,7} \circ \mathsf{ssdr}_{2,3}(v) = b48e5845,$$
$$\psi_2(v) = \mathsf{ssdr}_{7,8} \circ \mathsf{ssdr}_{3,4}(v) = b26e5625,$$
$$\psi_3(v) = \mathsf{ssdr}_{6,7} \circ \mathsf{ssdr}_{3,4}(v) = b28e5825,$$
$$\psi_4(v) = \mathsf{ssdr}_{2,3} \circ \mathsf{ssdr}_{7,8}(v) = b46e5645.$$

Note that $\psi_1(v) \equiv \psi_2(v) \equiv \psi_3(v) \equiv \psi_4(v)$.

References

1. Cavalcanti, A., Clarke, T.H., Landweber, L.: MDS_IES_DB: a database of macronuclear and micronuclear genes in spirotrichous ciliates. Nucleic Acids Research 33, 396–398 (2005)
2. Ehrenfeucht, A., Harju, T., Petre, I., Prescott, D.M., Rozenberg, G.: Computation in Living Cells: Gene Assembly in Ciliates. Springer, Heidelberg (2003)
3. Ehrenfeucht, A., Prescott, D.M., Rozenberg, G.: Computational aspects of gene (un)scrambling in ciliates. In: Landweber, L.F., Winfree, E. (eds.) Evolution as Computation, pp. 216–256. Springer, New York (2001)
4. Harju, T., Li, C., Petre, I., Rozenberg, G.: Parallelism in gene assembly. Natural Computing (2006)
5. Harju, T., Li, C., Petre, I., Rozenberg, G.: Complexity Measures for Gene Assembly. In: Tuyls, K., Westra, R., Saeys, Y., Nowé, A. (eds.) KDECB 2006. LNCS (LNBI), vol. 4366. Springer, Heidelberg (2007)
6. Harju, T., Li, C., Petre, I., Rozenberg, G.: Modelling simple operations for gene assembly. In: Chen, J., Jonoska, N., Rozenberg, G. (eds.) Nanotechnology: Science and Computation, pp. 361–376. Springer, Heidelberg (2006)
7. Landweber, L.F., Kari, L.: The evolution of cellular computing: Nature's solution to a computational problem. In: Proceedings of the 4th DIMACS Meeting on DNA-Based Computers, Philadelphia, PA, pp. 3–15 (1998)
8. Landweber, L.F., Kari, L.: Universal molecular computation in ciliates. In: Landweber, L.F., Winfree, E. (eds.) Evolution as Computation. Springer, New York (2002)
9. Langille, M., Petre, I.: Simple gene assembly is deterministic. Fundamenta Informaticae 72, 1–12 (2006)
10. Prescott, D.M., Ehrenfeucht, A., Rozenberg, G.: Molecular operations for DNA processing in hypotrichous ciliates. Europ. J. Protistology 37, 241–260 (2001)

On the Computational Power of Genetic Gates with Interleaving Semantics: The Power of Inhibition and Degradation

Nadia Busi[1] and Claudio Zandron[2]

[1] Dipartimento di Scienze dell'Informazione, Università di Bologna,
Mura A. Zamboni 7, I-40127 Bologna, Italy
busi@cs.unibo.it
[2] DISCo, Università di Milano-Bicocca,
via Bicocca degli Arcimboldi 8, I-20126, Milano, Italy
zandron@disco.unimib.it

Abstract. Genetic Systems are a formalism inspired by genetic regulatory networks, suitable for modeling the interactions between genes and proteins, acting as regulatory products. The evolution is driven by genetic gates: a new object (representing a protein) is produced when all activator objects are available in the system, and no inhibitor object is present. Activators are not consumed by the application of such a rule. Objects disappear because of degradation: each object is equipped with a lifetime, and the object decays when such a lifetime expires.

We investigate the computational expressiveness of Genetic Systems with interleaving semantics (a single action is executed in a computational step): we show that they are Turing equivalent by providing a deterministic encoding of Random Access Machines in Genetic Systems. We also show that the computational power strictly decreases when moving to Genetic Systems where either the degradation or the inhibition mechanism are absent.

1 Introduction

Most biological processes are regulated by networks of interactions between regulatory products and genes. To investigate the dynamical properties of these genetic regulatory networks, various formal approaches, ranging from discrete to stochastic and to continuous models, have been proposed (see [8] for a review).

In [6] we introduced *Genetic Systems*, a simple discrete formalism for the modeling of genetic networks, and we started an investigation of the ability of such a formalism to act as a computational device. We continue here along the lines of [6], by investigating the computational expressiveness obtained when equipping *Genetic Systems* with a different semantics w.r.t. the one used in [6].

Genetic Systems are based on the following ingredients: *genetic gates*, that are rules modeling the behaviour of genes, and *objects*, that represent proteins. Proteins both regulate the activity of a gene – by activating or inhibiting transcription – and represent the product of the activity of a gene.

E. Csuhaj-Varjú and Z. Ésik (Eds.): FCT 2007, LNCS 4639, pp. 173–186, 2007.
© Springer-Verlag Berlin Heidelberg 2007

A genetic gate is essentially a *contextual rewriting rule* consisting of three components: the set of activators, the set of inhibitors and the transcription product. A genetic gate is activated if the activator objects are present in the system, and all inhibitor objects are absent. The result of the application of a genetic gate rule is the production of a new object (without removing the activator objects from the system).

In biological systems, proteins can disappear in (at least) two ways (see, e.g., [1]): a protein can either decay because its lifetime is elapsed, or because it is neutralized by a repressor protein. To model the decaying process, we equip objects with a lifetime, which is decremented at each computational step. When the lifetime of an object becomes zero, the object disappears. In our model we represent both decaying and persistent objects: while the lifetime of a decaying object is a natural number, persistent objects are equipped with an infinite lifetime. The behaviour of repressor proteins is modeled through *repressor rules*, consisting of two components: the repressor object and the object to be destroyed. When both objects are present in the system, the rule is applied: the object to be destroyed disappears, while the repressor is not removed.

In [6] we investigated the computational power of Genetic Systems with the *maximal parallelism semantics* – all the rules that *can be* applied simultaneously, *must be* applied in the same computational step – and we gave a deterministic encoding of Random Access Machines (RAMs) [16], a well-known, deterministic Turing equivalent formalism. As the maximal parallelism semantics is a very powerful synchronization mechanism, we consider the semantics at the opposite side of the spectrum, namely, the *interleaving* (or *sequential*) *semantics*, where a single rule is applied in each computational step. The biological intuition behind this choice is that the cell contains a finite number of RNA polymerases, that are the enzymes that catalyze the transcription of genes: the case of interleaving semantics corresponds to the presence of a single RNA polymerase enzyme. Universality is not affected when moving to the interleaving semantics, as shown in [5], where a variant of the RAM encoding of [6] is provided. The encoding is deterministic and a RAM terminates if and only if its encoding terminates (i.e. no additional divergent or failed computations are added in the encoding). Hence, both existential termination (i.e., the existence of a terminated, or deadlocked, computation) and universal termination (i.e., all computations terminate) are undecidable in this case. Such an universality result is obtained by using several ingredients: persistent and decaying objects, repressor rules, positive and negative regulation. Here we investigate what happens when some of these ingredients are removed; we obtain that both decaying objects and negative regulation are needed to obtain deterministic encodings of RAMs. We also show that the existential termination is decidable for Genetic Systems without decaying objects. Hence, there exists no encoding of RAMs in Genetic Systems without decaying objects. This is shown by means of a safe Petri net with contextual (i.e., inhibitor and read) arcs, that has the same behaviour w.r.t. termination. Then, we show that universal termination is decidable for Genetic Systems without negative regulation; the proof is based on the theory of

Well-Structured Transition Systems [9]. Thus, there exists no deterministic encoding of RAMs in Genetic Systems without negative regulation.

The paper is organized as follows. In Section 2 we give basic definitions that will be used throughout the paper. The syntax and the semantics of Genetic Systems is presented in Section 3; we also show that Genetic Systems with interleaving semantics are Turing equivalent, by providing an encoding of RAMs (the details are reported in [5]). Then, we give some results concerning weaker Genetic Systems: in Section 4 we consider systems without decaying objects, and in section 5 we consider systems without inhibitors. Section 6 reports some conclusive remarks.

2 Basic Definitions

In this section we provide some basic definitions that will be used throughout the paper. With $I\!N$ we denote the set of natural numbers, whereas $I\!N_\infty$ denotes $I\!N \cup \{\infty\}$. We start with the definition of multisets and multiset operations.

Definition 1. *Given a set S, a* finite multiset *over S is a function $m : S \to I\!N$ such that the set $dom(m) = \{s \in S \,|\, m(s) \neq 0\}$ is finite. The* multiplicity *of an element s in m is given by the natural number $m(s)$. The set of all finite multisets over S, denoted by $\mathcal{M}_{fin}(S)$, is ranged over by m. A multiset m such that $dom(m) = \emptyset$ is called* empty. *The empty multiset is denoted by $\emptyset$. Given the multiset m and m', we write $m \subseteq m'$ if $m(s) \leq m'(s)$ for all $s \in S$ while $\oplus$ denotes their* multiset union: *$m \oplus m'(s) = m(s) + m'(s)$. The operator $\setminus$ denotes* multiset difference: *$(m \setminus m')(s) = if\ m(s) \geq m'(s)\ then\ m(s) - m'(s)\ else\ 0$.*

The set of parts of a set S is defined as $\mathcal{P}(S) = \{X \mid X \subseteq S\}$. Given a set $X \subseteq S$, with abuse of notation we use X to denote also the multiset

$$m_X(s) = \begin{cases} 1 & \text{if } s \in X \\ 0 & \text{otherwise} \end{cases}$$

We provide some basic definitions on strings, cartesian products and relations.

Definition 2. *A* string *over S is a finite (possibly empty) sequence of elements in S. Given a string $u = x_1 \ldots x_n$, the* length *of u is the number of occurrences of elements contained in u and is defined as follows: $|u| = n$. The empty string is denoted by λ. With S^* we denote the set of strings over S, and $u, v, w, \ldots$ range over S^*. Given $n \geq 0$, with S^n we denote the set of strings of length n over S. Given a string $u = x_1 \ldots x_n$, the multiset corresponding to u is defined as follows: for all $s \in S$, $m_u(s) = |\{i \mid x_i = s \land 1 \leq i \leq n\}|$. With abuse of notation, we use u to denote also m_u[1].*

Definition 3. *With $S \times T$ we denote the cartesian product of sets S and T, with $\times_n S$, $n \geq 1$, we denote the cartesian product of n copies of set S and with*

---

[1] In some cases we denote a multiset by one of its corresponding strings, because this permits to define functions on multisets in a more insightful way.

$\times_{i=1}^{n} S_i$ we denote the cartesian product of sets $S_1, \ldots, S_n$, i.e., $S_1 \times \ldots \times S_n$. The ith projection of $x = (x_1, \ldots, x_n) \in \times_{i=1}^{n} S_i$ is defined as $\pi_i(x) = x_i$, and lifted to subsets $X \subseteq \times_{i=1}^{n} S_i$ as follows: $\pi_i(X) = \{\pi_i(x) \mid x \in X\}$.

Given a binary relation R over a set S, with R^n we denote the composition of n instances of R, with R^+ we denote the transitive closure of R, and with R^* we denote the reflexive and transitive closure of R.

3   Genetic Systems

In this section, we present the definition of Genetic Systems (G Systems for short) and the definitions which we need to describe their functioning. To this aim, given a set X, we define $\mathcal{R}_X = \mathcal{P}(X) \times \mathcal{P}(X) \times (X \times I\!N_\infty)$.

Definition 4. *A* Genetic System *is a tuple* $G = (V, GR, RR, w_0)$ *where*

1. V *is a finite alphabet whose elements are called* objects;
2. GR *is a finite multiset[2] over* $\mathcal{R}_V$ *of genetic gates over* V; *these gates are of the forms* $u_{act}, \neg u_{inh} :\rightarrow (b, t)$ *where* $u_{act} \cap u_{inh} = \emptyset$. $u_{act} \subseteq V$ *is the* positive regulation (activation)[3], $u_{inh} \subseteq V$ *the* negative regulation (inhibition), $b \in V$ *the transcription of the gate[4] and* $t \in I\!N_\infty$ *the duration of object* b;
3. RR *is a finite set[5] of* repressor rules *of the form* $(rep : b \rightarrow)$ *where* $rep, b \in V$ *and* $rep \neq b$;
4. w_0 *is a string over* $V \times I\!N_\infty$, *representing the multiset of objects contained in the system at the beginning of the computation. The objects are of the form* (a, t), *where* a *is a symbol of the alphabet* V *and* $t > 0$ *represents the decay time of that object.*

We say that a gate is *unary* if $|u_{act} \oplus u_{inh}| = 1$. The multiset represented by w_0 constitutes the *initial state* of the system.

A transition between states is governed by an application of the transcription rules (specified by the genetic gates) and of the repressor rules. The gate $u_{act}, \neg u_{inh} :\rightarrow (b, t)$ can be activated if the current state of the system contains enough free activators and no free inhibitors. If the gate is activated, the regulation objects (activators) in the set u_{act} are bound to such a gate, and they cannot be used for activating any other gate in the same maximal parallelism evolution step. In other words, the gate $u_{act}, \neg u_{inh} :\rightarrow (b, t)$ in a state formed by a multiset of (not yet bound) objects w can be activated if u_{act} is contained in w and no object in u_{inh} appears in w; if the gate performs the transcription,

---

[2] Here we use multisets of rules, instead of sets, for compatibility with the definition in [6].

[3] We consider sets of activators, meaning that a genetic gate is never activated by more than one instance of the same protein.

[4] Usually the expression of a genetic gate consists of a single protein.

[5] We use sets of rules, instead of multisets, because each repressor rule denotes a chemical law; hence, a repressor rule can be applied for an unbounded number of times in each computational step.

then a new object (b, t) is produced. Note that the objects in u_{act} and u_{inh} are not consumed by the transcription operation, but will be released at the end of the operation and (if they do not disappear because of the decay process) they can be used in the next evolution step. Each object starts with a decay number, which specify the number of steps after which this object disappears. The decay number is decreased after each parallel step; when it reaches the value zero, the object disappears. If the decay number of an object is equal to ∞, then the object is persistent and it never disappears. The repressor rule $(rep : b \rightarrow)$ is activated when both the repressor rep and the object b are present, and the repressor rep destroys the object b. We adopt the following notation for gates. The activation and inhibition sets are denoted by one of the corresponding strings, i,e, $a, b, \neg c :\rightarrow (c, 5)$ denotes the gate $\{a, b\}, \neg\{c\} :\rightarrow (c, 5)$. If either the activation or the inhibition is empty then we omit the corresponding set, i.e., $a :\rightarrow (b, 3)$ is a shorthand for the gate $\{a\}, \neg\emptyset :\rightarrow (b, 3)$. The *nullary gate* $\emptyset, \neg\emptyset :\rightarrow (b, 2)$ is written as $:\rightarrow (b, 2)$.

3.1   Configurations, Reaction Relation and Interleaving Computational Step

Once defined Genetic Systems, we are ready to describe their functioning. A transition between two states of the system is governed by an application of the transcription rules (specified by the genetic gates) and of the repressor rules. Different semantics can be adopted, depending on the number of rules that are applied in each computational step, and on the way in which the set of rules to be applied is chosen. In [6] we adopted the so-called *maximal parallelism semantics*: all the rules that *can be* applied simultaneously, *must be* applied in the same computational step. In this paper we consider the semantics at the opposite side of the spectrum, i.e., the *interleaving* (or *sequential*) *semantics*. In this case, at each computational step only a single rule is applied. In particular, for deterministic systems, at each computational step there is at most one rule that can be applied. For non–deterministic systems, at each computational step one rule, among all applicable ones, is chosen to be applied.

We give now the definitions for configuration, reaction relation, and heating and decaying function. A configuration represents the current state of the system, consisting in the (multi)set of objects currently present in the system.

Definition 5. *Let* $G = (V, GR, RR, w_0)$ *be a Genetic System. A* configuration *of* G *is a multiset* $w \in \mathcal{M}_{fin}(V \times I\!N_\infty)$. *The* initial configuration *of* G *is the multiset* w_0.

The activation of a genetic gate is formalized by the notion of reaction relation. In order to give a formal definition we need the function $obj : (V \times I\!N_\infty)^* \rightarrow V^*$, defined as follows. Assume that $(a, t) \in (V \times (I\!N_\infty))$ and $w \subseteq (V \times (I\!N_\infty))^*$. Then, $obj(\lambda) = \lambda$ and $obj((a, t)w) = a \, obj(w)$.

We also need to define a function DecrTime which is used to decrement the decay time of objects, destroying the objects which reached their time limit.

Definition 6. *The function* $DecrTime : (V \times I\!N_\infty)^* \to (V \times I\!N_\infty)^*$ *is defined as follows:* $DecrTime(\lambda) = \lambda$ *and*

$$DecrTime((a,t)w) = \begin{cases} (a,t-1)DecrTime(w) & \text{if } t > 1 \\ DecrTime(w) & \text{if } t = 1 \end{cases}$$

We are now ready to give the notion of *reaction relation*.

Definition 7. *Let* $G = (V, GR, RR, w_0)$ *be a Genetic System. The reaction relation* $\mapsto$ *over* $\mathcal{M}_{fin}(V \times I\!N_\infty) \times \mathcal{M}_{fin}(V \times I\!N_\infty)$ *is defined as follows:* $w \mapsto w'$ *iff one of the following holds:*

- *there exist* $u_{act}, \neg u_{inh} :\to (b,t) \in GR$ *and* $w_{act} \subseteq w$ *such that*
  - $u_{inh} \cap dom(obj(w)) = \emptyset$
  - $m_{obj(w_{act})} = m_{u_{act}}$[6]
  - $w' = DecrTime(w) \oplus \{(b,t)\}$
- *there exists* $(rep : b \to) \in RR$ *such that*
  - *there exist* $t_{rep}, t_b \in I\!N_\infty$ *such that* $\{(rep, t_{rep}), (b, t_b)\} \subseteq w$
  - $w' = DecrTime(w) \setminus \{(rep, t_{rep}), (b, t_b)\} \oplus DecrTime((rep, t_{rep}))$

Now we are ready to define the interleaving computational step $\rightsquigarrow$:

Definition 8. *Let* $G = (V, GR, RR, w_0)$ *be a Genetic System. The* interleaving computational step $\rightsquigarrow$ *over configurations of* G *is defined as follows:* $w_1 \rightsquigarrow w_2$ *iff one of the following holds:*

- *either* $w_1 \mapsto w_2$, *or*
- $w_1 \not\mapsto$ *and there exists* $(a,t) \in w_1$ *such that* $t \neq \infty$ *and* $w_2 = DecrTime(w_1)$.

We say that a configuration w *is* terminated *if no interleaving step can be performed, i.e.,* $w \not\rightsquigarrow$. *The set of configurations* reachable *from a given configuration* w *is defined as* $Reach(w) = \{w' \mid w \rightsquigarrow^* w'\}$. *The set of* reachable configurations *in* G *is* $Reach(w_0)$.

Note that a computational step can consist either in the application of a rule, or in the passing of one time unit, in case the system is deadlocked (i.e., no rule can be applied). We also need computational steps of the second kind, as it may happen that the computation restarts after some object – acting as inhibitor for some rule – decays. Consider, e.g., a system with a negative gate $\neg b :\to (a, 3)$ and a positive gate $a :\to (a, 3)$, reaching a configuration containing only the object $(b, 2)$. The system cannot evolve until 2 time units have elapsed; then, an object $(a, 3)$ is produced and, because of the positive gate, the system will never terminate.

To illustrate the difference between the interleaving and the maximal parallelism semantics, consider a system with gates $\neg d, a :\to b$ and $\neg b, c :\to d$ and with initial state ac. According to the maximal parallelism semantics only the

---

[6] We recall that $m_{u_{act}}$ is the multiset containing exactly one occurrence of each object in the set u_{act}. Hence, the operator $=$ is intended here to be the equality operator on multisets.

following step can be performed: $ac \Rightarrow acbd \not\Rightarrow$, where $\Rightarrow$ is a maximal parallelism computational step. On the other hand, according to the interleaving semantics also the following sequence of steps can be performed: $ac \rightsquigarrow acb \rightsquigarrow acbb \rightsquigarrow acbbb \rightsquigarrow \ldots$.

Now we introduce some notions concerned with divergence and termination that will be useful in the following.

Definition 9. *Let* $G = (V, GR, RR, w_0)$ *be a Genetic System. We say that a configuration* w *has a divergent computation (or infinite computation) if there exists an infinite sequence of configurations* $w_1, \ldots, w_i, \ldots$ *such that* $w = w_1$ *and* $\forall i \geq 1 : w_i \rightsquigarrow w_{i+1}$. *We say that a configuration* w *universally terminates if it has no divergent computations. We say that* w *is deterministic iff for all* w', w'': *if* $w \rightsquigarrow w'$ *and* $w \rightsquigarrow w''$ *then* $w' = w''$. *We say that* w *has a terminating computation (or a deadlock) if there exists* w' *such that* $w \rightsquigarrow^* w'$ *and* $w \not\rightsquigarrow$. *The system* G *satisfies the universal termination property if* w_0 *has no divergent computations. The system* G *satisfies the existential termination property if* w_0 *has a deadlock. Note that the existential termination and the universal termination properties are equivalent on deterministic systems.*

3.2 Expressiveness of Genetic Systems with Interleaving Semantics

In [6] we showed that Genetic Systems with maximal parallelism semantics are Turing powerful. We strengthen the result by showing that interleaving semantics is enough to get Turing equivalence. As maximal parallelism semantics turns out to be a very powerful synchronization mechanism, often the expressiveness of a formalism is increased when moving from the interleaving to the maximal parallelism semantics; see, e.g., [10] for some examples of this phenomenon in Membrane Systems [13] and [3] for a fragment of the Brane Calculus [7].

In order to show that Genetic Systems with interleaving semantics are Turing powerful, we provide a variant of the encoding of Random Access Machines (RAMs) [16] proposed in [6]. Such an encoding is deterministic and it enjoys the following property: a RAM terminates if and only if its encoding terminates (i.e. no additional divergent or failed computations are added in the encoding).

The detailed description of the encoding, preceded by a description of RAMs are given in the extended version of this paper [5].

4 Genetic Systems Without Decaying Objects

In the result we presented in the previous section, we proved Turing equivalence of Genetic Systems by using several ingredients: persistent and decaying objects, repressor rules, positive and negative regulation. Are all such ingredients needed in order to obtain such a result? In this section, we show that existential termination is decidable for Genetic Systems without decaying objects. As a consequence, there exist no (deterministic or weak) encoding of RAMs in Genetic systems without decaying objects. This is shown by constructing a safe Petri net with contextual (i.e., inhibitor and read) arcs, that has the same behaviour w.r.t. termination.

4.1 Contextual P/T Nets

We recall Place/Transition nets (see, e.g., [15]) extended with contextual (i.e., inhibitor and read) arcs (see, e.g., [12,2]). Read (resp. inhibitor) arcs do not modify the contents of the places of the net, but permit to test respectively for presence (resp. absence) of a token in a place for a transition to fire. Here we provide a characterization of this model which is convenient for our aims.

Definition 10. *A Contextual P/T net is a pair* (S, T) *where* S *is the set of places and* $T \subseteq \mathcal{M}_{fin}(S) \times \mathcal{P}(S) \times \mathcal{P}(S) \times \mathcal{M}_{fin}(S)$ *is the set of transitions.*

Finite multisets over the set S *of places are called* markings. *Given a marking* m *and a place* s, *we say that the place* s *contains* $m(s)$ *tokens. A Contextual P/T net is finite if both* S *and* T *are finite.*

A Contextual P/T system is a triple $N = (S, T, m_0)$ *where* (S, T) *is a P/T net and* m_0 *is the* initial marking.

A transition $t = (Pre, Read, Inhib, Post)$ *is usually written in the following form:* $(Pre; Read; Inhib) \rightarrow Post$. *The marking* Pre, *denoted by* $^{\bullet}t$, *is called the* preset *of* t *and represents the tokens to be* consumed; *the marking* $Post$, *denoted by* $t^{\bullet}$, *is called the* postset *of* t *and represents the tokens to be* produced. *The set* $Read$, *denoted with* $\hat{t}$, *is called the* contextual set *of* t *and represents the tokens to be* tested for presence; *the set* $Inhib$, *denoted with* $^{\circ}t$, *is called the* inhibitor set *of* t *and represents the tokens to be* tested for absence.

A transition t *is enabled at* m *if* $^{\bullet}t \oplus \hat{t} \subseteq m$ *and* $dom(m) \cap {}^{\circ}t = \emptyset$.

The execution of a transition t *enabled at* m *produces the marking* $m' = (m \setminus {}^{\bullet}t) \oplus t^{\bullet}$. *This is written as* $m[t\rangle m'$ *or simply* $m[\rangle m'$ *when the transition* t *is not relevant. We use* σ, τ *to range over sequences of transitions; the empty sequence is denoted by* ε; *let* $\sigma = t_1, \dots, t_n$, *we write* $m[\sigma\rangle m'$ *to mean the firing sequence* $m[t_1\rangle \cdots [t_n\rangle m'$.

A marking m' *is* reachable *from a marking* m *if there exists a sequence of transitions* σ *such that* $m[\sigma\rangle m'$. *The set of markings reachable from* m *is denoted by* $[m\rangle$. *We say that the marking* m *is reachable in the Contextual P/T system* N *if* $m \in [m_0\rangle$ *(i.e., if* m *is reachable from the initial marking).*

A deadlock *is a marking* m *such that, for all* $t \in T$, $\neg m[t\rangle$. *We say that the Contextual P/T system* N *has a deadlock if there exists a deadlock* $m \in [m_0\rangle$.

A Contextual P/T system is safe if, in all reachable markings, each place contains at most one token.

Definition 11. *A Contextual P/T system* $N = (S, T, m_0)$ *is* safe *if, for all* $m \in [m_0\rangle$ *and for all* $s \in S$, $m(s) \leq 1$.

Clearly, for safe, finite systems the set $[m_0\rangle$ of reachable markings is finite; hence, both existential and universal termination are decidable for such a class of nets.

4.2 Mapping Genetic Systems Without Degradation on Safe Contextual P/T Systems

Given a Genetic System without degradation, we show how to construct a corresponding P/T system with the same behaviour w.r.t. existential termination.

More precisely, given a Genetic System G, we construct a safe Contextual P/T system $Net(G)$ satisfying the following property: G has a deadlock if and only if $Net(G)$ has a deadlock.

The net has a place a for each object (a, ∞) of the system; place a contains a token iff there is at least one occurrence of object (a, ∞) in the system. Given, e.g., a genetic gate $a, \neg b :\to (c, \infty)$, the net has the following transitions: $(\emptyset; a; bc) \to c$ and $(\emptyset; ac; b) \to \emptyset$. For each repressor rule $repr : c \to$ there is the following transition: $(c; repr; emptyset) \to \emptyset$. The idea is the following: if a transition corresponding to a repressor rule $(c; repr; emptyset) \to \emptyset$ fires, this corresponds to the fact that all the occurrences of c are destroyed. The function dec maps configurations of the system to markings of the net, and is defined as $dec(w) = \pi_1(dom(w))$. The idea is the following: a place a in the net contains one token iff the corresponding configuration contains at least one occurrence of object a; on the other hand, if no object of kind a is contained in the configuration, then the place a is empty in the corresponding marking.

Definition 12. *Let $G = (V, GR, RR, w_0)$ be a Genetic System. The function $dec : \mathcal{M}_{fin}(V \times I\!N_\infty) \to \mathcal{P}(V)$ is defined as follows: $dec(w) = \pi_1(dom(w))$ for all configurations w.*

The Contextual P/T system $Net(G) = (S, T, m_0)$ is defined as follows:
$S = V$
$T = \{(\emptyset; u_{act}; u_{inhib} \cup \{b\}) \to \{b\} \mid (u_{act}, \neg u_{inh} :\to (b, \infty)) \in GR\} \cup$
$\qquad \{(\emptyset; u_{act} \cup \{b\}; u_{inhib}) \to \emptyset \mid (u_{act}, \neg u_{inh} :\to (b, \infty)) \in GR\} \cup$
$\qquad \{(b; rep; \emptyset) \to \emptyset \mid (rep : b \to) \in RR\}$
$m_0 = dec(w_0)$

Proposition 1. *Let $G = (V, GR, RR, w_0)$ be a Genetic System. Then the Contextual P/T system $Net(G)$ is safe and finite.*

Lemma 1. *Let $G = (V, GR, RR, w_0)$ be a Genetic System and w, w' be configurations of G. If $w \rightsquigarrow w'$ then either $dec(w) = dec(w')$ or $dec(w)[t\rangle dec(w')$.*

Lemma 2. *Let $G = (V, GR, RR, w_0)$ be a Genetic System, w a configuration of G and m a marking of $Net(G)$. If $dec(w)[t\rangle m'$ then there exists w' such that $w \rightsquigarrow^+ w'$ and $m = dec(w')$.*

Theorem 1. *Let $G = (V, GR, RR, w_0)$ be a Genetic System and w, w' be configurations of G. If $w \rightsquigarrow^* w' \not\rightsquigarrow$ then there exists a sequence of transitions σ such that $dec(w)[\sigma\rangle dec(w')$ and $dec(w')$ is a deadlock.*

Theorem 2. *Let $G = (V, GR, RR, w_0)$ be a Genetic System, w a configuration of G and m' a marking of $Net(G)$. If $dec(w)[\sigma\rangle m'$ and m' is a deadlock then there exists a configuration w' such that $w \rightsquigarrow^* w' \not\rightsquigarrow$ and $dec(w') = m'$.*

Corollary 1. *Let $G = (V, GR, RR, w_0)$ be a Genetic System. G has a deadlock if and only if $Net(G)$ has a deadlock.*

We notice that this approach cannot be trivially extended to the maximal parallelism semantics, because in this case the number of occurrences of an object does matter. Consider, e.g., the system with rules $\neg d, a :\to b$ and $\neg b, a :\to d$. In this case, the initial states a and aa lead to different evolutions: from a an infinite computation can start, but from aa only a deadlocked state can be reached.

5   Genetic Systems Without Inhibitors

In this section we show that universal termination is decidable for Genetic Systems without inhibitors. The decidability proof is based on the theory of Well-Structured Transition Systems [9]: the existence of an infinite computation starting from a given state is decidable for finitely branching transition systems, provided that the set of states can be equipped with a well-quasi-ordering, i.e., a quasi-ordering relation which is compatible with the transition relation and such that each infinite sequence of states admits an increasing subsequence. After recalling the part of theory of Well-Structured Transition Systems needed for our aim, we show that the condition above is fulfilled by Genetic systems.

5.1   Well-Structured Transition Systems

We start by recalling some basic definitions and results from [9], concerning well-structured transition systems, that will be used in the following.

A *quasi-ordering* (qo) is a reflexive and transitive relation.

Definition 13. *A well-quasi-ordering (wqo) is a quasi-ordering $\leq$ over a set X such that, for any infinite sequence $x_0, x_1, x_2, \ldots$ in X, there exist indexes $i < j$ such that $x_i \leq x_j$.*

Note that, if $\leq$ is a wqo, then any infinite sequence $x_0, x_1, x_2, \ldots$ contains an infinite increasing subsequence $x_{i_0}, x_{i_1}, x_{i_2}, \ldots$ (with $i_0 < i_1 < i_2 < \ldots$).

Transition systems can be formally defined as follows.

Definition 14. *A transition system is a structure $TS = (S, \to)$, where S is a set of states and $\to \subseteq S \times S$ is a set of transitions. We write $Succ_\to(s)$ to denote the set $\{s' \in S \mid s \to s'\}$ of immediate successors of $s \in S$.*

TS is finitely branching if $\forall s \in S : Succ(s)$ is finite. We restrict to finitely branching transition systems.

Well-structured transition systems, defined as follows, provide the key tool to decide properties of computations.

Definition 15. *A well-structured transition system (with strong compatibility) is a transition system $TS = (S, \to)$, equipped with a quasi-ordering $\leq$ on S, also written $TS = (S, \to, \leq)$, such that the following two conditions hold:*

1. **well-quasi-ordering:** $\leq$ *is a well-quasi-ordering, and*
2. **strong compatibility:** $\leq$ *is (upward) compatible with $\to$, i.e., for all $s_1 \leq t_1$ and all transitions $s_1 \to s_2$, there exists a state t_2 such that $t_1 \to t_2$ and $s_2 \leq t_2$.*

The following theorem (a special case of a result in [9]) will be used to obtain our decidability result.

Theorem 3. *Let $TS = (S, \rightarrow, \leq)$ be a finitely branching, well-structured transition system with decidable $\leq$ and computable $Succ$. The existence of an infinite computation starting from a state $s \in S$ is decidable.*

To show that the quasi-ordering relation we will define on Genetic systems is a well-quasi-ordering we need the following result on well-quasi-ordering relations for multisets.

Lemma 3. *Let S be a finite set. The relation $\subseteq$ is a wqo over $\mathcal{M}_{fin}(S)$.*

5.2   Decidability of Universal Termination for Genetic Systems Without Inhibition

We show that it is possible to equip the set of reachable configurations of a Genetic System with a well-quasi-ordering relation, compatible with $\rightsquigarrow$, that permits to fulfill the conditions of Theorem 3. The idea is to use Lemma 3 to show that the multiset inclusion $\subseteq$ is a wqo over (a superset of) the set of reachable configurations. Two problems need to be solved:

- The relation $\subseteq$ is not compatible with $\rightsquigarrow$. Consider, e.g., the Genetic System containing only the repressor rule $(rep : a \rightarrow)$, and the two related configurations $(a, 5) \subseteq (a, 5)(rep, 2)$. The only move that can be performed by the first configuration is the passing of time, i.e., $(a, 5) \rightsquigarrow (a, 4)$, whereas for the second configuration the only move is the application of the repressor rule, i.e., $(a, 5)(rep, 2) \rightsquigarrow (rep, 1)$, and $(a, 4) \not\subseteq (rep, 1)$. However, we note that for Genetic Systems with positive regulation only the passing of time does not influence the overall behaviour of the system. Hence, the interleaving computational step $\rightsquigarrow$ is equivalent to the reaction relation $\mapsto$ w.r.t. the terminating behaviour of the system.
- We cannot take $\mathcal{M}_{fin}(V \times I\!N_\infty)$ as a superset of the reachable configurations of a Genetic System, because the set $V \times I\!N_\infty$ is infinite and the Lemma cannot be applied. However, we show that there exists an upper bound for the decay time of all nonpersistent objects belonging to a reachable configuration. This fact permits to define a finite set, such that all reachable configurations of a Genetic System turn out to be finite multisets over such a finite set.

The following definition turns out to be useful to solve the above problems. The maximum decay time of decaying objects belonging to a configuration, produced by a genetic gate and belonging to a Genetic System is defined as follows:

Definition 16. *Let $G = (V, GR, RR, w_0)$ be a Genetic System and w be a configuration of G. The function $maxtime : (V \times I\!N_\infty) \rightarrow I\!N$ is defined on configurations as follows:*

$$maxtime(\varepsilon) \quad = 0$$
$$maxtime(w(a, t)) = \begin{cases} maxtime(w) & \text{if } t = \infty \\ max\{t, maxtime(w) & \text{otherwise} \end{cases}$$

The function maxtime is defined on a genetic gate $gg = u_{act}, \neg u_{inib} :\rightarrow (b, t)$ as follows:

$$maxtime(u_{act}, \neg u_{inib} :\rightarrow (b, t)) = \begin{cases} 0 & \text{if } t = \infty \\ t & \text{otherwise} \end{cases}$$

The function maxtime is defined on the Genetic System G as follows:

$$maxtime(G) = max\{maxtime(w_0), max\{maxtime(gg) \mid gg \in GR\}\}$$

The first problem is solved by considering the relation $\mapsto$ instead of $\rightsquigarrow$ for Genetic Systems with positive regulation. The following proposition is useful to show that $\mapsto$ and $\rightsquigarrow$ are equivalent w.r.t. the termination of the system.

Proposition 2. *Let $G = (V, GR, RR, w_0)$ be a Genetic System and w, w' configurations of G. If $w \not\mapsto$ and $w \rightsquigarrow w'$ then $w' \not\mapsto$. If $w \not\mapsto$ and $w \rightsquigarrow^n w'$ then $n \leq maxtime(w)$.*

Corollary 2. *Let $G = (V, GR, RR, w_0)$ be a Genetic System. There exist $\{w_i \mid 1 \leq i \leq n\}$ such that $w_0 \rightsquigarrow w_1 \rightsquigarrow \ldots w_i \rightsquigarrow \ldots$ if and only if there exist $\{w'_i \mid 1 \leq i \leq n\}$ such that $w_0 \mapsto w'_1 \mapsto \ldots w'_i \mapsto \ldots$.*

Now we tackle the second problem. The decay time of the nonpersistent objects in a reachable configuration is bounded by the maximum decay time of a Genetic System:

Proposition 3. *Let $G = (V, GR, RR, w_0)$ be a Genetic System. For all reachable configurations w, for all $(a, t) \in dom(w)$, if $t \neq \infty$ then $t \leq maxtime(G)$.*

Hence, all the reachable configurations of a Genetic System turn out to be finite multisets over the finite set $V \times \{0, \ldots, maxtime(G), \infty\}$

Corollary 3. *Let $G = (V, GR, RR, w_0)$ be a Genetic System. For all configurations w, $w_0 \mapsto^* w$ then $w \in \mathcal{M}_{fin}(V \times \{0, \ldots, maxtime(G), \infty\})$.*

We are ready to build the transition system corresponding to a Genetic System:

Definition 17. *The transition system corresponding to a Genetic System $G = (V, GR, RR, w_0)$ is $TS(G) = (\mathcal{M}_{fin}(V \times \{0, \ldots, maxtime(G), \infty\}), \mapsto, \subseteq)$.*

To apply Theorem 3 we need to show that $TS(G)$ is finitely branching:

Proposition 4. *Let $G = (V, GR, RR, w_0)$ be a Genetic System and w be a configuration of G. The set $\{w' \mid w \mapsto w'\}$ is finite.*

Now we are ready to show that the conditions of Theorem 3 are fulfilled:

Theorem 4. *Let $G = (V, GR, RR, w_0)$ be a Genetic System. Then $TS(G)$ is a finitely branching, well-structured transition system with decidable $\leq$ and computable Succ.*

Thus, universal termination is decidable for Genetic Systems without inhibitors. Note that, if we add inhibition, the above result does not hold because the ordering relation $\subseteq$ cannot be compatible with $\leadsto$. (Consider, e.g., the system with genetic gate $a, \neg b :\rightarrow (c, \infty)$ and the two related configurations $(a, \infty) \subseteq (a, \infty)(b, \infty)$).

We showed that universal termination is decidable, hence there exist no deterministic encoding of RAMs in Genetic Systems without inhibitors. We leave as an open problem the existence of a nondeterministic encoding of RAMs. Note that the results of Section 4 cannot be applied. In Section 4 we consider systems without decaying objects, and such systems satisfy the following property (P): given a repressor rule $(rep : b \rightarrow)$, once the repressor object rep is produced, it can be used to remove all the occurrences of (b, ∞) in the system. Clearly, such a rule is not satisfied by systems with decaying objects, because the repressor object rep can decay before it has consumed all occurrences of b (consider, e.g., by the configuration $(rep, 1)(b, \infty)(b, \infty)$. We mimicked the behavior of a Genetic System without decaying objects with a safe net, where each repressor rule $(rep : b \rightarrow)$ is either mimicked by a transition that removes the token in place b (corresponding to the fact that *all* the occurrences of (b, ∞) in the corresponding configuration of the system are removed, by a sequence of computational steps), or by no transition (corresponding to the fact that at least one occurrence of (b, ∞) is left in the system after the application of hte repressor rule). In presence of decaying objects, the transition of the net – that mimicks the fact that all objects of kind (b, t) in the system are removed – may be not reproducible in the corresponding Genetic System, because property (P) may not hold.

6 Conclusion

We investigated the computational expressiveness of Genetic Systems, a formalism modeling the interactions occurring between genes and regulatory products.

A study of the expressiveness of rewriting rules inspired by genetic networks have been carried out in [4], in the context of Membrane Systems [13]. The result presented in [4] is incomparable with the result presented in this paper, because the semantics of the rules are different (in this paper, the modeling of repressors is more faithful to the biological reality, a more abstract semantics for genetic gates is used, and interleaving semantics instead of maximal parallelism is adopted), and because the result in [4] crucially depends on the use of membranes, permitting to localize to a specific compartment the objects and the rules, and of rules modeling the movement of objects across membranes. While both approaches are inspired by DNA, Genetic Systems turn out to be different from *DNA computing* (see, e.g., [14] for a survey), where the basic ingredients are strings, representing DNA strands, that evolve through the splicing operation.

In the result we presented in section 4, several ingredients are used to achieve Turing equivalence: the use of both persistent and decaying objects, repressor rules, positive and negative regulation. We showed that both Genetic Systems with persistent objects only and Genetic Systems with positive regulation cannot provide a deterministic encoding of RAMs. Regarding systems without repressor rules, we conjecture that existential termination is decidable, hence it is not possible to provide an encoding of RAMs in such a class of systems. The idea is to provide a mapping of Genetic Systems without repressor rules on safe Contextual Petri Net, which preserves the existence of a deadlock, using a technique similar to the one employed in Section 4.

References

1. Blossey, R., Cardelli, L., Phillips, A.: A Compositional Approach to the Stochastic Dynamics of Gene Networks. In: Priami, C., Cardelli, L., Emmott, S. (eds.) Transactions on Computational Systems Biology IV. LNCS (LNBI), vol. 3939. Springer, Heidelberg (2006)
2. Busi, N., Pinna, G.M.: A Causal Semantics for Contextual P/T nets. In: Proc. ICTCS'95, pp. 311–325. World Scientific, Singapore (1995)
3. Busi, N.: On the Computational Power of the Mate/Bud/Drip Brane Calculus: Interleaving vs. Maximal Parallelism. In: Freund, R., Păun, G., Rozenberg, G., Salomaa, A. (eds.) WMC 2005. LNCS, vol. 3850. Springer, Heidelberg (2006)
4. Busi, N., Zandron, C.: Computing with Genetic Gates, Proteins and Membranes. In: Hoogeboom, H.J., Păun, G., Rozenberg, G., Salomaa, A. (eds.) WMC 2006. LNCS, vol. 4361. Springer, Heidelberg (2006)
5. Busi, N., Zandron, C.: On the computational power of Genetic Gates with interleaving semantics: The power of inhibition and degradation, www.cs.unibo.it/%7Ebusi/FCT07long.pdf
6. Busi, N., Zandron, C.: Computing with Genetic Gates. In: Proc. Third Conference on Computability in Europe. Computation and Logic in the Real World (CiE 2007). LNCS, vol. 4497. Springer, Heidelberg 2007 (to appear)
7. Cardelli, L.: Brane Calculi - Interactions of Biological Membranes. In: Danos, V., Schachter, V. (eds.) CMSB 2004. LNCS (LNBI), vol. 3082, pp. 257–280. Springer, Heidelberg (2005)
8. De Jong, H.: Modeling and Simulation of Genetic Regulatory Systems: A Literature Review. Journal of Computatonal Biology 9, 67–103 (2002)
9. Finkel, A., Schnoebelen, Ph.: Well-Structured Transition Systems Everywhere! Theoretical Computer Science 256, 63–92 (2001)
10. Freund, R.: Asynchronous P Systems and P Systems Working in the Sequential Mode. In: Mauri, G., Păun, G., Pérez-Jiménez, M.J., Rozenberg, G., Salomaa, A. (eds.) WMC 2004. LNCS, vol. 3365, pp. 36–62. Springer, Heidelberg (2005)
11. Minsky, M.L.: Computation: Finite and Infinite Machines. Prentice-Hall, Englewood Cliffs (1967)
12. Montanari, U., Rossi, F.: Contextual Nets. Acta Inform. 32(6), 545–596 (1995)
13. Păun, G.: Membrane Computing. An Introduction. Springer, Heidelberg (2002)
14. Păun, G., Rozenberg, G., Salomaa, A.: DNA Computing. New computing paradigms. Springer, Heidelberg (1998)
15. Reisig, W.: Petri nets: An Introduction. EATCS Monographs in Computer Science. Springer, Heidelberg (1985)
16. Shepherdson, J.C., Sturgis, J.E.: Computability of Recursive Functions. Journal of the ACM 10, 217–255 (1963)

On Block-Wise Symmetric Signatures for Matchgates

Jin-Yi Cai[1] and Pinyan Lu[2,*]

[1] Computer Sciences Department, University of Wisconsin
Madison, WI 53706, USA
jyc@cs.wisc.edu
[2] Department of Computer Science and Technology, Tsinghua University
Beijing, 100084, P.R. China
lpy@mails.tsinghua.edu.cn

Abstract. We give a classification of block-wise symmetric signatures in the theory of matchgate computations. The main proof technique is matchgate identities, a.k.a. useful Grassmann-Plücker identities.

1 Introduction

The most fundamental question in computational complexity theory is what differentiate between polynomial time and exponential time problems. On the one hand, we have many completeness results and conjectured separations of complexity classes. On the other hand we have precious few unconditional separations. In fact, the most spectacular advances in the field in the past 20 years have been *upper bounds*, i.e., surprising ways to do computation efficiently. Valiant's theory of matchgate and holographic algorithms [11,13] is one such methodology.

The basic idea in matchgate computations is to encode 0-1 bits of a computation in terms of *perfect matchings*. The complexity of graph matching is very interesting in its own right, having inspired the notion of P in the first place [5]. While a brute force attempt at graph matching seems to take exponential time, it turns out that the decision problem is in P. More relevant, counting perfect matchings is known to be in P for planar graphs by the FKT method [7,8,10]. (Counting all, not necessarily perfect, matchings for planar graphs is #P-complete, as is counting perfect matchings for general graphs [6].) So one can say that graph matching is right at the border of polynomial time and (probably) exponential time. Valiant's theory of matchgate computations uses the FKT method as the starting point.

To give a flavor of this methodology, let's consider the problem #$_7$Pl-Rtw-Mon-3CNF. Given a planar read-twice monotone 3CNF formula, this problem asks for the number of satisfying assignments modulo 7. Without the modulo 7, it is #P-complete even for such restricted formulae [14]. Furthermore, counting

---

* Supported by NSF CCR-0511679 and by the National Natural Science Foundation of China Grant 60553001 and the National Basic Research Program of China Grant 2007CB807900, 2007CB807901.

E. Csuhaj-Varjú and Z. Ésik (Eds.): FCT 2007, LNCS 4639, pp. 187–198, 2007.

mod 2, denoted as $\#_2$Pl-Rtw-Mon-3CNF, is $\oplus$P-complete (hence NP-hard). But, using matchgates Valiant showed that $\#_7$Pl-Rtw-Mon-3CNF $\in$ P [14].

A matchgate is a weighted planar graph with some external nodes. E.g., let π be a path of length 3: all 3 edges have weight 1, and the two end vertices are external nodes. If we remove exactly one of the two external nodes we have 3 vertices left and therefore there is no perfect matching. If we remove either both or none of the two external nodes we get a unique perfect matching with weight 1 (the product of weights of matching edges). We can record this information as $(1, 0, 0, 1)^{\mathrm{T}}$, indexed by $00, 01, 10, 11$; this is called the (standard) *signature* of π. One can use this gadget to replace a Boolean variable x in a planar formula φ, and $00, 01, 10, 11$ will naturally correspond to truth values of x to be fanned-out to the 2 clauses of φ in which x appears (recall it is read-twice). Then the signature $(1, 0, 0, 1)^{\mathrm{T}}$ indicates consistency of this truth assignment on x.

Now for each clause in φ we wish to find a matchgate with 3 external nodes having signature $(0, 1, 1, 1, 1, 1, 1, 1)^{\mathrm{T}}$, indexed by $000, 001, \ldots, 111$. This signature corresponds to a Boolean OR. One can replace each clause by such a gadget, and connect its 3 external nodes to the gadgets of its 3 variables. Then the total number of perfect matchings of the resulting planar graph is exactly the number of satisfying assignments of φ. This can be computed by the FKT method, which would imply $\mathrm{P}^{\#\mathrm{P}} = \mathrm{P}$.

It turns out that a matchgate with the *standard* signature $(0, 1, 1, 1, 1, 1, 1, 1)^{\mathrm{T}}$ does not exist. However, using a basis transformation a (non-standard) signature in the form $(0, 1, 1, 1, 1, 1, 1, 1)^{\mathrm{T}}$ *is realizable* over the field $\mathbf{Z}_7$ (but not $\mathbf{Q}$). This gives the result that $\#_7$Pl-Rtw-Mon-3CNF $\in$ P. (In this paper we will not be concerned with non-standard signatures.)

The signatures $(1, 0, 0, 1)^{\mathrm{T}}$ and $(0, 1, 1, 1, 1, 1, 1, 1)^{\mathrm{T}}$ are called symmetric signatures, since their values only depend on the Hamming weight of the index. Symmetric signatures have natural combinatorial meanings (such as two equal bits or the Boolean OR). Therefore the study of symmetric signatures is of foremost importance in order to understand the power of these exotic algorithms. To this end, we have achieved a complete classification of bit-wise symmetric signatures [3].

In Valiant's surprising algorithm for $\#_7$Pl-Rtw-Mon-3CNF he took another innovative step in the use of matchgates. In his algorithm, the matchgates have external nodes grouped in blocks of 2 each (called "2-rail" in [14]). This naturally raises the question of classification of block-wise symmetric signatures. This paper is concerned with this classification.

The classification theorem of block-wise symmetric signatures is more difficult compared to that of bit-wise symmetric signatures. The main reason for this is that matchgate signatures are characterized by a set of parity requirements (due to consideration of perfect matchings) and an exponential sized set of algebraic constraints called *Matchgate Identities* (MGI) a.k.a. the *useful Grassman-Plücker Identities* [9,12,2,1]. These MGI are non-linear, and are more subtle compared to parity requirements. They come about due to an equivalence between the perfect matching polynomial PerfMatch and the *Pfaffian* [2,1]. For bit-wise symmetric signatures, these MGI degenerate into something more read-ily treatable. This paper is the first time one is able to mount a successful and systematic attack on these MGI. We find proofs on MGI technically challenging, with almost every step a struggle (at least to the authors).

At a higher level, the new theory of matchgate and holographic algorithms represents a *novel* algorithm design methodology by Valiant, with its ultimate reach unknown. Will the new theory lead to a collapse of complexity classes? We don't know. Only a systematic study will (hopefully) tell. To get a classification theorem for block-wise symmetric signatures seems a useful step.

2   Background

Let $G = (V, E, W)$ be a weighted undirected planar graph. A *matchgate* Γ is a tuple (G, X) where $X \subseteq V$ is a set of external nodes, ordered counterclockwise on the external face. Γ is called an odd (resp. even) matchgate if it has an odd (resp. even) number of nodes.

Each matchgate Γ with n external nodes is assigned a (standard) *signature* $(\Gamma^\alpha)_{\alpha \in \{0,1\}^n}$ with 2^n entries,

$$\Gamma^{i_1 i_2 \ldots i_n} = \text{PerfMatch}(G - Z) = \sum_M \prod_{(i,j) \in M} w_{ij},$$

where the sum is over all perfect matchings M of $G - Z$, and $Z \subseteq X$ is the subset of external nodes having the characteristic sequence $\chi_Z = i_1 i_2 \ldots i_n$.

An entry Γ^α is called an even (resp. odd) entry if the Hamming weight $\text{wt}(\alpha)$ is even (resp. odd). It was proved in [1,2] that standard signatures are characterized by the following two sets of conditions. (1) The parity requirements: either all even entries are 0 or all odd entries are 0. This is due to perfect matchings. (2) A set of Matchgate Identities (MGI) defined as follows: A pattern α is an n-bit string, i.e., $\alpha \in \{0,1\}^n$. A position vector $P = \{p_i\}, i \in [l]$, is a subsequence of $\{1, 2, \ldots, n\}$, i.e., $p_i \in [n]$ and $p_1 < p_2 < \cdots < p_l$. We also use p to denote the pattern, whose $(p_1, p_2, \ldots, p_l)$-th bits are 1 and others are 0. Let $e_i \in \{0,1\}^n$ be the pattern with 1 in the i-th bit and 0 elsewhere. Let $\alpha + \beta$ be the bitwise XOR of α and β. Then for any pattern $\alpha \in \{0,1\}^n$ and any position vector $P = \{p_i\}, i \in [l]$,

$$\sum_{i=1}^{l} (-1)^i \Gamma^{\alpha + e_{p_i}} \Gamma^{\alpha + p + e_{p_i}} = 0. \tag{1}$$

The use of MGI will be central in this paper. These MGI come from the Grassmann-Plücker identities valid for Pfaffians. In fact initially Valiant introduced *two* theories of matchgate computation: The first is the matchcircuit theory with general (non-planar) matchgates [11]. These matchgates have *characters* which are defined in terms of Pfaffians. The second is the theory of matchgrid/holographic algorithms [13]. These use planar matchgates with signatures defined by PerfMatch. In [2] it was proved that MGI characterize (general) matchgate characters. In [1] an equivalence theorem of characters and signatures was established, and thus MGI also characterize planar matchgate signatures. The dual forms of the theory have been useful in both ways: some times it is easier to reason and construct planar gadgets, other times the algebraic Pfaffian setup seems essential. A case in point is symmetric signatures.

A signature Γ is (bit-wise) symmetric if Γ^α only depends on $\mathrm{wt}(\alpha)$. A bit-wise symmetric signature can be denoted as $[z_0, z_1, \ldots, z_n]$, where $\Gamma^\alpha = z_{\mathrm{wt}(\alpha)}$. It was proved in [2] that for even matchgates, a signature $[z_0, z_1, \ldots, z_n]$ is realizable iff for all odd i, $z_i = 0$, and there exist constants r_1, r_2 and λ, such that $z_{2i} = \lambda \cdot (r_1)^{\lfloor n/2 \rfloor - i} \cdot (r_2)^i$, for $0 \leq i \leq \lfloor \frac{n}{2} \rfloor$. Similar results hold for odd matchgates. These are proved via MGI and Pfaffians. It is interesting to note that the only construction for a planar matchgate realizing this signature is through a non-planar matchgate Γ and its character theory. There is no known direct construction.

A tensor (Γ^α) on index $\alpha = \alpha_1 \ldots \alpha_n$, where each $\alpha_i \in \{0, 1\}^k$, is *block-wise symmetric* if Γ^α only depends on the number of k-bit patterns of α_i, i.e., $\Gamma^{\cdots \alpha_i \cdots \alpha_j \cdots} = \Gamma^{\cdots \alpha_j \cdots \alpha_i \cdots}$, for all $1 \leq i < j \leq n$.

For an even (resp. odd) matchgate Γ with arity n, the *condensed signature* (g^α) of Γ is a tensor of arity $n - 1$, and $g^\alpha = \Gamma^{\alpha b}$ (resp. $g^\alpha = \Gamma^{\alpha \bar{b}}$), where $\alpha \in \{0, 1\}^{n-1}$ and $b = p(\alpha)$ is the parity of $\mathrm{wt}(\alpha)$.

3 Decomposition Theory for Block-Wise Symmetric Signatures

Theorem 1. *Let (Γ^α) be a block-wise symmetric tensor with block size k and arity nk. Assume $n \geq 4$ and $\Gamma^{00\cdots 0} \neq 0$ (or $\Gamma^{e_1 0 \cdots 0} \neq 0$). Then Γ is realizable by a matchgate iff there exist a matchgate Γ_0 with arity $k + 1$ and condensed signature $(g^\alpha)_{\alpha \in \{0,1\}^k}$, and a symmetric matchgate Γ_s such that*

$$\Gamma^{\alpha_1 \alpha_2 \cdots \alpha_n} = \Gamma_s^{p(\alpha_1) p(\alpha_2) \cdots p(\alpha_n)} g^{\alpha_1} g^{\alpha_2} \cdots g^{\alpha_n}. \tag{2}$$

We only prove the case $\Gamma^{00\cdots 0} \neq 0$ here, So it must be an even matchgate. For odd matchgates (the case $\Gamma^{e_1 00 \cdots 0} \neq 0$), the proof is slightly more complicated but along similar lines. Due to space limitation, the proof of this case is omitted here and can be found in the full paper[4].

Proof. We prove "$\Leftarrow$" by a direct construction. In Figure 1, we extend every external node of Γ_s by a copy of the matchgate with condensed signature g, and view the remaining k external nodes of each copy as external. This gives us a new matchgate with nk external nodes, whose signature is given by (2). Therefore every signature which has form (2) is realizable.

Now we prove "$\Rightarrow$": Since $\Gamma^{00\cdots 0} \neq 0$, by adding an extra isolated edge with weight $1/\Gamma^{00\cdots 0}$ we can assume $\Gamma^{00\cdots 0} = 1$. First we assume $r_1 = \Gamma^{e_1 e_1 00 \cdots 0} \neq 0$ (where for convenience we consider $e_1 \in \{0, 1\}^k$), and prove the theorem under this assumption. We take Γ_s to be an even symmetric matchgate with signature $z_{2i} = (r_1)^{-i}$. By [2] this Γ_s exists. Since the given (Γ^α) is realizable, it can be

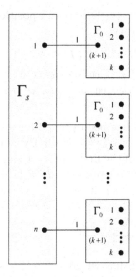

Fig. 1. Block-wise symmetric signature

realized by a matchgate Γ with nk external nodes. View its first $k + 1$ external nodes still as external nodes and the other nodes as internal, we have a matchgate with $k + 1$ external nodes. This is our Γ_0. By definition its condensed signature is

$$g^\alpha = \begin{cases} \Gamma^{\alpha 00 \cdots 0} & \text{when wt}(\alpha) \text{ is even,} \\ \Gamma^{\alpha e_1 0 \cdots 0} & \text{when wt}(\alpha) \text{ is odd.} \end{cases}$$

Note that $g^0 = 1$ and $g^{e_1} = r_1$. We prove (2) by induction on $\text{wt}(\alpha_1 \alpha_2 \cdots \alpha_n) \geq 0$ and $\text{wt}(\alpha_1 \alpha_2 \cdots \alpha_n)$ is even.

If $\text{wt}(\alpha_1 \alpha_2 \cdots \alpha_n) = 0$, we have the only case that $\alpha_1 \alpha_2 \cdots \alpha_n = 00 \cdots 0$. In this case (2) is obvious.

If $\text{wt}(\alpha_1 \alpha_2 \cdots \alpha_n) = 2$, we have two cases depending on whether the two 1s are in the same block or not. If they are in the same block, we can assume it is in the first block since Γ is block symmetric, then $\Gamma^{\alpha_1 \alpha_2 \cdots \alpha_n} = \Gamma^{\alpha_1 00 \cdots 0} = g^{\alpha_1}$ and (2) is satisfied. If they are not in the same block, by symmetry, we may assume $\alpha_1 \alpha_2 \cdots \alpha_n$ has the form $e_i e_j 00 \cdots 0$. When 0 appears in the sup index of Γ, sup index of g, a pattern or positions used by a MGI for Γ, it means a block of all zero. Using the pattern $0 e_j e_1 e_1 00 \cdots 0$ and positions $e_i e_j e_1 e_1 00 \cdots 0$, from (1) we have the following matchgate identity (applying block-wise symmetry):

$$\Gamma^{e_i e_j e_1 e_1 0 \cdots 0} \Gamma^{00 \cdots 0} - \Gamma^{e_1 e_1 0 \cdots 0} \Gamma^{e_i e_j 0 \cdots 0} + \Gamma^{e_j e_1 0 \cdots 0} \Gamma^{e_i e_1 0 \cdots 0} - \Gamma^{e_j e_1 0 \cdots 0} \Gamma^{e_i e_1 0 \cdots 0} = 0.$$

The last two terms cancel out, we get:

$$\Gamma^{e_i e_j e_1 e_1 00 \cdots 0} = \Gamma^{e_i e_j 00 \cdots 0} \Gamma^{e_1 e_1 00 \cdots 0}. \tag{3}$$

Next, using the pattern $0e_ie_je_i00\cdots0$ and positions $e_ie_1e_je_i00\cdots0$, we have the following matchgate identity:

$$\Gamma^{e_ie_je_1e_10\cdots0}\Gamma^{00\cdots0}-\Gamma^{e_je_10\cdots0}\Gamma^{e_ie_10\cdots0}+\Gamma^{e_1e_10\cdots0}\Gamma^{e_ie_j0\cdots0}-\Gamma^{e_je_10\cdots0}\Gamma^{e_ie_10\cdots0}=0.$$

Together with (3), we have $\Gamma^{e_ie_j00\cdots0}\Gamma^{e_1e_100\cdots0} = \Gamma^{e_ie_100\cdots0}\Gamma^{e_je_100\cdots0}$. Since $\Gamma^{e_1e_100\cdots0} = r_1 \neq 0$, we have $\Gamma^{e_ie_j00\cdots0} = \Gamma^{e_ie_100\cdots0}\Gamma^{e_je_100\cdots0}/r_1 = (r_1)^{-1}g^{e_i}g^{e_j}$. So (2) is satisfied.

Inductively we assume (2) has been proved for all $\mathrm{wt}(\alpha_1\alpha_2\cdots\alpha_n) \leq 2(i-1)$, for some $i \geq 2$. Now $\mathrm{wt}(\alpha_1\alpha_2\cdots\alpha_n) = 2i > 0$. By symmetry, we can assume $\alpha_1 \neq 00\cdots0$. Let t be the position of the first 1 in α_1. Using the pattern $\alpha_1\alpha_2\cdots\alpha_n+e_t$ and positions $\alpha_1\alpha_2\cdots\alpha_n$ (we denote it as $P = \{p_j\}$ where $j = 1, 2, \ldots, 2i$), we have the following matchgate identity:

$$\Gamma^{\alpha_1\alpha_2\cdots\alpha_n} = \sum_{j=2}^{2i}(-1)^j\Gamma^{\alpha_1\alpha_2\cdots\alpha_n+e_t+e_{p_j}}\Gamma^{e_t+e_{p_j}}. \tag{4}$$

Since every Γ^β in the RHS has $\mathrm{wt}(\beta) \leq 2i-2$, we can apply (2) to them.

Now we do the summation of the RHS in (4) block by block; the sum of the r-th block is denoted as S_r. Let $w_r = \mathrm{wt}(\alpha_r)$. Let $2q$ be the number of odd w_r, i.e., the number of blocks among $\alpha_1, \alpha_2, \ldots, \alpha_n$ with odd weight. Note that this number is even.

For the first block, if $w_1 = 1$, then $S_1 = 0$, being an empty sum. Assume $w_1 > 1$. In the notation below we consider $e_t, e_{p_j} \in \{0,1\}^k$ for convenience.

$$S_1 = \sum_{j=2}^{w_1}(-1)^j\Gamma^{(\alpha_1+e_t+e_{p_j})\alpha_2\cdots\alpha_n}\Gamma^{(e_t+e_{p_j})00\cdots0} \tag{5}$$

$$= r_1^{-q}g^{\alpha_2}\cdots g^{\alpha_n}\sum_{j=2}^{w_1}(-1)^j g^{\alpha_1+e_t+e_{p_j}}g^{e_t+e_{p_j}}. \tag{6}$$

Note that the exponent q in r_1^{-q} comes from the fact that the number of blocks with odd weight among $\alpha_1 + e_t + e_{p_j}, \alpha_2, \ldots, \alpha_n$ is $2q$.

If w_1 is odd, using the pattern $(\alpha_1 + e_t)1$ and positions $\alpha_1 1$, we have the following matchgate identity for Γ_0:

$$-g^{\alpha_1} + \sum_{j=2}^{w_1}(-1)^j g^{\alpha_1+e_t+e_{p_j}}g^{e_t+e_{p_j}} + g^{\alpha_1+e_t}g^{e_t} = 0.$$

Substituting this in (6), we have:

$$S_1 = r_1^{-q}g^{\alpha_2}\cdots g^{\alpha_n}(g^{\alpha_1} - g^{\alpha_1+e_t}g^{e_t}). \tag{7}$$

We note that this is also valid for $w_1 = 1$.

If w_1 is even, using the pattern $(\alpha_1 + e_t)0$ and positions $\alpha_1 0$, we have the following matchgate identity for Γ_0:

$$-g^{\alpha_1} + \sum_{j=2}^{w_1}(-1)^j g^{\alpha_1+e_t+e_{p_j}}g^{e_t+e_{p_j}} = 0.$$

Substituting this in (6), we have:

$$S_1 = r_1^{-q} g^{\alpha_1} g^{\alpha_2} \cdots g^{\alpha_n}. \tag{8}$$

If all S_r are empty block-wise sums for $r > 1$ (i.e., $w_r = 0$ for all $r > 1$), then w_1 must be even, and we are done. Now suppose there are non-empty block-wise sums S_r, for $r > 1$. For the r-th block, let v_r be the number of 1s in the first $r - 1$ blocks, and p_j^r ($j \in [w_r]$) be the position of the j-th 1 in α_r. Then

$$S_r = (-1)^{v_r} \sum_{j=1}^{w_r} (-1)^j \Gamma^{(\alpha_1+e_t)\alpha_2\cdots(\alpha_r+e_{p_j^r})\cdots\alpha_n} \Gamma^{(e_t)00\cdots(e_{p_j^r})\cdots 0} \tag{9}$$

$$= (-1)^{v_r} r_1^{-q'} g^{e_t} g^{\alpha_1+e_t} g^{\alpha_2} \cdots \widehat{g^{\alpha_r}} \cdots g^{\alpha_n} \sum_{j=1}^{w_r} (-1)^j g^{\alpha_r+e_{p_j^r}} g^{e_{p_j^r}}, \tag{10}$$

where $\widehat{g^{\alpha_r}}$ denotes a missing factor, and $2q'$ is the total number of odd blocks in $\alpha_1+e_t, \alpha_2, \ldots, \alpha_r+e_{p_j^r}, \ldots, \alpha_n$ from the first factor Γ and in $(e_t)00\cdots(e_{p_j^r})\cdots 0$ from the second factor Γ. If w_r is even, using the pattern $\alpha_r 1$ and positions $\alpha_r 0$, we have the following matchgate identity for Γ_0:

$$\sum_{j=1}^{w_r} (-1)^j g^{\alpha_r+e_{p_j^r}} g^{e_{p_j^r}} = 0.$$

Substituting this in (10), we have $S_r = 0$.

Therefore, among block sums S_r, for $r > 1$, we need only consider blocks with odd w_r. Assume w_r is odd now, we have $q' = q$ if w_1 is odd, and $q' = q + 1$ if w_1 is even. Using the pattern $\alpha_r 0$ and positions $\alpha_r 1$, we have the following MGI for Γ_0:

$$\sum_{j=1}^{w_r} (-1)^j g^{\alpha_r+e_{p_j^r}} g^{e_{p_j^r}} + g^{\alpha_r} = 0.$$

Substituting this in (10), we have $S_r = -(-1)^{v_r} r_1^{-q'} g^{e_t} g^{\alpha_1+e_t} g^{\alpha_2} \cdots g^{\alpha_r} \cdots g^{\alpha_n}$.

To summarize, after the first block sum S_1, every even block will be zero, and every odd block will alternatingly contribute a $\pm r_1^{-q'} g^{e_t} g^{\alpha_1+e_t} g^{\alpha_2} \cdots g^{\alpha_n}$. If S_1 is an even block sum, then this alternating sum has an even number of such terms, and they all cancel out. This leaves us with the desired result $\Gamma^{\alpha_1\alpha_2\cdots\alpha_n} = S_1 = r_1^{-q} g^{\alpha_1} g^{\alpha_2} \cdots g^{\alpha_n}$ from (8). If the first block is odd, then $q' = q$, and there are an odd number of alternating S_r for $r > 1$ and w_r odd, starting with the sign $-(-1)^{v_2} = +1$. These will cancel out pairwise except one $r_1^{-q} g^{e_t} g^{\alpha_1+e_t} g^{\alpha_2} \cdots g^{\alpha_n}$ left, which cancels the $-r_1^{-q} g^{e_t} g^{\alpha_1+e_t} g^{\alpha_2} \cdots g^{\alpha_n}$ in S_1 from (7). Finally in either cases, we have $\Gamma^{\alpha_1\alpha_2\cdots\alpha_n} = r_1^{-q} g^{\alpha_1} g^{\alpha_2} \cdots g^{\alpha_n}$. This is precisely (2).

Now we consider the case $\Gamma^{e_1 e_1 00\cdots 0} = 0$. If there exists any $i \in [k]$ such that $\Gamma^{e_i e_i 00\cdots 0} \neq 0$, the above proof can go through similarly. Therefore we assume for all $i \in [k]$, $\Gamma^{e_i e_i 00\cdots 0} = 0$.

Consider any $1 \leq i, j, s, t \leq k$ (not necessarily distinct). Using the pattern $0e_je_se_t00\cdots0$ and positions $e_ie_je_se_t00\cdots0$ we get (applying block symmetry),

$$\Gamma^{e_ie_je_se_t0\cdots0}\Gamma^{00\cdots0} - \Gamma^{e_se_t0\cdots0}\Gamma^{e_ie_j0\cdots0} + \Gamma^{e_ie_s0\cdots0}\Gamma^{e_je_t0\cdots0} - \Gamma^{e_je_s0\cdots0}\Gamma^{e_ie_t0\cdots0} = 0.$$

Also use the pattern $0e_se_je_t00\cdots0$ and positions $e_ie_se_je_t00\cdots0$ we get

$$\Gamma^{e_ie_se_je_t0\cdots0}\Gamma^{00\cdots0} - \Gamma^{e_je_t0\cdots0}\Gamma^{e_ie_s0\cdots0} + \Gamma^{e_se_t0\cdots0}\Gamma^{e_ie_j0\cdots0} - \Gamma^{e_se_j0\cdots0}\Gamma^{e_ie_t0\cdots0} = 0.$$

Adding the two, we get $\Gamma^{e_ie_se_je_t00\cdots0} = \Gamma^{e_se_j00\cdots0}\Gamma^{e_ie_t00\cdots0}$.
From this we have

$$(\Gamma^{e_ie_j00\cdots0})^2 = \Gamma^{e_ie_je_ie_j00\cdots0} = \Gamma^{e_ie_je_je_i00\cdots0} = \Gamma^{e_ie_i00\cdots0}\Gamma^{e_je_j00\cdots0} = 0.$$

Therefore for all $i, j \in [k]$, we have $\Gamma^{e_ie_j00\cdots0} = 0$. Now we define $g^\alpha = \Gamma^{\alpha00\cdots0}$ when $\mathrm{wt}(\alpha)$ is even, and $g^\alpha = 0$ when $\mathrm{wt}(\alpha)$ is odd, and inductively prove (2) similarly as before. (g^α) is the condensed signature of a realizable matchgate Γ_0 of arity $k + 1$ obtained from Γ as follows: View its first k external nodes (in the first block) still as external and the rest as internal, add a new isolated edge with weight 1, and one end as the $(k + 1)$-st external node and the other end an internal node. We will still arrive at (4). Now all block sums $S_r = 0$, for $r > 1$, since it involves a $\Gamma^{e_t+e_{p_j}}$, and e_t appears in the first block.

Consider the first block sum S_1. Suppose $q > 0$, i.e., there are some odd w_r. Then there are at least two odd blocks. Only the first block has a changed index in the sum, so some odd block among $\alpha_2, \ldots, \alpha_n$ remains in $\Gamma^{\alpha_1\alpha_2\cdots\alpha_n+e_t+e_{p_j}}$. Thus, by induction it is 0, since the corresponding $g^{\alpha_i} = 0$. Now suppose $q = 0$, i.e., all blocks are even. By induction we get

$$\Gamma^{\alpha_1\alpha_2\cdots\alpha_n} = g^{\alpha_2}\cdots g^{\alpha_n}\sum_{j=2}^{w_1}(-1)^j g^{\alpha_1+e_t+e_{p_j}}g^{e_t+e_{p_j}}.$$

Using the pattern $(\alpha_1 + e_t)0$ and positions α_10 on Γ_0, we have MGI,

$$-g^{\alpha_1} + \sum_{j=2}^{w_1}(-1)^j g^{\alpha_1+e_t+e_{p_j}}g^{e_t+e_{p_j}} = 0,$$

This gives $\Gamma^{\alpha_1\alpha_2\cdots\alpha_n} = g^{\alpha_1}g^{\alpha_2}\cdots g^{\alpha_n}$ proving (2). □

These theorems give an elegant decomposition structure of block-wise symmetric signatures. There is an underlying bit-wise symmetric signature Γ_s, whose structure is very clear to us. Therefore, the realizability condition is within each block.

4   Characterization of Block-Wise Symmetric Signature with Block Size 2

In Theorem 1, we have two assumptions $n \geq 4$ and $\Gamma^{00\cdots0} \neq 0$. $n \geq 4$ is necessary for some boundary reason. The assumption $\Gamma^{00\cdots0} \neq 0$ is more technical but we are not able to bypass it in general. However, in this section we show that this assumption is not necessary for block size $k = 2$.

Theorem 2. *If Γ is a block-wise symmetric signature for some matchgate, whose block size is 2 and arity $2n$ where $n \geq 4$. Then there exist four numbers $g^{00}, g^{01}, g^{10}, g^{11}$ and a realizable bit-wise symmetric signature Γ_s such that*

$$\Gamma^{\alpha_1\alpha_2\cdots\alpha_n} = \Gamma_s^{p(\alpha_1)p(\alpha_2)\cdots p(\alpha_n)} g^{\alpha_1} g^{\alpha_2} \cdots g^{\alpha_n}. \tag{11}$$

We only prove it for even matchgates here; the proof is similar for odd matchgates. If $\Gamma^{00,00,\dots,00} \neq 0$ or $\Gamma^{11,11,\dots,11} \neq 0$ (we use "," to separate blocks), we are done by Theorem 1. Note that flipping all bits preserves block-symmetry. Now we assume Γ is an even matchgate, $n \geq 4$, and $\Gamma^{00,00,\dots,00} = \Gamma^{11,11,\dots,11} = 0$. This assumption is made for all the following Claims.

Claim 1. *For any $\alpha \in \{00, 01, 10, 11\}^{n-4}$, we have*

$$\Gamma^{01,01,01,01,\alpha} \Gamma^{00,00,00,00,\alpha} = (\Gamma^{01,01,00,00,\alpha})^2.$$

$$\Gamma^{10,10,10,10,\alpha} \Gamma^{00,00,00,00,\alpha} = (\Gamma^{10,10,00,00,\alpha})^2.$$

$$\Gamma^{01,01,10,10,\alpha} \Gamma^{00,00,00,00,\alpha} = \Gamma^{01,01,00,00,\alpha} G^{10,10,00,00,\alpha} = (\Gamma^{01,10,00,00,\alpha})^2.$$

Proof. All three equations follow from MGI. The α part is not involved in the MGI. This means that the pattern for these bits is exactly α and the position vector bits for these bit locations are all 0. For convenience, we only list below the pattern and positions for the other bits, which are really involved in the MGI. We also use this simplified notation in the following Claims.

This Claim is quite direct from MGI. We only list the pattern and positions used, and omit the actual MGI. The first equation uses the pattern $00, 01, 01, 01$ and positions $01, 01, 01, 01$. The second equation uses the pattern $00, 10, 10, 10$ and positions $10, 10, 10, 10$. The last equation is from two MGI: one uses the pattern $00, 01, 10, 10$ and positions $01, 01, 10, 10$, the other uses the pattern $00, 10, 01, 10$ and positions $01, 10, 01, 10$. □

Claim 2

$$\Gamma^{00,00,\{00,01,10\}^{n-2}} = 0.$$

$$\Gamma^{11,11,\{11,01,10\}^{n-2}} = 0.$$

Proof. We only prove $\Gamma^{00,00,\{00,01,10\}^{n-2}} = 0$; the second equation can be obtained for the first by flipping all the bits. For $\alpha \in \{00, 01, 10\}^{n-2}$, we prove it by induction on $\mathrm{wt}(\alpha) \geq 0$ and $\mathrm{wt}(\alpha)$ is even. The case $\mathrm{wt}(\alpha) = 0$ is by assumption. We use Claim 1 to go from weight i to weight $i+2$. □

Claim 3. *For any $\alpha \in \{00, 01, 10, 11\}^{n-3}$,*

$$\Gamma^{00,00,00,\alpha} = 0.$$

$$\Gamma^{11,11,11,\alpha} = 0.$$

Proof. We also only need to prove $\Gamma^{00,00,00,\alpha} = 0$. For $\alpha \in \{00, 01, 10, 11\}^{n-3}$, we prove it by induction on the number of non-"00" blocks in α. (We denote this number by $N_0(\alpha)$.)

If every block in α is 00, then it is by assumption. Inductively we assume it has been proved for all $N_0(\alpha) < i$. Now $N_0(\alpha) = i$. If α does not have any block "11", it has been proved by Claim 2. Otherwise, we can assume $\alpha = 11, \alpha'$ by block-symmetry. Since $N_0(00, \alpha') = i - 1$, we have $\Gamma^{00,00,00,00,\alpha'} = 0$.

Using the pattern $00, 00, 01, 11$ and positions $00, 00, 11, 11$, we have MGI: (Note that we omit the α' part, and also we omit the symbol Γ in the MGI.)

$$
\begin{aligned}
0 &= (00, 00, 11, 11)(00, 00, 00, 00) \\
&- (00, 00, 00, 11)(00, 00, 11, 00) \\
&+ (00, 00, 01, 01)(00, 00, 10, 10) \\
&- (00, 00, 01, 10)(00, 00, 10, 01)
\end{aligned}
$$

The first term is 0, and by Claim 1, the last two terms cancel out. It follows that $\Gamma^{00,00,00,11,\alpha'} \Gamma^{00,00,11,00,\alpha'} = 0$, which is exactly $\Gamma^{00,00,00,\alpha} = 0$. $\square$

From Claim 3 and Claim 1, we have

Claim 4. *For any* $\alpha \in \{00, 01, 10, 11\}^{n-4}$,

$$
\Gamma^{01,10,00,00,\alpha} = \Gamma^{01,01,00,00,\alpha} = \Gamma^{10,10,00,00,\alpha} = 0.
$$

Claim 5. *For any* $\alpha \in \{00, 01, 10, 11\}^{n-2}$, *the following are all valid,*

$$
\Gamma^{00,00,\alpha} = 0, \qquad \Gamma^{11,11,\alpha} = 0, \qquad \Gamma^{00,11,\alpha} = 0.
$$

Claim 5 says that every non-zero entry Γ^α can have at most one even block. This is an important step in the proof. However, due to space limitation, the proof is omitted here, and is presented in the full paper[4]. The proof is by repeated applications of MGI (*death by a thousand cuts, an ancient Chinese disgrace; unfortunately we cannot find a* coup de grâce.)

Claim 6. *For any* $\alpha \in \{00, 01, 10, 11\}^{n-2}$, *we have*

$$
\Gamma^{01,01,\alpha} \Gamma^{10,10,\alpha} = (\Gamma^{01,10,\alpha})^2.
$$

Proof. Using the pattern $00, 01$ and positions $11, 11$ (omitting α), we have MGI:

$$
0 = (10, 01)(01, 10) - (01, 01)(10, 10) + (00, 11)(11, 00) - (00, 00)(11, 11).
$$

From Claim 5, we know the last two terms are both 0. So we have

$$
\Gamma^{01,01,\alpha} \Gamma^{10,10,\alpha} = (\Gamma^{01,10,\alpha})^2.
$$ $\square$

Claim 7. *For* $n \geq 4$, $k = 2$, *if* n *is even and* $\Gamma^{00,00,\dots,00} = \Gamma^{11,11,\dots,11} = 0$, *Theorem 2 holds.*

Proof. Suppose $\Gamma^{\alpha_1,\alpha_2,\ldots,\alpha_n} \neq 0$, we show each $\alpha_i \in \{01, 10\}$. Since n is even and we have an even matchgate, the number of odd blocks must be even, so that if it has any even block it has at least two even blocks. Then by Claim 5 it is 0.

If $\Gamma^{01,01,\ldots,01} \neq 0$, w.l.o.g, we assume $\Gamma^{01,01,\ldots,01} = 1$. Let Γ_s be the matchgate having symmetric signature $[0, 0, \ldots, 0, 1]$ (in the notation for bit-wise symmetric signatures), let $g^{01} = 1$ and $g^{10} = \Gamma^{10,01,01,\ldots,01}/\Gamma^{01,01,\ldots,01} = \Gamma^{10,01,01,\ldots,01}$. From Claim 6, we can verify that (11) is satisfied. This is seen as follows: Claim 6 allows one to "exchange" one block of 10 for one block of 01, incurring a factor of g^{10}. This works as long as $g^{10} \neq 0$. If $g^{10} = 0$, we can instead use Claim 6 to show that $\Gamma^{01,10,\alpha} = 0$, for all $\alpha \in \{01, 10\}^{n-2}$. Moreover we want to show that $\Gamma^{10,10,\ldots,10} = 0$ as well. For this purpose, we use MGI with the pattern $00, 10, 10, \ldots, 10$ and all positions, and get

$$0 = (10, 10, \ldots, 10)(01, 01, \ldots, 01) - (01, 10, \ldots, 10)(10, 01, \ldots, 01) + \ldots$$

The remaining terms (omitted) all have a 00 block in its first factor, and so they are all 0. The second term is also 0 as $g^{10} = 0$. Yet $(01, 01, 01, \ldots, 01) = 1$, so $(10, 10, 10, \ldots, 10) = 0$. This proves the Claim when $\Gamma^{01,01,\ldots,01} \neq 0$.

If $\Gamma^{01,01,\ldots,01} = 0$, again from the "exchange" argument by Claim 6, the only possible non-zero entry of Γ is $\Gamma^{10,10,\ldots,10}$. Let $g^{00} = g^{11} = g^{01} = 0$ and $g^{10} = \sqrt[n]{\Gamma^{10,10,\ldots,10}}$. Then (11) is satisfied. (This may require us to go to an algebraic extension field.) $\qquad \square$

Claim 8. *For $n \geq 4$, $k = 2$, n is odd and $\Gamma^{00,00,\ldots,00} = \Gamma^{11,11,\ldots,11} = 0$, Theorem 2 holds.*

Proof. Since n is odd and Γ is an even matchgate, from Claim 5, we know that if $\Gamma^{\alpha_1\alpha_2\cdots\alpha_n} \neq 0$, then there is exactly one $\alpha_i \in \{00, 11\}$ and all other $\alpha_j \in \{01, 10\}$. By block-symmetry, we assume $\alpha_1 \in \{00, 11\}$ and $\alpha_i \in \{01, 10\}$ (where $i = 2, 3, \ldots, n$).

If $\Gamma^{00,01,01,\ldots,01} \neq 0$, w.l.o.g, we assume $\Gamma^{00,01,01,\ldots,01} = 1$. Let $g^{00} = g^{01} = 1$. Using the pattern $10, 01, 01, \ldots, 01$ and the first four bits as positions, we have

$$(00, 01)(11, 10) - (11, 01)(00, 10) + (10, 11)(01, 00) - (10, 00)(01, 11) = 0.$$

By block-symmetry, the first and the last two terms are equal. So we have

$$\Gamma^{00,01,01,\ldots,01}\Gamma^{11,10,01,\ldots,01} - \Gamma^{11,01,01,\ldots,01}\Gamma^{00,10,01,\ldots,01} = 0. \tag{12}$$

Since $\Gamma^{00,01,01,\ldots,01} = 1$, let $g^{11} = \Gamma^{11,01,01,\ldots,01}$ and $g^{10} = \Gamma^{00,10,01,\ldots,01}$, we have $\Gamma^{11,10,01,\ldots,01} = g^{11}g^{10}$. And let Γ_s be the matchgate with symmetric signature $[0, 0, \ldots, 1, 0]$. The proof is similar with Claim 7. Degenerate cases happen when $g^{10} = 0$, or $g^{11} = 0$, or both. In particular, when $g^{10} = 0$, we need to prove $\Gamma^{00,10,10\ldots10} = 0$, which goes beyond Claim 6. This is shown by the MGI using the pattern $00, 00, 10, \ldots, 10$ and positions $00, 11, 11, \ldots, 11$ (all the bits except the first two). We also need to prove $\Gamma^{11,10,10,\ldots,10} = 0$ when $g^{10} = 0$ or $g^{11} = 0$ or both. This can be shown by the MGI using the pattern $10, 01, 01, \ldots, 01$ and all positions.

If $\Gamma^{11,01,01,\ldots,01} \neq 0$, we have a similar proof.

198   J-Y. Cai and P. Lu

Finally assume $\Gamma^{00,01,01,\ldots,01} = \Gamma^{11,01,01,\ldots,01} = 0$. From Claim 6 and the "exchange" argument, the only two possible non-zero entries of Γ are $\Gamma^{00,10,10,\ldots,10}$ and $\Gamma^{11,10,10,\ldots,10}$. If they are both 0, then Γ is trivial. Otherwise w.l.o.g. we assume $\Gamma^{00,10,10,\ldots,10} = 1$. Let $g^{01} = 0$, $g^{00} = g^{10} = 1$ and $g^{11} = \Gamma^{11,10,10,\ldots,10}$. And let Γ_s be the matchgate with symmetric signature $[0, 0, \ldots, 1, 0]$, we can verify that (11) is satisfied.   □

Together with Claim 7 and Claim 8, we have a complete proof for Theorem 2.

This paper presents an elegant decomposition theorem on the structure of block-wise symmetric signatures for matchgates. The main tool is Matchgate Identities. However the statement of Theorem 2 for $k > 2$ without any non-zero conditions is open. It would also be interesting to simplify the proofs.

References

1. Cai, J.-Y., Choudhary, V.: Some Results on Matchgates and Holographic Algorithms. In: Bugliesi, M., Preneel, B., Sassone, V., Wegener, I. (eds.) ICALP 2006. LNCS, vol. 4051, pp. 703–714. Springer, Heidelberg (2006)
2. Cai, J.-Y., Choudhary, V., Lu, P.: On the Theory of Matchgate Computations. In: IEEE Conference on Computational Complexity 2007 (to appear)
3. Cai, J.-Y., Lu, P.: On Symmetric Signatures in Holographic Algorithms. In: Thomas, W., Weil, P. (eds.) STACS 2007. LNCS, vol. 4393, pp. 429–440. Springer, Heidelberg (2007)
4. Cai, J.-Y., Lu, P.: On Block-wise Symmetric Signatures for Matchgates. Available at ECCC Report TR07-019
5. Edmonds, J.: Minimum partition of a matroid into independent subsets. J. Res. Nat. Bur. Standards Sect. B 69, 67–72 (1965)
6. Jerrum, M.: Two-dimensional monomer-dimer systems are computationally intractable. J. Stat. Phys. 48, 121–134 (1987) erratum, 59, 1087–1088 (1990)
7. Kasteleyn, P.W.: The statistics of dimers on a lattice. Physica 27, 1209–1225 (1961)
8. Kasteleyn, P.W.: Graph Theory and Crystal Physics. In: Harary, F. (ed.) Graph Theory and Theoretical Physics, pp. 43–110. Academic Press, London (1967)
9. Murota, K.: Matrices and Matroids for Systems Analysis. Springer, Heidelberg (2000)
10. Temperley, H.N.V., Fisher, M.E.: Dimer problem in statistical mechanics – an exact result. Philosophical Magazine 6, 1061–1063 (1961)
11. Valiant, L.G.: Quantum circuits that can be simulated classically in polynomial time. SIAM Journal of Computing 31(4), 1229–1254 (2002)
12. Valiant, L.G.: Expressiveness of Matchgates. Theoretical Computer Science 281(1), 457–471 (2002)
13. Valiant, L.G.: Holographic Algorithms (Extended Abstract). In: Proc. 45th IEEE Symposium on Foundations of Computer Science, pp. 306–315 (2004). A more detailed version appeared in ECCC Report TR05-099.
14. Valiant, L.G.: Accidental Algorithms. In: Proc. 47th Annual IEEE Symposium on Foundations of Computer Science, pp. 509–517 (2006)

Path Algorithms on Regular Graphs

Didier Caucal and Dinh Trong Hieu

IGM–CNRS, 5 bd Descartes, 77454 Marne-la-Vallée, France
caucal@univ-mlv.fr, dinh@univ-mlv.fr

Abstract. We consider standard algorithms of finite graph theory, like for instance shortest path algorithms. We present two general methods to polynomially extend these algorithms to infinite graphs generated by deterministic graph grammars.

1 Introduction

The regularity of infinite graphs was first considered by Muller and Schupp [6]. They studied the transition graphs of pushdown automata, called pushdown graphs, and showed that their connected components are the connected graphs of finite degree whose decomposition by distance from a(ny) vertex yields finitely many non-isomorphic connected components. More generally, a graph is regular if it admits a finite decomposition (not necessarily by distance) or, equivalently, if it can be generated by a deterministic graph grammar. Regular graphs have been defined by Courcelle and called hyperedge replacement equational graphs [3]. Any connected regular graph is finitely decomposable by distance, hence the connected components of pushdown graphs coincide with the connected regular graphs of finite degree. This identity was also generalized to non connected graphs: the regular restrictions of pushdown graphs are the regular graphs of finite degree [1]. A regular graph may be seen as an infinite automaton [9] : it recognizes the set of path labels between two given finite vertex sets. Even though a regular graph can have vertices of infinite degree, regular graphs recognize exactly the family of context-free languages. Many publications focus on finite graphs and their applications, but few deal with regular graphs. This paper is a first step towards developing an algorithmic theory of regular graphs.

We consider a set of edge labels forming an idempotent and continuous semiring. For any graph, we define the value of a path to be the product of its successive labels, that we extend by summation to any set of paths between vertex sets. The value (or the computation) of a grammar is then a vector whose components are the values of the graphs generated from each of the left hand sides of the grammar. In the case of deterministic graph grammars in a restricted form, we show the equivalence between the algebraic and operational semantics: the value of the grammar is the least upper bound of the sequence obtained by iteratively applying the interpretation of the grammar from the least element (Theorem 2.5). We then present two methods to compute the value of a grammar over a commutative (idempotent and continuous) semiring. The first method

E. Csuhaj-Varjú and Z. Ésik (Eds.): FCT 2007, LNCS 4639, pp. 199–212, 2007.

applies to linear semirings having a greatest element reached by any increasing Kleene sequence. In this case, the grammar value is determined by applying the grammar interpretation a number of times (Theorem 3.1). The second algorithm works with any commutative semiring and is a simple graph generalization of the Hopkins-Kozen method [5] developped for context-free grammars (cf. Corollary 3.5). Finally we give a polynomial transformation of any grammar into an equivalent grammar in restricted form (Proposition 4.3). We end up with a polynomial complexity algorithm to solve the shortest path problem on regular graphs using graph grammars (Corollary 4.4).

2   Computations with Graph Grammars

We consider a graph as a denumerable set of labelled arcs. The arc labels are elements of an idempotent and continuous semiring. The value of a path is the product of its successive labels. The value of a graph from initial vertices to final vertices is the sum of the values of its paths from an initial vertex to a final vertex.

We want to compute values of graphs generated by a deterministic graph grammar. In this section, we restrict ourselves to grammars in which each left hand side is a labelled arc from vertex 1 to vertex 2, and no right hand side has an arc from 2 or an arc to 1. The value of any left hand side is the value from 1 to 2 of its generated graph. The value of a grammar is the vector of its left hand side values. By considering a grammar as a function from and into the semiring (to the power of the rule number), its iterative application from the least element gives by least upper bound the value of the grammar (Theorem 2.5).

Recall that an algebra $(K, +, \cdot, 0, 1)$ is a *semiring* if $(K, +, 0)$ is a commutative monoid, $(K, \cdot, 1)$ is a monoid, *multiplication* $\cdot$ distributes over *addition* $+$, and 0 annihilates K: $0 \cdot a = a \cdot 0 = 0$ for any $a \in K$.

We say that a semiring K is *complete* if for the following relation:

$$a \leq b \quad \text{if} \quad a + c = b \quad \text{for some} \quad c \in K,$$

any increasing sequence $a_0 \leq a_1 \leq \ldots \leq a_n \leq \ldots$ has a least upper bound $\bigvee_n a_n$. This implies that $\leq$ is a partial order, 0 is the least element, and the operations $+$ and $\cdot$ are monotonous:

$$a \leq b \ \wedge \ a' \leq b' \implies a + a' \leq b + b' \ \wedge \ a \cdot a' \leq b \cdot b'.$$

This also permits to define the sum of any sequence $(a_n)_{n \geq 0}$ in K by

$$\sum_{n \geq 0} a_n := \bigvee_{n \geq 0} \left(\sum_{i=0}^{n} a_i \right).$$

We say that a semiring K is *idempotent* if $+$ is idempotent:

$$a + a = a \quad \text{for any} \ a \in K \quad \text{or equivalently} \quad 1 + 1 = 1.$$

For K complete and idempotent, we have $\sum_{n \geq 0} a_n = \sum_{n \geq 0} a_{\pi(n)}$ for any permutation π, and this sum is also denoted $\sum_{n \geq 0} \{ a_n \mid n \geq 0 \}$.

We say that a complete semiring K is *continuous* if $+$ and $\cdot$ are continuous *i.e.* commute with $\bigvee$: for any increasing sequences $(a_n)_{n \geq 0}$ and $(b_n)_{n \geq 0}$,

$$\left(\bigvee_{n \geq 0} a_n \right) * \left(\bigvee_{n \geq 0} b_n \right) = \bigvee_{n \geq 0} (a_n * b_n) \quad \text{where} \ * \ \text{stands for} \ + \ \text{or} \ \cdot$$

So $\left(\sum_{n \geq 0} a_n \right) + \left(\sum_{n \geq 0} b_n \right) = \sum_{n \geq 0} (a_n + b_n)$ for any $a_n, b_n \in K$.

For K continuous and idempotent and for any sequences $(a_n)_{n \geq 0}$ and $(b_n)_{n \geq 0}$

$$\left(\textstyle\sum_{n\geq0} a_n\right) \cdot \left(\textstyle\sum_{n\geq0} b_n\right) \;=\; \textstyle\sum\{\, a_i \cdot b_j \mid i, j \geq 0 \,\}.$$

From now on K will denote a complete and idempotent semiring.

Recall that $(K^*, ., \varepsilon)$ is the free monoid generated by the set K, i.e. $K^* = \bigcup_{n\geq0} K^n$ is the set of tuples of elements of K where . is the tuple concatenation and the neutral element ε is the 0-tuple (). Any tuple $(a_1, \ldots, a_p)$ is written $a_1 \ldots a_p$. The identity relation on K is extended to the morphism $[\ ]$ from $(K^*, ., \varepsilon)$ into the monoid $(K, \cdot, 1)$:

$$[\varepsilon] = 1 \quad ; \quad \forall\, a \in K, \;\; [a] = a \quad ; \quad \forall\, u, v \in K^*, \;\; [u.v] \;=\; [u] \cdot [v].$$

We extend by summation the *value* mapping $[\ ]$ to any language:

$$[U] \;:=\; \textstyle\sum\{\, [u] \mid u \in U \,\} \quad \text{for any } U \subseteq K^*.$$

Let L be an arbitrary set. Here a L-graph is just a set of arcs labelled in L. Precisely, a *L-graph* G is a subset of $V \times L \times V$ where V is an arbitrary set such that the *vertex* set of G defined by

$$V_G \;:=\; \{\, s \mid \exists\, a, t,\; (s, a, t) \in G \,\vee\, (t, a, s) \in G \,\}$$

is finite or countable, and its *label* set

$$L_G \;:=\; \{\, a \in L \mid \exists\, s, t,\; (s, a, t) \in G \,\} \quad \text{is finite.}$$

Any (s, a, t) of G is a *labelled arc* of *source* s, of *goal* (or target) t, with label a, and is identified with the labelled transition $s \xrightarrow[G]{a} t$, or directly $s \xrightarrow{a} t$ if G is understood. The transformation of a graph G by a function h from V_G into a set V is the graph

$$h(G) \;:=\; \{\, h(s) \xrightarrow{a} h(t) \mid s \xrightarrow[G]{a} t \,\wedge\, s, t \in Dom(h) \,\}.$$

An *isomorphism* h from a graph G to a graph H is a bijection from V_G to V_H such that $h(G) = H$. The *language recognized* by G from a vertex set I to a vertex set F is the set of labels of the paths from I to F:

$$L(G, I, F) \;:=\; \{\, a_1 \ldots a_n \mid n \geq 0 \,\wedge\, \exists\, s_0, \ldots, s_n,\; s_0 \in I \,\wedge\, s_n \in F \,\wedge$$
$$s_0 \xrightarrow{a_1} s_1 \ldots s_{n-1} \xrightarrow{a_n} s_n \,\}.$$

In particular $\varepsilon \in L(G, I, F)$ for $I \cap F \neq \emptyset$.

We also write $s \xrightarrow[G]{\;*\;} t$ for $L(G, s, t) \neq \emptyset$, and $s \xRightarrow[G]{u} t$ for $u \in L(G, s, t)$. The *graph value* on a K-graph G from $I \subseteq V_G$ to $F \subseteq V_G$ is defined as the value of its recognized language:

$$[G, I, F] \;:=\; [L(G, I, F)].$$

For any graph G and by identifying any label $a \in L_G$ with $\{a\}$, we can take the semiring $(2^{L_G^*}, \cup, ., \emptyset, \{\varepsilon\})$ of languages in which $[u] = \{u\}$ for any $u \in L_G^*$.

For this semiring, the values of G are the recognized languages:

$$[G, I, F] \;=\; L(G, I, F) \quad \text{for any } I, F \subseteq L_G^*.$$

Example 2.1. *We consider the following graph G:*

a) *By identifying a with $\{a\}$ and b with $\{b\}$, the value $[G, 1, 2]$ of G from vertex 1 to vertex 2 for the semiring $(2^{\{a,b\}^*}, \cup, ., \emptyset, \{\varepsilon\})$ is the Lukasiewicz language.*

b) *Now taking the semiring* $(2^{\mathbb{N} \times \mathbb{N}}, \cup, +, \emptyset, (0,0))$ *with* $a = \{(1,0)\}$ *and* $b = \{(0,1)\}$, *the value* $[\![G, 1, 2]\!]$ *is the Parikh image* $\{ (n, n+1) \mid n \geq 0 \}$ *of the Lukasiewicz language.*

c) *Taking* $a, b \in \mathbb{R}$ *and the semiring* $(\mathbb{R} \cup \{-\omega, \omega\}, Min, +, \omega, 0)$, *the value*
$$[\![G, 1, 2]\!] = \begin{cases} b & \text{if } a + b \geq 0, \\ -\omega & \text{otherwise,} \end{cases}$$
is the smallest value labelling the paths from 1 *to* 2.

d) *Finally having* $a, b \in \mathbb{R}_+$ *the set of non negative real numbers and for the semiring* $(\mathbb{R}_+ \cup \{\omega\}, Max, \times, 0, 1)$, *the value from* 1 *to* 2 *is*
$$[\![G, 1, 2]\!] = \begin{cases} b & \text{if } a \times b \leq 1, \\ \omega & \text{otherwise.} \end{cases}$$

For continuous and idempotent semirings, we want to extend algorithms computing the values of finite graphs to the graphs generated by graph grammars.

In this section, we restrict ourselves to grammars in which each left hand side is a labelled arc from vertex 1 to vertex 2, and no right hand side has an arc of goal 1 or of source 2.

Precisely a *2-grammar* R is a finite set of rules of the form:
$$(1, A, 2) \ \longrightarrow \ H$$
where $(1, A, 2)$ is an arc labelled by A from vertex 1 to vertex 2, and H is a finite graph.

The labels of the left hand sides form the set N_R of *non-terminals* of R:
$$N_R := \{ A \mid (1, A, 2) \in Dom(R) \}$$
and the remaining labels in R form the set T_R of *terminals*:
$$T_R := \{ A \notin N_R \mid \exists \, H \in Im(R), \ \exists \, s, t, \ (s, A, t) \in H \}.$$
Furthermore we require that each right hand side has no arc of goal 1 or of source 2:
$$\forall \, H \in Im(R), \ \forall \, (s, a, t) \in H, \ s \neq 2 \wedge t \neq 1.$$
We say that R is an *acyclic grammar* if its right hand sides are acyclic graphs.

Starting from any graph, we want a graph grammar to generate a unique graph up to isomorphism. We thus restrict ourselves to *deterministic* 2-grammars in which two rules have distinct left hand sides:
$$((1, A, 2), H), ((1, A, 2), K) \in R \implies H = K.$$
An example is given below.

Fig. 2.2. A deterministic graph grammar

Starting from a graph, this grammar generates a unique infinite graph obtained by applying indefinitely parallel rewritings. Precisely and for any 2-grammar R, the *rewriting* $\underset{R}{\longrightarrow}$ is the binary relation between graphs defined

by $M \xrightarrow{R} N$ if we can choose a non-terminal arc $X = (s, A, t)$ in M and a right hand side H of A in R to replace X by H in M :
$$N = (M - \{X\}) \cup h(H)$$
for some function h mapping vertex 1 to s, vertex 2 to t, and the other vertices of H injectively to vertices outside of M; this rewriting is denoted by $M \xrightarrow{R, X} N$. The rewriting $\xrightarrow{R, X}$ of a non-terminal arc X is extended in an obvious way to the rewriting $\xrightarrow{R, E}$ of any subset E of non-terminal arcs. A *complete parallel rewriting* $\underset{R}{\Longrightarrow}$ is the rewriting according to the set of all non-terminal arcs: $M \underset{R}{\Longrightarrow} N$ if $M \xrightarrow{R, E} N$ where E is the set of all non-terminal arcs of M.

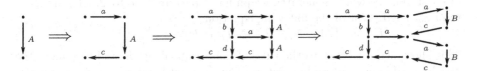

Fig. 2.3. Parallel rewritings according to the grammar of Figure 2.2

Due to the two non-terminal arcs in the right hand side of B for the grammar of Figure 2.2, we get after n parallel rewritings from $H_0 = \{ 1 \xrightarrow{A} 2 \}$, a graph H_n having an exponential number $|H_n|$ of arcs.

The *derivation* $\underset{R}{\Longrightarrow}^*$ is the reflexive and transitive closure for the composition of the parallel rewriting $\underset{R}{\Longrightarrow}$ *i.e.* $G \underset{R}{\Longrightarrow}^* H$ if H is obtained from G by a consecutive sequence of parallel rewritings. We denote by $[M]$ the set of terminal arcs of M :
$$[M] := M \cap V_M \times T_R \times V_M .$$
We now assume that any 2-grammar is deterministic. A *2-grammar over* K is a 2-grammar R such that $T_R \subseteq K$.

A graph G is *generated by a 2-grammar* R from a graph H if G is isomorphic to a graph in the following set $R^\omega(H)$ of isomorphic graphs:
$$R^\omega(H) := \{ \bigcup_{n \geq 0} [H_n] \mid H_0 = H \wedge \forall\, n \geq 0,\ H_n \underset{R}{\Longrightarrow} H_{n+1} \} .$$
For instance by iterating indefinitely the derivation of Figure 2.3, we get the infinite graph depicted below.

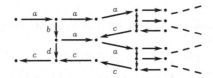

Fig. 2.4. Graph generated by the grammar of Figure 2.2

For any 2-grammar R and any non-terminal $A \in N_R$,
its right hand side in R is $R(\{(1, A, 2)\})$, also denoted $R(A)$,

the generated graph by R from A is $R^\omega(\{(1, A, 2)\})$, also denoted $R^\omega(A)$, if $T_R \subseteq K$, the value of A by R is $[R^\omega(A), 1, 2]$ also denoted $[R^\omega(A)]$.

Given a 2-grammar R over K, we want to compute $[R^\omega(A)]$ for any non-terminal A. First we put R in the following *reduced* form S:
$$N_S = \{ A \in N_R \mid L(R^\omega(A), 1, 2) \neq \emptyset \} \quad \text{and} \quad 0 \notin T_S$$
and for every $A \in N_S$, we have
$$[S^\omega(A)] = [R^\omega(A)] \quad \text{and} \quad \forall\, s \in V_{S(A)},\ 1 \xrightarrow[S(A)]{\ *\ } s \xrightarrow[S(A)]{\ *\ } 2.$$
We begin by removing the arcs labelled by 0 in the right hand sides of R. Then we compute the set
$$E = \{ A \in N_R \mid L(R^\omega(A), 1, 2) \neq \emptyset \}$$
of non-terminals whose generated graph has a path from 1 to 2. This set E is the least fixed point of the following equation:
$$E = \{ A \in N_R \mid \exists\, u \in (T_R \cup E)^*,\ 1 \underset{R(A)}{\overset{u}{\Longrightarrow}} 2 \}.$$
This allows us to restrict to the rules of non-terminals in E and to remove the arcs labelled by a non-terminal not in E in the right hand sides of the grammar. Finally we get S by restricting each right hand side to the vertices accessible from 1 and co-accessible from 2. The overall time complexity is quadratic according to the description length of R (due to the computation of E).

Henceforth we assume that any 2-grammar R is in reduced form. In that case,
$$R \text{ is acyclic} \iff R^\omega(A) \text{ is acyclic for all } A \in N_R.$$

Given a 2-grammar R over K, we order the set $N_R = \{A_1, \ldots, A_p\}$ of its p non-terminals. We want to determine the *grammar value* $[R^\omega]$ of R as being the following tuple in K^p:
$$[R^\omega] := ([R^\omega(A_1)], \ldots, [R^\omega(A_p)]).$$
A standard semantic way is to define the *interpretation* $[R]$ of R as being the mapping:
$$[R]: \qquad K^p \longrightarrow K^p$$
$$(a_1, \ldots, a_p) \longmapsto ([R(A_1)[a_1, \ldots, a_p]], \ldots, [R(A_p)[a_1, \ldots, a_p]])$$
where $G[a_1, \ldots, a_p]$ is the graph obtained from graph G by replacing each label A_i by a_i for every $1 \leq i \leq p$. This interpretation $[R]$ is a continuous mapping.

As K^p is a complete set for the product order whose least element is $0 = (0, \ldots, 0)$, we can apply the Knaster-Tarsky theorem: the least upper bound of the iterative application of $[R]$ from 0, is the least fixed point of $[R]$ denoted $\mu[R]$. A first result is that this least upper bound is also the value $[R^\omega]$ of R.

Theorem 2.5. *For any continuous and idempotent semiring K and for any 2-grammar R over K,*
$$[R^\omega] = \bigvee_{n \geq 0} [R]^n(0) = \mu[R].$$
The first equality holds for K complete and idempotent when R is acyclic.

Note that the first equality of Theorem 2.5 can be false if we allow in a right hand side of the grammar an arc of goal 1 or of source 2. For instance, taking the semiring $(2^{\{a,b,c\}^*}, \cup, ., \emptyset, \{\varepsilon\})$ and the grammar R reduced to the rule:

we have $\bigvee_{n\geq 0}[R]^n(\emptyset) = \emptyset \neq [R^\omega]$.

3   Computation Algorithms

We present two general algorithms to compute the value of any 2-grammar R for idempotent and continuous semirings which are commutative. The first algorithm compares the differences between the $|R|$-th ($|R|$ is the number of rules) approximant $[R]^{|R|}(0)$ with the next one in order to detect increments strictly greater than 1 in the value of the generated graphs (cf. Theorem 3.1). The second algorithm is a simple graph generalization of the Hopkins-Kozen method [5] developped for context-free grammars (cf. Corollary 3.5). Although the first algorithm is more efficient than the second one, it works with semirings whose order is linear and admits a greatest element.

Given a semiring and an algorithm to compute $[G, I, F]$ for any finite graph G (and vertex sets I, F) of time complexity $C_{G,I,F}$, we want to use this algorithm to compute $[R^\omega]$ for any 2-grammar R. The complexity will be expressed with the following parameters:

$|R| = |N_R|$ the number p of rules of R (its cardinality),

$\ell_R := \sum\{\, |R(A)| \mid A \in N_R \,\}$ the *length* of the description of R,

$C_R := \sum\{\, C_{R(A)[0],1,2} \mid A \in N_R \,\}$ the time complexity to compute $[R](0)$.

Theorem 2.5 requires continuous and idempotent semirings. Our algorithms require that these semirings are commutative: the multiplication $\cdot$ is commutative. A *cci-semiring* means a commutative continuous idempotent semiring.

The first algorithm also needs that the semirings be *linear*, meaning that the relation $\leq$ is a linear (total) order. Let us describe this first algorithm.

For any 2-grammar R, Theorem 2.5 gives a standard way to compute $[R^\omega]$: if there exists n such that $[R]^n(0) = [R]^{n+1}(0)$ then $[R^\omega] = [R]^n(0)$. This is true for any *bounded semiring* i.e. with no infinite increasing sequence. But the sequence $([R]^n(0))_n$ is strict in general. We compare the vectors $[R]^p(0)$ and $[R]^{p+1}(0)$ (with $p = |R|$) and determine the ranks for which they differ:

$$E = \{\, A_i \mid ([R]^p(0))_i \neq ([R]^{p+1}(0))_i \,\}.$$

By a classical pumping argument and for each $A \in E$, we can find an increment > 1 in the value of its generated graph. Assuming that we have a greatest element $\top$ reached by any strict increasing sequence $(a^n)_{n\geq 0}$, we get $[R^\omega(A)] = \top$.

This is also true for any non-terminal having A in its right hand side. We complete E into the set $\overline{E}$ which is the least fixed point of the following equation:

$$\overline{E} = E \cup \{\, A \in N_R \mid L_{R(A)} \cap \overline{E} \neq \emptyset \,\}.$$

We deduce that for any rank $1 \leq i \leq p$,

$$([\,R^\omega\,])_i \;=\; \begin{cases} \top & \text{if } A_i \in \overline{E}, \\ ([\,R\,]^P(0))_i & \text{otherwise.} \end{cases}$$

The time complexity is the complexity to compute $[\,R\,]^P(0)$.

Theorem 3.1. *For any linear cci-semiring K having a greatest element $\top$ such that $\bigvee_n a^n = \top$ for every $a > 1$ and $a\cdot\top = \top$ for every $a \neq 0$, we can compute $[\,R^\omega\,]$ in time $O(|R|\,C_R)$ for any 2-grammar R over K.*

We can apply Theorem 3.1 to the semiring $(\mathbb{R} \cup \{-\omega, \omega\}, Min, +, \omega, 0)$ of Example 2.1 (c). This allows us to solve the shortest path problem on any graph generated by a 2-grammar. We take Floyd's algorithm (with negative cycle test) on finite graphs. For any 2-grammar R, we have C_R in $O(\ell_R^3)$ thus the complexity to compute $[\,R^\omega\,]$ is $O(|R|\,\ell_R^3)$.

When R is acyclic, we can take Bellman's algorithm on finite graphs; so C_R is $O(\ell_R)$ and the complexity for $[\,R^\omega\,]$ is $O(|R|\,\ell_R)$, hence quadratic time. For instance taking the following 2-grammar R:

we have the following approximants:

n		0	1	2	3	4
$([\,R\,]^n(0))$	$_A$	ω	3	3	2	2
$([\,R\,]^n(0))$	$_B$	ω	ω	3	3	2
$([\,R\,]^n(0))$	$_C$	ω	1	1	1	1

giving by Theorem 3.1 the value $[\,R^\omega\,] = (-\omega, -\omega, 1)$ *i.e.*
$$[\,R^\omega(A)\,] = [\,R^\omega(B)\,] = -\omega \quad \text{and} \quad [\,R^\omega(C)\,] = 1.$$
Note that we can apply Theorem 3.1 to the semiring $(\mathbb{R}_+ \cup \{\omega\}, Max, \times, 0, 1)$ of Example 2.1 (d).

For the second algorithm, we need to show that (graph) 2-grammars are language-equivalent to cf-grammars (on words), 'cf' is short for context-free.
Recall that a *context-free grammar* P is a finite binary relation on words in which each left hand side is a letter called a non-terminal, and the remaining letters of P are terminals. By denoting N_P and T_P the respectives sets of non-terminals and terminals of P, the rewriting $\underset{P}{\longrightarrow}$ according to P is the binary relation on $(N_P \cup T_P)^*$ defined by
$$UAV \underset{P}{\longrightarrow} UWV \quad \text{if } (A, W) \in P \text{ and } U, V \in (N_P \cup T_P)^*.$$
The derivation $\underset{P}{\longrightarrow}^*$ is the reflexive and transitive closure of $\underset{P}{\longrightarrow}$ with respect to composition. The language $L(P, U)$ generated by P from any $U \in (N_P \cup T_P)^*$ is the set of terminal words deriving from U:
$$L(P, U) := \{\, u \in T_P^* \mid U \underset{P}{\longrightarrow}^* u \,\}.$$
2-grammars and cf-grammars are language-equivalent with linear time translations.

Proposition 3.2. a) *We can transform in linear time any 2-grammar* R *into a cf-grammar* P *such that* $L(R^\omega(A), 1, 2) = L(P, A)$ *for any* $A \in N_R$.

b) *We can transform in linear time any cf-grammar* P *into an acyclic 2-grammar* R *such that* $L(P, A) = L(R^\omega(A), 1, 2)$ *for any* $A \in N_P$.

The first transformation is analogous to the translation of any finite automaton into an equivalent right linear grammar. For each non-terminal $A \in N_R$, let h_A be a vertex renaming of $R(A)$ such that $h_A(1) = A$ and $h_A(2) = \varepsilon$, and the image $Im(h_A) - \{\varepsilon\}$ is a set of symbols with $Im(h_A) \cap Im(h_B) = \{\varepsilon\}$ for any $B \in N_R - \{A\}$. We define:

$$P := \{ (h_A(s), a h_A(t)) \mid \exists A \in N_R,\ s \xrightarrow[R(A)]{a} t \}.$$

Note that each right hand side of P is a word of length at most 2, and the number of non-terminals of P depends on the description length ℓ_R of R:

$$|N_P| = \left(\sum_{A \in N_R} |V_{R(A)}| \right) - |N_R|.$$

For the second transformation, we have $N_R = N_P$ and for each $A \in N_P$, its right hand side in R is the set of distinct paths from 1 to 2 labelled by the right hand sides of A in P. Note that by using the two transformations of Proposition 3.2, we can transform in linear time any 2-grammar into a language equivalent acyclic 2-grammar. Then we can apply Theorem 3.1 with the Bellman's algorithm for the shortest path problem.

Corollary 3.3. *For the semiring* $(\mathbb{R} \cup \{-\omega, \omega\}, Min, +, \omega, 0)$, *any 2-grammar* R *and any* $A \in N_R$, *the shortest path problem* $[R^\omega(A12), 1, 2]$ *can be solved in* $O(\ell_R^2)$, *and in* $O(|R|\ell_R)$ *when* $R^\omega(A12)$ *is acyclic.*

In particular for any context-free grammar P, the shortest path problem can be solved in $O(|P|\ell_P)$.

Starting from any pushdown automaton R, it is easy to transform R into a pushdown automaton R' recognizing the same language (by final states and/or empty stack) such that each right hand side is a state followed by at most two stack letters; the number of rules $|R'|$ hence its length of description $\ell_{R'}$ are in $O(|R|.\ell_R^2)$. Then we apply to R' the usual transformation to get an equivalent cf-grammar P: $|P|$ and ℓ_P are in $O(|R'|^3)$. Thus for any pushdown automaton R, the shortest path problem can be solved in $O((|R|.\ell_R^2)^6) = O(|R|^6.\ell_R^{12})$.

In next section, Corollary 3.3 will be extended to Corollary 4.4 starting from any generalized graph grammar.

Note also that by the two transformations of Proposition 3.2 and by the first equality of Theorem 2.5, we can compute the value of any 2-grammar for non continuous (but complete and idempotent) semirings.

A cf-grammar over K is a context-free grammar P such that $T_P \subseteq K$. By ordering the non-terminal set: $N_P = \{A_1, \ldots, A_p\}$, the *cf-grammar value* is

$$[L(P)] := ([L(P, A_1)], \ldots, [L(P, A_p)]).$$

For the semiring $(2^{\mathbb{N}^q}, \cup, +, \emptyset, (0, \ldots, 0))$ in Example 2.1 *(b)* of commutative languages over q terminals of the form $\{(0, \ldots, 0, 1, 0, \ldots, 0)\}$, a first solution to determine $[L(P)]$ was given by Parikh [7]. This method was refined in [8] and generalized in [5], and works for any *cci*-semiring. This last method is presented briefly below.

The *derivative* of any language $E \subseteq (N_P \cup T_P)^*$ by $A \in N_P$ is the following language:

$$\frac{\partial E}{\partial A} := \{ UV \mid UAV \in E \}$$

and as $+$ is idempotent, we can restrict to remove only the first occurrence of A i.e. $|U|_A = 0$. The *Jacobian matrix* of P is

$$P' := \left(\frac{\partial P(A_i)}{\partial A_j} \right)_{1 \le i,j \le p}$$

where $P(A) = \{ U \mid (A,U) \in P \}$ is the image of $A \in N_P$ by P.

The *interpretation* $[\![P']\!]$ of P' is the mapping:

$$[\![P']\!] : \qquad K^p \longrightarrow K^{p \times p}$$

$$(a_1, \ldots, a_p) \longmapsto \left([\![\tfrac{\partial P(A_i)}{\partial A_j} [a_1, \ldots, a_p]]\!] \right)_{1 \le i,j \le p}$$

where $E[a_1, \ldots, a_p]$ is the language over K obtained from $E \subseteq (N_P \cup T_P)^*$ by replacing in the words of E each label A_i by a_i for every $1 \le i \le p$.

The *Hopkins-Kozen transformation* is the mapping:

$$\mathrm{HK} : \qquad K^p \longrightarrow K^p$$

$$(a_1, \ldots, a_p) \longmapsto \left([\![P']\!](a_1, \ldots, a_p) \right)^* \times (a_1, \ldots, a_p)$$

where $M \times \vec{v} := \left(M \cdot (\vec{v})^t \right)^t$ with ^{t} for vector transposition, $\cdot$ for matrix multiplication and $*$ for its Kleene closure.

By applying iteratively this transformation from $([\![P(A_1)[0]]\!], \ldots, [\![P(A_p)[0]]\!])$, we get an increasing sequence which reaches its least upper bound $[\![L(P)]\!]$ after a finite number of iterations [5], and even after p iterations [4].

Theorem 3.4. [5] [4] *For any cci-semiring K and any cf-grammar P over K,*

$$[\![L(P)]\!] = \mathrm{HK}^{|P|} ([\![P(A_1) \cap T^*]\!], \ldots, [\![P(A_p) \cap T^*]\!]) .$$

Theorem 3.4 remains true even if the languages $P(A_1), \ldots, P(A_p)$ are infinite [4].

By applying the transformation of Proposition 3.2 (a) to any 2-grammar R over K, we get in linear time a cf-grammar $\widehat{R}$ over K such that $[\![R^\omega]\!] = [\![L(\widehat{R})]\!]$ that we can compute by Theorem 3.4 for any cci-semiring K. However the non-terminal set $N_{\widehat{R}}$ depends on ℓ_R and not on $|R|$. So it is more efficient to extend Theorem 3.4 directly to 2-grammars.

Let R be any graph grammar over K with non-terminal set $N_R = \{A_1, \ldots, A_p\}$.
Let us extend the derivative to any $(K \cup N_R)$-graph G with $1, 2, \in V_G$.

Let $A \in N_R$. We take a new symbol A_0 and we define the synchronization product

$$H := \left(G \cup \{ s \xrightarrow{A_0} t \mid s \xrightarrow[G]{A} t \} \right) \times S_A$$

$$\text{with } S_A := \{ 1 \xrightarrow{a} 1 \mid a \in K \cup N_P - \{A\} \} \cup \{1 \xrightarrow{A_0} 2\}$$

$$\cup \{ 2 \xrightarrow{a} 2 \mid a \in K \cup N_P \}.$$

Taking the vertex renaming h of H defined by $h(1,1) = 1$, $h(2,2) = 2$, and $h(x) = x$ for any $x \in V_H - \{(1,1), (2,2)\}$, and by restricting $h(H)$ to the graph $\overline{H}$ with vertices accessible from 1 and co-accessible from 2, the *derivative* of graph G by A is the graph:

$$\tfrac{\partial G}{\partial A} \; := \; \left(\overline{H} - V_{\overline{H}} \times A_0 \times V_{\overline{H}}\right) \cup \{ \, s \xrightarrow{1} t \mid s \xrightarrow{A_0} t \, \}.$$

The *Jacobian matrix* of R is $R' := \left(\tfrac{\partial R(A_i)}{\partial A_j}\right)_{1 \le i,j \le p}$ and its interpretation is

$$[\,R'\,](a_1,\ldots,a_p) \; := \; \left([\,\tfrac{\partial R(A_i)}{\partial A_j}\,[a_1,\ldots,a_p]\,]\right)_{1 \le i,j \le p} \quad \text{for any } a_1,\ldots,a_p \in K.$$

So $[\,R'\,] = [\,P'\,]$ for the infinite cf-grammar $P \colon P(A) = L(R(A),1,2)$ for any $A \in N_R$.

Corollary 3.5. *For any cci-semiring K and any graph grammar R over K,*
$$[\,R^\omega\,] \;=\; HK^{|R|}([\,R\,](0)).$$
For $+, \cdot, *$ in K in $O(1)$, the complexity is in $O(|R|^4 + |R|^2 C_R)$.

Contrary to Theorem 3.1, this corollary can be used for any *cci*-semiring. The computation of the Jacobian matrix R' is $O(\sum_{A \in N_R} |R||R(A)|) = O(|R|\ell_R)$. We compute $[\,R'\,](a_1,\ldots,a_p)$ in $O(\sum_{A \in N_R} |R| C_{R(A)[a_1,\ldots,a_p],1,2}) = O(|R| C_R)$. Having a square matrix $M \in K^{p \times p}$ and a vector $\vec{a} \in K^p$, the value of $\vec{b} = M^*.\vec{a}$ corresponds to solving the linear system $\vec{b} = \vec{a} + M.\vec{b}$ of p equations with p variables $\{b_1,\ldots,b_p\}$. By the Gauss elimination method, we use $O(p^3)$ operations $+$ and $\cdot$ operations, and $O(p)$ $*$ operations.

For the semiring $(\mathbb{R} \cup \{-\omega, \omega\}, Min, +, \omega, 0)$ of Example 2.1 (c) and using Floyd's algorithm over finite graphs, we get a complexity in $O(|R|^2 \, \ell_R^3)$ hence in $O(\ell_R^5)$ for computing $[\,R^\omega\,]$ for any 2-grammar R over K. This is a greater complexity than for Theorem 3.1.

4 Computations on Regular Graphs

The family of regular graphs contains the pushdown graphs. Precisely the regular graphs of finite degree are the regular restrictions of pushdown graphs. They are the graphs generated by deterministic graph grammars allowing non-terminal hyperarcs of arity greater than 2. By splitting these hyperacs into arcs, we provide a polynomial transformation of any graph grammar to a language-equivalent 2-grammar (Proposition 4.3). Then we can apply the previous results to obtain polynomial algorithms for solving the shortest path problem on regular graphs (Corollary 4.4).

We denote by a word $a s_1 \ldots s_{\varrho(a)}$ a *hyperarc* of label a with an *arity* $\varrho(a) \ge 1$, linking in order the vertices $s_1,\ldots,s_{\varrho(a)}$. A hyperarc $a s t$ of arity $\varrho(a) = 2$ is an arc (s,a,t). A *hypergraph* G is a set of hyperarcs, and its *length* is $\ell_G := \sum \{ \, |u| \mid u \in G \, \}$ for G finite.

A *graph grammar* R is a finite set of rules of the form:
$$A 1 \ldots \varrho(A) \; \longrightarrow \; H$$
where $A 1 \ldots \varrho(A)$ is a hyperarc labelled by A linking the vertices $1,\ldots,\varrho(A)$, and H is a finite hypergraph. Again the labels of the left hand sides form the

set N_R of non-terminals, and the remaining labels in the right hand sides form the set T_R of terminals. As we only want to generate graphs, we assume that any terminal is of arity 2.

We extend in a natural way to any graph grammar R the notions of rewriting $\underset{R}{\longrightarrow}$ and of parallel rewriting $\underset{R}{\Longrightarrow}$.

Henceforth a graph grammar R is *deterministic*: two rules have distinct left hand sides, and like for 2-grammars, we define the graph $R^\omega(H)$ generated from any hypergraph H. We give below a graph grammar reduced to a unique rule, and its generated graph.

Fig. 4.1. Graph grammar and generated graph

A graph grammar R over K means that $T_R \subseteq K$.
For any graph grammar R and for any non-terminal $A \in N_R$,
  its right hand side in R is $R(A 1 \ldots \varrho(A))$, also denoted $R(A)$,
  the graph generated by R from A is $R^\omega(A 1 \ldots \varrho(A))$, also denoted $R^\omega(A)$,
  $|R| = |N_R|$ is the number of rules of R (its cardinality),
  $\ell_R := \sum \{ |R(A)| + \varrho(A) \mid A \in N_R \}$ is the *length* of R,
  $\varrho_R := \sum \{ \varrho(A) \mid A \in N_R \}$ is the *arity* of R.
A *regular graph* is a graph generated by a graph grammar from a non-terminal.

We transform any graph grammar R into a language-equivalent 2-grammar by splitting any non-terminal hyperarc into all the possible arcs.
We assume that 0 is not a vertex of R and we take a new set of symbols
$$\{ A_{i,j} \mid A \in N_R \ \wedge \ 1 \le i,j \le \varrho(A) \}.$$
We define the *splitting* $\prec G \succ$ of any $(T_R \cup N_R)$-hypergraph G by the graph:

$$\prec G \succ := \{ s \xrightarrow{a} t \mid ast \in G \ \wedge \ a \in T_R \}$$
$$\cup \ \{ s \xrightarrow{A_{i,j}} t \mid A \in N_R \wedge 1 \le i,j \le \varrho(A) \ \wedge$$
$$\exists \, s_1, \ldots, s_{\varrho(A)}, \ As_1 \ldots s_{\varrho(A)} \in H \ \wedge \ s = s_i \ \wedge \ t = s_j \}.$$

This allows us to define the splitting of R by the 2-grammar:

$$\prec R \succ := \{ (1, A_{i,j}, 2) \longrightarrow h_{i,j}(R(A)_{i,j}) \mid A \in N_R \ \wedge \ 1 \le i,j \le \varrho(A) \}$$

where for $i \ne j$,
$$R(A)_{i,j} := \{ s \xrightarrow[\prec R(A) \succ]{a} t \mid s \ne j \wedge t \ne i \ \wedge \ s,t \notin \{1, \ldots, \varrho(A)\} - \{i,j\} \}$$

with $h_{i,j}$ the vertex renaming of $R(A)_{i,j}$ defined by
$$h_{i,j}(i) = 1, \ h_{i,j}(j) = 2, \ h_{i,j}(x) = x \ \text{otherwise},$$

and $R(A)_{i,i} := \{ s \xrightarrow[\prec R(A) \succ]{a} t \mid t \ne i \ \wedge \ s,t \notin \{1, \ldots, \varrho(A)\} - \{i\} \}$
$$\cup \ \{ s \xrightarrow{a} 0 \mid s \xrightarrow[\prec R(A) \succ]{a} i \}$$

Fig. 4.2. Splitting in reduced form of the graph grammar of Figure 4.1

with $h_{i,i}$ the vertex renaming of $R(A)_{i,i}$ defined by
$$h_{i,i}(i) = 1,\ h_{i,i}(0)=2,\ h_{i,i}(2)=0,\ h_{i,i}(x) = x\ \text{otherwise.}$$

We then put $\prec R\succ$ into the reduced form of Section 2.
Note that $|\prec R\succ| = \sum_{A\in N_R}\varrho(A)^2 \leq \varrho_R^2$

and $\ell_{\prec R\succ} \leq \sum_{A\in N_R}\varrho(A)^2\,|R(A)|^2 \leq \varrho_R^2\,\ell_R^2\,.$

So the splitting transformation is polynomial and is also language equivalent.

Proposition 4.3. *For any graph grammar R, any $(T_R \cup N_R)$-hypergraph H and any $s,t \in V_H$, we have $L(R^\omega(H),s,t) = L(\prec R\succ^\omega(\prec H\succ),s,t)$.*

Proposition 4.3 and Proposition 3.2 imply the well-known fact that the languages recognized (from and to a vertex) by regular graphs are exactly the context-free languages. But the judiciousness of Proposition 4.3 is that we can apply Theorem 3.1 and Corollary 3.5 to compute values of regular graphs. Let us extend Corollary 3.3.

Corollary 4.4. *For the semiring $(\mathbb{R} \cup \{-\omega,\omega\}, Min, +, \omega, 0)$, any graph grammar R, any left hand side X and any vertices $i,j \in V_X$, the shortest path problem $[R^\omega(X),i,j]$ can be solved in $O(\varrho_R^4\,\ell_R^4)$, and in $O(\varrho_R^4\,\ell_R^2)$ when $R^\omega(X)$ is acyclic.*

Corollary 4.4 is just a particular path problem which can be solved with polynomial complexity by using Proposition 4.3 with Theorem 3.1 or Corollary 3.5. To summarize the shortest path problem from devices generating context-free languages, we got the complexity
$O(n^2)$ for any context-free grammar,
$O(n^2)$ for any 2-grammar,
$O(n^6)$ for any graph grammar generating an acyclic graph,
$O(n^8)$ for any graph grammar,
$O(n^{18})$ for any pushdown automaton.
These algorithms result from a pumping argument: Theorem 3.1.

Of course, the complexity of the shortest path problem for pushdown automata would be improved. In this paper, we focus on path algorithms starting from graph grammars over semirings. We thank an 'anonymous' referee for this conclusion.

Note that the semiring approach cannot be used for computing the throughput value of cf-languages; a polynomial solution to this particular problem has been given in [2].

Many thanks to Antoine Meyer for his help in the writing of this paper.

References

1. Caucal, D.: On the regular structure of prefix rewriting. Theoretical Computer Science 106, 61–86 (1992)
2. Caucal, D., Czyzowicz, J., Fraczak, W., Rytter, W.: Efficient computation of throughput values of context-free languages. In: 12th CIAA, LNCS 2007 (to appear)
3. Courcelle, B.: Infinite graphs of bounded width. Mathematical Systems Theory 21(4), 187–221 (1989)
4. Esparza, J., Kiefer, S., Luttenberger, M.: On fixed point equations over commutative semirings. In: Thomas, W., Weil, P. (eds.) STACS 2007. LNCS, vol. 4393, pp. 296–307. Springer, Heidelberg (2007)
5. Hopkins, M., Kozen, D.: Parikh's theorem in commutative Kleene algebra. In: Longo, G. (ed.) 14th LICS, pp. 394–401. IEEE, Los Alamitos (1999)
6. Muller, D., Schupp, P.: The theory of ends, pushdown automata, and second-order logic. Theoretical Computer Science 37, 51–75 (1985)
7. Parikh, R.: On context-free languages. JACM 13(4), 570–581 (1966)
8. Pilling, D.: Commutative regular equations and Parikh's theorem. J. London Math. Soc. 6(4), 663–666 (1973)
9. Thomas, W.: A short introduction to infinite automata. In: Kuich, W., Rozenberg, G., Salomaa, A. (eds.) DLT 2001. LNCS, vol. 2295, pp. 130–144. Springer, Heidelberg (2002)

Factorization of Fuzzy Automata[*]

Miroslav Ćirić[1], Aleksandar Stamenković[1], Jelena Ignjatović[1],
and Tatjana Petković[2]

[1] Faculty of Sciences and Mathematics, University of Niš,
Višegradska 33, 18000 Niš, Serbia
ciricm@bankerinter.net, aca@pmf.ni.ac.yu, jejaign@yahoo.com
[2] Nokia, Joensuunkatu 7, FIN-24100 Salo, Finland
tatjana.petkovic@nokia.com

Abstract. We show that the size reduction problem for fuzzy automata
is related to the problem of solving a particular system of fuzzy relation
equations. This system consists of infinitely many equations, and finding
its general solution is a very difficult task, so we first consider one of its
special cases, a finite system whose solutions, called right invariant fuzzy
equivalences, are common generalizations of recently studied right invariant or well-behaved equivalences on NFAs, and congruences on fuzzy automata. We give a procedure for constructing the greatest right invariant
fuzzy equivalence contained in a given fuzzy equivalence, which work if the
underlying structure of truth values is a locally finite residuated lattice.

1 Introduction

Unlike deterministic automata, whose minimization is efficiently possible, it is
well-known that the state minimization of non-deterministic automata (NFA) is
computationally hard. For that reason, many researchers aimed their attention to
efficient NFA size reduction methods which do not necessarily give a minimal one.
The basic idea of reducing the size of NFAs by computing and merging indistinguishable states resembles the minimization algorithm for deterministic automata, but it is more complicated. That led to the concept of a right invariant
equivalence on an NFA, studied first by Ilie and Yu in [9], and then in [5, 6, 10,
11, 12]. From another aspect, the same concept was studied by Calude at all.
in [4], under the name well-behaved equivalence.

Fuzzy automata are generalizations of NFAs, and the mentioned problems concerning NFAs still exist in work with fuzzy automata. Size reduction algorithms
for fuzzy automata given in [1,7,13,15,16,17] are also based on the idea of computing and merging indistinguishable states, and the term minimization that we
meet there does not mean the usual construction of the minimal one in the set of
all fuzzy automata recognizing a given fuzzy language, but just the procedure of
computing and merging indistinguishable states.

---

[*] Research supported by Ministry of Science and Environmental Protection, Republic
of Serbia, Grant No. 144011.

E. Csuhaj-Varjú and Z. Ésik (Eds.): FCT 2007, LNCS 4639, pp. 213–225, 2007.

In this paper we consider size reduction of fuzzy automata from another point of view. We start from a fuzzy equivalence E on a set of states A of a fuzzy automaton $\mathscr{A}$, and without any restriction on E we turn the transition function on A into a related transition function on the factor set A_E, what results in the factor fuzzy automaton $\mathscr{A}_E$. However, fuzzy automata $\mathscr{A}_E$ and $\mathscr{A}$ are not necessarily compatible. We show that they are compatible if and only if E is a solution of a particular system of fuzzy relation equations including E, as an unknown fuzzy relation, and the transition relations on $\mathscr{A}$. If $\mathscr{A}$ is a fuzzy recognizer, it also includes fuzzy sets of initial and final states. The system consists of infinitely many equations, and finding its general solution is a very difficult task, so we point to certain special cases consisting of finitely many equations. One of them is a system whose solutions, called right invariant fuzzy equivalences, are common generalizations of the above mentioned right invariant or well-behaved equivalences, and of congruences of fuzzy automata, studied by the fourth author in [17]. We prove that any fuzzy equivalence E on $\mathscr{A}$ contains the greatest right invariant one, and we give a procedure for its construction, which works if the underlying structure $\mathscr{L}$ of truth values is a locally finite residuated lattice, but it does not necessarily work if $\mathscr{L}$ is not locally finite. This fact is not surprising if we have in mind recent results by Bĕlohlávek [2] and Li and Pedrycz [14], who found out that any finite fuzzy recognizer over $\mathscr{L}$ is equivalent to a deterministic fuzzy recognizer if and only if the semiring reduct $\mathscr{S} = (L, \vee, \otimes, 0, 1)$ of $\mathscr{L}$ is locally finite.

2   Preliminaries

In this paper we will use complete residuated lattices as the structures of truth values. A *residuated lattice* is an algebra $\mathscr{L} = (L, \wedge, \vee, \otimes, \rightarrow, 0, 1)$ such that

(L1) $(L, \wedge, \vee, 0, 1)$ is a lattice with the least element 0 and the greatest element 1,
(L2) $(L, \otimes, 1)$ is a commutative monoid with the unit 1,
(L3) $\otimes$ and $\rightarrow$ form an *adjoint pair*, i.e., they satisfy the *adjunction property*: for all $x, y, z \in L$,

$$x \otimes y \leqslant z \iff x \leqslant y \rightarrow z. \tag{1}$$

If, in addition, $(L, \wedge, \vee, 0, 1)$ is a complete lattice, then we call $\mathscr{L}$ a *complete residuated lattice*. The operations $\otimes$ (*multiplication*) and $\rightarrow$ (*residuum*) are intended for modeling the conjunction and implication of the corresponding logical calculus, supremum ($\bigvee$) and infimum ($\bigwedge$) are used for modeling the existential and general quantifier, and an operation $\leftrightarrow$ defined by $x \leftrightarrow y = (x \rightarrow y) \wedge (y \rightarrow x)$, called a *biresiduum* (or a *biimplication*), is used for modeling the equivalence of truth values. Emphasizing their monoidal structure, in some sources residuated lattices are called integral, commutative, residuated l-monoids.

It can be easily verified that with respect to $\leqslant$, $\otimes$ is isotonic in both arguments, $\rightarrow$ is isotonic in the second and antitonic in the first argument, and for any $x, y, z \in L$ and any $\{x_i\}_{i \in I}, \{y_i\}_{i \in I} \subseteq L$, the following hold:

$$x \leftrightarrow y \leqslant x \otimes z \leftrightarrow y \otimes z, \tag{2}$$

$$\left(\bigvee_{i \in I} x_i \right) \otimes x = \bigvee_{i \in I} (x_i \otimes x), \tag{3}$$

$$\bigwedge_{i \in I} (x_i \leftrightarrow y_i) \leqslant \left(\bigvee_{i \in I} x_i \right) \leftrightarrow \left(\bigvee_{i \in I} y_i \right). \tag{4}$$

For other properties of complete residuated lattices we refer to [2].

The most studied and applied structures of truth values, defined on the real unit interval $[0,1]$ with $x \wedge y = \min(x,y)$, $x \vee y = \max(x,y)$, are: *Łukasiewicz structure* ($x \otimes y = \max(x+y-1,0)$, $x \rightarrow y = \min(1-x+y,1)$), *product structure* ($x \otimes y = x \cdot y$, $x \rightarrow y = 1$ if $x \leqslant y$ and $= y/x$ otherwise) and *Gödel structure* ($x \otimes y = \min(x,y)$, $x \rightarrow y = 1$ if $x \leqslant y$ and $= y$ otherwise).

More generally, an algebra $([0,1], \wedge, \vee, \otimes, \rightarrow, 0, 1)$ is a complete residuated lattice if and only if $\otimes$ is a left-continuous t-norm and the residuum is defined by $x \rightarrow y = \bigvee \{ u \in [0,1] \mid u \otimes x \leqslant y \}$ (cf. [2]). Another important set of truth values is $\{a_0, a_1, \ldots, a_n\}$, $0 = a_0 < \cdots < a_n = 1$, with $a_k \otimes a_l = a_{\max(k+l-n,0)}$ and $a_k \rightarrow a_l = a_{\min(n-k+l,n)}$. A special case of the latter algebras is the two-element Boolean algebra of classical logic with the support $\{0,1\}$. The only adjoint pair on the two-element Boolean algebra consist of the classical conjunction and implication operations. A residuated lattice $\mathscr{L}$ satisfying $x \otimes y = x \wedge y$ is called a *Heyting algebra*, whereas a Heyting algebra satisfying the prelinearity axiom $(x \rightarrow y) \vee (y \rightarrow x) = 1$ is called a *Gödel algebra*. If any finitelly generated subalgebra of $\mathscr{L}$ is finite, then $\mathscr{L}$ is called *locally finite*. For example, every Gödel algbera, and hence, the Gödel structure, is locally finite, whereas the product structure is not locally finite.

In the further text $\mathscr{L}$ will be a complete residuated lattice. A *fuzzy subset* of a set A *over* $\mathscr{L}$, or simply a *fuzzy subset* of A, is any mapping of A into L. Let f and g be two fuzzy subsets of A. The *equality* of f and g is defined as the usual equality of mappings, i.e., $f = g$ if and only if $f(a) = g(a)$, for every $a \in A$. The *inclusion* $f \leqslant g$ is also defined pointwise: $f \leqslant g$ if and only if $f(a) \leqslant g(a)$, for every $a \in A$. Endowed with this partial order the set $\mathscr{F}(A)$ of all fuzzy subsets of A forms a completely distributive lattice, in which the meet (intersection) $\bigwedge_{i \in I} f_i$ and the join (union) $\bigvee_{i \in I} f_i$ of an arbitrary family $\{f_i\}_{i \in I}$ of fuzzy subsets of A are mappings from A into L defined by

$$\left(\bigwedge_{i \in I} f_i \right)(a) = \bigwedge_{i \in I} f_i(a), \qquad \left(\bigvee_{i \in I} f_i \right)(a) = \bigvee_{i \in I} f_i(a).$$

The *crisp part* of a fuzzy subset f of A, in notation $\widehat{f}$, is a crisp subset of A defined by $\widehat{f} = \{ a \in A \mid f(a) = 1 \}$. We will also consider $\widehat{f}$ as a mapping $\widehat{f} : A \rightarrow L$ defined by $\widehat{f}(a) = 1$, if $f(a) = 1$, and $\widehat{f}(a) = 0$, if $f(a) < 1$.

A *fuzzy relation* on A is any mapping from $A \times A$ into L, that is to say, any fuzzy subset of $A \times A$, and the equality, inclusion, joins, meets and ordering of fuzzy relations are defined as for fuzzy sets. For fuzzy relations R and S on A, their *composition* $R \circ S$ is a fuzzy relation on A defined by

$$(R \circ S)(a,b) = \bigvee_{c \in A} R(a,c) \otimes S(c,b), \tag{5}$$

for all $a,b \in A$, and for a fuzzy subset f of A and a fuzzy relation R on A, the *compositions* $f \circ R$ and $R \circ f$ are fuzzy subsets of A defined, for any $a \in A$, by

$$(f \circ R)(a) = \bigvee_{b \in A} f(b) \otimes R(b,a), \quad (R \circ f)(a) = \bigvee_{b \in A} R(a,b) \otimes f(b). \tag{6}$$

If R is symmetric, then $f \circ R = R \circ f$. For fuzzy subsets $f,g \in \mathscr{F}(A)$ we write

$$f \circ g = \bigvee_{a \in A} f(a) \otimes g(a). \tag{7}$$

We know that the composition of fuzzy relations is associative, and also

$$(f \circ R) \circ S = f \circ (R \circ S), \quad (f \circ R) \circ g = f \circ (R \circ g), \tag{8}$$

for arbitrary fuzzy subsets f and g of A, and fuzzy relations R and S on A, and hence, the parentheses in (8) can be omitted. Note also that if A is a finite set with n elements, then R and S can be treated as $n \times n$ fuzzy matrices over $\mathscr{L}$ and $R \circ S$ is the matrix product, whereas $f \circ R$ can be treated as the product of a $1 \times n$ matrix f and an $n \times n$ matrix R, and $R \circ f$ as the product of an $n \times n$ matrix R and an $n \times 1$ matrix f^t (the transpose of f).

A fuzzy relation E on A is

(R) *reflexive* if $E(a,a) = 1$, for every $a \in A$;

(S) *symmetric* if $E(a,b) = E(b,a)$, for all $a,b \in A$;

(T) *transitive* if $E(a,b) \otimes E(b,c) \leqslant E(a,c)$, for all $a,b,c \in A$.

If E is reflexive and transitive, then $E \circ E = E$. A fuzzy relation on A which is reflexive, symmetric and transitive is called a *fuzzy equivalence*. With respect to the ordering of fuzzy relations, the set $\mathscr{E}(A)$ of all fuzzy equivalence relations on a set A is a complete lattice, in which the meet coincide with the ordinary intersection of fuzzy relations, but in the general case, the join in $\mathscr{E}(A)$ does not coincide with the ordinary union of fuzzy relations. For a fuzzy equivalence E on A and $a \in A$ we define a fuzzy subset $E_a \in \mathscr{F}(A)$ by:

$$E_a(x) = E(a,x), \quad \text{for every } x \in A.$$

We call E_a an *equivalence class* of E determined by a. The set $A_E = \{E_a \mid a \in A\}$ is called the *factor set* of A w.r.t. E (cf. [2,8]). Cardinality of the factor set A_E, in notation $\mathrm{ind}(E)$, is called the *index* of E. To any fuzzy subset f of A we assign a fuzzy equivalence E_f on A defined by $E_f(a,b) = f(a) \leftrightarrow f(b)$, for all $a,b \in A$.

3   Fuzzy Automata

By a *fuzzy automaton over* $\mathscr{L}$, or simply a *fuzzy automaton*, we mean a triple $\mathscr{A} = (A, X, \delta)$, where A and X are sets, called respectively a *set of states* and an *input alphabet*, and $\delta : A \times X \times A \to L$ is a fuzzy subset of $A \times X \times A$, called a *fuzzy transition function*. The input alphabet X will be always finite, but from methodological reasons we will allow the set of states A to be infinite. A fuzzy automaton whose set of states is finite is called a *finite fuzzy automaton*.

Let X^* denote the free monoid over the alphabet X, and let $e \in X^*$ be the empty word. Then δ can be extended up to a mapping $\delta^* : A \times X^* \times A \to L$ in the following way: If $a, b \in A$, then $\delta^*(a, e, b) = 1$, for $a = b$, and $\delta^*(a, e, b) = 0$, for $a \neq b$, and if $a, b \in A$, $u \in X^*$ and $x \in X$, then

$$\delta^*(a, ux, b) = \bigvee_{c \in A} \delta^*(a, u, c) \otimes \delta(c, x, b). \tag{9}$$

Without danger of confusion we shall write just δ instead of δ^*.

By (3) and Theorem 3.1 [14], we have that

$$\delta(a, uv, b) = \bigvee_{c \in A} \delta(a, u, c) \otimes \delta(c, v, b), \tag{10}$$

for all $a, b \in A$ and $u, v \in X^*$, i.e., if $w = x_1 \cdots x_n$, for $x_1, \ldots, x_n \in X$, then

$$\delta(a, w, b) = \bigvee_{(c_1, \ldots, c_{n-1}) \in A^{n-1}} \delta(a, x_1, c_1) \otimes \delta(c_1, x_2, c_2) \otimes \cdots \otimes \delta(c_{n-1}, x_n, b). \tag{11}$$

If for any $u \in X^*$ we define a fuzzy relation $\delta_u \in \mathscr{F}(A \times A)$, called the *transition relation* determined by u, by $\delta_u(a, b) = \delta(a, u, b)$, for all $a, b \in A$, then for all $u, v \in X^*$, the equality (10) can be written as $\delta_{uv} = \delta_u \circ \delta_v$.

A *fuzzy recognizer* is a five-tuple $\mathscr{A} = (A, \sigma, X, \delta, \tau)$, where (A, X, δ) is a fuzzy automaton, and σ and τ are fuzzy subsets of A, called respectively a fuzzy set of *initial states*, and a fuzzy set of *terminal states*.

A *fuzzy language* in X^* over $\mathscr{L}$, or briefly a *fuzzy language*, is any fuzzy subset of X^*, i.e., any mapping from X^* into L. A fuzzy recognizer $\mathscr{A} = (A, \sigma, X, \delta, \tau)$ *recognizes* a fuzzy language $f \in \mathscr{F}(X^*)$ if for any $u \in X^*$ we have

$$f(u) = \bigvee_{a, b \in A} \sigma(a) \otimes \delta(a, u, b) \otimes \tau(b). \tag{12}$$

Using notation from (6), and the second equality in (8), we can state (12) as

$$f(u) = \sigma \circ \delta_u \circ \tau. \tag{13}$$

The fuzzy language recognized by a fuzzy recognizer $\mathscr{A}$ is denoted by $L(\mathscr{A})$, and called the *fuzzy language of* $\mathscr{A}$.

If δ is a crisp subset of $A \times X \times A$, i.e., $\delta : A \times X \times A \to \{0,1\}$, then $\mathscr{A}$ is an ordinary crisp non-deterministic automaton, and if δ is a mapping of $A \times X$ into A, then $\mathscr{A}$ is an ordinary deterministic automaton. Evidently, in these two cases we have that δ is also a crisp subset of $A \times X^* \times A$, and a mapping of $A \times X^*$ into A, respectively.

For undefined notions and notations we refer to [2] and [16].

4   Factor Fuzzy Automata and Fuzzy Relation Equations

Let $\mathscr{A} = (A, X, \delta)$ be a fuzzy automaton and let E be a fuzzy equivalence on A. Without any restriction on the fuzzy equivalence E, we can define a fuzzy transition function $\delta_E : A_E \times X \times A_E \to L$ by

$$\delta_E(E_a, x, E_b) = \bigvee_{a',b' \in A} E(a, a') \otimes \delta(a', x, b') \otimes E(b', b) \qquad (14)$$

or equivalently,

$$\delta_E(E_a, x, E_b) = (E \circ \delta_x \circ E)(a, b) = E_a \circ \delta_x \circ E_b, \qquad (15)$$

for any $a, b \in A$ and $x \in X$. Evidently, δ_E is well-defined, and $\mathscr{A}_E = (A_E, X, \delta_E)$ is a fuzzy automaton, called the *factor fuzzy automaton* of $\mathscr{A}$ w.r.t. E.

If, in addition, $\mathscr{A} = (A, \sigma, X, \delta, \tau)$ is a fuzzy recognizer, then without any restriction on E, we can also define a fuzzy set $\sigma_E \in \mathscr{F}(A_E)$ of initial states and a fuzzy set $\tau_E \in \mathscr{F}(A_E)$ of terminal states by

$$\sigma_E(E_a) = \bigvee_{a' \in A} \sigma(a') \otimes E(a', a) = (\sigma \circ E)(a) = \sigma \circ E_a, \qquad (16)$$

$$\tau_E(E_a) = \bigvee_{a' \in A} \tau(a') \otimes E(a', a) = (\tau \circ E)(a) = \tau \circ E_a, \qquad (17)$$

for any $a \in A$. Clearly, σ_E and τ_E are well-defined and $\mathscr{A}_E = (A_E, \sigma_E, X, \delta_E, \tau_E)$ is a fuzzy recognizer, called the *factor fuzzy recognizer* of $\mathscr{A}$ w.r.t. E.

The factor automaton $\mathscr{A}_E$ is not necessarily compatible with $\mathscr{A}$, i.e., for any $a, b \in A$ and $u = x_1 x_2 \cdots x_n \in X^*$ we have that

$$\delta_E(E_a, u, E_b) = (E \circ \delta_{x_1} \circ E \circ \delta_{x_2} \circ E \circ \cdots \circ \delta_{x_n} \circ E)(a, b), \qquad (18)$$

and, in the general case, $\delta_E(E_a, u, E_b)$ is not necessarily equal to $(E \circ \delta_u \circ E)(a, b)$. If for all $a, b \in A$ and $u \in X^*$ we have that

$$\delta_E(E_a, u, E_b) = (E \circ \delta_u \circ E)(a, b), \qquad (19)$$

then we say that the factor fuzzy automaton $\mathscr{A}_E$ is *compatible* with $\mathscr{A}$. Thus, $\mathscr{A}_E$ is compatible with $\mathscr{A}$ if and only if E is a solution of a fuzzy relation equation

$$E \circ \delta_{x_1} \circ E \circ \delta_{x_2} \circ E \circ \cdots \circ \delta_{x_n} \circ E = E \circ \delta_{x_1} \circ \delta_{x_2} \circ \cdots \circ \delta_{x_n} \circ E, \qquad (20)$$

for every $n \in \mathbb{N}$ and $x_1, x_2, \ldots, x_n \in X$.

Furthermore, if $\mathscr{A}$ is a fuzzy recognizer, then the fuzzy language recognized by the factor fuzzy recognizer $\mathscr{A}_E$ is given by

$$L(\mathscr{A}_E)(u) = \sigma \circ E \circ \delta_{x_1} \circ E \circ \delta_{x_2} \circ E \circ \cdots \circ \delta_{x_n} \circ E \circ \tau, \qquad (21)$$

for any $u = x_1 x_2 \cdots x_n \in X^*$, and we have that $L(\mathscr{A}) = L(\mathscr{A}_E)$ if and only if E is a solution of a fuzzy relation equation

$$\sigma \circ E \circ \delta_{x_1} \circ E \circ \delta_{x_2} \circ E \circ \cdots \circ \delta_{x_n} \circ E \circ \tau = \sigma \circ \delta_{x_1} \circ \delta_{x_2} \circ \cdots \circ \delta_{x_n} \circ \tau \qquad (22)$$

for every $n \in \mathbb{N}$ and $x_1, x_2, \ldots, x_n \in X$. Clearly, the systems (20) and (22) have at least one solution in $\mathscr{E}(A)$, the equality relation on A, but to obtain the best possible reduction of $\mathscr{A}$ we have to find the greatest solution in $\mathscr{E}(A)$, if it exists, or to find as big a solution as possible.

The systems (20) and (22) consist of infinitely many equations, and finding their general solution is a very difficult task. However, imposing certain conditions on E, the number of equations can be considerably reduced. In particular, if we assume that E is a solution of a finite system

$$\delta_x \circ E \circ \delta_y \circ E = \delta_x \circ \delta_y \circ E, \quad x, y \in X, \tag{23}$$

by induction we can prove that E is also a solution of any equation of the form

$$\delta_{x_1} \circ E \circ \delta_{x_2} \circ E \circ \cdots \circ \delta_{x_n} \circ E = \delta_{x_1} \circ \delta_{x_2} \circ \cdots \circ \delta_{x_n} \circ E, \tag{24}$$

for any $n \in \mathbb{N}$, $x_1, \ldots, x_n \in X$, so it is a solution of the system (20). In this case we have the following system, whose any solution is also a solution of (22):

$$\begin{aligned} \delta_x \circ E \circ \delta_y \circ E &= \delta_x \circ \delta_y \circ E, \quad x, y \in X, \\ \sigma \circ E &= \sigma, \\ \tau \circ E &= \tau. \end{aligned} \tag{25}$$

Analogously, if we assume that E is a solution of a finite system

$$E \circ \delta_x \circ E \circ \delta_y = E \circ \delta_x \circ \delta_y, \quad x, y \in X, \tag{26}$$

we obtain the following system, whose any solution is also a solution of (22):

$$\begin{aligned} E \circ \delta_x \circ E \circ \delta_y &= E \circ \delta_x \circ \delta_y, \quad x, y \in X, \\ \sigma \circ E &= \sigma, \\ \tau \circ E &= \tau. \end{aligned} \tag{27}$$

Note that the equations $\sigma \circ E = \sigma$ and $\tau \circ E = \tau$ can be easily solved. It is well-known that E is a solution of $\sigma \circ E = \sigma$ (resp. $\tau \circ E = \tau$) if and only if $E \leqslant E_\sigma$ (resp. $E \leqslant E_\tau$), and hence, E_σ (resp. E_τ) is the greatest solution of this equation. Thus, solutions of the system (25) (resp. (27)) are exactly those solutions of the system (23) (resp. (26)) which are contained in $E_\sigma \wedge E_\tau$. The equations $\sigma \circ E = \sigma$ and $\tau \circ E = \tau$ also have a natural interpretation. Roughly speaking, $E(a, b) \leqslant E_\sigma(a, b)$ and $E(a, b) \leqslant E_\tau(a, b)$, for all $a, b \in A$, mean that E does not merge initial and noninitial, and terminal and nonterminal states.

5 Right Invariant Fuzzy Equivalences

We start study of the system (23) considering one of its most interesting special cases, whose solutions in $\mathscr{E}(A)$ are generalizations both of right invariant equivalences on non-deterministic automata, studied by Ilie, Yu and others [5, 6, 9, 10, 11, 12], or well-behaved equivalences, studied by Calude et al. [4], and of congruences on fuzzy automata, studied by the fourth author in [17].

Let $\mathscr{A} = (A, X, \delta)$ be a fuzzy automaton. A fuzzy equivalence E on A will be called *right invariant* if it is a solution of the system of fuzzy relation equations

$$E \circ \delta_x \circ E = \delta_x \circ E, \quad x \in X. \tag{28}$$

Using the reverse equations we define *left invariant* fuzzy equivalences.

Right invariant fuzzy equivalences can be characterized as follows:

Theorem 1. *Let $\mathscr{A} = (A, X, \delta)$ be a fuzzy automaton and E a fuzzy equivalence on A. Then the following conditions are equivalent:*

(i) *E is a right invariant fuzzy equivalence;*
(ii) *$E \circ \delta_x \leqslant \delta_x \circ E$, for every $x \in X$;*
(iii) *for every $a, b \in A$ we have*

$$E(a, b) \leqslant \bigwedge_{x \in X} \bigwedge_{c \in A} (\delta_x \circ E)(a, c) \leftrightarrow (\delta_x \circ E)(b, c). \tag{29}$$

Proof. (i)$\Leftrightarrow$(ii). Consider an arbitrary $x \in X$. If $E \circ \delta_x \circ E = \delta_x \circ E$, then we have that $E \circ \delta_x \leqslant E \circ \delta_x \circ E = \delta_x \circ E$. Conversely, if $E \circ \delta_x \leqslant \delta_x \circ E$ then $E \circ \delta_x \circ E \leqslant \delta_x \circ E \circ E = \delta_x \circ E$, and since the opposite inequality always hold, we conclude that $E \circ \delta_x \circ E = \delta_x \circ E$.

(i)$\Rightarrow$(iii). Let E be right invariant. Then for any $x \in X$ and $a, b, c \in A$ we have that $E(a, b) \otimes (\delta_x \circ E)(b, c) \leqslant (E \circ \delta_x \circ E)(a, c) = (\delta_x \circ E)(a, c)$, and by the adjunction property, we obtain that $E(a, b) \leqslant (\delta_x \circ E)(b, c) \rightarrow (\delta_x \circ E)(a, c)$. By symmetry, $E(a, b) = E(b, a) \leqslant (\delta_x \circ E)(a, c) \rightarrow (\delta_x \circ E)(b, c)$, and hence,

$$E(a, b) \leqslant (\delta_x \circ E)(a, c) \leftrightarrow (\delta_x \circ E)(b, c). \tag{30}$$

Since (30) is satisfied for every $c \in A$ and $x \in X$, we conclude that (29) holds.

(iii)$\Rightarrow$(i). If (iii) holds, then for arbitrary $x \in X$ and $a, b, c \in A$ we have that

$$E(a, b) \leqslant (\delta_x \circ E)(a, c) \leftrightarrow (\delta_x \circ E)(b, c) \leqslant (\delta_x \circ E)(b, c) \rightarrow (\delta_x \circ E)(a, c),$$

and by the adjunction property, $E(a, b) \otimes (\delta_x \circ E)(b, c) \leqslant (\delta_x \circ E)(a, c)$. Now,

$$(E \circ \delta_x \circ E)(a, c) = \bigvee_{b \in A} E(a, b) \otimes (\delta_x \circ E)(b, c) \leqslant (\delta_x \circ E)(a, c),$$

and hence, $E \circ \delta_x \circ E \leqslant \delta_x \circ E$. Since the opposite inequality always hold, we conclude that $E \circ \delta_x \circ E = \delta_x \circ E$, for each $x \in X$.   □

The next two remarks show that right invariant fuzzy equivalences are common generalizations of right invariant or well-behaved equivalences on non-deterministic automata and of congruences on fuzzy automata.

Remark 1. Let $\mathscr{A} = (A, X, \delta)$ be a crisp non-deterministic automaton and E an equivalence on A. It is easy to check that $E \circ \delta_x \subseteq \delta_x \circ E$ is equivalent to

(P_2) $(\forall a, b \in A)(\forall x \in X)((a, b) \in E \Rightarrow (\forall b' \in \delta(b, x))(\exists a' \in \delta(a, x))(a', b') \in E)$,

what is the second of two conditions by which Ilie, Navarro and Yu in [11] defined the notion of a right invariant equivalence on a non-deterministic automaton (see also [12,6]). The first one, which requires that terminal and non-terminal states are not E-equivalent, can be written in the fuzzy case as $\tau \circ E = \tau$. Here we have excluded this condition from the definition of a right invariant fuzzy equivalence, and it will be considered separately. Calude et al. in [4] called equivalences satisfying (P_2) well-behaved. Note also that an equivalent form of the condition (P_2), appearing in [9,10], correspond to our condition (iii) in Theorem 1.

Remark 2. Let $\mathscr{A} = (A, X, \delta)$ be a fuzzy automaton and E a crisp equivalence on A. By (iii) of Theorem 1 we have that E is a right invariant equivalence on $\mathscr{A}$ if and only if for any $a, a' \in A$, by $(a, a') \in E$ it follows $(\delta_x \circ E)(a, b) = (\delta_x \circ E)(a', b)$, for all $x \in X$ and $b \in A$. But, $(\delta_x \circ E)(a, b) = (\delta_x \circ E)(a', b)$ is equivalent to

$$\bigvee_{b' \in E_b} \delta(a, x, b') = \bigvee_{b' \in E_b} \delta(a', x, b'),$$

and hence, right invariant crisp equivalences on fuzzy automata are nothing else than congruences on fuzzy automata studied by the fourth author in [17] (or partitions with substitution property from [1]).

Let $\mathscr{A} = (A, X, \delta)$ be a fuzzy automaton and E a fuzzy equivalence on A. Let us define the fuzzy relations $E^x : A \times A \to L$, for any $x \in X$, and $E^r : A \times A \to L$ by

$$E^x(a, b) = \bigwedge_{c \in A} (\delta_x \circ E)(a, c) \leftrightarrow (\delta_x \circ E)(b, c), \quad E^r(a, b) = \bigwedge_{x \in X} E^x(a, b). \quad (31)$$

By the well-known Valverde's Representation Theorem [18] (see also [2,8]) we have that E^x, for each $x \in X$, and E^r are fuzzy equivalences.

Lemma 1. *Let $\mathscr{A} = (A, X, \delta)$ be a fuzzy automaton, and let E and F be fuzzy equivalences on A. If $E \leqslant F$, then $E^r \leqslant F^r$.*

Proof. Consider arbitrary $a, b \in A$ and $x \in X$. By $E \leqslant F$ it follows $E \circ F = F$, and by (2), for arbitrary $c, d \in A$ we have that

$$(\delta_x \circ E)(a, c) \leftrightarrow (\delta_x \circ E)(b, c) \leqslant (\delta_x \circ E)(a, c) \otimes F(c, d) \leftrightarrow (\delta_x \circ E)(b, c) \otimes F(c, d).$$

Now, by (4) we obtain that

$$E^r(a, b) \leqslant \bigwedge_{c \in A} (\delta_x \circ E)(a, c) \leftrightarrow (\delta_x \circ E)(b, c)$$

$$\leqslant \bigwedge_{c \in A} \left[(\delta_x \circ E)(a, c) \otimes F(c, d) \leftrightarrow (\delta_x \circ E)(b, c) \otimes F(c, d) \right]$$

$$\leqslant \left[\bigvee_{c \in A} (\delta_x \circ E)(a, c) \otimes F(c, d) \right] \leftrightarrow \left[\bigvee_{c \in A} (\delta_x \circ E)(b, c) \otimes F(c, d) \right]$$

$$= (\delta_x \circ E \circ F)(a, d) \leftrightarrow (\delta_x \circ E \circ F)(b, d) = (\delta_x \circ F)(a, d) \leftrightarrow (\delta_x \circ F)(b, d).$$

This holds for any $x \in X$ and $d \in A$, so we conclude that $E^r \leqslant F^r$. □

Theorem 2. *The set $\mathscr{E}^{\mathrm{ri}}(\mathscr{A})$ of all right invariant fuzzy equivalences on a fuzzy automaton $\mathscr{A} = (A, X, \delta)$ forms a complete lattice. This lattice is a complete join-subsemilattice of the lattice $\mathscr{E}(A)$ of all fuzzy equivalences on A.*

Proof. Since $\mathscr{E}^{\mathrm{ri}}(\mathscr{A})$ contains the least element of $\mathscr{E}(A)$, the equality relation on A, it is enough to prove that $\mathscr{E}^{\mathrm{ri}}(\mathscr{A})$ is a complete join-subsemilattice of $\mathscr{E}(A)$.

Let $\{E_i\}_{i \in I}$ be a family of right invariant fuzzy equivalences on $\mathscr{A}$, and E its join in $\mathscr{E}(A)$. For any $i \in I$, by $E_i \leqslant E$ and Lemma 1 we have that $E_i \leqslant E_i^r \leqslant E^r$, so $E \leqslant E^r$. Thus, by (iii) of Theorem 1 we obtain that E is right invariant. $\square$

By Theorem 2 it follows that for any fuzzy equivalence E on $\mathscr{A}$ there exists the greatest right invariant fuzzy equivalence contained in E, that will be denoted by $E^\flat$. In the next theorem we consider the problem how to construct it.

Theorem 3. *Let $\mathscr{A} = (A, X, \delta)$ be a fuzzy automaton, E a fuzzy equivalence on A and $E^\flat$ the greatest right invariant fuzzy equivalence contained in E.*

Define inductively a sequence $\{E_k\}_{k \in \mathbb{N}}$ of fuzzy equivalences on A as follows:

$$E_1 = E, \quad E_{k+1} = E_k \wedge E_k^r, \quad \text{for each } k \in \mathbb{N}.$$

Then

(a) $E^\flat \leqslant \cdots \leqslant E_{k+1} \leqslant E_k \leqslant \cdots \leqslant E_1 = E$;
(b) *If $E_k = E_{k+m}$, for some $k, m \in \mathbb{N}$, then $E_k = E_{k+1} = E^\flat$;*
(c) *If $\mathscr{A}$ is finite and $\mathscr{L}$ is locally finite, then $E_k = E^\flat$ for some $k \in \mathbb{N}$.*

Proof. (a) Evidently, $E_k \leqslant E_{k+1}$, for each $k \in \mathbb{N}$, and $E^\flat \leqslant E_1$. Suppose that $E^\flat \leqslant E_k$, for some $k \in \mathbb{N}$. Then $E^\flat \leqslant (E^\flat)^r \leqslant E_k^r$, so $E^\flat \leqslant E_k \wedge E_k^r = E_{k+1}$. Hence, by induction we obtain that $E^\flat \leqslant E_k$, for any $k \in \mathbb{N}$.

(b) Let $E_k = E_{k+m}$, for some $k, m \in \mathbb{N}$. Then $E_k = E_{k+m} \leqslant E_{k+1} = E_k \wedge E_k^r \leqslant E_k^r$, what means that E_k is a right invariant fuzzy equivalence. Since $E^\flat$ is the greatest right invariant fuzzy equivalence contained in E, we conclude that $E_k = E_{k+1} = E^\flat$.

(c) Let $\mathscr{A}$ be a finite fuzzy automaton and $\mathscr{L}$ be a locally finite algebra. Let the carrier of a subalgebra of $\mathscr{L}$ generated by the set $\delta(A \times X \times A) \cup E(A \times A)$ be denoted by $L_\mathscr{A}$. This generating set is finite, so $L_\mathscr{A}$ is also finite, and hence, the set $L_\mathscr{A}^{A \times A}$ of all fuzzy relations on A with values in $L_\mathscr{A}$ is finite. By definitions of fuzzy relations E_k and E_k^r we have that $E_k \in L_\mathscr{A}^{A \times A}$, which implies that there are $k, n \in \mathbb{N}$ such that $E_k = E_{k+m}$, and by (b) we conclude that $E_k = E^\flat$. $\square$

The above theorem gives a procedure for construction of the greatest right invariant fuzzy equivalence contained in a given fuzzy equivalence E on a finite fuzzy automaton, which works if $\mathscr{L}$ is locally finite, but it does not necessarily work if $\mathscr{L}$ is not locally finite, what the following example shows:

Example 1. Let $\mathscr{L}$ be the product structure, $\mathscr{A} = (A, X, \delta)$ a fuzzy automaton over $\mathscr{L}$ with $A = \{1, 2\}$, $X = \{x\}$, and a transition relation δ_x given as in (32), and E the universal relation on A. Applying to E the procedure from Theorem 3, we obtain a sequence $\{E_k\}_{k\in\mathbb{N}}$ of fuzzy equivalences (shown also in (32)) whose all members are different. We have that $E^\flat$ is the equality relation on A, since it is the only right invariant fuzzy equivalence on $\mathscr{A}$.

$$\delta_x = \begin{bmatrix} 0 & 1 \\ \frac{1}{2} & 0 \end{bmatrix}, \qquad E_k = \begin{bmatrix} 1 & \frac{1}{2^{k-1}} \\ \frac{1}{2^{k-1}} & 1 \end{bmatrix}, k\in\mathbb{N}, \qquad E^\flat = \begin{bmatrix} 1 & 0 \\ 0 & 1 \end{bmatrix}. \qquad (32)$$

Therefore, if the complete residuated lattice $\mathscr{L}$ is not locally finite, we know that there exists the greatest right invariant fuzzy equivalence $E^\flat$ contained in E, but the problem is how to construct it. In some cases, to reduce $\mathscr{A}$, we can try to find the greatest right invariant crisp equivalence E° contained in E (or in $\widehat{E}$), which can have the same index as $E^\flat$, so the factor fuzzy automata $\mathscr{A}_{E^\circ}$ and $\mathscr{A}_{E^\flat}$ would have the same number of states. This can be done using a procedure given by the fourth author in [17], which can be stated here as follows:

Theorem 4. [17] *Let* $\mathscr{A} = (A, X, \delta)$ *be a fuzzy automaton,* ϱ *an equivalence on* A *and* ϱ° *the greatest right invariant equivalence contained in* ϱ.

Define inductively a sequence $\{\varrho_k\}_{k\in\mathbb{N}}$ *of fuzzy equivalences on* A *as follows:*

$$\varrho_1 = \varrho, \quad \varrho_{k+1} = \varrho_k \cap \varrho_k^r, \ \text{for each } k\in\mathbb{N}, \ \text{where}$$

$$(a,b) \in \varrho_k^r \ \Leftrightarrow \ (\forall x\in X)(\forall c\in A)(\delta_x \circ \varrho_k)(a,c) = (\delta_x \circ \varrho_k)(b,c).$$

Then

(a) $\varrho^\circ \leqslant \cdots \leqslant \varrho_{k+1} \leqslant \varrho_k \leqslant \cdots \leqslant \varrho_1 = \varrho$;

(b) *If* $\varrho_k = \varrho_{k+m}$, *for some* $k, m \in \mathbb{N}$, *then* $\varrho_k = \varrho_{k+1} = \varrho^\circ$;

(c) *If* $\mathscr{A}$ *is finite, then* $\varrho_k = \varrho^\circ$ *for some* $k\in\mathbb{N}$. $\qquad\qquad$ □

However, E° can have the greather index than $E^\flat$, and hence, $\mathscr{A}_{E^\circ}$ can have more states than $\mathscr{A}_{E^\flat}$, as the following example shows:

Example 2. Let $\mathscr{L}$ be the Gödel structure, $\mathscr{A} = (A, X, \delta)$ a fuzzy automaton over $\mathscr{L}$ with $A = \{1, 2, 3, 4\}$, $X = \{x, y\}$, and transition relations given by

$$\delta_x = \begin{bmatrix} 1 & 0.8 & 0.6 & 0.8 \\ 0.8 & 1 & 0.8 & 0.6 \\ 0.2 & 0.3 & 0.8 & 0.9 \\ 0.2 & 0.3 & 0.8 & 0.9 \end{bmatrix}, \qquad \delta_y = \begin{bmatrix} 0.8 & 1 & 0.6 & 0.8 \\ 1 & 0.6 & 0.5 & 0.9 \\ 0.3 & 0.2 & 0.4 & 0.8 \\ 0.5 & 0.3 & 0.3 & 1 \end{bmatrix}.$$

and let E be the universal relation on A. Applying to E procedures from Theorems 3 and 4, we obtain that $E_2 = E_3 = E^\flat$ and $\varrho_3 = \varrho_4 = E^\circ$, where

$$E^\flat = \begin{bmatrix} 1 & 1 & 0.8 & 0.9 \\ 1 & 1 & 0.8 & 0.9 \\ 0.8 & 0.8 & 1 & 0.8 \\ 0.9 & 0.9 & 0.8 & 1 \end{bmatrix}, \qquad E^\circ = \begin{bmatrix} 1 & 0 & 0 & 0 \\ 0 & 1 & 0 & 0 \\ 0 & 0 & 1 & 0 \\ 0 & 0 & 0 & 1 \end{bmatrix}.$$

Therefore, E° does not make a reduction of $\mathscr{A}$, whereas $E^\flat$ has three classes, and reduces $\mathscr{A}$ to a fuzzy automaton $\mathscr{A}_{E^\flat}$ having three states and transition matrices

$$\delta_x^{E^\flat} = \begin{bmatrix} 1 & 0.8 & 0.9 \\ 0.9 & 0.8 & 0.9 \\ 0.9 & 0.8 & 0.9 \end{bmatrix}, \qquad \delta_y^{E^\flat} = \begin{bmatrix} 1 & 0.8 & 0.9 \\ 0.8 & 0.8 & 0.8 \\ 0.9 & 0.8 & 1 \end{bmatrix},$$

with entries taken from the matrices $\delta_x \circ E^\flat = \delta_x \circ E_2$ and $\delta_y \circ E^\flat = \delta_y \circ E_2$.

If $\mathscr{L}$ has a property that $\vee K = 1$ implies $1 \in K$, for any finite $K \subseteq L$, what is satisfied, for example, whenever $\mathscr{L}$ is linearly ordered, then for arbitrary fuzzy relations R and S on a finite set A we have that $\widehat{R \circ S} = \widehat{R} \circ \widehat{S}$. In this case, for any right invariant fuzzy equivalence E on a fuzzy automaton $\mathscr{A} = (A, X, \delta)$ we have that $\widehat{E}$ is a right invariant crisp equivalence on the non-deterministic automaton $\widehat{\mathscr{A}} = (A, X, \widehat{\delta})$, the crisp part of $\mathscr{A}$. Also, for any fuzzy equivalence F on $\mathscr{A}$, the crisp part of $F^\flat$ is the greatest right invariant equivalence on $\widehat{\mathscr{A}}$ contained in $\widehat{F}$.

Note that dual results for left invariant fuzzy equivalences can be obtained.

As a conclusion let us remark that this paper imposes a lot of open questions, like: Whether the systems (20) and (22) have greatest solutions in $\mathscr{E}(A)$, and what their special cases have such solutions? When such solutions exist, how to construct them? Especially, how to construct the greatest right invariant fuzzy equivalence in the case when $\mathscr{L}$ is not locally finite? As different special cases of the systems (20) and (22) give different types of fuzzy equivalence relations, how to achieve a better reduction combining various types of these fuzzy equivalence relations? All these questions will be considered in our further research.

References

1. Basak, N.C., Gupta, A.: On quotient machines of a fuzzy automation and the minimal machine. Fuzzy Sets and Systems 125, 223–229 (2002)
2. Bělohlávek, R.: Fuzzy Relational Systems: Foundations and Principles. Kluwer, New York (2002)
3. Bělohlávek, R.: Determinism of fuzzy automata. Information Sciences 143, 205–209 (2002)
4. Calude, C.S., Calude, E., Khoussainov, B.: Finite nondeterministic automata: Simulation and minimality. Theoretical Computer Science 242, 219–235 (2000)
5. Câmpeanu, C., Sântean, N., Yu, S.: Mergible states in large NFA. Theoretical Computer Science 330, 23–34 (2005)
6. Champarnaud, J.-M., Coulon, F.: NFA reduction algorithms by means of regular inequalities. Theoretical Computer Science 327, 241–253 (2004)
7. Cheng, W., Mo, Z.: Minimization algorithm of fuzzy finite automata. Fuzzy Sets and Systems 141, 439–448 (2004)
8. Ćirić, M., Ignjatović, J., Bogdanović, S.: Fuzzy equivalence relations and their equivalence classes. Fuzzy Sets and Systems 158, 1295–1313 (2007)
9. Ilie, L., Yu, S.: Algorithms for computing small NFAs. In: Diks, K., Rytter, W. (eds.) MFCS 2002. LNCS, vol. 2420, pp. 328–340. Springer, Heidelberg (2002)
10. Ilie, L., Yu, S.: Reducing NFAs by invariant equivalences. Theoretical Computer Science 306, 373–390 (2003)
11. Ilie, L., Navarro, G., Yu, S.: On NFA reductions. In: Karhumäki, J., Maurer, H., Păun, G., Rozenberg, G. (eds.) Theory Is Forever. LNCS, vol. 3113, pp. 112–124. Springer, Heidelberg (2004)
12. Ilie, L., Solis-Oba, R., Yu, S.: Reducing the size of NFAs by using equivalences and preorders. In: Apostolico, A., Crochemore, M., Park, K. (eds.) CPM 2005. LNCS, vol. 3537, pp. 310–321. Springer, Heidelberg (2005)

13. Lei, H., Li, Y.M.: Minimization of states in automata theory based on finite lattice-ordered monoids. Information Sciences 177, 1413–1421 (2007)
14. Li, Y.M., Pedrycz, W.: Fuzzy finite automata and fuzzy regular expressions with membership values in lattice ordered monoids. Fuzzy Sets and Systems 156, 68–92 (2005)
15. Malik, D.S., Mordeson, J.N., Sen, M.K.: Minimization of fuzzy finite automata. Information Sciences 113, 323–330 (1999)
16. Mordeson, J.N., Malik, D.S.: Fuzzy Automata and Languages: Theory and Applications. Chapman & Hall/CRC, Boca Raton, London (2002)
17. Petković, T.: Congruences and homomorphisms of fuzzy automata. Fuzzy Sets and Systems 157, 444–458 (2006)
18. Valverde, L.: On the structure of F-indistinguishability operators. Fuzzy Sets and Systems 17, 313–328 (1985)

Factorisation Forests for Infinite Words
Application to Countable Scattered Linear Orderings

Thomas Colcombet

CNRS/IRISA
thomas.colcombet@irisa.fr

Abstract. The theorem of *factorisation forests* shows the existence of nested factorisations — a la Ramsey — for finite words. This theorem has important applications in semigroup theory, and beyond.

We provide two improvements to the standard result. First we improve on all previously known bounds for the standard theorem. Second, we extend it to every 'complete linear ordering'. We use this variant in a simplified proof of complementation of automata over words of countable scattered domain.

Keywords: Formal languages, semigroups, infinite words, automata.

1 Introduction

Factorisation forests were introduced by Simon [15]. The associated theorem — which we call the theorem of factorisation forests below — states that for every semigroup morphism from words to a finite semigroup S, every word has a Ramseyan factorisation tree of height linearly bounded by $|S|$ (see below). An alternative presentation states that for every morphism φ from A^+ to some finite semigroup S, there exists a regular expression evaluating to A^+ in which the Kleene star L^* is allowed only when $\varphi(L) = \{e\}$ for some $e = e^2 \in S$; i.e. the Kleene star is allowed only if it produces a Ramseyan factorisation of the word.

The theorem of factorisation forests provides a very deep insight on the structure of finite semigroups, and has therefore many applications. Let us cite some of them. Distance automata are nondeterministic finite automata mapping words to naturals. An important question concerning them is the limitedness problem: decide whether this mapping is bounded or not. It has been shown decidable by Simon using the theorem of factorisation forests [15]. This theorem also allows a constructive proof of Brown's lemma on locally finite semigroups [2]. It is also used in the characterisation of subfamilies of the regular languages, for instance the polynomial closure of varieties in [11]. Or to give general characterisations of finite semigroups [10]. In the context of languages of infinite words indexed by ω, it has also been used in a complementation procedure [1] extending Buchi's lemma [4]. In [7], a deterministic variant of the theorem of factorisation forest is used for proving that every monadic second-order interpretation is equivalent over trees to the composition of a first-order interpretation and a monadic

E. Csuhaj-Varjú and Z. Ésik (Eds.): FCT 2007, LNCS 4639, pp. 226–237, 2007.

second-order marking. This itself provides new result in the theory of finitely presentable infinite structures.

The present paper aims first at advertising the theorem of factorisation forest which, though already used in many papers, is in fact known only to a quite limited community. The reason for this is that its proofs rely on the use of Green's relations: Green's relations form an important tool in semigroup theory, but are technical and uncomfortable to work with. The merit of the factorisation forest theorem is that it is usable without any significant knowledge of semigroup theory, while it encapsulates nontrivial parts of this theory. Furthermore, as briefly mentioned above, this theorem has natural applications in automata theory.

This paper contains three contributions. First, we provide a new proof of the original theorem improving on all previously known bounds in [15] and [6]. Second, we extend the result to the infinite case (i.e., to infinite words, though we use a different presentation). Third, we use this last extension in a simplified proof of complementation of automata on countable scattered linear orderings, a result known from Carton and Rispal [5].

The content of the paper is organised as follows. Section 2 is dedicated to definitions. Section 3 presents the original theorem of factorisation forests as well as a variant in terms of Ramseyan splits and its extension to the infinite case. In Section 4 we apply this last extension to the complementation of automata over countable scattered linear orderings.

2 Definitions

In this section, we successively present linear orderings, words indexed by them, semigroups and additive labellings.

2.1 Linear Orderings

A *linear ordering* $\alpha = (L, <)$ is a set L equipped with a total ordering relation $<$; i.e., an irreflexive, antisymmetric and transitive relation such that for every distinct elements x, y in L, either $x < y$ or $y < x$. Two linear orderings $\alpha = (L, <)$ and $\beta = (L', <')$ have same *order type* if there exists a bijection f from L onto L' such that for every x, y in L, $x < y$ iff $f(x) <' f(y)$. We denote by $\omega, -\omega, \zeta$ the order types of respectively $(\mathbb{N}, <)$, $(-\mathbb{N}, <)$ and $(\mathbb{Z}, <)$. Below, we do not distinguish between a linear ordering and its order type unless necessary. This is safe since all the constructions we perform are defined up to similar order type.

A *subordering* β of α is a subset of L equipped with the same ordering relation; i.e., $\beta = (L', <)$ with $L' \subseteq L$. We write $\beta \subseteq \alpha$. A *convex subset of* α is a subset S of α such that for all $x, y \in S$ and $x < z < y$, $z \in S$. We use the notations $[x, y], [x, y[,]x, y],]x, y[,]-\infty, y],]-\infty, y[, [x, +\infty[$ and $]x, +\infty[$ for denoting the usual *intervals*. Intervals are convex, but the converse does not hold in general if α is not complete (see below). Given two subsets X, Y of a linear ordering, $X < Y$ holds if for all $x \in X$ and $y \in Y$, $x < y$.

The *sum* of two linear orderings $\alpha_1 = (L_1, <_1)$ and $\alpha_2 = (L_2, <_2)$ (up to renaming, assume L_1 and L_2 disjoint), denoted $\alpha_1 + \alpha_2$, is the linear ordering $(L_1 \cup L_2, <)$ with $<$ coinciding with $<_1$ on L_1, with $<_2$ on L_2 and such that $L_1 < L_2$. More generally, given a linear ordering $\alpha = (L, <)$ and for each $x \in L$ a linear ordering $\beta_x = (K_x, <_x)$ (the K_x are assumed disjoint), we denote by $\sum_{x \in \alpha} \beta_x$ the linear $(\cup_{x \in L} K_x, <')$ with $x' <' y'$ if $x < y$ or $x = y$ and $x' <_x y'$, where $x' \in K_x$ and $y' \in K_y$.

A linear ordering α is *complete* if every nonempty subset of α with an upper bound has a least upper bound in α, and every nonempty subset of α with a lower bound has a greatest lower bound in α.

A (Dedekind) *cut* of a linear ordering $\alpha = (L, <)$ is a couple (E, F) where $\{E, F\}$ is a partition of L, and $E < F$. Cuts are totally ordered by $(E, F) < (E', F')$ if $E \subsetneq E'$. This order has a minimal element $\bot = (\emptyset, L)$ and a maximal element $\top = (L, \emptyset)$. We denote by $\overline{\alpha}$ the set of cuts of α, and we abbreviate by $\overline{\alpha}^{]}, \overline{\alpha}^{[[}, \overline{\alpha}^{]]}, \overline{\alpha}^{][}$ the sets $\overline{\alpha}, \overline{\alpha} \setminus \{\top\}, \overline{\alpha} \setminus \{\bot\}, \overline{\alpha} \setminus \{\bot, \top\}$ respectively. An important remark is that $\overline{\alpha}$ is a complete linear ordering. Cuts can be thought as new elements located between the elements of α: given $x \in \alpha$, $x^- = (]-\infty, x[, [x, +\infty[)$ represents the cut placed just before x, while $x^+ = (]-\infty, x],]x, +\infty[)$ is the cut placed just after x. We say in this case that x^+ is *the successor of x^- through x*. But not all cuts are successors or predecessors of another cut. A cut c is a *right limit* (resp. a *left limit*) if it is not the minimal element and not of the form x^+ for some x in α (resp. not the maximal element and not of the form x^-).

A linear ordering α is *dense* if for every $x < y$ in α, there exists z in $]x, y[$. A linear ordering is *scattered* if none of its subordering is dense. For instance $(\mathbb{Q}, <)$ and $(\mathbb{R}, <)$ are dense, while $(\mathbb{N}, <)$ and $(\mathbb{Z}, <)$ are scattered. Being scattered is preserved under taking a subordering. A scattered sum of scattered linear orderings also yields a scattered linear ordering. Every ordinal is scattered. Furthermore, if α is scattered, then $\overline{\alpha}$ is scattered. And if α is countable and scattered, then $\overline{\alpha}$ is also countable and scattered.

Additional material on linear orderings can be found in [13].

2.2   Words, Languages

We use a generalised version of words: words indexed by linear orderings. Given a linear ordering $\alpha = (L, <)$ and a finite alphabet A, an α-*word u over the alphabet A* is a mapping from L to A. We also say that α is the *domain* of the word u, or that u is a word *indexed* by α. Below we always consider word up to isomorphism of the domain, unless a specific presentation of the domain is required. Standard finite words are simply the words indexed by finite linear orderings. Given a word u of domain α and $\beta \subseteq \alpha$, we denote by $u|_\beta$ the word u restricted to its positions in β. Given an α-word u and a β-word v, uv represents the $(\alpha + \beta)$-word defined by $(uv)(x)$ is $u(x)$ if x belongs to α and $v(x)$ if x belongs to β. The product is extended to languages of words in a natural way. The product of words is naturally generalised to the infinite product $\prod_{i \in \alpha} u_i$, where α is an order type and u_i are linear β_i-words; the resulting being a $(\sum_{i \in \alpha} \beta_i)$-word.

For a language W and a linear ordering α, one defines W^α to be the language containing all the words $\prod_{i \in \alpha} u_i$, where $u_i \in W$ for all $i \in \alpha$.

Given an alphabet A, we denote by $A^\diamond$ the set of words indexed by a countable scattered linear ordering.

2.3   Semigroups and Additive Labellings

For a thorough introduction to semigroups, we refer the reader to [8,9]. A *semigroup* $(S,.)$ is a set S equipped with an associative binary operator written multiplicatively. Groups and monoids are particular instances of semigroups. The set of nonempty finite words A^+ over an alphabet A is a semigroup – it is the semigroup freely generated by A. A *morphism of semigroups* from a semigroup $(S,.)$ to a semigroup $(S',.')$ is a mapping φ from S to S' such that for all x, y in S, $\varphi(x.y) = \varphi(x).'\varphi(y)$. An *idempotent* in a semigroup is an element e such that $e^2 = e$.

Let α be a linear ordering and $(S,.)$ be a semigroup. A mapping σ from couples (x, y) with $x, y \in \alpha$ and $x < y$ to S is called an *additive labelling* if for every $x < y < z$ in α, $\sigma(x, y).\sigma(y, z) = \sigma(x, z)$.

Given a semigroup morphism φ from $(A^\diamond,.)$ to some semigroup $(S,.)$ and a word u in $A^\diamond$ of domain α, there is a natural way to construct an additive labelling φ_u from $\overline{\alpha}$ to $(S,.)$: for every two cuts $x < y$ in $\overline{\alpha}$, set $\varphi_u(x, y)$ to be $\varphi(u_{x,y})$, where $u_{x,y}$ is the word u restricted to its positions between x and y; i.e., $u_{x,y} = u|_{F \cap E'}$ for $x = (E, F)$ and $y = (E', F')$.

3   Factorisation Forest Theorems

In this section, we present various theorems of factorisation forest. We first give the original statement in Section 3.1. In Section 3.2, we introduce the notion of a split, and use it in a different presentation of the result. In Section 3.3, we state the extension to every complete linear ordering.

3.1   Factorisation Forest Theorem

Fix an alphabet A and a semigroup morphism φ from A^+ to a finite semigroup $(S,.)$. A *factorisation tree* is an ordered unranked tree in which each node is either a leaf labelled by a letter, or an internal node. The *value* of a node is

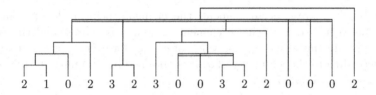

Fig. 1. A factorisation tree

the word obtained by reading the leaves below from left to right. A *factorisation tree* of a word $u \in A^+$ is a factorisation tree of value u. The *height* of the tree is defined as usual, with the convention that the height of a single leaf is 0. A factorisation tree is *Ramseyan* (for φ) if every node 1) is a leaf, or 2) has two children, or, 3) the values of its children are all mapped by φ to the same idempotent of S.

Example 1. Fix $A = \{0, 1, 2, 3, 4\}$, $(S, .) = (\mathbb{Z}/5\mathbb{Z}, +)$ and φ to be the only semigroup morphism from A^+ to $(S, .)$ mapping each letter to its value. Figure 1 presents a Ramseyan factorisation tree for the word $u = 210232300322002$. In this drawing, internal nodes appear as horizontal lines. Double lines correspond to case 3 in the description of Ramseyanity.

The theorem of factorisation forests is then the following.

Theorem 1 (factorisation forests, Simon [15]). *For every alphabet A, finite semigroup $(S, .)$, semigroup morphism φ from A^+ to S and word u in A^+, u has a Ramseyan factorisation tree of height at most $3|S|$.*

The original theorem is due to Simon [15], with a bound of $9|S|$. An improved bound of $7|S|$ is provided by Chalopin and Leung [6]. The value of $3|S|$ is a byproduct of the present work (see Theorem 2 below and subsequent comments).

3.2 A Variant Via Ramseyan Splits

The variant presented here of the factorisation forest theorem uses the notion of splits. We reuse this framework later on.

A *split of height* N of a linear ordering α is a mapping s from α to $[1, N]$. Given a split, two elements x and y in α such that $s(x) = s(y) = k$ are k-*neighbours* if $s(z) \geq k$ for all $z \in [x, y]$. k-neighbourhood is an equivalence relation over $s^{-1}(k)$. Fix an *additive labelling* from α to some finite semigroup S. A split of α is *Ramseyan* for σ — we also say a *Ramseyan split for* (α, σ) — if for every $k \in [1, N]$, every $x < y$ and $x' < y'$ such that all the elements x, y, x', y' are k-neighbours, then $\sigma(x, y) = \sigma(x', y') = (\sigma(x, y))^2$; Equivalently, for all k, every class of k-neighbourhood is mapped by σ to a single idempotent.

Example 2. Let S be $\mathbb{Z}/5\mathbb{Z}$ equipped with the addition $+$. Consider the linear ordering of 17 elements and the additive labelling σ defined by:

$$| 3 | 1 | 0 | 2 | 3 | 2 | 3 | 0 | 0 | 3 | 2 | 2 | 0 | 0 | 0 | 2 |$$

Each symbol '|' represents an element, the elements being ordered from left to right. Between two consecutive elements x and y is represented the value of $\sigma(x, y) \in S$. In this situation, the value of $\sigma(x, y)$ for every $x < y$ is uniquely defined according to the additivity of σ: it is obtained by summing all the values between x and y modulo 5.

A split s of height 3 is the following, where we have written above each element x the value of $s(x)$:

$$1 \quad 3 \quad 2 \quad 2 \quad 1 \quad 2 \quad 1 \quad 2 \quad 2 \quad 2 \quad 3 \quad 2 \quad 1 \quad 1 \quad 1 \quad 1 \quad 2$$
$$|\; 2\; |\; 1\; |\; 0\; |\; 2\; |\; 3\; |\; 2\; |\; 3\; |\; 0\; |\; 0\; |\; 3\; |\; 2\; |\; 2\; |\; 0\; |\; 0\; |\; 0\; |\; 2\; |$$

In particular, if you choose $x < y$ such that $s(x) = s(y) = 1$, then the sum of elements between them is 0 modulo 5. If you choose $x < y$ such that $s(x) = s(y) = 2$ but there is no element z in between with $s(z) = 1$ — i.e., x and y are 2-neighbours — the sum of values separating them is also 0 modulo 5. Finally, it is impossible to find two distinct 3-neighbours in our example.

Theorem 2. *For every finite linear ordering α, every finite semigroup $(S, .)$ and additive labelling σ from α to S, there exists a Ramseyan split for (α, σ) of height at most $|S|$.*

Let us state the link between Ramseyan splits and factorisation trees. Fix an alphabet A, a semigroup S, a morphism φ from A^+ to S and a word $u \in A^+$ of finite domain α. The following is easy to establish:

- every Ramseyan factorisation tree of height k of u can be turned into a Ramseyan split of height at most k of $(\overline{\alpha}^{][}, \varphi_u)$,
- every Ramseyan split of height k of $(\overline{\alpha}^{][}, \varphi_u)$ can be turned into a factorisation tree of height at most $3k$ of u.

Using this last argument and Theorem 2, we directly obtain a proof of Theorem 1 with the announced bound of $3|S|$.

3.3 Ramseyan Splits for Complete Linear Orderings

We generalise Theorem 2 to complete linear orderings as follows.

Theorem 3. *For every complete linear ordering α, every finite semigroup $(S, .)$ and additive labelling σ from α to S, there exists a Ramseyan split for (α, σ) of height at most $3|S|$ ($|S|$ if α is an ordinal).*

Compared to Theorem 2, we trade the finiteness — which is replaced by the completeness — for a bound of $3|S|$ — which replaces a bound of $|S|$. The special case of α being a finite ordinal yields Theorem 2.

The proof by itself follows the lines of [6]. This means using three different arguments according to three different situations arising in the decomposition of the semigroup by Green's relations. The first situation amounts to treat the case of S being a group. The second case is the one of a single $\mathcal{J}$-class ($\mathcal{J}$ is one of the Green's relation). And finally one performs an induction on the number of $\mathcal{J}$-classes. Examples 1 and 2 do only involve the first situation.

This rough sketch contains certain technicalities when the proof need be formalised. In particular one performs many gluing and nesting of splits. An explanation of the improvement of the bound in the finite case is that splits are more versatile in handling those details. E.g., the use of the 'border types' $[[,][,]], []$ allow to glue more easily pieces of Ramseyan splits together, while Ramseyan factorisation trees do correspond only to the case $][.$

4   Application to Countable Scattered Linear Orderings

In this section, we present the automata theoretic approach to regularity of languages over words of countable scattered domain. This notion has been developed in [3], in which a suitable family of automata is proposed. These automata are easily shown closed under union, intersection and projection, and their emptiness is decidable. The closure under complementation is more involved and is due to Carton and Rispal [5]. In this section we give a simplified proof to this result.

The properties of these automata result directly in the decidability of the monadic (second-order) theory of countable scattered linear orderings. This decidability result can be independently established using the famous theorem of Rabin [12] (see [16] for a modern presentation), and its consequence, the decidability of the monadic theory of $(\mathbb{Q}, <)$. But this technique is less informative and is not totally satisfying. More precisely, using the theorem of Rabin signifies the use of infinite trees, and also has to do with the theory of Müller/parity games and their determinacy. We believe that these subtle issues are not relevant when considering the theory of linear orderings, and thus are worth being avoided. Furthermore the approach using the theorem of Rabin does not help much for understanding the notions of regularity over linear orderings.

Another application of Theorem 3 – to some extent a variant of the application proposed here – is to give a compositional proof for the decidability of the monadic theory of countable scattered linear orderings. Generally speaking, the compositional method allows to devise automata-free proofs of decidability of monadic theories (or other logics). It was used by Shelah [14] in his seminal work on the monadic theory of linear orderings. But so far it could not be used in situations like scattered orderings by lack of the correct combinatorial result. Theorem 3 bridges this gap.

In this section we concentrate ourselves solely on the technical core of the theory: the closure under complementation of automata over countable scattered linear orderings. We present the suitable family of automata, then the corresponding semigroup, and finally the complementation proof itself.

4.1   Automata over Countable Scattered Linear Orderings

In this section, we define priority automata and show how they accept words indexed by countable scattered linear orderings. Those automaton were introduced in [3], but in their 'Muller' form, while here we adopt the 'parity-like' approach (to this respect, the results given below are new).

Definition 1. *A priority automaton* $\mathcal{A} = (Q, A, I, F, p, \delta)$ *consists of a finite set of* states Q, *a finite alphabet* A, *a set of* initial states I, *a set of final states* F,

a priority mapping $p : Q \mapsto [1, N]$ *(N being a natural) and a* transition relation
$\delta \subseteq (Q \times A \times Q) \uplus ([1, N] \times Q) \uplus (Q \times [1, N])$.

A *run* of the automaton $\mathcal{A}$ over an α-word u is a mapping ρ from $\overline{\alpha}$ to Q such
that for all cuts c, c':

- if c' is the successor of c through x, then $(\rho(c), u(x), \rho(c')) \in \delta$,
- if c is a right limit, then $(k, \rho(c)) \in \delta$ where $k = \max \bigcap_{c' < c} p(\rho(]c', c[))$,
- if c is a left limit, then $(\rho(c), k) \in \delta$ where $k = \max \bigcap_{c' > c} p(\rho(]c, c'[))$.

The first case corresponds to standard automata on finite words: a transition
links one state to another while reading a single letter in the word. The second
case verifies that the highest priority appearing infinitely close to the left of c
corresponds to a transition. The third case is symmetric. An α-word u is *accepted*
by $\mathcal{A}$ if there is a run ρ of $\mathcal{A}$ over u such that $\rho(\bot) \in I$ and $\rho(\top) \in F$.

Example 3. Consider the automaton with states $\{q, r\}$, alphabet $\{a\}$, initial
states $\{q, r\}$, final state q, priority mapping constant equal to 0 and transitions $\{(q, a, q), (q, a, r), (0, q), (r, 0)\}$). It accepts those words in $\{a\}^\diamond$ which have
a complete domain. For this, note that a linear ordering is complete iff no cut is
simultaneously a left and a right limit.

Consider a word $u \in \{a\}^\diamond$ which has a complete domain α. For $c \in \overline{\alpha}$, set $\rho(c)$
to be q if c is $\top$ or if c has a successor, else $\rho(c)$ is r. Under the hypothesis of
completeness, it is simple to verify that ρ is a run witnessing the acceptance of
the word. Conversely, assume that there is a run ρ over the α-word u with α
not complete. There is a cut $c \in \overline{\alpha}$ which is both a left and a right limit. If $\rho(c)$
is r, then, as c is a left limit, there is no corresponding transition; else if $\rho(c)$
is q the same argument can be applied to the right of c. In both cases there is a
contradiction.

It is easy to prove that the languages of $\diamond$-words accepted by priority automata are
closed under union, intersection, and projection [5]. It is also easy to establish the
decidability of their emptiness problem. Below, after introducing the necessary
semigroup, we show the more difficult closure under complementation.

4.2 Semigroup Structure

In order to use Theorem 3, we have to relate automata with semigroups. Let us
fix ourselves an automaton of states Q and priorities $[1, N]$. One equips

$$S = 2^{Q \times [1, N] \times Q}$$

of a semigroup structure as usual with

for $a, b \in S$, $a.b = \{(p, \max\{m, n\}, r) \ : \ (p, m, q) \in a, (q, n, r) \in b\}$.

This definitions naturally comes together with a semigroup morphism φ from $\diamond$-words to S such that for every word u, $\varphi(u)$ contains (p, n, q) iff there exists a run of the automaton reading u, starting from state p, finishing with state q, and of maximal priority n.

The semigroup defined so far does not entirely capture the semantic of the automaton. In particular it contains no limit passing features. We resolve this issue by defining the exponentiations under ω and $-\omega$ of idempotents of the semigroup. One defines e^ω (and symmetrically $e^{-\omega}$) for an idempotent e by:

$$e^\omega = e.\{(q, m, r) \ : \ (q, m, q) \in e, \ (\max(m, p(q)), r) \in \delta\} \ ,$$
$$\text{and} \quad e^{-\omega} = \{(r, m, q) \ : \ (q, m, q) \in e, \ (r, \max(m, p(q))) \in \delta\}.e \ .$$

One also defines e^ζ as $e^{-\omega}.e^\omega$.

The essential property of these exponentiations is the following. Given a sequence of words $(u_i)_\beta$ indexed by $\beta = \omega, -\omega, \zeta$, and such that for all i in β, $\varphi(u_i) = e$, then

$$\varphi(\prod_{i \in \beta} u_i) = e^\beta \ .$$

4.3   Complementation

We sketch now a short proof of the following theorem.

Theorem 4 (Carton and Rispal [5]). *Languages of countable scattered words accepted by priority automata are closed under complement.*

Let k be a natural number, a be in S, and ι among $[], [[,]],][$, set $S_k^\iota(a)$ to be the set of $\diamond$-words u such that $\varphi(u) = a$ and $(\overline{\alpha}^\iota, \varphi_u)$ admits a Ramseyan split of height k (by convention, ε does not belong to $S_k^{]\ [}(a)$). We prove by induction on k that for every a in S and $\iota = [], [[,]],][$, $S_k^\iota(a)$ is accepted by a priority automaton. Since by Theorem 3, $\varphi^{-1}(a) = S_{3|S|}^{[]}(a)$, we deduce that $\varphi^{-1}(a)$ would be accepted by a priority automaton. As the complement language we are aiming at is a finite union of such languages, it would also be accepted by a priority automaton. This argument concludes the proof. What remains to be done is to establish the induction.

The base case is obtained by remarking that the following languages are accepted by priority automata:

$$S_0^{]\ [}(a) = \varphi^{-1}(a) \cap A \,, \quad S_0^{[[}(a) = S_0^{]]}(a) = \varphi^{-1}(a) \cap \{\varepsilon\} \,, \quad \text{and} \quad S_0^{[]}(a) = \emptyset \,.$$

For all $k \geq 1$ and idempotent e, let $C_{e,k}$ be the set of $\diamond$-words u of domain α such that $\varphi(u) = e$, and there exists a split s of height k of $\overline{\alpha}$ such that $s(\perp) = s(\top) = 1$.

Our first step is to show how to construct an automaton accepting $C_{e,k+1}$ from automata accepting the languages $S_k^{\iota}(a)$. For this, consider the following languages:

$$M_{e,k} = S_k^{][}(e),\qquad\qquad M_{e,k}^{\leftarrow} = \sum_{ae^{-\omega}=e} S_k^{]]}(a),$$

$$M_{e,k}^{\rightarrow} = \sum_{e^{\omega}a=e} S_k^{[[}(a),\qquad \text{and } M_{e,k}^{\rightarrow\leftarrow} = \sum_{e^{\omega}ae^{-\omega}=e} S_k^{[]}(a).$$

By induction hypothesis, those languages are accepted by priority automata. Wlog, we choose them to use distinct priorities, and we set $n-1$ to be the maximal priority involved in those automata. We use them in the construction of the automaton $\mathcal{A}_{e,k+1}$ depicted Fig. 2.

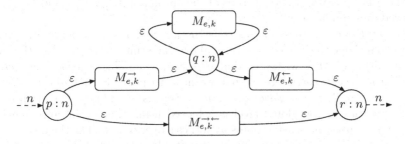

Fig. 2. The automaton $\mathcal{A}_{e,k+1}$

This construction makes use of ε-transitions. This is just a commodity of notation and can be removed using standard techniques. The automaton itself is made of disjoint copies of the automata accepting $M_{e,k}, M_{e,k}^{\rightarrow\leftarrow}, M_{e,k}^{\rightarrow}$, and $M_{e,k}^{\leftarrow}$, together with three new states p, q, r. Each ε-transition entering one of the subautomata represents in fact all possible ε-transitions with an initial state as destination; similarly, every ε-transition exiting a subautomaton represents all possible ε-transitions with as origin any of the final states of the automaton. The priority of the new state q is n, a priority unused elsewhere by construction. One chooses also p and r to have priority n (this is not of real importance since it is impossible to see infinitely often p or r in a run without seeing infinitely often q: the priority of q only matters). The two dashed arrows represent the two limit transitions (n, p) and (r, n).

Let $L_{e,k+1}[q_1, q_2]$ be the language accepted by this automaton with initial state q_1 and final state q_2 for q_1, q_2 among p, q, r.

The core of the proof is embedded in the following lemma.

Lemma 1. *For every idempotent e, $L_{e,k+1}[q, q] = C_{e,k+1}$.*

Proof. (sketch of the difficult inclusion: $L_{e,k+1}[q, q] \subseteq C_{e,k+1}$)

Let u be in $L_{e,k+1}[q, q]$, we have to construct a Ramseyan split s of height $k+1$ of $\varphi_u^{]}$ with $s(\bot) = s(\top) = 1$. Since $u \in L_{e,k}[q, q]$, there exists a corresponding run ρ of the automaton $\mathcal{A}_{e,k+1}$ from state q to state q. Let I be the set of cuts c such that $\rho(c) = q$.

Set $s(c) = 1$ for all c in I. Let now $J \subseteq \bar{\alpha}$ be a maximal interval not intersecting I. Let us define s over J. Let x be inf J and y be sup J, J is either $[x,y]$, $[x,y[, \ ]x,y]$ or $]x,y[$. Assume $J = [x,y[$. In this case, since $y \notin J$, $y \in I$ and hence $\rho(y) = q$. Furthermore, since $x \in J$, there exists an infinite sequence $x_1 < x_2 < \dots$ of length ω and limit x in I. As the priority of $\rho(x_i) = q$ is the maximal one, namely n, the only possible state for $\rho(x)$ compatible with limit transitions is p. Furthermore the state q is never visited by ρ in $[x,y[$ (by definition of J). By inspecting the automaton, we conclude that the only possibility is that ρ restricted to $[x,y]$ is in fact a run of the subautomaton $M_{e,k}^{\rightarrow}$. By induction hypothesis, since $M_{e,k}^{\rightarrow}$ is a union of languages $S_k^{[[}$, this means that there exists a split s_J of height k of J, Ramseyan for σ. We set s to coincide with $s_J + 1$ over J. For the other possibilities for J, runs of the automata $M_{e,k}^{\rightarrow}$, $M_{e,k}^{\leftarrow}$ and $M_{e,k}^{\rightarrow\leftarrow}$ are involved in a similar way.

Proving the correctness of this construction requires some more arguments. Let us come back to the case $J = [x,y[$ above. The run ρ over $[x,y]$ together with the definition of $M_{e,k}^{\rightarrow}$ witnesses that $e^\omega \sigma(x,y) = e$. This is a *local correctness property* for the construction. What we have to prove is that $\sigma(x,y) = e$ for every $x < y$ in I; i.e., a *global correctness conclusion*. This propagation of the local equalities to the global level is achieved using topological arguments. In particular, it uses the scatteredness hypothesis over α as well as the countability hypothesis. It also involves the use the countable axiom of choice. $\square$

We can derive from the last lemma the following.

Corollary 1. $L_{e,k}[q,p] = C_{e,k}^\omega$, $L_{e,k}[r,q] = C_{e,k}^{-\omega}$, and $L_{e,k}[r,p] = C_{e,k}^\zeta$.

And we terminate by remarking that, for $\iota = [], [[,]],][$ and $a \in S$, the language $S_k^\iota(a)$ can be written in terms of the S_{k-1} and the $C_{e,k}$ languages using finite sums, concatenation and ω, $-\omega$ and ζ exponentiations.

5   Conclusion and Future Work

We believe that the factorisation forest theorem cannot be improved further in the directions presented here. In particular, the bounds in Theorem 2 cannot be improved in general. And in Theorem 3, removing the completeness hypothesis makes the result fail.

Concerning automata over countable scattered linear orderings, our complementation proof has the advantage – with respect to the original one in [5] – to isolate the combinatorial part from the problems related to scatteredness itself. Our proof is in fact very resemblant to the original one of Buchi for ω-words [4] in which the theorem of Ramsey would be replaced by Theorem 3. Along the same lines, Theorem 3 can also be used in a compositional proof of the decidability of the monadic theory of countable scattered linear orderings.

The question is whether there are other applications for Theorem 3 since $(\mathbb{R}, <)$ does not have a decidable monadic theory [14]. We believe that it is the case, for instance for tackling the conjecture of Rabin that the monadic theory of $(\mathbb{R}, <)$ is decidable when monadic variables are interpreted over Borelian sets (let us remark that the theory of $(\mathbb{R}, <)$ with quantification over boolean combinations of opens sets is already known to be decidable from Rabin [12]). We are working in this direction.

Acknowledgement. I am very grateful to Olivier Carton for his numerous comments on this work. I also thank the anonymous referees who helped in improving this document.

References

1. Bojańczyk, M., Colcombet, T.: Bounds in omega-regularity. In: IEEE Symposium on Logic In Computer Science, pp. 285–296. IEEE Computer Society Press, Los Alamitos (2006)
2. Brown, T.C.: An interesting combinatorial method in the theory of locally finite semigroups. Pacific Journal of Mathematics 36(2), 277–294 (1971)
3. Bruyère, V., Carton, O.: Automata on linear orderings. In: Sgall, J., Pultr, A., Kolman, P. (eds.) MFCS 2001. LNCS, vol. 2136, pp. 236–247. Springer, Heidelberg (2001)
4. Büchi, J.R.: On a decision method in restricted second order arithmetic. In: Proceedings of the International Congress on Logic, Methodology and Philosophy of Science, pp. 1–11. Stanford University press (1960)
5. Carton, O., Rispal, C.: Complementation of rational sets on countable scattered linear orderings. Int. J. Found. Comput. Sci. 16(4), 767–786 (2005)
6. Chalopin, J., Leung, H.: On factorization forests of finite height. Theoretical Computer Science 310(1–3), 489–499 (2004)
7. Colcombet, T.: A combinatorial theorem for trees. In: ICALP'07. LNCS, Springer, Heidelberg (2007)
8. Lallement, G.: Semigroups and Combinatorial Applications. Wiley, New-York (1979)
9. Pin, J.: Varieties of formal languages. North Oxford, London and Plenum, New-York (1986)
10. Pin, J., le Saëc, B., Weil, P.: Semigroups with idempotent stabilizers and application to automata theory. Int. J. of Alg. and Comput. 1(3), 291–314 (1991)
11. Pin, J.-E., Weil, P.: Polynominal closure and unambiguous product. Theory Comput. Syst. 30(4), 383–422 (1997)
12. Rabin, M.: Decidability of second-order theories and automata on infinite trees. Trans. Amer. Math. soc. 141, 1–35 (1969)
13. Rosenstein, J.G.: Linear Orderings. Academic Press, New York (1982)
14. Shelah, S.: The monadic theory of order. Annals Math. 102, 379–419 (1975)
15. Simon, I.: Factorization forests of finite height. Theor. Comput. Sci. 72(1), 65–94 (1990)
16. Thomas, W.: Languages, automata, and logic. In: Rozenberg, G., Salomaa, A. (eds.) Handbook of Formal Language Theory, vol. III, pp. 389–455. Springer, Heidelberg (1997)

Marked Systems and Circular Splicing[*]

Clelia De Felice, Gabriele Fici, and Rosalba Zizza

Dipartimento di Informatica e Applicazioni
Università di Salerno
via Ponte Don Melillo
84084 Fisciano (SA), Italy
{defelice,fici,zizza}@dia.unisa.it

Abstract. Splicing systems are generative devices of formal languages, introduced by Head in 1987 to model biological phenomena on linear and circular DNA molecules. In this paper we introduce a special class of *finite circular splicing systems* named *marked systems*. We prove that a marked system S generates a *regular circular language* if and only if S satisfies a special (decidable) property. As a consequence, we show that we can decide whether a regular circular language is generated by a marked system and we characterize the structure of these regular circular languages.

1 Introduction

The notion of *splicing systems* was first introduced in [11], where Head modelled a recombinant behaviour of DNA molecules (under the action of restriction and ligase enzymes) as a particular operation between words. In short, two strands of DNA are cut at specified substrings (sites) by restriction enzymes that recognize a pattern inside the molecule and then the fragments are pasted by ligase enzymes. Since 1987, this basic idea has been formalized in terms of generative mechanisms for formal languages, the splicing systems. A splicing system is defined by giving an initial language I (initial set of DNA molecules) and a set of special words or rules R (enzymes). The set I is then transformed by repeated applications of the splicing operation. In nature, the DNA molecules are present in the form of a linear or a circular sequence and, correspondingly, there are three definitions of linear splicing systems and three definitions of circular splicing systems, given by Head, Paun and Pixton respectively [12].

While there have been many articles on linear splicing, relatively few works on circular splicing systems have been published [12,15]. In particular, some questions still unanswered are related to the computational power of the latter systems. Notice that in this context, at least three aspects should be considered.

---

[*] Partially supported by **MIUR** Project *"Automi e Linguaggi Formali: aspetti matematici e applicativi"* (2005), by 60% Project *"Linguaggi formali e codici: problemi classici e modelli innovativi"* (University of Salerno, 2005) and by 60% Project *"Linguaggi formali e codici a lunghezza variabile: proprietà strutturali e nuovi modelli di rappresentazione"* (University of Salerno, 2006).

E. Csuhaj-Varjú and Z. Ésik (Eds.): FCT 2007, LNCS 4639, pp. 238–249, 2007.

Indeed, this computational power depends on (a) whether additional hypotheses are taken into account (i.e., reflexive and symmetric set of rules, self-splicing, see Sections 2.2 and 5), (b) which of the three definitions (Head's, Paun's or Pixton's definition) is considered, (c) the level in the (circular) Chomsky hierarchy the initial set I and the set R of rules belong to. It is known that Paun circular splicing systems generate regular (resp. context-free, recursively enumerable) circular languages when I is a regular (resp. context-free, recursively enumerable) circular language and R is a finite set of rules which satisfies the additional hypotheses in (a) (see [12] for a more general result). Furthermore, the same result holds for Pixton splicing systems with a regular set of rules in place of a finite one [16,17]. However, as observed in [12], the problem of characterizing the corresponding generated languages remains open in all these cases.

The results presented in this paper fit into this framework. We mainly deal with finite Paun systems $SC_{PA} = (A, I, R)$, i.e., Paun circular splicing systems with both I and R finite sets, with no additional hypotheses, and with the corresponding class of generated languages, denoted $C(Fin, Fin)$. It is known that in contrast with the linear case, $C(Fin, Fin)$ is not intermediate between two classes of languages in the Chomsky hierarchy. Indeed, $C(Fin, Fin)$ contains context-free circular languages which are not regular (see [20]), context-sensitive circular languages which are not context-free (see [10]) and there exist regular circular languages which are not in $C(Fin, Fin)$ (see [5]). Furthermore, it has been claimed that $C(Fin, Fin)$ is contained in the class of context-sensitive circular languages (see [10]). However, the structure of regular circular languages in $C(Fin, Fin)$ is not completely understood. Thus, we focus our interest on the search for a characterization of these languages. Partial results are known (see [3,4,5,6]). In particular the above problem has been solved for languages over a one-letter alphabet in [5,6]. On the other hand, a special class of circular splicing systems (*Paun circular semi-simple splicing systems* or *CSSH systems*) has already been considered in [7,8,20]. Concerning the computational power of CSSH systems, in [10] the author claimed that the class of circular languages generated by circular semi-simple splicing systems is contained in the class of context-free circular languages. Once again, a characterization of the regular circular languages generated by circular semi-simple splicing systems is still lacking.

Let us briefly explain the results proved in this paper. In Section 3.1 we define *marked systems*, i.e., CSSH systems with I and R satisfying additional hypotheses. In the same section we also prove that the language L generated by a marked system S is a disjoint union of a finite number of languages L_g generated by special "subsystems" of S, defined by means of an equivalence relation on I. Therefore, we prove that L is a regular (resp. context-free) circular language if and only if each L_g is regular (resp. context-free). As a direct result, we prove that in order to characterize regular circular languages generated by marked systems, it suffices to characterize them for *transitive marked systems*, i.e., marked systems such that I is an equivalence class (Section 3.2).

As a main result, we prove that a (transitive) marked system S generates a regular circular language if and only if S satisfies a special (decidable) property

(Section 4). As a consequence, we show that we can decide whether a regular circular language is generated by a marked system and we characterize the structure of these regular circular languages (Section 4). We also give a classification of transitive marked systems S generating regular circular languages in terms of their *diameter* $d(S)$, a notion that will be defined in Section 3.3. Finally, we prove that each language generated by a marked system with self-splicing is a regular circular language (Section 5).

Let us briefly illustrate the organization of this paper. In Section 2 we gathered basics on circular words and circular splicing. Sections 3 and 4 are devoted to our main results on marked systems without self-splicing whereas Section 5 deals with marked systems with self-splicing. The main results of this paper were also presented at *AutoMathA 2007* (Palermo, Italy, 18-22 June 2007).

2   Basics

2.1   Circular Words and Languages

We denote by A^* the free monoid over a finite alphabet A and we set $A^+ = A^* \backslash 1$, where 1 is the empty word. For a word $w \in A^*$, $|w|$ is the length of w and for every $a \in A$, we denote by $|w|_a$ the number of occurrences of a in w. We also set $alph(w) = \{a \in A \mid |w|_a > 0\}$. For a subset X of A^*, $|X|$ is the cardinality of X. A word $x \in A^*$ is a *prefix* of $w \in A^*$ if $y \in A^*$ exist such that $w = xy$ and we set $Pref(w) = \{x \in A^+ \mid \exists y \in A^* : xy = w\}$. Furthermore, Fin (resp. Reg) denotes the class of finite (resp. regular) languages over A, at times represented by means of regular expressions.

Circular splicing deals with circular strings, a notion which has been intensively examined in formal language theory (see [2,9,14,19]). For a given word $w \in A^*$, a circular word $^\sim w$ is the equivalence class of w with respect to the *conjugacy* relation $\sim$ defined by $xy \sim yx$, for $x, y \in A^*$ (see [14]). The notations $|^\sim w|$, $alph(^\sim w)$ will be defined as $|w|$, $alph(w)$ for any representative w of $^\sim w$. When the context does not make it ambiguous, we will use the notation w for a circular word $^\sim w$. Prefixes of the words w' such that $w' \sim w$ are also needed, and so we set $Pref_c(w) = \{x \in A^+ \mid \exists w' \sim w : x \in Pref(w')\}$.

Let $^\sim A^*$ denote the set of all circular words over A, i.e., the quotient of A^* with respect to $\sim$. Given $L \subseteq A^*$, $^\sim L = \{^\sim w \mid w \in L\}$ is the *circularization* of L, i.e., the set of all circular words corresponding to elements of L. A subset C of $^\sim A^*$ is named a *circular language* and every language L such that $^\sim L = C$ is called a *linearization* of C. In particular, the *full linearization* $Lin(C)$ of C is the set of all the strings in A^* corresponding to the elements of C, i.e., $Lin(C) = \{w' \in A^* \mid \exists \, ^\sim w \in C : \, w' \sim w\}$. For simplicity of notation, we will use the same letter to designate a set of words (resp. circular words) of length one and its circularization (resp. full linearization). We will also often write $^\sim w$ instead of $\{^\sim w\}$ when no confusion arises.

Given a family of languages FA in the Chomsky hierarchy, $FA^\sim$ is the set of all those circular languages C which have some linearization in FA. In this paper we deal only with circular languages having a regular linearization, i.e., with $Reg^\sim = \{C \subseteq {}^\sim A^* \mid \exists L \in Reg : {}^\sim L = C\}$. It is classically known that given a regular language $L \subseteq A^*$, $Lin({}^\sim L)$ is regular (see Exercise 4.2.11 in [13]). As a result, given a circular language C, we have $C \in Reg^\sim$ if and only if its full linearization $Lin(C)$ is regular [12]. If $C \in Reg^\sim$ then C is a *regular circular language*. Analogously, we can define context-free (resp. context-sensitive, recursive, recursively enumerable) circular languages and, once again, a circular language C is context-free if and only if $Lin(C)$ is context-free [12].

2.2 Circular Splicing

As in the linear case, there are three definitions of the circular splicing operation. In this paper, we will mainly take into account a restricted version of Paun's definition reported below.

Paun's definition [12,18]. A *Paun circular splicing system* is a triple $SC_{PA} = (A, I, R)$, where A is a finite alphabet, I is the initial circular language, with $I \subseteq {}^\sim A^*$ and R is the set of rules, with $R \subseteq A^*\#A^*\$A^*\#A^*$ and $\#, \$ \notin A$. Then, given a rule $r = u_1\#u_2\$u_3\#u_4$ and two circular words $w' = {}^\sim u_2 h u_1$, $w'' = {}^\sim u_4 k u_3$, the rule r cuts and linearizes the two circular strings, obtaining $u_2 h u_1, u_4 k u_3$, pastes them and circularizes, obtaining ${}^\sim u_2 h u_1 u_4 k u_3$. We say that ${}^\sim u_2 h u_1 u_4 k u_3$ is generated starting with w', w'' and by using r. We also say that $u_1 u_2, u_3 u_4$ are *sites* of splicing and we denote $SITES(R)$ the set of sites of the rules in R.

We must note that in the original definition of circular splicing languages given by Paun, rules in R can be used in two different ways [12]. One way has been described above while we will be concerned with the other, known as *self-splicing*, in Section 5 only. Furthermore, as observed in [12], additional hypotheses can be added to the definition of circular splicing. Namely, we may assume that R is *reflexive* (i.e., for each $u_1\#u_2\$u_3\#u_4 \in R$, we have $u_1\#u_2\$u_1\#u_2$, $u_3\#u_4\$u_3\#u_4 \in R$) or R is *symmetric* (i.e., for each $u_1\#u_2\$u_3\#u_4 \in R$, we have $u_3\#u_4\$u_1\#u_2 \in R$).

Remark 1. In view of the definition of circular splicing, for every circular splicing system $SC_{PA} = (A, I, R)$, it is natural to assume that R is symmetric [12]. On the contrary, we do not assume that R is reflexive.

In the remainder of this paper, "splicing system" will be synonymous with "circular splicing system". Furthermore, we assume that $SC_{PA} = (A, I, R)$ is a *finite* splicing system, i.e., a circular splicing system with both I and R finite sets. We now give the definition of circular splicing languages. For a given splicing system SC_{PA}, we denote $(w', w'') \vdash_r w$ the fact that w is generated (or spliced) starting with w', w'' and by using a rule r. Furthermore, given a language $C \subseteq {}^\sim A^*$, we denote $\sigma'(C) = \{z \in {}^\sim A^* \mid \exists w', w'' \in C, \exists r \in R : (w', w'') \vdash_r z\}$. Then, we define $\sigma^0(C) = C$, $\sigma^{i+1}(C) = \sigma^i(C) \cup \sigma'(\sigma^i(C))$, $i \geq 0$, and $\sigma^*(C) = \bigcup_{i \geq 0} \sigma^i(C)$.

Definition 1. *Given a splicing system* $SC_{PA} = (A, I, R)$, *the circular language* $C(SC_{PA}) = \sigma^*(I)$ *is the language* generated *by* SC_{PA}.
A circular language C *is* C_{PA} generated *(or* C *is a* circular splicing language*) if a splicing system* SC_{PA} *exists such that* $C = C(SC_{PA})$.

2.3 Circular Semi-simple Splicing Systems

As already said, we are interested in finding a characterization of regular circular languages generated by splicing systems and, as a first step, we restrict our attention to a special class of systems already considered in [7,8,10,20].

Precisely, let us consider those splicing systems $SC_{PA} = (A, I, R)$, named *Paun circular semi-simple splicing systems* (or *CSSH systems*) in [7], such that, for each rule $r = u_1\#u_2\$u_3\#u_4 \in R$, we have $|u_1u_2| = |u_3u_4| = 1$. Thus, using the terminology of [7], an (i, j)-*CSSH system*, with $(i, j) \in \{(1, 3), (2, 4)\}$, (resp. a $(2, 3)$-*CSSH system*) is a CSSH system where for each $u_1\#u_2\$u_3\#u_4 \in R$ we have $u_i, u_j \in A$ (resp. $u_2, u_3 \in A$ or $u_1, u_4 \in A$). The special case $u_1u_2 = u_3u_4 \in A$ (*simple systems*) was first considered in [20] using Head's definition and then in [8] by taking into account Paun's definition. Given $(i, j) \in \{(1, 3), (2, 4), (2, 3)\}$, an (i, j)-*circular simple system* is an (i, j)-CSSH system which is simple [8]. In the former paper [20], the authors claim that these systems generate regular circular languages. In the latter paper [8], the authors compared the class of all circular languages generated by (i, j)-circular simple systems with that of all circular languages generated by (i', j')-circular simple systems, with $(i, j) \neq (i', j')$ and they gave a precise description of the relationship among these classes of languages along with some of their closure properties. An analogous viewpoint was adopted for Paun circular semi-simple splicing systems in [7] where the authors strengthened differences between circular simple and CSSH systems. Finally, in [10], the author claimed that the class of circular languages generated by semi-simple splicing systems is contained in the class of context-free circular languages. However, the following problem is still unsolved:

Problem 1. Find a characterization of the class of regular circular languages generated by finite circular semi-simple splicing systems.

In this paper we focus on a restricted version of Problem 1, namely we will only take into account $(1, 3)$-CSSH systems with I satisfying an additional hypothesis. While it is not difficult to extend our results to $(2, 4)$-CSSH systems, we do not yet know whether these results still hold when the hypothesis on the position of the letters is dropped.

Remark 2. Notice that in a $(1, 3)$-CSSH system, each rule r in R has the form $r = a_i\#1\$a_j\#1$, with $a_i, a_j \in A$. Furthermore, the circular splicing can be rephrased as follows: given a rule $a_i\#1\$a_j\#1$ and two circular words $^\sim ha_i, ^\sim ka_j$, the circular splicing yields as a result $^\sim ha_ika_j$.

3   Marked Systems Without Self-splicing

3.1   Marked Systems

In this section we define the circular splicing systems we deal with. We consider the special class of $(1,3)$-CSSH systems $SC_{PA} = (A, I, R)$ such that each word in I contains as a factor one occurrence of a site of R at most, i.e., for each $w \in I$, for each $a \in SITES(R)$, we have $|w|_a \leq 1$ (e.g. $SC_{PA} = (A, I, R)$, with $A = \{a, b, c\}$, $I = {}^{\sim}\{cca, b\}$, $R = \{a\#1\$b\#1\}$). It can be proved that in order to characterize the class of regular circular languages generated by these systems, we can adopt the definition that follows.

Definition 2 (Marked system). *A* marked system $SC_{PA} = (A, I, R)$ *is a* $(1,3)$-*CSSH system such that* $I = SITES(R) = A$.

To shorten notation, from now on $S = (I, R)$ will denote a marked system and $L(I, R)$ will be the corresponding generated language. Furthermore, it is understood that (a_i, a_j) is an abridged notation for a rule $r = a_i\#1\$a_j\#1$ in R. It is a simple matter to prove the proposition that follows.

Proposition 1. *For each marked system* $S = (I, R)$, *we have* $L(I, R) \subseteq {}^{\sim}I^*$. *Furthermore, for every* $a \in I$, *we have* $L(I, R) \cap {}^{\sim}a^+ a \neq \emptyset$ *if and only if* $(a, a) \in R$. *If* $(a, a) \in R$ *then* ${}^{\sim}a^+ \subseteq L(I, R)$.

We now introduce a relation $\approx$ in I that allows us to state a useful decomposition of the language generated by a marked system in Proposition 3.

Definition 3. *Let* $a_i, a_j \in I$. *Then* $a_i \approx a_j$ *if and only if* $b_1, \ldots, b_k \in I$ *exist such that* $b_1 = a_i$, $b_k = a_j$ *and* $(b_h, b_{h+1}) \in R$, *for all* $h \in \{1, \ldots, k-1\}$.

Example 1. Let $S = (I, R)$ be a marked system, with $I = \{a, b\}$, $R = \{(a, b)\}$. Then, $a \approx b$ (by taking $k = 2$ and $b_1 = a$, $b_2 = b$), $a \approx a$ (by taking $k = 3$ and $b_1 = a = b_3$, $b_2 = b$) and $b \approx b$ (by taking $k = 3$ and $b_1 = b = b_3$, $b_2 = a$).

Proposition 2. *The relation* $\approx$ *is an equivalence relation on* I.

Remark 3. We denote $I_1, \ldots, I_g$ the equivalence classes with respect to $\approx$. Obviously $\mathcal{I}_{\approx} = \{I_1, \ldots, I_g\}$ is a partition of I. Furthermore, notice that if $(a_i, a_j) \in R$ then $a_i \approx a_j$ and so, $\approx$ also defines a partition $\mathcal{R}_{\approx} = \{R_1, \ldots, R_g\}$ of R, where R_h is defined by $SITES(R_h) = I_h$, $1 \leq h \leq g$.

Given a marked system $S = (I, R)$, we now define some special "subsystems" of S by means of $\approx$.

Definition 4 (Canonical decomposition). *Let* $S = (I, R)$ *be a marked system. The* canonical decomposition *of* S *is the family* $\{(I_h, R_h) \mid 1 \leq h \leq g\}$ *of marked systems, with* $\mathcal{I}_{\approx} = \{I_1, \ldots, I_g\}$ *(resp.* $\mathcal{R}_{\approx} = \{R_1, \ldots, R_g\}$*) being the partition of* I *(resp.* R*) induced by* $\approx$.

We now state the already mentioned canonical decomposition of the language generated by a marked system.

Proposition 3. *Let $S = (I, R)$ be a marked system and let $\{(I_h, R_h) \mid 1 \leq h \leq g\}$ be the canonical decomposition of S. Then $L = L(I, R) = \bigcup_{h=1}^{g} L_h$, where $L_h = L(I_h, R_h)$, and $L_h \cap L_{h'} = \emptyset$, for $h, h' \in \{1, \ldots, g\}$.*

3.2 Transitive Marked Systems

As stated in Proposition 3, each circular language $L(I, R)$ generated by a marked system $S = (I, R)$ is a disjoint union of the circular languages $L(I_g, R_g)$ generated by means of the canonical decomposition of $S = (I, R)$, i.e., $L = L(I, R) = \bigcup_{h=1}^{g} L_h$. Consequently, $Lin(L) = \bigcup_{h=1}^{g} Lin(L_h)$. On the other hand, it is classically known that the class of regular (resp. context-free) languages is closed under union and intersection (resp. intersection with a regular language). As a result, L is regular (resp. context-free) circular if and only if each $Lin(L_h) = Lin(L) \cap I_h^+$ is regular (resp. context-free). The above arguments show that in order to characterize the structure of the regular languages generated by marked systems, we can assume that S is transitive, i.e., S satisfies the definition given below.

Definition 5. *Let $S = (I, R)$ be a marked system. S is transitive if $\{(I, R)\}$ is the canonical decomposition of S, i.e., for each $a_i, a_j \in I$ we have $a_i \approx a_j$.*

Lemma 1 will be mentioned in Section 3.4.

Lemma 1. *Let $L = L(I, R)$ be the circular language generated by a transitive marked system. Then $Pref_c(Lin(L)) = \cup_{w \in Lin(L)} Pref_c(w) = I^+$.*

3.3 Distance and Diameter

We introduce two notions, both given below: the *distance* between two letters and the *diameter* of a transitive marked system.

Definition 6. *Let $S = (I, R)$ be a transitive marked system. For each $a_i, a_j \in I$ the distance $d(a_i, a_j)$ between a_i and a_j is defined by $d(a_i, a_j) = min\{k \mid \exists b_1, \ldots, b_k \in I : (b_h, b_{h+1}) \in R, 1 \leq h \leq k-1, b_1 = a_i, b_k = a_j\}$.*

Definition 7. *Let $S = (I, R)$ be a transitive marked system. The diameter of S is $d(S) = \max\{d(a_i, a_j) \mid a_i, a_j \in I, a_i \neq a_j\}$ if $|I| \geq 2$, $d(S) = 2$ otherwise.*

Notice that if $S = (I, R)$ is a marked system with $I = \{a\}$, S is transitive if and only if $R = \{(a, a)\}$. In Section 4 we will see that the class of transitive marked systems S such that $d(S) = 3$ is a boundary between $Reg^\sim$ and its complement.

3.4   Forbidden Chains

The main result of this section shows that if a transitive marked system $S = (I, R)$ satisfies a special condition, then the corresponding generated language $L(I, R)$ is not regular (Proposition 6). As a matter of fact, in Section 4, we prove that this condition characterizes transitive marked systems generating regular languages.

In order to prove this result, until further notice we assume that $S = (I, R)$, where $I = \{a_1, a_2, a_3, a_4\}$ and $R = \{(a_1, a_2), (a_2, a_3), (a_3, a_4)\}$. We recall that, for each word $v \in I^+$ there exists $\sim w$ in $L(I, R)$ such that $v \in Pref_c(w)$ (Lemma 1). So, we define some special words v and we prove that for each w satisfying the above property we have $|w| \geq 2|v|$ (Definition 8, Lemma 3). An application of the classical pumping lemma allows us to state that $L(I, R)$ is not regular (Proposition 4). Then, by using classical closure properties of regular languages along with Proposition 5, we prove the more general result in Proposition 6. The special above-mentioned words are defined below.

Definition 8. *Let ϕ be the morphism from I^* to I^* defined by $\phi(a_1) = a_4$, $\phi(a_4) = a_1$, $\phi(a_2) = a_3$, $\phi(a_3) = a_2$. For every $n \geq 2$ we define:*

$$v_n = \phi(v'_n) = \begin{cases} (a_1 a_4)^{\frac{n}{2}} & \text{if } n \text{ is even,} \\ (a_1 a_4)^{\frac{n-1}{2}} a_1 & \text{if } n \text{ is odd.} \end{cases}$$

$$v'_n = \phi(v_n) = \begin{cases} (a_4 a_1)^{\frac{n}{2}} & \text{if } n \text{ is even,} \\ (a_4 a_1)^{\frac{n-1}{2}} a_4 & \text{if } n \text{ is odd.} \end{cases}$$

We also denote by $W_n = \{w_n \mid n \geq 1\}$ (resp. $W'_n = \{w'_n \mid n \geq 1\}$) the set of the words w_n (resp. w'_n) recursively defined as follows:

- $w_1 = a_1 a_2$, $w'_1 = a_4 a_3$.
- $\forall n \geq 1$ $w_n = a_1 w_{n-1} a_2$, $w'_n = a_4 w_{n-1} a_3$.

We consider the circular words $\sim w_n$ (resp. $\sim w'_n$) corresponding to the words w_n (resp. w'_n) and we will refer to w_n (resp. w'_n) as to the canonical linearization of $\sim w_n$ (resp. $\sim w'_n$). By abuse of notation, we set $W_n = \{\sim w_n \mid n \geq 1\}$ (resp. $W'_n = \{\sim w'_n \mid n \geq 1\}$). We also extend ϕ to the circular words and we set $\phi(\sim w_n) = \sim \phi(w_n)$. Notice that the $\phi(\sim w_n)$ does not depend on which representative in $\sim w_n$ we choose to define it. Finally, if $r = (x, y) \in R$, then we also have $\phi(r) = (\phi(x), \phi(y)) \in R$.

Example 2. We have $\sim w_4 = \sim a_1 a_4 a_1 a_4 a_3 a_2 a_3 a_2$ and $\sim w'_4 = \sim a_4 a_1 a_4 a_1 a_2 a_3 a_2 a_3$.

Special properties of the words in $W_n \cup W'_n$ are pointed out in Lemmas 2 and 3.

Lemma 2. $W_n \cup W'_n \subseteq L(I, R)$.

Lemma 3. *Let n be a positive integer such that $n \geq 2$. Then, for each $\sim w \in L(I, R)$ such that $\sim w = \sim v_n z$ (resp. $\sim w = \sim v'_n z$) we have $|\sim w| \geq 2n$. Furthermore, there exists a unique shortest word $\sim w \in L(I, R)$ (of length $2n$) such that $\sim w = \sim v_n z$ (resp. $\sim w = \sim v'_n z$), namely $\sim w = \sim w_n$ (resp. $\sim w = \sim w'_n$).*

In view of Lemma 3, Proposition 4 is a consequence of a particular version of Pumping Lemma, also reported in [1] (*Ogden's Iteration Lemma for Regular Languages*).

Proposition 4. $L(I, R)$ *is not a regular language.*

Obviously Proposition 4 still holds if $R = \{(a_i, a_j), (a_j, a_h), (a_h, a_k)\}$, where $\{a_i, a_j, a_h, a_k\} = \{a_1, a_2, a_3, a_4\}$. We now assume that S is a transitive marked system with no additional hypotheses.

Proposition 5. *Let* $S = (I, R)$ *be a transitive marked system. Let* $J \subseteq I$ *and let* $R^J = R \cap (J \times J) = \{(a_i, a_j) \in R \mid a_i, a_j \in J\}$. *Then* $L(I, R) \cap {}^\sim J^+ = L(J, R^J)$. *Consequently, we have* $Lin(L(J, R^J)) = Lin(L(I, R) \cap {}^\sim J^+) = Lin(L(I, R)) \cap J^+$.

By using Proposition 5 we prove the result below which is an extension of Proposition 4 to the general case.

Proposition 6. *Let* $S = (I, R)$ *be a transitive marked system. Assume that there exists a subset* $J = \{a_1, a_2, a_3, a_4\}$ *of* I *such that* $R^J = \{(a_1, a_2), (a_2, a_3), (a_3, a_4)\}$. *Then* $L(I, R)$ *is not a regular circular language.*

4   A Classification of Marked Systems

In this section we give results which allow us to give a description of the behaviour of a transitive marked system S and in this description the diameter of S intervenes. We begin with a result concerning marked systems S with $d(S) \neq 3$.

Proposition 7. *Let* $S = (I, R)$ *be a transitive marked system. If* $d(S) = 2$ *then* $L(I, R) = {}^\sim I^+ \setminus \cup_{a \in I, (a,a) \notin R} {}^\sim a^+ a$ *is a regular circular language. If* $d(S) \geq 4$ *then* $L(I, R)$ *is not a regular circular language.*

The main difficulty in carrying out a classification of transitive marked systems is to handle the case $d(S) = 3$. Indeed, it is easy to give an example of a marked system $S = (I, R)$ with $L(I, R)$ being regular and $d(S) = 3$ (e.g., all transitive marked systems $S = (I, R)$ with $|I| = 3$) but we can also exhibit a non regular circular language generated by a transitive marked system having the same diameter 3. A characterization of regular circular languages generated by transitive marked systems (with diameter 3) will be given in Theorem 1.

In the remainder of this section we assume $S = (I, R)$ to be a transitive marked system such that $|I| \geq 4$ (otherwise $L(I, R)$ is a regular circular language) and $d(S) = 3$ (otherwise Proposition 7 applies). Furthermore, in the property that follows, we say that a subset J of I is *transitive* if $S^J = (J, R^J)$ is a transitive marked system.

Property 1. For every transitive subset $J = \{a_1, a_2, a_3, a_4\}$ of I we have $R^J \neq \{(a_i, a_j), (a_j, a_h), (a_h, a_k)\}$, where $\{a_i, a_j, a_h, a_k\} = \{a_1, a_2, a_3, a_4\}$.

Propositions 8 and 10 may be summarized by saying that $L(I, R)$ is regular if and only if $S = (I, R)$ satisfies Property 1.

Proposition 8. *If $L(I, R)$ is regular then $S = (I, R)$ satisfies Property 1.*

In Proposition 10, we state that if $S = (I, R)$ satisfies Property 1 then $L(I, R)$ is regular. In order to do so, Proposition 9 is a preliminary step.

Proposition 9. *For each $\sim w$, with $\sim w \in L(I, R)$ and $|w| \geq 2$, the set $J = alph(w)$ is a transitive subset of I and $\sim w \in L(J, R^J)$. Consequently, we have $L(I, R) \subseteq I \cup \bigcup_{J \subseteq I, \, J \text{ transitive}} {}^{\sim}(\cap_{a_i \in J} J^* a_i J^*)$.*

Proposition 10. *Assume that $S = (I, R)$ satisfies Property 1 and let $\sim w$ be such that $alph(w)$ is a transitive subset of I. Then $\sim w \in L(I, R)$. Consequently, $L(I, R) = I \cup \bigcup_{J \subseteq I, \, J \text{ transitive}} {}^{\sim}(\cap_{a_i \in J} J^* a_i J^*)$ is a regular circular language.*

Theorem 1. *The following conditions are equivalent:*

1) *$L(I, R)$ is a regular circular language.*
2) *$S = (I, R)$ satisfies Property 1.*
3) *$L(I, R) = I \cup \{w \in {}^{\sim}I^+ \mid alph(w) \text{ is transitive }\}$.*
4) *$L(I, R) = I \cup \bigcup_{J \subseteq I, \, J \text{ transitive}} {}^{\sim}(\cap_{a_i \in J} J^* a_i J^*)$.*

Proposition 11. *Given a marked splicing system S, we can decide whether $L(I, R)$ is a regular circular language. Given a regular circular language C we can decide whether a marked splicing system $S = (I, R)$ exists such that $C = L(I, R)$.*

Remark 4. In this paper we have restricted our attention to marked systems. Attempting to extend the corresponding results to the more general case of CSSH systems is a natural research direction which arose. The following example shows that this attempt fails and actually underlines the difference between marked and CSSH systems. Indeed, let $S = (I, R)$ be a marked system, where $R = \{(a, b)\}$. Then, we necessarily have $I = {}^{\sim}\{a, b\}$ and, in view of Proposition 7, $L(I, R)$ is regular. On the contrary, we can prove that $C(SC_{PA})$ is not a regular circular language, where $SC_{PA} = (A, I, R)$ is the $(1, 3)$-CSSH system defined by $A = \{a, b\}$, $I = {}^{\sim}\{ab\}$, $R = \{(a, b)\}$.

5   Marked Systems with Self-splicing

As we have already said, in the original definition of circular splicing languages given by Paun in [12], rules in R can be used in two different ways: one way was described in Section 2.2, the other, called *self-splicing*, will be reported below. In the remainder of this paper we assume that both splicing and self-splicing are allowed and the results proved in the previous sections are correspondingly reviewed. For better readability, we continue to use the same terminology (finite splicing systems, CSSH systems, (i, j)-CSSH systems, marked systems) even if it is understood that we refer to this different setting.

We recall that given a circular word $w = {}^\sim h u_1 u_2 k u_3 u_4$ and a rule $r = u_1 \# u_2 \$ u_3 \# u_4$, the self-splicing produces $z' = {}^\sim h u_1 u_2$, $z'' = {}^\sim k u_3 u_4$ [12]. The notation $w \vdash_r (z', z'')$ means that z', z'' are obtained by application of self-splicing, starting with w and by using r. In particular, for $(1, 3)$-CSSH systems we have $u_2 = u_4 = 1$ and $w = {}^\sim h u_1 k u_3 \vdash_r ({}^\sim h u_1, {}^\sim k u_3)$.

The definition of the circular language generated by a splicing system is given below.

Definition 9. *Let $SC_{PA} = (A, I, R)$ be a splicing system and let $C \subseteq {}^\sim A^*$. We denote: $\tau'(C) = \{z \in {}^\sim A^* \mid \exists w', w'' \in C, \exists r \in R \ (w', w'') \vdash_r z\} \cup \{z', z'' \in {}^\sim A^* \mid \exists w \in C, \exists r \in R \ w \vdash_r (z', z'')\}$. Thus, we define: $\tau^0(C) = C$, $\tau^{i+1}(C) = \tau^i(C) \cup \tau'(\tau^i(C))$, $i \geq 0$, $\tau^*(C) = \bigcup_{i \geq 0} \tau^i(C)$. We say that $C(SC_{PA}) = \tau^*(I)$ is the language generated by SC_{PA}.*

As in the first part of this paper, three models of splicing systems exist according to which of the three definitions (Head's, Paun's or Pixton's definition) is considered. When we refer to the computational power of these systems, one of the most interesting results deals with Pixton systems with additional hypotheses (see Section 2.2). This result, which is partially reported below, generalizes a similar theorem proved for linear splicing [17].

Theorem 2. *[16,17] Let $SC_{PI} = (A, I, R)$ be a Pixton circular splicing system, where I is a regular circular language and R is a regular, reflexive and symmetric set of rules. Then $\tau^*(I)$ is a regular circular language.*

It can be proved that Proposition 3 and Lemma 1 are still true if we assume that self-splicing is also taken into account. Finally, we state below that each marked system generates a regular circular language. We recall that in the special case of CSSH systems, each finite Paun circular splicing system S may be transformed into a finite Pixton circular splicing system which generates the same language as S (see Remark 3.1 in [5]). However, Proposition 12 is not a consequence of Theorem 2 since in a marked system the set of rules is not necessarily assumed to be reflexive.

Proposition 12. *Let $S = (I, R)$ be a marked system and let $\overline{L} = \tau^*(I)$. Then we have $\overline{L} = \bigcup_{h=1}^{g} {}^\sim I_h^+$ and $\overline{L}$ is a regular circular language. In particular, if S is transitive then we have $\overline{L} = {}^\sim I^+$.*

Acknowledgments. The authors wish to thank the referees for a number of helpful comments.

References

1. Berstel, J.: Transductions and Context-free Languages. B.G. Teubner, Stuttgart (1979)
2. Berstel, J., Restivo, A.: Codes et sousmonoides fermes par conjugaison. Sem. LITP 81(45), 10 pages (1981)

3. Bonizzoni, P., De Felice, C., Mauri, G., Zizza, R.: DNA and Circular Splicing. In: Condon, A., Rozenberg, G. (eds.) DNA 2000. LNCS, vol. 2054, pp. 117–129. Springer, Heidelberg (2001)

4. Bonizzoni, P., De Felice, C., Mauri, G., Zizza, R.: Decision Problems on Linear and Circular Splicing. In: Ito, M., Toyama, M. (eds.) DLT 2002. LNCS, vol. 2450, pp. 78–92. Springer, Heidelberg (2003)

5. Bonizzoni, P., De Felice, C., Mauri, G., Zizza, R.: Circular splicing and regularity. Theoretical Informatics and Applications 38, 189–228 (2003)

6. Bonizzoni, P., De Felice, C., Mauri, G., Zizza, R.: On the power of circular splicing. Discrete Applied Mathematics 150, 51–66 (2005)

7. Ceterchi, R., Martin-Vide, C., Subramanian, K.G.: On Some classes of splicing languages. In: Jonoska, N., Păun, G., Rozenberg, G. (eds.) Aspects of Molecular Computing. LNCS, vol. 2950, pp. 83–104. Springer, Heidelberg (2003)

8. Ceterchi, R., Subramanian, K.G.: Simple circular splicing systems. Romanian Journal of Information Science and Technology 6, 121–134 (2003)

9. Choffrut, C., Karhumaki, J.: Combinatorics on words. In: Rozenberg, G., Salomaa, A. (eds.) Handbook of Formal Languages, vol. 1, pp. 329–438. Springer, New York (1996)

10. Fagnot, I.: Simple circular splicing systems. Preproc. of Dixième Journées Montoises d'Informatique Théorique, Liege (2004)

11. Head, T.: Formal Language Theory and DNA: an analysis of the generative capacity of specific recombinant behaviours. Bulletin of Mathematical Biology 49, 737–759 (1987)

12. Head, T., Paun, G., Pixton, D.: Language theory and molecular genetics. Generative mechanisms suggested by DNA recombination. In: Rozenberg, G., Salomaa, A. (eds.) Handbook of Formal Languages, vol. 2, pp. 295–360. Springer, New York (1996)

13. Hopcroft, J.E., Motwani, R., Ullman, J.D.: Introduction to Automata Theory, Languages, and Computation. Addison-Wesley, Reading, Mass (2001)

14. Lothaire, M.: Combinatorics on Words. Encyclopedia of Mathematics and its Applications. Addison Wesley, Reading (1983)

15. Paun, G., Rozenberg, G., Salomaa, A.: DNA computing, New Computing Paradigms. Springer, New York (1998)

16. Pixton, D.: Linear and Circular Splicing Systems. In: Proceedings of the 1st International Symposium on Intelligence in Neural and Biological Systems, pp. 181–188. IEEE Computer Society Press, Los Alamitos (1995)

17. Pixton, D.: Regularity of splicing languages. Discrete Applied Mathematics 69, 101–124 (1996)

18. Pixton, D.: Splicing in abstract families of languages. Theoretical Computer Science 234, 135–166 (2000)

19. Reis, C., Thierren, G.: Reflective star languages and codes. Information and Control 42, 1–9 (1979)

20. Siromoney, R., Subramanian, K.G., Dare, A.: Circular DNA and splicing systems. In: Nakamura, A., Saoudi, A., Inoue, K., Wang, P.S.P., Nivat, M. (eds.) ICPIA 1992. LNCS, vol. 654, pp. 260–273. Springer, Heidelberg (1992)

The Quantum Query Complexity
of Algebraic Properties

Sebastian Dörn[1] and Thomas Thierauf[2]

[1] Institut für Theoretische Informatik, Universität Ulm
Sebastian.Doern@Uni-Ulm.de
[2] Fakultät Elektronik und Informatik, HTW Aalen
Thomas.Thierauf@HTW-Aalen.de

Abstract. We present quantum query complexity bounds for testing algebraic properties. For a set S and a binary operation on S, we consider the decision problem whether S is a semigroup or has an identity element. If S is a monoid, we want to decide whether S is a group.

We present quantum algorithms for these problems that improve the best known classical complexity bounds. In particular, we give the first application of the new quantum random walk technique by Magniez, Nayak, Roland, and Santha [18] that improves the previous bounds by Ambainis [3] and Szegedy [23]. We also present several lower bounds for testing algebraic properties.

1 Introduction

Quantum algorithms have the potential to demonstrate that for some problems quantum computation is more efficient than classical computation. A goal of quantum computing is to determine whether quantum computers are faster than classical computers.

In search problems, the access to the input is done via an oracle. This motivates the definition of the query complexity, which measures the number of accesses to the oracle. Here we study the quantum query complexity, which is the number of quantum queries to the oracle. For some problems the quantum query complexity can be exponentially smaller than the classical one; an example is the Simon algorithm [22].

Quantum query algorithms have been presented for several problems, see [8,3, 13,17,19,10,11,12]. These algorithms use search techniques like Grover search [14], amplitude amplification [9] and quantum random walk [3,23].

In this paper we study the quantum query complexity for testing algebraic properties. Our input is a multiplication table for a set S of size $n \times n$. In Section 3 we consider the *semigroup problem*, that is, whether the operation on S is associative. Rajagopalan and Schulman [21] developed a randomized algorithm for this problem that runs in time $O(n^2)$. As an additional parameter, we consider the binary operation $\circ : S \times S \to M$, where $M \subseteq S$. We construct a quantum algorithm for this problem whose query complexity is $O(n^{5/4})$, if the size of M is constant. Our algorithm is the first application of the new quantum random

E. Csuhaj-Varjú and Z. Ésik (Eds.): FCT 2007, LNCS 4639, pp. 250–260, 2007.

walk search scheme by Magniez, Nayak, Roland, and Santha [18]. With the quantum random walk of Ambainis [3] and Szegedy [23], the query complexity of our algorithm would not improve the obvious Grover search algorithm for this problem. We show a quantum query lower bound for the semigroup problem of $\Omega(n)$ in Section 5.

In Section 4 we consider the *group problem*, that is, whether the monoid M given by its multiplication table is a group. We present a randomized algorithm that solves the problem with $O(n^{\frac{3}{2}})$ classical queries to the multiplication table. This improves the naive $O(n^2)$ algorithm that searches for an inverse in the multiplication table for every element. Then we show that on a quantum computer the query complexity can be improved to $\widetilde{O}(n^{\frac{11}{14}})$, where the $\widetilde{O}$-notation hides a logarithmic factor.

In Section 5, we show linear lower bounds for the semigroup problem and the identity problem. In the latter problem we have given a multiplication table of a set S and have to decide whether S has an identity element. As an upper bound, the identity problem can be solved with linearly many quantum queries, which matches the lower bound. Finally we show linear lower bounds for the quasigroup and the loop problem, where one has to decide whether a multiplication table is a quasi group or a loop, respectively.

2 Preliminaries

2.1 Quantum Query Model

In the query model, the input $x_1, \ldots, x_N$ is contained in a black box or oracle and can be accessed by queries to the black box. As a query we give i as input to the black box and the black box outputs x_i. The goal is to compute a Boolean function $f : \{0,1\}^N \rightarrow \{0,1\}$ on the input bits $x = (x_1, \ldots, x_N)$ minimizing the number of queries. The classical version of this model is known as decision tree.

The quantum query model was explicitly introduced by Beals *et al.* [6]. In this model we pay for accessing the oracle, but unlike the classical case, we use the power of quantum parallelism to make queries in superposition. The state of the computation is represented by $|i, b, z\rangle$, where i is the query register, b is the answer register, and z is the working register.

A quantum computation with k queries is a sequence of unitary transformations

$$U_0 \rightarrow O_x \rightarrow U_1 \rightarrow O_x \rightarrow \ldots \rightarrow U_{k-1} \rightarrow O_x \rightarrow U_k,$$

where each U_j is a unitary transformation that does not depend on the input x, and O_x are query (oracle) transformations. The oracle transformation O_x can be defined as $O_x : |i, b, z\rangle \rightarrow |i, b \oplus x_i, z\rangle$.

The computation consists of the following three steps:

1. Go into the initial state $|0\rangle$.
2. Apply the transformation $U_T O_x \cdots O_x U_0$.
3. Measure the final state.

The result of the computation is the rightmost bit of the state obtained by the measurement.

The quantum computation determines f with bounded error, if for every x, the probability that the result of the computation equals $f(x_1, \ldots, x_N)$ is at least $1 - \epsilon$, for some fixed $\epsilon < 1/2$. In the query model of computation each query adds one to the query complexity of an algorithm, but all other computations are free.

2.2   Tools for Quantum Algorithms

For the basic notation on quantum computing, we refer the reader to the textbook by Nielsen and Chuang [20]. Here, we give three tools for the construction of our quantum algorithms.

Quantum Search. A search problem is a subset $P \subseteq \{1, \ldots, N\}$ of the search space $\{1, \ldots, N\}$. With P we associate its characteristic function $f_P : \{1, \ldots, N\} \to \{0, 1\}$ with

$$f_P(x) = \begin{cases} 1, & \text{if } x \in P, \\ 0, & \text{otherwise.} \end{cases}$$

Any $x \in P$ is called a solution to the search problem. Let $k = |P|$ be the number of solutions of P.

Theorem 1. [14,7] *For $k > 0$, the expected quantum query complexity for finding one solution of P is $O(\sqrt{N/k})$, and for finding all solutions, it is $O(\sqrt{kN})$. Futhermore, whether $k > 0$ can be decided in $O(\sqrt{N})$ quantum queries to f_P.*

Amplitude Amplification. Let $\mathcal{A}$ be an algorithm for a problem with small success probability at least ϵ. Classically, we need $\Theta(1/\epsilon)$ repetitions of $\mathcal{A}$ to increase its success probability from ϵ to a constant, for example $2/3$. The corresponding technique in the quantum case is called amplitude amplification.

Theorem 2. [9] *Let $\mathcal{A}$ be a quantum algorithm with one-sided error and success probability at least ϵ. Then there is a quantum algorithm $\mathcal{B}$ that solves $\mathcal{A}$ with success probability $2/3$ by $O(\frac{1}{\sqrt{\epsilon}})$ invocations of $\mathcal{A}$.*

Quantum Walk. Quantum walks are the quantum counterpart of Markov chains and random walks. The quantum walk search provide a promising source for new quantum algorithms, like quantum walk search algorithm [16], element distinctness algorithm [3], triangle finding [19], testing group commutativity [17], and matrix verification [10].

Let $P = (p_{xy})$ be the transition matrix of an ergodic symmetric Markov chain on the state space X. Let $M \subseteq X$ be a set of marked states. Assume that the search algorithms use a data structure D that associates some data $D(x)$ with every state $x \in X$. From $D(x)$, we would like to determine if $x \in M$. When operating on D, we consider the following three types of cost:

Setup cost s: The worst case cost to compute $D(x)$, for $x \in X$.

Update cost u: The worst case cost for transition from x to y, and update $D(x)$ to $D(y)$.

Checking cost c: The worst case cost for checking if $x \in M$ by using $D(x)$.

Magniez *et al.* [18] developed a new scheme for quantum search, based on any ergodic Markov chain. Their work generalizes previous results by Ambainis [3] and Szegedy [23]. They extend the class of possible Markov chains and improve the query complexity as follows.

Theorem 3. [18] *Let $\delta > 0$ be the eigenvalue gap of a ergodic Markov chain P and let $\frac{|M|}{|X|} \geq \epsilon$. Then there is a quantum algorithm that determines if M is empty or finds an element of M with cost*

$$s + \frac{1}{\sqrt{\epsilon}} \left(\frac{1}{\sqrt{\delta}} u + c \right).$$

In the most practical application (see [3, 19]) the quantum walk takes place on the Johnson graph $J(n,r)$, which is defined as follows: the vertices are subsets of $\{1, \ldots, n\}$ of size r and two vertices are connected iff they differ in exactly one number. It is well known, that the spectral gap δ of $J(n,r)$ is $1/r$.

2.3   Tool for Quantum Query Lower Bounds

In this paper, we use the following special case of a method by Ambainis [1] to prove lower bounds for the quantum query complexity.

Theorem 4. [1] *Let $A \subset \{0,1\}^n, B \subset \{0,1\}^n$ and $f : \{0,1\}^n \to \{0,1\}$ such that $f(x) = 1$ for all $x \in A$, and $f(y) = 0$ for all $y \in B$. Let m and m' be numbers such that*

1. *for every $(x_1, \ldots, x_n) \in A$ there are at least m values $i \in \{1, \ldots, n\}$ such that $(x_1, \ldots, x_{i-1}, 1 - x_i, x_{i+1}, \ldots, x_n) \in B$,*
2. *for every $(x_1, \ldots, x_n) \in B$ there are at least m' values $i \in \{1, \ldots, n\}$ such that $(x_1, \ldots, x_{i-1}, 1 - x_i, x_{i+1}, \ldots, x_n) \in A$.*

Then every bounded-error quantum algorithm that computes f has quantum query complexity $\Omega(\sqrt{m \cdot m'})$.

3   The Semigroup Problem

In the semigroup problem we have given two sets S and $M \subseteq S$ and a binary operation $\circ : S \times S \to M$ represented by a table. We denote with n the size of the set S. One has to decide whether S is a semigroup, that is, whether the operation on S is associative.

The complexity of this problem was first considered by Rajagopalan and Schulman [21], who gave a randomized algorithm with time complexity of

$O(n^2 \log \frac{1}{\delta})$, where δ is the error probability. They also showed a lower bound of $\Omega(n^2)$. The previously best known algorithm was the naive $\Omega(n^3)$-algorithm that checks all triples.

In the quantum setting, one can do a Grover search over all triples $(a, b, c) \in S^3$ and check whether the triple is associative. The quantum query complexity of the search is $O(n^{3/2})$. We construct a quantum algorithm for the semigroup problem that has query complexity $O(n^{5/4})$, if the size of M is constant. In Section 5 we give a quantum query lower bound of $\Omega(n)$ for this problem.

Our algorithm is the first application of the recent quantum random walk search scheme by Magniez et al. [18]. The quantum random walk of Ambainis [3] and Szegedy [23] doesn't suffice to get an improvement of the Grover search mentioned above.

Theorem 5. *Let $k = n^\alpha$ be the size of M with $0 \le \alpha \le 1$. The quantum query complexity of the semigroup problem is*

$$\begin{cases} O(n^{\frac{5}{4}+\frac{\alpha}{2}}), & \text{for } 0 < \alpha \le \frac{1}{6}, \\ O(n^{\frac{6}{5}+\frac{4}{5}\alpha}), & \text{for } \frac{1}{6} < \alpha \le \frac{3}{8}, \\ O(n^{\frac{3}{2}}), & \text{for } \frac{3}{8} < \alpha \le 1. \end{cases}$$

Proof. We use the quantum walk search scheme of Theorem 3. To do so, we construct a Markov chain and a database for checking if a vertex of the chain is marked.

Let A and B two subsets of S of size r that are disjoint from M. We will determine r later. The database is the set

$$D(A, B) = \{ (a, b, a \circ b) \mid a \in A \cup M \text{ and } b \in B \cup M \}.$$

Our quantum walk is done on the categorical graph product of two Johnson graphs $G_J = J(n - k, r) \times J(n - k, r)$. The marked vertices of G_J correspond to pairs (A, B) with $(A \circ B) \circ S \ne A \circ (B \circ S)$. In every step of the walk, we exchange one row and one column of A and B.

Now we compute the quantum query costs for the setup, update and checking. The setup cost for the database $D(A, B)$ is $(r + k)^2$ and the update cost is $r + k$. To check whether a pair (A, B) is marked, we search for a pair $(b, c) \in B \times S$ with $(A \circ b) \circ c \ne A \circ (b \circ c)$. The quantum query cost to check this inequality is $O(k)$, by using our database. Therefore, by applying Grover search, the checking cost is $O(k\sqrt{nr})$. The spectral gap of the walk on G_J is $\delta = O(1/r)$ for $1 \le r \le \frac{n}{2}$, see [10]. If there is a triple (a, b, c) with $(a \circ b) \circ c \ne a \circ (b \circ c)$, then there are at least $\binom{n-k-1}{r-k-1}^2$ marked sets (A, B). Therefore we have

$$\epsilon \ge \frac{|M|}{|X|} \ge \left(\frac{\binom{n-k-1}{r-k-1}}{\binom{n-k}{r-k}} \right)^2 \ge \left(\frac{r-k}{n-k} \right)^2.$$

Let $r = n^\beta$, for $0 < \beta < 1$. Assuming $r > 2k$ we have

$$\frac{1}{\sqrt{\epsilon}} \le \frac{n-k}{r-k} \le \frac{n}{r/2} = \frac{2n}{r}.$$

Then the quantum query complexity of the semigroup problem is

$$O\left(r^2 + \frac{n}{r}\left(\sqrt{r} \cdot r + \sqrt{nr} \cdot k\right)\right) = O\left(n^{2\beta} + n^{1+\frac{\beta}{2}} + n^{\frac{3}{2}+\alpha-\frac{\beta}{2}}\right).$$

Now we choose β depending on α such that this expression is minimal. A straight forward calculation gives the bounds claimed in the theorem. □

For the special case that $\alpha = 0$, i.e., only a constant number of elements occurs in the multiplication table, we get

Corollary 1. *The quantum query complexity of the semigroup problem is $O(n^{\frac{5}{4}})$, if M has constant size.*

Note that the time complexity of our algorithm is $O(n^{1.5} \log n)$.

4   Group Problems

In this section we consider the problem whether a given finite monoid M is in fact a group. That is, we have to check whether every element of M has an inverse. The monoid M has n elements and is given by its multiplication table and the identity element e.

To the best of our knowledge, the group problem has not been studied before. The naive approach for the problem checks for every element $a \in M$, whether e occurs in a's row in the multiplication table. The query complexity is $O(n^2)$. We develop a (classical) randomized algorithm that solves the problem with $O(n^{\frac{3}{2}})$ queries to the multiplication table. Then we show that on a quantum computer the query complexity can be improved to $\widetilde{O}(n^{\frac{11}{14}})$.

Theorem 6. *Whether a given monoid with n elements is a group can be decided with query complexity*

1. *$O(n^{\frac{3}{2}})$ by a randomized algorithm with probability $\geq 1/2$,*
2. *$O(n^{\frac{11}{14}} \log n)$ by a quantum query algorithm.*

Proof. Let $a \in M$. We consider the sequence of powers $a, a^2, a^3, \ldots$. Since M is finite, there will be a repetition at some point. We define the *order of a* as the smallest power t, such that $a^t = a^s$, for some $s < t$. Clearly, if a has an inverse, s must be zero.

Lemma 1. *Let $a \in M$ of order t. Then a has an inverse iff $a^t = e$.*

Hence the powers of a will tell us at some point whether a has an inverse. On the other hand, if a has no inverse, the powers of a provide more elements with no inverse as well.

Lemma 2. *Let $a \in M$. If a has no inverse, then a^k has no inverse, for all $k \geq 1$.*

Our algorithm has two phases. In phase 1, it computes the powers of every element up to certain number r. That is, we consider the sequences $S_r(a) = (a, a^2, \ldots, a^r)$, for all $a \in M$. If $e \in S_r(a)$ then a has an inverse by Lemma 1. Otherwise, if we find a repetition in the sequence $S_r(a)$, then, again by Lemma 1, a has no inverse and we are done.

If we are not already done by phase 1, i.e. there are some sequences $S_r(a)$ left such that $e \notin S_r(a)$ and $S_r(a)$ has pairwise different elements, then the algorithm proceeds to phase 2. It selects some $a \in M$ uniformly at random and checks whether a has an inverse by searching for e in the row of a in the multiplication table. This step is repeated n/r times.

The query complexity $t(n)$ of the algorithm is bounded by nr in phase 1 and by n^2/r in phase 2. That is $t(n) \leq nr + n^2/r$, which is minimized for $r = n^{\frac{1}{2}}$. Hence we have $t(n) \leq 2n^{\frac{3}{2}}$.

For the correctness observe that the algorithm accepts with probability 1 if M is a group. Now assume that M is not a group. Assume further that the algorithm does not already detect this in phase 1. Let a be some element without an inverse. By Lemma 1, the sequence $S_r(a)$ has r pairwise different elements which don't have inverses too by Lemma 2. Therefore in phase 2, the algorithm picks an element without an inverse with probability at least r/n. By standard arguments, the probability that at least one out of n/r many randomly chosen elements has no inverse is constant.

For the quantum query complexity we use Grover search and amplitude amplification. In phase 1, we search for an $a \in M$, such that the sequence $S_r(a)$ has r pairwise different entries different from e. This property can be checked by first searching $S_r(a)$ for an occurance of e by a Grover search with $\sqrt{r} \log r$ queries. Then, if e doesn't occur in $S_r(a)$, we check whether there is an element in $S_r(a)$ that occurs more than once. This is the element distinctness problem and can be solved with $r^{2/3} \log r$ queries, see [3]. Therefore the quantum query complexity of phase 1 is bounded by $\sqrt{n} \cdot r^{2/3} \log r$.

In phase 2 we search for an $a \in M$ such that a has no inverse. In phase 2 we actually search the row of a in the multiplication table. Hence this takes $\sqrt{n}$ queries. Since at least r of the a's don't have an inverse, by amplitude amplification we get $\sqrt{n}\sqrt{n/r} = n/\sqrt{r}$ queries in phase 2.

In summary, the quantum query complexity is $\sqrt{n} \cdot r^{2/3} \log r + n/\sqrt{r}$, which is minimized for $r = n^{\frac{3}{7}}$. Hence we have a $O(n^{\frac{11}{14}} \log n)$ quantum query algorithm. $\qquad \square$

5   Lower Bounds

Theorem 7. *The semigroup problem requires $\Omega(n)$ quantum queries.*

Proof. Let S be a set of size n and $\circ : S \times S \rightarrow \{0, 1\}$ a binary operation represented by a table. We apply Theorem 4. The set A consists of all $n \times n$ matrices, where the entry of position $(1, 1), (1, c), (c, 1)$ and (c, c) is 1, for $c \in S - \{0, 1\}$, and zero otherwise. It is easy to see, that the multiplication tables of A are associative, since $(x \circ y) \circ z = x \circ (y \circ z) = 1$ for all $x, y, z \in \{1, c\}$ and zero otherwise.

The set B consists of all $n \times n$ matrices, where the entry of position $(1, 1)$, $(1, c)$, $(c, 1)$, (c, c) and (a, b) is 1, for fixed $a, b, c \in S - \{0, 1\}$ with $a, b \neq c$, and zero otherwise. Then $(a \circ b) \circ c = 1$ and $a \circ (b \circ c) = 0$. Therefore the multiplication tables of B are not associative.

From each $T \in A$, we can obtain $T' \in B$ by replacing the entry 0 of T at (a, b) by 1, for any $a, b \notin \{0, 1, c\}$. Hence we have $m = \Omega(n^2)$. From each $T' \in B$, we can obtain $T \in A$ by replacing the entry 1 of T' at position (a, b) by 0, for $a, b \notin \{0, 1, c\}$. Then we have $m' = 1$. By Theorem 4, the quantum query complexity is $\Omega(\sqrt{m \cdot m'}) = \Omega(n)$. □

Next, we consider the *identity problem*: given the multiplication table on a set S, decide whether there is an identity element.[1] We show that the identity problem requires linearly many quantum queries. We start by considering the *1-column problem*: given a 0-1-matrix of order n, decide whether it contains a column that is all 1.

Lemma 3. *The 1-column problem requires $\Omega(n)$ quantum queries.*

Proof. We use Theorem 4. The set A consists of all matrices, where in $n - 1$ columns there is exactly one entry with value 0, and the other entries of the matrix are 1. The set B consists of all matrices, where in every column there is exactly one entry with value 0, and the other entries of the matrix are 1. From each matrix $T \in A$, we can obtain $T' \in B$ by changing one entry in the 1-column from 1 to 0. Then we have $m = n$. From each matrix $T' \in B$, we can obtain $T \in A$ by changing one entry from 0 to 1. Then we have $m' = n$. By Theorem 4, the quantum query complexity is $\Omega(n)$. □

Theorem 8. *The identity problem requires $\Omega(n)$ quantum queries.*

Proof. We reduce the 1-column problem to the identity problem. Given a 0-1-matrix $M = (m_{i,j})$ of order n. We define $S = \{0, 1, \ldots, n\}$ and a multiplication table $T = (t_{i,j})$ with $0 \leq i, j \leq n$ for S as follows:

$$t_{i,j} = \begin{cases} 0, & \text{if } m_{i,j} = 0, \\ i, & \text{if } m_{i,j} = 1, \end{cases}$$

and $t_{0,j} = t_{i,0} = 0$. Then M has a 1-column iff T has an identity element. □

Finding an identity element is simple. We choose an element $a \in S$ and then we test if a is the identity element by using Grover search in $O(\sqrt{n})$ quantum queries. The success probability of this procedure is $\frac{1}{n}$. By using the amplitude amplification we get an $O(n)$ quantum query algorithm for finding an identity element (if there is one). Since the upper and the lower bound match, we have determined the precise complexity of the identity problem.

---

[1] Here we consider right identity, the case of left identity is analogous.

Corollary 2. *The identity problem has quantum query complexity* $\Theta(n)$.

In the *quasigroup problem* we have given a set S and a binary operation on S represented by a table. One has to decide whether S is a quasigroup, that is, whether all equations $a \circ x = b$ and $x \circ a = b$ have unique solutions. In the *loop problem*, one has to decide whether S is a loop. A loop is a quasigroup with an identity element e such that $a \circ e = a = e \circ a$ for all $a \in S$.

   In the multiplication table of a quasigroup, every row and column is a permutation of the elements of S. In a loop, there must occur the identity permutation in some row and some column. We have already seen how to determine an identity element with $O(n)$ quantum queries. A row or column is a permutation, if no element appears twice. Therefore one can use the element distinctness quantum algorithm by Ambainis [3] to search for a row or column with two equal elements. The quantum query complexity of the search is $O(\sqrt{n} \cdot n^{\frac{2}{3}}) = O(n^{\frac{7}{6}})$. We show in the following theorem an $\Omega(n)$ lower bound for these problems.

Theorem 9. *The quasigroup problem and the loop problem require* $\Omega(n)$ *quantum queries.*

Proof. We reduce the *identity matrix problem* to the loop problem. Given a 0-1-matrix $M = (m_{i,j})$ of order n, decide whether M is the identity matrix. It is not hard to see that the identity matrix problem requires $\Omega(n)$ quantum queries (similar as for the 1-column problem).

   We define $S = \{0, 1, \ldots, n-1\}$ and a multiplication table $T = (t_{i,j})$ for S. For convenience, we take indices $0 \le i, j \le n-1$ for M and T. The entries of the second diagonal are

$$t_{i,n-1-i} = \begin{cases} n-1, & \text{if } m_{i,i} = 1, \\ 0, & \text{otherwise.} \end{cases}$$

For $j \ne n - 1 - i$ we define

$$t_{i,j} = \begin{cases} (i+j) \bmod n, & \text{if } m_{i,n-1-j} = 0, \\ 0, & \text{if } m_{i,n-1-j} = 1 \text{ and } (i+j) \bmod n \ne 0, \\ 1, & \text{otherwise.} \end{cases}$$

If M is the identity matrix, then T is a circular permutation matrix

$$T = \begin{pmatrix} 0 & 1 \cdots n - 2 & n - 1 \\ 1 & 2 \cdots n - 1 & 0 \\ \vdots & \vdots \quad\quad \vdots & \vdots \\ n - 1 & 0 \cdots n - 3 & n - 2 \end{pmatrix}.$$

Hence S is a loop with identity 0.

   Suppose M is not the identity matrix. If M has a 0 on the main diagonal, say at position (i,i), then the value $n-1$ doesn't occur in row i in T, and hence, row i

is not a permutation of S. If M has a 1 off the main diagonal, say at position $(i, n-1-j)$, then there will be a 0 or 1 at position (i,j) in T, which is different from $(i+j)$ mod n. Hence, either there will be two 0's or two 1's in row i in T, in which case row i is not a permutation of S, or a 0 or 1 changes to 1 or 0. In the latter case, the i-th row of T can be a permutation only if correspondingly the other 0 or 1 changes as well to 1 or 0, respectively. But then this carries over to the other rows and columns of T. That is, there must be more 1's in M off the main diagonal, so that all 0's and 1's in T switch their place with respect to their position when M is the identity matrix. However, then there is no identity element in T and hence, T is not a loop.

The reduction to the quasigroup problem can be done with similar arguments. $\qquad\square$

6   Conclusion and Open Problems

In this paper we present quantum query complexity bounds of algebraic properties. We construct a quantum algorithm for the semigroup problem whose query complexity is $O(n^{5/4})$, if the size of M is constant. Then we consider the group problem, and presented a randomized algorithm that solves this problem with $O(n^{\frac{3}{2}})$ classical queries and $\tilde{O}(n^{\frac{11}{14}})$ quantum queries to the multiplication table. Finally we show linear lower bounds for the semigroup, identity, quasigroup and loop problem.

Some questions remain open: Is there a quantum algorithm for the semigroup problem which is better then the Grover search bound of $O(n^{\frac{3}{2}})$ for $|M| \geq \frac{3}{8}$. It is not clear, whether we can apply the technique of the randomized associative algorithm by Rajagopalan and Schulman [21] in connection with the quantum walk search schema of Magniez et al. [18].

Some quantum query lower bound remain open. Are we able to prove a nontrivial lower bound for the group problem. Our upper bound for this problem is $\tilde{O}(n^{\frac{11}{14}})$. It would also be very interesting to close the gap between the $\Omega(n)$ lower bound and the $O(n^{7/6})$ upper bound for the quasigroup and the loop problem.

References

1. Ambainis, A.: Quantum Lower Bounds by Quantum Arguments. Journal of Computer and System Sciences 64, 750–767 (2002)
2. Ambainis, A.: Quantum walks and their algorithmic applications. International Journal of Quantum Information 1, 507–518 (2003)
3. Ambainis, A.: Quantum walk algorithm for element distinctness. In: Proceedings of FOCS'04, pp. 22–31 (2004)
4. Ambainis, A.: Quantum Search Algorithms, Technical Report arXiv:quant-ph/0504012 (2005)
5. Ambainis, A., Špalek, R.: Quantum Algorithms for Matching and Network Flows. In: Durand, B., Thomas, W. (eds.) STACS 2006. LNCS, vol. 3884. Springer, Heidelberg (2006)

6. Beals, R., Buhrman, H., Cleve, R., Mosca, M., de Wolf, R.: Quantum lower bounds by polynomials. Journal of ACM 48, 778–797 (2001)
7. Boyer, M., Brassard, G., Høyer, P., Tapp, A.: Tight bounds on quantum searching. Fortschritte Der Physik 46(4-5), 493–505 (1998)
8. Buhrman, H., Dürr, C., Heiligman, M., Høyer, P., Magniez, F., Santha, M., de Wolf, R.: Quantum Algorithms for Element Distinctness. In: Proceedings of CCC'01, pp. 131–137 (2001)
9. Brassard, G., Hóyer, P., Mosca, M., Tapp, A.: Quantum amplitude amplification and estimation. In: Quantum Computation and Quantum Information: A Millennium Volume. AMS Contemporary Mathematics Series (2000)
10. Buhrman, H., Špalek, R.: Quantum Verification of Matrix Products. In: Proceedings of SODA'06, pp. 880–889 (2006)
11. Dörn, S.: Quantum Complexity Bounds of Independent Set Problems. In: van Leeuwen, J., Italiano, G.F., van der Hoek, W., Meinel, C., Sack, H., Plášil, F. (eds.) SOFSEM 2007. LNCS, vol. 4362, pp. 25–36. Springer, Heidelberg (2007)
12. Dörn, S.: Quantum Algorithms for Graph Traversals and Related Problems. In: Proceedings of CIE'07 (2007)
13. Dürr, C., Heiligman, M., Høyer, P., Mhalla, M.: Quantum query complexity of some graph problems. In: Díaz, J., Karhumäki, J., Lepistö, A., Sannella, D. (eds.) ICALP 2004. LNCS, vol. 3142, pp. 481–493. Springer, Heidelberg (2004)
14. Grover, L.: A fast mechanical algorithm for database search. In: Proceedings of STOC'96, pp. 212–219 (1996)
15. Kavitha, T.: Efficient Algorithms for Abelian Group Isomorphism and Related Problems. In: Pandya, P.K., Radhakrishnan, J. (eds.) FST TCS 2003: Foundations of Software Technology and Theoretical Computer Science. LNCS, vol. 2914, pp. 277–288. Springer, Heidelberg (2003)
16. Kempe, J., Shenvi, N., Whaley, K.B.: Quantum Random-Walk Search Algorithm. Physical Review Letters A 67(5) (2003)
17. Magniez, F., Nayak, A.: Quantum complexity of testing group commutativity. In: Caires, L., Italiano, G.F., Monteiro, L., Palamidessi, C., Yung, M. (eds.) ICALP 2005. LNCS, vol. 3580, pp. 1312–1324. Springer, Heidelberg (2005)
18. Magniez, F., Nayak, A., Roland, J., Santha, M.: Search via Quantum Walk. In: Proceedings of STOC'07 (2007)
19. Magniez, F., Santha, M., Szegedy, M.: Quantum Algorithms for the Triangle Problem. In: Proceedings of SODA'05, pp. 1109–1117 (2005)
20. Nielsen, M.A., Chuang, I.L.: Quantum Computation and Quantum Information. Cambridge University Press, Cambridge (2003)
21. Rajagopalan, S., Schulman, L.J.: Verification of identities. SIAM J. Computing 29(4), 1155–1163 (2000)
22. Simon, D.R.: On the power of quantum computation. In: Proceedings of FOCS'94, pp. 116–123 (1994)
23. Szegedy, M.: Quantum speed-up of markov chain based algorithms. In: Proceedings of FOCS'04, pp. 32–41 (2004)

On the Topological Complexity of Weakly Recognizable Tree Languages

Jacques Duparc[1] and Filip Murlak[2,*]

[1] Université de Lausanne, Switzerland
jduparc@unil.ch
[2] Warsaw University, Poland
fmurlak@mimuw.edu.pl

Abstract. We show that the family of tree languages recognized by weak alternating automata is closed by three set theoretic operations that correspond to sum, multiplication by ordinals $< \omega^\omega$, and pseudo-exponentiation with the base ω_1 of the Wadge degrees. In consequence, the Wadge hierarchy of weakly recognizable tree languages has the height of at least ε_0, that is the least fixed point of the exponentiation with the base ω.

1 Introduction

Topological hierarchies stormed into the theory of formal languages with Klaus Wagner's fundamental works on regular ω-languages [16,17]. The incredible coincidence of the Wagde hierarchy and the index hierarchy for these languages encouraged further investigation of the Wadge hierarchies of wider classes of ω-languages, corresponding to more powerful recognizing devices: push-down automata and Turing machines [3,4,13]. It was only a matter of time before the same questions were asked for tree languages. Deterministic languages, an acclaimed "easy" subclass, were considered first. Albeit more complex, they admitted a number of techniques developed for ω-languages. Soon, the Borel hierarchy, the Wadge hierarchy, and the index hierarchy of deterministic tree languages were described and proved decidable [7,8,9,10].

The real challenge seems to be nondeterminism. The power it gives to tree automata makes them extremely difficult to tackle. Therefore, the investigation has basically concentrated on a very special sub-case – weakly recognizable languages. This class is the intersection of Büchi and co-Büchi languages [6,12], so it is a rather small subclass of all regular tree languages. In fact, it does not even contain all deterministic languages. On the other hand, it captures some real nondeterminism, as it contains a lot of languages that cannot be recognized by deterministic automata: Skurczyński showed that weakly recognizable languages can have any finite Borel rank [14], while deterministic languages are either Π_1^1-complete or are in Π_3^0 [9].

* The second author was supported by Polish government grant no. N206 005 31/0881. A part of this work was done during the author's visit at the University of Lausanne, Switzerland, financed by AutoMathA (ESF Short Visit Grant 1410).

E. Csuhaj-Varjú and Z. Ésik (Eds.): FCT 2007, LNCS 4639, pp. 261–273, 2007.

More precisely, Skurczyński gave examples of Π_n^0 and Σ_n^0-complete languages recognized by weak alternating automata using ranks $[0, n]$ and $[1, n+1]$ accordingly. In this paper we extend this result by showing that weak automata using ranks $[0, n]$ can only recognize Π_n^0 languages, and dually $[1, n + 1]$-automata can only recognize Σ_n^0 languages. (One may conjecture that the converse also holds, i. e., every weakly recognizable Π_n^0 language can be recognized by a $[0, n]$-automaton, and dually for the additive classes.) We actually prove a bit stronger result. We consider so called weak game languages $W_{[\iota, \kappa]}$, to which all languages recognized by weak $[\iota, \kappa]$-automata can be reduced. We show that $W_{[0,n]} \in \Pi_n^0$ and $W_{[1,n+1]} \in \Sigma_n^0$ (by Skurczyński's results, they are hard for these classes). The languages $W_{[\iota, \kappa]}$ are natural weak counterparts of strong game languages considered lately by Arnold and Niwiński. The strong game languages also form a strict hierarchy, but they are all non-Borel [1].

The main result of this paper is a lower bound for the Wadge hierarchy of weakly recognizable languages. We show that weakly recognizable languages are closed by three set-theoretic operations corresponding to the sum, multiplication by ordinals $< \omega^\omega$ and pseudo-exponentiation with the base ω_1 of the Wadge degrees. As a consequence, the hierarchy has the height of at least ε_0, which is the least fixpoint of the exponentiation with the base ω. Again, this should be contrasted with the height of the hierarchy of deterministic tree languages, which is as low as $(\omega^\omega)^3 + 3$.

2   Weak Alternating Automata

A *tree over* Σ is a partial function $t : X^* \longrightarrow \Sigma$ with a prefix closed domain. For the purpose of this paper we call such trees *conciliatory*. We do that to remind the reader that we are working with the trees that may have infinite and finite branches. A tree t is *full* if $\operatorname{dom} t = X^*$. A tree is called *binary* if $X = \{0, 1\}$. Let T_Σ denote the set of full binary trees over Σ, and let $\tilde{T}_\Sigma$ be the set of all conciliatory binary trees over Σ. By $t.v$ we denote the subtree of t rooted in $v \in \operatorname{dom} t$.

A *weak alternating automaton* $\mathcal{A} = \langle \Sigma, Q_\exists, Q_\forall, q_0, \delta, \operatorname{rank} \rangle$ consists of a finite input alphabet Σ, a finite set of states Q partitioned into existential states $Q_\exists$ and universal states $Q_\forall$, an initial state q_0, a transition relation $\delta \subseteq Q \times \Sigma \times \{0, 1, \varepsilon\} \times Q$, and a priority function $\operatorname{rank} : Q \to [\iota, \kappa]$, where $[\iota, \kappa]$ stands for $\{\iota, \iota+1, \ldots, \kappa\}$. The transitions of the automaton are usually written as $p \xrightarrow{\sigma, d} q$, instead of $(p, \sigma, d, q) \in \delta$.

The *run* of the automaton $\mathcal{A}$ on a conciliatory input tree $t \in \tilde{T}_\Sigma$ is a finitely branching conciliatory tree ρ_t labeled with $Q \times \{0, 1, \varepsilon\}$. The root of the tree is labeled with (q_0, ε). Suppose we have already labeled a node X of ρ_t. Let (p_1, d_1), (p_2, d_2), $\ldots$, (p_m, d_m) be the sequence of labels on the unique path leading form the root to X. Let $v = d_1 d_2 \ldots d_m$, where the ε's occurring in the sequence $d_1 d_2 \ldots d_m$ are interpreted as empty words. If $v \notin \operatorname{dom} t$, then X is a leaf in ρ_t. Otherwise, for each transition $p_m \xrightarrow{t(v), d} q$, add a child Y to the node X and label it with (q, d). Note that the number of children of each node can by bounded by $3|Q|$.

The reader should not be puzzled by the fact that leaves of ρ_t do not correspond to any nodes of t. This is a notorious inconvenience in automata on finite objects: the number of states visited always exceeds by one the number of letters read. Let us imagine that cutting off a subtree produces a stub, and this is where the leaves of ρ_t dwell.

The accepting runs are defined by means of a modified weak parity game. Let ρ be a run of $\mathcal{A}$. The game G_ρ is played by Adam and Eve on the tree ρ. They move a token along the edges of the tree, starting from the root. The move is always made by the owner of the node: if the current node is labeled with a state from $Q_\exists$, it is Eve who moves the token to the next node, otherwise it is Adam. The play is infinite, unless it reaches a leaf. A play is won by Eve if the maximum of the ranks of states seen on the labels of visited nodes is even. Note that classically, when a play is finite, the owner of the last position looses. Here, we give no special rules for finite plays: the highest rank decides.

A run ρ is *accepting* if Eve has a winning strategy in the game G_ρ. A tree t is *accepted by the automaton* if ρ_t is accepting. The *language recognized by the automaton*, $L(\mathcal{A})$, is the set of accepted trees. A language is *weakly recognizable* if it is recognized by a weak alternating automaton.

While our automata work on conciliatory trees, the classical automata work on full binary trees. Instead of $L(\mathcal{A})$ one considers $L^\omega(\mathcal{A}) = L(\mathcal{A}) \cap T_\Sigma$. In order to relate the two versions we have to disguise conciliatory trees to make them look full.

Consider $T_{\Sigma \cup \{s\}}$, where s stands for "skip". For a tree $t \in T_{\Sigma \cup \{s\}}$ we will define a conciliatory tree $u(t)$, called the *undressing* of t. Informally, we want to omit the skips in a top-down manner. Suppose we are in a node v such that $t(v) = s$. We would like to ignore this node and replace it with the next one, but in case of trees we have two nodes to choose from: $v0$ and $v1$. Let us always choose $v0$. Another problem is that we may encounter an infinite sequence of s's. This would keep us replacing the current node with its left child, and never get to a symbol different from s. In that case, the tree $u(t)$ simply does not contain this node. Now, let us see a formal definition. Let v be the first node *not* labeled with s on the leftmost path of t (if there is no such node, $u(t)$ is empty). For each $w \in \{0,1\}^*$ consider two possibly infinite sequences:

- $v_0 = v$, $w_0 = w$,
- $v_{i+1} = v_i b$, $w_{i+1} = w'_i$ if $w_i = bw'_i$ and $t(v_i b) \neq s$,
- $v_{i+1} = v_i 0$, $w_{i+1} = w_i$ if $w_i = bw'_i$ and $t(v_i b) = s$.

If $w_n = \varepsilon$ for some n, then $w \in \mathrm{dom}\, u(t)$ and $u(t)(w) = t(v_n)$. Otherwise, $w \notin \mathrm{dom}\, u(t)$. For a conciliatory language L, define L_s as the set of trees that belong to L after undressing, i. e., $L_s = \{t \in T_{\Sigma \cup \{s\}} : u(t) \in L\}$.

An automaton $\mathcal{A}$ can be transformed easily into $\mathcal{A}'$ such that $L^\omega(\mathcal{A}') = (L(\mathcal{A}))_s$. Simply, whenever you see a node labeled with s, move deterministically to the left without changing the state. In other words, it is enough to add $\{q \xrightarrow{s,0} q : q \in Q\}$ to the transition relation of $\mathcal{A}$.

3   Games, Hierarchies, and Topology

Let us start this section with the definition of a *conciliatory version of the Wadge game* (see [2]). For any pair of conciliatory tree languages L, M the game $G_C(L, M)$ is played by Spoiler and Duplicator. Each player builds a tree, t_S and t_D respectively. In every round, first Spoiler adds a finite number of nodes to t_S and then Duplicator adds a finite number of nodes to t_D. Nodes added by Duplicator and Spoiler must be children of nodes previously added. Both players are allowed to skip, i. e., add no nodes to their trees. Duplicator wins the game if $t_S \in L \iff t_D \in M$. Note that the resulting trees are conciliatory: they may contain finite branches, or even be finite.

For conciliatory languages L, M we use the notation $L \leq_C M$ iff Duplicator has a winning strategy in the game $G_C(L, M)$. If $L \leq_C M$ and $M \leq_C L$, we will write $L \equiv_C M$. The *conciliatory hierarchy* is the order induced by $\leq_C$ on the $\equiv_C$ classes of conciliatory languages.

The *classical Wadge game* $G_W(L, M)$ is defined for languages of full infinite trees, therefore a restriction on the moves is needed. The players must add both child nodes under each node they had put in the previous round. Only Duplicator is allowed to skip, and he must not skip forever. He must make infinitely many real moves, so that the tree he constructs is full.

The classical Wadge games provide a well-known criterion for continuous reducibility. T_Σ and the space of ω-words over Σ are equipped with the standard Cantor-like topology. For trees it is induced by the metric

$$d(s, t) = \begin{cases} 2^{-\min\{|x| \,:\, x \in \{0,1\}^*, \, s(x) \neq t(x)\}} & \text{if } s \neq t, \\ 0 & \text{if } s = t. \end{cases}$$

L is *continuously reducible* (or *Wadge reducible*) to M, if there exists a continuous function φ such that $L = \varphi^{-1}(M)$. We will write $L \leq_W M$, if L is Wadge reducible to M. Similarly we define $\equiv_W$ and $<_W$. The *Wadge hierarchy* is the order induced on $\equiv_W$ classes of languages.

Lemma 1 (Wadge). *For $L, M \subseteq T_\Sigma$, $L \leq_W M$ iff Duplicator has a winning strategy in $G_W(L, M)$.*

The conciliatory hierarchy embeds naturally into the Wadge hierarchy by the mapping $L \mapsto L_s$. A strategy in one game can be translated easily to a strategy in the other: arbitrary skipping in $G_C(L, M)$ gives the same power as the s labels in $G_W(L_s, M_s)$.

Lemma 2. *For all conciliatory languages L and M, $L \leq_C M \iff L_s \leq_W M_s$.*

Recall that a language L is called *self dual* if it is equivalent to its complement L^C. The conciliatory hierarchy does not contain self dual languages: a strategy for Spoiler in $G_C(L, L^C)$ is to skip in the first round, and then copy Duplicator's moves. By the lemma above, L_s is non self dual in terms of ordinary Wadge reducibility. Altogether, this shows that the conciliatory languages correspond to certain non self dual languages. Which ones? For sets of infinite words of the finite Borel ranks – all of them.

A conciliatory word language is simply $L \subseteq \Sigma^{\leq \omega} = \Sigma^* \cup \Sigma^\omega$, i. e. a set of finite or infinite words. As for trees, we define L_s as the set of words over $\Sigma \cup \{s\}$, such that when we ignore all the s we obtain a word (finite or infinite) from L. Obviously, Lemma 2 holds also for words, but – as we have already disclosed – we get much more than that.

Theorem 1. *(Duparc [2]) For every* $L \subseteq \Sigma^\omega$ *of finite Borel rank,* L *is non self dual iff there exists* $F \subseteq \Sigma^*$ *such that* $L \equiv_W (F \cup L)_s$.

In particular, an ω-language of finite Borel rank is non self dual iff it is Wadge equivalent to a disguised conciliatory set. From the theorem above it follows that this also holds for tree languages.

Corollary 1. *For every* $L \subseteq T_\Sigma$ *of finite Borel rank,* L *is non self dual iff* $L \equiv_W C_s$ *for some conciliatory language* C.

Proof. First, observe that L is Wadge equivalent to L_w, which is the set of sequences obtained by writing down the trees from L level by level from left to right. The "writing down" and its inverse are suitable continuous reductions. By Theorem 1, L_w is equivalent to $(L_w \cup F)_S$ for some set of finite words F. Now, we need a conciliatory tree language C, equivalent to $L_w \cup F$. For a conciliatory tree t let fixed(t) denote the sequence obtained by writing down the tree level by level from left to right until the first missing node is found. Note that fixed(t) is infinite iff t is a full tree. Let $C = \{t \in \tilde{T}_\Sigma : \text{fixed}(t) \in L_w \cup F\}$. The identity function reduces L to C_s, so $C_s \geq_W L \equiv_W L_w \cup F$. Let us prove the converse inequality.

We will consider a mixed game $G(C_s, (L_w \cup F)_s)$. (Formally, instead of $(L_w \cup F)_s$ one can take a Wadge equivalent language $T(L_w \cup F)$, consisting of trees which have the leftmost path in $(L_w \cup F)_s$.) A winning strategy for Duplicator is to undress on-line the tree t_S constructed by Spoiler and write it down level by level, from left to right. When Duplicator finds a missing node, he plays s until Spoiler plays the missing node. At the end of the play, fixed$(u(t_S)) = u(w_D)$. Hence, $t_S \in C_s \iff w_D \in (L_w \cup F)_s$. □

Let us end this section by recalling the notion of the *Wadge degree*. Since the Wadge ordering is well-founded [5], one may proceed by induction:

- $d_W(\emptyset) = d_W(\emptyset^C) = 1$,
- $d_W(L) = \sup\{d_W(M) + 1 : M \text{ is non self dual}, M <_W L\}$ for $L >_W \emptyset$.

The *conciliatory degree* of a language is defined analogously:

- $d_C(\emptyset) = d_C(\emptyset^C) = 1$,
- $d_C(L) = \sup\{d_C(M) + 1 : M <_C L\}$ for $L >_C \emptyset$.

By Corollary 1, for conciliatory L such that L_s has finite Borel rank, $d_C(L) = d_W(L_s)$. This observation lets us work with the conciliatory hierarchy instead of the Wadge hierarchy, as long as we restrict ourselves to non self dual sets of finite Borel ranks.

4   Up and Down the Hierarchy

In this section we present a handful of set-theoretical operations. Four of them will be the main tools in the remaining of the paper.

First, note that the choice of the alphabet Σ is of no importance. Let Σ and Σ' be finite alphabets containing at least two letters. For any language L over Σ (conciliatory or not), one can find an equivalent language L' over Σ'. Furthermore, if L is recognized by an automaton, so can be L', and the construction of the new automaton is effective. Therefore, without loss of generality we may assume $\Sigma = \{a, b\}$.

For $L, M \subseteq \tilde{T}_\Sigma$ define $M + L$ as the set of trees $t \in \tilde{T}_\Sigma$ satisfying any of the following conditions:

- $t.0 \in L$ and $t(10^n) = a$ for all n,
- 10^n is the first node on the path 10^* labeled with b and either $t(10^n 0) = a$ and $t.10^n 00 \in M$ or $t(10^n 0) = b$ and $t.10^n 00 \in M^\complement$.

When playing a conciliatory Wadge game, being in charge of the language $M + L$ is like being in charge of L with one extra move that erases everything played so far and replaces L with M or $M^\complement$. This move can be played only once during the play, and is executed by playing b on the path 10^* for the first time. By choosing the next letter on this path we make choice between M and $M^\complement$.

The next operation is a generalization of the previous one. It lets a player choose from a countable collection of languages. Let $L_n \subseteq \tilde{T}_\Sigma$ for $n < \omega$. Define $\sup_{n<\omega} L_n$ as the set of trees $t \in \tilde{T}_\Sigma$ satisfying one of the following conditions:

- $t(1^n) = a$ for all n,
- 1^n is the first node on 1^* labeled with b and $t.1^n 0 \in L_n$.

Note that $\sup_{n<\omega} L_n$ is conciliatory, even if the languages L_n contain only full trees.

The multiplication by countable ordinals is defined as an iterated sum:

- $L \cdot 1 = L$,
- $L \cdot (\alpha + 1) = L \cdot \alpha + L$,
- $L \cdot \lambda = \sup_{\gamma < \lambda} L \cdot \gamma$ for limit ordinals λ.

Finally let us define the pseudo-exponentiation. Let $L \subseteq \tilde{T}_\Sigma$. For $t \in \tilde{T}_\Sigma$ let

$$i(t)(a_1 a_2 \ldots a_n) = \begin{cases} t(a_1 0 a_2 \ldots 0 a_n 0) & \text{if } \forall_k \ t(a_1 0 a_2 \ldots 0 a_n 1^k) = a \\ s & \text{if } \exists_k \ t(a_1 0 a_2 \ldots 0 a_n 1^k) = b \end{cases}.$$

Intuitively, the rightmost path starting in $a_1 0 a_2 \ldots 0 a_n$ tells us whether to skip the node $a_1 0 a_2 \ldots 0 a_n 0$ or not. Let

$$\exp L = \{t \in \tilde{T}_\Sigma : \ u(i(t)) \in L\}.$$

A player in charge of $\exp L$ is like a player in charge of L with an extra possibility to decide that a chosen node labeled in the past (and the subtree rooted in its right child) is to be ignored.

The names of the operations and the notation used make the following theorem rather expected.

Theorem 2 (Duparc [2]). *For $L, M \subseteq \tilde{T}_\Sigma$, L_s, M_s Borel of finite rank, and a countable ordinal α it holds that*

$$d_C(L + M) = d_C(L) + d_C(M),$$
$$d_C(L \cdot \alpha) = d_C(L) \cdot \alpha,$$
$$d_C(\exp L) = \omega_1^{d_C(L)+\varepsilon},$$

where

$$\varepsilon = \begin{cases} -1 & \text{if } d_C(L) < \omega \\ 0 & \text{if } d_C(L) = \beta + n \text{ and } \operatorname{cof}\beta = \omega_1 \\ +1 & \text{if } d_C(L) = \beta + n \text{ and } \operatorname{cof}\beta = \omega \end{cases}.$$

A kind of inverse operation for $\exp L$ was introduced in [2]. The operation relies on auxiliary moves and involves games where players must, along with playing letters as usual, answer questions about the future of the play. One may easily imagine that when the opponent asks questions like this, it may be much more difficult to win. For convenience, we describe this operation on infinite words, but it can easily be modified to apply to infinite trees (via the usual correspondence between infinite sequences and trees with finite branching).

Let us define the space in which the player evolves when answering questions about the future of the play. We call such a space a *question tree*. We will be working with non-labeled trees, which are simply prefix closed subsets of X^* for a fixed set X. A tree is *pruned* if it has no finite branches. For a finite or infinite word $x = x_0 x_1 x_2 \ldots$ we write $\frac{x}{2}$ for the word $x_0 x_2 x_4 \ldots = \langle x_{2i} : i < |x|/2 \rangle$.

Definition 1. *A question tree related to an alphabet Σ is a non-labeled non-empty pruned tree $T \subseteq (\Sigma')^\omega$, with $\Sigma' = \Sigma \cup \{\langle !w \rangle : w \in \Sigma\} \cup \{\langle ? \rangle\}$, satisfying for every node $u \in T$:*

if $|u|$ is even, *then $u = v\sigma$ for some $\sigma \in \Sigma$ (these nodes correspond to the main play),*

if $|u|$ is odd, *then it is an auxiliary move with two different kinds of options:*

$\langle ! \rangle$ *$u = v \langle !w \rangle$ for some $w \in \Sigma^*$ extending $\frac{u}{2}$. In this case we demand that any node $u' \in T$ extending u satisfies $\frac{u'}{2} \sqsubseteq w$ or $w \sqsubseteq \frac{u'}{2}$. Moreover, we require that for any $v \langle !w' \rangle \in T$ either $w' = w$ or $w' \not\sqsubseteq w \wedge w \not\sqsubseteq w'$.*

$\langle ? \rangle$ *$u = v \langle ? \rangle$. This is an option to avoid all positions of the form $v \langle !w \rangle$. Formally, for each $u' \in T$ extending $v \langle ? \rangle$ and each w such that $v \langle !w \rangle \in T$, $w \not\sqsubseteq \frac{u'}{2}$.*

Recall that $[T]$ denotes the set of branches of T. Notice that for every infinite word $x \in \Sigma^\omega$ there exists a unique infinite branch $y \in [T]$ such that $\frac{y}{2} = x$.

In even moves a player moving along a question tree simply plays letters from Σ, just like in an ordinary play. In odd moves, everything looks like his opponent were asking him questions of the form: "Do you intend to stay forever in this closed subset of Σ^ω? If you are willing to exit, let me know which of the positions I submitted to you, you intend to reach." So, the player can choose $\langle !v' \rangle$ and say: "OK, I'm going to reach this position v", or taking the option $\langle ? \rangle$ he may answer: "No, I won't reach any of these, I will rather stay out of the open set formed by the union of the basic open sets $v\Sigma^\omega$ for all positions v you are asking me about." Taking the latter option means remaining in the complement of this open set, hence in a closed subset of Σ^ω.

Definition 2. Let $L \subseteq \Sigma^\omega$ and $\mathcal{T}$ be a question tree on Σ' related to Σ. $L^{\mathcal{T}}$ consists of $x \in (\Sigma')^\omega$ such that $x \in [\mathcal{T}] \wedge \frac{x}{2} \in L$ or $x \notin [\mathcal{T}] \wedge \exists x'$ $(wx' \in [\mathcal{T}] \wedge \frac{wx'}{2} \in L)$ where w is the longest prefix of x that belongs to $\mathcal{T}$.

The definition of $L^{\mathcal{T}}$ may seem a bit awkward. We defined it this way so that Duplicator still has a winning strategy in the restricted version of the Wadge game $G_W(L^{\mathcal{T}}, L^{\mathcal{T}})$, where Duplicator is not allowed to exit the question tree $\mathcal{T}$, while Spoiler can play anything he wants. Hence, a player in charge of $L^{\mathcal{T}}$ in a Wadge game has a winning strategy if and only if he has a winning strategy that always remains inside $\mathcal{T}$. Therefore in the sequel we always assume that a strategy involving a set of the form $L^{\mathcal{T}}$ remains in the underlying question tree; in other words it restricts its moves to the legal ones.

For any $L \subseteq \Sigma^\omega$ and any question tree $\mathcal{T}$ related to Σ, $L^{\mathcal{T}} \leq_W L$. In particular, whenever L is Borel, $L^{\mathcal{T}}$ is Borel too.

Definition 3. Let $L \subseteq \Sigma^\omega$ be Borel, $\log L$ stands for a $<_W$-minimal element of the form $L^{\mathcal{T}}$ where $\mathcal{T}$ ranges over all question trees related to Σ.

In [2] it is proved that for a fixed L all $<_W$-minimal sets of the form $L^{\mathcal{T}}$ are Wadge equivalent. However, $L \mapsto \log L$ is an operation only if we make it functional, which requires the full Axiom of Choice. Since this functional character is not needed in the proofs, we insist on the fact that $\log L$ is just a notational convenience and none of the foregoing proofs involving it requires the Axiom of Choice.

The operation log preserves the Wadge ordering.

Proposition 1. For $L, M \subseteq \Sigma^\omega$ Borel, $L \leq_W M \implies \log L \leq_W \log M$.

For the present application the key property of the operation log is the following.

Proposition 2. Let $L \subseteq \Sigma^\omega$,

$$L \text{ is } \Pi^0_{n+1}\text{-complete} \iff \log L \text{ is } \Pi^0_n\text{-complete},$$
$$L \text{ is } \Sigma^0_{n+1}\text{-complete} \iff \log L \text{ is } \Sigma^0_n\text{-complete}.$$

Actually, the operation is even finer: it is a precise counterpart of the pseudo--exponentiation. We state the following result for tree languages despite the fact it was proved for word languages in [2]. The extension to tree languages is straightforward.

Proposition 3. *Given L a (full) tree language, and M a conciliatory tree language, with L, M_s both Borel,*

$$\log L \leq_W M_s \iff L \leq_W (\exp M)_s,$$
$$M_s \leq_W \log L \iff (\exp M)_s \leq_W L.$$

5 Weak Index Vs. Borel Rank

Fix a natural number N. Let us call a game-tree, a N-ary tree T whose nodes are boxes or diamonds equipped with ranks. For $\iota = 0, 1$ and $\kappa > \iota$ let $W_{[\iota,\kappa]}$ be the set of all game-trees T with ranks inside $[\iota, \kappa]$ and such that Eve has a winning strategy in the underlying *weak* parity game.

Theorem 3. *For each n, $W_{[0,n]} \in \Pi_n^0$ and $W_{[1,n+1]} \in \Sigma_n^0$.*

Proof. By the determinacy of parity games, $(W_{[0,n]})^{\complement} \equiv_W W_{[1,n+1]}$. Hence, it is enough to prove the claim for ranks inside $[0, n]$. The proof goes by induction with respect to n.

Let us first see that $W_{[0,1]}$ is closed. Let t_k denote the restriction of a tree t to its k first levels. By Weak König Lemma, Adam has a winning strategy in t iff for some k, Adam has a winning strategy in t_k. This means that if a tree t does not belong to $W_{[0,1]}$, then one already knows it after looking at some finite initial part of it. This is precisely the condition that defines closed sets.

Let $\mathcal{T}$ be a question tree defined as follows. Given any game tree t with ranks inside $[0, n + 1]$, for each node u in t which is the first node on this branch with priority n, $\mathcal{T}$ asks the question whether Adam or Eve would have a winning strategy if the game were to start from this particular node. By what we have already proved, if n is even, "Eve has a winning strategy" is a closed condition and if n is odd, "Adam has a winning strategy" is a closed condition. Hence, the question above is legal.

Fix a tree T over $\Sigma \cup \{\langle !w \rangle \colon w \in \Sigma\} \cup \{\langle ? \rangle\}$, where $\Sigma = \{\Diamond, \Box\} \times [0, n + 1]$, which answers questions asked by $\mathcal{T}$ (strictly speaking this means the one and only play inside $\mathcal{T}$ which corresponds to T). Let us construct a tree t' over $\{\Diamond, \Box\} \times [0, n]$ which is exactly the same as T except that for each node u as above: if the answer is "Adam has a winning strategy", every node in the subtree rooted in u receives the rank n for odd n's and $n-1$ for even n's, and if the answer is "Eve has a winning strategy" – the other way round. This gives a continuous reduction of $W_{[0,n+1]}^{\mathcal{T}}$ to $W_{[0,n]}$.

Hence, $\lg W_{[0,n+1]} \leq_W W_{[0,n+1]}^{\mathcal{T}} \leq_W W_{[0,n]}$. By induction hypothesis we have $W_{[0,n]} \in \Pi_n^0$, and in consequence $W_{[0,n+1]} \in \Pi_{n+1}^0$.   □

As a corollary we get the promised improvement of Skurczyński's result [14].

Corollary 2. *For every weak alternating automaton with ranks inside $[0, n]$ (resp. $[1, n + 1]$) it holds that $L^\omega(\mathcal{A}) \in \Pi_n^0$ (resp. $L^\omega(\mathcal{A}) \in \Sigma_n^0$).*

Proof. Let $\mathcal{A}$ be an automaton with priorities inside $[\iota, \kappa]$. For sufficiently large N we may assume without loss of generality that the runs of the automaton are N-ary trees. By assigning to an input tree the run of $\mathcal{A}$, one obtains a continuous function reducing $L^\omega(\mathcal{A})$ to $W_{(\iota,\kappa)}$. Hence, the claim follows from the theorem above.   □

The results described in this section give yet another argument to one of the opposing parties in the everlasting dispute between the big-endians and the little-endians of game theory. Had we defined a play to be winning for Eve if the *lowest* rank was even, the correspondence between the indices and Borel classes would be rather ugly.

6   Three Simple Constructions

Let $\mathcal{A}$, $\mathcal{B}$ be weak alternating tree automata. As it was explained in the beginning of Sect. 4, we may assume without loss of generality that the automata have the same input alphabet $\Sigma = \{a, b\}$. We will construct automata recognizing languages equivalent (in the conciliatory sense) to $L(\mathcal{B}) + L(\mathcal{A})$, $L(\mathcal{A}) \cdot \omega$, and $\exp L(\mathcal{A})$.

Sum. Consider the automaton $\mathcal{B} + \mathcal{A}$ defined on Fig. 1. The diamond states are existential and the box states are universal. The circle states can be treated as existential, but in fact they give no choice to either player. The transitions leading to $\mathcal{A}$, $\mathcal{B}$ and $\mathcal{B}^{\complement}$ should be understood as transitions to the initial states of the according automata. The priority functions of $\mathcal{B}$ and $\mathcal{B}^{\complement}$ might need shifting up, so that they were not using the value 0. It is easy to check that $L(\mathcal{B} + \mathcal{A}) = L(\mathcal{B}) + L(\mathcal{A})$.

Multiplication by ω. The automaton $\mathcal{A} \cdot \omega$ is shown on Fig. 1. The language recognized by $\mathcal{A} \cdot \omega$ consists of trees having no b's on the path 1^* or satisfying the following conditions for some $0 < i \le k$ and n:

- 1^k is the first node labeled with b on the path 1^*,
- i is minimal such that for all $i < j \le k$ the path $1^j 0^+$ contains no b's,
- $1^i 0^n$ is the first node labeled with b lying on the path $1^i 0^+$,
- either $t(1^i 0^n 0) = a$ and $t.1^i 0^n 00 \in L(\mathcal{A})$ or $t(1^i 0^n 0) = b$ and $t.1^i 0^n 00 \in L(\mathcal{A})^{\complement}$.

Let L_k denote the set of trees satisfying the four conditions above for a fixed k. Observe that $L_k \equiv_W \emptyset + L(\mathcal{A}) \cdot k$. Intuitively, we cannot use the subtrees rooted in $10, 110, \ldots, 1^k 0$ together, because making the first nontrivial move in the subtree rooted in $1^i 0$ (putting the first b on the path $1^i 0^+$) makes the subtrees rooted in $10, 110, \ldots, 1^{i-1} 0$ irrelevant. The best we can do is to use them one by one, and this gives exactly the power of $\emptyset + L(\mathcal{A}) \cdot k$.

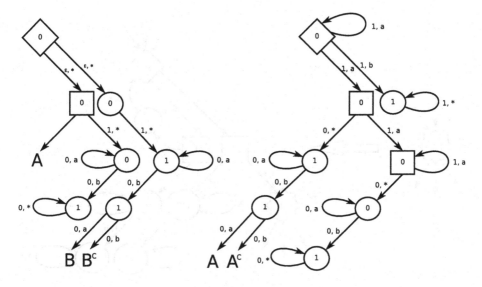

Fig. 1. The automata $\mathcal{B} + \mathcal{A}$ and $\mathcal{A} \cdot \omega$

Now, consider $G_C(L(\mathcal{A}) \cdot \omega, L(\mathcal{A} \cdot \omega))$. By definition, $L(\mathcal{A}) \cdot \omega$ consists of trees having no b on the rightmost path, or such that 1^k is the first node on this path labeled with b, and $t.1^k 0 \in L(\mathcal{A}) \cdot k$. Consider the following strategy for Duplicator. First, only observe the rightmost path of Spoiler's tree t_S. While Spoiler plays a's, keep playing a's in t_D (Duplicator's tree). If Spoiler never plays a b, Duplicator wins. Suppose Spoiler plays his first b in the node 1^k. Duplicator should also play b in the node 1^k. Now, the result of the play depends only on whether $t_S.1^k 0 \in L(\mathcal{A}) \cdot k \iff t_D \in L_k$, and Duplicator should simply use the strategy from $G_C(L(\mathcal{A}) \cdot k, L_k)$.

In the game $G_C(L(\mathcal{A} \cdot \omega), L(\mathcal{A}) \cdot \omega))$ the only difference is that Duplicator should play one more a: if Spoiler plays the first b on the rightmost path in the node 1^k, then Duplicator should put his first b in 1^{k+1}, so that he can later use the strategy from $G_C(L_k, L(\mathcal{A}) \cdot (k+1))$.

Pseudo-exponentiation. Both previous constructions were performed by combining two or three automata with a particularly chosen gadget. The automaton $\exp \mathcal{A}$ is a bit more tricky. This time, we have to change the whole structure of the automaton. Instead of adding one gadget, we replace each state of $\mathcal{A}$ by a different gadget.

The gadget for a state p is shown on Fig. 2. By replacing p with the gadget we mean that all the transitions ending in p should now end in p' and all the transitions starting in p should start in p''. Note that the state p'' is the place where the original transition is chosen, so p'' should be existential iff p is existential. It is not difficult to see that $\exp \mathcal{A}$ recognizes exactly $\exp L(\mathcal{A})$.

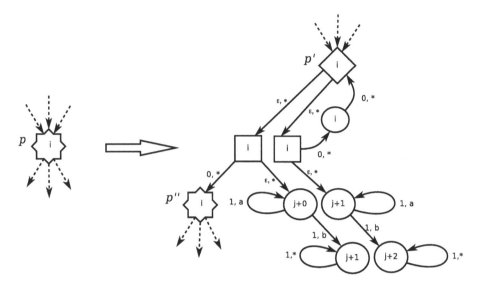

Fig. 2. The gadget to replace p in the construction of $\exp \mathcal{A}$. The state p'' is existential iff p is existential, $i = \operatorname{rank} p$, and j is the least even number greater or equal to i.

7   A Lower Bound

In the previous section we have shown that weakly recognizable languages are closed by sum, multiplication by ω, and pseudo-exponentiation with the base ω_1. By iterating finitely many times sum and multiplication by ω we obtain closure by multiplication by ordinals of the form $\omega^n k_n + \ldots + \omega k_1 + k_0$, i.e., all ordinals less then ω^ω. In other words, we can find a weakly recognizable language of any conciliatory degree from the closure of $\{1\}$ by ordinal sum, multiplication by ordinals $< \omega^\omega$ and pseudo-exponentiation with the base ω_1. It is easy to see that the order type of this set is not changed if we replace pseudo-exponentiation with ordinary exponentiation $\alpha \mapsto \omega_1^\alpha$. This in turn is isomorphic with the closure of $\{1\}$ by ordinal sum, multiplication by ordinals $< \omega^\omega$, and exponentiation with the base ω^ω. This last set is obviously ε_0, the least fixpoint of the exponentiation with the base ω.

By Lemma 2 and the final remark of Sect. 2 it follows that the mapping $L \mapsto L_s$ embeds the conciliatory hierarchy of weakly recognizable languages into the Wadge hierarchy of weakly recognizable languages of full trees. Hence, we obtain the main result of this paper.

Theorem 4. *The Wadge hierarchy of weakly recognizable tree languages has the height of at least ε_0.*

Our intuition tells us the bound is tight, but we have no evidence for that. The question of the exact height of the hierarchy for weak automata remains open.

Acknowledgments

The first author would like to express his gratitude to David Janin who incidentally made him initiate this study of the hierarchy of weak alternating tree automata. We also thank Damian Niwiński and the anonymous referees for very helpful comments.

References

1. Arnold, A., Niwiński, D.: Continuous separation of game languages. Manuscript, 2006 (submitted)
2. Duparc, J.: Wadge hierarchy and Veblen hierarchy. Part I: Borel sets of finite rank. The Journal of Symbolic Logic 66 (2001)
3. Duparc, J.: A hierarchy of deterministic context-free ω-languages. Theoret. Comput. Sci. 290, 1253–1300 (2003)
4. Finkel, O.: Wadge Hierarchy of Omega Context Free Languages. Theoret. Comput. Sci. 269, 283–315 (2001)
5. Kechris, A.S.: Classical Descriptive Set Theory. Graduate Texts in Mathematics 156 (1995)
6. Muller, D.E., Saoudi, A., Schupp, P.E.: Alternating automata. The weak monadic theory of the tree, and its complexity. In: Kott, L. (ed.) Automata, Languages and Programming. LNCS, vol. 226, pp. 275–283. Springer, Heidelberg (1986)
7. Murlak, F.: On deciding topological classes of deterministic tree languages. In: Ong, L. (ed.) CSL 2005. LNCS, vol. 3634, pp. 428–441. Springer, Heidelberg (2005)
8. Murlak, F.: The Wadge hierarchy of deterministic tree languages. In: Bugliesi, M., Preneel, B., Sassone, V., Wegener, I. (eds.) ICALP 2006. LNCS, vol. 4052, pp. 408–419. Springer, Heidelberg (2006)
9. Niwiński, D., Walukiewicz, I.: A gap property of deterministic tree languages. Theoret. Comput. Sci. 303, 215–231 (2003)
10. Niwiński, D., Walukiewicz, I.: Deciding nondeterministic hierarchy of deterministic tree automata. In: Proc. WoLLiC 2004. Electronic Notes in Theoret. Comp. Sci., pp. 195–208 (2005)
11. Perrin, D., Pin, J.-E.: Infinite Words. Automata, Semigroups, Logic and Games. Pure and Applied Mathematics, vol. 141. Elsevier, Amsterdam (2004)
12. Rabin, M.O.: Weakly definable relations and special automata. In: Mathematical Logic and Foundations of Set Theory, North-Holland, pp. 1–70 (1970)
13. Selivanov, V.: Wadge Degrees of ω-languages of deterministic Turing machines. Theoret. Informatics Appl. 37, 67–83 (2003)
14. Skurczyński, J.: The Borel hierarchy is infinite in the class of regular sets of trees. Theoret. Comput. Sci. 112, 413–418 (1993)
15. Urbański, T.F.: On deciding if deterministic Rabin language is in Büchi class. In: Welzl, E., Montanari, U., Rolim, J.D.P. (eds.) ICALP 2000. LNCS, vol. 1853, pp. 663–674. Springer, Heidelberg (2000)
16. Wagner, K.: Eine topologische Charakterisierung einiger Klassen regulärer Folgenmengen. J. Inf. Process. Cybern. EIK 13, 473–487 (1977)
17. Wagner, K.: On ω-regular sets. Inform. and Control 43, 123–177 (1979)

Productivity of Stream Definitions[*]

Jörg Endrullis[1], Clemens Grabmayer[3], Dimitri Hendriks[1],
Ariya Isihara[1], and Jan Willem Klop[1,2]

[1] Vrije Universiteit Amsterdam, Department of Computer Science,
De Boelelaan 1081a, 1081 HV Amsterdam, The Netherlands
{ariya,joerg}@few.vu.nl, {diem,klop}@cs.vu.nl
[2] Radboud Universiteit Nijmegen, Department of Computer Science,
Toernooiveld 1, 6525 ED Nijmegen, The Netherlands
[3] Universiteit Utrecht, Department of Philosophy,
Heidelberglaan 8, 3584 CS Utrecht, The Netherlands
clemens@phil.uu.nl

Abstract. We give an algorithm for deciding productivity of a large
and natural class of recursive stream definitions. A stream definition is
called 'productive' if it can be evaluated continuously in such a way that
a uniquely determined stream is obtained as the limit. Whereas productivity is undecidable for stream definitions in general, we show that it can
be decided for 'pure' stream definitions. For every pure stream definition
the process of its evaluation can be modelled by the dataflow of abstract
stream elements, called 'pebbles', in a finite 'pebbleflow net(work)'. And
the production of a pebbleflow net associated with a pure stream definition, that is, the amount of pebbles the net is able to produce at its
output port, can be calculated by reducing nets to trivial nets.

1 Introduction

In functional programming, term rewriting and λ-calculus, there is a wide arsenal
of methods for proving termination such as recursive path orders, dependency
pairs (for term rewriting systems, [15]) and the method of computability (for
λ-calculus, [13]). All of these methods pertain to finite data only. In the last two
decades interest has grown towards infinite data, as witnessed by the application
of type theory to infinite objects [2], and the emergence of coalgebraic techniques
for infinite data types like streams [11]. While termination cannot be expected
when infinite data are processed, infinitary notions of termination become relevant. For example, in formal frameworks for the manipulation of infinite objects
such as infinitary rewriting [7] and infinitary λ-calculus [8], basic notions are the
properties WN^∞ of infinitary weak normalisation and SN^∞ of infinitary strong
normalisation [9].

In the functional programming literature the notion of 'productivity' has
arisen, initially in the pioneering work of Sijtsma [12], as a natural strengthening

---

[*] This research has been partially funded by the Netherlands Organisation for Scientific Research (NWO) under FOCUS/BRICKS grant number 642.000.502.

E. Csuhaj-Varjú and Z. Ésik (Eds.): FCT 2007, LNCS 4639, pp. 274–287, 2007.
© Springer-Verlag Berlin Heidelberg 2007

of the property WN^∞. A recursive stream definition is called productive if not only can the definition be evaluated continuously to build up an infinite normal form, but the resulting infinite expression is also meaningful in the sense that it is a constructor normal form which allows to read off consecutively individual elements of the stream. Since productivity of stream definitions is undecidable in general, the challenge is to find increasingly larger classes of stream definitions significant to programming practice for which productivity is decidable, or for which at least a powerful method for proving productivity exists.

Contribution and Overview. We show that productivity is decidable for a rich class of recursive stream definitions that hitherto could not be handled automatically. In Section 2 we define 'pure stream constant specifications' (SCSs) as orthogonal term rewriting systems, which are based on 'weakly guarded stream function specifications' (SFSs). In Section 3 we develop a 'pebbleflow calculus' as a tool for computing the 'degree of definedness' of SCSs. The idea is that a stream element is modelled by an abstract 'pebble', a stream definition by a finite 'pebbleflow net', and the process of evaluating a definition by the dataflow of pebbles in the associated net. More precisely, we give a translation of SCSs into 'rational' pebbleflow nets, and prove that this translation is production preserving. Finally in Section 4, we show that the production of a 'rational' pebbleflow net, that is, the amount of pebbles such a net is able to produce at its output port, can be calculated by an algorithm that reduces nets to trivial nets. We obtain that productivity is decidable for pure SCSs. We believe our approach is natural because it is based on building a pebbleflow net corresponding to an SCS as a model that is able to reflect the local consumption/production steps during the evaluation of the definition in a quantitatively precise manner.

We follow [12] in describing the quantitative input/output behaviour of a stream function f by a non-decreasing 'production function' $\beta_f : (\overline{\mathbb{N}})^r \to \overline{\mathbb{N}}$ such that the first $\beta_f(n_1, \ldots, n_r)$ elements of $f(t_1, \ldots, t_r)$ can be computed whenever the first n_i elements of t_i are defined. More specifically, we employ 'rational' production functions $\beta : (\overline{\mathbb{N}})^r \to \overline{\mathbb{N}}$ that, for $r = 1$, have eventually periodic difference functions $\Delta_\beta(n) := \beta(n+1) - \beta(n)$, that is $\exists n, p \in \mathbb{N}. \forall m \geq n. \Delta_\beta(m) = \Delta_\beta(m+p)$. This class is effectively closed under composition, and allows to calculate fixed points of unary functions. Rational production functions generalise those employed by [16], [5], [2], and [14], and enable us to precisely capture the consumption/production behaviour of a large class of stream functions.

Related Work. It is well-known that networks are devices for computing least fixed points of systems of equations [6]. The notion of 'productivity' (sometimes also referred to as 'liveness') was first mentioned by Dijkstra [3]. Since then several papers [16,12,2,5,14,1] have been devoted to criteria ensuring productivity. The common essence of these approaches is a quantitative analysis. In [16], Wadge uses dataflow networks to model fixed points of equations. He devises a so-called *cyclic sum test*, using production functions of the form $\beta_f(n_1, \ldots, n_r) = \min(n_1 + a_{f,1}, \ldots, n_r + a_{f,r})$ with $a_{f,i} \in \mathbb{Z}$, i.e. the output *leads* or *lags* the input by a fixed value $a_{f,i}$. Sijtsma [12] points out that this class of production functions is too restrictive to capture the behaviour of commonly used stream operations like even, dup, zip and so forth. Therefore he develops an approach allowing arbitrary production functions $\beta_f : \mathbb{N}^r \to \mathbb{N}$, having the only drawback of not being automatable in full generality. Coquand [2] defines a syntactic criterion called 'guardedness' for ensuring productivity. This criterion is too restrictive for programming practice, because it disallows function applications to recursive calls. Telford and Turner [14] extend the notion of guardedness with a method in the flavour of Wadge. However, their approach does not overcome Sijtsma's criticism. Hughes, Pareto and Sabry [5] introduce a type system using production functions with the property that $\beta_f(a \cdot x + b) = c \cdot x + d$ for some $a, b, c, d \in \mathbb{N}$. This class of functions is not closed under composition, leading to the need of approximations and a loss of power. Moreover their typing system rejects definitions like $\mathsf{M} = a : b : \mathsf{tail}(\mathsf{M})$, where ':' is the infix stream constructor, because tail is applied to the recursive call. Buchholz [1] presents a formal type system for proving productivity, whose basic ingredients are, closely connected to [12], unrestricted production functions $\beta_f : \mathbb{N}^r \to \mathbb{N}$. In order to obtain an automatable method, Buchholz also devises a syntactic criterion to ensure productivity. This criterion easily handles all the examples of [14], but fails to deal with functions that have a negative effect 'worse than tail'.

2   Recursive Stream Specifications

In this section the concepts of 'stream constant specification' (SCS) and 'stream function specification' (SFS) are introduced. We use a two-layered set-up, which is illustrated by the well-known definition $\mathsf{M} = 0{:}1{:}\mathsf{zip}(\mathsf{tail}(\mathsf{M}), \mathsf{inv}(\mathsf{tail}(\mathsf{M})))$ of the Thue–Morse sequence. This corecursive definition employs separate definitions of the stream functions zip and tail, contained in Ex. 1 below, and of the definition $\mathsf{inv}(x{:}\sigma) = (1-x){:}\mathsf{inv}(\sigma)$ of the stream function inv. Stream constants are written using uppercase letters, stream and data functions are written lowercase.

In order to distinguish between *data terms* and *streams* we use the framework of many-sorted term rewriting. Let S be a finite set of sorts. An S-sorted set A is a family of sets $(A_s)_{s \in S}$. An S-sorted signature Σ is a set of function symbols, each having a fixed arity $\mathsf{ar}(f) \in S^* \times S$. Let X be an S-sorted set of variables. The S-sorted set of terms $Ter(\Sigma, X)$ is inductively defined by: $X_s \subseteq Ter(\Sigma, X)_s$ for all $s \in S$ and $f(t_1, \ldots, t_n) \in Ter(\Sigma, X)_s$ whenever $f \in \Sigma$ with arity $\langle s_1 \cdots s_n, s \rangle$ and $t_i \in Ter(\Sigma, X)_{s_i}$. An S-sorted term rewriting system (TRS) over an S-sorted signature Σ is an S-sorted set R where $R_s \subseteq Ter(\Sigma, X)_s \times Ter(\Sigma, X)_s$ for all $s \in S$, satisfying the standard TRS requirements for rules. An S-sorted TRS is called *finite* if both its signature and the set of all of its rules are finite.

In the sequel let $S = \{d, s\}$ where d is the sort of *data terms* and s is the sort of *streams*. We say that a $\{d, s\}$-sorted TRS $\langle \Sigma, R \rangle$ is a *stream TRS* if there exists a partition of the signature $\Sigma = \Sigma_d \uplus \Sigma_{sf} \uplus \Sigma_{sc} \uplus \{:\}$ such that the arity of the symbols from Σ_d is in $\langle d^*, d \rangle$, for Σ_{sf} in $\langle \{s, d\}^*, s \rangle$, for Σ_{sc} in $\langle \epsilon, s \rangle$ and ':' has arity $\langle ds, s \rangle$. Accordingly, the symbols in Σ_d are referred to as the *data symbols*, ':' as the *stream constructor symbol*, the symbols in Σ_{sf} as the *stream function symbols* and the symbols in Σ_{sc} as the *stream constant symbols*.

Without loss of generality we assume that for all $f \in \Sigma_{sf}$ the stream arguments are in front. That is, f has arity $\langle s^{r_s} d^{r_d}, s \rangle$ for some $r_s, r_d \in \mathbb{N}$; we say that f has arity $\langle r_s, r_d \rangle$ for short.

Definition 1. Let $\mathcal{T} = \langle \Sigma, R \rangle$ be a finite stream TRS with $\Sigma = \Sigma_d \uplus \Sigma_{sf} \uplus \{:\}$ and a partition $R = R_d \uplus R_{sf}$ of its set of rules. Then $\mathcal{T}$ (together with these partitions) is called a *weakly guarded stream function specification (SFS)* if:

(i) $\mathcal{T}$ is orthogonal (i.e. left-linear, non-overlapping redex patterns, see [15]).
(ii) $\langle \Sigma_d, R_d \rangle$ is a strongly normalising TRS.
(iii) For every stream function symbol $f \in \Sigma_{sf}$ there is precisely one rule in R_{sf}, denoted by ρ_f, the *defining rule* for f. Furthermore, for all $f \in \Sigma_{sf}$ with arity $\langle r_s, r_d \rangle$, the rule $\rho_f \in R_{sf}$ has the form:

$$f((\boldsymbol{x}_1 : \sigma_1), \ldots, (\boldsymbol{x}_{r_s} : \sigma_{r_s}), y_1, \ldots, y_{r_d}) \to u$$

where $\boldsymbol{x}_i : \sigma_i$ stands for $x_{i,1} : \ldots : x_{i,n_i} : \sigma_i$, and u is one of the following forms:

$$u \equiv t_1 : \ldots : t_{m_f} : g(\sigma_{\pi_f(1)}, \ldots, \sigma_{\pi_f(r'_s)}, t'_1, \ldots, t'_{r'_d}), \qquad (1)$$

$$u \equiv t_1 : \ldots : t_{m_f} : \sigma_i \qquad (2)$$

Here, the terms $t_1, \ldots, t_{m_f} \in Ter(\Sigma_d)$ are called *guards* of f. Furthermore, $g \in \Sigma_{sf}$ with arity $\langle r'_s, r'_d \rangle$, $\pi_f : \{1, \ldots, r'_s\} \to \{1, \ldots, r_s\}$ is an injection used to permute stream arguments, $n_1, \ldots, n_{r_s}, m_f \in \mathbb{N}$, and $1 \le i \le r_s$. In case (1) we write $f \rightsquigarrow g$, and say f 'depends on' g.
(iv) Every stream function symbol $f \in \Sigma_{sf}$ is *weakly guarded* in $\mathcal{T}$, i.e. on every dependency cycle $f \rightsquigarrow g \rightsquigarrow \cdots \rightsquigarrow f$ there is at least one guard.

It is easy to show that every function symbol $f \in \Sigma_{sf}$ in an SFS defines a unique function that maps stream arguments and data arguments to a stream, which can be computed, for given infinite stream terms $u_1, \ldots, u_{r_s}$ in constructor normal form (that is, being of the form $s_0 : s_1 : s_2 : \ldots$) and data terms $t_1, \ldots, t_{r_d}$, by infinitary rewriting as the infinite normal form of the term $f(u_1, \ldots, u_{r_s}, t_1, \ldots, t_{r_d})$. Note that the definition covers a large class of stream functions including tail, even, odd, zip, add. However, the function head defined by $head(x : \sigma) = x$, possibly creating 'look-ahead' as in the well-defined example $S = 0 : head(tail^2(S)) : S$ from [12], is not included.

Now we are ready to define the concept of 'stream constant specification'.

Definition 2. Let $\mathcal{T} = \langle \Sigma, R \rangle$ be a finite stream TRS with a partition $\Sigma = \Sigma_d \uplus \Sigma_{sf} \uplus \Sigma_{sc} \uplus \{:\}$ of its signature and a partition $R = R_d \uplus R_{sf} \uplus R_{sc}$ of its set of rules. Then $\mathcal{T}$ (together with these partitions) is called a *pure stream constant specification (SCS)* if the following conditions hold:

(i) $\langle \Sigma_d \uplus \Sigma_{sf} \uplus \{:\}, R_d \uplus R_{sf} \rangle$ is an SFS.
(ii) $\Sigma_{sc} = \{M_1, \ldots, M_n\}$ is a non-empty set of constant symbols, and $R_{sc} = \{M_i \to rhs_{M_i} \mid 1 \le i \le n, rhs_{M_i} \in Ter(\Sigma)_s\}$. The rule $\rho_{M_i} := M_i \to rhs_{M_i}$ is called *the defining rule for* M_i *in* $\mathcal{T}$.

Note that an SCS $\mathcal{T}$ is orthogonal as a consequence of (i) and (ii).

An SCS is called *productive* if every $\mathsf{M} \in \Sigma_{sc}$ has a stream of data terms as infinite normal form (an infinite *constructor normal form*). Note that orthogonality implies that infinite normal forms are unique.

Example 1. Let $\mathcal{T}_\mathsf{D} = \langle \Sigma_d \uplus \Sigma_{sf} \uplus \Sigma_{sc} \uplus \{:\}, R_d \uplus R_{sf} \uplus R_{sc} \rangle$ be the SCS with $\Sigma_d = \{\mathsf{s}, 0, \mathsf{a}\}$, $\Sigma_{sf} = \{\mathsf{tail}, \mathsf{even}, \mathsf{odd}, \mathsf{zip}, \mathsf{add}\}$, $\Sigma_{sc} = \{\mathsf{D}\}$, and R_{sc} consists of

$$\mathsf{D} \to 0 : 1 : 1 : \mathsf{zip}(\mathsf{add}(\mathsf{tail}(\mathsf{D}), \mathsf{tail}(\mathsf{tail}(\mathsf{D}))), \mathsf{even}(\mathsf{tail}(\mathsf{D}))),$$

R_{sf} consists of the rules

$$\mathsf{tail}(x : \sigma) \to \sigma \qquad \mathsf{even}(x : \sigma) \to x : \mathsf{odd}(\sigma) \qquad \mathsf{odd}(x : \sigma) \to \mathsf{even}(\sigma)$$

$$\mathsf{zip}(x : \sigma, \tau) \to x : \mathsf{zip}(\tau, \sigma) \qquad \mathsf{add}(x : \sigma, y : \tau) \to \mathsf{a}(x, y) : \mathsf{add}(\sigma, \tau)$$

and $R_d = \{\mathsf{a}(x, \mathsf{s}(y)) \to \mathsf{s}(\mathsf{a}(x, y)), \ \mathsf{a}(x, 0) \to x\}$. Note that D has the infinite constructor normal form $0:1:1:2:1:3:2:3:3:4:3:5:4:5:5:6:5:7:6:7:7:\ldots$, and hence is productive in $\mathcal{T}_\mathsf{D}$.

Example 2. Consider the rule $\mathsf{J} \to 0 : 1 : \mathsf{even}(\mathsf{J})$ together with Σ, R_d, R_{sf} as in Ex. 1. The infinite normal form of J is $0:1:0:0:\mathsf{even}(\mathsf{even}(\ldots))$, which is not a constructor normal form. Hence J is WN^∞ (in fact SN^∞), but not productive.

3   Modelling with Nets

We introduce nets as a means to model SCSs and to visualise the *flow* of stream elements. As our focus is on productivity of SCSs, we are interested in the *production* of such a net, that is, the number of stream elements produced by a net. Therefore, stream elements are abstracted from in favour of occurrences of the symbol •, which we call *pebble*. The nets we study are called *pebbleflow nets*; they are inspired by interaction nets [10], and could be implemented in the framework of interaction nets with little effort.

First we give an operational description of pebbleflow nets, explaining what the components of nets are and the way how the components process pebbles. To ease manipulation of and reasoning about nets, we employ term representations. Term constructs corresponding to net components, as well as the rules governing the flow of pebbles through a net, are given on the fly. Their formal definitions are given in Subsec. 3.2. Finally, in Subsec. 3.3, we define a production preserving translation of pure stream specifications into rational nets.

We denote the set of *coinductive natural numbers* by $\overline{\mathbb{N}} = \mathbb{N} \cup \{\infty\}$ and the *numerals* representing the elements of $\overline{\mathbb{N}}$ by $\underline{n} = \mathsf{s}^n(0)$ for $n \in \mathbb{N}$, and $\underline{\infty} = \mathsf{s}^\omega$.

3.1   Nets

Wires. The directed edges of a net, along which pebbles travel, are called *wires*. Wires are idealised in the sense that there is no upper bound on the number of pebbles they can store; arbitrarily long queues are allowed. Wires have

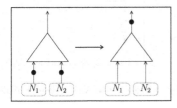   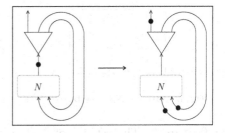

Fig. 1. $\triangle(\bullet(N_1), \bullet(N_2)) \to \bullet(\triangle(N_1, N_2))$ **Fig. 2.** $\mu x.\bullet(N(x)) \to \bullet(\mu x.N(\bullet(x)))$

no counterpart on the term level; in this sense they are akin to the edges of a term tree. Wires connect *boxes*, *meets*, *fans*, and *sources*, that we describe next.

Meets. A *meet* is waiting for a pebble at each of its input ports and only then produces one pebble at its output port, see Fig. 1. Put differently, the number of pebbles a meet produces equals the minimum of the numbers of pebbles available at each of its input ports. Meets enable explicit branching; they are used to model stream functions of arity > 1, as will be explained in the part "Boxes and gates" below. A meet with an arbitrary number $n \geq 1$ of input ports is implemented by using a single wire in case $n = 1$, and if $n = k + 1$ with $k \geq 1$, by connecting the output port of a 'k-ary meet' to one of the input ports of a (binary) meet.

Fans. The behaviour of a *fan* is dual to that of a meet: a pebble at its input port is duplicated along its output ports. A fan can be seen as an explicit sharing device, and thus enables the construction of cyclic nets. More specifically, we use fans only to implement feedback when drawing nets; there is no explicit term representation for the fan in our term calculus. In Fig. 2 a pebble is sent over the output wire of the net and at the same time is fed back to the 'recursion wire(s)'. Turning a cyclic net into a term (tree) means to introduce a notion of binding; certain nodes need to be labelled by a name (μx) so that a wire pointing to that node is replaced by a name (x) *referring* to the labelled node.

Sources. A *source* has an output port only, contains a number $k \in \overline{\mathbb{N}}$ of pebbles, and can fire if $k > 0$. In Sec. 4 we show how to reduce 'closed' nets to sources.

Boxes and Gates. A *box* consumes pebbles at its input port and produces pebbles at its output port, controlled by an infinite sequence $\sigma \in \{+, -\}^\omega$ associated with the box. This consumption/production behaviour of the box is then also be expressed by the 'production function' $\beta_\sigma : \overline{\mathbb{N}} \to \overline{\mathbb{N}}$ of the box, see Fig. 5. For example, consider the unary stream function dup, defined as follows, and its corresponding 'I/O sequence':

$$\mathsf{dup}(x : \sigma) = x : x : \mathsf{dup}(\sigma) \qquad\qquad -++-++-++\ldots$$

which is to be thought of as: *for* dup *to produce two outputs, it first has to consume one input, and this process repeats indefinitely.* Intuitively, the symbol $-$ represents a requirement for an input pebble, and $+$ represents a ready state for an output pebble. Pebbleflow through boxes is visualised in Figs. 3 and 4.

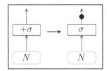

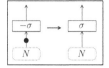

Fig. 3. $\mathsf{box}(+\sigma, N) \to \bullet(\mathsf{box}(\sigma, N))$ **Fig. 4.** $\mathsf{box}(-\sigma, \bullet(N)) \to \mathsf{box}(\sigma, N)$

Definition 3. The set $\pm^\omega$ of *I/O sequences* is defined as the set of infinite sequences over the alphabet $\{+, -\}$ that contain an infinite number of $+$'s:

$$\pm^\omega := \{\sigma \in \{+, -\}^\omega \mid \forall n. \exists m. \sigma(n+m) = +\}$$

Further, we define the set $\pm^\omega_{rat} \subseteq \pm^\omega$ of *rational I/O sequences*. A sequence $\sigma \in \pm^\omega$ is called *rational* if there exist lists $\alpha, \gamma \in \{+, -\}^*$ such that $\sigma = \alpha\overline{\gamma}$, where γ is not the empty list and $\overline{\gamma}$ denotes the infinite sequence $\gamma\gamma\gamma\ldots$. The pair $\langle \alpha, \gamma \rangle$ is called a *rational representation* of σ.

To model stream functions of arbitrary arity, we introduce *gates*. Gates are compounded of meets and boxes, as depicted in Fig. 6. The precise construction of a gate corresponding to a given stream function is described in Subsec. 3.3.

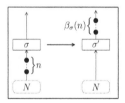

Fig. 5. $\mathsf{box}(\sigma, \bullet^n(N)) \to \bullet^{\beta_\sigma(n)}(\mathsf{box}(\sigma', N))$ **Fig. 6.** A gate for modelling r_s-ary stream functions

Definition 4. The *production function* $\beta_\sigma : \overline{\mathbb{N}} \to \overline{\mathbb{N}}$ of (a box containing) a sequence $\sigma \in \pm^\omega$ is corecursively defined, for all $n \in \overline{\mathbb{N}}$, by $\beta_\sigma(n) := \beta(\sigma, \underline{n})$:

$$\beta(+\sigma, n) = \mathsf{s}(\beta(\sigma, n)) \qquad \beta(-\sigma, 0) = 0 \qquad \beta(-\sigma, \mathsf{s}(n)) = \beta(\sigma, n)$$

Intuitively, $\beta_\sigma(n)$ is the number of outputs of a box containing sequence σ when fed n inputs. Note that production functions are well-defined due to our requirement on I/O sequences.

3.2   A Rewrite System for Pebbleflow

We define terms representing nets, and a rewrite system to model pebbleflow.

Definition 5. Let $\mathcal{V}$ be a set of variables. The set $\mathcal{N}$ of *terms for pebbleflow nets* is generated by:

$$N ::= \mathsf{src}(\underline{k}) \mid x \mid \bullet(N) \mid \mathsf{box}(\sigma, N) \mid \mu x.N \mid \triangle(N, N)$$

where $k \in \bar{\mathbb{N}}$, $x \in V$, and $\sigma \in \pm^{\omega}$. Furthermore, the set $\mathcal{N}_{rat}$ of *terms for rational pebbleflow nets* is defined by the same inductive clauses, but now with the restriction $\sigma \in \pm^{\omega}_{rat}$.

The importance of identifying the subset of rational nets will become evident in Sec. 4, where we introduce a rewrite system for reducing nets to trivial nets (pebble sources). That system will be terminating for rational nets, and will enable us to determine the total production of a rational net.

The rules that govern pebbleflow are listed in Def. 6.

Definition 6. The *pebbleflow rewrite relation* $\rightarrow_{\mathsf{P}}$ is defined as the compatible closure of the union of the following rules:

$$\triangle(\bullet(N_1), \bullet(N_2)) \rightarrow \bullet(\triangle(N_1, N_2)) \tag{P1}$$

$$\mu x.\bullet(N(x)) \rightarrow \bullet(\mu x.N(\bullet(x))) \tag{P2}$$

$$\mathsf{box}((+\sigma), N) \rightarrow \bullet(\mathsf{box}(\sigma, N)) \tag{P3}$$

$$\mathsf{box}((-\sigma), \bullet(N)) \rightarrow \mathsf{box}(\sigma, N) \tag{P4}$$

$$\mathsf{src}(\mathsf{s}(\underline{k})) \rightarrow \bullet(\mathsf{src}(\underline{k})) \tag{P5}$$

The first four rewrite rules in the definition above are visualised in Figures 1, 2, 3, and 4, respectively. In rule (P2) the feedback of pebbles along the recursion wire(s) of the net N is accomplished by substituting $\bullet(x)$ for all free occurrences x of N. Observe that $\rightarrow_{\mathsf{P}}$ constitutes an orthogonal CRS [15], hence:

Theorem 1. *The relation* $\rightarrow_{\mathsf{P}}$ *is confluent.*

3.3   Translating Pure Stream Specifications

First we give a translation of the stream function symbols in an SFS into rational gates (gates with boxes containing rational I/O sequences) that precisely model their quantitative consumption/production behaviour. The idea is to define, for a stream function symbol f, a rational gate by keeping track of the 'production' (sequence of guards encountered) and the 'consumption' of the rules applied, during the finite or eventually periodic dependency sequence on f.

Definition 7. Let $\mathcal{T} = \langle \Sigma_d \uplus \Sigma_{sf} \uplus \{:\}, R_d \uplus R_{sf} \rangle$ be an SFS. Then, for each $\mathsf{f} \in \Sigma_{sf}$ with arity $\langle r_s, r_d \rangle$ the *translation* of f is a rational gate $[\mathsf{f}] : \mathcal{N}^{r_s} \rightarrow \mathcal{N}$ as defined by:

$$[\mathsf{f}](N_1, \ldots, N_{r_s}) = \triangle_{r_s}(\mathsf{box}([\mathsf{f}]_1, N_1), \ldots, \mathsf{box}([\mathsf{f}]_{r_s}, N_{r_s}))$$

where $[\mathsf{f}]_i \in \pm^{\omega}_{rat}$ is defined as follows. We distinguish the two formats a rule $\rho_{\mathsf{f}} \in R_{sf}$ can have. Let $\boldsymbol{x_i} : \sigma_i$ stand for $x_{i,1} : \ldots : x_{i,n_i} : \sigma_i$. If ρ_{f} has the form: $\mathsf{f}(\boldsymbol{x_1} : \sigma_1, \ldots, \boldsymbol{x_{r_s}} : \sigma_{r_s}, y_1, \ldots, y_{r_d}) \rightarrow t_1 : \ldots : t_{m_{\mathsf{f}}} : u$, where:

(a) $u \equiv \mathsf{g}(\sigma_{\pi_{\mathsf{f}}(1)}, \ldots, \sigma_{\pi_{\mathsf{f}}(r'_s)}, t'_1, \ldots, t'_{r'_d})$, then   (b) $u \equiv \sigma_j$, then

$$[\mathsf{f}]_i = \begin{cases} -^{n_i}+^{m_{\mathsf{f}}}[\mathsf{g}]_j & \text{if } \pi_{\mathsf{f}}(j) = i \\ -^{n_i}\mp & \text{if } \neg\exists j. \ \pi_{\mathsf{f}}(j) = i \end{cases} \qquad [\mathsf{f}]_i = \begin{cases} -^{n_i}+^{m_{\mathsf{f}}}\overline{-+} & \text{if } i = j \\ -^{n_i}\mp & \text{if } i \neq j \end{cases}$$

In the second step, we define a translation of the stream constants in an SCS into rational nets. Here the idea is that the recursive definition of a stream constant M is unfolded step by step; the terms thus arising are translated according to their structure by making use of the translation of the stream function symbols encountered; whenever a stream constant is met that has been unfolded before, the translation stops after establishing a binding to a μ-binder created earlier.

Definition 8. Let $T = \langle \Sigma_d \uplus \Sigma_{sf} \uplus \Sigma_{sc} \uplus \{:\}, R_d \uplus R_{sf} \uplus R_{sc}\rangle$ be an SCS. Then, for each $M \in \Sigma_{sc}$ with rule $\rho_M \equiv M \to rhs_M$ the translation $[M] := [M]_\varnothing$ of M to a pebbleflow net is recursively defined by (α a set of stream constant symbols):

$$[M]_\alpha = \begin{cases} \mu M.[rhs_M]_{\alpha \cup \{M\}} & \text{if } M \notin \alpha \\ M & \text{if } M \in \alpha \end{cases}$$

$$[t:u]_\alpha = \bullet([u]_\alpha)$$

$$[f(u_1,\ldots,u_{r_s},t_1,\ldots,t_{r_d})]_\alpha = [f]([u_1]_\alpha,\ldots,[u_{r_s}]_\alpha)$$

Example 3. Reconsider the SCS defined in Example 1. The translation of the stream function symbols $tail, zip \in \Sigma_{sf}$ is carried out as follows:

$$[tail](N) = \triangle_1(box([tail]_1, N)) \qquad [zip](N_1, N_2) = \triangle_2(box([zip]_1, N_1), box([zip]_2, N_2))$$
$$= box([tail]_1, N) \qquad\qquad [zip]_1 = -+[zip]_2 = -++[zip]_1 = \overline{-++}$$
$$[tail]_1 = -\overline{-+} \qquad\qquad [zip]_2 = +[zip]_1 = +-+[zip]_2 = \overline{+-+}$$

(Note that to obtain rational representations of the translated stream functions we use loop checking on top of Def. 7.) Then, the stream constant D is translated to the following pebbleflow net, depicted in Fig. 7:

$$[D] = \mu D.\bullet(\bullet(\bullet([zip]([add]([tail](D), [tail]([tail](D))), [even]([tail](D)))))) .$$

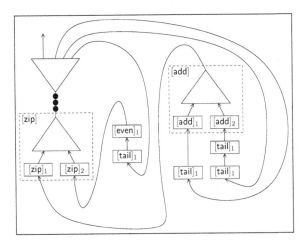

Fig. 7. The pebbleflow net [D] corresponding to the stream D

The theorem below is the basis of our decision algorithm. It states that the translation is 'production preserving', based on the following terminology: The *production* $\pi(N)$ of a pebbleflow net N is the supremum of the number of pebbles the net can 'produce': $\pi(N) := \sup\{n \in \mathbb{N} \mid N \twoheadrightarrow_{\mathsf{P}} \bullet^n(N')\}$, where $\twoheadrightarrow_{\mathsf{P}}$ denotes the reflexive–transitive closure of $\rightarrow_{\mathsf{P}}$. Likewise for an SCS $\mathcal{T} = \langle \Sigma, R \rangle$ the *production* $\pi_{\mathcal{T}}(t)$ of a term $t \in Ter(\Sigma)$ is the supremum of the number of data elements t can 'produce': $\pi_{\mathcal{T}}(t) := \sup\{n \in \mathbb{N} \mid t \twoheadrightarrow s_1 : \ldots : s_n : t'\}$.

Theorem 2. *Let $\mathcal{T}$ be a pure SCS. Then, $\pi([\mathsf{M}]) = \pi_{\mathcal{T}}(\mathsf{M})$ for all $\mathsf{M} \in \Sigma_{sc}$.*

4   Deciding Productivity

We define a rewriting system for pebbleflow nets that, for every net N, allows to reduce N to a single source while preserving the production of N.

Definition 9. We define *the net reduction relation* $\rightarrow_{\mathsf{R}}$ *on closed pebbleflow nets* by the compatible closure of the following rule schemata:

$$\bullet(N) \rightarrow \mathsf{box}((+\overline{-+}), N) \qquad\qquad (R1)$$

$$\mathsf{box}(\sigma, \mathsf{box}(\tau, N)) \rightarrow \mathsf{box}((\sigma \cdot \tau), N) \qquad\qquad (R2)$$

$$\mathsf{box}(\sigma, \triangle(N_1, N_2)) \rightarrow \triangle(\mathsf{box}(\sigma, N_1), \mathsf{box}(\sigma, N_2)) \qquad\qquad (R3)$$

$$\mu x.\triangle(N_1, N_2) \rightarrow \triangle(\mu x.N_1, \mu x.N_2) \qquad\qquad (R4)$$

$$\mu x.N \rightarrow N \qquad\qquad \text{if } x \notin \mathsf{FV}(N) \quad (R5)$$

$$\mu x.\mathsf{box}(\sigma, x) \rightarrow \mathsf{src}(\underline{\mathsf{fix}(\sigma)}) \qquad\qquad (R6)$$

$$\triangle(\mathsf{src}(\underline{k_1}), \mathsf{src}(\underline{k_2})) \rightarrow \mathsf{src}(\underline{\min(k_1, k_2)}) \qquad\qquad (R7)$$

$$\mathsf{box}(\sigma, \mathsf{src}(\underline{k})) \rightarrow \mathsf{src}(\underline{\beta_\sigma(k)}) \qquad\qquad (R8)$$

$$\mu x.x \rightarrow \mathsf{src}(0) \qquad\qquad (R9)$$

where $\sigma, \tau \in \pm^\omega$, $k, k_1, k_2 \in \overline{\mathbb{N}}$, and $\underline{\min(n, m)}$, $\underline{\beta_\sigma(k)}$, $\underline{\sigma \cdot \tau}$ (see Def. 10) and $\underline{\mathsf{fix}(\sigma)}$ (see Def. 11) are term representations of operation results.

Definition 10. The operation *composition* $\cdot : \pm^\omega \times \pm^\omega \rightarrow \pm^\omega$, $\langle \sigma, \tau \rangle \mapsto \sigma \cdot \tau$ of *I/O sequences* is defined corecursively by the following equations:

$$(+\sigma) \cdot \tau = +(\sigma \cdot \tau) \qquad (-\sigma) \cdot (+\tau) = \sigma \cdot \tau \qquad (-\sigma) \cdot (-\tau) = -((-\sigma) \cdot \tau)$$

Composition of sequences $\sigma \cdot \tau \in \pm^\omega$ exhibits analogous properties as composition of functions over natural numbers: it is associative, but not commutative. Furthermore, for all $\sigma, \tau \in \pm^\omega, n \in \overline{\mathbb{N}}$ we have $\beta_{\sigma \cdot \tau}(n) = \beta_\sigma(\beta_\tau(n))$. Because we formalised the I/O behaviour of boxes by sequences and because we are interested in (dis)proving productivity, for the formalisation of the pebbleflow rewrite relation in Def. 6 the choice has been made to give output priority over input. This becomes apparent in the definition of composition above: the net $\mathsf{box}(+\overline{-+}, \mathsf{box}(-\overline{-+}, x))$ is able to consume an input pebble at its free input port x as well as to produce an output pebble, whereas the result $\mathsf{box}(+-\overline{-+}, x)$ of the composition can only consume input *after* having fired.

The fixed point of a box is the production of the box when fed its own output.

Definition 11. The operations *fixed point* fix : $\pm^\omega \to \overline{\mathbb{N}}$ and *requirement removal* $\delta : \pm^\omega \to \pm^\omega$ on *I/O sequences* are corecursively defined as follows:

$$\underline{\mathsf{fix}(+\sigma)} = \mathsf{s}(\underline{\mathsf{fix}(\delta(\sigma))}) \qquad\qquad \delta(+\sigma) = +\delta(\sigma)$$
$$\underline{\mathsf{fix}(-\sigma)} = 0 \qquad\qquad\qquad \delta(-\sigma) = \sigma$$

For all $\sigma \in \pm^\omega$, we have $\beta_\sigma(\mathsf{fix}(\sigma)) = \mathsf{fix}(\sigma)$. Moreover, $\mathsf{fix}(\sigma)$ is the *least* fixed point of β_σ. Observe that $\beta_{\sigma \cdot \sigma \cdot \sigma \cdots} = \beta_\sigma(\beta_\sigma(\beta_\sigma(\ldots))) = \mathsf{fix}(\sigma)$. Therefore, the infinite self-composition $\mathsf{box}(\sigma, \mathsf{box}(\sigma, \mathsf{box}(\sigma, \ldots)))$ is 'production equivalent' to $\mathsf{src}(\mathsf{fix}(\sigma))$.

Lemma 1. *The net reduction relation $\to_R$ is production preserving, that is, $N \to_R N'$ implies $\pi(N) = \pi(N')$, for all nets $N, N' \in \mathcal{N}$. Furthermore, $\to_R$ is terminating and every closed net normalises to a unique normal form, a source.*

Observe that net reduction employs infinitary rewriting for fixed point computation and composition (Def. 10 and 11). To compute normal forms in finite time we make use of finite representations of rational sequences and exchange the numeral s^ω with a constant ∞. The reader may confer [4] for further details.

Lemma 2. *There is an algorithm that, if $N \in \mathcal{N}_{rat}$ and rational representations of the sequences $\sigma \in \pm^\omega_{rat}$ in N are given, computes the $\to_R$-normal form of N.*

Proof (Hint). Note that composition preserves rationality, that is, $\sigma \cdot \tau \in \pm^\omega_{rat}$ whenever $\sigma, \tau \in \pm^\omega_{rat}$. Similarly, it is straightforward to show that for sequences $\sigma, \tau \in \pm^\omega_{rat}$ with given rational representations the fixed point $\mathsf{fix}(\sigma)$ and a rational representation of the composition $\sigma \cdot \tau$ can be computed in finite time. □

Theorem 3. *Productivity is decidable for pure stream constant specifications.*

Proof. The following steps describe a decision algorithm for productivity of a stream constant M in an SCS $\mathcal{T}$: First, the translation [M] of M into a pebbleflow net is built according to Def. 8. It is easy to verify that [M] is in fact a rational net. Second, by the algorithm stated by Lem. 2, [M] is collapsed to a source $\mathsf{src}(n)$ with $n \in \overline{\mathbb{N}}$. By Thm. 2 it follows that [M] has the same production as M in $\mathcal{T}$, and by Lem. 1 that the production of [M] is n. Consequently, $\pi_\mathcal{T}(M) = n$. Hence the answers "$\mathcal{T}$ is productive for M" and "$\mathcal{T}$ is not productive for M" are obtained if $n = \infty$ and if $n \in \mathbb{N}$, respectively. □

5   Examples

We give three examples to show how our algorithm decides productivity of SCSs. First we recognise our running example (Ex. 1) to be productive. Next, we give a simple example of an SCS that is not productive. Finally, we illustrate that productivity is sensitive to the precise definitions of the stream functions used.

Example 4. We revisit Ex. 3 where we calculated the pebbleflow net [D] for D and show the last five steps of the reduction to $\to_R$-normal form.

$$[D] \to_R \triangle(\triangle(\mu D.\mathsf{box}(+++-\overline{-++}, D), \mu D.\mathsf{box}(+++-\overline{-++}, \mathsf{box}(-\overline{-++}, D))), \mathsf{src}(\infty))$$
$$\to_{R6} \triangle(\triangle(\mathsf{src}(\infty), \mu D.\mathsf{box}(+++-\overline{-++}, \mathsf{box}(-\overline{-++}, D))), \mathsf{src}(\infty))$$
$$\to_{R2} \triangle(\triangle(\mathsf{src}(\infty), \mu D.\mathsf{box}(+++--\overline{-++}, D)), \mathsf{src}(\infty))$$
$$\to_{R6} \triangle(\triangle(\mathsf{src}(\infty), \mathsf{src}(\infty)), \mathsf{src}(\infty)) \to_{R7} \triangle(\mathsf{src}(\infty), \mathsf{src}(\infty)) \to_{R7} \mathsf{src}(\infty) .$$

Hence D is productive in the SCS of Ex. 1.

Example 5. For the definition of J from Ex. 2 we get:

$$[J] = \mu J.\bullet(\bullet(\mathsf{box}(\overline{-+-}, J))) \to_{R1}^2 \mu J.\mathsf{box}(+\overline{-+}, \mathsf{box}(+\overline{-+}, \mathsf{box}(\overline{-+-}, J)))$$
$$\to_{R2} \mu J.\mathsf{box}(++\overline{-+}, \mathsf{box}(\overline{-+-}, J)) \to_{R2} \mu J.\mathsf{box}(++\overline{-+-}, J) \to_{R6} \mathsf{src}(\underline{4}) ,$$

proving that J is not productive (only 4 elements can be evaluated).

Example 6. Let $T = \langle \Sigma_d \uplus \Sigma_{sf} \uplus \Sigma_{sc} \uplus \{:\}, R_d \uplus R_{sf} \uplus R_{sc}\rangle$ be an SCS where $\Sigma_d = \{0\}$, $\Sigma_{sf} = \{\mathsf{zip}, \mathsf{tail}, \mathsf{even}, \mathsf{odd}\}$, $\Sigma_{sc} = \{\mathsf{C}\}$, $R_d = \varnothing$, R_{sc} consists of:

$$\mathsf{C} \to 0 : \mathsf{zip}(\mathsf{C}, \mathsf{even}(\mathsf{tail}(\mathsf{C}))) ,$$

and R_{sf} consists of the rules:

$$\mathsf{tail}(x : \sigma) \to \sigma \qquad\qquad \mathsf{zip}(x : \sigma, \tau) \to x : \mathsf{zip}(\tau, \sigma)$$
$$\mathsf{even}(x : \sigma) \to x : \mathsf{odd}(\sigma) \qquad\qquad \mathsf{odd}(x : \sigma) \to \mathsf{even}(\sigma) .$$

Then, we obtain the following translations:

$$[\mathsf{zip}](N_1, N_2) = \triangle_2(\mathsf{box}(\overline{-++}, N_1), \mathsf{box}(\overline{+-+}, N_2))$$
$$[\mathsf{even}](N) = \mathsf{box}(\overline{-+-}, N)$$
$$[\mathsf{tail}](N) = \mathsf{box}(-\overline{-+}, N)$$
$$[\mathsf{C}] = \mu C.\bullet(\triangle(\mathsf{box}(\overline{-++}, C), \mathsf{box}(\overline{+-+}, \mathsf{box}(\overline{-+-}, \mathsf{box}(-\overline{-+}, C))))) .$$

Now by rewriting [C] with parallel outermost rewriting (except that composition of boxes is preferred to reduce the size of the terms) according to $\to_R$ we get:

$$[C] \to_{R2} \mu C.\bullet(\triangle(\mathsf{box}(\overline{-++}, C), \mathsf{box}(+\overline{-++-}, \mathsf{box}(-\overline{-+}, C))))$$
$$\to_{R2} \mu C.\bullet(\triangle(\mathsf{box}(\overline{-++}, C), \mathsf{box}(\overline{+--+}, C)))$$
$$\to_{R1} \mu C.\mathsf{box}(+\overline{-+}, \triangle(\mathsf{box}(\overline{-++}, C), \mathsf{box}(\overline{+--+}, C)))$$
$$\to_{R3} \mu C.\triangle(\mathsf{box}(+\overline{-+}, \mathsf{box}(\overline{-++}, C)), \mathsf{box}(+\overline{-+}, \mathsf{box}(\overline{+--+}, C)))$$
$$\to_{R2}^2 \mu C.\triangle(\mathsf{box}(\overline{+-+}, C), \mathsf{box}(++\overline{--}, C))$$
$$\to_{R4} \triangle(\mu C.\mathsf{box}(\overline{+-+}, C), \mu C.\mathsf{box}(++\overline{--}, C))$$
$$\to_{R6}^2 \triangle(\mathsf{src}(\infty), \mathsf{src}(\infty))$$
$$\to_{R7} \mathsf{src}(\infty)$$

witnessing productivity of C in T. Note that the 'fine' definitions of zip and even are crucial in this setting. If we replace the definition of zip in T by the 'coarser' one: $\mathsf{zip}^*(x : \sigma, y : \tau) \to x : y : \mathsf{zip}^*(\sigma, \tau)$, we obtain an SCS T^* where:

$$[\mathsf{zip}^*](N_1, N_2) = \triangle_2(\mathsf{box}(\overline{-++}, N_1), \mathsf{box}(\overline{-++}, N_2))$$

$$[\mathsf{C}] = \mu C.\bullet(\triangle(\mathsf{box}(\overline{-++}, C), \mathsf{box}(\overline{-++}, \mathsf{box}(\overline{-+-}, \mathsf{box}(\overline{--+}, C)))))$$
$$\to_{\mathsf{R2}} \mu C.\bullet(\triangle(\mathsf{box}(\overline{-++}, C), \mathsf{box}(\overline{-++-}, \mathsf{box}(\overline{--+}, C))))$$
$$\to_{\mathsf{R2}} \mu C.\bullet(\triangle(\mathsf{box}(\overline{-++}, C), \mathsf{box}(\overline{--++}, C)))$$
$$\to_{\mathsf{R1}} \mu C.\mathsf{box}(+\overline{-+}, \triangle(\mathsf{box}(\overline{-++}, C), \mathsf{box}(\overline{--++}, C)))$$
$$\to_{\mathsf{R3}} \mu C.\triangle(\mathsf{box}(+\overline{-+}, \mathsf{box}(\overline{-++}, C)), \mathsf{box}(+\overline{-+}, \mathsf{box}(\overline{--++}, C)))$$
$$\to^2_{\mathsf{R2}} \mu C.\triangle(\mathsf{box}(\overline{+-+}, C), \mathsf{box}(\overline{+--+}, C))$$
$$\to_{\mathsf{R4}} \triangle(\mu C.\mathsf{box}(\overline{+-+}, C), \mu C.\mathsf{box}(\overline{+--+}, C))$$
$$\to^2_{\mathsf{R6}} \triangle(\mathsf{src}(\infty), \mathsf{src}(\underline{1}))$$
$$\to_{\mathsf{R7}} \mathsf{src}(\underline{1}) \ .$$

Hence C is not productive in $\mathcal{T}^*$ (here it produces only one element).

Similarly, if we change the definition of even to $\mathsf{even}(x : y : \sigma) \to x : \mathsf{even}(\sigma)$, giving rise to the translation $[\mathsf{even}](N) = \mathsf{box}(\overline{--+}, N)$, then only the first two elements of C can be evaluated.

6   Conclusion and Ongoing Research

We have shown that productivity is decidable for stream definitions that belong to the format of SCSs. The class of SCSs contains definitions that cannot be recognised automatically to be productive by the methods of [16,12,2,5,14,1] (e.g. the stream constant definition in Ex. 1). These previous approaches established criteria for productivity that are not applicable for disproving productivity; furthermore, these methods are either applicable to general stream definitions, but cannot be mechanised fully, or can be automated, but give a 'productive'/'don't know' answer only for a very restricted subclass. Our approach combines the features of being automatable and of obtaining a definite 'productive'/'not productive' decision for a rich class of stream definitions.

Note that we obtain decidability of productivity by restricting only the stream function definition part of a stream definition (formalised as an orthogonal TRS), while imposing no conditions on how the stream constant definition part makes use of the stream functions. The restriction to weakly guarded stream function definitions in SCSs is motivated by the wish to formulate an effectively recognisable format of stream definitions for which productivity is decidable. More general recognisable formats to which our method can be applied are possible. If the requirement of a recognisable format is dropped, our approach allows to show decidability of productivity for stream definitions that are based on stream function specifications which can (quantitatively) faithfully be described by 'rational' I/O sequences. Finally, also lower and upper 'rational' bounds on the production of stream functions can be considered to obtain computable criteria for productivity and its complement. This will allow us to deal with stream functions that depend quantitatively on the value of stream elements and data parameters. All of these extensions of the result presented here are the subject of ongoing research (see also [4]).

The reader may want to visit http://infinity.few.vu.nl/productivity/ for additional material. There, an implementation of the decision algorithm for productivity of SCSs as well as an animation tool for pebbleflow nets can be found. We have tested the usefulness and feasibility of the implementation of our decision algorithm on various SCSs from the literature, and so far have not encountered excessive run-times. However, a precise analysis of the run-time complexity of our algorithm remains to be carried out.

Acknowledgement. For useful discussions we want to thank Clemens Kupke, Milad Niqui, Vincent van Oostrom, Femke van Raamsdonk, and Jan Rutten. Also, we would like to thank the anonymous referees for their encouraging comments.

References

1. Buchholz, W.: A term calculus for (co-)recursive definitions on streamlike data structures. Annals of Pure and Applied Logic 136(1-2), 75–90 (2005)
2. Coquand, Th.: Infinite Objects in Type Theory. In: Barendregt, H., Nipkow, T. (eds.) TYPES 1993. LNCS, vol. 806, pp. 62–78. Springer, Heidelberg (1994)
3. Dijkstra, E.W.: On the productivity of recursive definitions, EWD749 (1980)
4. Endrullis, J., Grabmayer, C., Hendriks, D.: Productivity of stream definitions. Technical report, Vrije Universiteit Amsterdam (2007), available via http://infinity.few.vu.nl/productivity/
5. Hughes, J., Pareto, L., Sabry, A.: Proving the correctness of reactive systems using sized types. In: POPL '96, pp. 410–423 (1996)
6. Kahn, G.: The semantics of a simple language for parallel programming. Information Processing, 471–475 (1974)
7. Kennaway, R., Klop, J.W., Sleep, M.R., de Vries, F.-J.: Transfinite reductions in orthogonal term rewriting systems. Inf. and Comput. 119(1), 18–38 (1995)
8. Kennaway, R., Klop, J.W., Sleep, M.R., de Vries, F.-J.: Infinitary lambda calculus. TCS 175(1), 93–125 (1997)
9. Klop, J.W., de Vrijer, R.: Infinitary normalization. In: We Will Show Them: Essays in Honour of Dov Gabbay (2). College Publications, pp. 169–192 (2005), Item 95 at http://web.mac.com/janwillemklop/iWeb/Site/Bibliography.html
10. Lafont, Y.: Interaction nets. In: POPL '90, pp. 95–108. ACM Press, New York (1990)
11. Rutten, J.J.M.M.: Behavioural differential equations: a coinductive calculus of streams, automata, and power series. TCS 308(1-3), 1–53 (2003)
12. Sijtsma, B.A.: On the productivity of recursive list definitions. ACM Transactions on Programming Languages and Systems 11(4), 633–649 (1989)
13. Tait, W.W.: Intentional interpretations of functionals of finite type I. Journal of Symbolic Logic 32(2) (1967)
14. Telford, A., Turner, D.: Ensuring the Productivity of infinite structures. Technical Report 14-97, The Computing Laboratory, Univ. of Kent at Canterbury (1997)
15. Terese: Term Rewriting Systems. Cambridge Tracts in Theoretical Computer Science, vol. 55. Cambridge University Press, Cambridge (2003)
16. Wadge, W.W.: An extensional treatment of dataflow deadlock. TCS 13, 3–15 (1981)

Multi-dimensional Packing with Conflicts

Leah Epstein[1], Asaf Levin[2], and Rob van Stee[3],[*]

[1] Department of Mathematics, University of Haifa, 31905 Haifa, Israel
lea@math.haifa.ac.il
[2] Department of Statistics, The Hebrew University, Jerusalem 91905, Israel
levinas@mscc.huji.ac.il
[3] Department of Computer Science, University of Karlsruhe,
D-76128 Karlsruhe, Germany
vanstee@ira.uka.de

Abstract. We study the multi-dimensional version of the bin packing problem with conflicts. We are given a set of squares $V = \{1, 2, \ldots, n\}$ with sides $s_1, s_2, \ldots, s_n \in [0, 1]$ and a conflict graph $G = (V, E)$. We seek to find a partition of the items into independent sets of G, where each independent set can be packed into a unit square bin, such that no two squares packed together in one bin overlap. The goal is to minimize the number of independent sets in the partition.

This problem generalizes the square packing problem (in which we have $E = \emptyset$) and the graph coloring problem (in which $s_i = 0$ for all $i = 1, 2, \ldots, n$). It is well known that coloring problems on general graphs are hard to approximate. Following previous work on the one-dimensional problem, we study the problem on specific graph classes, namely, bipartite graphs and perfect graphs.

We design a $2 + \varepsilon$-approximation for bipartite graphs, which is almost best possible (unless $P = NP$). For perfect graphs, we design a 3.2744-approximation.

1 Introduction

Two-dimensional packing of squares is a well-known problem, with applications in stock cutting and other fields. In the basic problem, the input consists of a set of squares of given sides. The goal is to pack the input into bins, which are unit squares. A packed item receives a location in the bin so that no pair of squares have an overlap. The goal is to minimize the number of used bins.

However, in computer related applications, items often represent processes. These processes may have conflicts due to efficiency, fault tolerance or security reasons. In such cases, the input set of items is accompanied with a conflict graph where each item corresponds to a vertex. A pair of items that cannot share a bin are represented by an edge in the conflict graph between the two corresponding vertices.

Formally, the problem is defined as follows. We are given a set of squares $V = \{1, 2, \ldots, n\}$ whose sides are denoted by $s_1, s_2 \ldots, s_n$ and satisfy $s_i \in [0, 1]$

[*] Research supported by Alexander von Humboldt Foundation.

for all $1 \leq i \leq n$. We are also given a conflict graph $G = (V, E)$. A valid output is a partition of the items into independent sets of G, together with a packing of the squares of each set into a unit square bin. The packing of a bin is valid if no two squares that are packed together in this bin overlap. The goal is to find such a packing with a minimum number of independent sets.

This problem is a generalization of the square packing problem [1], where $E = \emptyset$, and of the graph coloring problem, where $s_i = 0$ for all $i = 1, 2, \ldots, n$. It is well known that coloring problems on general graphs are hard to approximate. Following previous work on the one-dimensional problem, we study the problem on specific graph classes, namely, bipartite graphs and perfect graphs.

For an algorithm $\mathcal{A}$, we denote its cost on an input I by $\mathcal{A}(I)$, and simply by $\mathcal{A}$, if I is clear from the context. An optimal algorithm that uses a minimum number of bins is denoted by OPT. We consider the (absolute) approximation ratio that is defined as follows. The (absolute) approximation ratio of $\mathcal{A}$ is the infimum $\mathcal{R}$ such that for any input I, $\mathcal{A}(I) \leq \mathcal{R} \cdot \text{OPT}(I)$. We restrict ourselves to algorithms that run in polynomial time. The asymptotic approximation ratio is defined as be $\limsup\limits_{n \to \infty} \sup\limits_{I} \left\{ \frac{\mathcal{A}(I)}{\text{OPT}(I)} | \text{OPT}(I) = n \right\}$.

The one dimensional problem (where items are one-dimensional rather than squares) was studied on these graph classes. Jansen and Öhring [12] introduced the problem and designed approximation algorithms which work in two phases. The first phase is a coloring phase, where the graph is colored using a minimum number of colors. In the second phase, each independent set (which corresponds to a color class) is packed using a bin packing algorithm. Using this method, they obtained a 2-approximation algorithm for bipartite graphs and a 2.7-approximation algorithm for perfect graphs.

In [6], improved algorithms were designed. It was shown that the approximation ratio of the algorithm of [12] for perfect graphs is actually approximately 2.691, and a 2.5-approximation algorithm was designed. The algorithm applies a matching phase in which some pairs of relatively large items are packed in dedicated bins, and applies the methods of [12] as above on the remaining subgraph. An improved 1.75-approximation for bipartite conflict graphs was achieved by applying the algorithm of [12] on inputs with large enough values of OPT, while finding better solutions for inputs with small values of OPT.

Several papers [12,11,6] contain further results for additional graph classes. The paper [12] considered a class of graphs, on which the PRECOLORING EXTENSION problem (see [10,16,17]) can be solved in polynomial time. In this problem a graph is to be colored using a minimum number of colors with the constraint that some vertices already have given colors (a different color to each such vertex). This class contains chordal graphs, interval graphs, forests, split graphs, complements of bipartite graphs, cographs, partial K-trees and complements of Meyniel graphs. For these graphs, they designed a 2.5-approximation algorithm which is based on solving the PRECOLORING EXTENSION problem on the graph (where the items of size larger than $\frac{1}{2}$ are pre-colored each with a different color). In [6] an improved $\frac{7}{3}$-approximation algorithm, which is based on a pre-processing phase in which subsets of at most three items are packed into dedicated bins, was designed.

For all $\varepsilon > 0$, Jansen and Öhring [12] also presented a $(2 + \varepsilon)$-approximation algorithm for one-dimensional packing with conflicts on cographs and partial K-trees. Jansen [11] showed an asymptotic fully polynomial time approximation scheme for the one-dimensional problem on d-inductive graphs, where d is a constant. A d-inductive graph has the property that the vertices can be assigned distinct numbers $1, \ldots, n$ such that each vertex is adjacent to at most d lower numbered vertices. This includes the cases of trees, grid graphs, planar graphs and graphs with constant tree-width. Additional papers [20,18] studied the one-dimensional problem on graphs that are unions of cliques, but their results are inferior to work of Jansen and Öhring [12].

The inapproximability results known for the two-dimensional and one-dimensional packing problems are as follows. Since standard bin packing (two-dimensional packing of squares and one-dimensional packing, respectively), is a special case of the problems with conflicts, the same inapproximability results holds for them as well. This means that the one-dimensional problem cannot be approximated up to a factor smaller than $\frac{3}{2}$, unless $P = NP$, (due to a simple reduction from the PARTITION problem, see problem SP12 in [8]). Also, the two-dimensional problem cannot be approximated up to a factor smaller than 2, unless $P = NP$, since it was shown in [15] that given a set of squares, it is NP-hard to check whether these squares can be packed into one bin. These results hold for the graph classes we consider since an empty graph (i.e., a graph with an empty edge set) is both bipartite and perfect.

Square packing was studied in many variants. An algorithm of approximation 2 (best possible unless $P = NP$) was shown in [22]. Unlike coloring problems, bin packing is often studied with respect to the asymptotic approximation ratio. An asymptotic approximation scheme was given by Bansal et al. [1,2,5]. This was the last result after a sequence of improvements [4,13,3,14,21,7].

Our results. In this paper we design approximation algorithms for bipartite graphs and perfect graphs. For bipartite graphs, we give an algorithm of approximation ratio $2 + \varepsilon$ for any $\varepsilon > 0$. Note that unlike the one-dimensional case, this is almost best possible unless $P = NP$. The algorithm chooses the best solution out of several algorithms, which are designed for various values of OPT. For perfect graphs we design algorithms which have clever pre-processing phases. We analyze an algorithm which chooses the best solution out of the outputs of all the algorithms we design. This results in an algorithm of approximation ratio at most 3.2744.

2   Bipartite Graphs

In this section, we present an algorithm and analysis for the case where the conflict graph is bipartite. This algorithm will use the well-known square packing algorithm NEXT FIT DECREASING (NFD) [19] and a natural variant of it, FIRST FIT DECREASING (FFD), as subroutines. We begin by giving some properties of these two algorithms in Section 2.1. In Section 2.2, we introduce a new algorithm called SixEleven, which is a variation of FFD which packs items differently in one special, crucial case. This helps to get a better area guarantee in a bin packed with SixEleven. We then describe our main algorithm for the cases OPT = 1 and OPT = 2 (Section 2.3), OPT = 3 (Section 2.4), OPT is a constant $k > 3$ (Section 2.5) and finally the case where OPT is not constant (Section 2.6). Since the value of OPT is unknown to the algorithm, the algorithm needs to apply all these possibilities and among these that output a valid solution, choose the one with the smallest cost. We will therefore assume that OPT is known to the algorithm (but make sure that the number of different algorithms applied is constant).

2.1 NFD and FFD

NFD packs items in slices, which are rectangular regions of the bin of width 1 that are stacked on top of each other starting from the bottom of the bin. The height of a slice is defined as the side of the first item packed into it. Each item is packed immediately to the right of the previously packed item, or in the next slice in case it does not fit in the current slice. When a new slice does not fit in the current bin, a new bin is opened for it. FFD works the same, but tries to put each new item in each slice that has been opened so far (to the right of the last item in the slice) instead of only trying the last slice or a new one. Regarding NFD and FFD, we have the following results.

Lemma 1 (Meir & Moser [19]). *Let L be a list of squares with sides $x_1 \geq x_2 \geq \ldots$ Then L can be packed in a rectangle of height $a \geq x_1$ and width $b \geq x_1$ using* NEXT FIT DECREASING *if one of the following conditions is satisfied:*

- *the total area of items in L is at most $x_1^2 + (a - x_1)(b - x_1)$.*
- *the total area of items in L is at most $ab/2$.*

In the following, we will abuse notation and use x_i to denote both the ith item in the input and its side, i.e., the length of one of its sides.

Lemma 2 (van Stee [22]). *Consider a bin that is packed by NFD, and suppose the largest item in this bin has side at most $1/3$. If after packing this bin, there are still unpacked items with side at most $\frac{1}{3}$ left, then the total area of the items in the bin is at least 9/16.*

2.2 Algorithm SixEleven

Algorithm SixEleven is displayed in Figure 1. It has the following properties.

Lemma 3. *Consider a set of squares where the largest item has side strictly more than $1/3$. If the two largest items have total side (i.e., sum of sides) at most 1, and the largest item that remains unpacked has side at most $1/5$, then SixEleven packs at least a total area of 6/11 in this bin, unless it runs out of items.*

Define a *large item* to be an item with side more than $1/88$. An item that is not large is said to be a *small item*. A large item is *huge* if its side is more than $1/3$.

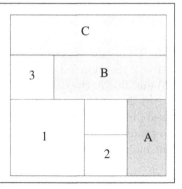

Input: A list of squares $\{x_1, \ldots, x_n\}$, sorted in order of nonincreasing side

Output: A packing of the input or a prefix of it in a single bin.

1. If $x_1 + x_2 + x_3 > 1$, but $x_1 + x_2 + x_4 \leq 1$, pack the three largest items as shown on the right. Pack the area A using NFD starting from the fourth item, then continue in area B with NFD (considering this to be a single slice), and finally pack area C using NFD.
2. Else, use FFD.

Fig. 1. Algorithm SixEleven

Definition 1. *A* good *set of squares is a set S with at least one of the following properties:*

1. *The two largest items in S have total side at most 1, and the total area of the large items is at most 6/11.*
2. *S contains only one large item.*

Theorem 1. *For any input set S which is good, SixEleven either packs S in one bin, or packs at least an area of 6/11 in the first bin.*

This Theorem implies that when SixEleven packs a good set, all the large items in the set are packed in the first bin.

2.3   The Algorithm for OPT = 1 and OPT = 2

Recall that the conflict graph is bipartite. Thus, it is 2-colorable in all cases. If OPT $= 1$, we get that all items can be packed into a single bin, and therefore the conflict graph is empty. We can apply the 2-approximation from [22].

If OPT $= 2$, we act as follows. There are at most 18 huge items. Consider all partitions (a constant number) of the huge items into two sets L_1 and L_2. For the analysis it suffices to consider the iteration of the correct guess. So each such set of huge items can be packed with one bin (and we can find such a packing using the algorithm from Bansal et al. [1], which gives a constant time algorithm to pack a constant number of squares into a bin, is possible), and the coloring of the huge items (where the color of an item is determined by the set it is in) can be extended to a 2-coloring of the entire input as explained below.

For each connected component that contains a huge item the 2-coloring is defined uniquely (unless it contains at least two huge items and we get that it is impossible to extend the coloring accordingly, in this case the partition of the huge items is incorrect), and it remains to decide on the 2-coloring of the connected components of the remaining items. For this problem we apply a similar idea to the one in [6] on the 1-dimensional case, only the partition into two sets must be done more carefully here. For each connected component we find its 2-coloring and we need to decide which color is red and which color is blue (in each of the connected components). We see the problem of balancing the area of blue items and red items as a load balancing problem. Let t be the number of connected components. For each connected component i, let c_i and d_i be the areas of items of the two colors in component i, we define $p_i = \max\{c_i, d_i\}$ and $\Delta(i) = \min\{c_i, d_i\}$. Clearly, each color has in total an area of at least $\sum_{i=1}^{t} \Delta(i)$. We define a load balancing problem on the residual area, i.e., we would like to balance the loads $p_i - \Delta(i)$ between two "machines", where assigning "job" i to machine 1 means that in component i, the color class of larger area got red color, and assigning "job" i to machine 2 means that in component i, the color class of larger area got blue color. Some "jobs" are pre-assigned to a machine if the coloring of this component is determined by the huge items. Therefore, we have a restricted assignment problem. This is a special case of load balancing on two unrelated machines, which admits an FPTAS, see [9].

Consider an optimal solution to the original bin packing problem. The total size of the items that are packed with L_i for $i = 1, 2$ is at most 1. Since we are using an FPTAS, where some area may be removed, the totals remain at most 1 and the total size of the larger set of items is at most 1.006 (for $\varepsilon = 0.006$).

Next we show that we can apply an algorithm based on SixEleven for each color class, which uses at most two bins (and four in total). First consider the case where the set of the huge items in this color has size at least $4/9$. Then the huge items use at most one bin (using the packing of the algorithm from [1]), and for the other items, if by packing them using SixEleven, we need at least two bins, then we have an area guarantee of at least $9/16$ in the second bin by Lemma 2, and this is a contradiction as $4/9 + 9/16 > 1.006$.

On the other hand, if the total area of the huge items is at most $4/9$, then we use SixEleven on the complete color class. We would like to show that the area guarantee of the first packed bin is at least $4/9$, if there is a second bin. If there is a single huge item and it has side at least $2/3$ we are done. Otherwise, the huge item can fit next to any other item. If there are at least two huge items, since the huge items can fit into one bin, the sum of sides of the largest two items in at most 1. We get from the proof of Lemma 3 that if an item does not fit into the first bin, then the area guarantee is $6/11$ in cases 1,2,4, no matter what the size of the next item is, and a guarantee of $1/2$ in case 3, unless the bin contains exactly four items. Since the next item had side of at most $1/3$, we get a guarantee of $\frac{4}{9}$ in this case (similarly to the proof for the case that this item is bounded by $1/5$). So if there is a second bin, the first one has an area guarantee of $\frac{4}{9}$. If we are using three bins, then the second bin again has an area guarantee of $9/16$ by Lemma 2, which again leads to a contradiction.

2.4 The Algorithm for OPT $= 3$

We call items with side in $(1/3, 1/2]$ items of type 2, and larger items are type 1. In this section, items with side at most $1/88$ are called *small*, and the others are *large*. If OPT $= 3$, there are at most $3 \cdot 87^2$ large items. In constant time, find

- A two-coloring of these items that can be extended to a valid coloring for the entire input. This can be done by standard methods. We color the entire conflict graph ignoring the sizes of items.
- A packing of these items in at most three bins. This can be again done by checking all possible partitions of large items into three sets, and application of the algorithm of [1] on each set to pack it into a bin.

Note that the two results are unrelated and we do not require the packing to be consistent with the two-coloring. There are two cases. First, if the total area of the small items is at most $2 \cdot (\frac{87}{88})^2 \approx 1.9548$, do the following.

1. Use an arbitrary valid two-coloring for the small items.
2. Pack the largest set of small items in 2 bins, and the smallest set in at most 1 bin, using NFD.
3. Pack the large items in at most three bins according to the packing found above.

To see that Step 2 can indeed by applied, note that the smallest set has area at most $(\frac{87}{88})^2$, and the largest set has area at most twice this. The first bin packed for the largest set has area packed at least $(\frac{87}{88})^2$ by Lemma 1, leaving at most the same amount for the second bin, which can be packed there using NFD again by Lemma 1.

If the total area of the small items is more than $2 \cdot (\frac{87}{88})^2$, consider the packing for the large items (in at most 3 bins) that we have found. This packing gives us (at most) three sets, denoted by L_1, L_2, L_3. Each set may contain items of both colors. The total area of these items is at most 1.0452. In total, there are at most three items with side more than $1/2$, since all items can be packed in three bins.

We are going to *repack* these items so that each bin contains only items of one color. In this way we ensure that we do not pack conflicting items together. We next show the following auxiliary claim.

Claim. All large items can be packed in at most four bins. For any color, if not all items of that color are packed with large items, then the bins with large items have area guarantee of at least $6/11$.

We now have two cases. First, one of the colors (say blue) might be good. This means that if we pack all blue items using SixEleven, by Theorem 1 SixEleven packs an area of at least $6/11$ in the first blue bin (unless perhaps if it needs only one bin for *all* blue items). By Lemma 1, the area guarantee of any other bin for this color (except, always, the last one) is $(\frac{87}{88})^2 > 0.977$. This gives us the following area guarantees.

Bins needed for blue items	1	2	3	4
Total area guarantee of blue items packed	6/11	1.5228	2.5002	3
Maximum possible area of red items	3	2.4546	1.4772	0.4998
Packed in red bin 1, 2, 3	6/11	6/11	6/11	1/2
Packed in red bin 4, 5 (if needed)	0.977	0.977	-	-

The area guarantees for the red bins follow from Claim 2.4. Using this table, it is easy to verify that in this case (i.e., if the set of blue items is good) we never need more than six bins.

We give one example of such a verification. Suppose the total area of the blue items is 1.6, and the set of blue items is good. Then by the above table, we need at most three bins for the blue items. Since the total area guarantee for the first three red bins is $18/11 > 1.4 = 3 - 1.6$, we need at most three bins for the red items as well, so at most six bins in total.

If neither color is good, the large blue items are packed into *two* bins (either by SixEleven, or in some other way). In this case by Claim 2.4, we can pack the red items with area guarantees of $6/11$ in the first two bins. Therefore, all large red items are packed in the first two bins since $12/11 > 1.0452$. Therefore, any further red bin that is packed using SixEleven (which uses FFD in this case) will again have an area guarantee of $(\frac{87}{88})^2 > 0.977$ by Lemma 1. Overall we find the following table.

Bins needed for blue items	1	2	3	4
Total area guarantee of blue items packed	6/11	12/11	2.068	3
Maximum possible area of red items	3	2.4546	1.909	0.932
Packed in red bin 1, 2	6/11	6/11	6/11	6/11
Packed in red bin 3, 4 (if needed)	0.977	0.977	0.977	-

Again, it can be verified that this is sufficient to pack all items in at most six bins in all cases.

2.5   The Algorithm for OPT $= k > 3$

For any constant value k of OPT, we can find using Lemma 1 a value ε such that the area guarantee for NFD on items of side at most ε is at least $(k - 1.0452)/(k - 1) = 1 - \frac{0.0452}{k-1}$. Then, if the small items have total area at most $k - 1.0452$, we can pack them into at most k bins using NFD, and find an optimal packing for the items with side larger than ε using complete enumeration.

Else, the items with side at least ε have total area at most 1.0452. The proof of Claim 1 shows that *in case there are at most three items of type 1 we need at most four bins for all large items.* We now show that we need at most $2k$ bins for all the items. If SixEleven needs more bins for both colors, this follows because the area guarantee in the four bins with large items is $24/11$, so a total area of at most $k - \frac{24}{11}$ remains to be packed, and we have

$$k - \frac{24}{11} < (2k - 6)\left(1 - \frac{0.0452}{k-1}\right) \qquad \text{for } k \geq 4. \qquad (1)$$

So we need at most $2k - 5$ bins for the small items of both colors: we lose (at most) one bin compared to (1) because there are two colors. (If there are less than four bins with large items, the area guarantee of the remaining bins improves.)

If SixEleven has already packed one color, then the small items of the other color have total area at most $\min(k, k - \frac{6}{11}(j - 2))$ where $j \leq 4$ is the number of bins packed so far (there may be two almost empty bins that contain large items, since we have two colors). These items can be packed in at most $2k - j$ bins for $k \geq 4$, since

$$\min(k, k - \frac{6}{11}(j-2)) < (2k-j)\left(1 - \frac{0.0452}{k-1}\right) \qquad \text{for } k \geq 4, j = 0, \dots, 4. \quad (2)$$

The only case that is not covered yet is the case where there are **four** items of type 1 (since there cannot be more than four such items because the total size of items with side at least ε is at most 1.0452). If all these items are red (say), the blue items are good, and we pack the large red items in four bins. In case we need more bins for both colors, we now have five bins with area guarantee 6/11, and we can pack the remaining items in at most $2k - 5$ bins since $k - \frac{30}{11} < (2k-6)(1 - \frac{0.0452}{k-1})$ for $k \geq 4$. If one color is already packed, we can pack the remaining items into at most $2k - 6$ bins by (1) if we packed five bins so far, and into at most $2k - j$ bins by (2) if we packed $j < 5$ bins so far.

If only one item of type 1 is blue, the blue items are still good. In this case the red items are also good if we exclude the two largest red items, so we need only four bins for all large items (again packing the red items as in Case 1A). Finally, if there are two blue items of type 1, we can pack the large items of each color into two bins, since removing the largest item of either color leaves a good set.

2.6   The Algorithm for Large OPT

Consider a fixed value $\varepsilon > 0$. There are two cases: if $\varepsilon \cdot \text{OPT} > 2$, color the items with two colors, and on each of them apply the APTAS of [1] for square packing. Since the minimum number of bins required to pack each color class is no larger than OPT, it needs only at most $2((1 + \varepsilon)\text{OPT} + 1) \leq (2 + 3\varepsilon)\text{OPT}$ bins. Else, $\text{OPT} \leq 2/\varepsilon$ which is a constant, so use the method from the previous section and use at most 2OPT bins. Note that for the case $\varepsilon \cdot \text{OPT} > 2$, we run just one algorithm, so in total we run at most $2/\varepsilon + 1$ polynomial-time algorithms and take the one that gives the best output.

3   An Algorithm for Perfect Graphs

3.1   An Algorithm for Independent Sets

Given an independent set of items, we use the following packing algorithm.

Algorithm Pack Independent Set (PackIS):

1. As long as there exists an item of side in $(\frac{1}{2}, 1]$, pack such an item in a bin.
2. As long as the number of items of side in $(\frac{1}{3}, \frac{1}{2}]$ is at least four, pack four such items in a bin.
3. As long as the number of items of side in $(\frac{1}{4}, \frac{1}{3}]$ is at least 9, pack 9 such items in a bin.
4. As long as the number of items of side in $(\frac{1}{5}, \frac{1}{4}]$ is at least 16, pack 16 such items in a bin.

5. If there are no items of side in $(\frac{1}{3}, \frac{1}{2}]$ left, pack the remaining items using NFD and halt.

6. Pack all items of side in $(0, \frac{1}{3}]$ using NFD. Call the resulting set of bins S, and let $m = |S|$. Let s_a be the side of the first item of bin m of S.

   Take bin m of S and remove all items from it. Pack its contents together with the remaining larger items (of side in $(\frac{1}{3}, \frac{1}{2}]$), possibly using a second bin, by applying algorithm SixEleven on the first bin, and NFD on the second bin. The items packed in the second adapted bin are those which did not fit into the first adapted bin.

   If a second bin is needed for the adapted packing and $s_a \leq \frac{1}{5}$, keep the first adapted bin packed with the items of side in $(\frac{1}{3}, \frac{1}{2}]$. Re-pack all other items (the ones in S plus the ones in the second adapted bin) once again with NFD. Note that this may affect the packing of bin $m-1$. Otherwise, the current packing (S without bin m together with one or two adapted bins) is given as output.

Note that there is at most one bin packed in the last step whose first packed item has side in the interval $(\frac{1}{k+1}, \frac{1}{k}]$, for $k = 2, 3, 4, 5$. To analyze our algorithm, we use three parameters, $\frac{8}{5} \leq r \leq \frac{16}{9}$, $\frac{1}{4} \leq \mu \leq \frac{2}{7}$ and $\frac{1}{9} \leq \nu \leq \frac{1}{7}$. These bounds imply

$$\nu \geq \frac{r}{16} \quad \text{and} \quad \mu \geq \frac{r}{9}. \tag{3}$$

We moreover require

$$r\frac{43}{99} + \mu \geq 1, \quad r\frac{5}{16} + 4\nu \geq 1, \quad r\frac{331}{648} + \nu \geq 1. \tag{4}$$

We assign weights as follows.

side	$(\frac{1}{2}, 1]$	$(\frac{1}{3}, \frac{1}{2}]$	$(\frac{1}{4}, \frac{1}{3}]$	$(0, \frac{1}{4}]$
weight	1	$\mu + r(x^2 - \frac{1}{9})$	$\nu + r(x^2 - \frac{1}{16})$	$r \cdot x^2$
expansion	1	$r + (\mu - \frac{r}{9})/x^2$	$r + (\nu - \frac{r}{16})/x^2$	r

Expansion is defined as the minimum ratio of weight over size of an item. By (3), it can be seen that the expansion of any item of side at most $\frac{1}{2}$ is at least r, so it is at least $\frac{8}{5}$.

Claim. Let ℓ be the number of bins created by Algorithm PackIS applied on a given color class. The sum of weights of items in this color class is at least $\ell - 1$.

3.2   The General Algorithm

Algorithm Matching Preprocessing (PM):

1. Define the following auxiliary bipartite graph. One set of vertices consists of all items of side in $(\frac{1}{2}, 1]$. The other set of vertices consists of items of side in $(\frac{1}{4}, \frac{1}{2}]$. An edge (a, b) between vertices of items of sides $s_a > \frac{1}{2}$ and $s_b \leq \frac{1}{2}$ occurs if both following conditions hold.

    (a) $s_a + s_b \leq 1$.

    (b) $(a, b) \notin E(G)$.

    That is, if these two items can be placed in a bin together. If this edge occurs, we give it the cost μ if $s_b \geq \frac{1}{3}$ and ν otherwise.

2. Find a maximum cost matching in the bipartite graph.
3. Each pair of matched vertices is removed from G and packed into a bin together.
4. Let G' denote the induced subgraph over the items that were not packed in the preprocessing (i.e., during Steps 1,2,3).
5. Compute a feasible coloring of G' using $\chi(G')$ colors.
6. For each color class, apply the PackIS algorithm described above .

We analyze algorithm PM using weighting functions. Denote the weight function defined in the analysis of Algorithm PackIS for independent sets by w_1. We define the weight function for items packed into bins which are created in the preprocessing to be $1 - \mu$ for an item of side in $(\frac{1}{2}, 1]$ which is packed with an item of side in $(\frac{1}{3}, \frac{1}{2}]$, and $1 - \nu$ otherwise (i.e., if it is packed with an item of side in $\left(\frac{1}{4}, \frac{1}{3}\right]$).

We define a second weight function w_2 which is based on an optimal packing OPT of the entire input which we fix now. This weight function is defined differently from w_1 only for items of side in $(\frac{1}{2}, 1]$. Specifically, for a given such item x, consider the bin in which OPT packs x. If all items in this bin are of side in $(0, \frac{1}{4}]$, we define $w_2(x) = 1$. If the bin contains at least one other item of side larger than $\frac{1}{3}$, we define $w_2(x) = 1 - \mu$ and otherwise $w_2(x) = 1 - \nu$. Note that matching each item of side in $(\frac{1}{2}, 1]$, which got a weight strictly smaller than 1 with respect to w_2, with the largest item that shares its bin in OPT, gives a valid matching in the auxiliary bipartite graph. Therefore, if W_i denotes the total weight of all items with respect to the weight function w_i, then we have $W_1 \leq W_2$.

Theorem 2. *The approximation ratio of PM is at most 3.277344.*

Running an alternative algorithm which combines five possible preprocessing steps instead of just one improves the upper bound on the approximation ratio to 3.2743938. Details for this are omitted due to space constraints.

References

1. Bansal, N., Correa, J., Kenyon, C., Sviridenko, M.: Bin packing in multiple dimensions: Inapproximability results and approximation schemes. Mathematics of Operations Research 31(1), 31–49 (2006)
2. Bansal, N., Sviridenko, M.: New approximability and inapproximability results for 2-dimensional packing. In: Proceedings of the 15th Annual Symposium on Discrete Algorithms, pp. 189–196. ACM/SIAM (2004)
3. Caprara, A.: Packing 2-dimensional bins in harmony. In: Proc. 43rd Annual Symposium on Foundations of Computer Science, pp. 490–499 (2002)

4. Chung, F.R.K., Garey, M.R., Johnson, D.S.: On packing two-dimensional bins. SIAM Journal on Algebraic and Discrete Methods 3, 66–76 (1982)
5. Correa, J., Kenyon, C.: Approximation schemes for multidimensional packing. In: Proceedings of the 15th ACM/SIAM Symposium on Discrete Algorithms, pp. 179–188. ACM/SIAM (2004)
6. Epstein, L., Levin, A.: On bin packing with conflicts. In: Proc. of the 4th Workshop on Approximation and online Algorithms (WAOA2006), pp. 160–173 (2006)
7. Epstein, L., van Stee, R.: Optimal online bounded space multidimensional packing. In: Proc. of 15th Annual ACM-SIAM Symposium on Discrete Algorithms (SODA'04), pp. 207–216. ACM Press, New York (2004)
8. Garey, M.R., Johnson, D.S.: Computers and intractability. W. H. Freeman and Company, New York (1979)
9. Horowitz, E., Sahni, S.: Exact and approximate algorithms for scheduling nonidentical processors. Journal of the ACM 23(2), 317–327 (1976)
10. Hujter, M., Tuza, Z.: Precoloring extension, III: Classes of perfect graphs. Combinatorics, Probability and Computing 5, 35–56 (1996)
11. Jansen, K.: An approximation scheme for bin packing with conflicts. Journal of Combinatorial Optimization 3(4), 363–377 (1999)
12. Jansen, K., Öhring, S.: Approximation algorithms for time constrained scheduling. Information and Computation 132, 85–108 (1997)
13. Kenyon, C., Rémila, E.: A near optimal solution to a two-dimensional cutting stock problem. Mathematics of Operations Research 25(4), 645–656 (2000)
14. Kohayakawa, Y., Miyazawa, F.K., Raghavan, P., Wakabayashi, Y.: Multidimensional cube packing. Algorithmica 40(3), 173–187 (2004)
15. Leung, J.Y.-T., Tam, T.W., Wong, C.S., Young, G.H., Chin, F.Y.L.: Packing squares into a square. Journal on Parallel and Distributed Computing 10, 271–275 (1990)
16. Marx, D.: Precoloring extension, http://www.cs.bme.hu/dmarx/prext.html
17. Marx, D.: Precoloring extension on chordal graphs. In: Graph Theory in Paris. Proceedings of a Conference in Memory of Claude Berge, Trends in Mathematics, pp. 255–270. Birkhäuser (2007)
18. McCloskey, B., Shankar, A.: Approaches to bin packing with clique-graph conflicts. Technical Report UCB/CSD-05-1378, EECS Department, University of California, Berkeley (2005)
19. Meir, A., Moser, L.: On packing of squares and cubes. J. Combinatorial Theory Ser. A 5, 126–134 (1968)
20. Oh, Y., Son, S.H.: On a constrained bin-packing problem. Technical Report CS-95-14, Department of Computer Science, University of Virginia (1995)
21. Seiden, S.S., van Stee, R.: New bounds for multi-dimensional packing. Algorithmica 36(3), 261–293 (2003)
22. van Stee, R.: An approximation algorithm for square packing. Operations Research Letters 32(6), 535–539 (2004)

On Approximating Optimal Weighted Lobbying, and Frequency of Correctness Versus Average-Case Polynomial Time*

Gábor Erdélyi[1], Lane A. Hemaspaandra[2], Jörg Rothe[1], and Holger Spakowski[1]

[1] Institut für Informatik, Universität Düsseldorf, 40225 Düsseldorf, Germany
{erdelyi,rothe,spakowski}@ccc.cs.uni-duesseldorf.de
[2] Department of Computer Science, University of Rochester, Rochester,
NY 14627, USA
www.cs.rochester.edu/u/www/u/lane/

Abstract. We investigate issues regarding two hard problems related to voting, the optimal weighted lobbying problem and the winner problem for Dodgson elections. Regarding the former, Christian et al. [2] showed that optimal lobbying is intractable in the sense of parameterized complexity. We provide an efficient greedy algorithm that achieves a logarithmic approximation ratio for this problem and even for a more general variant—optimal weighted lobbying. We prove that essentially no better approximation ratio than ours can be proven for this greedy algorithm.

The problem of determining Dodgson winners is known to be complete for parallel access to NP [11]. Homan and Hemaspaandra [10] proposed an efficient greedy heuristic for finding Dodgson winners with a guaranteed frequency of success, and their heuristic is a "frequently self-knowingly correct algorithm." We prove that every distributional problem solvable in polynomial time on the average with respect to the uniform distribution has a frequently self-knowingly correct polynomial-time algorithm. Furthermore, we study some features of probability weight of correctness with respect to Procaccia and Rosenschein's junta distributions [15].

1 Introduction

Preference aggregation and election systems have been studied for centuries in social choice theory, political science, and economics. Recently, these topics have become the focus of attention in various areas of computer science as well, such as artificial intelligence (especially with regard to distributed AI in multiagent settings), systems (e.g., for spam filtering), and computational complexity.

* Supported in part by DFG grants RO 1202/9-1 and RO 1202/9-3, NSF grant CCF-0426761, the Alexander von Humboldt Foundation's TransCoop program, and a Friedrich Wilhelm Bessel Research Award. Work done in part while the second author was visiting Heinrich-Heine-Universität Düsseldorf. Some of the results of Section 3 of this paper were presented at the *First International Workshop on Computational Social Choice*, December 2006.

E. Csuhaj-Varjú and Z. Ésik (Eds.): FCT 2007, LNCS 4639, pp. 300–311, 2007.
© Springer-Verlag Berlin Heidelberg 2007

This paper's topic is motivated by two hard problems that both are related to voting, the optimal weighted lobbying problem and the winner problem for Dodgson elections. Regarding the former problem, Christian et al. [2] defined its unweighted variant as follows: Given a 0-1 matrix that represents the No/Yes votes for multiple referenda in the context of direct democracy, a positive integer k, and a target vector (of the outcome of the referenda) of an external actor ("The Lobby"), is it possible for The Lobby to reach its target by changing the votes of at most k voters? They proved the optimal lobbying problem complete for the complexity class W[2], thus providing strong evidence that it is intractable even for small values of the parameter k. However, The Lobby might still try to find an approximate solution efficiently. We propose an efficient greedy algorithm that establishes the first approximation result for the weighted version of this problem in which each voter has a price for changing his or her 0-1 vector to The Lobby's specification. Our approximation result applies to Christian et al.'s original optimal lobbying problem (in which each voter has unit price), and also provides the first approximation result for that problem. In particular, we achieve logarithmic approximation ratios for both these problems.

The Dodgson winner problem was shown NP-hard by Bartholdi, Tovey, and Trick [1]. Hemaspaandra, Hemaspaandra, and Rothe [11] optimally improved this result by showing that the Dodgson winner problem is complete for $P_{\parallel}^{NP}$, the class of problems solvable via parallel access to NP. Since these hardness results are in the worst-case complexity model, it is natural to wonder if one at least can find a heuristic algorithm solving the problem efficiently for "most of the inputs occurring in practice." Homan and Hemaspaandra [10] proposed a heuristic, called Greedy-Winner, for finding Dodgson winners. They proved that if the number of voters greatly exceeds the number of candidates (which in many real-world cases is a very plausible assumption), then their heuristic is a *frequently self-knowingly correct algorithm*, a notion they introduced to formally capture a strong notion of the property of "guaranteed success frequency" [10]. We study this notion in relation with average-case complexity. We also investigate Procaccia and Rosenschein's notion of deterministic heuristic polynomial time for their so-called junta distributions, a notion they introduced in their study of the "average-case complexity of manipulating elections" [15]. We show that under the junta definition, when stripped to its basic three properties, every NP-hard set is $\leq_m^p$-reducible to a set in deterministic heuristic polynomial time relative to some junta distribution and we also show a very broad class of sets (including many NP-complete sets) to be in deterministic heuristic polynomial time relative to some junta distribution. We note (see also [17]) that the "average-case complexity" results of [15] are not really average-case complexity results (in the sense of being about some sort of averaging of running times), but rather are frequency of correctness—or, to be more precise, probability weight of correctness—results (as are also the results of Homan and Hemaspaandra).

This paper is organized as follows. In Section 2, we propose and analyze an efficient greedy algorithm for approximating the optimal weighted lobbying problem. In Section 3, we show that every problem solvable in average-case polynomial time with respect to the uniform distribution has a frequently self-knowingly correct polynomial-time algorithm, and we study Procaccia and Rosenschein's junta distributions. The heuristic Greedy-Score on which Greedy-Winner is based [10], some technical definitions from average-case complexity theory [13,9,18], and the proofs omitted due to space constraints can be found in the full version of this paper [6].

2 Approximating Optimal Weighted Lobbying

2.1 Optimal Lobbying and Its Weighted Version

Christian et al. [2] introduced and studied the following problem. Suppose there are m voters who vote on n referenda, and there is an external actor, which is referred to as "The Lobby" and seeks to influence the outcome of these referenda by making voters change their votes. It is assumed that The Lobby has complete information about the voters' original votes, and that The Lobby's budget allows for influencing the votes of a certain number, say k, of voters. Formally, the Optimal-Lobbying problem is defined as follows: Given an $m \times n$ 0-1 matrix V (whose rows represent the voters, whose columns represent the referenda, and whose 0-1 entries represent No/Yes votes), a positive integer $k \le m$, and a target vector $x \in \{0,1\}^n$, is there a choice of k rows in V such that by changing the entries of these rows the resulting matrix has the property that, for each j, $1 \le j \le n$, the jth column has a strict majority of ones (respectively, zeros) if and only if the jth entry of the target vector x of The Lobby is one (respectively, zero) [2]?

Christian et al. [2] showed that Optimal-Lobbying (with respect to parameter k, the number of voters influenced by The Lobby) is complete for the complexity class W[2]; see, e.g., Downey and Fellows [4] and Flum and Grohe [7] for background on the theory of parameterized complexity and in particular for the definition of W[2].

This result is considered strong evidence that Optimal-Lobbying is intractable, even for small values of the parameter k. However, even though the optimal goal of The Lobby cannot be achieved efficiently, it might be approximable within some factor. That is, given an $m \times n$ 0-1 matrix V and a target vector $x \in \{0,1\}^n$, The Lobby might try to reach its target by changing the votes of as few voters as possible.

We consider the more general problem Optimal-Weighted-Lobbying, where we assume that influencing the 0-1 vector of each voter v_i exacts some price, $price(v_i) \in \mathbb{Q}$, where $\mathbb{Q}$ denotes the set of nonnegative rational numbers. In this scenario, The Lobby seeks to minimize the amount of money spent to reach its goal. The problem Optimal-Lobbying (redefined as an optimization problem rather than a parameterized problem) is the unit-prices special case of Optimal-Weighted-Lobbying, i.e., where $price(v_i) = 1$ for each voter v_i. It follows that Optimal-Weighted-Lobbying (redefined as a parameterized rather than an optimization problem, where the parameter is The Lobby's budget of money to be spent) inherits the W[2]-hardness lower bound from its special case

Optimal-Lobbying, and that the logarithmic approximation algorithm we build for Optimal-Weighted-Lobbying will provide the same approximation ratio for Optimal-Lobbying.

In the remainder of this section, we describe and analyze an efficient greedy algorithm for approximating Optimal-Weighted-Lobbying.

2.2   A Greedy Algorithm for Optimal Weighted Lobbying

Let a matrix $V \in \{0,1\}^{m \times n}$ be given, where the columns $r_1, r_2, \ldots, r_n$ of V represent the referenda and the rows $v_1, v_2, \ldots, v_m$ of V represent the voters. Without loss of generality, we may assume that The Lobby's target vector is of the form $x = 1^n$ (and thus may be dropped from the problem instance), since if there is a zero in x at position j, we can simply flip this zero to one and also flip the corresponding zeros and ones in column r_j.

For each column r_j, define the *deficit* d_j to be the minimum number of zeros that need to be flipped to ones such that there are strictly more ones than zeros in this column. Let $D_0 = \sum_{j=1}^n d_j$ be the sum of all initial deficits.

Figure 1 gives the greedy algorithm, which proceeds by iteratively choosing a most "cost-effective" row of V and flipping to ones all those zeros in this row that belong to columns with a positive deficit, until the deficits in all columns have decreased to zero. We assume that ties between rows with equally good cost-effectiveness are broken in any simple way, e.g., in favor of the tied v_i with lowest i.

Let R be the set of columns of V whose deficits have already vanished at the beginning of an iteration, i.e., all columns in R already have a strict majority of ones. Let $v_{i \upharpoonright R^c}$ denote the entries of v_i restricted to those columns not in R, and let $\#_0(v_{i \upharpoonright R^c})$ denote the number of zeros in $v_{i \upharpoonright R^c}$. (For i such that $\#_0(v_{i \upharpoonright R^c}) = 0$, we consider $price(v_i)/\#_0(v_{i \upharpoonright R^c})$ to be $+\infty$.) During an iteration, the *cost per flipped entry in row* v_i (for decreasing the deficits in new columns by flipping v_i's zeros to ones) is $price(v_i)/\#_0(v_{i \upharpoonright R^c})$. We say a voter v_i is *more cost-effective* than a voter v_j if v_i's cost per flipped entry is less than v_j's. When our algorithm chooses to alter a row v_i, we will think of its price being distributed equally among the new columns with decreased deficit, and at that instant will permanently associate with every flipped entry, e_k, in that row its portion of the cost, i.e., $cost(e_k) = price(v_i)/\#_0(v_{i \upharpoonright R^c})$.

Clearly, the greedy algorithm in Figure 1 always stops, and its running time is polynomial, since the while loop requires only linear (in the input size) time and has to be executed at most $D_0 = \sum_{j=1}^n d_j \leq n \cdot \lceil (m+1)/2 \rceil$ times (note that at most $\lceil (m+1)/2 \rceil$ flips are needed in each column to achieve victory for The Lobby's position).

Now, enumerate the D_0 entries of V that have been flipped in the order in which they were flipped by the algorithm. Let $e_1, e_2, \ldots, e_{D_0}$ be the resulting enumeration. Let OPT be the money that would be spent by The Lobby for an optimal choice of voters such that its target is reached.

Lemma 1. *For each $k \in \{1, 2, \ldots, D_0\}$, we have $cost(e_k) \leq \text{OPT}/(D_0 - k + 1)$.*

The proof of Lemma 1 can be found in the full version of this paper [6].

1. **Input:** A matrix $V \in \{0,1\}^{m \times n}$.
2. **Initialize:**
   Compute the deficits d_j, $1 \le j \le n$.
   $D \leftarrow \sum_{j=1}^{n} d_j$.   /* Initially, $D = D_0$. */
   $X \leftarrow \emptyset$.
3. **While** $D \ne 0$ **do**
   Let R be the set of columns r_j with $d_j = 0$.
   Find a voter whose cost-effectiveness is greatest, say v_i.
   Let $\gamma_i = price(v_i)/\#_0(v_{i\rceil R^c})$.
   Choose v_i and flip all zeros in $v_{i\rceil R^c}$ to ones.
   For each flipped entry e in v_i, let $cost(e) = \gamma_i$.
                                                                       /* $cost(e)$ will be used in our analysis. */
   $X \leftarrow X \cup \{i\}$.
   $d_j \leftarrow d_j - 1$, for each column r_j for which a zero was flipped.
   $D \leftarrow \sum_{j=1}^{n} d_j$.
4. **Output:** X.

Fig. 1. Greedy algorithm for Optimal-Weighted-Lobbying

Theorem 1. *The greedy algorithm presented in Figure 1 approximates the problem Optimal-Weighted-Lobbying with approximation ratio at most*

$$\sum_{i=1}^{D_0} \frac{1}{i} \le 1 + \ln D_0 \le 1 + \ln\left(n \left\lceil \frac{m+1}{2} \right\rceil\right).$$

Proof. The total price of the set of voters X picked by the greedy algorithm is the sum of the costs of those entries flipped. That is, $price(X) = \sum_{i \in X} price(v_i) = \sum_{k=1}^{D_0} cost(e_k) \le \left(1 + \frac{1}{2} + \cdots + \frac{1}{D_0}\right) \cdot \text{OPT}$, where the last inequality follows from Lemma 1.   ❑

Since the input size is lower-bounded by $m \cdot n$, Theorem 1 establishes a logarithmic approximation ratio for Optimal-Weighted-Lobbying (and also for Optimal-Lobbying). Note that the proof of Theorem 1 establishes an approximation ratio bound that is (sometimes nonstrictly) stronger than $\sum_{i=1}^{D_0} 1/i$. In particular, if the number of zeros flipped in successive iterations of the algorithm's while loop are $\ell_1, \ell_2, \ldots, \ell_p$, where $\ell_1 + \ell_2 + \cdots + \ell_p = D_0$, then the proof gives a bound on the approximation ratio of

$$\frac{\ell_1}{D_0} + \frac{\ell_2}{D_0 - \ell_1} + \cdots + \frac{\ell_p}{D_0 - (\ell_1 + \cdots + \ell_{p-1})} = \sum_{j=1}^{p} \frac{\ell_j}{D_0 - \sum_{k=1}^{j-1} \ell_k}.$$

This is strictly better than $\sum_{i=1}^{D_0} 1/i$ except in the case that each ℓ_j equals 1. And this explains why, in the example we are about to give that shows that

Table 1. A tight example for the greedy algorithm in Figure 1

	r_1	r_2	r_3	$\cdots$	r_n	$price(v_i)$
v_1	0	1	1	$\cdots$	1	1
v_2	1	0	1	$\cdots$	1	$1/2$
v_3	1	1	0	$\cdots$	1	$1/3$
$\vdots$	$\vdots$	$\vdots$	$\vdots$	$\ddots$	$\vdots$	$\vdots$
v_n	1	1	1	$\cdots$	0	$1/n$
v_{n+1}	0	0	0	$\cdots$	0	$1+\epsilon$
v_{n+2}	1	0	0	$\cdots$	0	2
v_{n+3}	0	1	0	$\cdots$	0	2
v_{n+4}	0	0	1	$\cdots$	0	2
$\vdots$	$\vdots$	$\vdots$	$\vdots$	$\ddots$	$\vdots$	$\vdots$
v_{2n+1}	0	0	0	$\cdots$	1	2

the algorithm can at times yield a result with ratio essentially no better than $\sum_{i=1}^{D_0} 1/i$, each ℓ_j will equal 1.

Now, we show that the $\sum_{i=1}^{D_0} 1/i$ approximation ratio stated in Theorem 1 is essentially the best possible that can be stated for the greedy algorithm of Figure 1. Consider the example given in Table 1. The prices for changing the voters' 0-1 vectors are shown in the right-most column of Table 1: Set $price(v_i) = 1/i$ for each $i \in \{1, 2, \ldots, n\}$, set $price(v_i) = 2$ for each $i \in \{n+2, n+3, \ldots, 2n+1\}$, and set $price(v_{n+1}) = 1 + \epsilon$, where $\epsilon > 0$ is a fixed constant that can be set arbitrarily small. Note that, for each j, $1 \leq j \leq n$, we have $d_j = 1$, and hence $D_0 = n$.

When run on this input, our greedy algorithm sequentially flips, for $i = n$, $n - 1, \ldots, 1$, the single zero-entry of voter v_i to a one. Thus the total money spent is $1 + 1/2 + \cdots + 1/n = 1 + 1/2 + \cdots + 1/D_0$. On the other hand, the optimal choice consists of influencing just voter v_{n+1} by flipping all of v_{n+1}'s entries to ones, which costs only $1 + \epsilon$.

3   Frequency of Correctness Versus Average-Case Polynomial Time

3.1   A Motivation: How to Find Dodgson Winners Frequently

An election (C, V) is given by a set C of candidates and a set V of voters, where each vote is specified by a preference order on all candidates and the underlying preference relation is strict (i.e., irreflexive and antisymmetric), transitive, and complete. A Condorcet winner of an election is a candidate i such that for each candidate $j \neq i$, a strict majority of the voters prefer i to j. Not all elections have a Condorcet winner, but when a Condorcet winner exists, he or she is unique. In 1876, Dodgson [5] proposed an election system that is based on a combinatorial optimization problem: An election is won by those candidates who are "closest" to being a Condorcet winner. More precisely, given a Dodgson election (C,V), every candidate c in C is assigned a score, denoted by DodgsonScore(C,V,c), which gives the smallest number of sequential exchanges of adjacent preferences in the voters' preference orders needed to make c a Condorcet winner with respect to the resulting preference orders. Whoever has the lowest Dodgson score wins.

The problem Dodgson-Winner is defined as follows: Given an election (C,V) and a designated candidate c in C, is c a Dodgson winner in (C,V)? (The search version of this decision problem can easily be stated.) As mentioned earlier, Hemaspaandra et al. [11] have shown that this problem is $P_{\parallel}^{NP}$-complete.

It certainly is not desirable to have an election system whose winner problem is hard, as only systems that can be evaluated efficiently are actually used in practice. Fortunately, there are a number of positive results on Dodgson elections and related systems as well (see, e.g., [1,8,16,14]). One of these positive results is due to Homan and Hemaspaandra [10] who proposed a greedy heuristic that finds Dodgson winners with a "guaranteed high frequency of success." To capture a strengthened version of this property formally, they introduced the notion of a "frequently self-knowingly correct algorithm."

Definition 1 ([10]). *Let $f : S \to T$ be a function, where S and T are sets. We say an algorithm $\mathcal{A} : S \to T \times \{$ "definitely", "maybe"$\}$ is self-knowingly correct for f if, for each $s \in S$ and $t \in T$, whenever $\mathcal{A}$ on input s outputs $(t,$ "definitely") then $f(s) = t$. An algorithm $\mathcal{A}$ that is self-knowingly correct for $g : \Sigma^* \to T$ is said to be* frequently self-knowingly correct *for g if*

$$\lim_{n \to \infty} \frac{\|\{x \in \Sigma^n \mid \mathcal{A}(x) \in T \times \{ \text{"maybe"}\}\}\|}{\|\Sigma^n\|} = 0.$$

3.2   On AvgP and Frequently Self-knowingly Correct Algorithms

The theory of average-case complexity was initiated by Levin [13]. A problem's average-case complexity can be viewed as a more significant measure than its worst-case complexity in many cases, for example in cryptographic applications. For an excellent introduction to this theory, we refer to Goldreich [9] and Wang [18]. Formal definitions can be found there and in the full version of this paper [6]. An alternative view of the definition of Levin's class average polynomial time (AvgP) was provided by Impagliazzo [12].

Definition 2 ([12]). *An algorithm computes a function f with* benign faults *if it either outputs an element of the image of f or "?," and if it outputs anything other than "?" it is correct. For any distribution μ on Σ^*, let $\mu_{\leq n}$ denote the restriction of μ to strings of length at most n. A* polynomial-time benign algorithm scheme *for a function f on μ is an algorithm $\mathcal{A}(x,\delta)$ such that:*

1. *$\mathcal{A}$ runs in time polynomial in $|x|$ and $1/\delta$.*
2. *$\mathcal{A}$ computes f with benign faults.*
3. *For each δ, $0 < \delta < 1$, and for each $n \in \mathbb{N}^+$, $\text{Prob}_{\mu_{\leq n}}[\mathcal{A}(x,\delta) = \ ?] \leq \delta$.*

Our main result in this section is that every distributional problem that has a polynomial-time benign algorithm scheme with respect to the uniform distribution must also have a frequently self-knowingly correct polynomial-time algorithm. It follows that all uniformly distributed AvgP problems have a frequently self-knowingly correct polynomial-time algorithm. The proofs of Theorem 2 and Proposition 1 (which says that the converse implication of that of Corollary 1 below is not true) can be found in the full version of this paper [6].

Theorem 2. *Suppose that $\mathcal{A}(x, \delta)$ is a polynomial-time benign algorithm scheme for a distributional problem f on the standard uniform distribution. Then there is a frequently self-knowingly correct polynomial-time algorithm $\mathcal{A}'$ for f.*

Theorem 2 and Proposition 2 in [12] establish the following corollary.

Corollary 1. *Every distributional problem that under the standard uniform distribution is in AvgP has a frequently self-knowingly correct polynomial-time algorithm.*

Proposition 1. *There exist (distributional) problems with a frequently self-knowingly correct polynomial-time algorithm that are not in AvgP under the standard uniform distribution.*

3.3   A Basic Junta Distribution for SAT

Procaccia and Rosenschein [15] introduced "junta distributions" in their study of NP-hard manipulation problems for elections. The goal of a junta is to be such a hard distribution (that is, to focus so much weight on hard instances) that, loosely put, if a problem is easy relative to a junta then it will be easy relative to any reasonable distribution (such as the uniform distribution). This is a goal, not (currently) a theorem; Procaccia and Rosenschein [15] do not formally establish this, but rather seek to give a junta definition that might satisfy this. Their paper in effect encourages others to weigh in and study the suitability of the notion of a junta and the notion built on top of it, heuristic polynomial time. Furthermore, they repeatedly describe their theory as one of average-case complexity. In the full version of this paper [6] we suggest that it is potentially confusion-inducing to describe their theory as one of average-case complexity. Their theory adds to the study of frequency of correctness the notion of probability weight of correctness. This is a very valuable direction, but we point out (see also [17]) that it is neither explicitly about, nor does it seem to implicitly yield claims about, average-case complexity. Their paper states that work of Conitzer and Sandholm [3] is also about average-case complexity but, similarly, we mention that that work is not about average-case complexity; it is about (and carefully and correctly frames itself as being about) frequency of correctness. We do not mean this as a weakness: We feel that frequency of (or probability weight of) correctness, most especially when as in the work of Homan and Hemaspaandra [10] the algorithm is "self-knowingly" correct a guaranteed large portion of the time, is an interesting and important direction.

Regarding Procaccia and Rosenschein's notion of juntas, they state three "basic" conditions for a junta, and then give two additional ones that are tailored specifically to the needs of NP-hard voting manipulation problems. They state their hope that their scheme will extend more generally, using the three basic conditions and potentially additional conditions, to other mechanism problems. One might naturally wonder whether their junta/heuristic polynomial-time/susceptibility approach applies more generally to studying the probability weight of correctness for NP-hard problems, since their framework in effect (aside from the two "additional" junta conditions just about voting manipulation) is a general one relating problems to probability weight of correctness. We first carefully note that in asking this we are taking their notion beyond the realm for which it was explicitly designed, and so we do not claim to be refuting any claim of their paper. What we will do, however, is show that the three basic conditions for a junta are sufficiently weak that one can construct a junta relative to which the standard NP-complete problem SAT—and a similar attack can be carried out on a wide range of natural NP-complete problems—has a deterministic heuristic polynomial-time algorithm. So if one had faith in the analog of their approach, as applied to SAT, one would have to believe that under essentially every natural distribution SAT is easy (in the sense that there is an algorithm with a high probability weight of correctness under that distribution). Since the latter is not widely believed, we suggest that the right conclusion to draw from the main result of this section is simply that if one were to hope to effectively use on typical NP-complete sets the notion of juntas and of heuristic polynomial time w.r.t. juntas, one would almost certainly have to go beyond the basic three conditions and add additional conditions. Again, we stress that Procaccia and Rosenschein didn't focus on examples this far afield, and even within the world of mechanisms implied that unspecified additional conditions beyond the core three might be needed when studying problems other than voting manipulation problems. This section's contribution is to give a construction indicating that the core three junta conditions, standing on their own, seem too weak.

Since we will use the Procaccia–Rosenschein junta notion in a more general setting than merely manipulation problems, we to avoid any chance of confusion will use the term "basic junta" to denote that we have removed the word "manipulation" and that we are using their three "basic" properties, and not the two additional properties that are specific to voting manipulation. Our definition of "deterministic heuristic polynomial-time algorithm" is identical to theirs, except we have replaced the word "junta" with "basic junta"—and so again we are allowing their notion to be extended beyond just manipulation and mechanism problems.

Definition 3 (see [15]). *Let $\mu = \{\mu_n\}_{n\in\mathbb{N}}$ be a distribution over the possible instances of an NP-hard problem L. (In this model, each μ_n sums to 1 over all length n instances.) We say μ is a* basic junta distribution *if and only if μ has the following properties:*

1. **Hardness:** *The restriction of L to μ is the problem whose possible instances are only $\bigcup_{n\in\mathbb{N}}\{x \mid |x| = n \text{ and } \mu_n(x) > 0\}$. Deciding this restricted problem is still NP-hard.*

2. **Balance:** *There exist constants $c > 1$ and $N \in \mathbb{N}$ such that for all $n \geq N$ and for all instances x, $|x| = n$, we have $1/c \leq \mathrm{Prob}_{\mu_n}[x \in L] \leq 1 - 1/c$.*
3. **Dichotomy:** *There exists some polynomial p such that for all n and for all instances x, $|x| = n$, either $\mu_n(x) \geq 2^{-p(n)}$ or $\mu_n(x) = 0$.*

Let (L, μ) be a distributional decision problem (see, e.g., [6, Definition B.1]). An algorithm $\mathcal{A}$ is said to be a deterministic heuristic polynomial-time algorithm for (L, μ) *if $\mathcal{A}$ is a deterministic polynomial-time algorithm and there exist a polynomial q and $N \in \mathbb{N}$ such that for each $n \geq N$, $\mathrm{Prob}_{\mu_n}[x \notin L \Longleftrightarrow \mathcal{A}$ accepts $x] < \frac{1}{q(n)}$.*

We now explore their notion of deterministic heuristic polynomial time and their notion of junta, both however viewed for general NP problems and using the "basic" three conditions. We will note that the notion in such a setting is in some senses not restrictive enough and in other senses is too restrictive. Let us start with the former. We need a definition.

Definition 4. *We will say that a set L is* well-pierced *(respectively,* uniquely well-pierced*) if there exist sets $Pos \in P$ and $Neg \in P$ such that $Pos \subseteq L$, $Neg \subseteq \overline{L}$, and there is some $N \in \mathbb{N}$ such that at each length $n \geq N$, each of Pos and Neg has at least one string at length n (respectively, each of Pos and Neg has exactly one string at length n).*

Each uniquely well-pierced set is well-pierced. Note that, under quite natural encodings, such NP-complete sets as, for example, SAT certainly are well-pierced and uniquely well-pierced. (All this says is that, except for a finite number of exceptional lengths, there is one special string at each length that can easily, uniformly be recognized as in the set and one that can easily, uniformly be recognized as not in the set.) Indeed, under quite natural encodings, undecidable problems such as the halting problem are uniquely well-pierced.

Recall that juntas are defined in relation to an infinite list of distributions, one per length (so $\mu = \{\mu_n\}_{n \in \mathbb{N}}$). The Procaccia and Rosenschein definition of junta does not explicitly put computability or uniformity requirements on such distributions in the definition of junta, but it is useful to be able to make claims about that. So let us say that such a distribution is *uniformly computable in polynomial time* (respectively, is *uniformly computable in exponential time*) if there is a polynomial-time function (respectively, an exponential-time function) f such that for each i and each x, $f(i, x)$ outputs the value of $\mu_i(x)$ (say, as a rational number—if a distribution takes on other values, it simply will not be able to satisfy our notion of good uniform time).

Theorem 3. *Let A be any NP-hard set that is well-pierced. Then there exists a basic junta distribution relative to which A has a deterministic heuristic polynomial-time algorithm (indeed, it even has a deterministic heuristic polynomial-time algorithm whose error weight is bounded not merely by $1/poly$ as the definition requires, but is even bounded by $1/2^{n^2 - n}$). Moreover, the junta is uniformly computable in exponential time, and if we in addition assume that A is uniquely well-pierced, the junta is uniformly computable in polynomial time.*

The proof of Theorem 3, additional results, and extensive related discussions on the junta approach can be found in the full version of this paper [6].

4   Conclusions

Christian et al. [2] introduced the optimal lobbying problem and showed it complete for W[2], and so generally viewed as intractable in the sense of parameterized complexity. In Section 2, we proposed an efficient greedy algorithm for approximating the optimal solution of this problem, even if generalized by assigning prices to voters. This greedy algorithm achieves a logarithmic approximation ratio and we prove that that is essentially the best approximation ratio that can be proven for this algorithm. We mention as an interesting open issue whether more elaborate algorithms can achieve better approximation ratios.

Section 3 studied relationships between average-case polynomial time, benign algorithm schemes, and frequency (and probability weight) of correctness. We showed that all problems having benign algorithm schemes relative to the uniform distribution (and thus all sets in average-case polynomial time relative to the uniform distribution) have frequently self-knowingly correct algorithms. We also studied, when limited to the "basic" three junta conditions, the notion of junta distributions and of deterministic heuristic polynomial time, and we showed that they admit some extreme behaviors. We argued that deterministic heuristic polynomial time should not be viewed as a model of average-case complexity.

Acknowledgments. We are deeply grateful to Chris Homan for his interest in this work and for many inspiring discussions on computational issues related to voting. We also thank the anonymous FCT 2007 and COMSOC 2006 workshop referees for their helpful comments.

References

1. Bartholdi III, J., Tovey, C., Trick, M.: Voting schemes for which it can be difficult to tell who won the election. Social Choice and Welfare 6(2), 157–165 (1989)
2. Christian, R., Fellows, M., Rosamond, F., Slinko, A.: On complexity of lobbying in multiple referenda. In: Endriss, U., Lang, J. (eds.) First International Workshop on Computational Social Choice (COMSOC 2006), pp. 87–96 (workshop notes). Universiteit van Amsterdam (December 2006)
3. Conitzer, V., Sandholm, T.: Nonexistence of voting rules that are usually hard to manipulate. In: Proceedings of the 21st National Conference on Artificial Intelligence. AAAI Press, Stanford, California, USA (2006)
4. Downey, R., Fellows, M.: Parameterized Complexity. Springer, New York (1999)
5. Dodgson, C.: A method of taking votes on more than two issues. Pamphlet printed by the Clarendon Press, Oxford (1876)
6. Erdélyi, G., Hemaspaandra, L., Rothe, J., Spakowski, H.: On approximating optimal weighted lobbying, and frequency of correctness versus average-case polynomial time. Technical Report TR-914, Department of Computer Science, University of Rochester, Rochester, NY (March 2007)

7. Flum, J., Grohe, M.: Parameterized Complexity Theory. EATCS Texts in Theoretical Computer Science. Springer, Heidelberg (2006)
8. Fishburn, P.: Condorcet social choice functions. SIAM Journal on Applied Mathematics 33(3), 469–489 (1977)
9. Goldreich, O.: Note on Levin's theory of average-case complexity. Technical Report TR97-058, Electronic Colloquium on Computational Complexity (November 1997)
10. Homan, C., Hemaspaandra, L.: Guarantees for the success frequency of an algorithm for finding Dodgson-election winners. In: Královič, R., Urzyczyn, P. (eds.) MFCS 2006. LNCS, vol. 4162, pp. 528–539. Springer, Heidelberg (2006)
11. Hemaspaandra, E., Hemaspaandra, L., Rothe, J.: Exact analysis of Dodgson elections: Lewis Carroll's 1876 voting system is complete for parallel access to NP. Journal of the ACM 44(6), 806–825 (1997)
12. Impagliazzo, R.: A personal view of average-case complexity. In: Proceedings of the 10th Structure in Complexity Theory Conference, pp. 134–147. IEEE Computer Society Press, Los Alamitos (1995)
13. Levin, L.: Average case complete problems. SIAM Journal on Computing 15(1), 285–286 (1986)
14. McCabe-Dansted, J., Pritchard, G., Slinko, A.: Approximability of Dodgson's rule. In: Endriss, U., Lang, J. (eds.) First International Workshop on Computational Social Choice (COMSOC 2006), pp. 331–344 (workshop notes). Universiteit van Amsterdam (December 2006)
15. Procaccia, A., Rosenschein, J.: Junta distributions and the average-case complexity of manipulating elections. Journal of Artificial Intelligence Research 28, 157–181 (2007)
16. Rothe, J., Spakowski, H., Vogel, J.: Exact complexity of the winner problem for Young elections. Theory of Computing Systems 36(4), 375–386 (2003)
17. Trevisan, L.: Lecture notes on computational complexity (Lecture 12) (2002), www.cs.berkeley.edu/~luca/notes/complexitynotes02.pdf
18. Wang, J.: Average-case computational complexity theory. In: Hemaspaandra, L., Selman, A. (eds.) Complexity Theory Retrospective II, pp. 295–328. Springer, Heidelberg (1997)

Efficient Parameterized Preprocessing for Cluster Editing

Michael Fellows[1,2,*], Michael Langston[3,**], Frances Rosamond[1,***], and Peter Shaw[1]

[1] University of Newcastle, Callaghan NSW 2308, Australia
{michael.fellows,frances.rosamond,peter.shaw}@newcastle.edu.au
[2] Durham University, Institute of Advanced Study,
Durham DH1 3RL, United Kingdom
[3] University of Tennessee, Knoxville, Tennessee 37996-3450, U.S.A.
langston@cs.utk.edu

Abstract. In the CLUSTER EDITING problem, a graph is to be changed to a disjoint union of cliques by at most k operations of *edge insertion* or *edge deletion*. Improving on the best previously known quadratic-size polynomial-time kernelization, we describe how a crown-type structural reduction rule can be used to obtain a $6k$ kernelization bound.

1 Introduction

The CLUSTER EDITING problem takes as input an undirected graph G, and asks whether k *edge changes* are sufficient to transform G into a graph G' that is a disjoint union of complete subgraphs. Such a graph G' is called a *cluster graph*. The problem was first introduced by Bansal, Blum and Chawla [2] (where it is called CORRELATION CLUSTERING) in the context of machine learning, and by Shamir, Sharan and Tsur [25] in the context of bioinformatics applications such as the analysis of gene expression data. Chen, Jiang and Lin [9] and Damaschke [12] have described applications in phylogenetics. An implementation with target applications in gene regulatory network analysis has been described in [15].

* Research supported by the Australian Research Council through the Australian Centre in Bioinformatics, by the University of Newcastle Parameterized Complexity Research Unit under the auspices of the Deputy Vice-Chancellor for Research, by a Fellowship to the Durham University Institute for Advanced Studies and by a William Best Fellowship at Grey College, Durham, while the paper was in preparation.
** Research supported in part by the U.S. National Institutes of Health under grants 1-P01-DA-015027-01, 5-U01-AA-013512 and 1-R01-MH-074460-01, by the U.S. Department of Energy under the EPSCoR Laboratory Partnership Program, by the European Commission under the Sixth Framework Programme, and by the Australian Research Council under grants to Griffith University and the University of Newcastle.
*** Research supported by the Australian Centre in Bioinformatics.

E. Csuhaj-Varjú and Z. Ésik (Eds.): FCT 2007, LNCS 4639, pp. 312–321, 2007.
© Springer-Verlag Berlin Heidelberg 2007

In the latter application, the vertices represent genes, and edges join co-regulated genes belonging to functional groups represented by the complete sub-graphs. The observed graph G might not be a cluster graph, due to experimental errors, noisy data and other reasons. A reasonable approach to formulating a par-simonious hypothesis concerning a hidden clustering is to determine a minimum number of edge changes that can transform the observed graph into a cluster graph.

1.1   Previous Work

CLUSTER EDITING is NP-hard [20] and does not admit a PTAS unless $P = NP$ [8]. A polynomial-time 4-approximation algorithm for CLUSTER EDITING is de-scribed in [8]. The problem is easily seen to be in FPT by a search tree algorithm that runs in time $O^*(3^k)$, based on the observation that the problem is equivalent to destroying (by means of the allowed operations) all occurences of an induced P_3 (a vertex-induced path consisting of three vertices). (It can also be classified as FPT using general results of Cai [5].) This was improved by a more sophis-ticated search tree strategy by Gramm, Guo, Hüffner and Niedermeier [18] to $O^*(2.27^k)$, and then further improved to $O^*(1.92^k)$ based on automated search tree generation and analysis [17]. In realistic applications, the enumeration of all possible solutions, for a given G and k, may be important, and Damaschke has described practical FPT algorithms for this [10]. Damaschke has also shown that a number of nontrivial and applications-relevant generalizations of CLUS-TER EDITING are fixed-parameter tractable [11]. (These generalizations study situations where the clusters may have limited overlap, rather than be com-pletely disjoint, a matter of importance in many data-clustering applications.) For general background on parameterized complexity, see [13, 16, 21].

The best known FPT kernelization for CLUSTER EDITING, to a graph on $O(k^2)$ vertices having $O(k^3)$ edges, is shown in [18] (exposited in [21]) and has been further improved by a constant factor by Damaschke [10]. Subsequent to, and extending the work reported here, a polynomial-time kernelization to a graph on at most $4k$ vertices has been announced [19].

1.2   Our Results

We describe a many:1 polynomial time kernelization to a problem kernel graph on $O(k)$ vertices, based on a crown-type reduction rule. Crown-type reduction rules have proved to be a surprisingly powerful method in FPT kernelization, applicable to a wide variety of problems [7, 14, 24, 23]. In particular, we obtain a kernelization to a graph on at most $6k$ vertices. Our result is roughly analogous to the $2k$ kernelization for the VERTEX COVER problem due to Nemhauser and Trotter [22]. In the case of VERTEX COVER, linear kernelization can be achieved more than one way. In particular, a $2k$ kernelization for the VERTEX COVER problem can be achieved by a crown reduction rule (see [21] for an exposition), that resembles the reduction rule for CLUSTER EDITING that we employ here. The main idea of a crown-type reduction rule is to identify cutsets that separate off a subgraph with homogeneous structure, allowing the input to be simplified.

Reduction rules often cascade and can have great power in practical settings. Although parameterization allows the efficiency of reduction rules to be measured, it is not necessary for the parameter to be small in order for this output of the study of parameterized algorithmics and complexity to be useful, because the reduction rules can be applied in polynomial time, and hence are of use even when the parameter is not guaranteed to be small.

2   A Crown Reduction Rule

Suppose that (G, k) is a yes-instance of the CLUSTER EDITING problem. Figure 1 shows a depiction of a solution, where the cliques that result from the editing are represented by the boxes C_i, $i = 1, ..., m$. The depiction shows only the *edits* of the solution. Define a vertex to be *of type A* if it is involved in an edge addition. Define a vertex to be *of type B* if it is not of type A, and is involved in an edge deletion. Define a vertex to be *of type C* if it is not of type A and not of type B.

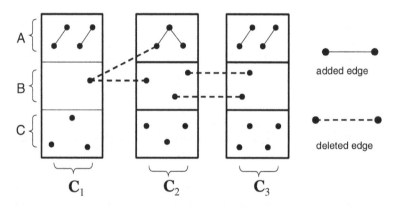

Fig. 1. A depiction of a solution showing (only) the edits

The following lemma is trivial.

Lemma 1. *There are at most $2k$ vertices of type A or B in a solution S.*

It follows from the lemma that if G is "large", then almost all of the vertices of the graph are of type C. Consider two vertices u and v of type C that belong to the clique C_i. Then u and v are adjacent, and adjacent to every vertex of type A or type B in C_i, and are not adjacent to any vertex of a clique C_j for $j \neq i$. In other words, we know everything there is to know about u and v, merely because they are of type C. We might expect that if the number of vertices of type C, for a given clique C_i is sufficiently large, then some reduction rule might apply. We identify such a reduction rule, based on the following notion of a structural decomposition applicable to this problem.

Definition 1. *A cluster crown decomposition of a graph $G = (V, E)$ is a partition of the vertices of V into four sets (C, H, N, X) satisfying the following conditions:*

1. *C is a clique.*
2. *Every vertex of C is adjacent to every vertex of H.*
3. *H is a cutset, in the sense that there are no edges between C and $N \cup X$.*
4. *$N = \{v \in V - C - H : \exists u \in H, uv \in E\}$.*

Our kernelization algorithm is based on the following reduction rule.

The Cluster Crown Reduction Rule. If (G, k) is an instance of the CLUSTER EDITING problem, and G admits a cluster crown decomposition (C, H, N, X) where

$$|C| \geq |H| + |N| - 1$$

then replace (G, k) with (G', k') where $G' = G - C - H$ and $k' = k - e - f$, where e is the number of edges that need to be added between vertices of H in order to make $C \cup H$ into a clique, and f is the number of edges between H and N.

3   A Linear Kernelization Bound

We defer the discussion of soundness for the Cluster Crown Reduction Rule, as well as a proof that it can be exhaustively applied in polynomial time, to §4 and §5.

We next argue that this rule gives us a linear kernelization for the problem.

Theorem 1. *Suppose that (G, k) is a yes-instance that does not admit a cluster crown decomposition to which the Cluster Crown Reduction Rule applies. Then G has less than $6k$ vertices.*

Proof. In order to discuss the situation, we introduce some notation. Suppose $\mathcal{S}$ is a solution for the instance (G, k). Let $\mathcal{C}_i$ denote the cliques formed by $\mathcal{S}$, $i = 1, ..., m$. Corresponding to each $\mathcal{C}_i$ is a cluster crown decomposition (C_i, H_i, N_i, X_i) of G. Let $c_i = |C_i|$, $h_i = |H_i|$ and $n_i = |N_i|$.

By the assumption that the Cluster Crown Reduction Rule does not apply, we have that for $\forall i$:

$$c_i \leq h_i + n_i - 2$$

It follows that

$$\sum_{i=1}^{m} c_i \leq \sum_{i=1}^{m} h_i + \sum_{i=1}^{m} n_i - 2m$$

Lemma 1 shows that $\sum_{i=1}^{m} h_i \leq 2k$, and we also have the bound $\sum_{i=1}^{m} n_i \leq 2k$ because the solution $\mathcal{S}$ involves deleting at most k edges between the sets H_i. These edges are the only source of neighbors in the sets N_i, and each such edge is counted twice in the sum $\sum_{i=1}^{m} n_i$. Therefore $\sum_{i=1}^{m} c_i \leq 4k - 4$, and G therefore has at most $6k - 4$ vertices, noting that $|V| = \sum_{i}(c_i + h_i)$.   □

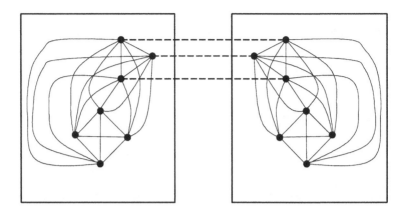

Fig. 2. Irreducible "yes" instance for $k = 3$ with $6k - 4$ vertices

Figure 2 shows that the bound of Theorem 1 cannot be improved: shown is a construction (for $k = 3$, this easily generalizes) of an irreducible yes-instance having $6k - 4$ vertices.

4   Soundness of the Reduction Rule

In this section we prove the soundness of the Cluster Crown Reduction Rule.

Lemma 2. *The Cluster Crown Reduction Rule is sound. That is, suppose we have a cluster crown decomposition (C, H, N, X) where $|C| \geq |H| + |N| - 1$ for an instance (G, k), and that (G', k') is the reduced instance. Then (G, k) is a YES-instance for* CLUSTER EDITING *if and only if (G', k') is a YES-instance.*

Proof. Any solution to the problem may be viewed as a partition of the vertex set of G, with associated costs: (1) the number of edges that must be *added* within classes of the partition, (2) the number of edges that must be *deleted* between classes of the partition, where the total cost is minimized. If π is a partition of V, then we will use $\Gamma(\pi)$ to denote the total editing cost of making the vertex classes of π into the disjoint cliques of a solution.

Consider a partition π of the vertex set V that minimizes the total editing cost, where π partitions V into m classes V_i, $i = 1, ..., m$. In this situation, we use C_i to denote $C \cap V_i$, H_i to denote $H \cap V_i$, N_i to denote $N \cap V_i$ and X_i to denote $X \cap V_i$. Let $c_i = |C_i|$, $h_i = |H_i|$, $n_i = |N_i|$ and $x_i = |X_i|$. If U and W are two disjoint sets of vertices, we write $e(U, W)$ to denote the number of adjacent pairs of vertices, $u \in U$ and $w \in W$, and we will write $\bar{e}(U, W)$ to denote the number of nonadjacent pairs of vertices, $u \in U$ and $w \in W$.

We argue that the partition π' that subtracts all vertices of $C \cup H$ from the classes of π and makes a new class consisting of $C \cup H$, has total editing cost no worse than that of π. That is, we will argue that $\Gamma(\pi) - \Gamma(\pi') \geq 0$. It is sufficient to show that

$$\sum_{i<j} c_i c_j + \sum_{i<j} (c_i h_j + c_j h_i) + \sum_i c_i n_i + \sum_i h_i x_i \tag{1}$$

$$+ \sum_{i<j} e(H_i, H_j) - \sum_{i<j} \bar{e}(H_i, H_j) \tag{2}$$

$$+ \sum_i \bar{e}(H_i, N_i) - \sum_i e(H_i, N_i) \tag{3}$$

$$\geq 0 \tag{4}$$

The terms in line (2) are greater than or equal to $-\sum_{i<j} h_i h_j$, and the terms in line (3) are greater than or equal to $-\sum_i h_i n_i$. Therefore it is enough to show that the following inequality holds.

$$\sum_{i<j} c_i c_j + \sum_{i<j} (c_i h_j + c_j h_i) + \sum_i c_i n_i - \sum_{i<j} h_i h_j - \sum_i h_i n_i \geq 0 \tag{5}$$

Let $c = |C| = \sum_i c_i$, $h = |H| = \sum_i h_i$ and $n = |N| = \sum_i n_i$. We can think of the inequality (5) as simply a statement to be proved about 3 by m matrices of non-negative integers, where the first row is $c_1...c_m$, the second row is $h_1...h_m$ and the third row is $n_1...n_m$. The statement is to hold for any such matrix, so long as $c \geq h + n - 1$.

Observation. An inspection of (5) shows that if there is any counterexample, then there is a counterexample where $c = h + n - 1$. Call such a matrix M *balanced* if for all i, $c_i \geq h_i$. It is straightforward to verify that (5) holds for all balanced matrices; the first positive sum dominates the first negative sum, and the third positive sum dominates the second negative sum.

The inequality (5) also holds if $m = 1$, since the hypothesis that $c \geq h + n - 1$ implies that $cn \geq hn + n^2 - n$ so that $cn \geq hn$ (which is what we must show in this simple case). We have now established the necessary base cases for an induction. Suppose the matrix M is not balanced and that M has l columns. Then M has a column j where $c_j < h_j$. The truth of the inequality (5) is unaffected by permutations of the columns, so we can assume that $j = l$. Write $\alpha(M)$ to denote the value of the lefthand side of (5) for a matrix M. Let M' be M with column l deleted. Certainly M' satisfies $c' \geq h' + n' - 1$, where c', h' and n' denote the row sums for M'. However, while we assume that $c = h + n - 1$ for M, the analogous equality does not hold for M', but this is not a problem, because, by the observation above, and our induction on l, $\alpha(M') \geq 0$ for *any* M' with $l - 1$ columns. It suffices to argue that $\Delta(M) = \alpha(M) - \alpha(M') \geq 0$. Elaborating, we must show that:

$$\Delta(M) = \sum_{i=1}^{l-1} c_i c_l + \sum_{i=1}^{l-1} (h_i c_l + c_i h_l) - \sum_{i=1}^{l-1} h_i h_l + c_l n_l - h_l n_l \geq 0$$

Factoring some of these terms, what we must show is:

$$c_l \left(\sum_i c_i \right) + c_l \left(\sum_i h_i \right) + h_l \left(\sum_i c_i \right) - h_l \left(\sum_i h_i \right) + c_l n_l - h_l n_l \geq 0$$

The last inequality holds if

$$c_l h' + h_l c' - h_l h' + c_l n_l - h_l n_l \ \geq \ 0$$

and this holds if

$$c_l h' + h_l c - c_l h_l - h_l h' + c_l n_l - h_l n_l \ \geq \ 0$$

Replacing c with $h + n - 1$, it is enough to show

$$c_l h' + h_l \left(\sum_{i=1}^{l} (h_i + n_i) - 1 \right) - c_l h_l - h_l h' + c_l n_l - h_l n_l \ \geq \ 0$$

which can be rewritten as

$$c_l h' + (h_l h' + h_l n' + h_l^2 + h_l n_l - h_l) - c_l h_l - h_l h' + c_l n_l - h_l n_l \ \geq \ 0$$

Cancelling and gathering terms, our task is to show

$$c_l h' + h_l n' + h_l (h_l - c_l - 1) + c_l n_l \ \geq \ 0$$

which is true, because $c_l < h_l$.   □

5   Efficiently Applying the Reduction Rule

In this section we describe how to compute in polynomial time whether the input graph G admits a cluster crown decomposition to which the Cluster Crown Reduction Rule can be applied.

Definition 2. *Distinct vertices u, v of a graph G are termed twins if:*
(1) u and v are adjacent, and
(2) $N[u] = N[v]$.
For vertices u and v that are twins, we will denote this by $u \sim v$.

Observe that if (C, H, N, X) is a cluster crown decomposition for a graph G, then every pair of vertices in C are twins. The next lemma shows that in some sense we can restrict our attention to cluster crown decompositions that have a kind of "maximality" property.

Lemma 3. *If G admits any cluster crown decomposition (C, H, N, X) satisfying:*
(1) $|C| \geq |H| + |N| - 1$
then G admits a cluster crown decomposition (C', H', N', X') that satisfies (1) and also the further condition:
(2) $\forall u \in C' \ \forall v \in H : \neg(u \sim v)$.

Proof. Suppose (C, H, N, X) is a cluster crown decomposition satisfying (1), and that there are vertices $u \in C$ and $v \in H$ with $u \sim v$. This implies that v has no neighbors in $N \cup X$. If we take $C' = C \cup \{v\}$ and $H' = H - \{v\}$ then we also have a cluster crown decomposition satisfying (1).   □

A cluster crown decomposition that satisfies the two conditions of the lemma above is termed *maximal*.

Definition 3. *The twin graph $\tau(G)$ of a graph $G = (V, E)$ has the same vertex set V, and two vertices u, v are adjacent in $\tau(G)$ if and only if $u \sim v$.*

We say that a subgraph H of a graph G is an *isolated clique* if H is a complete subgraph, and the vertices of H are adjacent in G only to vertices of H. (In other words, G consists of a disjoint union of H and the subgraph $G - H$.)

Lemma 4. *(1) If (C, H, N, X) is a maximal cluster crown decomposition of a graph $G = (V, E)$, then the vertices of C form an isolated clique in $\tau(G)$.*
(2) Conversely, an isolated clique in $\tau(G)$ corresponds to a maximal cluster crown decomposition where the vertices of the set C are the vertices of the isolated clique.

Proof. (1) follows from the observation that every pair of vertices in C are twins, and the definition of maximality. To see that (2) holds, let C denote the set of vertices of an isolated clique in $\tau(G)$. Take

$$H = (\cup_{v \in C} N(v)) - C$$
$$N = N(H) - C$$
$$X = V - C - H - N$$

It is easy to check that the conditions for a cluster crown decomposition are satisfied.   □

Our algorithm is described as follows.

Kernelization Algorithm.
On input (G, k):
Step 1. Compute the twin graph $\tau(G)$.
Step 2. Identify the isolated cliques in $\tau(G)$, and the corresponding maximal cluster crown decompositions. Apply the Reduction Rule if an opportunity is found.
      The algorithm can clearly be implemented in polynomial time.

6   Discussion and Open Problems

We have shown that the CLUSTER EDITING problem admits a polynomial time many:1 kernelization to a graph on at most $6k$ vertices. Based in part on the ideas introduced here, this kernelization bound has recently been improved to $4k$ [19].
      Linear kernelization for the CLUSTER EDITING problem could be viewed as an analog result, for a problem somewhat related to the VERTEX COVER problem, of the first linear kernelization result for that problem due to Nemhauser and Trotter [22]. Later (recently) a different combinatorial route to linear kernelization

for VERTEX COVER was discovered [7, 1], based on so-called "crown-type" reduction rules. This has proved useful in practical applications [3, 4]. Both approaches yield a $2k$ kernelization for VERTEX COVER. The importance of kernelization is that pre-processing is a nearly universal practical strategy for coping with hard problems. Since kernelization rules can be applied in polynomial time, the overall situation is that parameterization allows us to measure their *efficiency*, but their *relevance* and practical significance is not tied to situations where the parameter is small: the main outcome, as seen from the practical side, is just a "smart" preprocessing subroutine that can be deployed in *any* algorithmic approach to the NP-hard CLUSTER EDITING problem, including heuristic approaches.

The VERTEX COVER problem, the CLUSTER EDITING problem, and (for example) the generalizations studied by Damaschke [11], are all, roughly speaking, concerned with editing a graph to one or more (or a specified number) of clusters (possibly with limited overlap). We may roughly conceptualize here a class of "graph editing" problems, parameterized by the number of edit operations. Suitably formalized: are all such problems fixed-parameter tractable? Do they all admit linear kernels? Are general results possible concerning this area of investigation?

References

1. Abu-Khzam, F., Collins, R., Fellows, M., Langston, M., Suters, W.H., Symons, C.: Kernelization algorithms for the vertex cover problem: theory and experiments. In: Proceedings of ALENEX'04, pp. 62–69. ACM-SIAM Publications (2004)
2. Bansal, N., Blum, A., Chawla, S.: Correlation clustering. Machine Learning 56, 89–113 (2004) (Preliminary version In: Proceedings 43rd IEEE FOCS 2002, pp. 238–247)
3. Baldwin, N.E., Chesler, E.J., Kirov, S., Langston, M.A., Snoddy, J.R., Williams, R.W., Zhang, B.: Computational, integrative and comparative methods for the elucidation of gene regulatory networks. J. Biomedicine and Biotechnology 2, 172–180 (2005)
4. Baldwin, N.E., Collins, R.L., Langston, M.A., Leuze, M.R., Symons, C.T., Voy, B.H.: High performance computational tools for motif discovery. In: Proceedings of the IEEE International Workshop on High Performance Computational Biology (HiCOMB). IEEE Computer Society Press, Los Alamitos (2004)
5. Cai, L.: Fixed-parameter tractability of graph modification problems for hereditary properties. Information Processing Letters 58, 171–176 (1996)
6. Chlebík, M., Chlebíková, J.: Improvement of Nemhauser-Trotter theorem and its applications in parameterized complexity. In: Hagerup, T., Katajainen, J. (eds.) SWAT 2004. LNCS, vol. 3111, pp. 174–186. Springer, Heidelberg (2004)
7. Chor, B., Fellows, M., Juedes, D.: Linear kernels in linear time. In: Hromkovič, J., Nagl, M., Westfechtel, B. (eds.) WG 2004. LNCS, vol. 3353, pp. 36–53. Springer, Heidelberg (2004)
8. Charikar, M., Guruswami, V., Wirth, A.: Clustering with qualitative information. Journal of Computer and System Sciences 71, 360–383 (2005)
9. Chen, Z.Z., Jiang, T., Lin, G.: Computing phylogenetic roots with bounded degrees and errors. SIAM J. Computing 32, 864–879 (2003)

10. Damaschke, P.: On the fixed-parameter enumerability of cluster editing. In: Kratsch, D. (ed.) WG 2005. LNCS, vol. 3787, pp. 283–294. Springer, Heidelberg (2005)
11. Damaschke, P.: Fixed-parameter tractable generalizations of cluster editing. In: Calamoneri, T., Finocchi, I., Italiano, G.F. (eds.) CIAC 2006. LNCS, vol. 3998, pp. 321–332. Springer, Heidelberg (2006)
12. Damaschke, P.: Parameterized enumeration, transversals, and imperfect phylogeny reconstruction. Theoretical Computer Science 351, 337–350 (2006)
13. Downey, R.G., Fellows, M.R.: Parameterized Complexity. Springer, Heidelberg (1999)
14. Dehne, F., Fellows, M., Rosamond, F., Shaw, P.: Greedy localization, iterative compression and modeled crown reductions: new FPT techniques, an improved algorithm for set splitting and a novel $2k$ kernelization for vertex cover. In: Downey, R.G., Fellows, M.R., Dehne, F. (eds.) IWPEC 2004. LNCS, vol. 3162, pp. 271–280. Springer, Heidelberg (2004)
15. Dehne, F., Langston, M.A., Luo, X., Pitre, S., Shaw, P., Zhang, Y.: The Cluster Editing problem: implementations and experiments. In: Bodlaender, H.L., Langston, M.A. (eds.) IWPEC 2006. LNCS, vol. 4169, pp. 13–24. Springer, Heidelberg (2006)
16. Flum, J., Grohe, M.: Parameterized Complexity Theory. Springer, Heidelberg (2006)
17. Gramm, J., Guo, J., Hüffner, F., Niedermeier, R.: Automated generation of search tree algorithms for hard graph modification problems. Algorithmica 39, 321–347 (2004)
18. Gramm, J., Guo, J., Hüffner, F., Niedermeier, R.: Graph-modeled data clustering: exact algorithms for clique generation. Theory of Computing Systems 38, 373–392 (2005) (preliminary version In: Proceedings of the 5th Italian Conference on Algorithms and Complexity (CIAC '03). Lecture Notes in Computer Science 2653, pp. 108–119. Springer-Verlag (2003))
19. Guo, J.: Manuscript (2007)
20. Krivanek, M., Moravek, J.: NP-hard problems in hierarchical tree clustering. Acta Informatica 23, 311–323 (1986)
21. Niedermeier, R.: Invitation to Fixed Parameter Algorithms. Oxford University Press, Oxford (2006)
22. Nemhauser, G., Trotter, L.: Vertex packings: structural properties and algorithms. Mathematical Programming 8, 232–248 (1975)
23. Prieto-Rodriguez, E.: Systematic kernelization in FPT algorithm design. Ph.D. Thesis, School of EE&CS, University of Newcastle, Australia (2005)
24. Prieto, E., Sloper, C.: Looking at the stars. In: Downey, R.G., Fellows, M.R., Dehne, F. (eds.) IWPEC 2004. LNCS, vol. 3162, pp. 138–148. Springer, Heidelberg (2004)
25. Shamir, R., Sharan, R., Tsur, D.: Cluster graph modification problems. Discrete Applied Mathematics 144, 173–182 (2004) (Preliminary version In: 28th WG 2002, LNCS 2573, pp. 379–390 (2002))

Representing the Boolean OR Function by Quadratic Polynomials Modulo 6

Gyula Győr

Eötvös Loránd University, Budapest
Pázmány P. sétány 1/C, H-1117, Budapest, Hungary
angelofd.gy@gmail.com, angelofd@inf.elte.hu

Abstract. We give an answer to a question of Barrington, Beigel and Rudich, asked in 1992, concerning the largest n such that the OR function in n variables can be weakly represented by a quadratic polynomial modulo 6. More specifically, we show that no 11-variable quadratic polynomial exists that is congruent to zero modulo 6 if all arguments are 0, and non-zero modulo 6 on the set $\{0,1\}$, otherwise.

1 Introduction

In this paper we answer an open question of Barrington, Beigel and Rudich [1] asked in 1992. The polynomial P weakly represents the N-variable OR function modulo m if $P(0,0,\cdots,0) \equiv 0 \pmod{m}$ and for all $x \in \{0,1\}^N$, where $x \neq (0,0,\ldots,0)$ it holds that $P(x) \not\equiv 0 \pmod{m}$. It is known that if the polynomial P weakly represents the N-variable Boolean OR function modulo p, where p is a prime, then its degree is at least $\left\lceil \frac{N}{p-1} \right\rceil$ [4]. On the other hand Barrington, Beigel and Rudich proved that there exists a degree $O(\sqrt[r]{N})$ polynomial that weakly represents the N-variable OR function modulo m, where r is the number of distinct prime divisors of m. So polynomials modulo 6 are more powerful than polynomials modulo a prime power. This construction uses only symmetric polynomials where the bounds are matching. Barrington, Beigel and Rudich asked what is the largest N such that the N-variable Boolean OR function can be weakly represented by a quadratic polynomial modulo 6? It is known [1] that for symmetric polynomials the answer is 8, but it is not hard to construct polynomials showing that $N \geq 10$. As a consequence of their theorem, Tardos and Barrington[5] showed that $N \leq 18$.

Supposing that P is a multilinear polynomial containing no constant monomials, deciding this question by a brute force algorithm is hopeless, because we need to check $6^{\binom{11}{2}+11} = 6^{66}$ polynomials. To check one polynomial we need $2^{11} = 2048$ operations, so it will take $1.48 \cdot 10^{38}$ years to complete the task at the speed of 10^9 operations/sec.

Our algorithm does exhaustive search in approximately 800 2.2GHz-CPU hours.

Ramsey's theorem shows that each graph with 2^n vertices has either a clique or an independent set of $\frac{n}{2}$ vertices. Erdős showed with the probabilistic method

E. Csuhaj-Varjú and Z. Ésik (Eds.): FCT 2007, LNCS 4639, pp. 322–327, 2007.

[2] that there exists a graph with 2^n vertices with no clique and independent set of size $2 \cdot n$. Constructive bounds to date are far from the upper bound. The connection between these problems was shown by Grolmusz [3]. He used polynomial representations of the OR function to construct Ramsey graphs and showed that lower degree representation leads to better Ramsey graphs. Because the best lower bound for these type of representations is only $\Omega(\log n)$ it may be possible to achieve better graph constructions this way.

2   The Testing Algorithm

Definition 1 ([1]). *We say that the n-variable polynomial P over the ring of integers modulo m* weakly represents *$f : \{0,1\}^n \to \{0,1\}$ if and only if*

$$\forall x, y \in \{0,1\}^n : (f(x) \neq f(y)) \text{ implies } (P(x) \neq P(y)).$$

Let $P^{*(n,m)}$ denote the set of n-variable multilinear polynomials over the ring of integers modulo m that do not contain constant monomials. For any polynomial that weakly represents the n-variable OR, there exists a multilinear polynomial that weakly represents the n-variable OR and does not contain constant monomials. The proof is immediate from the fact that $\left|OR^{(-1)}(0)\right| = 1$ and $x_i = x_i^2$ on the set $\{0,1\}$.

Definition 2. *Let $H^{*(n,m)} \subseteq P^{*(n,m)}$ be defined as*

$$H^{*(n,m)} := \left\{ h \in P^{*(n,m)} : h \text{ weakly represents the n-variable OR function.} \right\} \tag{1}$$

Let us call the elements of $H^{*(n,m)}$ well representing.

Definition 3 ([5]). *We call a Boolean function g a* strict restriction *of the Boolean function f if g can be obtained from f by setting some variables of f to 0 and identifying some variables. The number of variables of g is therefore the number of equivalence classes of the nonzero variables of f. We call polynomial Q a* strict restriction *of polynomial P if we can obtain Q from P via this kind of restriction.*

Definition 4. *Let $p \in P^{*(n,m)}$. Let x be the first variable of p for some fixed total ordering of the variables and z the vector of all other variables. p can be written in the following unique form: $p = (a+L(z))x+Q(z)$ where a is a number, L is linear, Q is quadratic, and L and Q do not contain variable x and constant monomials. Let*

$$C(p) := a; L(p) := L(z) \quad \text{and} \quad Q(p) := Q(z).$$

The following function maps three well representing n-variable polynomials to an $(n + 1)$-variable polynomial.

Definition 5. Let $\Phi : \left(H^{*(n,m)} \right)^3 \to P^{*(n+1,m)}$. We define

$$\Phi(h_1, h_2, h_3) := (C(h_1) + L(h_1))x + (C(h_2) + L(h_2))y$$
$$+ (C(h_3) - C(h_1) - C(h_2))xy + Q(h_1)$$

where x and y are new variables.

The following main lemma shows a method to build all $(n + 1)$-variable well representing polynomials from the n-variable ones. The main idea is to split the first variable.

Lemma 1

$$H^{*(n+1,m)} = \{ \Phi(h_1, h_2, h_3) : \exists h_1, h_2, h_3 \in H^{*(n,m)} : L(h_1) + L(h_2) \equiv$$
$$\equiv L(h_3) \pmod{m} \quad \text{and} \quad Q(h_1) = Q(h_2) = Q(h_3) \}.$$

Proof. Let H' denote the set on the RHS of the equation.

Fact 1. $H^{*(n+1,m)} \subseteq H'$.
Let $h(x, y, z) \in H^{*(n+1,m)}$ where x is the first, y is the second variable and z is the array of the other variables. We will show that $h \in H'$. Let

$$h_1 := h(x, 0, z), \ h_2 := h(0, x, z), \ h_3 := h(x, x, z)$$

Because any strict restriction of OR is also an OR function, $h_1, h_2, h_3 \in H^{*(n,m)}$, and these satisfy $L(h_1) + L(h_2) \equiv L(h_3) \pmod{m}$ and $Q(h_1) = Q(h_2) = Q(h_3)$.

Fact 2. $H' \subseteq H^{*(n+1,m)}$.
Let $h = \Phi(h_1, h_2, h_3) \in H'$. It is enough to show that h is well representing. If $x = 0$ then h behaves exactly like h_2. If $y = 0$ then h behaves exactly like h_1. If $x = y$ then h behaves exactly like h_3. Because h_1, h_2 and h_3 are well representing, h is also well representing. $\qquad\square$

Remark 1. It is easy to show that if $H_1, H_2 \in H^{*(n,m)}$ for some n and $Q(H_1) \neq Q(H_2)$, then for every H_1^* that is a descendant of H_1 and for every H_2^* that is a descendant of $H_2 : Q(H_1^*) \neq Q(H_2^*)$. So function Q partitions $H^{*(n,m)}$ into classes, therefore we should search separately in these classes. We can thus divide the problem into disjoint subproblems and solve these subproblems with reduced computational resources.

Remark 2. $H^{*(1,m)} = \{ x, 2x, \cdots, (m-1)x \}$.

The basic algorithm builds all $(n+1)$-variable well representing polynomials from the n-varaible ones by induction, using Lemma 1, Remark 1 and Remark 2.

3 Permutation Filtering

We say that two polynomials are *equivalent* if and only if there exists a permutation of their variables that makes them equal. Clearly, this relation partitions the polynomials.

Lemma 2. *Let polynomials P_1 and P_2 be in the same class. Then P_1 represents the OR function if and only if P_2 does.*

Proof. Since OR is a symmetric function the proof is obvious. □

The previous algorithm does not filter such permutations. Our goal is to check one representative in each class. This is a hard task, but we can define the following property of our polynomials, that helps us to check only a small number of polynomials in each class.

Let $h \in P^{*(n,m)}$. h can be written in the following form:

$$h(x) = \sum_{i=1}^{n} \left(a_i x_i + \sum_{i>j} a_{i,j} x_i x_j \right).$$

We say that $\tau(h)$ is true exactly when the following conditions are satisfied:

- $a_2 \leq a_3 \leq \cdots \leq a_n$.
- If $a_i = a_j$ for some $1 < i < j$, then $(a_{n,i}, a_{n-1,i}, \cdots, a_{j+1,i})$ is lexicographically not greater than $(a_{n,j}, a_{n-1,j}, \cdots, a_{j+1,j})$.

The search is performed only for polynomials satisfying τ.

Remark 3. There are no conditions for the first variable.

Lemma 3. *For all polynomial classes there is at least one polynomial that satisfies τ.*

Proof. We will show that there exists at least one permutation of variables for any polynomial in $P^{*(n,m)}$ that satisfies τ.

Let $h \in P^{*(n,m)}$ be given in the above form. In this part we represent the ring of integers modulo m by the numbers $0, 1, \cdots, m-1$ and we use the natural ordering on these numbers. Let $\pi : \{1, 2, \cdots, n\} \to \{1, 2, \cdots, n\}$. We construct the inverse of the promised permutation. Let π_n satisfy the following: $\forall i \in \{1, 2, \cdots, n\} : a_i \leq a_{\pi_n}$ We assume that $\pi_{s+1}, \cdots, \pi_n$ are already defined. Now we construct π_s. Let relation R be defined on the set of non-chosen variables, where x_i is non-chosen if $\pi^{(-1)}(i) = \emptyset$.

Let $x_i R x_j$ (where x_i, x_j are non-chosen) if and only if

$$(a_i < a_j) \text{ or } (a_i = a_j \text{ and } \exists t \in \{s+1, s+2, \cdots, n\} \forall t' \in \{t+1, t+2, \cdots, n\} :$$
$$a_{\max\{i, \pi_{t'}\}, \min\{i, \pi_{t'}\}} = a_{\max\{j, \pi_{t'}\}, \min\{j, \pi_{t'}\}} \wedge$$
$$\wedge a_{\max\{i, \pi_t\}, \min\{i, \pi_t\}} < a_{\max\{j, \pi_t\}, \min\{j, \pi_t\}})$$

Relation R is antisymmetric and transitive, so it is a partial order on the set of non-chosen variables. Let π_s be the index of one of the maximal variables with respect to R.

Now π is a one-to-one mapping and $\tau \left(h(x_{\pi_1^{-1}}, x_{\pi_2^{-1}}, \cdots, x_{\pi_n^{-1}}) \right)$ is true. □

Definition 6. *We define*

$$H^{**(n,m)} := \left\{ h \in H^{*(n,m)} : \tau(h) \right\}.$$

The following lemma corresponds to Lemma 1, but now the sets contain only polynomials that satisfy τ.

Lemma 4

$$H^{**(n+1,m)} = \{ \Phi(h_1, h_2, h_3) : \tau(\Phi(h_1, h_2, h_3)) \wedge \exists h_1, h_2, h_3 \in H^{**(n,m)} :$$

$$L(h_1) + L(h_2) \equiv L(h_3) \pmod{m} \quad \text{and} \quad Q(h_1) = Q(h_2) = Q(h_3) \}.$$

Proof. The proof follows from Lemma 1, Remark 3, and the definition of H^{**}.   □

Remark 4. $\tau(\Phi(h_1, h_2, h_3))$ depends only on h_2.

Algorithm 1 *Final*

```
H:={[x],[2x],[3x],[4x],[5x]}
for level:=2 to 11 do
begin
  Hnew:=empty;
  forall H' Q-class of H do
    forall h2 in H' do
      if isGoodPermutation(h2) then
        forall h1,h3 in H' do
          if (L(h1)+L(h2)-L(h3) mod 6=0) then
            Hnew:=Hnew+[PHI(h1,h2,h3)];
  H:=Hnew;
end;
if H=empty then
  print('There is no quadratic polynomial '+
        'representing the 11 variables OR.')
else
  print('There is quadratic polynomial '+
        'representing the 11 variables OR.');
```

4   The Result

We have used the above algorithm with minor technical modifications and implemented it in C++. It uses STL data structures (sets) to speed up the search. It did not find any 11 variable polynomials, so the largest n such that the OR function of n variable can be weakly represented by a quadratic polynomials modulo 6, is 10.

The computation took approximately 800 2.2GHz-CPU hours on 6 computers.

Acknowledgements. The author wishes to thank Vince Grolmusz for his help in this work.

References

1. Mix Barrington, D.A., Beigel, R., Rudich, S.: Representing boolean functions as polynomials modulo composite numbers (extended abstract). In: STOC, pp. 455–461. ACM Press, New York (1992)
2. Erdős, P.: Some remarks on the theory of graphs. Bull. Am. Math. Soc. 53, 292–294 (1947)
3. Grolmusz, V.: Superpolynomial size set-systems with restricted intersections mod 6 and explicit ramsey graphs. Combinatorica 20(1), 71–86 (2000)
4. Smolensky, R.: Algebraic methods in the theory of lower bounds for boolean circuit complexity. In: STOC, pp. 77–82. ACM, New York (1987)
5. Tardos, G., Mix Barrington, D.A.: A lower bound on the mod 6 degree of the or function. In: ISTCS, pp. 52–56 (1995)

On the Complexity of Kings

Edith Hemaspaandra[1,*], Lane A. Hemaspaandra[2,**], Till Tantau[3],
and Osamu Watanabe[4,***]

[1] Dept. of Comput. Science, Rochester Institute of Technology, Rochester, NY, USA
[2] Dept. of Comput. Science, University of Rochester, Rochester, NY, USA
[3] Inst. of Theoretical Comput. Science, Universität zu Lübeck, Germany
[4] Dept. of Math. & Comput. Sciences, Tokyo Institute of Technology, Japan

Abstract. A k-king in a directed graph is a node from which each node in the graph can be reached via paths of length at most k. Recently, kings have proven useful in theoretical computer science, in particular in the study of the complexity of reachability problems and semifeasible sets. In this paper, we study the complexity of recognizing k-kings. For each succinctly specified family of tournaments (completely oriented digraphs), the k-king problem is easily seen to belong to Π_2^p. We prove that the complexity of kingship problems is a rich enough vocabulary to pinpoint every nontrivial many-one degree in Π_2^p. That is, we show that for every $k \geq 2$ every set in Π_2^p other than $\emptyset$ and Σ^* is equivalent to a k-king problem under $\leq_m^p$-reductions. The equivalence can be instantiated via a simple padding function. Our results can be used to show that the radius problem for arbitrary succinctly represented graphs is Σ_3^p-complete. In contrast, the diameter problem for arbitrary succinctly represented graphs (or even tournaments) is Π_2^p-complete.

1 Introduction

We study the complexity of recognizing k-kings in graphs. A vertex of a graph is said to be a *k-king* if every vertex of the graph can be reached from it via a path of length at most k. For the *k-kingship problem* we are given a graph and a vertex as inputs and would like to tell whether the vertex is a k-king. We can vary the problem by allowing different ways of encoding graphs (the more succinctly, the harder the problem) and by allowing different kinds of input graphs (the more restricted, the easier the problem).

Much is known about the *existence* of kings in graphs. For example, in the 1950s Landau [11] discovered the simple but lovely result that every tournament

* Supported in part by grant NSF-CCR-0311021. Work done in part while on sabbatical at the University of Rochester and while visiting the Tokyo Institute of Technology.
** Supported in part by grant NSF-CCF-0426761, JSPS Invitational Fellowship S-05022, a TransCoop grant, and a Friedrich Wilhelm Bessel Research Award. Work done in part while visiting the Tokyo Institute of Technology.
*** Supported in part by "New Horizons in Computing" (2004–2006), an MEXT Grant-in-Aid for Scientific Research on Priority Areas.

E. Csuhaj-Varjú and Z. Ésik (Eds.): FCT 2007, LNCS 4639, pp. 328–340, 2007.
© Springer-Verlag Berlin Heidelberg 2007

has a 2-king. A tournament is a directed graph G such that for each pair u and v of distinct vertices exactly one of the directed edges $u \to v$ or $v \to u$ is present in the graph and such that there are no loops. A well-known (see [21]) way to easily see that Landau's result holds is to note that every vertex with maximum degree must be a 2-king. More recently, similar results were proven for generalizations of tournaments, such as multipartite tournaments ([5,14], see also [1] and the references therein).

When graphs are specified explicitly in the natural way (say, via an adjacency matrix), it is not hard to see that the k-kingship problem is first-order definable and thus very simple from a computational point of view. However, when we specify graphs *succinctly*, k-kingship problems (provably) get harder. There are different ways of specifying graphs succinctly, ranging from the general Galperin–Wigderson model to the polynomial-time uniform tournament family specifiers that arise in the study of semifeasible sets.

In the Galperin–Wigderson model, input graphs are specified as follows: A directed graph G with a vertex set $\{0,1\}^n$ is specified using a circuit C with $2n$ input gates and one output gate. For any two vertices $x, y \in \{0,1\}^n$ there is an edge $x \to y$ in G if and only if $C(xy) = 1$. (This definition does allow the possibility of self-loops.) Note that a circuit whose size is polynomial in n can encode a graph whose vertex set has size 2^n, which is exponential in n. For this model, the k-kingship problem can be formalized as follows: SUCCINCT-k-KINGS $=$ $\{\langle \text{code}(C), x \rangle \mid C$ specifies (in the manner specified above) a graph G in which x is a k-king$\}$. Here, $\text{code}(C)$ denotes a standard binary encoding of the circuit C. Furthermore, $\langle \text{code}(C), x \rangle$ is a standard binary encoding of the circuit C paired with a bitstring x.

In the tournament family specifier model, input graphs are specified using *polynomial-time computable, commutative selector functions*. A *selector function* f gets two words u and v as inputs and outputs one of them, thereby telling us where the edge between u and v heads. More formally, a selector function $f : \{0,1\}^* \times \{0,1\}^* \to \{0,1\}^*$ defines an (infinite) graph with the vertex set $\{0,1\}^*$ where there is an edge from u to v if and only if $f(u, v) = v$ and $u \ne v$. The graph will be a tournament if f is commutative, that is, if for each u and v it holds that $f(u, v) = f(v, u)$. Polynomial-time selector functions were originally introduced by Selman [17, 18, 19] in his study of so-called P-selective sets: P-selective set can be defined as the sets of vertices in tournaments specified by polynomial-time computable, commutative selector functions that are closed under reachability.

Instead of using a selector function to describe a single infinite tournament, we can also use them to describe one tournament per word length: Given a commutative selector function f and a word length n, we say that f describes the length-n tournament whose vertex set is $\{0,1\}^n$ and whose edge set is defined as above: There is an edge from $u \in \{0,1\}^n$ to $v \in \{0,1\}^n$ if $f(u, v) = v$ and $u \ne v$. In this model we can also consider the k-kingship problem:

$$k\text{-Kings}_f = \{x \in \{0,1\}^* \mid x \text{ is a } k\text{-king in the length-}|x| \text{ tournament}$$
$$\text{specified by } f\}.$$

A language $L \subseteq \{0,1\}^*$ for which there exists a commutative polynomial-time selector function f such that $L = k\text{-Kings}_f$ will be called a P-k-*king language*. Our main interest in this paper is to study which languages are P-k-king languages.

One can view the k-Kings_f problems (one for each f) as very restricted cases of the more general SUCCINCT-k-KINGS problem: For the k-Kings_f we must check k-kingship for *a single tournament per word length*. In complexity terms, this tremendous uniformity of specification—a polynomial-time computable function specifying for us a single tournament at each length—will naturally tend to tie our hands in terms of showing hardness for higher levels of the polynomial hierarchy. Nonetheless, what we will actually show in this paper is that we can free our hands from those cords. Our main result is that, for each $k \geq 2$, every language in $\Pi_2^p - \{\emptyset, \Sigma^*\}$ is many-one equivalent to a P-k-king language. Informally put, this shows that k-king languages are comprehensively descriptive in terms of naming the complexity of the nontrivial Π_2^p many-one degrees.

We obtain this main result via an even stronger main tool, which shows something about the uniformity and simplicity of a set of reductions that can instantiate the above equivalences. Namely, we show that, for every $k \geq 2$, a language L is in Π_2^p if and only if $\text{pad}'_j(L)$ is a P-k-king language for some j. Here, pad'_j is a padding operator whose exact definition will be given later.

Motivations for studying P-k-king languages. Our study of P-k-king languages is motivated from several contrasting directions.

Relationship to the radius problem. Kings are closely related to the *radius problem* for graphs. A ball of radius r around a vertex v is the set of vertices that can be reached from v in r steps. The radius of a graph is smallest radius of a ball that covers the whole graph. This means that the radius of a graph is at most r if and only if there *exists* an r-king in the graph. (Note, in contrast, that the k-king problems focus on whether a given node, which is explicitly stated as part of the input, is a k-king.) We use our results on P-k-king languages to give a short proof that radius problems for succinctly specified graphs (using the Galperin–Wigderson model) are complete for the *third* level of the polynomial hierarchy [12, 20], i.e., are complete for $\Sigma_3^p = \text{NP}^{\text{NP}^{\text{NP}}}$. This result is interesting in its own right. While for the first level of the polynomial hierarchy (NP) countless natural complete problems are known, for higher levels the collection of such problems is less extensive (see also Section 1's comments on complete sets for such classes). The succinct radius problem is a new and fairly natural problem that is complete for Σ_3^p.

Relationship to the diameter problem. Kings are also closely related to the *diameter problem* for graphs. The diameter of a graph is the maximum over all ordered vertex pairs of the shortest distance (via a directed path) from the first vertex of the pair to the second vertex of the pair. (If the second node isn't reachable from the first, this distances is ∞.) This means that a graph has diameter at most d if and only if *every* vertex of the graph is a d-king of the graph. Based on this relationship we show that diameter problems for succinctly represented graphs are complete for the *second* level of the polynomial hierarchy, i.e., are complete for $\Pi_2^P = \text{coNP}^{\text{NP}}$.

Relationship to P-selective sets. P-2-king languages are closely related to P-selective languages. For a P-selective language A, for each n, within the length-n graph specified by the selector function it always holds that the reachability closure of the length n words of A is precisely the length n words of A. For a P-2-king language, the words in the language of length n are the 2-kings in the length-n tournament specified by the selector function. This means that for a P-selective set the 2-kings of the tournaments induced by a selector are (speaking very informally) the "least likely" words to be contained in the language. More precisely, unless all words of a given word length are in the language, none of the 2-kings of the tournament specified by the selector for this word length is in the language. This observation can be used to show that P-selective sets cannot be $\Pi_2^P/1$-immune [10].

Relationship to the second level of the polynomial hierarchy. Despite the close relationship of P-2-king languages and P-selective languages, there are fundamental differences. For example, it is easy to see that all P-k-king languages are in Π_2^P, see [9] for a detailed proof, but can be "arbitrarily complex" in a sense that can be crisply formalized. This encourages us to investigate which languages are P-k-king languages. Many languages in Π_2^P are not P-k-king languages—for example, since every tournament has a 2-king, a P-k-king language contains at least one word for every word length. However, the tool underpinning our main result shows that for every $k \geq 2$ and every language $L \in \Pi_2^P$ a certain padded version of L is a P-k-king language. Thus, although not every language in Π_2^P is a P-k-king language, for every such language a very closely related language is a P-k-king language. And from this we have our main result, which is that every Π_2^P (many-one) degree, except those of $\emptyset$ and Σ^*, contains a P-k-king language.

Relationship to quantifier characterizations. By the quantifier characterization of the polynomial hierarchy a language L is in Π_2^P if and only if there exist a polynomial p and a ternary polynomial-time decidable relation R such that $x \in L \iff (\forall y \in \{0,1\}^{p(|x|)})(\exists z \in \{0,1\}^{p(|x|)})[R(x,y,z)]$. For P-2-king languages a more restrictive characterization is possible: A language L is a P-2-king language if and only if there exists a binary polynomial-time decidable relation S such that $x \in L \iff (\forall y \in \{0,1\}^{|x|})(\exists z \in \{0,1\}^{|x|})[S(x,y) \wedge S(y,z)]$ and such that for all distinct $x, y \in \{0,1\}^*$ we have $S(x,y) \leftrightarrow \neg S(y,x)$.

Related work. The work most closely related to that of this paper is the work of Nickelsen and Tantau on the complexity of reachability problems [13], the path-breaking modeling and complexity work of Galperin and Wigderson [3], and the existing work on the complexity of kings and in particular their use in the study of the semifeasible sets [8, 10, 9].

It is well worth mentioning that without the work of Landau [11], which showed that 2-kings always exist in tournaments, it is unlikely that the notion of 2-kings would even be available for study. And Landau's work has led to a rich (though, naturally, not complexity-theoretic) body of work on the existence of k-kings in a variety graph-theoretic structures (for example, for the case of multipartite tournaments see [5, 14, 1] and the references therein).

For reasons of focus and coherence, all tournaments in this paper follow the typical notion of a tournament. However, one central result of this paper, our Π_2^p-completeness result for the k-kings problem, has been studied for the case of j-partite tournaments (though in a more circuit-focused model) in [6], where a dichotomy theorem is given that completely characterizes what happens in that case, namely, for the boundary case of "1-kingship" one gets P algorithms and for all other cases Π_2^p-completeness holds.

In this paper, we will show problems to be complete for classes at the second and third levels of the polynomial hierarchy. These levels have nothing resembling the range and number of known, natural complete problems that NP has (see, for example, the famous compendium of Garey and Johnson [4]). Nonetheless, these levels do have a larger range and number of known, natural complete problems than many people realize. Schaefer and Umans have provided a very nice "Garey and Johnson" for classes at levels of the polynomial hierarchy beyond the first [15, 16].

Organization of this paper. Section 2 provides notations and definitions. Section 3 studies the complexity of the diameter problem, and shows that it is complete for the Π_2^p level of the polynomial hierarchy. That section also introduces tools that will be used in subsequent proofs in the paper. Section 4 proves our main result, namely, that k-kings problems have the descriptive flexibility to name every nontrivial Π_2^p degree. We do so by showing that for each $k \geq 2$ it holds that via a certain family of padding functions each Π_2^p language can be turned into a P-k-king language. Section 5 studies the complexity of the radius problem, and shows that it is complete for the Σ_3^p level of the polynomial hierarchy, which provides an interesting contrast with Section 3's Π_2^p-completeness result for the diameter problem. Due to lack of space, this extended abstract contains no proofs and some technical definitions have been replaced by descriptions of the key ideas; we refer the interested reader to the technical report version of this paper [6] for the complete proofs and technical definitions.

2   Basic Definitions and Tools

Alphabets, padding, reductions. Throughout this paper $\Sigma = \{0, 1\}$. We refer to elements of $\{0, 1\}^* = \Sigma^*$ as bitstrings. The *length* of a bitstring b is denoted $|b|$. We define a pairing function $\langle ., . \rangle \colon \Sigma^* \times \Sigma^* \to \Sigma^*$ as follows (unlike some other papers, we will not require our pairing function to be a surjective function): For every two bitstrings $x, y \in \Sigma^*$ where the individual bits of x are x_1 to x_n, let $\langle x_1 x_2 \cdots x_n, y \rangle = 0 x_1 0 x_2 \cdots 0 x_n 1 y$. This function, which clearly is injective, has a number useful properties: It is polynomial-time computable, polynomial-time invertible, when $|x| = |x'|$ and $|y| = |y'|$ then $|\langle x, y \rangle| = |\langle x', y' \rangle|$, and no word pair is mapped to an element of $\{0\}^*$. For a tuple $(x_1, \ldots, x_n)$ of words, $n \geq 1$, let $\langle x_1, \ldots, x_n \rangle = \langle x_1, \langle x_2, \ldots, \langle x_{n-1}, x_n \rangle \cdots \rangle \rangle$.

For a positive integer j, we define a padding function $\mathrm{pad}_j \colon \Sigma^* \to \Sigma^*$ by $\mathrm{pad}_j(x) = x0^{|x|^j + j + 3}$. Thus, we add $|x|^j + j + 3$ zeros after x. The reason for the slightly startling "+ 3" will become clear later on. Note that for every word $y \in \Sigma^*$ there can be at most one word x such that $\mathrm{pad}_j(x) = y$ and, if such an x exists, it is easy to compute.

We define two padded versions of languages. The "usual" way to define a padded version of a language L is to consider the image of L under the padding function pad_j, that is, for a given language $L \subseteq \Sigma^*$ let $\mathrm{pad}_j(L) = \{\mathrm{pad}_j(x) \mid x \in L\}$. The "interesting" words in a padded language are those in $\mathrm{pad}_j(\Sigma^*)$. Words outside $\mathrm{pad}_j(\Sigma^*)$ are not in L. For the second padded version of L we change this latter property: The membership of the words in $\mathrm{pad}_j(\Sigma^*)$ is the same, but (almost) all other words are *in* the second padded version. Formally, for a language $L \subseteq \Sigma^*$ we define

$$\mathrm{pad}'_j(L) = \mathrm{pad}_j(L) \cup \left(\Sigma^* - \mathrm{pad}_j(\Sigma^*) - \{1, 11\} \right).$$

Once more, there is a startling part of the definition, namely the "$- \{1, 11\}$" and, once more, this will be explained later on. The padded versions of a language L has exactly the typical properties that we expect from a padded language.

Let $A \leq^P_m B$ denote that A is many-to-one polynomial-time reducible to B.

Graphs and tournaments. A *directed graph* is a pair (V, E) consisting of a nonempty vertex set V together with an edge set $E \subseteq V \times V$ and we write $u \to v$ if $(u, v) \in E$. A *path of length l* in a graph is sequence $(v_0, v_1, \ldots, v_l)$ of distinct vertices such that $v_{i-1} \to v_i$ holds for all $i \in \{1, \ldots, l\}$. For positive integers k, a *k-king* of a graph is a vertex v such that there is a path of length at most k from v to every other vertex. The *diameter* of a graph is the smallest number d such that for every pair $u, v \in V$ of vertices there is a path from u to v of length at most d. Note that the diameter of a graph is exactly the smallest number k such that every vertex of the graph is a k-king. If a graph has more than one strongly connected component, its diameter is ∞. The *radius* of a graph is the smallest number r such there exists a vertex v from which there are paths of length at most r to all other vertices. Note that the radius of a graph is exactly the smallest number k such that there exists a k-king in the graph. It is possible for a graph (though, as we will see, not a tournament) to have a radius of ∞.

A *tournament* is a directed graph such that (a) there are no self-loops, that is, the edge relation E is irreflexive and (b) for every pair $u, v \in V$ of distinct vertices we have either $(u, v) \in E$ or $(v, u) \in E$, but not both. It is well known that any vertex of maximal out-degree in a tournament is a 2-king of the tournament. In particular, every tournament has a 2-king. Except for the tournament consisting of a single vertex or of no vertices, a tournament obviously cannot have a diameter strictly less than 2. In the following, we will often need to construct tournaments that have diameter exactly 2. The following lemmas show when and how this can be done.

Lemma 2.1. *Let n be a positive integer. Then there exists an n-vertex tournament of diameter 2 if and only if $n \notin \{1, 2, 4\}$.*

Succinct representations of general graphs. By *circuit* we refer to combinatorial circuits containing input-, output-, negation-, and-, and or-gates. The fan-in of each gate is at most 2. Fan-out is not restricted. For a circuit C with n input gates and m output gates, we in a slight overloading of notation also use C to denote the function computed by the circuit C. This function, C, maps elements of Σ^n to Σ^m. For a circuit C we use $\mathrm{code}(C)$ to denote a standard binary encoding of the circuit. The exact details of such a coding will not be important, but note that for n-input and m-output circuits C the coding will have length at least $n + m$.

We use circuits to define graphs succinctly. For positive integers n, given a $2n$-input, 1-output circuit C, we say that it *specifies the graph* G whose vertex set is $V = \Sigma^n$ and whose edge set is defined as follows: There is an edge from $x \in V$ to $y \in V$ if and only if $C(xy) = 1$. We say that C is a *succinct representation of* G and write $G(C)$ for G. Note that a graph G has many succinct representations. We formalize the radius, diameter, and k-kingship problems for succinctly specified graphs for fixed positive integers k as follows:

$$\textsc{succinct-}k\textsc{-radius} = \{\mathrm{code}(C) \mid G(C) \text{ has radius} \leq k\}.$$
$$\textsc{succinct-}k\textsc{-diameter} = \{\mathrm{code}(C) \mid G(C) \text{ has diameter} \leq k\}.$$
$$\textsc{succinct-}k\textsc{-king} = \{\langle \mathrm{code}(C), x \rangle \mid x \text{ is a } k\text{-king in } G(C)\}.$$

Tournament family specifiers. We can use circuits as introduced above to describe tournaments succinctly. A second way of specifying tournaments, which we sketched already in the introduction, is more computationally uniform, but less flexible: A *tournament family specifier* is a function $f \colon \Sigma^* \times \Sigma^* \to \Sigma^*$ such that

1. f is a polynomial-time computable function.
2. f is commutative, that is, for all $x, y \in \Sigma^*$ we have $f(x, y) = f(y, x)$.
3. f is a selector, that is, for all $x, y \in \Sigma^*$ we have $f(x, y) \in \{x, y\}$.

We interpret this as specifying, in the following way, a family of tournaments, one per length. At each length n, the nodes in the length-n tournament specified by f will be the bitstrings in Σ^n. For each two distinct nodes among these, x and y, the edge between them will be $x \to y$ if $f(x, y) = y$ and it will be $y \to x$ if $f(x, y) = x$. There will be no self-loops. Since our function f always chooses one of its inputs and is commutative, this indeed yields a family of tournaments. We call the tournament just described *the length-n tournament induced by f.*

Recall from the introduction that k-Kings$_f$ is the set of all k-kings in the tournaments specified by f and a language L is called a P-k-*king language* if there exists a tournament family specifier f with k-Kings$_f = L$.

3   The Complexity of the Diameter Problem

In this section we prove that the succinct diameter problem is complete for Π_2^p, see Theorem 3.6 for the exact claim. Indeed, we show that the problem is already hard when we restrict ourselves to tournaments. The tools that we introduce for the proof of this result will be important in the following sections.

Definition 3.1. *An ℓ-layer tournament is a tournament whose vertex set is the disjoint union of ℓ nonempty sets $L_1, \ldots, L_\ell$ such that the following holds: For any vertex $u \in L_i$ and $v \in L_j$ with $i < j - 1$, there is an edge $v \rightarrow u$.*

Lemma 3.2. *Let T be an ℓ-layer tournament, let $u \in L_1$, and let $v \in L_\ell$. Then the shortest path from u to v has length at least $\ell - 1$.*

Our next task is the definition of a rather complex tournament that will be used in later proofs. Recall that we want to show that every problem in Π_2^p reduces to the succinct diameter problem. For this, we construct a tournament in which all vertices are k-kings, except possibly for one vertex, which will be a k-king exactly if a certain "for all ... exists ..." property is true.

For the definition of the tournament and also for the definition of even more complicated tournaments later on, we proceed in three steps. First, we describe the structure of the tournament. This means that we explain how many vertices are present, which names we are going to use to abstractly refer to these vertices, and how these vertices are connected by edges. However, in this first step we do not yet fix which words will later on be used as being associated with these vertices. Rather, the description of the structure of the tournament is given only in terms of the names of the vertices. Second, we prove important properties of the tournament, such as the property that all vertices except possibly for one vertex are k-kings. Here, we still argue in terms of the names of the vertices of the tournament and are not interested which words are represented by the names. Third, we fix which words we are going to use. That is, for each named vertex we present a unique word that is represented by that name.

The following definition of the tournament $T^k(R, J)$ just explains the basic idea behind the construction.

Definition 3.3. *Let a word length n, an integer k, a relation $R \subseteq \Sigma^n \times \Sigma^n$, and a set J of even cardinality be given. The vertices in J are called the junk vertices, which are mainly needed to ensure later on that tournaments have a size that is a power of 2. We define a tournament $T^k(R, J)$ as follows: It has $k + 1$ layers $L_1, \ldots, L_{k+1}$. The first layer, called the potential king layer, and the next $k - 2$ layers, called the antenna layers, all contain a single vertex. The vertex in the potential king layer is called p, the vertex in antenna layer L_i is called a_i. Layer L_k is called the z-layer and contains one vertex named β_z for each $z \in \Sigma^n$ plus two vertices z_1, z_2 and possibly a vertex z_3 so that the whole tournament has even size. Let Z denote the set $\{\beta_z \mid z \in \Sigma^n\}$. Layer L_{k+1} is called the y-layer and contains one vertex named α_y for each $y \in \Sigma^n$ plus six special vertices $\{c_0, c_1, c_2, d_0, d_1, d_2\}$ and all the junk vertices. Let Y denote the* set $\{\alpha_y \mid y \in \Sigma^n\}$. *The edges between the vertices are directed as follows: Between vertices in nonadjacent layers, the direction is already fixed by the fact that the tournament is layered. For the potential king layer and the antenna layers, there is an edge to every vertex on the next layer. For the z- and y-layers, the most important connections are those between a vertex β_z and a vertex α_y: The edge goes from β_z to α_y iff $(y, z) \in R$. The remaining edges (to and from the junk vertices and the special vertices z_i, c_i, and d_i) are setup appropriately so that all vertices except possibly p are k-kings of the tournament.*

We will often need to talk about "an arbitrary vertex in some layer L_i." We will generally use the variables l_i, l_i' and so on to denote such vertices.

Lemma 3.4. *Let $k \geq 2$, let $n \geq 3$, let $R \subseteq \Sigma^n \times \Sigma^n$ be a relation, and let J be a set of even size. Then the tournament $T^k(R, J)$ has the following properties:*

1. *The vertex p in the potential k-king layer is a k-king if and only if for every $y \in \Sigma^n$ there exists a $z \in \Sigma^n$ such that $(y, z) \in R$ holds.*
2. *All other vertices are k-kings of the tournament.*

In the definition of the tournament $T^k(R, J)$ we left open which vertices are to be used as vertices p or a_i or α_y. We only needed that all these vertices are different. Since our aim is to prove something about succinctly specified tournaments, we will have to use the set Σ^m for some m as the set of vertices. This means that we have to use bitstrings for the vertices p, a_i, and so on. The following definition fixes which bitstrings we are going to use.

Definition 3.5. *Let $k \geq 2$ and $n \geq 3$ be integers, let $R \subseteq \Sigma^n \times \Sigma^n$, and let m be an integer such that $2^m > k + 8 + 2 \cdot 2^n$. We define a tournament $T^k_{\Sigma^m}(R)$ as follows. Its vertex set is Σ^m. Let r be the size of the tournament $T^k(R, \emptyset)$ from Definition 3.3. Let σ_i denote the lexicographically ith element of the set Σ^m, starting with $i = 1$. Thus, σ_1 is the all-0 bitstring of length m, while σ_{2^m} is the all-1 bitstrings. Let $J = \{\sigma_i \mid r - 5 \leq i \leq 2^m - 6\}$ and note that J has even cardinality. The tournament $T^k_{\Sigma^m}(R)$ is the tournament $T^k(R, J)$ with the named vertices assigned to bitstrings in the manner described in the following.*

1. *The vertex p is mapped to σ_1.*
2. *The vertices $a_2, \ldots, a_{k-2}$ in the antenna layers are mapped to $\sigma_2, \ldots, \sigma_{k-2}$, respectively.*
3. *The vertices in the z-layer are mapped to the vertices $\sigma_{k-1}, \ldots, \sigma_{k+2^n-1}$ for even k and to $\sigma_{k-1}, \ldots, \sigma_{k+2^n}$ for odd k. The ordering of the vertices of the z-layer is the same as the one described in the Definition 3.3: The vertices $\beta_z \in Z$ come first, in their lexicographic ordering, followed by z_1, z_2, and possibly z_3.*
4. *The vertices in the y-layer are mapped to the remaining vertices as follows. The vertices α_y are mapped, in lexicographical order, to the vertices $\sigma_{r-2^n-6}, \ldots, \sigma_{r-6}$. The junk vertices inside the y-layer are simply mapped to themselves, namely to σ_{r-5} to σ_{2^m-6}. The six special vertices c_i and d_j are mapped to σ_{2^m-5} and σ_{2^m}.*

The above establishes a one-to-one correspondence between the vertices mentioned in the definition of $T^k(R, J)$ and the elements of Σ^m. This concludes the description of the tournament $T^k_{\Sigma^m}(R)$.

We now introduce the restricted version of the succinct diameter problem and prove its hardness. Let SUCCINCT-k-DIAMETER-TOURNAMENT denote the set $\{\text{code}(C) \mid$ the graph specified by C is a tournament of diameter at most $k\}$.

Theorem 3.6. *Let $k \geq 2$. Then* SUCCINCT-k-DIAMETER-TOURNAMENT *is $\leq^p_m$-complete for Π^p_2.*

A simple corollary of the above theorem is that SUCCINCT-k-DIAMETER is also $\leq^p_m$-complete for Π^p_2 for $k \geq 2$. Note that SUCCINCT-1-DIAMETER is easily seen to be $\leq^p_m$-complete for coNP while SUCCINCT-1-TOURNAMENT-DIAMETER is the empty set (graphs specified by circuits have size at least 2).

4   The Complexity of P-k-King Languages

In this section we establish the following result, which is the main result of this paper.

Theorem 4.1. *Let $k \geq 2$. Each language in $\Pi^p_2 - \{\emptyset, \Sigma^*\}$ is $\leq^p_m$-equivalent to a P-k-king language.*

This result says that, excluding from our attention the trivial singleton degrees of the empty set and of Σ^*, every many-one degree can be named by a k-kings problem (for each k). That is, k-kings problems are so flexible that they take on every possible nontrivial Π^p_2 complexity level.

Note in particular that the theorem applies to the complete Π^p_2 degree. Thus we have the following corollary, which shows that the result of [9] that P-2-king languages are all in Π^p_2 is optimal.

Corollary 4.2. *For each $k \geq 2$, there is a Π^p_2-complete P-k-king language.*

We prove Theorem 4.1 via showing a result, Theorem 4.3, that is even stronger.

Theorem 4.3. *Let L be a language and let $k \geq 2$. Then $L \in \Pi^p_2$ if and only if there exists a positive integer j such that $\text{pad}'_j(L)$ is a P-k-king language.*

Theorem 4.3 says that each set in Π^p_2 has a padded version of itself that is a P-k-king language. In light of the particular class of padding functions we use, it is easy to see that Theorem 4.1 follows easily from this, as each padded version is clearly many-one equivalent to the underlying set it is a padding of when that underlying set is neither the empty set nor Σ^*. Indeed, a nontrivial language and its padded versions are even equivalent under first-order reductions (and even under even more restrictive reductions), which are the same as DLOGTIME-uniform many-one AC^0-reductions, see Barrington, Immerman, and Straubing [2]. This means that the $\leq^p_m$-equivalence in Theorem 4.1 and the (implicit) $\leq^p_m$-completeness in Corollary 4.2 can be respectively replaced by first-order equivalence and completeness.

To prove Theorem 4.3, we introduce new tournaments and, as in the previous section, we do so in three steps: First, we explain how these tournaments are structured, but do not fix which words we are going to use as vertices. Second, we prove that the tournament has desirable properties. Third, we explain which words we are going to use for the vertices. The core idea behind the definition of the tournaments is to "weave together" multiple $T^k_{\Sigma^m}(R)$ tournaments.

Definition 4.4. *Let $k \geq 2$, $n \geq 1$, $n' \geq 3$, and $m > \log_2(k + 8 + 2 \cdot 2^{n'})$ be integers, let $R \subseteq \Sigma^n \times \Sigma^{n'} \times \Sigma^{n'}$, and let F (called the fill-up vertices) be a set whose cardinality is neither 0, 2, nor 4. The woven tournament $W^k(R, F, m)$ is obtained as follows: Each $x \in \Sigma^n$ induces a relation $R_x \subseteq \Sigma^{n'} \times \Sigma^{n'}$, namely the set of all pairs (y, z) such that $(x, y, z) \in R$. For each $x \in \Sigma^n$ we build a new version of the tournament $T^k_{\Sigma^m}(R_x)$ by tagging each vertex $v \in T^k_{\Sigma^m}(R_x)$ with x and we write v^x for this tagged vertex. The vertices of the woven tournament are all these tagged vertices, that is, all vertices v^x for $x \in \Sigma^n$ and $v \in T^k_{\Sigma^m}(R_x)$, plus the vertices in the set F. The woven tournament is also a layered tournament and tagging does not change the layer, that is, if $v \in T^k_{\Sigma^m}(R_x)$ lies in layer L_i, then v^x also lies on layer L_i of the woven tournament. The fill-up vertices lie on layer L_k. The edges between the vertices of the woven tournament are directed as follows: We inherit all edges from the individual tournaments $T^k_{\Sigma^m}(R_x)$ and the property that the woven tournament is layered fixes the direction of edges between nonadjacent layers. For adjacent layers, edges generally point from the vertex in the layer with the larger index to the layer with the smaller index; but there are two exceptions: First, edges point from the vertices in layer L_{k-1} to the fill-up vertex. Second, for every vertex v^x in layer L_k and every vertex $u^{x'}$ in layer L_{k+1} with $x \neq x'$ there is an edge from v^x to $u^{x'}$. Inside each layer, edges that are not yet fixed are setup appropriately to ensure that all vertices except possibly the vertices in the potential king layer are k-kings of the woven tournament.*

Lemma 4.5. *Let $k \geq 2$, $n \geq 1$, and $n' \geq 3$ be integers, let $R \subseteq \Sigma^n \times \Sigma^{n'} \times \Sigma^{n'}$ be a ternary relation, let m be an integer with $2^m > k + 8 + 2 \cdot 2^{n'}$, and let F be a set of vertices whose cardinality is not 0, 2, or 4. Then the woven tournament $W^k(R, F, m)$ has the following properties:*

1. *For every $x \in \Sigma^n$ the vertex p^x is a k-king of the woven tournament if and only if for every $y \in \Sigma^{n'}$ there exists a $z \in \Sigma^{n'}$ such that $(x, y, z) \in R$ holds.*
2. *All other vertices (vertices other than the p^x) are k-kings of the woven tournament.*

The next step is to fix how the vertices of the tournament $W^k(R, F, m)$ are coded. Lemma 4.7 then tells us that the resulting tournament has all the properties needed to prove the main result, Theorem 4.3.

Definition 4.6. *Let $k \geq 2$, $n \geq 0$, and $n' \geq 3$ be integers, let $R \subseteq \Sigma^n \times \Sigma^{n'} \times \Sigma^{n'}$ be a ternary relation, and let m be an integer such that $2^m > k + 8 + 2 \cdot 2^{n'}$*

and let $l = n + m + 3$. The tournament $W_{\Sigma^l}^k(R)$ has the vertex set $V = \Sigma^l$. The set F is the set of all elements of V that do not end with 000. Note that this set does not have size 0, 2, or 4. The vertices of the different $T_{\Sigma^m}^k(R_x)$ are encoded as follows: A vertex u^x is mapped to the bitstring $xu000$. Thus, we prefix the vertices of $T_{\Sigma^m}^k(R_x)$ with x and add 000 at the end.

Lemma 4.7. Let $k \geq 2$, $n \geq 0$, and $n' \geq 3$ be integers, let $R \subseteq \Sigma^n \times \Sigma^{n'} \times \Sigma^{n'}$ be a ternary relation. Then the woven tournament $W_{\Sigma^l}^k(R)$ has the following properties:

1. For every $x \in \Sigma^n$, the bitstring $x0^{l-n}$ is a k-king of the woven tournament if and only if for every $y \in \Sigma^{n'}$ there exists a $z \in \Sigma^{n'}$ such that $(x, y, z) \in R$ holds.
2. All bitstrings that do not end with 0^{l-n} are k-kings of the woven tournament.

5 The Complexity of the Radius Problem

The results on P-king languages can be directly used to prove that the succinct radius problem for directed graphs is complete for Σ_3^p:

Theorem 5.1. Let $k \geq 2$. Then SUCCINCT-k-RADIUS is $\leq_m^p$-complete for Σ_3^p.

The proof of the theorem uses Theorem 4.3 instead of just the statement that every nontrivial language in Π_2^p is equivalent to a P-k-king language. We remark that it is also possible to prove Theorem 5.1 directly, but the proof then has to redo several of the constructions of the proof of Theorem 4.3.

6 Conclusion

Kings problems have tremendous flexibility, and in fact can be used as a naming scheme for the nontrivial Π_2^p many-one degrees. We found that king, radius, and diameter problems for succinctly specified graphs are complete for classes at the second or third level of the polynomial hierarchy. However, defining languages in terms of k-kings does not always lead to sets that are complete for different levels of the polynomial hierarchy. In the technical report version [6] of this paper we study *initial component languages*, which are languages of the form $\bigcup_k k\text{-Kings}_f$, and show that such languages always lie in Π_2^p, but cannot be NP-hard unless $P = NP$.

References

1. Bang-Jensen, J., Gutin, G.: Generalizations of tournaments: A survey. J. Graph Theory 28(4), 171–202 (1998)
2. Barrington, D., Immerman, N., Straubing, H.: On uniformity within NC^1. J. Comput. Syst. Sci. 41(3), 274–306 (1990)

3. Gál, A., Wigderson, A.: Boolean complexity classes versus their arithmetic analogs. Random Structures and Algorithms 9(1–2), 99–111 (1996)
4. Garey, M., Johnson, D.: Computers and Intractability: A Guide to the Theory of NP-Completeness. W. H. Freeman and Company, New York (1979)
5. Gutin, G.: The radii of n-partite tournaments. Math. Notes 40, 414–417 (1986)
6. Hemaspaandra, E., Hemaspaandra, L., Tantau, T., Watanabe, O.: On the complexity of kings. Technical Report URCS-TR905, Computer Science Dept., Univ. Rochester (2006)
7. Hemaspaandra, E., Hemaspaandra, L., Watanabe, O.: The complexity of kings. Technical Report URCS-TR870, Computer Science Dept., Univ. Rochester (2005)
8. Hemaspaandra, L., Nasipak, C., Parkins, K.: A note on linear-nondeterminism, linear-sized, Karp–Lipton advice for the P-selective sets. J. Universal Comput. Sci. 4(8), 670–674 (1998)
9. Hemaspaandra, L., Ogihara, M., Zaki, M., Zimand, M.: The complexity of finding top-Toda-equivalence-class members. Theory of Comput. Sys. 39(5), 669–684 (2006)
10. Hemaspaandra, L., Torenvliet, L.: P-selectivity, immunity, and the power of one bit. In: Gibet, S., Courty, N., Kamp, J.-F. (eds.) GW 2005. LNCS (LNAI), vol. 3881, pp. 323–331. Springer, Heidelberg (2006)
11. Landau, H.: On dominance relations and the structure of animal societies, III: The condition for score structure. Bulletin of Math. Biophysics 15(2), 143–148 (1953)
12. Meyer, A., Stockmeyer, L.: The equivalence problem for regular expressions with squaring requires exponential space. In: Proc. 13th IEEE Symposium on Switching and Automata Theory, pp. 125–129. IEEE Press, Los Alamitos (1972)
13. Nickelsen, A., Tantau, T.: The complexity of finding paths in graphs with bounded independence number. SIAM J. Comput. 34(5), 1176–1195 (2005)
14. Petrovic, V., Thomassen, C.: Kings in k-partite tournaments. Discrete Math. 98, 237–238 (1991)
15. Schaefer, M., Umans, C.: Completeness in the polynomial-time hierarchy: Part I: A compendium. SIGACT News 33(3) (2002)
16. Schaefer, M., Umans, C.: Completeness in the polynomial-time hierarchy: Part II. SIGACT News 33(4) (2002)
17. Selman, A.: P-selective sets, tally languages, and the behavior of polynomial time reducibilities on NP. Math. Syst. Theory 13(1), 55–65 (1979)
18. Selman, A.: Some observations on NP real numbers and P-selective sets. J. Comput. Syst. Sci. 23(3), 326–332 (1981)
19. Selman, A.: Reductions on NP and P-selective sets. Theoret. Comput. Sci. 19(3), 287–304 (1982)
20. Stockmeyer, L.: The polynomial-time hierarchy. Theoret. Comput. Sci. 3(1), 1–22 (1976)
21. West, D.: Introduction to Graph Theory, 2nd edn. Prentice-Hall, Englewood Cliffs (2001)

Notions of Hyperbolicity in Monoids

Michael Hoffmann and Richard M. Thomas*

Department of Computer Science, University of Leicester, Leicester LE1 7RH, UK
mh55@mcs.le.ac.uk, rmt@mcs.le.ac.uk

Abstract. We introduce a notion of hyperbolicity in monoids which is
a restriction of that suggested by Duncan and Gilman. One advantage
is that the notion gives rise to efficient algorithms for dealing with certain questions; for example, the word problem can be solved in time
$\mathcal{O}(n \log n)$. We also introduce a new way of defining automatic monoids
which provides a uniform framework for the discussion of these concepts.

1 Introduction

The notions of hyperbolic [7] and automatic [4] groups have played a fundamental
role in computational group theory in recent years. It has been noted (see [12,13]
for example) that the definition of automaticity generalizes naturally from groups
to semigroups and an exploration of the basic properties of automatic semigroups
was undertaken in [2]. There are some issues with this generalization; see [9], for
example, where it was shown that the idea generalizes in several non-equivalent
ways. Notwithstanding this, a coherent theory of automatic semigroups has been
developed, with some fundamental properties (such as the solution of the word
problem in quadratic time) generalizing to semigroups.

Whilst the usual definition of automatic lends itself naturally to such a generalization, this has not been the case for hyperbolic. There were several equivalent ways known of defining hyperbolic groups (see [1] for example) but none of
these really apply to semigroups. The situation changed with Gilman's elegant
characterization of hyperbolic groups in [6] using pushdown automata; this new
condition generalizes naturally to the semigroup setting. As a result, Duncan
and Gilman [3] proposed this as the definition of a hyperbolic semigroup.

Their definition is entirely natural. One issue, however, is the absence (so far)
of efficient algorithms for dealing with hyperbolic semigroups and monoids. It is
well known that the word problem for hyperbolic groups can be solved in linear
time (even in real time [11]) but the best known algorithm for the word problem
in a hyperbolic monoid is exponential [8]. Other questions (such as the conjugacy

---

* This paper was written whilst the authors were on study leave from the University
of Leicester and the support of the University in this regard is much appreciated. In
addition, the paper was completed whilst the authors were visiting Friedrich Otto
in Kassel; they are very grateful to him for his hospitality and for his constructive
comments on the paper. The authors would also like to thank Chen-Hui Chiu and
Hilary Craig for all their help and encouragement.

E. Csuhaj-Varjú and Z. Ésik (Eds.): FCT 2007, LNCS 4639, pp. 341–352, 2007.

problem, which can be solved efficiently in hyperbolic groups [5]) are still open as far as hyperbolic monoids are concerned, even as regards decidability.

The purpose of this paper is to show how a restriction of the definition used by Duncan and Gilman in [3] does lead to efficient algorithms. An analysis of Gilman's proof in [6] shows that one can impose restrictions on the push-down automata used in the definition; we describe these in Section 3. These new definitions are also natural; we point out that it is possible to define automaticity and biautomaticity in terms of pushdown automata (see Remark 2), and these new notions of hyperbolicity arise directly from this observation. An essential part of all this is the definition of a special type of context-free language which we term "sync linear" (see Definition 5). This gives rise to new perspectives on the relationship between hyperbolic and automatic monoids; it enables us to view hyperbolic monoids (at least, in the sense presented here) and automatic monoids in a more uniform fashion than has previously been the case.

A particular aspect of this is the following. It is known that a hyperbolic group is necessarily automatic [4], but this does not generalize to monoids [8]. With the notions of hyperbolicity given here, we recapture this connection; in fact, a monoid satisfying one of these new notions is (as is the case in groups) necessarily biautomatic (see Theorem 3); this means that, for example, the conjugacy problem is solvable in such monoids.

In general, the algorithmic properties of this new class of monoids suggest that they are worthy of further study. Whilst the new definitions are equivalent to the previous one in the group setting (see Theorem 1), they allow us to develop efficient algorithms for monoids; for example, the word problem can be solved in time $\mathcal{O}(n \log n)$ (see Theorem 2). Further work is in progress, and it seems that the techniques described here give rise to a number of efficient algorithms for other problems. One interesting feature (which mirrors the developments in automatic monoids) is that, when developing algorithms, the techniques involve formal languages and automata (as opposed to the situation in groups, where the techniques have been more geometric).

We conclude this section by mentioning some notation we will use. For any $k \in \mathbb{N}$, let A^k denote the set of all words α in A^* with $|\alpha| = k$ and $A^{\leqslant k}$ the set of all words α with $|\alpha| \leqslant k$. Let α^{rev} denote the reversal of the word α. If M is a monoid and $A \subseteq M$ a set of generators of M, then there is a homomorphism $\theta : A^* \to M$ where each α in A^* is mapped to the corresponding element of M. We will be concerned with finite sets A, so that M is finitely generated. In this context, if α and β are elements of A^*, we write $\alpha \equiv \beta$ if α and β are identical as words, and $\alpha = \beta$ if α and β represent the same element of M (i.e. if $\alpha\theta = \beta\theta$).

2   Synchronously Regular Languages

In this section we describe some aspects of synchronous two-tape finite automata that will be used in our algorithms; we also define some notions of biautomaticity in monoids that we will need later in the paper.

If $\alpha \equiv a_1 a_2 \ldots a_n$ and $\beta \equiv b_1 b_2 \ldots b_m$, we have an FSA (finite state automaton) with input alphabet $A \times A$ and reading pairs (a_1, b_1), (a_2, b_2), and so on. To deal with the case where $n \neq m$, we introduce a *padding symbol* \$. More formally, we define a mapping $\delta^R : A^* \times A^* \to A(2, \$)^*$, where $\$ \notin A$ and $A(2, \$) = (A \cup \{\$\}) \times (A \cup \{\$\}) - \{(\$, \$)\}$, by

$$(\alpha, \beta)\delta^R = \begin{cases} (a_1, b_1) \ldots (a_n, b_n) & \text{if } n = m \\ (a_1, b_1) \ldots (a_n, b_n)(\$, b_{n+1}) \ldots (\$, b_m) & \text{if } n < m \\ (a_1, b_1) \ldots (a_m, b_m)(a_{m+1}, \$) \ldots (a_n, \$) & \text{if } n > m. \end{cases}$$

We have a map that inserts paddings on the left instead of the right; we define $\delta^L : A^* \times A^* \to A(2, \$)^*$ by $(\alpha, \beta)\delta^L = ((\alpha^{rev}, \beta^{rev})\delta^R)^{rev}$. If α and β have the same length then $(\alpha, \beta)\delta^L$ and $(\alpha, \beta)\delta^R$ coincide; we sometimes just write $(\alpha, \beta)\delta$ in this case. The following facts about regular languages will be useful:

Lemma 1. *If* $J \subseteq (A^* \times B^*)\delta^X$ *and* $K \subseteq (B^* \times C^*)\delta^X$ *are regular with* $X \in \{L, R\}$, *then* $\{(\alpha, \gamma)\delta^X : \exists \beta \in B \text{ with } (\alpha, \beta)\delta^X \in J, (\beta, \gamma)\delta^X \in K\}$ *is regular.*

Lemma 2. *Suppose that* $K \subseteq (A^* \times A^*)\delta^X$ *is regular and* $X \in \{L, R\}$. *Let* α *be a word in* A^*. *If there exists* $\beta \in A^*$ *such that* $(\alpha, \beta)\delta^X \in K$, *then such a word* β *can be found in time* $\mathcal{O}(|\alpha|)$.

Lemma 3. *Let* $M = (Q, A(2, \$), \tau, q_0, F)$ *be an FSA and* $X \in \{L, R\}$. *For any* $\alpha \equiv a_1 a_2 \ldots a_n \in A^*$ *we can create the following collection of sets in time* $\mathcal{O}(n)$:

$$D_i = \{q \in Q : \text{there exists } \beta \in A^* \text{ with } q_0 \xrightarrow{(a_1 \ldots a_i, \beta)\delta^X} q\}.$$

Fundamental in the notion of automatic and biautomatic monoids is the concept of "padded" pairs of words. If M is a monoid generated by a finite set A, L is a regular subset of A^* and $a \in A \cup \{\epsilon\}$, then we define:

$$\begin{aligned} {}^{\$}_a L &= \{(\alpha, \beta)\delta^L : \alpha, \beta \in L, a\alpha = \beta\}; & {}^{\$}L_a &= \{(\alpha, \beta)\delta^L : \alpha, \beta \in L, \alpha a = \beta\}; \\ {}_a L^{\$} &= \{(\alpha, \beta)\delta^R : \alpha, \beta \in L, a\alpha = \beta\}; & L^{\$}_a &= \{(\alpha, \beta)\delta^R : \alpha, \beta \in L, \alpha a = \beta\}. \end{aligned}$$

Recall that $\alpha a = \beta$ means that αa and β represent the same element of M, not that αa and β are identical as words. Given this, we now recall some notions of biautomaticity in monoids (see [10]):

Definition 1. *Let* M *be a monoid generated by a finite set* A *and suppose that* L *is a regular language over* A *that maps onto* M. *Let* $\bar{A}$ *represent* $A \cup \{\epsilon\}$. *The pair* (A, L) *is said to be*

- *a left-biautomatic structure if* ${}^{\$}_a L$ *and* ${}^{\$}L_a$ *are regular for* $a \in \bar{A}$;
- *a right-biautomatic structure if* ${}_a L^{\$}$ *and* $L^{\$}_a$ *are regular for* $a \in \bar{A}$.

A monoid M *is said to be* left-biautomatic *if it has a left-biautomatic structure and* right-biautomatic *if it has a right-biautomatic structure.*

3   Types of Context-Free Languages

Throughout this paper, if $P = (Q, \Sigma, \Gamma, \tau, q_0, F)$ is a PDA (pushdown automaton), we assume there is a special symbol $\perp$ (where $\perp \notin \Gamma$) on the bottom of the stack. This symbol is present at the start of the computation (i.e. the stack only contains $\perp$ initially), is never deleted nor appears anywhere else on the stack. We accept by accept state but our machines all have an "empty stack" (i.e. a stack only containing $\perp$) when a word is accepted. If $(r, \omega') \in \tau((q, \omega), \alpha)$ we have a transition from state q to state r reading α where the stack contents change from ω to ω'; we write $(q, \omega) \overset{\alpha}{\to} (r, \omega')$. When we refer to the stack contents, we omit $\perp$ unless the stack is empty (in which case we denote the contents by $\perp$).

As we mentioned above, we consider restrictions to PDA's. To do this, we want to define types of sequences of moves. As above, let $\bar{A}$ denote $A \cup \{\epsilon\}$. Let $O = \{h, y, p\}$ represent the possible operations push, stay, pop to the stack .

Definition 2. *Let $(q, \omega_1) \overset{\alpha}{\to} (r, \omega_2)$ be a transition in a PDA P. We say that $(a_1, o_1)(a_2, o_2) \dots (a_n, o_n)$, where $a_i \in \bar{A}$ and $o_i \in O$, is a* trace *of the transition if $\alpha \equiv a_1 \dots a_n$ and there exists a computation path from (q, ω_1) to (r, ω_2) such that the ith step reads a_i and performs the stack operation o_i. (We may have that $a_i = \epsilon$ for some values of i.)*

If $\alpha \in L(P)$ we say α has a trace $t = (a_1, o_1)(a_2, o_2) \dots (a_n, o_n)$ if there exists $q \in F$ such that there is a transition $(q_0, \perp) \overset{\alpha}{\to} (q, \perp)$ with trace t.

Note that, when defining traces for words, we are are only doing so for words accepted by P. We have an analogous concept for a language:

Definition 3. *Let $T \subseteq (\bar{A} \times O)^*$. A PDA is of* type T *if every word accepted has a trace in T; a language L is of* type T *if there is a PDA of type T accepting L.*

To reduce the notation needed for types and traces we use the following:

Definition 4. *If $B \subseteq \bar{A}$ and $o \in \{h, y, p\}$ we write B_o for $B \times \{o\}$ and b_o for $\{b\} \times \{o\}$. Let $(p, \omega) \overset{\alpha}{\underset{T}{\to}} (q, \omega')$ denote a transition with trace in T which starts in state p with stack contents ω, reads α, and ends in state q with stack contents ω'.*

We now introduce a certain kind of CFL (context-free language):

Definition 5. *A CFG (context-free grammar) $G = (N, A \cup \{\#\}, R, S)$ (where $\# \notin A$) is said to be* sync linear *if each production rule in R is of the form $X \to aYb$ or $X \to \#$ where $X, Y \in N$ and $a, b \in A$. A language K is said to be* sync linear *if it is generated by a sync linear grammar.*

Remark 1. If K is a sync linear language (as in Definition 5), then K must be a subset of $\{\alpha\#\beta : \alpha, \beta \in A^*, |\alpha| = |\beta|\}$.   □

The following result will be used throughout the paper:

Lemma 4. *If $\Sigma = A \cup \{\#\}$ and $K \subseteq \Sigma^*$, then the following are equivalent:*

(i) K is sync linear.
(ii) K^{rev} is sync linear.

(iii) $K \subseteq \{\alpha\#\beta : |\alpha| = |\beta|\}$ *and* $\{(\alpha,\beta)\delta : \alpha^{rev}\#\beta \in K\}$ *is regular.*
(iv) $K \subseteq \{\alpha\#\beta : |\alpha| = |\beta|\}$ *and* $\{(\alpha,\beta)\delta : \alpha\#\beta^{rev} \in K\}$ *is regular.*
(v) K *is a CFL of type* $A_{\mathrm{h}}^* \#_{\mathrm{y}} A_{\mathrm{p}}^*$.

Remark 2. An advantage of Lemma 4 is that it lets us consider automaticity and biautomaticity in terms of CFLs. For example, if $L \subseteq A^*$ and we consider $L_a^\$ = \{(\alpha,\beta)\delta^R : \alpha,\beta \in L, \alpha a = \beta\}$ (as in Definition 1), the regularity of $L_a^\$$ is equivalent to $K = \{\tilde\alpha\#\tilde\beta^{rev} : (\alpha,\beta)\delta^R \in L_a^\$\}$ being sync linear, where $\tilde\alpha$ and $\tilde\beta$ are obtained from α and β by padding the shorter of the two words on the right by symbols \$ to make them of the same length. Another way of saying that $L_a^\$$ is regular is to say that K is a CFL of type $B_{\mathrm{h}}^*\#_{\mathrm{y}}B_{\mathrm{p}}^*$ where $B = A \cup \{\$\}$.   □

The next result is straightforward; the proof via PDA's that CFL's are closed under concatenation and union still works if we restrict the types of machine (given that, in our machines, the stack is always empty when accepting a word).

Lemma 5. *If L and K are CFL's of type T_L and T_K respectively then LK is a CFL of type $T_L T_K$ and $L \cup K$ is a CFL of type $T_L \cup T_K$.*

It is also known that one can insert a CFL into another to yield a CFL; modifying the proof slightly gives the following result:

Lemma 6. *Suppose that $L \subseteq A^*\{\#\}A^*$ (where $\# \notin A$) is a CFL of type $W_1\#_{\mathrm{y}}W_2$ with $W_1, W_2 \subseteq (\bar{A} \times \{h,y,p\})^*$ and that K is a CFL of type T_K. Then $L' = \{\alpha\beta\gamma : \alpha\#\gamma \in L, \beta \in K\}$ is a CFL of type $W_1 T_K W_2$.*

Another result in a similar vein is the following:

Lemma 7. *If $L \subseteq A^*$ is a CFL of type T and $K \subseteq A^*$ is a regular language then $L \cap K$ is a CFL of type T.*

The following result, which allows us to change the type of a language, is a little technical but will be useful in what follows:

Lemma 8. *If $B \subseteq A$, $W_1, W_2 \in (\{A \cup \{\epsilon\}\} \times \{h,y,p\} - B_y)^*$ and $L \subseteq A^*$ is a CFL of type $W_1 B_y W_2$, then L is also of type $W_1 B_h W_2 \epsilon_{\mathrm{p}}$.*

4 Hyperbolic Structures

In this section we introduce our notions of hyperbolicity; as we explained in Section 1, these are obtained by following the definition given in [3] but imposing constraints on the type of the PDA. The three types we will consider are:

$$T_1 = A_{\mathrm{h}}^* \#_{\mathrm{y}} A_{\mathrm{p}}^* A_{\mathrm{y}}^{\leqslant 1} A_{\mathrm{h}}^* \#_{\mathrm{y}} A_{\mathrm{p}}^*;$$
$$T_2 = A_{\mathrm{h}}^* \#_{\mathrm{y}} A_{\mathrm{p}}^* A_{\mathrm{h}}^* \#_{\mathrm{y}} A_{\mathrm{p}}^* (\epsilon_{\mathrm{p}}^* \cup A_{\mathrm{y}});$$
$$T_3 = A_{\mathrm{h}}^* \#_{\mathrm{y}} A_{\mathrm{p}}^* A_{\mathrm{h}}^* \#_{\mathrm{y}} (\epsilon_{\mathrm{p}}^* \cup A_{\mathrm{y}}) A_{\mathrm{p}}^*.$$

Given this, we now make the following definition:

Definition 6. *A monoid M is called T_i-hyperbolic if M has a hyperbolic structure (A, L) such that $\{\alpha \# \beta \# \gamma^{rev} : \alpha, \beta, \gamma \in L, \alpha\beta = \gamma\}$ is of type T_i; (A, L) is then a T_i-hyperbolic structure for M.*

We will refer to the language $\{\alpha \# \beta \# \gamma^{rev} : \alpha, \beta, \gamma \in L, \alpha\beta = \gamma\}$ as L_{hyp} for the remainder of this paper.

Not all monoids which are hyperbolic in these new ways are close to being groups; for example, consider $M = \langle a, b, x : xa^i x = xb^i x \text{ for } i > 0 \rangle$. M is neither finitely presented nor cancellative; however we can show that M is T_1-hyperbolic. Let $A = \{a, b, x\}$ and $L = A^* - A^*\{x\}\{b\}^*\{x\}A^*$; then, for all $\alpha, \beta \in L$, either $\alpha\beta \equiv \gamma \in L$ or $\alpha \equiv \alpha_1 x b^i$ and $\beta \equiv b^j x \beta_2$ for some $i + j > 0$ and $\alpha_1, \beta_2 \in A^*$. In the latter case $\alpha\beta = \alpha_1 x a^{i+j} x \beta_2 \equiv \gamma \in L$. A PDA that pushes all the elements of α and β onto the stack can verify that γ is of the required form while reading γ^{rev} and popping a symbol off the stack for each symbol of γ^{rev}.

The following result follows directly from Lemma 8:

Lemma 9. *If M is a T_1-hyperbolic monoid then M is also T_2-hyperbolic.*

Remark 3. In a similar fashion (given an analogue of Lemma 8) one can show that, if M is a T_1-hyperbolic monoid, then M is also T_3-hyperbolic. In fact, using the techniques developed in this paper, one can show that a monoid M is T_3-hyperbolic if and only if M^{rev} is T_2-hyperbolic.   □

Given Remark 3, we will focus on T_2-hyperbolic. As we explained in the introduction, part of the motivation for these notions springs from the following:

Theorem 1. *If M is a group then M is hyperbolic if and only if M is T_i-hyperbolic for $1 \leqslant i \leqslant 3$.*

Proof. "⇒": Let M be a hyperbolic group generated by a set A; given Lemma 9 and Remark 3, it is sufficient to show that M is T_1-hyperbolic.

Each element in M is represented by several words in A^*; we are only interested in, for any given element, the representatives of minimum length. Let L be the set of all such words (so that, if $\alpha \in L$, $\beta \in A^*$ and $\alpha = \beta$, then $|\alpha| \leqslant |\beta|$; such words α label geodesics in the Cayley graph of M). It is well known that, for a hyperbolic group, this set L is regular. In addition, Gilman's characterization of hyperbolic groups in [6] shows that the language L_{hyp} is context-free. His proof proceeds via a CFG G which can be taken to be of the following form. The set of non-terminals N is the disjoint union of sets X (which contains the sentence symbol), Y and Z; the production rules are of the form:

$$X_i \rightarrow aX_j b; \quad X_i \rightarrow Y_k Z_l; \quad X_i \rightarrow Y_k c Z_l;$$
$$Y_i \rightarrow aY_j b; \quad Y_i \rightarrow \#; \quad Z_i \rightarrow aZ_j b; \quad Z_i \rightarrow \#,$$

where $a, b, c \in A$, $X_\ell \in X$, $Y_\ell \in Y$ and $Z_\ell \in Z$. We will build $L = L(G)$ out of smaller components of particular types; we then assemble these components and show that L_{hyp} is of type T_1.

Let $L_{Y_i} = \{\eta \in A^* : Y_i \overset{*}{\Rightarrow} \eta\}$ and $L_{Z_i} = \{\eta \in A^* : Z_i \overset{*}{\Rightarrow} \eta\}$; these are sync linear and, by Lemma 4, are of type $A_h^* \#_y A_p^*$. By Lemma 5, for any Y_i and Z_j and any $c \in A$, the languages $L_{Y_i} L_{Z_j}$ and $L_{Y_i} c L_{Z_j}$ are of type $A_h^* \#_y A_p^* A_y^{\leqslant 1} A_h^* \#_y A_p^*$.

Let G' be the CFG with non-terminals X, the same starting symbol as G and the following transitions:

$$X_k \to aX_lb \quad \text{if } X_k \to aX_lb \text{ is a transition in } G;$$
$$X_k \to \#_{i,c,j} \quad \text{if } X_k \to Y_i c Z_j \text{ is a transition in } G;$$
$$X_k \to \#_{i,j} \quad \text{if } X_k \to Y_i Z_j \text{ is a transition in } G.$$

Let $L_{i,c,j} = L(G') \cap A^* \{\#_{i,c,j}\} A^*$ and $L_{i,j} = L(G') \cap A^* \{\#_{i,j}\} A^*$. By Lemmas 4 and 7, each of $L_{i,j}$ and $L_{i,c,j}$ is of type $A_h^* B_y A_p^*$ with B the set of all the symbols $\#_{i,j}$ and $\#_{i,c,j}$. If we replace $\#_{i,c,j}$ in $L_{i,c,j}$ with $L_{Z_i} \{c\} L_{Y_j}$ we get $K_{i,c,j}$, and $K_{i,j}$ is obtained in a similar fashion:

$$K_{i,c,j} = \{\eta \xi \zeta : \eta \#_{i,c,j} \zeta \in L_{i,c,j}, \xi \in L_{Z_i} \{c\} L_{Y_j}\},$$
$$K_{i,j} = \{\eta \xi \zeta : \eta \#_{i,j} \zeta \in L_{Z_i} L_{Y_j}, \xi \in L_{i,j}\}.$$

By Lemma 6, $K_{i,c,j}$ and $K_{i,j}$ are of type T_1. Since, by construction, $L = L(G) = \bigcup K_{i,c,j} \cup \bigcup K_{i,j}$, we have, by Lemma 5, that L is of type T_1 as required.

"$\Leftarrow$": If M is a group with T_i-hyperbolic structure (A, L) then the set L_{hyp} is a CFL; so M is hyperbolic by [6].   □

5   Word Problem of T_i-Hyperbolic Monoids

Given a monoid with a T_2-hyperbolic structure (A, L) we will show that the word problem is solvable in time $\mathcal{O}(n \log(n))$. Our first aim is perform a "multiplication" of two words in L into a word in L in linear time.

Let $P = (Q, A \cup \{\#\}, \Gamma, \tau, q_o, F)$ be a PDA of type T_2 with $L(P) = L_{hyp}$. We are particular interested at what happens when we read the $\#$ symbols; we think of a triangle with sides labelled by α, β and γ^{rev}, and talk about the "corners" of the triangle. Let Γ' denote $\Gamma \cup \{\bot\}$. For $\mu, \nu \in A^*$ and $\omega \in \Gamma^*$ let

$$T_{\mu,\nu,\omega} = \{(p,q,t) \in Q \times Q \times \Gamma' : (p, t\omega) \xrightarrow[A_h^* \#_y A_p^*]{\mu \# \nu} (q, t\omega)\}.$$

If $t = \bot$ we must have $\omega \equiv \epsilon$ (this convention applies to similar situations in the remainder of the paper). Here μ represents a suffix of α and ν a prefix of β.

The trace of the transition specifies that elements will be pushed on the stack, followed by a stay operation, and then elements will be popped off the stack. Since the end configuration has the same stack as the initial one, P has performed the same number of pushes as pops. The element t will never removed from the stack during the transition and therefore the set is independent of ω; so, from now on, we will omit ω and denote this set by $T_{\mu,\nu}$. Each such set $T_{\mu,\nu}$ is a subset of $Q \times Q \times \Gamma'$ and is therefore bounded in size by the choice of P.

When dealing with these sets we want to be able to construct $T_{a\mu,\nu b}$ out of $T_{\mu,\nu}$; this can be done in the following way:

$$T_{a\mu,\nu b} = \{(p,q,t) \in Q \times Q \times \Gamma' : \text{there exists } (p',q',t') \in T_{\mu,\nu}$$
$$\text{with } (p,t\omega) \xrightarrow[A_h]{a} (p',t't\omega) \text{ and } (q',t't\omega) \xrightarrow[A_p]{b} (q,t\omega)\}.$$

If $t = \perp$, then we have $(p,t\omega) = (p,\perp) \xrightarrow[A_h]{a} (p',t')$ and $(q',t') \xrightarrow[A_p]{b} (q,\perp)$; again, we adopt a similar convention for the remainder of the paper. This enables us to create a complete deterministic FSA M_T where each state corresponds to a subset of $Q \times Q \times \Gamma'$; the input alphabet is $A \times A$ and $\tau_T(s_T,(\mu^{rev},\nu)\delta) = T_{\mu,\nu}$.

For any given $p,q \in Q$ and $t \in \Gamma'$ we can choose the accept states of M to be all states which contain (p,q,t). Hence the set

$$C_{p,q,t} = \{(\mu^{rev},\nu)\delta : (p,t\omega) \xrightarrow[A_h^* \#_y A_p^*]{\mu \# \nu} (q,t\omega)\} = \{(\mu^{rev},\nu) : (p,q,t) \in T_{\mu,\nu}$$

is regular. In terms of our triangle, the sets $C_{p,q,t}$ are relevant when considering the corner between α and β. We will now give similar arguments to define a deterministic complete FSA and regular set for each of the other two corners.

First consider the corner between β and γ. For $\mu,\nu \in A^*$ and $\omega \in \Gamma^*$ let

$$V_{\mu,\nu,\omega} = \{(p,q,t) \in Q \times Q \times \Gamma' : (p,t\omega) \xrightarrow[A_h^* \#_y A_p^* \epsilon_p^*]{\mu \# \nu} (q,t\omega)\}.$$

Again these sets are independent of ω and we can build the sets up. Here μ represents a suffix of β and ν a prefix of γ^{rev}. Since μ could be longer than ν, we have to distinguish between the following two cases:

for $|\mu| = |\nu|$: $V_{b\mu,\nu c} = \{(p,q,t) \in Q \times Q \times \Gamma' : \exists(p',q',t') \in V_{\mu,\nu}$
$\text{with } (p,t\omega) \xrightarrow[A_h]{b} (p',t't\omega), (q',t't\omega) \xrightarrow[A_p]{c} (q,t\omega)\};$

for any μ,ν: $V_{b\mu,\nu} = \{(p,q,t) \in Q \times Q \times \Gamma' : \exists(p',q',t') \in V_{\mu,\nu}$
$(|\mu| \geq |\nu|)$ $\text{with } (p,t\omega) \xrightarrow[A_h]{b} (p',t't\omega), (q',t't\omega) \xrightarrow[\epsilon_p]{\epsilon} (q,t\omega)\}.$

This leads to a complete deterministic FSA M_V with two sorts of transition depending whether or not a padding symbol has already been used. The states of are subsets of $Q \times Q \times \Gamma'$, the alphabet is $A(2,\$)$ and $\tau_V(s_V,(\mu^{rev},\nu)\delta^R) = V_{\mu,\nu}$.

As before, for any $p,q \in Q$ and $t \in \Gamma'$, we can set the accept states in M_V to be all states that contain (p,q,t); so the following set is regular:

$$E_{p,q,t} = \{(\mu^{rev},\nu)\delta^R : (p,q,t) \in V_{\mu,\nu}\}.$$

We now use similar arguments for the corner between α and γ. For $\mu,\nu \in A^*$ let

$$U_{\mu,\nu} = \{(p,q,t) \in Q \times Q \times \Gamma' : (q_0,\perp) \xrightarrow[A_h^*]{\mu} (p,t\omega),$$
$$(q,t\omega) \xrightarrow[A_p^*(A_y^* \cup \epsilon_p^*)]{\nu^{rev}} (q_f,\perp) \text{ for some } q_f \in F \text{ for some } \omega \in \Gamma^*\}.$$

Again we can build the sets up. Here μ represents a prefix of α and ν a prefix of γ. However due to the fact that we can either clear the stack with empty moves or else read the rest of γ^{rev} whilst the stack is empty, we have to distinguish three cases. Let $\mu, \nu \in A^*$ and $a, c \in A$; then:

$$U_{\mu a, \nu c} = \{(p, q, t) : \exists (p', q', t') \in U_{\mu, \nu} \text{ with } (p', t'\omega) \xrightarrow[A_h]{a} (p, tt'\omega),$$
$$(q', tt'\omega) \xrightarrow[A_p]{c} (q, t'\omega)\};$$

$$U_{\mu a, \epsilon} = \{(p, q, t) : \exists (p', q', t') \in U_{\mu, \epsilon} \text{ with } (p', t'\omega) \xrightarrow[A_h]{a} (p, tt'\omega),$$
$$(q', tt'\omega) \xrightarrow[\epsilon_p]{\epsilon} (q, t'\omega)\};$$

$$U_{\epsilon, \nu c} = \{(q_0, q, \perp) : \exists (q_0, q', \perp) \in U_{\epsilon, \nu} \text{ with } (q', \perp) \xrightarrow[A_y]{c} (q, \perp)\}.$$

This leads to a deterministic complete FSA M_U over the alphabet $A(2, \$)$ with states $Q \times Q \times \Gamma'$ and transitions $\tau_U(s_U, (\mu, \nu)\delta^L) = U_{\mu, \nu}$. As before, for any $p, q \in Q$ and $t \in \Gamma'$, we can set the accept states to be all states that contain (p, q, t); therefore the following set is regular:

$$D_{p,q,t} = \{(\mu, \nu)\delta^L : (p, q, t) \in U_{\mu, \nu}\}.$$

The main step in solving the word problem is now the following result:

Lemma 10. *Let M be a monoid with a T_2-hyperbolic structure (A, L). Given $\alpha, \beta \in L$, a word $\gamma \in L$ with $\alpha\beta = \gamma$ can be constructed in time $\mathcal{O}(|\alpha| + |\beta|)$.*

Proof. Let $P = (Q, A \cup \{\#\}, \Gamma, \tau, q_0, F)$ be a PDA of type T_2 accepting L_{hyp}. Assume that $\alpha \equiv a_1 a_2 \ldots a_n$ and $\beta \equiv b_1 b_2 \ldots b_m$ are given; we want to construct $\gamma \in L$ with $\gamma = \alpha\beta$. The algorithm will work in two steps; the figure below indicates some of the notation used.

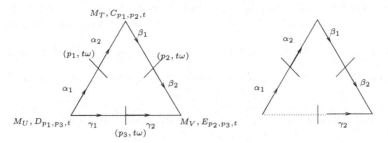

Goal 1: Find i, p_1, p_2, p_3 and t such that there exist $\gamma_1, \gamma_2 \in A^*$, $\omega \in \Gamma^*$ and $q_f \in F$ with either:

$$(q_0, \perp) \xrightarrow[A_h^*]{\alpha_1} (p_1, t\omega) \xrightarrow[A_h^* \#_y A_p^*]{\alpha_2 \# \beta_1} (p_2, t\omega) \xrightarrow[A_h^* \#_y A_p^*]{\beta_2 \# \gamma_2^{rev}} (p_3, t\omega) \xrightarrow[A_p^* (\epsilon_p^* \cup A_y^*)]{\gamma_1^{rev}} (q_f, \perp)$$

$$\text{or:} \quad (q_0, \perp) \xrightarrow[A_h^*]{\alpha_1} (p_1, t\omega) \xrightarrow[A_h^* \#_y A_p^*]{\alpha_2 \# \beta_1} (p_2, t\omega) \xrightarrow[A_h^* \#_y A_p^* \epsilon_p^+]{\beta_2 \# \gamma_2^{rev}} (p_3, t\omega) \xrightarrow[\epsilon_p]{\epsilon} (q_f, \perp)$$

where $\alpha \equiv \alpha_1 \alpha_2$ with $|\alpha_2| = i$ and $\beta \equiv \beta_1 \beta_2$ with $|\beta_1| = i$.

Goal 2: Create $\gamma \in L$ with $\gamma = \alpha\beta$.

We now describe the steps of our algorithm that allow us to achieve these goals.

Step 1: Let G_j be the set of states which M_U can be in for any input of the form $(a_1 a_2 \ldots a_j, \gamma_1)\delta^L$ with $\gamma_1 \in A^*$. and I_k the set of states that M_V can be in for any input $(b_m \ldots b_{m-k+1}, \gamma_2)\delta^R$ with $\gamma_2 \in A^*$. Let H_i be $T_{a_n \ldots a_{n-i+1}, b_1 \ldots b_i}$.

By Lemma 3 all of these sets can be created in time $\mathcal{O}(|\alpha|+|\beta|)$. The algorithm will now find the least i such that there exists $p_1, p_2, p_3 \in Q$, $t \in \Gamma$, $I \in I_{m-i}$, $G \in G_{n-i}$ and $H \in H_i$ such that $(p_1, p_2, t) \in H$, $(p_2, p_3, t) \in I$ and $(p_1, p_3, t) \in G$. Since the sizes of all sets H_i, G_j and I_k can be uniformly bounded in terms of P, the check for any particular i is done in constant time; hence we obtain i in time $\mathcal{O}(|\alpha| + |\beta|)$.

Step 2: From Step 1 we have determined i, p_1, p_2, p_3 and t. Let $\alpha_1 \alpha_2 \equiv \alpha$ with $|\alpha_2| = i$ and $\beta_1 \beta_2 \equiv \beta$ with $|\beta_1| = i$.

Since $D_{p_1, p_3, t}$ is regular we can find γ_1 such that $(\alpha_1, \gamma_1)\delta \in D_{p_1, p_3, t}$ in time $\mathcal{O}(n - i)$ by Lemma 2. Similarly, $E_{p_2, p_3, t}$ is regular and we can find γ_2 such that $(\beta_2^{rev}, \gamma_2^{rev})\delta^L \in E_{p_2, p_3, t}$ in time $\mathcal{O}(m - i)$ by Lemma 2. So we have that either:

$$(q_0, \bot) \xrightarrow[A_h^*]{\alpha_1} (p_1, tw) \xrightarrow[A_h^* \#_y A_p^*]{\alpha_2 \# \beta_1} (p_2, tw) \xrightarrow[A_h^* \#_y A_p^*]{\beta_2 \# \gamma_2^{rev}} (p_3, tw) \xrightarrow[A_p^*(\epsilon_p^* \cup A_y^*)]{\gamma_1^{rev}} (q_f, \bot)$$

$$\text{or:} \quad (q_0, \bot) \xrightarrow[A_h^*]{\alpha_1} (p_1, tw) \xrightarrow[A_h^* \#_y A_p^*]{\alpha_2 \# \beta_1} (p_2, tw) \xrightarrow[A_h^* \#_y A_p^* \epsilon_p^+]{\beta_2 \# \gamma_2^{rev}} (p_3, tw) \xrightarrow[\epsilon_p^*]{\epsilon} (q_f, \bot)$$

for some $q_f \in F$ and $\omega \in \Gamma^*$. Hence we have that $\alpha \# \beta \#(\gamma_1 \gamma_2)^{rev} \in L(P)$ and $\gamma \equiv \gamma_1 \gamma_2 = \alpha\beta$ with so $\gamma \in L$ as required. $\qquad\square$

Given this, we now have the following:

Lemma 11. *Let (A, L) be a T_2-hyperbolic structure of a monoid M. If $\zeta \in A^*$ with $|\zeta| = n$, then $\lambda \in L$ with $\lambda = \zeta$ can be calculated in time $\mathcal{O}(n \log n)$.*

Proof. We split ζ into two words ζ_1 and ζ_2 of length at most $\lceil |\zeta|/2 \rceil$ and construct $\lambda_1, \lambda_2 \in L$ with $\lambda_1 = \zeta_1$ and $\lambda_2 = \zeta_2$ recursively. By Lemma 10 we can construct λ from λ_1 and λ_2 in time $O(|\lambda_1| + |\lambda_2|)$, and hence find λ in time $\mathcal{O}(n \log n)$. $\qquad\square$

The last step in solving the word problem is given by the following result:

Lemma 12. *Let (A, L) be a T_2-hyperbolic structure for a monoid M and $\beta \in L$; then the set $\{(\alpha, \gamma)\delta^L : \alpha, \gamma \in L, \alpha\beta = \gamma\}$ is regular.*

Proof. We will continue the notation used above. We are interested in the set S of all $(\mu, \nu, p, q, t) \in A^* \times A^* \times Q \times Q \times \Gamma'$ such that

$$(p, tw) \xrightarrow[A_h^k \#_y A_p^k A_h^l \#_y A_p^{l-m} \epsilon_p^m]{\mu \# \beta \# \nu^{rev}} (q, tw)$$

is a transition in P for some $k, l, m \in \mathbb{N}$ and some $\omega \in \Gamma^*$.

Since β is fixed and $k, l, m \leqslant |\beta|$, the set S is finite. As described above the FSA M_U reads words over $A(2, \$)$ and $\tau(s_U, (\alpha_1, \gamma_1)\delta^L) = U_{\alpha_1, \gamma_1}$. We construct a new FSA M' by adding a state f (the only accept state of M') and transitions

$$\left\{ x \xrightarrow{(\mu, \nu)\delta^R} f : \text{there exists } (p, q, t) \in x \text{ and } (\mu, \nu, p, q, t) \in S \right\}.$$

Note that words accepted by M' are generally padded at the left, but a bounded number of padding symbols can appear on the right (due to the transitions to f). Let $(a_1, c_1) \ldots (a_n, c_n) \in L(M')$; let α be the word resulting from $a_1 \ldots a_n$ after removing all the padding symbols and let γ be the analogous word for $c_1 \ldots c_n$. By the construction of M' there must exist $\alpha_1, \alpha_2, \gamma_1, \gamma_2 \in A^*$, $p, q \in Q$ and $t \in \Gamma'$ with $\alpha_1 \alpha_2 \equiv \alpha$, $\gamma_1 \gamma_2 \equiv \gamma$, $(p, q, t) \in U_{(\alpha_1, \gamma_1) \delta^L}$ and

$$(q_0, \bot) \xrightarrow{\alpha_1} (p, tw) \xrightarrow{\alpha_2 \# \beta \# \gamma_2^{rev}} (q, tw) \xrightarrow{\gamma_1^{rev}} (q_f, \bot)$$

for some $q_f \in F$, $w \in \Gamma^*$. Hence $\alpha \# \beta \# \gamma^{rev} \in L(P)$ and $\alpha \beta = \gamma$ with $\alpha, \gamma \in L$.

Conversely, if $\alpha, \gamma \in L$ with $\alpha \beta = \gamma$, then $\alpha \# \beta \# \gamma^{rev}$ is in $L(P)$. So there exist $\alpha_1, \alpha_2, \gamma_1, \gamma_2 \in A^*$ with $\alpha_1 \alpha_2 \equiv \alpha$, $\gamma_1 \gamma_2 \equiv \gamma$, $|\alpha_2| = k$, $|\gamma_2| = l$, and either:

$$(q_0, \bot) \xrightarrow[A_h^*]{\alpha_1} (p, tw) \xrightarrow[A_h^k \#_y A_p^k A_h^l \# A_p^l]{\alpha_2 \# \beta \# \gamma_2^{rev}} (q, tw) \xrightarrow[A_p^*(\epsilon_p^* \cup A_y^*)]{\gamma_1^{rev}} (q_f, \bot)$$

or: $(q_0, \bot) \xrightarrow[A_h^*]{\alpha_1} (p, tw) \xrightarrow[A_h^k \#_y A_p^k A_h^l \# A_p^{l-m} \epsilon_p^m]{\alpha_2 \# \beta \# \gamma_2^{rev}} (q, tw) \xrightarrow[\epsilon_p^*]{\gamma_1^{rev}} (q_f, \bot)$

for some $m > 0$, $p, q \in Q$, $t \in \Gamma'$, $q_f \in F$ and $w \in \Gamma^*$; note that γ_1 must be the empty word in the second case. By the construction of M' the word $(\alpha_1, \gamma_1) \delta^L (\alpha_2, \gamma_2) \delta^R$ is accepted by M'. Using Lemma 4.1 from [9] we see that the set $\{(\alpha, \gamma) \delta^L : \alpha, \gamma \in L, \alpha \beta = \gamma\}$ is regular. □

In a similar vein, one can prove:

Lemma 13. *Let (A, L) be a T_2-hyperbolic structure for a monoid M and $\alpha \in L$; then the set $\{(\beta, \gamma) \delta^L : \beta, \gamma \in L, \alpha \beta = \gamma\}$ is regular.*

Given Lemma 12, we can check whether two given words in L represent the same element of M in linear time by choosing β to represent the identity element. Given Lemma 11, we now have the following result:

Theorem 2. *The word problem of a T_2-hyperbolic monoid is solvable in time $\mathcal{O}(n \log n)$.*

Given Remark 3 the word problem of a T_3-hyperbolic monoid is solvable in time $\mathcal{O}(n \log n)$; by Lemma 9, this is also true for T_1-hyperbolic monoids.

6   Connections with Biautomaticity

Lemmas 12 and 13 give that, for any given element m in a monoid M with a T_2-hyperbolic structure (A, L), the sets ${}_m^{\$} L = \{(\alpha, \beta) \delta^L : m\alpha = \beta\}$ and ${}^{\$} L_m = \{(\alpha, \beta) \delta^L : \alpha m = \beta\}$ are regular. Since, by the definition of a T_2-hyperblic structure, L is regular and L maps onto M, we have the following.

Theorem 3. *If M is a monoid with a T_2-hyperbolic structure (A, L) then (A, L) is also a left-biautomatic structure for M.*

We note that, by Remark 3, a T_3-hyperbolic structure for a monoid M is also a right-biautomatic structure for M. Given Lemma 9, we see that a T_1-hyperbolic structure is both a left-biautomatic and a right-biautomatic structure.

We finish with the following observation:

Lemma 14. *If (A, L) is a T_2-hyperbolic structure for a monoid M then there exists $K \subseteq L$ such that (A, K) is a T_2-hyperbolic structure for M and K maps bijectively to M.*

Proof. (A, L) is also a left-biautomatic structure for M by Theorem 3. By Corollary 5.5 in [2] there exists a regular language $K \subseteq L$ such that K maps bijectively to M. The set $K\{\#\}K\{\#\}K^{rev}$ is clearly regular; by Lemma 7 the set

$$K\{\#\}K\{\#\}K^{rev} \cap L_{hyp} = \{(\alpha\#\beta\#\gamma^{rev} : \alpha\beta = \gamma, \alpha, \beta, \gamma \in K\} = K_{hyp}$$

is also a CFL of type T_2; hence (A, K) is also a T_2-hyperbolic structure for M and K maps bijectively to M as required.   □

We have a similar result to Lemma 14 for T_1-hyperbolic and T_3-hyperbolic structures.

References

1. Alonso, J.M, Brady, T., Cooper, D., Ferlini, V., Lustig, M., Michalik, M., Shapiro, M., Short, H.: Notes on word hyperbolic groups. In: Ghys, E., Haefliger, A., Verjovsky, A. (eds.) Group Theory from a Geometric Viewpoint, pp. 3–63. World Scientific, Singapore (1991)
2. Campbell, C.M., Robertson, E.F., Ruškuc, N., Thomas, R.M.: Automatic semigroups. Theoret. Comp. Sci. 250, 365–391 (2001)
3. Duncan, A., Gilman, R.H.: Word hyperbolic semigroups. Math. Proc. Cambridge Philos. Soc. 136, 513–524 (2004)
4. Epstein, D.B.A., Cannon, J.W., Holt, D.F., Levy, S., Paterson, M.S., Thurston, W.: Word Processing in Groups. Jones & Barlett (1992)
5. Epstein, D.B.A., Holt, D.F.: The linearity of the conjugacy problem in word-hyperbolic groups. Internat. J. Algebra Comput. 16, 287–305 (2006)
6. Gilman, R.H.: On the definition of word hyperbolic groups. Math. Z. 242, 529–541 (2002)
7. Gromov, M.: Hyperbolic groups. In: Gersten, S.M. (ed.) Essays in Group Theory, vol. 8, pp. 75–263 (MSRI Publ., Springer-Verlag) (1987)
8. Hoffmann, M., Kuske, D., Otto, F., Thomas, R.M.: Some relatives of automatic and hyperbolic groups. In: Gomes, G.M.S., Pin, J.-E., Silva, P.V. (eds.) Semigroups, Algorithms, Automata and Languages, pp. 379–406. World Scientific, Singapore (2002)
9. Hoffmann, M., Thomas, R.M.: Notions of automaticity in semigroups. Semigroup Forum 66, 337–367 (2003)
10. Hoffmann, M., Thomas, R.M.: Biautomatic semigroups. In: Liśkiewicz, M., Reischuk, R. (eds.) FCT 2005. LNCS, vol. 3623, pp. 56–67. Springer, Heidelberg (2005)
11. Holt, D.F.: Word-hyperbolic groups have real-time word problem. Internat. J. Algebra Comput. 10, 221–227 (2000)
12. Hudson, J.F.P.: Regular rewrite systems and automatic structures. In: Almeida, J., Gomes, G.M.S., Silva, P.V. (eds.) Semigroups, Automata and Languages, pp. 145–152. World Scientific, Singapore (1998)
13. Otto, F., Sattler-Klein, A., Madlener, K.: Automatic monoids versus monoids with finite convergent presentations. In: Nipkow, T. (ed.) Rewriting Techniques and Applications. LNCS, vol. 1379, pp. 32–46. Springer, Heidelberg (1998)

P Systems with Adjoining Controlled Communication Rules

Mihai Ionescu[1] and Dragoş Sburlan[2,3]

[1] Rovira i Virgili University,
Research Group on Mathematical Linguistics,
Tarragona, Spain
armandmihai.ionescu@urv.cat
[2] Ovidius University
Faculty of Mathematics and Informatics
Constantza, Romania
dsburlan@univ-ovidius.ro
[3] Research Group on Natural Computing,
University of Seville, Spain

Abstract. This paper proposes a new model of P systems where the rules are activated by the presence/absence of certain objects in the neighboring regions. We obtain the computational completeness considering only two membranes, external inhibitors, and carriers. Leaving the carriers apart we obtain equivalence with ET0L systems in terms of number sets.

1 Introduction

Having as inspiration the way living cells are divided by membranes into cellular compartments where various biochemical processes take place, *P systems* (also known as *membrane systems*) area grew rapidly since Gheorghe Păun proposed the first model in 1998 ([5]). A complete bibliography of P systems can be found on the P system webpage ([9]).

Within the living cell there are several energy consuming activities. Among them there is the transport activity which is of three types: *diffusion, facilitated diffusion*, and *active transport*. Simple diffusion means that the molecules can pass directly through the membrane, always down a concentration gradient, while in the case of facilitated diffusion and active transport, molecules can pass both down an up the concentration gradient. In the facilitated diffusion membrane protein channels are used to allow charged molecules (which otherwise could not diffuse across the cell membrane) to freely diffuse the cell, while active transport requires the expenditure of energy to transport the molecule from one side of the membrane to the other.

Hence, living cells get/expel from/to their environment many substances and for this aim they have developed specific transport systems across membranes, even against a concentration gradient. Often enough this necessity of the living cell

E. Csuhaj-Varjú and Z. Ésik (Eds.): FCT 2007, LNCS 4639, pp. 353–364, 2007.

to expel or attract various molecules is triggered by the presence or the absence of certain chemicals in the immediate neighboring (inner or outer) regions.

Here we deal with P systems where the rules from a given region are activated precisely by the presence or the absence of certain symbols in the neighboring regions. This model has a biological counterpart and it is inspired by the chemicals that pass through the membranes of the cell, from one region to another, in the sense of polarization gradient. In this case, the electrical charge plays the role of the promoter. Attempts to formalize these biological phenomena were done also in [1].

Before going into the definition of the new model and its computational power under certain restrictions (Section 3) we briefly remind the reader some basic notions and notations (Section 2). Section 4 is dedicated to the conclusions and challenges for further research.

2   Preliminaries and Definitions

We assume familiarity with the basics of formal language theory (see [7]), as well as with the basics of membrane computing (see [6]).

An *alphabet* is a finite set of symbols (letters), and a word (string) over an alphabet Σ is a finite sequence of letters from Σ. We denote the empty word by λ, the length of a word w by $|w|$, and the number of occurrences of a symbol a in w by $|w|_a$. The (con)catenation of two words x and y is denoted by xy.

A *language* over Σ is a (possibly infinite) set of words over Σ. The language consisting of all words over Σ is denoted by Σ^*, and $\Sigma^+ = \Sigma^* \backslash \{\lambda\}$. We denote by REG, CF, ETOL, CS, RE the families of languages generated by regular, context-free, extended tabled interactionless Lindemayer systems, context-sensitive, and of arbitrary Chomsky grammars, respectively (RE stands for the class of recursively enumerable languages). The following strict inclusions hold: REG $\subset$ CF $\subset$ ETOL $\subset$ CS $\subset$ RE.

For a family FL of languages, NFL denotes the family of length sets of languages in FL. The following relations hold: NREG = NCF $\subset$ NETOL $\subset$ NCS $\subset$ NRE.

The multisets over a given finite support (alphabet) are represented by strings of symbols. The order of symbols does not matter, because the number of copies of an object in a multiset is given by the number of occurrences of the corresponding symbol in the string (see [2] for other ways to specify multisets).

3   The Model

Based on the biological observations mentioned in the introductory section we introduce the following new class of P systems.

3.1   Defining the Model

Definition 1. *A P system with adjoining controlled communication rules (called in short, a PACC system) is a construct*

$$\Pi = (V, C, \mu, w_1, \ldots, w_m, R_1, \ldots, R_m, i_0),$$

where:

- V is the alphabet of objects;
- $C \subseteq V$ is the set of carriers;
- μ is a membrane structure with m membranes (labeled in a one-to-one manner by $1, \ldots, m$);
- $w_1, \ldots, w_m$ are the multisets of objects initially present in the regions of Π;
- $R_1, \ldots, R_m$ are finite sets of communication rules associated to membranes, that are of the following types:
  ◇ simple rules:
  $[A]_i \longrightarrow [\ ]_i\alpha$ or $A[\ ]_i \longrightarrow [\alpha]_i$, for $A \in V \setminus C$, $\alpha \in (V \setminus C)^*$,
  ◇ promoted simple rules:
  $[A]_iB \longrightarrow [\ ]_i\alpha$ or $A[B]_i \longrightarrow [\alpha]_i$, for $A, B \in V \setminus C$, $\alpha \in (V \setminus C)^*$,
  ◇ inhibited simple rules:
  $[A]_i\neg B \longrightarrow [\ ]_i\alpha$ or $A[\neg B]_i \longrightarrow [\alpha]_i$, for $A, B \in V \setminus C$, $\alpha \in (V \setminus C)^*$,
  ◇ carrier rules:
  pairs of rules $[cA]_i \longrightarrow [\ ]_ic\alpha$ *and* $c[\ ]_i \longrightarrow [c]_i$, *for* $A \in V \setminus C$, $c \in C$, $\alpha \in (V \setminus C)^*$, *or*
  pairs of rules $cA[\ ]_i \longrightarrow [c\alpha]_i$ *and* $[c]_i \longrightarrow [\ ]_ic$ *for* $A \in V \setminus C$, $c \in C$, $\alpha \in (V \setminus C)^*$,
  ◇ promoted carrier rules:
  pairs of rules $[cA]_iB \longrightarrow [\ ]_ic\alpha$ *and* $c[\ ]_i \longrightarrow [c]_i$, *for* $A, B \in V \setminus C$, $c \in C$, $\alpha \in (V \setminus C)^*$, *or*
  pairs of rules $cA[B]_i \longrightarrow [c\alpha]_i$ *and* $[c]_i \longrightarrow [\ ]_ic$, *for* $A, B \in V \setminus C$, $c \in C$, $\alpha \in (V \setminus C)^*$;
  ◇ inhibited carrier rules:
  pairs of rules $[cA]_i\neg B \longrightarrow [\ ]_ic\alpha$ *and* $c[\ ]_i \longrightarrow [c]_i$, *for* $A, B \in V \setminus C$, $c \in C$, $\alpha \in (V \setminus C)^*$, *or*
  pairs of rules $cA[\neg B]_i \longrightarrow [c\alpha]_i$ *and* $[c]_i \longrightarrow [\ ]_ic$, *for* $A, B \in V \setminus C$, $c \in C$, $\alpha \in (V \setminus C)^*$;
- $i_0 \in \{1, \ldots, m\}$ is an elementary membrane of μ (the output membrane).

In a *simple rule* an object is rewritten in a multiset of objects, in the inner or in the outer region with respect to the initial position of the object. A *promoted simple rule/inhibited simple rule* has the same action as a simple rule but it can be applied only in the presence/absence of certain objects (chemicals) called promoters/inhibitors. To be more precise we take as example the rule $[A]_iB \longrightarrow [\ ]_i\alpha$, which implies that object A is rewritten in α in the outer membrane only if *promoter* B is present there. If we replace B with $\neg B$, the object plays the role of the *inhibitor*, and only by its presence it blocks the execution of the rule.

In a *carrier rule* the objects can be rewritten only if they are guided by an object, the *carrier*. Note that the carrier is not actively participating in the reaction. Its role is to "accompany" the reaction and to inhibit the parallelism. As an example, by rule $[cA]_i \longrightarrow [\ ]_ic\alpha$ we mean that object A evolves to α (in the outer region of object A) iff there is an object c that helps A to be rewritten.

Promoted/Inhibited carrier rules can be applied if besides the carrier there is also a promoter/inhibitor which triggers/blocks the reaction.

As usual in membrane computing, the rules are used in a nondeterministic maximally parallel manner starting from an initial configuration. In this way, we obtain transitions between the configurations of the system. A configuration is described by the m-tuple of the multisets of objects present in the m regions of the system. The initial configuration is $(w_1, \ldots, w_m)$.

A sequence of transitions between configurations of the system constitutes a *computation*; a computation is successful if it *halts*, i.e., it reaches a configuration (the halting configuration) where no rule can be applied to any of the objects.

The *result* of a successful computation is the number of objects present within the membrane with the label i_o in the halting configuration. A computation which never halts yields no result.

We use the notation $NPACC_m(\alpha, \beta)$, where $\alpha \in \{smp\} \cup \{car_k \mid k \geq 0\}$, $\beta \in \{proR_i, inhR_i\}$ to denote the family of sets of natural numbers generated by P systems with adjoining controlled communication rules having at most m membranes, communication rules that can be simple $\alpha = smp$, or carrier $\alpha = car_k$, using at most k carriers, and external promoters $\beta = proR_i$ or external inhibitors $\beta = inhR_i$ of weight i at the level of rules.

3.2   An Example

Let us now exemplify the functioning of the model defined above throughout an **example**. Here it shown how such machines can be used to compute functions.

Consider the following system

$$\Pi_1 = (\{A, B, D\}, C = \{c\}, [\ [\ ]_2]_1, w_1 = A^n, w_2 = c, R_1, R_2, 2),$$

where:

- $R_1 = \emptyset$, $R_2 = \{A[\ ]_2 \longrightarrow [ABD]_2, [B]_2 \longrightarrow [\ ]_2B, [cD]_2 \longrightarrow [\ ]_2c,$
  $B[D]_2 \longrightarrow [AB]_2, c[\ ]_2 \longrightarrow [c]_2\}.$

The system Π is fed with $n \geq 1$ copies of object A in region 1 and when it halts, the contents of the output region contains n^2 copies of A.

The functioning of the system is rather simple. The only rule we can apply in the initial configuration is the one which rewrites object A in ABD in the inner region, hence in the second step of the computation we will have all the objects of the system (n copies of A, n copies of B, n copies of D and the object initially present here, carrier c) in region 2. Then, we expel all objects B in region 1 and we start consuming objects D by applying the rule $[cD]_2 \longrightarrow [\ ]_2c$, hence object D is sent outside membrane 2 and is rewritten to λ having carrier c accompanying the reaction.

Note that object D plays the role of the counter and each time a copy of D is deleted (for example in step i of the computation), n more copies of A are produced (in step $i + 2$ of the computation). One by one, the n-th copies of D are consumed, adding for each of them n copies to object A. In the rule $B[D]_2 \longrightarrow [AB]_2$, the object D plays also the role of promoter and the object

B can be rewritten into AB only in its presence. The computation ends with n^2 copies of A in region 2, hence the system computes the number-theoretic function $f(n) = n^2$, $n \geq 1$.

3.3 The Results

In what follows we will prove that the class of sets of numbers generated by P systems with external inhibitors equals the class of sets of numbers generated by P systems with external inhibitors and only two membranes.

Lemma 1. $NPACC_m(smp, inhR_1) = NPACC_2(smp, inhR_1), m \geq 2$.

Proof. Obviously, $NPACC_m(smp, inhR_1) \supseteq NPACC_2(smp, inhR_1)$. For the opposite inclusion we have to show that for any P system with external inhibitors $\overline{\Pi} = (\overline{V}, \overline{C}, \overline{\mu}, \overline{R}, \overline{i_0})$ generating a set of natural numbers, there exists an equivalent P system with external inhibitors $\Pi = (V, C, \mu, R, i_0)$ with only 2 membranes.

To this aim, we simulate the computation of $\overline{\Pi}$, with the system Π defined as follows.

Let us denote by $\mathcal{L} = \{1, 2, \ldots, m\}$ the set of labels of the regions in $\overline{\Pi_m}$. In addition, assume that $\overline{R} = \{R_1, \ldots, R_m\}$, and each $\overline{R_i} \in \overline{R}$, $1 \leq i \leq m$, contains all the rules whose objects "cross" membrane i. Then, we define:
- $V = \{a_i \mid a \in \overline{V}, i \in \mathcal{L}\}$;
- $C = \overline{C} = \emptyset$;

Let $h : \overline{V}^* \times \mathcal{L} \to V^*$ be a mapping such that
1) $h(a, i) = a_i, a \in \overline{V}, i \in \mathcal{L}$,
2) $h(\lambda, j) = \lambda, j \in \mathcal{L}$,
3) $h(x_1 x_2, j) = h(x_1, j)h(x_2, j), x_1, x_2 \in \overline{V}^*, j \in \mathcal{L}$,

- denote $w = h(\overline{w_1}, 1)h(\overline{w_2}, 2) \ldots h(\overline{w_m}, m)$, where $\overline{w_i}$ is the multiset present in region $i \in \mathcal{L}$ of $\overline{\Pi_m}$ at the beginning of the computation.
- R is defined as follows.

For each rule $A[\ ]_i \longrightarrow [\alpha]_i \in R_i$, $A \in \overline{V}$, $\alpha \in \overline{V}^*$, $i \in \mathcal{L}$, we add to R the rule $h(A, j)[\ ]_1 \longrightarrow [h(\alpha', i)]_1$, providing that j is the label of the outer membrane of membrane i.

For each rule $A[\neg B]_i \longrightarrow [\alpha]_i \in R_i$, $A, B \in \overline{V}$, $\alpha \in \overline{V}^*$, $i \in \mathcal{L}$, we add to R the rule $h(A, j)[\neg h(B, i)]_1 \longrightarrow [h(\alpha', 2)]_1$, providing that j is the label of the outer membrane of membrane i.

For each rule $[A]_i \longrightarrow [\ ]_i \alpha \in R_i$, $A, B \in \overline{V}$, $\alpha \in \overline{V}^*$, $i \in \mathcal{L}$, we add to R the rule $[h(A, i)]_1 \longrightarrow [\ ]_1 h(\alpha', j)$ providing that j is the outer membrane of membrane i.

For each rule $[A]_i \neg B \longrightarrow [\ ]_i \alpha \in R_i$, $A, B \in \overline{V}$, $\alpha \in \overline{V}^*$, $i \in \mathcal{L}$, we add to R the rule $[h(A, i)]_1 \neg h(B, j) \longrightarrow [\ ]_1 h(\alpha', j)$ providing that j is the outer membrane of membrane i.

Generally speaking, the purpose of membranes is to keep private the interior rules and objects from the neighboring ones and vice-versa. However, in our case we can express the passage of certain symbol through the membranes by using new symbols that we add to vocabulary and that encode both the crossed membrane label and the symbols from where they derive. In this way we can rewrite the rules, using the new symbols that perfectly describe the passage of objects in the membrane structure; consequently, in our case, we can shrink an arbitrarily membrane structure to only two membranes. The morphism used by the above construction accomplishes the encoding procedure.

The system Π simulates all the moves of $\overline{\Pi}$ and it stops whenever $\overline{\Pi}$ stops. However, in the halting configuration, in the designated output region of Π, there could be some objects representing the encoded version of the objects present in the regions of $\overline{\Pi}$. Therefore, we have to modify the above set of rules such that Π eliminates all these objects in order to generate the same set of numbers as $\overline{\Pi}$. This can be accomplish by producing an object D whenever a rule of $\overline{\Pi}$ is simulated (by adding the object D at the right hand side of each above rule), deleting it at each step (we add to R rules of type $D[\ ]_1 \longrightarrow [\lambda]_1$ and $[D]_1 \longrightarrow [\ ]_1\lambda$). Finally, if $\overline{\Pi}$ stops, then Π will not produce the object D anymore, hence the absence of this object can trigger an inhibited rule that deletes all the unnecessary objects. Consequently, we have that $NPACC_m(smp, inhR_1) = NPACC_2(smp, inhR_1), m \geq 2$.   □

Here we will prove that the family of sets of numbers generated by P systems with external inhibitors equals the family of sets of numbers generated by ET0L systems.

Theorem 1. $NPACC_2(smp, inhR_1) = NET0L.$

Proof. We will prove the result by showing that P systems with external inhibitors are equivalent with P systems with inhibitors, which at their turn, generates the same class of sets of numbers as the Parikh image of $ET0L$ as shown in [8]. Let $NP_1(smp, inhR_1)$ be the family of sets of numbers generated by P systems with inhibitors.

The proof of the inclusion $NP_1(smp, inhR_1) \supseteq NPACC_2(smp, inhR_1)$ is rather simple and is based on a similar encoding of regions into new objects as was presented above.

For the inclusion $NP_1(smp, inhR_1) \subseteq NPACC_2(smp, inhR_1)$ we will simulate the computation of a P system with one region $\Pi_{inh} = (V, C, \mu, w, R, i_0)$. We assume that the set of rules R contains rules of type $A \rightarrow \alpha$ or $A \rightarrow \alpha|_{\neg B}$, $A, B \in V$, $\alpha \in V^*$.

Let us consider the sets $\widetilde{V} = \{\widetilde{A} \mid A \in V\}$ and $\dot{V} = \{\dot{A} \mid A \in V\}$. In addition, let us define the morphisms:

$$h_1 : V^* \rightarrow \widetilde{V}^*, \text{ such that } h_1(A) = \widetilde{A}, \text{ for all } A \in V;$$
$$h_2 : V^* \rightarrow \dot{V}^*, \text{ such that } h_3(A) = \dot{A}, \text{ for all } A \in V.$$

We construct a P system $\Pi_{cc} = (\overline{V}, \overline{C}, \overline{\mu}, \overline{R}, \overline{i_0})$, simulating Π_{inh}, defined as follows:

$$\overline{V} = V \cup \widetilde{V} \cup \dot{V} \cup \{F\}; \quad w_1 = w;$$
$$C = \emptyset; \quad w_2 = w;$$
$$\mu = [\ [\ ]_2\ ]_1; \quad i_0 = 1.$$

The set of rules R is defined as follows[1]:

step i $A[\neg B] \longrightarrow [h_1(\alpha)h_2(\alpha)]$, for all rules $A \to \alpha|_{\neg B} \in R_{inh}$,

step i $A[\ ] \longrightarrow [h_1(\alpha)h_2(\alpha)]$, for all rules $A \to \alpha \in R_{inh}$,

step i $[A] \longrightarrow [\ ]F$, if exists $A \to \alpha \in R_{inh}$,

step i $[A]\neg B \longrightarrow [\ ]F$, if exists $A \to \alpha|_{\neg B} \in R_{inh}$,

step $i+1$ $F[\ ] \longrightarrow [\lambda]$,

step $i+1$ $[h_1(A)] \longrightarrow [\ ]h_1(A)$, for all objects $A \in V$,

step $i+2$ $h_1(A)[\ ] \longrightarrow [A]$, for all objects $A \in V$,

step $i+2$ $[h_2(A)]\neg F \longrightarrow [\ ]A$, for all $A \in V$.

Here is how the system Π_{cc} simulates the computation of Π_{inh}. First, remark that in order to correctly simulate the moves of Π_{inh}, we will maintain during the computation in both regions of Π_{cc} a copy of the multiset w – the multiset that represent the current configuration of Π_{inh}. This is especially useful when trying to simulate rules of type $A \to \alpha|_{\neg B} \in R_{inh}$ because we have to know whether or not the external inhibitor is present.

We assume that the system is in a configuration given by the strings $w_1 = w_2 = w$. The system attempts to execute simultaneously the rules of type

step i $A[\neg B] \longrightarrow [h_1(\alpha)h_2(\alpha)]$, for all rules $A \to \alpha|_{\neg B} \in R_{inh}$,

step i $A[\ ] \longrightarrow [h_1(\alpha)h_2(\alpha)]$, for all rules $A \to \alpha \in R_{inh}$,

step i $[A] \longrightarrow [\ ]F$, if exists $A \to \alpha \in R_{inh}$,

step i $[A]\neg B \longrightarrow [\ ]F$, if exists $A \to \alpha|_{\neg B} \in R_{inh}$.

Remark that the rules of first two types are used to generate inside the inner region, two copies of multiset α (represented by $h_1(\alpha)$ and $h_2(\alpha)$). In the same time, the rules of second type delete from region 2 the objects that were within the scope of rules of first type. In addition, remark that there are no other rules that can be applied in this step. Moreover, they produce in region 1 objects R; these objects will be used later for synchronizing the moments when multiset α appears in both regions.

Next, are executed the rules of type:

step $i+1$ $F[\ ] \longrightarrow [\lambda]$,

step $i+1$ $[h_1(A)] \longrightarrow [\ ]h_1(A)$, for all objects $A \in V$.

Observe that the presence of object(s) F in this computational step inhibits the executions of rules of type $[h_2(A)]\neg F \longrightarrow [\ ]A$, for all $A \in V$. Hence, in the third step, the rules of type

---

[1] For the present proof, we will simplify the notation by not including the membrane labels into the syntax of the rules; this is possible here since we have only two membranes and we do not allow the interaction with the environment. In addition, we have specified on their left hand side the moment of their executions during the simulation of one computational step in Π_{inh}.

$$h_1(A)[\ ] \longrightarrow [A], \text{ for all objects } A \in V,$$
$$[h_2(A)]\neg F \longrightarrow [\ ]A, \text{ for all } A \in V,$$

will be executed. The new objects appear at the same time in both regions of the system Π_{cc} and the simulation of the next computational step of Π_{inh} can start. Finally, if the system Π_{inh} stops because there are no rules to be applied, then also Π_{cc} halts.

Before we conclude, remark that the maximal parallelism as well as the universal clock is fundamental for the construction.

Consequently we have proved that the computation of an arbitrary P system with inhibitors can be simulated by a P system with external inhibitors, hence we have $NP_1(smp, inhR_1) \subseteq NPACC_2(smp, inhR_1)$. Therefore we have that $NP_1(smp, inhR_1) = NPACC_2(smp, inhR_1) = NET0L$. $\square$

The following theorem shows that P systems with external inhibitors and carriers are computationally complete.

Theorem 2. $NPACC_2(car_1, inhR_1) = NRE$.

Proof. The inclusion $NPACC_2(car_1, inhR_1) \subseteq NRE$ is assumed true by invoking the Turing-Church thesis.

For the inclusion $NPACC_2(car_1, inhR_1) \supseteq NRE$ we will simulate the computation of an arbitrary non-deterministic register machine $M = (n, \mathcal{P}, l_0, l_h)$. Such register machines are computational universal if $n \geq 3$.

We construct $\Pi = (V, C, \mu, w_1, w_2, R_1, i_0)$ as follows:

$$V = \{a_i, A_i, S_i \mid 1 \leq i \leq n\} \cup \{l, \bar{l}, \bar{\bar{l}}, \tilde{l}, \tilde{\tilde{l}}, L \mid l \in Lab(\mathcal{P})\} \cup \{c\}$$
$$\cup \{K, \overline{K}, \overline{\overline{K}}, \overline{\overline{\overline{K}}}, T_0, T_1, X, \overline{X}\};$$
$$C = \{c\};$$
$$\mu = [\ [\ ]_2\ ]_1;$$
$$w_1 = l_0 L_0 a_1^{k_1} \dots a_n^{k_n} c;$$
$$w_2 = A_1 \dots A_n S_1 \dots S_n;$$
$$i_0 = 1.$$

The set of rules R is defined as follows:
- for each instruction $(l_1 : ADD(j), l_2, l_3) \in \mathcal{P}$, the set R contains the rules:

$$l_1[\ ] \longrightarrow [A_1 \dots A_{j-1}A_{j+1} \dots A_n S_1 \dots S_n a_j l_2], \ l_1 \neq l_h,$$
$$l_1[\ ] \longrightarrow [A_1 \dots A_{j-1}A_{j+1} \dots A_n S_1 \dots S_n a_j l_3], \ l_1 \neq l_h,$$
$$L_1[\neg A_j] \longrightarrow [A_1 \dots A_n S_1 \dots S_n],$$
$$[l_2] \longrightarrow [\ ]l_2,$$
$$[l_3] \longrightarrow [\ ]l_3,$$
$$[a_j] \longrightarrow [\ ]a_j,$$
$$[A_i] \longrightarrow [\ ]\lambda, 1 \leq i \leq n,$$
$$[S_i] \longrightarrow [\ ]\lambda, 1 \leq i \leq n;$$

- for each instruction $(l_1 : \mathrm{SUB}(r), l_2, l_3) \in \mathcal{P}$, the set R contains the rules:

$l_1[\ ] \longrightarrow [A_1 \ldots A_n S_1 \ldots S_{j-1} S_{j+1} \ldots S_n \overline{l_1}], l_1 \neq l_h,$

$ca_j[\neg S_j] \longrightarrow [A_1 \ldots A_n S_1 \ldots S_n X],$

$L_1[\neg S_j] \longrightarrow [A_1 \ldots A_n S_1 \ldots S_n K],$

$[\overline{l_1}] \longrightarrow [\ ]\overline{\overline{l_1}},$

$[X] \longrightarrow [\ ]\overline{X},$

$\overline{\overline{l_1}}[\ ] \longrightarrow [\overline{\overline{l_1}} T_0 A_1 \ldots A_n S_1 \ldots S_n],$

$[K] \longrightarrow [\ ]\overline{K},$

$\overline{\overline{l_1}}\neg X \longrightarrow [\ ]\widetilde{l_3},$

$\overline{X}[\neg T_0] \longrightarrow [l_2],$

$\overline{K}[\ ] \longrightarrow [A_1 \ldots A_n S_1 \ldots S_n \overline{\overline{K}}],$

$[T_0] \longrightarrow [\ ]T_1,$

$\overline{\overline{l_1}}\neg\overline{K} \longrightarrow [\ ]\lambda,$

$[l_2] \longrightarrow [\ ]l_2 L_2,$

$T_1[\ ] \longrightarrow [A_1 \ldots A_n S_1 \ldots S_n],$

$\widetilde{l_3}[\ ] \longrightarrow [\widetilde{\widetilde{l_3}}],$

$[\widetilde{\widetilde{l_3}}] \longrightarrow [\ ]l_3 L_3,$

$[\overline{\overline{K}}] \longrightarrow [\ ]\overline{\overline{\overline{K}}},$

$\overline{\overline{\overline{K}}}[\ ] \longrightarrow [A_1 \ldots A_n S_1 \ldots S_n],$

$[A_i] \longrightarrow [\ ]\lambda, 1 \leq i \leq n,$

$[S_i] \longrightarrow [\ ]\lambda, 1 \leq i \leq n.$

Here is how the P system Π simulates the computation of the register machine M. Observe for the beginning that in the P system Π we will represent the number stored into register j of M as the multiplicity of the object a_j. In addition, remark that objects $A_j, S_j, 1 \leq j \leq n$, stand for the addition/subtraction command over register j – both in the simulation of an ADD or SUB instruction, the absence of symbol A_j or S_j allows the addition or deletion of one occurrence of object a_j. Objects $A_j, S_j, 1 \leq j \leq n$, are produced all the time during the computation except the moment when we actually want to increment or subtract one occurrence of object a_j from the multiset; at that moment we generate all objects $A_i, S_i, 1 \leq i \leq n$, such that $i \neq j$.

Let us see in more details how the simulation of the addition instruction $(l_1 : \mathrm{ADD}(j), l_2, l_3) \in \mathcal{P}$ works. Assume that at a certain moment during the computation, the current multisets in regions 1 and 2 are represented by the strings $w_1 = l_1 L_1 a_1^{k_1} \ldots a_n^{k_n} c$ and $w_2 = A_1 \ldots A_n S_1 \ldots S_n$, respectively. Then, the rules that can be executed are:

$l_1[\;] \longrightarrow [A_1 \ldots A_{j-1}A_{j+1} \ldots A_n S_1 \ldots S_n a_j l_2]$ or the rule involving l_3,

$[A_i] \longrightarrow [\;]\lambda, 1 \leq i \leq n,$

$[S_i] \longrightarrow [\;]\lambda, 1 \leq i \leq n.$

As a consequence of executing the above rules the next configuration will be described by $w_1 = L_1 a_1^{k_1} \ldots a_n^{k_n} c$ and $w_2 = A_1 \ldots A_{j-1}A_{j+1} \ldots A_n S_1 \ldots S_n a_j l_2$. Now, since in region 2 the object A_j is missing, then the rule

$L_1[\neg A_j] \longrightarrow [A_1 \ldots A_n S_1 \ldots S_n]$

can be executed; its role is to reestablish the initial configuration in region 2. Simultaneously, the system runs the rules

$[l_2] \longrightarrow [\;]l_2,$

$[a_j] \longrightarrow [\;]a_j,$

$[A_i] \longrightarrow [\;]\lambda, 1 \leq i \leq n,$

$[S_i] \longrightarrow [\;]\lambda, 1 \leq i \leq n.$

The rule $[l_2] \longrightarrow [\;]l_2$ produces in region 1 the object l_2 that corresponds to register machine label l_2. In addition, by the execution of the rule $[a_j] \longrightarrow [\;]a_j$, the number of objects a_j in region 1 (that corresponds to the number stored in register j of M) is incremented.

Concerning the simulation of the subtract instruction $(l_1 : \mathrm{SUB}(j), l_2, l_3) \in \mathcal{P}$, the system Π, being in a configuration represented by $w_1 = l_1 L_1 a_1^{k_1} \ldots a_n^{k_n} c$ and $w_2 = A_1 \ldots A_n S_1 \ldots S_n$, executes first the rules:

$l_1[\;] \longrightarrow [A_1 \ldots A_n S_1 \ldots S_{j-1}S_{j+1} \ldots S_n \overline{l_1}],$

$[A_i] \longrightarrow [\;]\lambda, 1 \leq i \leq n,$

$[S_i] \longrightarrow [\;]\lambda, 1 \leq i \leq n.$

In a similar manner as presented in the addition simulation, the rule $l_1[\;] \longrightarrow [A_1 \ldots A_n S_1 \ldots S_{j-1}S_{j+1} \ldots S_n \overline{l_1}]$ creates the context required for starting the simulation. The absence of object S_j in region 2 allows, in the second step, the (possible) execution of the rules

$ca_j[\neg S_j] \longrightarrow [A_1 \ldots A_n S_1 \ldots S_n X],$

$L_1[\neg S_j] \longrightarrow [A_1 \ldots A_n S_1 \ldots S_n K].$

Observe that in case there exists an object a_j in region 1, both rules are executed, while if there is not, only the rule $L_1[\neg S_j] \longrightarrow [A_1 \ldots A_n S_1 \ldots S_n K]$ will be executed.

In the same step, the rule $[\overline{l_1}] \longrightarrow [\;]\overline{\overline{l_1}}$ performs. As we will see, the objects derived from object l_1 will be used later to check whether or not the rule $ca_j[\neg S_j] \longrightarrow [A_1 \ldots A_n S_1 \ldots S_n X]$ was executed. Moreover, they will be also used to introduce in region 2 the objects $A_1, \ldots, A_n, S_1, \ldots, S_n$, that forbid a new addition or subtraction of objects a_j.

Let us consider the first case, i.e., the region 1 contains at least one object a_j. Then, as a consequence of executing the above rules we will have the multisets $w_1 = a_1^{k_1} \ldots a_j^{k_j - 1} \ldots a_n^{k_n} c$ and $w_2 = A_1^2 \ldots A_n^2 S_1^2 \ldots S_n^2 X K$. The following rules will be further applied:

$$\overline{\overline{l_1}}[\ ] \longrightarrow [\overline{\overline{l_1}}T_0A_1 \ldots A_n S_1 \ldots S_n],$$
$$[K] \longrightarrow [\ ]\overline{K},$$
and possibly the rule
$$[X] \longrightarrow [\ ]\overline{X}.$$

Remark that the objects derived from $\overline{l_1}$ are within the scope of rules that introduce at each odd step the objects $A_1, \cdots, A_n, S_1, \ldots, S_n$ (or the objects $A_1, \ldots, A_{j-1}, A_{j+1}, \ldots, A_n, S_1, \ldots, S_n$, in the first step). In a similar manner the objects derived from K are within the scope of rules that introduce at each even step the objects $A_1, \ldots, A_n, S_1, \ldots, S_n$. Anyway, at each step we delete by the rules $A_i \to \lambda$ and $S_i \to \lambda$, $1 \le i \le n$, all the objects A_i and S_i.

Now, since in the third step an object $\overline{X}$ was introduced in region 1 then, in the fourth step, the rule $[\overline{\overline{l_1}}]\neg X \longrightarrow [\ ]\widetilde{l_3}$ cannot be executed. Moreover, because in region 2 exists an object T_0 also the rule $\overline{X}[\neg T_0] \longrightarrow [l_2]$ cannot be executed. However, in the fourth step the rule $[T_0] \longrightarrow [\ ]T_1$ runs and it will allow, in the fifth step, the execution of the rule $\overline{X}[\neg T_0] \longrightarrow [l_2]$. In the same time, rule $[\overline{\overline{l_1}}]\neg \overline{K} \longrightarrow [\ ]\lambda$ is executed and so there will be no way to rewrite $\overline{\overline{l_1}}$ into $\widetilde{l_3}$ and furthermore into l_3. Finally, by rule $[l_2] \longrightarrow [\ ]l_2L_2$ the label of the new register machine instruction to be simulated is generated.

Now let us see what how the simulation is done when the system Π attempts to simulate the instruction $(l_1 : \mathrm{SUB}(j), l_2, l_3) \in \mathcal{P}$ in the case when the register j is empty. Then, the simulation works in a similar manner as in the above presented case with the main difference being that in the fourth step the rule $[\overline{\overline{l_1}}]\neg X \longrightarrow [\ ]\widetilde{l_3}$ is executed because the object $\overline{X}$ was not produced (the rules $ca_j[\neg S_j] \longrightarrow [A_1 \ldots A_n S_1 \ldots S_n X]$ and $[X] \longrightarrow [\ ]\overline{X}$ were not ran since the object a_j was missing from the initial multiset). So, the following rules are executed in sequence $\widetilde{l_3}[\ ] \longrightarrow [\widetilde{\widetilde{l_3}}]$, $[\widetilde{\widetilde{l_3}}] \longrightarrow [\ ]l_3L_3$. As a consequence, the symbol that corresponds to the next instruction to be simulated is generated.

If l_h is generated then the computation stops, having in the output region a number of objects a_i, $1 \le i \le n$, equals with the contents of register i of M. In this way the execution of the entire register machine program is simulated.

Since one can easily construct a register machine, equivalent with M, that in a successful computation clears its registers except a special designated one (the output register) we have that $NPACC_2(car, inhR_1) \supseteq NRE$.

Therefore, we have proved the equality $NPACC_2(car_1, inhR_1) = NRE$. □

4 Conclusions and Further Research

The model we introduced is based on the observation that various chemical reactions within a compartment of a living cell are activated from the neighboring compartments of the cell. We have proved that the family of sets of vectors of numbers generated by P systems with adjoining controlled communication rules when only simple inhibited rules are used equals the family of sets of numbers generated by ET0L systems. We have also proved the computational completeness if, in addition, carriers are used. In addition, we conjecture that similar results can be obtained if, instead of inhibited simple rules, promoted ones are considered.

Trying to get more "realistic", we believe that it is worthwhile to investigate the power of the above systems to whom we add execution times for the rules and to study their properties (for more details we refer to [3]). Another possible line for further research is to investigate the power of the systems not considering the family of sets of vectors of numbers generated as we have done here, but considering the family of Parikh images generated by such systems.

References

1. Bernardini, F., Manca, V.: P Systems with Boundary Rules. In: Păun, Gh., Rozenberg, G., Salomaa, A., Zandron, C. (eds.) Membrane Computing. LNCS, vol. 2597. Springer, Heidelberg (2003)
2. Calude, C., Păun, Gh., Rozenberg, G., Salomaa, A.: Multiset Processing. LNCS, vol. 2235. Springer, Heidelberg (2001)
3. Cavaliere, M., Sburlan, D.: Time and Synchronization in Membrane Systems. Fundamenta Informaticae 64(1-4), 65–77 (2005)
4. Ionescu, M., Sburlan, D.: On P Systems with Promoters/Inhibitors. Journal of Universal Computer Science 10(5), 581–599 (2004)
5. Păun, Gh.: Computing with Membranes. Journal of Computer and System Sciences 618(1), 108–143 (2000)
6. Păun, Gh.: Membrane Computing. An Introduction. Springer, Heidelberg (2002)
7. Salomaa, A., Rozenberg, G. (eds.): Handbook of Formal Languages. Springer, Heidelberg (1997)
8. Sburlan, D.: Further Results on P Systems with Promoters/Inhibitors. International Journal of Foundations of Computer Science 17(1), 205–221 (2006)
9. The P Systems Web Page: http://psystems.disco.unimib.it/

The Simplest Language Where
Equivalence of Finite Substitutions
Is Undecidable

Michal Kunc*

Department of Mathematics and Statistics, Masaryk University,
Janáčkovo nám. 2a, 602 00 Brno, Czech Republic
kunc@math.muni.cz
http://www.math.muni.cz/~kunc/

Abstract. We show that it is undecidable whether two finite substitutions agree on the binary language a^*b. This in particular means that equivalence of nondeterministic finite transducers is undecidable even for two-state transducers with unary input alphabet and whose all transitions start from the initial state.

1 Introduction

Existence of solutions of equations over words was proved decidable in a breakthrough paper of Makanin [12]. It is now also well known that solvability of word equations is decidable even for infinite rational systems of equations [2,1,5]. However, if we consider instead of equations over words equations over languages where the only operation is concatenation, the solvability problem becomes much more complicated.

If constants in equations are allowed to be any regular languages, existence of arbitrary solutions becomes undecidable already for very simple systems of equations [9]. But there is no such result about equations with only finite constants, and we also have virtually no knowledge about the solvability of finite systems of equations over finite or regular languages, i.e. where only finite or regular solutions are allowed. On the other hand, it is known that already for a very simple fixed rational system of such equations, it is even undecidable whether given finite languages form its solution. This can be equivalently formulated as undecidability of equivalence of two finite substitutions on a fixed regular language. Such a result was first proved for the regular language $a\{b,c\}^*d$ by Lisovik [11], and later improved to the language ab^*c in [8]. In this paper we prove that the same undecidability result actually holds even for the simplest language where the problem is not trivially decidable, namely for the language a^*b (note that for each language over a one-letter alphabet it is always sufficient to perform a certain fixed finite number of tests).

These results about finite substitutions can be also interpreted as undecidability of the equivalence problem for very restricted classes of finite transducers,

* Supported by the Grant no. 201/06/0936 of the Grant Agency of the Czech Republic.

continuing the long lasting search for such restrictions initiated in 1968 by the undecidability result of Griffiths [4] for general transducers (see [8] for a more comprehensive overview of related results). From this point of view, our result corresponds to transducers over unary input alphabet having only two states and no transitions starting from the final state. Therefore the result provides in this direction the smallest class where the equivalence problem is undecidable.

We assume the reader to be familiar with basic notions of formal language theory, which can be found for instance in [13]. As usual, we denote by X^* the set of all finite words obtained by concatenating elements of X together with the empty word ε. The length of a word $w \in A^*$ over an alphabet A is written as $|w|$. The concatenation of two words $u, v \in A^*$ is denoted by uv. The operation of concatenation is extended to languages by the rule $KL = \{ uv \mid u \in K, \ v \in L \}$. If some words $u, v, w \in A^*$ satisfy $uv = w$, then u and v are called a prefix and a suffix of w, respectively, and we write $u = wv^{-1}$ and $v = u^{-1}w$. We will also use the notation $A^{-1}L = \{ w \in A^* \mid Aw \cap L \neq \emptyset \}$. And if $uvw = z$ for some $u, v, w, z \in A^*$, then v is called a factor of z.

2   Main Result

We are going to prove our result in the following form:

Theorem 1. *It is undecidable whether given three finite languages K, L, M satisfy $K^n M = L^n M$ for every non-negative integer n. In particular, the system of language equations $\{ X^n Z = Y^n Z \mid n \in \mathbb{N}_0 \}$ is not equivalent to any finite system.*

This result can be easily translated to undecidability results for the problems of equivalence of finite substitutions on a^*b and equivalence of finite transducers. If we consider for finite languages K, L, M the substitutions $\varphi, \psi \colon \{a, b\}^* \to A^*$ defined by $\varphi(a) = K$, $\psi(a) = L$ and $\varphi(b) = \psi(b) = M$, then Theorem 1 can be directly reformulated in terms of deciding equivalence of finite substitutions as follows:

Corollary 1. *Equivalence of finite substitutions on the binary language a^*b is undecidable.*

As in [8], the language a^*b can be replaced by a two-state automaton with outputs of loops in the initial state defined according to K or L and outputs of transitions leading to the final state defined according to M; this way our result immediately implies undecidability of equivalence of finite transducers of a very special form.

Corollary 2. *It is undecidable whether given two nondeterministic two-state finite transducers with unary input alphabet, whose all transitions start from the initial state, are equivalent.*

3   Proof of the Result

The rest of the paper is devoted to proving Theorem 1. The proof generally follows the idea used to prove the analogous result in [8] for the equality $NK^nM = NL^nM$. We encode into our problem the universality problem for blind counter automata with all states final; these automata were first studied by Greibach [3], and the universality problem for the restricted class of blind counter automata we consider was proved undecidable by Lisovik [10] (see also [6]). A blind counter automaton consists of a nondeterministic finite automaton and one counter that can assume arbitrary integer values. Each transition of the automaton reads a letter from the input and possibly modifies the counter by either adding or subtracting one. No information about the current value of the counter is available to the automaton. The automaton accepts a given word if, starting from the initial state and zero-valued counter, it can read the word so that at the end the counter returns back to zero. We can assume that the automaton works over a binary alphabet $\{a, b\}$ (see [7, Corollary 1.2]).

Formally, such an automaton is a triple $\mathcal{S} = (S, 1, \delta)$, where $S = \{1, \ldots, s\}$ is its set of states, that we denote simply by numbers, 1 is the initial state, and $\delta = \{\mathbf{t}_1, \ldots, \mathbf{t}_t\} \subseteq S \times S \times \{a, b\} \times \{-1, 0, 1\}$ is the set of transitions, where a transition $\mathbf{t}_q = (i, j, x, k)$ starting from the state i and leading to the state j is labelled by x and has value k. Without loss of generality, we additionally assume that the initial state 1 is not the target of any transition of $\mathcal{S}$ and that all transitions starting from the initial state have value zero (since we are interested only in universality of the automaton, we can add to it a new initial state together with a zero-valued transition to the original initial state for both labels a and b). The set of all states reachable from a state i by a path labelled by a word $v \in \{a, b\}^*$ and having the sum of values of its transitions p will be denoted $\delta^*(i, v, p)$. The automaton accepts a word $v \in \{a, b\}^*$ if $\delta^*(1, v, 0) \neq \emptyset$.

In order to prove Theorem 1, we construct for each such automaton $\mathcal{S}$ finite languages K, L, M such that the automaton accepts all words of length n if and only if these languages satisfy $K^nM = L^nM$. This is achieved by encoding into both languages K^nM and L^nM all computations of the automaton $\mathcal{S}$ and by adding into L^nM all words representing zero-valued computations of length n.

The main difference from the construction in [8] is that we have to encode into the languages K and L not only transitions of the automaton and their labels and values, but also the start of the computation. In the case of substitutions on ab^*c, all the additional words in the second substitution are placed into the image of a, so that they start the composition of words corresponding to zero-valued computations. However, in our case these additional initial words have to be placed into L, so they can occur also in the middle of concatenations from L^nM, and we have to ensure that all incorrect words produced in L^nM by using the additional words in the middle belong also to K^nM.

The words in K^nM encoding computations of $\mathcal{S}$ consist of many concatenated copies of the words $(wa)^3$ and $(wb)^3$, where w is a certain word containing only auxiliary letters distinct from a and b, which can be cut in different places to represent the current state of the computation or the currently performed transition. A word read by the automaton $\mathcal{S}$ represented by a given element of $K^n M$ is obtained by replacing every copy of $(wa)^3$ by a and every copy of $(wb)^3$ by b.

The values of transitions are encoded as the number of words from K needed to construct the corresponding copy of $(wa)^3$ or $(wb)^3$. For zero-valued transitions, the corresponding copy is constructed using one element of K; for decrementing transitions, already one half of an element of K builds the whole copy; and for incrementing transitions, one and a half element of K is needed. This means that the number of copies of $(wa)^3$ and $(wb)^3$ in a word from $K^n M$ corresponding to a correct computation of $\mathcal{S}$ is equal to n only if the final value of the counter is zero. If the value is positive, less than n copies are produced, and if it is negative, then we obtain more than n copies.

In order to achieve this, all words in K and L will be in fact constructed by concatenating two words from a certain language L_1 of basic building blocks, and each of these blocks will represent one half of a word from K or L.

The language $L \supseteq K$ is constructed so that $L^n M$ contains in addition to the words of $K^n M$ exactly words obtained by concatenating n copies of words $(wa)^3$ and $(wb)^3$, so it allows to count the number of performed transitions and to compare it with the values obtained by accumulating the numbers during the production of a word in $K^n M$. These numbers are equal if and only if the sum of transition values of the computation corresponding to the word from $K^n M$ is zero. Therefore the equality $K^n M = L^n M$ becomes true precisely if all words over $\{a, b\}$ of length n correspond to some computation of $\mathcal{S}$ that resets the counter to zero.

This is achieved by constructing L by adding to the language K two words (the sublanguage J below) that start this counting. Then counting proceeds by building each copy of $(wa)^3$ and $(wb)^3$ using exactly two basic blocks, which corresponds to zero-valued transitions. Since the words initiating counting belong to L, they can actually occur anywhere in the power L^n, and therefore we need many auxiliary words in K for building words obtained in L^n by using words from J inside the product.

Finally, the language M serves for stopping the computation at any state by completing the currently unfinished copy of w.

The languages K, L, M will be defined over the alphabet

$$A = \{a, b, \#, \$, a_1, \ldots, a_s, b_1, \ldots, b_t, c_1, \ldots, c_t, e, \bar{e}, f, \bar{f}, g, \bar{g}\}.$$

First, consider the word $w = \#a_1 \cdots a_s \$ e\bar{e} f \bar{f} g \bar{g} b_1 c_1 \cdots b_t c_t$, and note that every letter occurs only once in w. We denote by ${}_x w$ the suffix of w starting with $x \in A$. Let $w_x = w({}_x w)^{-1}$ be the corresponding prefix, and let ${}_x w_y = (w_x)^{-1} w({}_y w)^{-1}$ be the factor of w determined by letters $x, y \in A$. The fact that the automaton $\mathcal{S}$ is in a state i will be represented by cutting the word w right before the corresponding letter a_i. Similarly, we will cut w before b_q or c_q if the automaton is just performing the transition $\mathbf{t}_q$. Both letters b_q and c_q are needed to deal with transitions incrementing the counter (recall that the corresponding copy of $(wa)^3$ or $(wb)^3$ should consist of three basic blocks; the first cut will occur right before b_q, and the second cut right before c_q). Only the first cutting point is needed for zero-valued transitions, which are produced from two blocks, and none for decrementing transitions. Finally, the letters e, f and g will be used for the same purpose during counting in $L^n M$: building a new copy of $(wa)^3$ or $(wb)^3$ will start with the letter g, and we will cut before e or f to separate the two blocks producing a copy of $(wa)^3$ or $(wb)^3$, respectively.

Now we take the two words responsible for starting counting in L^n:

$$J = \{(wa)^2 w_e, (wb)^2 w_f\}$$

Then we define an auxiliary language L_0, which is a union of several languages defined below. The first part of L_0 consists of the two words from J and some additional words used to compensate these two words in $K^n M$ whenever they occur in $L^n M$ somewhere else than at the very beginning:

$$\{\#^{-1}, \varepsilon, \#\}(wa)^2 w_e \{\$^{-1}, \varepsilon, \$, \bar{e}, \bar{f}\} \cup$$
$$\{\#^{-1}, \varepsilon, \#\}(wb)^2 w_f \{\bar{e}^{-1}, \varepsilon, \$, \bar{e}, \bar{f}\} \subseteq L_0$$

Note that these additional words differ from the words of J by at most one additional or removed letter on each side. This allows us to deal with words from J placed in the middle of an element of $L^n M$ as follows. We add to L_1 words that differ from words already in L_1 by only one letter at the beginning or at the end. Incorrectly placed words from J can then be modified either by adding one letter from the neighbouring word or by removing one letter and shifting it to the neighbouring word; in this way the word from J is replaced by another word belonging to L_1, and so a word from L containing it is replaced by a word from K. Let us point out that these modified words can again appear in $L^n M$ as words neighbouring with words from J. This is why we have to properly alternate addition and removal of letters in order to prevent words in L_1 from becoming too long. Actually, it turns out that we do not need to cut more than one letter and to add more than two letters on each side.

The rest of counting is performed by means of the words

$$\{_e waw_g, {}_f wbw_g, {}_g wawaw_e, {}_g wbwbw_f\} \subseteq L_0. \tag{1}$$

These words will be compensated by

$$_\$ waw_g \{\bar{f}^{-1}, \$, \bar{e}, \bar{f}\} \cup {}_{\bar{e}} wbw_g \{\bar{f}^{-1}, \$, \bar{e}, \bar{f}\} \cup$$
$$_{\bar{f}} wawaw_e \{\$^{-1}, \$, \bar{e}, \bar{f}\} \cup {}_{\bar{f}} wbwbw_f \{\bar{e}^{-1}, \$, \bar{e}, \bar{f}\} \subseteq L_0.$$

Now we can see why letters e, f and g are doubled in w. The only words we would like to have in $L^n M$ in addition to elements of $K^n M$ are those starting with a word from J and continuing by correctly appending words from (1) up to a word from M. Therefore, if an element of $L^n M$ starts with a word from J followed by several correctly placed words from (1), but it is not of the desired form, we still need to find this element also in $K^n M$. This is achieved by adding to L_1 words from (1) shifted by one letter to the left (for instance, $_\$ waw_{\bar{f}}$ arises from shifting $_ewaw_g$). Then we can produce this element of L^nM in K^nM by shortening the initial word by one letter and shifting all the correctly placed words from (1).

Now we add to L_0 some words for every transition $\mathbf{t}_q = (i, j, x, k)$. If it is an initial transition, i.e. $i = 1$, then we add the words $(wx)^2w_{b_q}$ and $_{b_q}wxw_{a_j}$. If this transition decrements the counter, i.e. $k = -1$, then we add only the word $_{a_i}wx(wx)^2w_{a_j}$. If $k = 0$ and $i \neq 1$, we add to L_0 the words $_{a_i}wxwxw_{b_q}$ and $_{b_q}wxw_{a_j}$. And finally, if $k = 1$ then we add the words $_{a_i}wxw_{b_q}$, $_{b_q}wxw_{c_q}$ and $_{c_q}wxw_{a_j}$. This concludes the definition of L_0.

To complete the definition of basic blocks, it remains to say which words created by adding letters to the beginning or to the end of words from L_0 should also belong to L_1. The main restriction on adding new words is that we have to ensure that there is no way of using these new words to produce in L^nM some word corresponding to a correct computation of an automaton. As every word representing a computation is a factor of some word from the language $\{wa, wb\}^*$, the easiest way to achieve this is to require that no new words are such factors. Let us denote the language consisting of all factors of words from $\{wa, wb\}^*$ by F.

Let L_1 consist of

$$L_1' = \{_\$waw_g\#, _{\bar{e}}wbw_g\#, _{\bar{f}}wawaw_e\#, _{\bar{f}}wbwbw_f\#\}$$

and of all words of the form xuy, where $x \in \{\varepsilon, \$, \bar{e}, \bar{f}\}$, $u \in L_0$ and $y \in \{\varepsilon, \#\}$, such that

$$x \neq \varepsilon \implies xu \notin F \quad \text{and} \quad y \neq \varepsilon \implies uy \notin F.$$

In this definition, we require both words xu and uy not to belong to F since then removal of one of the letters x and y from xuy produces again a word satisfying this condition and therefore belonging to L_1. This property is useful when we need to move one of the letters x and y to a word neighbouring with xuy in some element of L^nM.

In the following, we will also use the notations $K_0 = L_0 \setminus J$ and $K_1 = L_1 \setminus J$. Note that in particular $L_0 \subseteq L_1$ and $K_0 \subseteq K_1$.

Finally, words in K and L should be formed by concatenating arbitrary two basic blocks, and so we define $K = K_1^2$, $L = L_1^2$ and

$$M = \{\varepsilon\} \cup \bigcup_{i \neq 1}\{_{a_i}w_g, _{\bar{f}a_i}w_g\}.$$

Claim 1. *For every $n \geq 1$, the languages K, L, M satisfy*

$$L^nM = K^nM \cup \{(wa)^3, (wb)^3\}^nw_g.$$

Proof. Since $L \subseteq K$, in order to prove the claim, it is enough to verify the inclusions

$$\{(wa)^3, (wb)^3\}^nw_g \subseteq L^nM, \tag{2}$$

$$L^nM \subseteq K^nM \cup \{(wa)^3, (wb)^3\}^nw_g. \tag{3}$$

The inclusion (2) follows by induction on n using the empty word from M and the formulas

$$(wa)^3 w_g = (wa)^2 w_e \cdot {}_e waw_g \in L \,,$$
$$(wb)^3 w_g = (wb)^2 w_f \cdot {}_f wbw_g \in L$$

for the basis of the induction and

$$\{(wa)^3, (wb)^3\}^n (wa)^3 w_g = \{(wa)^3, (wb)^3\}^n w_g \cdot {}_g wawaw_e \cdot {}_e waw_g \subseteq L^{n+1} \,,$$
$$\{(wa)^3, (wb)^3\}^n (wb)^3 w_g = \{(wa)^3, (wb)^3\}^n w_g \cdot {}_g wbwbw_f \cdot {}_f wbw_g \subseteq L^{n+1}$$

for the induction step.

In order to verify (3), we take any $2n + 1$-element sequence σ belonging to

$$\underbrace{L_1 \times \cdots \times L_1}_{2n \text{ times}} \times M$$

whose concatenation does not belong to $\{(wa)^3, (wb)^3\}^n w_g$, and successively modify σ to replace all words from J in this sequence by words from K_1 without changing neither the length of the sequence nor the resulting concatenation. We distinguish several cases according to the word in σ directly preceding the word from J. When writing parts of σ, we separate neighbouring words in the sequence by means of the multiplication sign.

Let us start with the case of a word from J preceded by a word from L_1'. Then the word from J can be replaced using one of the following rules:

$$_\$waw_g \# \cdot J = {}_\$waw_{\bar{f}} \cdot \bar{f} \# J \subseteq K_0 \cdot K_1$$
$$_{\bar{e}}wbw_g \# \cdot J = {}_{\bar{e}}wbw_{\bar{f}} \cdot \bar{f} \# J \subseteq K_0 \cdot K_1$$
$$_{\bar{f}}wawaw_e \# \cdot J = {}_{\bar{f}}wawaw_\$ \cdot \$ \# J \subseteq K_0 \cdot K_1$$
$$_{\bar{f}}wbwbw_f \# \cdot J = {}_{\bar{f}}wbwbw_{\bar{e}} \cdot \bar{e} \# J \subseteq K_0 \cdot K_1 \,.$$

Next we deal with words from L_1 with $y = \varepsilon$, i.e. of the form xu, where $x \in \{\varepsilon, \$, \bar{e}, \bar{f}\}$ and $u \in L_0$. First, observe that no word from L_0 ends on a or b, and therefore $xu\# \in K_1$. This allows us to replace any word from J following xu by means of the inclusion

$$xu \cdot J = xu\# \cdot \#^{-1} J \subseteq K_1 \cdot K_0 \,.$$

Now consider the other words from L_1, i.e. of the form $xu\#$, where $x \in \{\varepsilon, \$, \bar{e}, \bar{f}\}$ and $u \in L_0$. In this case we have $xu \in L_1$. If $xu \in K_1$ then we can use

$$xu\# \cdot J \subseteq K_1 \cdot K_0 \,.$$

And if $xu \in J$ then $xu \in K_0\{\$, \bar{e}\}$ and from the inclusion $\#J \subseteq L_0 \setminus F$ we obtain

$$xu\# \cdot J \subseteq K_0\{\$, \bar{e}\}(L_0 \setminus F) \subseteq K_0 \cdot K_1 \,.$$

It remains to deal with a word from J that is the first word of σ. We can assume that it is already the only word from J in σ. Let us take the longest initial part ρ of σ which is also an initial part of some of the sequences of the form

$$\{(wa)^2 w_e \cdot {}_e waw_g, (wb)^2 w_f \cdot {}_f wbw_g\} \cdot \{{}_g wawaw_e \cdot {}_e waw_g, {}_g wbwbw_f \cdot {}_f wbw_g\}^* .$$

If the length of ρ is greater than one, we replace every word belonging to ρ except for the last one according to the following rules: $(wa)^2 w_e$ by $(wa)^2 w_\$ \in K_0$, $(wb)^2 w_f$ by $(wb)^2 w_{\bar{e}} \in K_0$, ${}_e waw_g$ by ${}_\$ waw_{\bar{f}} \in K_0$, ${}_f wbw_g$ by ${}_{\bar{e}} wbw_{\bar{f}} \in K_0$, ${}_g wawaw_e$ by ${}_{\bar{f}} wawaw_\$ \in K_0$ and ${}_g wbwbw_f$ by ${}_{\bar{f}} wbwbw_{\bar{e}} \in K_0$. Observe that this modification does not affect the concatenation of these words of ρ except for losing one letter at the end, which is one of $\$$, $\bar{e}$ and $\bar{f}$, depending on what the last word of ρ is. This letter together with the last word of ρ forms one of the following words, that remain to be dealt with: ${}_\$ waw_g$, ${}_{\bar{e}} wbw_g$, ${}_{\bar{f}} wawaw_e$ and ${}_{\bar{f}} wbwbw_f$. If the sequence ρ consists of only one word, then we have to deal with this sole word, which belongs to J. Let us denote this remainder of ρ in both cases by r, and let v be the word that follows r in σ.

First assume that the length of ρ is maximal possible, i.e. equal to $2n$. Then $v \in M$ is the last word of σ and the remainder r is either ${}_\$ waw_g$ or ${}_{\bar{e}} wbw_g$. The last word in σ cannot be the empty word $\varepsilon \in M$, since then the concatenation of σ would belong to $\{(wa)^3, (wb)^3\}^n w_g$. If the last word in σ is ${}_{a_i} w_g \in M$, then it is sufficient to add to v the redundant letter $\bar{f}$ of r to obtain the word $\bar{f}_{a_i} w_g \in M$. And if the last word in σ is $\bar{f}_{a_i} w_g \in M$, then we replace r by one of the words ${}_\$ waw_g \bar{f} \in K_0$ and ${}_{\bar{e}} wbw_g \bar{f} \in K_0$, and v by the word ${}_{a_i} w_g \in M$.

Now assume that the length of ρ is smaller than $2n$. If $v \in L_1'$ then $r \cdot v$ can be replaced by a sequence from $r\{\$, \bar{e}, \bar{f}\} \cdot A^{-1} L_1' \subseteq K_0 \cdot K_1$. Otherwise, we have $v = xuy \in K_1$. If $x \in \{\$, \bar{e}, \bar{f}\}$ then $uy \in L_1$ and we distinguish two cases. For $uy \in K_1$, we replace $r \cdot v$ by $rx \cdot uy \in K_0 \cdot K_1$. And for $uy \in J$, we replace it by $rx\# \cdot \#^{-1} uy \in K_1 \cdot K_0$.

It remains to consider words v where $x = \varepsilon$. In this case we replace r in the same way as the other words of ρ and denote by z the last letter of r, which is again one of $\$$, $\bar{e}$ and $\bar{f}$. It remains to verify that the word zv belongs to K_1, so that we can use it instead of v. This is certainly true if zv does not start with one of the pairs $\$e$, $\bar{e}f$ and $\bar{f}g$, since then we can take the letter z for x. So assume that zv starts with one of these pairs. Observe that v cannot be one of the words ${}_e waw_g$, ${}_f wbw_g$, ${}_g wawaw_e$ and ${}_g wbwbw_f$ because of the maximality of ρ. Therefore we have

$$v \in \{{}_e waw_g \#, {}_f wbw_g \#, {}_g wawaw_e \#, {}_g wbwbw_f \#\},$$

which means that $zv \in L_1' \subseteq K_1$. $\square$

To describe which words from $\{(wa)^3, (wb)^3\}^n w_g$ belong to $K^n M$ we will use the injective homomorphism $\varphi : \{a, b\}^* \to A^*$ defined by the rule $\varphi(x) = (wx)^3$ for $x \in \{a, b\}$.

Claim 2. *For every $n \geq 1$, the languages K, M satisfy*

$$K^n M \cap \{(wa)^3, (wb)^3\}^n w_g = \{\varphi(v)w_g \mid v \in \{a,b\}^n, \ \delta^*(1,v,0) \neq \emptyset\}.$$

Proof. In order to prove the converse inclusion, we show by induction with respect to the length of a word $v \in \{a,b\}^+$ that if $i \in \delta^*(1,v,p)$ for some state i and some integer p, then

$$\varphi(v)w_{a_i} \in K_0^{2|v|+p}. \tag{4}$$

Starting with words of length one, take any $x \in \{a,b\}$ and any transition from the initial state labelled by x, which is by our initial assumption of the form $\mathbf{t}_q = (1,j,x,0)$. Then

$$\varphi(x)w_{a_j} = (wx)^2 w_{b_q} \cdot {}_{b_q} wxw_{a_j} \in K_0^2.$$

Now assume that (4) is true for some v, i and p, and consider any transition starting from i, i.e. of the form $\mathbf{t}_q = (i,j,x,k)$ with $x \in \{a,b\}$ and $k \in \{-1,0,1\}$. If $k = -1$ then

$$\varphi(vx)w_{a_j} = \varphi(v)w_{a_i} \cdot {}_{a_i} wx(wx)^2 w_{a_j} \in K_0^{2|vx|+(p-1)}.$$

For $k = 0$, we have $i \neq 1$ since the initial state is not reachable by v, and therefore

$$\varphi(vx)w_{a_j} = \varphi(v)w_{a_i} \cdot {}_{a_i} wxwxw_{b_q} \cdot {}_{b_q} wxw_{a_j} \in K_0^{2|vx|+p}.$$

And finally, if $k = 1$ then

$$\varphi(vx)w_{a_j} = \varphi(v)w_{a_i} \cdot {}_{a_i} wxw_{b_q} \cdot {}_{b_q} wxw_{c_q} \cdot {}_{c_q} wxw_{a_j} \in K_0^{2|vx|+(p+1)}.$$

This proves the induction step.

If we now take $v \in \{a,b\}^n$ such that $i \in \delta^*(1,v,0)$, then (4) gives us $\varphi(v)w_{a_i} \in K^n$, which implies

$$\varphi(v)w_g = \varphi(v)w_{a_i} \cdot {}_{a_i} w_g \in K^n \cdot M,$$

as required.

Now we turn to the direct inclusion. Since all factors of words from the set $\{(wa)^3, (wb)^3\}^n w_g$ are in F, it is enough to consider only words from K and M which belong to F. Observe that all words in $K \cap F$ clearly belong to $(K_0 \cap F)^2$, and that the language $K_0 \cap F$ consists of the words

$$a_1 wawaw_\$, \ a_1 wawaw_e, \ (wa)^2 w_\$, \ a_1 wbwbw_{\bar{e}}, \ a_1 wbwbw_f, \ (wb)^2 w_{\bar{e}},$$
$$_e waw_g, \ _f wbw_g, \ _g wawaw_e, \ _g wbwbw_f, \ _\$ waw_{\bar{f}}, \ _{\bar{e}} wbw_{\bar{f}}, \ _{\bar{f}} wawaw_\$, \ _{\bar{f}} wbwbw_{\bar{e}}$$

and additionally all words corresponding to the transitions of the automaton $\mathcal{S}$. It is easy to verify by induction on m that every word $u \in K_1^m$, which is a prefix of a word from $\{wa, wb\}^*$, actually belongs to one of the following sets:

$$\{wa, wb\}^*\{w_\$, w_{\bar{e}}, w_{\bar{f}}\}$$
$$\{\varphi(v)w_{a_i} \mid i \in \delta^*(1, v, p),\ m = 2|v| + p\}$$
$$\{\varphi(v)(wx)^2 w_{b_q} \mid i \in \delta^*(1, v, p),\ \mathbf{t}_q = (i, j, x, 0) \in \delta,\ m = 2|v| + p + 1\}$$
$$\{\varphi(v)wxw_{b_q} \mid i \in \delta^*(1, v, p),\ \mathbf{t}_q = (i, j, x, 1) \in \delta,\ m = 2|v| + p + 1\}$$
$$\{\varphi(v)(wx)^2 w_{c_q} \mid i \in \delta^*(1, v, p),\ \mathbf{t}_q = (i, j, x, 1) \in \delta,\ m = 2|v| + p + 2\}$$

This in particular shows that u cannot belong to the language $\{(wa)^3, (wb)^3\}^n w_g$. Since $M \cap F$ contains, apart from the empty word, only words $_{a_i} w_g$, every element of $K^n M \cap \{wa, wb\}^* w_g$ belongs to the set

$$\{\varphi(v)w_{a_i} \cdot_{a_i} w_g \mid i \in \delta^*(1, v, p),\ 2n = 2|v| + p\}.$$

This finally gives

$$K^n M \cap \{(wa)^3, (wb)^3\}^n w_g \subseteq \{\varphi(v)w_g \mid \delta^*(1, v, p) \neq \emptyset,\ 2n = 2|v| + p,\ |v| = n\}$$
$$\subseteq \{\varphi(v)w_g \mid v \in \{a, b\}^n,\ \delta^*(1, v, 0) \neq \emptyset\},$$

which concludes the proof of the claim. $\qquad\square$

From these two claims we can now easily prove the main result.

Proof (of Theorem 1). It is sufficient to show that $\mathcal{S}$ accepts all words from $\{a, b\}^+$ if and only if the languages K, L, M constructed from $\mathcal{S}$ satisfy $K^n M = L^n M$ for every n. First assume that $\mathcal{S}$ accepts all words. Then Claim 2 gives us

$$K^n M \cap \{(wa)^3, (wb)^3\}^n w_g = \varphi(\{a, b\}^n) w_g = \{(wa)^3, (wb)^3\}^n w_g,$$

and therefore $K^n M = L^n M$ by Claim 1. Conversely, if $K^n M = L^n M$ then Claim 1 implies $\{(wa)^3, (wb)^3\}^n w_g \subseteq K^n M$, which turns Claim 2 into

$$\{\varphi(v)w_g \mid v \in \{a, b\}^n,\ \delta^*(1, v, 0) \neq \emptyset\} = \{(wa)^3, (wb)^3\}^n w_g,$$

showing $\delta^*(1, v, 0) \neq \emptyset$ for every word $v \in \{a, b\}^n$. This proves that every word in $\{a, b\}^+$ is accepted by $\mathcal{S}$. $\qquad\square$

References

1. Albert, M.H., Lawrence, J.: A proof of Ehrenfeucht's conjecture. Theoret. Comput. Sci. 41(1), 121–123 (1985)
2. Culik II, K., Karhumäki, J.: Systems of equations over a free monoid and Ehrenfeucht's conjecture. Discrete Math. 43(2-3), 139–153 (1983)
3. Greibach, S.A.: Remarks on blind and partially blind one-way multicounter machines. Theoret. Comput. Sci. 7(3), 311–324 (1978)
4. Griffiths, T.V.: The unsolvability of the equivalence problem for Λ-free nondeterministic generalized machines. J. Assoc. Comput. Mach. 15, 409–413 (1968)
5. Guba, V.S.: Equivalence of infinite systems of equations in free groups and semigroups to finite subsystems. Mat. Zametki 40(3), 321–324 (1986)

6. Halava, V., Harju, T.: Undecidability in integer weighted finite automata. Fund. Inform. 38(1-2), 189–200 (1999)
7. Halava, V., Harju, T.: Undecidability of the equivalence of finite substitutions on regular language. Theor. Inform. Appl. 33, 117–124 (1999)
8. Karhumäki, J., Lisovik, L.P.: The equivalence problem of finite substitutions on ab^*c, with applications. Internat. J. Found. Comput. Sci. 14(4), 699–710 (2003)
9. Kunc, M.: The power of commuting with finite sets of words. Theory Comput. Syst. 40(4), 521–551 (2007)
10. Lisovik, L.P.: An undecidable problem for countable Markov chains. Kibernetika 2, 1–6 (1991)
11. Lisovik, L.P.: The equivalence problem for finite substitutions on regular languages. Dokl. Akad. Nauk 357(3), 299–301 (1997)
12. Makanin, G.S.: The problem of solvability of equations in a free semigroup. Mat. Sb. 103(2), 147–236 (1977)
13. Rozenberg, G., Salomaa, A. (eds.): Handbook of Formal Languages. Springer, Heidelberg (1997)

Real-Time Reversible Iterative Arrays

Martin Kutrib[1] and Andreas Malcher[2]

[1] Institut für Informatik, Universität Giessen,
Arndtstr. 2, 35392 Giessen, Germany
kutrib@informatik.uni-giessen.de
[2] Institut für Informatik,
Johann Wolfgang Goethe-Universität Frankfurt,
60054 Frankfurt am Main, Germany
a.malcher@em.uni-frankfurt.de

Abstract. Iterative arrays are one-dimensional arrays of interconnected interacting finite automata. The cell at the origin is equipped with a one-way read-only input tape. We investigate iterative arrays as acceptors for formal languages. In particular, we consider real-time devices which are reversible on the core of computation, i.e., from initial configuration to the configuration given by the time complexity. This property is called real-time reversibility. It is shown that real-time reversible iterative arrays can simulate restricted variants of stacks and queues. It turns out that real-time reversible iterative arrays are strictly weaker than real-time reversible cellular automata. On the other hand, a non-semilinear language is accepted. We show that real-time reversibility itself is not even semidecidable, which extends the undecidability for cellular automata and contrasts the general case, where reversibility is decidable for one-dimensional devices. Moreover, we prove the non-semidecidability of several other properties. The closure under Boolean operations is also derived.

1 Introduction

Reversibility in the context of computing devices means that deterministic computations are also backward deterministic. Roughly speaking, in a reversible device no information is lost and every configuration occurring in any computation has at most one predecessor. Many different formal models have been studied in connection with reversibility. For example, reversible Turing machines have been introduced in [3], where it is shown that any irreversible Turing machine can be simulated by a reversible one. With respect to the number of tapes and tape symbols the result is significantly improved in [20]. On the opposite end of the automata hierarchy, reversibility in very simple devices, namely deterministic finite automata, has been studied in [2] and [25]. Here we study linear arrays of identical copies of deterministic finite automata. The single nodes, except the node at the origin, are homogeneously connected to its both immediate neighbors. Moreover, they work synchronously at discrete time steps. The distinguished cell at the origin, the communication cell, is equipped with a one-way

E. Csuhaj-Varjú and Z. Ésik (Eds.): FCT 2007, LNCS 4639, pp. 376–387, 2007.
© Springer-Verlag Berlin Heidelberg 2007

read-only input tape. Such devices are commonly called *iterative arrays* (IA). In connection with formal language recognition IAs have been introduced in [7]. In [9] a real-time acceptor for prime numbers has been constructed. A characterization of various types of IAs in terms of restricted Turing machines and several results, especially speed-up theorems, are given in [11]. Some recent results concern infinite hierarchies beyond linear time [13] and between real time and linear time [6], hierarchies depending on the amount of nondeterminism [5], and descriptional complexity issues [18].

Closely related to iterative arrays are *cellular automata* (CA). Basically, the difference is that cellular automata receive their input in parallel. That is, in our setting the input is fed to the cells 0 to $n-1$ in terms of states during a pre-initial step. There is no extra input tape. It is well known that conventional *real-time* cellular automata are strictly more powerful than *real-time* iterative arrays [26]. Our particular interest lies in reversible iterative arrays as acceptors for formal languages. An early result on general reversible CAs is the possibility to make any CA, possibly irreversible, reversible by increasing the dimension. In detail, in [27] it is shown that any k-dimensional CA can be embedded into a $(k + 1)$-dimensional reversible CA. Again, this result has significantly been improved by showing how to make irreversible one-dimensional CAs reversible without increasing the dimension [23]. A solution is presented which preserves the neighborhood but increases time ($O(n^2)$ time for input length n). Furthermore, it is known that even reversible one-dimensional one-way CAs are computationally universal [19,21]. Once a reversible computing device is under consideration, the natural question arises whether reversibility is decidable. For example, reversibility of a given deterministic finite automaton or of a given regular language is decidable [25]. For cellular automata, injectivity of the global transition function is equivalent to the reversibility of the automaton. It is shown in [1] that global reversibility is decidable for one-dimensional CAs, whereas the problem is undecidable for higher dimensions [14]. Additional information on some aspects of CAs may be found in [15]. All these results concern cellular automata with unbounded configurations. Moreover, in order to obtain a reversible device the neighborhood as well as the time complexity may be increased. In [8] it is shown that the neighborhood of a reverse CA is at most $n-1$ when the given reversible CA has n states. Additionally, this upper bound is shown to be tight. In connection with pattern recognition reversible two-dimensional partitioned cellular automata have been investigated in [22,24].

Here we consider iterative arrays that are reversible on the core of computation, i.e., from initial configuration to the configuration given by the time complexity. Our main interest is in fast computations, i.e., real-time computations. Consequently, we call such devices real-time reversible. Recently, cellular automata have been investigated under this aspect [17]. Here we continue this work. In particular, we want to know whether for a given real-time IA there exists a reverse real-time IA with the same neighborhood. At first glance, such a setting should simplify matters. But quite the contrary, we prove that real-time reversibility is not even semidecidable, which extends the undecidability for cellular automata and contrasts the general case, where reversibility is decidable for one-dimensional devices. Moreover, in Section 4 we prove the non-semidecidability of several other properties. Section 3 is devoted to the simulation of restricted variants of stacks and queues by real-time reversible iterative arrays. It turns out that real-time reversible iterative arrays are strictly weaker than real-time reversible cellular automata. In the following section we present some basic notions and definitions. The particularities in connection with reversibility are identified by an example which deals with a non-semilinear language. The closure under Boolean operations is also derived.

2   Real-Time Reversible Iterative Arrays

We denote the set of non-negative integers by $\mathbb{N}$. The empty word is denoted by λ, and the reversal of a word w by w^R. For the length of w we write $|w|$. We use $\subseteq$ for inclusions and $\subset$ for strict inclusions. An iterative array is a semi-infinite array of deterministic finite automata, sometimes called cells. Except for the leftmost automaton each one is connected to its both nearest neighbors. For convenience we identify the cells by their coordinates, i.e., by non-negative integers. The distinguished leftmost cell at the origin is connected to its right neighbor and, additionally, equipped with a one-way read-only input tape. At the outset of a computation the input is written with an infinite number of end-of-input symbols to the right on the input tape, and all cells are in the so-called quiescent state. The finite automata work synchronously at discrete time steps. The state transition of all cells but the communication cell depends on the current state of the cell itself and the current states of its neighbors. The state transition of the communication cell additionally depends on the current input symbol. The head of the one-way input tape is moved at any step to the right. With an eye towards language recognition the machines have no extra output tape but the states are partitioned in accepting and rejecting states.

Definition 1. *An* iterative array *(IA) is a system* $\langle S, A, F, s_0, \lhd, \delta, \delta_0 \rangle$, *where S is the finite, nonempty set of* cell states, *A is the finite, nonempty set of* input symbols, *$F \subseteq S$ is the set of* accepting states, *$s_0 \in S$ is the* quiescent state, *$\lhd \notin A$ is the* end-of-input symbol, *$\delta : S^3 \to S$ is the* local transition function for non-communication cells *satisfying $\delta(s_0, s_0, s_0) = s_0$, $\delta_0 : (A \cup \{\lhd\}) \times S^2 \to S$ is the* local transition function for the communication cell.

Let $\mathcal{M}$ be an IA. A configuration of $\mathcal{M}$ at some time $t \geq 0$ is a description of its global state which is a pair (w_t, c_t), where $w_t \in A^*$ is the remaining input sequence and $c_t : \mathbb{N} \to S$ is a mapping that maps the single cells to their current states. The configuration (w_0, c_0) at time 0 is defined by the input word w_0 and the mapping $c_0(i) = s_0$, $i \geq 0$, while subsequent configurations are chosen according to the global transition function Δ: Let (w_t, c_t), $t \geq 0$, be a configuration, then its successor configuration (w_{t+1}, c_{t+1}) is as follows:

$$(w_{t+1}, c_{t+1}) = \Delta\big((w_t, c_t)\big) \iff \begin{cases} c_{t+1}(i) = \delta\big(c_t(i-1), c_t(i), c_t(i+1)\big), i \geq 1, \\ c_{t+1}(0) = \delta_0\big(a, c_t(0), c_t(1)\big) \end{cases},$$

where $a = \lhd$, $w_{t+1} = \lambda$ if $w_t = \lambda$, and $a = a_1$, $w_{t+1} = a_2 \cdots a_n$ if $w_t = a_1 \cdots a_n$. Thus, the global transition function Δ is induced by δ and δ_0. A word is accepted by an IA if at some time during its course of computation the communication cell becomes accepting.

Let $\mathcal{M} = \langle S, A, F, s_0, \lhd, \delta, \delta_0 \rangle$ be an IA. A word $w \in A^*$ is *accepted* by $\mathcal{M}$, if there exists a time step $i \geq 1$ such that $c_i(0) \in F$. $L(\mathcal{M}) = \{w \in A^* \mid w$ is accepted by $\mathcal{M}\}$ is the *language accepted* by $\mathcal{M}$. Let $t : \mathbb{N} \to \mathbb{N}$, $t(n) \geq n+1$, be a mapping. If all $w \in L(\mathcal{M})$ are accepted with at most $t(|w|)$ time steps, then L is said to be of *time complexity* t.

Now we turn to iterative arrays that are reversible on the core of computation, i.e., from initial configuration to the configuration given by the time complexity. Consequently, we call them t-time reversible, if the time complexity t is obeyed. Reversibility is meant with respect to the possibility of stepping the computation back and forth. Due to the domain S^3 and the range S, obviously, the local transition function cannot be injective in general. But for reverse computation steps we may utilize the information which is available for the cells, that is, the states of their neighbors, respectively. So, we have to provide a reverse local transition function.

For some mapping $t : \mathbb{N} \to \mathbb{N}$ let $\mathcal{M} = \langle S, A, F, s_0, \lhd, \delta, \delta_0 \rangle$ be a t-time iterative array. Then $\mathcal{M}$ is defined to be t-*reversible* (REV-IA), if there exist reverse local transition functions δ_R and δ_0^R such that $\Delta_R(\Delta(c_i)) = c_i$, for all configurations c_i of $\mathcal{M}$, $0 \leq i \leq t(n) - 1$. The global transition functions Δ and Δ_R are induced by δ, δ_0 and δ_R, δ_0^R, respectively. For distinctness, we denote $\langle S, A, F, s_0, \lhd, \delta_R, \delta_0^R \rangle$ by $\mathcal{M}_R$. The head of the input tape is always moved to the *left* when the reverse transition function is applied.

The family of all languages that are accepted by some REV-IA with time complexity t is denoted by $\mathscr{L}_t(\text{REV-IA})$. If t equals the function $n+1$, acceptance is said to be in *real time*, and we write $\mathscr{L}_{rt}(\text{REV-IA})$.

In order to introduce some of the particularities in connection with reversible language recognition we continue with an example. The goal is to define a real-time REV-IA that accepts the non-context-free language $\{a^{n^2} b^{2n-1} \mid n \geq 1\}$, which is not even semilinear. We start with a conventional iterative array. Basically, the idea is to recognize time steps which are square numbers. To this end, assume k cells of the array are marked at time k^2. Then a signal can be emitted by the communication cell. The signal moves through the marked area, extends it by one cell, and moves back again. So, the signal arrives at the communication cell at time $(k+1)^2$. Finally, the number of bs is checked by sending another signal through the marked area and back. Figure 1 (left) shows an accepting computation. But the transition functions have to be extended in order to reject words not belonging to the language. To this end, we consider possible errors and observe that all errors can be detected by the communication cell. We identify the following errors: (1) the first input symbol is a b, (2) an a follows b, (3) the number of as is not a square number, (4) the number of bs is insufficient, or (5) there are too many bs. Accordingly, we provide rules to cope with the situations. An example of a rejecting computation is given in Figure 1 (middle).

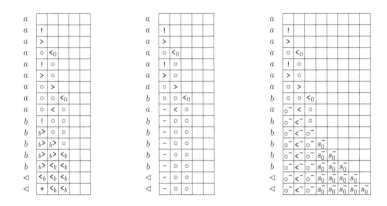

Fig. 1. Real-time IA accepting $\{a^{n^2}b^{2n-1} \mid n \geq 1\}$ (left), not being reversible (middle), rejecting reversibly (right). Cells in quiescent state are left blank.

Moreover, in our current construction the whole computation may get frozen before time step $n + 1$, for inputs not belonging to the language. Clearly, this implies non-reversibility. One reason is that for conventional computations we do not care about rejecting computations, except for keeping them rejecting. Nor do we care about the part of the computation that cannot influence the overall result, that is, the computation of cell $i \geq 1$ after time step $n + 1 - i$, i.e., the area below the diagonal starting from the lower left corner of the space-time diagram.

For reversible computations we do have to care about rejecting computations as well as for computations in the mentioned area. The idea of our construction is to send a signal from left to right which freezes the computation, whereby each cell passed through has to remember its current state. Clearly, this idea does not work in general. Sometimes much more complicated computations are necessary in order to obtain reversible computations. Next, we present the complete transition functions of a REV-IA accepting the language $\{a^{n^2}b^{2n-1} \mid n \geq 1\}$. For convenience, $\delta(p,q,r) = s$ is written as $pqr \rightarrow s$, and the same holds for δ_0. By x, z we denote arbitrary states.

δ_0	δ	δ_0	δ
$a\ s_0\ s_0 \rightarrow\ !$	$>\ s_0\ s_0 \rightarrow\ <_0$	$b\ s_0\ s_0 \rightarrow\ s_0^-$	$>^-\ s_0\ s_0 \rightarrow\ s_0^-$
$a\ !\ z \rightarrow\ >$	$>\ \circ\ y \rightarrow\ >$	$a\ {}_b{>}\ z \rightarrow\ {}_b{>}^-$	$>^-\ \circ\ y \rightarrow\ \circ^-$
$a\ >\ z \rightarrow\ \circ$	$x\ \circ\ <_0 \rightarrow\ <$	$a\ <_b\ z \rightarrow\ <_b^-$	$x^-\ \circ\ <_0 \rightarrow\ \circ^-$
$a\ \circ\ < \rightarrow\ !$	$x\ \circ\ < \rightarrow\ <$	$b\ \circ\ z \rightarrow\ \circ^-$	$x^-\ \circ\ < \rightarrow\ \circ^-$
$a\ \circ\ <_0 \rightarrow\ !$	$x\ >\ z \rightarrow\ \circ$	$b\ >\ z \rightarrow\ >^-$	$x^-\ >\ z \rightarrow\ >^-$
$b\ !\ s_0 \rightarrow\ <_b$	$x\ <_0\ s_0 \rightarrow\ \circ$	$b\ <_b\ z \rightarrow\ <_b^-$	$x^-\ <_0\ s_0 \rightarrow\ <_0^-$
$b\ !\ \circ \rightarrow\ {}_b{>}$	$x\ <\ z \rightarrow\ \circ$	$\triangleleft\ s_0\ z \rightarrow\ s_0^-$	$x^-\ <\ z \rightarrow\ <^-$
$b\ {}_b{>}\ <_b \rightarrow\ <_b$	${}_b{>}\ \circ\ \circ \rightarrow\ {}_b{>}$	$\triangleleft\ !\ z \rightarrow\ !^-$	${}_b{>}^-\ \circ\ \circ \rightarrow\ \circ^-$
$\triangleleft\ <_b\ z \rightarrow\ +$	${}_b{>}\ \circ\ s_0 \rightarrow\ <_b$	$\triangleleft\ >\ z \rightarrow\ >^-$	${}_b{>}^-\ \circ\ s_0 \rightarrow\ \circ^-$
	$x\ {}_b{>}\ <_b \rightarrow\ <_b$	$\triangleleft\ \circ\ z \rightarrow\ \circ^-$	$x^-\ {}_b{>}\ <_b \rightarrow\ {}_b{>}^-$
		$\triangleleft\ {}_b{>}\ z \rightarrow\ {}_b{>}^-$	$x^-\ s_0\ s_0 \rightarrow\ s_0^-$

The two blocks of transition rules at the left are for accepting computations. The third block provides rules for detecting that the input is of wrong format. The rules of the fourth block are for the freezing error signal. An example for a reversible rejecting computation is given in Figure 1 (right). It is not hard to obtain the reverse local transition functions δ_0^R and δ_R.

The technique to send a signal that freezes the computation in order to maintain reversibility in certain situations, yields the closure of the family in question under Boolean operations. A family of languages is said to be *effectively* closed under some operation, if the result of the operation can be constructed from the given language(s).

Lemma 2. *The family $\mathscr{L}_{rt}(REV\text{-}IA)$ is effectively closed under the Boolean operations complementation, union, and intersection.*

Proof. A real-time REV-IA accepts an input, if and only if the communication cell becomes accepting at any time during the computation. Once this happens, a freezing signal can be sent. So, the communication cell remembers forever that it has accepted. Next, we provide a copy of the non-accepting states, and modify the transition function δ_0 such that it changes to the corresponding new state if and only if the end-of-input symbol appears and the computation is not accepting. The real-time REV-IA is still reversible. Moreover, it accepts if and only if the state of the communication cell is accepting at time $n + 1$, and it rejects if and only if the state of the communication cell is a copied one at time $n + 1$. Simply defining the copied states to be accepting states shows the effective closure under complementation.

The closure under union and intersection follows by the well-known two-track technique. When two reversible computations are performed separately on different tracks, clearly, the whole computation is reversible, too. The interpretation of the states of the communication cell on both tracks at time $n + 1$ yields the closures.   □

3   Reversible Simulation of Data Structures

We next want to explore the computational capacity of real-time REV-IAs. To this end, we first consider the data structures *stack* and *queue*, and show that REV-IAs can simulate special variants thereof. We start with the stack. In detail, we consider real-time deterministic pushdown automata accepting linear context-free languages. Moreover, the stack behavior is restricted in such a way that in every step exactly one symbol is pushed on or popped from the stack. For convenience, we denote the family of languages accepted by such automata by DLR.

Theorem 3. *Every language from DLR belongs to the family $\mathscr{L}_{rt}(REV\text{-}IA)$.*

Proof. The principal idea is to simulate a stack by using the three register technique described in [4]. The content of the stack is stored in the first two registers and the third register is used as a buffer. Due to the fact that the given push-down automaton accepts a linear language, we know that there is at most one change between increasing and decreasing the stack. Thus, the stack behavior can be described as a sequence of push operations followed by a sequence of pop operations in which exactly one stack symbol is pushed or popped, respectively. An example of a computation is shown in Figure 2. Cells which represent an increasing stack or a decreasing stack are marked with the symbol ↑ or ↓, respectively. When the stack changes from increasing to decreasing, a signal → is sent to the right. With an eye towards reversibility, the communication cell stores a popped symbol on an additional track, and this information is shifted to the right. Thus, the history of the stack content is stored in the cells and can be reconstructed. Finally, the communication cell also simulates the state transition of the pushdown automaton, and stores the states on an additional track which also is shifted to the right.

We now have to show that a computation as described above is reversible. Obviously, shifting to the right can be made reversible by shifting to the left. The first phase of the computation (increasing stack height) is reversible, since in one step exactly one symbol has to be shifted backwards through the three registers in all cells. The second phase of the computation (decreasing stack height) consists of shifting the first register of each cell to the left. Thus, we obtain reversibility by shifting the first register of each cell to the right. The signal → is sent to the right and forces each cell to switch from increasing to decreasing. To achieve reversibility here, we send the signal to the left and observe that it meets the first entry of the stack (which is marked suitably) in the rightmost cell, which carries stack symbols in its first registers. In the next time step, cells with increasing stack height can be reconstructed. Finally,

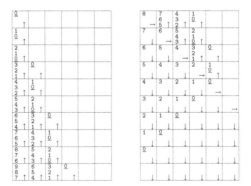

Fig. 2. Pushing (left) and popping (right) of ten pushdown symbols in real time. The left half of each cell contains the three registers for simulating the stack. The first two registers of the right half are used to store the current state of the communication cell and the last popped stack symbol, respectively. The last register indicates whether the stack is increasing (↑), decreasing (↓), or a switch takes place (→). The first entry of the stack is marked by underlining.

due the history of states and stack contents, the communication cell can always compute its predecessor state.   □

Now we can utilize the simulation in order to derive particular languages belonging to the family $\mathscr{L}_{rt}(\text{REV-IA})$.

Example 4. Every regular language as well as the languages $\{a^n b^n \mid n \geq 1\}$ and $\{wccw^R \mid w \in \{a, b\}^+\}$ are in DLR and, thus, belong to $\mathscr{L}_{rt}(\text{REV-IA})$.

Similar to the restricted stack behavior, we can show that real-time REV-IAs can simulate queues under certain conditions.

Lemma 5. *Let $\mathcal{Q}$ be an empty queue which is filled by a number of **in** operations, and then emptied by a sequence of **out** operations. Moreover, in every time step exactly one **in** or **out** operation is performed. Then $\mathcal{Q}$ can be simulated by a real-time REV-IA.*

Example 6. Consider the language $L = \{wcwc \mid w \in \{a, b\}^+\}$. First, the input prefix wc is inserted into a queue. Then, the content of the queue is removed step by step, whereby it is matched against the remaining input wc. Due to Lemma 5, language L belongs to the family $\mathscr{L}_{rt}(\text{REV-IA})$.

The restrictions on the data structures seem to be very natural, since in [16] it has been shown that the deterministic, linear context-free language $L = \{\$x_k\$ \cdots \$x_1 \# y_1 \$ \cdots \$ y_k \$ \mid x_i^R = y_i z_i, x_i, y_i, z_i \in \{a, b\}^*\}$ is not accepted by any conventional real-time iterative array. Related to iterative arrays are cellular automata. In [17] reversible language recognition by cellular automata is investigated. It is not hard to construct a real-time reversible cellular automaton which accepts L. So, we obtain a reversible relationship equal to the relationship in the conventional case.

Theorem 7. *The family $\mathscr{L}_{rt}(REV\text{-}IA)$ is properly included in the family of languages accepted by real-time reversible cellular automata.*

4   Decidability Questions

Now we turn to explore undecidable properties for real-time REV-IAs. To this end, we consider *valid computations of Turing machines* [10]. Roughly speaking, these are histories of accepting Turing machine computations. It suffices to consider deterministic Turing machines with a single tape and a single read-write head. Without loss of generality and for technical reasons, one can assume that any accepting computation has at least three and, in general, an odd number of steps. Therefore, it is represented by an even number of configurations. Moreover, it is assumed that the Turing machine cannot print blanks, and that a configuration is halting if and only if it is accepting. The language accepted by some machine $\mathcal{M}$ is denoted by $L(\mathcal{M})$.

Let S be the state set of some Turing machine M, where s_0 is the initial state, $T \cap S = \emptyset$ is the tape alphabet containing the blank symbol, $A \subset T$ is the input alphabet, and $F \subseteq S$ is the set of accepting states. Then a configuration of M can be written as a word of the form T^*ST^* such that $t_1 \cdots t_i s t_{i+1} \cdots t_n$ is used to express that M is in state s, scanning tape symbol t_{i+1}, and t_1 to t_n is the support of the tape inscription. The set of valid computations VALC($\mathcal{M}$) is now defined to be the set of words of the form $w_1 \texttt{\#\#\#\#} w_2 \texttt{\#\#\#\#} \cdots \texttt{\#\#\#\#} w_{2m} \texttt{\#\#\#\#}$, where $m \geq 2$, $\texttt{\#} \notin T \cup S$, $w_i \in T^*ST^*$ are configurations of M, w_1 is an initial configuration of the form $s_0 A^*$, w_{2m} is an accepting configuration of the form T^*FT^*, and w_{i+1} is the successor configuration of w_i, with $0 \leq i \leq 2m-1$. The set of *invalid computations* INVALC(M) is the complement of VALC(M) with respect to the alphabet $\{\texttt{\#}\} \cup T \cup S$.

The following lemma is the key tool to prove undecidability properties for real-time REV-IAs.

Lemma 8. *Let $\mathcal{M}$ be a Turing machine. Then the set* VALC[$\mathcal{M}$] *can be represented as the intersection of two languages from $\mathscr{L}_{rt}$(REV-IA).*

Proof. Let $L_3 = \{y \texttt{\#\#\#\#} z \texttt{\#\#\#\#} \mid z$ is successor of $y\}$. Then VALC($\mathcal{M}$) equals the intersection $L_1 \cap L_2$, where $L_1 = L_3^+$ and $L_2 = s_0 A^* \texttt{\#\#\#\#} L_3^* T^* FT^* \texttt{\#\#\#\#}$. We first describe how L_3 can be accepted by a real-time REV-IA. The principal idea is to read y, to compute its successor configuration y', and to store y' in a queue $\mathcal{Q}$. Then y' is matched against the input z. If $y' = z$, then the input is accepted, and otherwise rejected. To compute y' from y, we consider the four possible steps of $\mathcal{M}$: (1) ZqX is replaced by pZY, if $\mathcal{M}$ writes Y, and moves the head to the left, (2) $Zq\texttt{\#}$ is replaced by $pZY\texttt{\#}$, if $\mathcal{M}$ writes Y extending the support of the configuration at the right, and moves the head to the left, (3) qX is replaced by Yp, if $\mathcal{M}$ writes Y, and moves the head to the right, (4) $q\texttt{\#}$ is replaced by $Yp\texttt{\#}$, if $\mathcal{M}$ writes Y extending the support of the configuration at the left, and moves the head to the right. Thus, a string of length at most 3 is replaced by some string of length at most 4. Since $\mathcal{M}$ is deterministic, we know which of the above four cases applies. We add to the communication cell two buffers (`buffer1`, `buffer2`) of length 3 and 4, respectively. Now, the input is read and the first three input symbols are written into `buffer1`. Any next input symbol is written into the third place of `buffer1` and the contents of the third and second place are shifted to the second and first place, respectively. The content of the first place is inserted into the queue. If `buffer1` contains some triple to which δ can be applied, we write the result of the replacement into `buffer2`. While reading the next input symbols, the first place of `buffer2` is inserted into the queue, all other places are shifted to the left, and the last read input symbol is written into the rightmost place of `buffer2`. It should be remarked that the handling of `buffer2` is depending on the fact which case has to be simulated. If the fourth symbol $\texttt{\#}$ is read, we start to empty the queue and match the input with the queue. We observe that after filling `buffer1` within the first three time steps, there is exactly one `in` or `out` operation in $\mathcal{Q}$. Due to Lemma 5, we know that $\mathcal{Q}$ can be implemented reversibly. Finally, since the management of the buffers and the checking of the correct format take place in the communication cell only, the computation can be made reversible by storing a "protocol" of their states on an additional track similar to the construction in

Theorem 3. Thus, we obtain that $L_3 \in \mathscr{L}_{rt}(\text{REV-IA})$. By a simple extention of the queue simulation we obtain $L_1 = L_3^+ \in \mathscr{L}_{rt}(\text{REV-IA})$.

To accept L_2, we consider the above-constructed IA and check in the communication cell whether the input starts with a string of the form $s_0 A^*$####. If so, the simulation of the queue is started. Otherwise, the input is rejected. To check that the input has a suffix of the form $T^* F T^*$####, we implement a deterministic finite automaton $\mathcal{A}$ in the communication cell, which starts after every substring #### to check whether the next input is of the form $T^* F T^*$####. If the end-of-input symbol is read and $\mathcal{A}$ is in an accepting state, then the input is accepted, and otherwise rejected.   □

By Lemma 2 the family $\mathscr{L}_{rt}(\text{REV-IA})$ is closed under intersection. So, we obtain the following corollary.

Corollary 9. *Let $\mathcal{M}$ be a Turing machine. Then the set* VALC[$\mathcal{M}$] *belongs to the family $\mathscr{L}_{rt}(REV\text{-}IA)$.*

Theorem 10. *Emptiness, finiteness, infiniteness, universality, inclusion, equivalence, regularity, and context-freedom are not semidecidable for real-time REV-IAs.*

Proof. Let $\mathcal{M}$ be a Turing machine. By simple pumping arguments it can be shown that VALC($\mathcal{M}$) is context free if and only if $\mathcal{M}$ accepts a finite set. The finiteness problem of Turing machines is known to be not semidecidable. If, e.g., regularity were semidecidable for real-time REV-IAs, then we could semidecide whether a real-time REV-IA accepting VALC($\mathcal{M}$) accepts a regular language. Thus, we could semidecide the finiteness of Turing machines which is a contradiction. Similarly, the problems of emptiness, finiteness, inclusion, equivalence, and context-freedom can be proven to be not semidecidable for real-time REV-IAs. If VALC($\mathcal{M}$) is infinite, then $\mathcal{M}$ accepts an infinite set. Thus, infiniteness is also not semidecidable. Since $\mathscr{L}_{rt}(\text{REV-IA})$ is closed under complementation, universality is not semidecidable as well.   □

Theorem 11. *Let $\mathcal{M}$ be a real-time IA. It is not semidecidable whether or not $\mathcal{M}$ is real-time reversible.*

Proof. Let $\mathcal{M}'$ be a real-time REV-IA. We consider a real-time IA $\mathcal{M}''$ accepting $\{w\#\#va^{4(|w|+2)} \mid w \in L(\mathcal{M}'), v = \lambda$ if $|w|$ is even, and $v = \#a^4$ if $|w|$ is odd$\}$, where a and # are new alphabet symbols. We show that $\mathcal{M}''$ is reversible if and only if $L(\mathcal{M}')$ is empty. Since emptiness is not semidecidable for real-time reversible IAs, we obtain that reversibility is not semidecidable for real-time IAs.

The construction of $\mathcal{M}''$ may be sketched as follows. We consider four tracks. The correct input format is checked in the communication cell. On the first track the correct number of as is verified. To this end, all input symbols up to the first a are stored in a queue which uses $n = (|w| + 2)/2$ cells. In detail, we implement four copies of an empty queue (queue1, ..., queue4) and insert all incoming symbols into queue1. When reading the first a we start to empty queue1 and insert all deleted symbols from queue1 into queue2. When queue1 is empty, we start to empty `queue2` and insert their symbols into `queue3`. Then, `queue3` is copied into `queue4` and finally `queue4` is emptied. When reading the end-of-input symbol we know whether the number of as is correct, and can accept or reject. Clearly, the computation on the first track is reversible.

The second track is used to store the input $w = a_1 a_2 \ldots a_{|w|}$## in a stack. Observe that a_1 arrives in cell $n - 1$ at time $|w| + 2 + n$. Subsequently, the simulation of $\mathcal{M}'$ is started on the third track from right to left, i.e., cell $n - 1$ serves as the communication cell which gets its input from the stack stored on the second track. Additionally, two cells of $\mathcal{M}'$ are packed into one cell of the third track. Observe that the simulation takes time $|w| + 1$ and that we can decide after at most $2|w| + 3 + n$ time steps in cell $n - 1$ whether or not the input w is accepted in $\mathcal{M}'$. We want to achieve that cell $n - 1$ accepts or rejects after exactly $2(|w| + 2) + n$ time steps. This can be realized reversibly by using similar techniques as in Lemma 2. Thus, we can observe that the computation on the second and third track is reversible, because a stack can be simulated reversibly and $\mathcal{M}'$ is reversible.

If w is accepted, then we send a signal with speed $1/5$ to the left which causes each cell to enter some new permanent state g. Obviously, g erases any information from the cells. The communication cell changes to state g at time $5(|w| + 2)$. If the input has the correct format and the correct number of as, then we accept the input, and reject it in all other cases. Thus, $w \in L(\mathcal{M}')$ results in an accepting, non-reversible computation of $\mathcal{M}''$.

When the first a is read a reversible version of the FSSP according to the construction given in [12] is started on the fourth track. At time $2(|w| + 2) + n$ the cells $0, \ldots, n - 1$ are synchronized and change synchronously to some states which preserve the current contents of their second and third tracks. On the fourth track the reverse FSSP is started. So, the whole computation is reversible as long as no state g occurs. Thus, $w \notin L(\mathcal{M}')$ results in a non-accepting, reversible computation of $\mathcal{M}''$. Altogether, we obtain that $\mathcal{M}''$ is reversible if and only if $L(\mathcal{M}')$ is empty. □

References

1. Amoroso, S., Patt, Y.N.: Decision procedures for surjectivity and injectivity of parallel maps for tesselation structures. J. Comput. System Sci. 6, 448–464 (1972)
2. Angluin, D.: Inference of reversible languages. J. ACM 29, 741–765 (1982)
3. Bennet, C.H.: Logical reversibility of computation. IBM Journal of Research and Development 17, 525–532 (1973)
4. Buchholz, Th., Kutrib, M.: Some relations between massively parallel arrays. Parallel Comput. 23, 1643–1662 (1997)
5. Buchholz, Th., Klein, A., Kutrib, M.: Iterative arrays with limited nondeterministic communication cell. In: Words, Languages and Combinatorics III, pp. 73–87. World Scientific Publishing, Singapore (2003)
6. Buchholz, Th., Klein, A., Kutrib, M.: Iterative arrays with small time bounds. In: Nielsen, M., Rovan, B. (eds.) MFCS 2000. LNCS, vol. 1893, pp. 243–252. Springer, Heidelberg (2000)

7. Cole, S.N.: Real-time computation by n-dimensional iterative arrays of finite-state machines. IEEE Trans. Comput. C-18, 349–365 (1969)
8. Czeizler, E., Kari, J.: A tight linear bound on the neighborhood of inverse cellular automata. In: Caires, L., Italiano, G.F., Monteiro, L., Palamidessi, C., Yung, M. (eds.) ICALP 2005. LNCS, vol. 3580, pp. 410–420. Springer, Heidelberg (2005)
9. Fischer, P.C.: Generation of primes by a one-dimensional real-time iterative array. J. ACM 12, 388–394 (1965)
10. Hartmanis, J.: Context-free languages and Turing machine computations. Proc. Symposia in Applied Mathematics 19, 42–51 (1967)
11. Ibarra, O.H., Palis, M.A.: Some results concerning linear iterative (systolic) arrays. J. Parallel Distributed Comput. 2, 182–218 (1985)
12. Imai, K., Morita, K.: Firing squad synchronization problem in reversible cellular automata. Theoret. Comput. Sci. 165, 475–482 (1996)
13. Iwamoto, C., Hatsuyama, T., Morita, K., Imai, K.: Constructible functions in cellular automata and their applications to hierarchy results. Theoret. Comput. Sci. 270, 797–809 (2002)
14. Kari, J.: Reversibility and surjectivity problems of cellular automata. J. Comput. System Sci. 48, 149–182 (1994)
15. Kari, J.: Theory of cellular automata: a survey. Theoret. Comput. Sci. 334, 3–33 (2005)
16. Kutrib, M.: Automata arrays and context-free languages. In: Where Mathematics, Computer Science and Biology Meet, pp. 139–148. Kluwer Academic Publishers, Dordrecht (2001)
17. Kutrib, M., Malcher, A.: Fast reversible language recognition using cellular automata. In: Language and Automata Theory and Applications (LATA 2007). LNCS, Springer, Heidelberg 2007 (to appear)
18. Malcher, A.: On the descriptional complexity of iterative arrays. IEICE Trans. Inf. Syst. E87-D, 721–725 (2004)
19. Morita, K., Harao, M.: Computation universality of one dimensional reversible injective cellular automata. Trans. IEICE E72, 758–762 (1989)
20. Morita, K., Shirasaki, A., Gono, Y.: A 1-tape 2-symbol reversible Turing machine. Trans. of the IEICE E72, 223–228 (1989)
21. Morita, K.: Computation-universality of one-dimensional one-way reversible cellular automata. Inform. Process. Lett. 42, 325–329 (1992)
22. Morita, K., Ueno, S.: Parallel generation and parsing of array languages using reversible cellular automata. Int. J. Pattern Recog. and Artificial Intelligence 8, 543–561 (1994)
23. Morita, K.: Reversible simulation of one-dimensional irreversible cellular automata. Theoret. Comput. Sci. 148, 157–163 (1995)
24. Morita, K., Ueno, S., Imai, K.: Characterizing the ability of parallel array generators on reversible partitioned cellular automata. Int. J. Pattern Recog. and Artificial Intelligence 13, 523–538 (1999)
25. Pin, J.E.: On reversible automata. In: Simon, I. (ed.) LATIN 1992. LNCS, vol. 583, pp. 401–416. Springer, Heidelberg (1992)
26. Smith III, A.R.: Real-time language recognition by one-dimensional cellular automata. J. Comput. System Sci. 6, 233–253 (1972)
27. Toffoli, T.: Computation and construction universality of reversible cellular automata. J. Comput. System Sci. 15, 213–231 (1977)

The Computational Complexity of Monotonicity in Probabilistic Networks

Johan Kwisthout*

Department of Information and Computer Sciences, University of Utrecht,
P.O. Box 80.089, 3508 TB Utrecht, The Netherlands
johank@cs.uu.nl

Abstract. Many computational problems related to probabilistic networks are complete for complexity classes that have few 'real world' complete problems. For example, the decision variant of the inference problem (PR) is PP-complete, the MAP-problem is NP^{PP}-complete and deciding whether a network is monotone in mode or distribution is co-NP^{PP}-complete. We take a closer look at monotonicity; more specific, the computational complexity of determining whether the values of the variables in a probabilistic network can be ordered, such that the network is monotone. We prove that this problem – which is trivially co-NP^{PP}-hard – is complete for the class co-$NP^{NP^{PP}}$ in networks which allow implicit representation.

1 Introduction

Probabilistic networks [6] (also called Bayesian or belief networks) represent a joint probability distribution on a set of statistical variables. A probabilistic network is often described by a directed acyclic graph and a set of conditional probabilities. The nodes represent the statistical variables, the arcs (or lack of them) represent (in)dependencies induced by the joint probability distribution. Probabilistic networks are often used in decision support systems such as medical diagnosis systems (see e.g. [2] or [11]). Apart from their relevance in practical situations, they are interesting from a theoretical viewpoint as well.

Many problems related to probabilistic networks happen to be complete for complexity classes that have few 'real world' complete problems. For example, the decision variant of the inference problem PR (is the probability of a specific instantiation of a variable higher than p) is PP-complete [3], where the exact inference problem is #P-complete [7]. The problem of finding the most probable explanation (MPE), i.e., the most likely instantiation to all variables, has an NP-complete decision variant [8]. On the other hand, determining whether there is an instantiation to a subset of all variables (the so-called MAP variables), such that there exists an instantiation to the other variables with probability higher

---

* The work of this author was partially supported by the Netherlands Organisation for Scientific Research NWO.

E. Csuhaj-Varjú and Z. Ésik (Eds.): FCT 2007, LNCS 4639, pp. 388–399, 2007.

than p (the MAP problem) is NP^{PP}-complete [5]. Determining whether a network is monotone (in mode or in distribution) is co-NP^{PP}-complete [10].

Monotonicity is often studied in the context of probabilistic classification, where a network is constructed of evidence variables (like observable symptoms and test results), non-observable intermediate variables, and one or more classification variables. Informally, the conditional probability of a variable C given evidence variables E is monotone, if higher ordered instantiations to E always lead to higher values of C (isotone) or always lead to lower values of C (antitone). The question whether these relations are monotone is particularly relevant during the construction and verification of the network. Often, domain experts will declare that certain relations ought to be monotone, and the conditional probabilities in the network should then respect these assumptions. When a violation is found, the probabilities should be reconsidered, by elicitating better estimations or using more data to learn from.

While complexity results are known for the MONOTONICITY problem when all variables have fixed orderings, no such results have been obtained yet for the related problem where no such fixed order is presumed. Nevertheless, while variables sometimes have a trivial ordering (e.g., *always* > *sometimes* > *never*), such an ordering might be arbitrary, and determining a 'good' ordering might reduce the part of the network where monotonicity is violated. This problem is interesting from a theoretical viewpoint as well. If we can determine whether adding this extra 'degree of freedom' to the MONOTONICITY problem 'lifts' the complexity of the problem into a broader class, we gain some insight in the properties and power of these types of complexity classes.

In the remainder of this paper, some preliminaries are introduced in Section 2, and various monotonicity problem variants and their computational complexity are discussed in Section 3. In Section 4, we present an (alternative) proof for a restricted version of the MONOTONICITY problem as presented in [10]. This proof technique is then used in Section 5 to show that the MONOTONICITY problem with no fixed orderings, is indeed complete for the class $\text{co-NP}^{\text{NP}^{\text{PP}}}$ if we allow a (rather broad) implicit probability representation. Finally, in Section 6 these results are discussed and the paper is concluded.

2 Preliminaries

Before formalising the problems for which we want to determine their computational complexity, we first need to introduce some definitions and notations. Let $\mathbf{B} = (G, \Gamma)$ be a probabilistic or Bayesian network where Γ, the set of conditional probability distributions, is composed of rational probabilities, and let Pr be its joint probability distribution. The conditional probability distributions in Γ can be explicit, i.e., represented with look-up tables, or implicit, i.e., represented by a polynomial time computable function. If Γ consists only of explicit distributions then $\mathbf{B}$ will be denoted as an explicit network. Let $\Omega(V)$ denote the set of values that $V \in V(G)$ can take. Vertex A is denoted as a predecessor of B if $(A, B) \in A(G)$. For a node B with predecessors $A_1, \ldots, A_n$, the *configuration template* **A** is defined as $\Omega(A_1) \times \ldots \times \Omega(A_n)$; a particular instantiation of $A_1, \ldots,$ A_n will be denoted as a *configuration* of **A**.

Monotonicity can be defined as stochastical dominance (monotone in distribution) or in a modal sense (monotone in mode)[1], furthermore monotonicity can be defined on a global scale, or locally (only relations along the arcs of the network are considered). In this paper, we discuss global monotonicity in distribution only. We distinguish between weak and strong global monotonicity.

Definition 1 (global monotonicity [10]). *Let F_{Pr} be the cumulative distribution function for a node $V \in V(G)$, defined by $F_{Pr}(v) = Pr(V \leq v)$ for all $v \in \Omega(V)$. Let C be a variable of interest (e.g., the main classifier or output variable in the network), let E denote the set of observable variables, and let $\mathbf{E}$ be the configuration template of E. The network is strongly monotone in E, if either*

$$e \preceq e' \rightarrow F_{Pr}(c\,|\,e) \leq F_{Pr}(c\,|\,e') \text{ for all } c \in \Omega(C) \text{ and all } e, e' \in \mathbf{E}\,, \text{ or}$$
$$e \preceq e' \rightarrow F_{Pr}(c\,|\,e) \geq F_{Pr}(c\,|\,e') \text{ for all } c \in \Omega(C) \text{ and all } e, e' \in \mathbf{E}$$

The network is weakly monotone in E, if the network is strongly monotone in $\{E_i\}$, for all variables $E_i \in E$.

Note that all networks that are strongly monotone in E are also weakly monotone, but not vice versa: whereas the strong variant assumes a partial order on all configurations of E, the weak variant allows independent isotone or antitone effects for all variables in E. Put in another way: we could make a weakly monotone network also strongly monotone by reversing the order of the values of some variables in E, such that all effects are antitone or all effects are isotone.

The above notions of monotonicity assume an implicit *ordering* on the values of the variables involved. Such an ordering is often trivial (e.g., $x > \bar{x}$ and *always > sometimes > never*) but sometimes it is arbitrary, like an ordering of the organs that might be affected by a disease. Nevertheless, a certain ordering is necessary to determine whether the network is monotone, or to determine which parts of the network are violating monotonicity assumptions. Thus, for nodes where no a priori ordering is given, we want to order the values of these nodes in a way that maximises the number of monotone arcs. We define the notion of an *interpretation* of V to denote a certain ordering on $\Omega(V)$, the set of values of V. Note, that the number of distinct interpretations of a node with k values equals $k!$, the number of permutations of these values.

Definition 2 (interpretation). *An interpretation of $V \in V(G)$, denoted I_V, is a total ordering on $\Omega(V)$. We will often omit the subscript if no confusion is possible; for arbitrary interpretations we will often use σ and τ. The interpretation set $\mathbf{I_V}$ is defined as the set of all possible interpretations of V.*

---

[1] For variable set E, with value assignments e and e' ($e \prec e'$) and output C, the network is isotone in distribution if $\Pr(C\,|\,e)$ is stochastically dominant over $\Pr(C\,|\,e')$. The network is isotone in mode if the most probable $c \in C$ given assignment e is lower ordered than the most probable $c \in C$ given assignment e'.

In the remainder, we assume that the reader is familiar with basic concepts of computational complexity theory, such as the classes P, NP and co-NP, hardness, completeness, oracles, and the polynomial hierarchy (PH). For a thorough introduction to these subjects, we refer to textbooks like [1] and [4].

In addition to these concepts, we use the *counting hierarchy* (CH) [12,9]. The counting hierarchy closely resembles (in fact, *contains*) the polynomial hierarchy, but also involves the class PP (probabilistic polynomial time), i.e., the class that contains languages accepted by a non-deterministic Turing Machine where the *majority* of the paths accept a string if and only if it is in that language. Recall that the polynomial hierarchy can be characterised by alternating existential and universal operators applied to P, where $\exists^P P$ equals $\Sigma_1^p = \text{NP}$, $\forall^P P$ equals $\Pi_1^p = \text{co-NP}$, while $\forall^P \exists^P \forall^P \ldots P$ equals Π_k^p and $\exists^P \forall^P \exists^P \ldots P$ equals Σ_k^p, where k denotes the number of alternating quantifiers.

A convenient way to relate the counting hierarchy to the polynomial hierarchy is by introducing an additional operator $\mathcal{C}$, where $\mathcal{C}_0^p$ equals P, $\mathcal{C}_1^p$ equals PP, and in general $\mathcal{C}_{k+1}^p = \mathcal{C} \cdot \mathcal{C}_k^p = (\mathcal{C}_k^p)^{\text{PP}}$. Interesting complexity classes can be defined using these operators $\exists^P$, $\forall^P$ and $\mathcal{C}$ in various combinations. For example, $\exists^P \mathcal{C} P$ equals the class NP^{PP}, $\forall^P \mathcal{C} P$ equals co-NP^{PP} and $\exists^P \forall^P \mathcal{C} P$ equals $\text{NP}^{\text{NP}^{\text{PP}}}$. Default complete problems for these kind of complexity classes are defined by Wagner [12] using quantified satisfiability variants. In this paper we consider in particular the complete problems MAJSAT, E-MAJSAT, A-MAJSAT, EA-MAJSAT and AE-MAJSAT which will be used in the hardness proofs. These problems are proven complete by Wagner [12] for the classes PP, NP^{PP}, co-NP^{PP}, $\text{NP}^{\text{NP}^{\text{PP}}}$ and $\text{co-NP}^{\text{NP}^{\text{PP}}}$, respectively. In all problems, we consider a boolean formula ϕ with n variables X_i, with $1 \leq i \leq n$, and we introduce quantifiers to bound subsets of these variables.

MAJSAT
> **Instance:** Let $\mathbf{X}$ denote the configuration template for ϕ.
> **Question:** Does at least half of the instantiations of $\mathbf{X}$ satisfy ϕ?

E-MAJSAT
> **Instance:** Let $1 \leq k \leq n$, let $\mathbf{X_E}$ denote the configuration template for the variables X_1 to X_k and let $\mathbf{X_M}$ denote the configuration template for X_{k+1} to X_n.
> **Question:** Is there an instantiation to $\mathbf{X_E}$, such that at least half of the instantiations of $\mathbf{X_M}$ satisfy ϕ?

A-MAJSAT
> **Instance:** Sets $\mathbf{X_A}$ and $\mathbf{X_M}$ as in E-MAJSAT.
> **Question:** Does, for every possible instantiation to $\mathbf{X_A}$, at least half of the instantiations of $\mathbf{X_M}$ satisfy ϕ?

EA-MAJSAT
> **Instance:** Let $1 \leq k \leq l \leq n$, let $\mathbf{X_E}$, $\mathbf{X_A}$, and $\mathbf{X_M}$ denote the

configuration templates for the variables X_1 to X_k, X_{k+1} to X_l, and X_{l+1} to X_n, respectively.

Question: Is there an instantiation to $\mathbf{X_E}$, such that, for every possible instantiation of $\mathbf{X_A}$, at least half of the instantiations of $\mathbf{X_M}$ satisfy ϕ?

AE-Majsat

Instance: Sets $\mathbf{X_A}$, $\mathbf{X_E}$, and $\mathbf{X_M}$ as in EA-Majsat.

Question: Is there, for all instantiations to $\mathbf{X_A}$, a possible instantiation of $\mathbf{X_E}$, such that at least half of the instantiations of $\mathbf{X_M}$ satisfy ϕ?

In the remainder, we denote the complement of a problem P as NOT-P, with 'yes' and 'no' answers reversed with respect to the original problem P. Note that, by definition, if P is in complexity class C, then NOT-P is in co-C, and, likewise, if NOT-P is in C, then P is in co-C.

3   Monotonicity Variants and Their Complexity

In this paper, we study the computational complexity of various variants of global monotonicity. The following problems are defined on a probabilistic network $\mathbf{B} = (G, \Gamma)$, where $G = (V, A)$ is a directed acyclic graph.

1. The STRONG GLOBAL MONOTONICITY problem is the problem of testing whether $\mathbf{B}$ is strongly globally monotone, given an interpretation for V. This problem is co-NP^{PP}-complete [10] for explicit networks.
2. The WEAK GLOBAL MONOTONICITY problem is the problem of testing whether $\mathbf{B}$ is weakly globally monotone, given an interpretation for V.
3. The GLOBAL E-MONOTONICITY problem is the problem of testing whether there exists an interpretation to $\Omega(V)$, such that $\mathbf{B}$ is globally monotone.

Note that, if there exists an interpretation such that $\mathbf{B}$ is *weakly* monotone, there must also be an interpretation such that $\mathbf{B}$ is *strongly* monotone.

WEAK GLOBAL MONOTONICITY and GLOBAL E-MONOTONICITY will be discussed in Sections 4 and 5. In these sections, we use a proof technique introduced by Park and Darwiche [5] to construct a probabilistic network $\mathbf{B}_\phi$ from a given Boolean formula ϕ with n variables. For all variables $X_i (1 \leq i \leq n)$ in this formula, we create a variable X_i in G, with possible values T and F and uniform probability distribution. For each logical operator in ϕ, we create an additional variable, whose parents are the corresponding sub-formulas (or single variable in case of a negation operator) and whose conditional probability table is equal to the truth table of that operator. For example, the $\wedge$-operator would have a conditional probability of 1 if and only if both its parents have the value T, and 0 otherwise. Furthermore, we denote the top-level operator in ϕ with V_ϕ. In Figure 1 such a network is constructed for the formula $\neg(x_1 \vee x_2) \wedge \neg x_3$. Now, for any particular instantiation $\mathbf{x}$ of the set of all variables $\mathbf{X}$ in the formula we have that the probability of V_ϕ, given the corresponding configuration equals 1 if $\mathbf{x}$ satisfies ϕ, and 0 if $\mathbf{x}$ does not satisfy ϕ. Without any instantiation, the

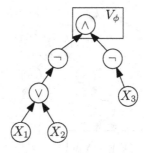

Fig. 1. The probabilistic network corresponding to $\neg(x_1 \vee x_2) \wedge \neg x_3$

probability of V_ϕ is $\frac{\#_q}{2^n}$, where $\#_q$ is the number of satisfying instantiations of $\mathbf{X}$. Using such constructs, Park and Darwiche proved that the decision variant of the MAP problem is NP^{PP}-complete; we will use this construct as a starting point to prove completeness results for WEAK GLOBAL MONOTONICITY and GLOBAL E-MONOTONICITY in the following Sections.

4 Weak Global Monotonicity

In this section, we present a proof for WEAK GLOBAL MONOTONICITY (with explicit representations) along the lines of Park and Darwiche. Note that STRONG GLOBAL MONOTONICITY has been proven to be co-NP^{PP}-complete in [10] using a reduction from the decision variant of the MAP-problem, and that hardness of the weak variant can be proven by restriction. We construct a reduction from A-MAJSAT, the relevant satisfiability variant discussed in Section 2, in order to facilitate our main result in the next section. First, we state the relevant decision problem:

WEAK GLOBAL MONOTONICITY
    Instance: Let $\mathbf{B} = (G, \Gamma)$ be a Bayesian network where Γ is composed of explicitly represented rational probabilities, and let Pr be its joint probability distribution. Let $C \in V(G)$ and $E \subseteq V(G) \setminus \{C\}$.
    Question: Is $\mathbf{B}$ weakly monotone in distribution in E?

We will see that any instance $(\phi, \mathbf{X_A}, \mathbf{X_M})$ of A-MAJSAT can be translated in a probabilistic network that is monotone, if and only if $(\phi, \mathbf{X_A}, \mathbf{X_M})$ is satisfiable. As an example, let us consider the formula $\phi = \neg(x_1 \wedge x_2) \vee \neg x_3$ (Figure 2), and let $\mathbf{X_A} = \{x_1, x_2\}$ and $\mathbf{X_M} = \{x_3\}$. This is a 'yes'-instance of A-MAJSAT because, for every configuration of $\mathbf{X_A}$, at least half of the configurations of $\mathbf{X_M}$ satisfies ϕ. From ϕ we construct a network $\mathbf{B}_\phi$ as described in the previous section. Furthermore, a node C ('classifier') and a node S ('select') is added, with arcs (S, C) and (V_ϕ, C), where V_ϕ is the top node in $\mathbf{B}_\phi$. S has values T and F with uniform distribution, and C has conditional probabilities as denoted in

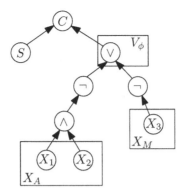

Fig. 2. Construct for hardness proof Monotonicity

Table 1. Conditional probability table for C

	c_1	c_2	c_3
$S = T \wedge V_\phi = T$	0.5	0.25	0.25
$S = T \wedge V_\phi = F$	0.5	0.25	0.25
$S = F \wedge V_\phi = T$	0.25	0.375	0.375
$S = F \wedge V_\phi = F$	0.375	0.5	0.125

Table 1. We claim, that $\Pr(C \mid S \wedge \mathbf{X_A})$ in the thus constructed network, is weakly monotone in distribution, if and only if the corresponding A-MAJSAT-instance $(\phi, \mathbf{X_A}, \mathbf{X_M})$ is satisfiable.

Theorem 1. WEAK GLOBAL MONOTONICITY *is co-*NP^{PP}*-complete*

Proof. To prove membership of co-NP^{PP}, we consider NOT-WEAK GLOBAL MONOTONICITY and prove membership of NP^{PP}. In this complement problem we want to know if there exist instantiations to the evidence variables E such that $\mathbf{B}$ is *not* monotone in distribution. This is clearly in NP^{PP}: we can non-deterministically choose instantiations $\mathbf{e_1} \preceq \mathbf{e_2}$ to E and values $c < c' \in \Omega(C)$, and verify that $F_{Pr}(c \mid e_1) \leq F_{Pr}(c' \mid e_1)$, but $F_{Pr}(c' \mid e_2) \leq F_{Pr}(c \mid e_2)$ since PR is PP-complete.

To prove co-NP^{PP}-hardness, we construct a transformation from A-MAJSAT. Let $(\phi, \mathbf{X_A}, \mathbf{X_M})$ be an instance of this problem, and let $\mathbf{B}_\phi$ be the network constructed from ϕ as described above. Given a particular configuration $\mathbf{x}$ of all n variables in $\mathbf{X_A} \cup \mathbf{X_M}$, $\Pr(V_\phi \mid \mathbf{x})$ equals 1 if $\mathbf{x}$ is a satisfying configuration and 0 if it is not, hence, for any configuration $\mathbf{X_A}$, $V_\phi \geq \frac{1}{2}$ if at least half of the instantiations to $\mathbf{X_M}$ satisfy ϕ. Since C is conditioned on V_ϕ, it follows from Table 1 that if any configuration of $\mathbf{X_A}$ leads to $\Pr(V_\phi) < \frac{1}{2}$, then C is no longer monotone in $S \wedge \mathbf{X_A}$, since $F_{Pr}(c_1 \mid S = T) > F_{Pr}(c_1 \mid S = F)$, but

$F_{Pr}(c_2 \mid S = T) < F_{Pr}(c_2 \mid S = F)$ as we can calculate[2] from the conditional probability table for C.

Thus, if we can decide whether $\mathbf{B}_\phi$ is weakly globally monotone in $S \cup \mathbf{X_A}$, we are able to decide $(\phi, \mathbf{X_A}, \mathbf{X_M})$. On the other hand, if $(\phi, \mathbf{X_A}, \mathbf{X_M})$ is a satisfying instantiation of A-MAJSAT, then $\Pr(V_\phi) \geq \frac{1}{2}$ and thus $\mathbf{B}_\phi$ is weakly globally monotone. Therefore WEAK GLOBAL MONOTONICITY is co-NP^{PP}-hard.   □

5 Global E-Monotonicity

We now use the proof technique from the previous section to prove that GLOBAL E-MONOTONICITY is co-NP$^{NP^{PP}}$-complete if we allow implicit representations for the conditional probability distributions, using a reduction from NOT-EA-MAJSAT, which is equivalent to AE-MAJSAT[3]. Again, we start with a formal definition of the relevant decision problem:

GLOBAL E-MONOTONICITY
  Instance: Let $\mathbf{B} = (G, \Gamma)$ be a Bayesian network where Γ is composed of rational probabilities, and let Pr be its joint probability distribution. Let $\Omega(V)$ denote the set of values that $V \in V(G)$ can take, and let $C \in V(G)$ and $E \subseteq V(G) \setminus \{C\}$.
  Question: Is there an interpretation I_V for all variables $V \in V(G)$, such that $\mathbf{B}$ is monotone in distribution in E?

We will see that any instance $(\phi, \mathbf{X_E}, \mathbf{X_A}, \overline{\mathbf{X_M}})$ of NOT-EA-MAJSAT can be translated in a probabilistic network for which exists an ordering of the values of its variables that makes the network monotone, if and only if $(\phi, \mathbf{X_E}, \mathbf{X_A}, \overline{\mathbf{X_M}})$ is *unsatisfiable*. As an example of the GLOBAL E-MONOTONICITY problem, let us consider the formula $\phi = \neg((x_1 \vee x_2) \wedge (x_3 \vee x_4)) \wedge x_5$ (Figure 3), let $\mathbf{X_E} = \{x_1, x_2\}$ and let $\mathbf{X_A} = \{x_3, x_4\}$ and $\overline{\mathbf{X_M}} = \{\neg x_5\}$. One can verify that this is indeed a 'yes'-instance of NOT-EA-MAJSAT: For every configuration of $\mathbf{X_E}$, the configuration $x_3 = x_4 = F$ ensures that at least half of the instantiations of $\overline{\mathbf{X_M}}$ satisfies ϕ. Thus, there does not exist an instantiation to $\mathbf{X_E}$, such that for *all* instantiations to $\mathbf{X_A}$ at least half of the instantiations of $\overline{\mathbf{X_M}}$ does *not* satisfy ϕ.

Again, we denote V_ϕ as the top node in $\mathbf{B}_\phi$. We now add three additional variables, C with values c_1, c_2, c_3, D with values d_1, d_2, and a variable ψ. This variable is implicitly defined and has (implicit) values w_0 to w_{2^m-1} $(m = \mid \mathbf{X_E} \mid)$ that correspond to configurations $\mathbf{x_E}$ of $\mathbf{X_E}$. These values are ordered by the

---

[2] $F_{Pr}(c_1 \mid S = T) = \Pr(c_1 \mid V_\phi = T \wedge S = T) \cdot \Pr(V_\phi = T) + \Pr(c_1 \mid V_\phi = F \wedge S = T) \cdot \Pr(V_\phi = F) = (0.5 + \epsilon) \cdot 0.5 + (0.5 - \epsilon) \cdot 0.5 = 0.5$. Likewise, $F_{Pr}(c_1 \mid S = F) = 0.25 \cdot (0.5 - \epsilon) + 0.375 \cdot (0.5 + \epsilon) = 0.3125 + 0.125\epsilon$. On the other hand, $F_{Pr}(c_2 \mid S = T) = \Pr(c_1 \mid S = T) + \Pr(c_2 \mid S = T) = 0.5 + 0.25 < F_{Pr}(c_2 \mid S = F) = \Pr(c_1 \mid S = F) + \Pr(c_2 \mid S = F) = (0.3125 + 0.125\epsilon) + (0.4375 + 0.125\epsilon) = 0.75 + 0.25\epsilon$.

[3] Thus, instead of $\forall^P \exists^P C$ we use the equivalent problem statement $\neg \exists^P \forall^P \neg C$. The reader can verify that this is an equivalent problem formulation.

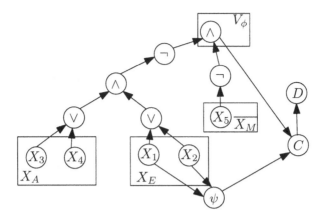

Fig. 3. Construct for hardness proof E-Monotonicity

binary representation of each configuration $\mathbf{x_E}$, e.g. for an instantiation $\mathbf{x_E} = X_1 = F, \ldots, X_{m-1} = F, X_m = T$ the binary representation would be $0 \ldots 01$ and therefore this particular configuration would correspond with w_1. Likewise, all possible configurations of $\mathbf{X_E}$ are mapped to values w_i of ψ. Furthermore, there are arcs (V_ϕ, C), (ψ, C), (C, D), and from every variable in $\mathbf{X_E}$ to ψ. The conditional probability $\Pr(C \mid V_\phi \wedge \psi)$ is defined in Table 2, where ϵ is a sufficiently small number, e.g. $\epsilon \leq \frac{1}{2^{m+3}}$. The conditional probabilities $\Pr(\psi \mid \mathbf{X_E})$ and $\Pr(D \mid C)$ are defined in Table 3. Note, that the conditional probability distribution of both ψ and C are defined implicitly.

Table 2. Conditional probability for C

$$\Pr(C = c_1 \mid V_\phi = T \wedge \psi = w_i) = \frac{1}{2} - \frac{i}{2^{m+1}} - \epsilon \quad \text{if } i = 0$$
$$\frac{1}{2} - \frac{i}{2^{m+1}} \quad \text{otherwise}$$
$$\Pr(C = c_2 \mid V_\phi = T \wedge \psi = w_i) = \frac{i+1}{2^m} - \frac{1}{2^{m+1}}$$
$$\Pr(C = c_3 \mid V_\phi = T \wedge \psi = w_i) = \frac{1}{2} - \frac{i+1}{2^{m+1}} + \epsilon \quad \text{if } i = 0$$
$$\frac{1}{2} - \frac{i+1}{2^{m+1}} \quad \text{otherwise}$$
$$\Pr(C = c_1 \mid V_\phi = F \wedge \psi = w_i) = 1 - \frac{i+1}{2^m}$$
$$\Pr(C = c_2 \mid V_\phi = F \wedge \psi = w_i) = \frac{i+1}{2^{m+1}}$$
$$\Pr(C = c_3 \mid V_\phi = F \wedge \psi = w_i) = \frac{i+1}{2^{m+1}}$$

Table 3. Conditional probabilities for $\Pr(D \mid C)$ and $\Pr(\psi \mid X)$

	d_1	d_2
c_1	0.20	0.80
c_2	0.40	0.60
c_3	0.60	0.40

$\Pr(\psi = w_i \mid \mathbf{x_E}) = 1$ if w_i corresponds to $\mathbf{x_E}$
$\phantom{\Pr(\psi = w_i \mid \mathbf{x_E}) =} 0 \quad$ otherwise

Table 4. Conditional probability for C in the example

	c_1	c_2	c_3		c_1	c_2	c_3
$\psi = w_0 \wedge V_\phi = T$	$0.5 - \epsilon$	0.125	$0.375 + \epsilon$	$\psi = w_0 \wedge V_\phi = F$	0.75	0.125	0.125
$\psi = w_1 \wedge V_\phi = T$	0.375	0.375	0.25	$\psi = w_1 \wedge V_\phi = F$	0.5	0.25	0.25
$\psi = w_2 \wedge V_\phi = T$	0.25	0.625	0.125	$\psi = w_2 \wedge V_\phi = F$	0.25	0.375	0.375
$\psi = w_3 \wedge V_\phi = T$	0.125	0.875	0	$\psi = w_3 \wedge V_\phi = F$	0	0.5	0.5

The conditional probabilities of D are chosen in such a way, that D is monotone in C if and only if $I_C = \{c_1 < c_2 < c_3\}$. We claim, that there is a possible interpretation I for all variables in $\mathbf{X_E} \cup \{\psi\}$ in the thus constructed network, such
that the network is globally monotone, if and only if the corresponding NOT-EA-
MAJSAT-instance is satisfiable. To support this claim, we take a closer look at the
example. The possible values of ψ are numbered as follows: $w_0 = \{X_1 = F, X_2 =
F\}, w_1 = \{X_1 = F, X_2 = T\}, w_2 = \{X_1 = T, X_2 = F\}, w_3 = \{X_1 = T, X_2 = T\}$.
For $i = 0 \ldots 3$, the conditional probability table $\Pr(C \mid V_\phi \wedge \psi = w_i)$ is defined as in Table 4. We have already seen that, for all configurations to $\mathbf{X_A}$, the
configuration $X_3 = X_4 = F$ of $\mathbf{X_E}$ ensures that the majority of the possible
configurations of $\mathbf{X_M}$ satisfies ϕ. Therefore, for all configurations of $\mathbf{X_A}$, there
is at least one configuration of ψ (namely, $\psi = w_0$) such that $V_\phi \geq \frac{1}{2}$. Since
C is conditioned on V_ϕ, we can calculate from the table that monotonicity is
violated: $F_{Pr}(c_1 \mid \psi = w_0) = 0.625 - 0.5\epsilon > F_{Pr}(c_1 \mid \psi = w_1) = 0.4375$ but
$F_{Pr}(c_2 \mid \psi = w_0) = 0.75 - 0.5\epsilon < F_{Pr}(c_2 \mid \psi = w_1) = 0.75$. Thus, independent of
the way the values of $\Omega(\psi)$ are ordered, there is always at least one violation of
monotonicity for any interpretation in $\mathbf{I}_\psi$ if $V_\phi \geq \frac{1}{2}$. If, on the other hand, there
does not exist such configuration to $\mathbf{X_E}$, then $V_\phi < \frac{1}{2}$ for all possible configurations of $\mathbf{X_E}$, and thus there is an ordering of the interpretations in $\mathbf{I}_\psi$ such that
$\Pr(C \mid \mathbf{X_A})$ is monotone. Note that we cannot assume an *a priori* ordering on the
values of ψ in this situation: although all configurations of $\mathbf{X_E}$ lead to $V_\phi < \frac{1}{2}$,
some may be closer to $\frac{1}{2}$ than others and thus, because of the conditioning on
V_ϕ, lead to higher values in C.

Theorem 2. GLOBAL E-MONOTONICITY *is* co-NP$^{\mathrm{NP}^{\mathrm{PP}}}$ *-complete*

Proof. For a membership proof we use NOT-WEAK GLOBAL MONOTONICITY as
an NP^{PP} oracle. With the aid of this oracle, an interpretation for the values of the
variables that violates monotonicity is an NP membership certificate for NOT-
GLOBAL E-MONOTONICITY , thus by definition the problem is in co-NP$^{\mathrm{NP}^{\mathrm{PP}}}$.

To prove co-NP$^{\mathrm{NP}^{\mathrm{PP}}}$-hardness, we construct a transformation from NOT-EA-
MAJSAT. Let $(\phi, \mathbf{X_E}, \mathbf{X_A}, \overline{\mathbf{X_M}})$ be an instance of this problem, and let $\mathbf{B}_\phi$ be
the network constructed from ϕ as described above. If $(\phi, \mathbf{X_E}, \mathbf{X_A}, \overline{\mathbf{X_M}})$ is *not*
satisfiable, then there exists an instantiation to ψ, such that $\Pr(V_\psi) \geq \frac{1}{2}$ and thus
– again, because of the conditioning of C on V_ψ – monotonicity is violated. But if
this is the case, then there exist $w_i, w_j \in \psi$ and $c < c' \in C$ such that $F_{Pr}(c \mid \psi =
w_i) \leq F_{Pr}(c' \mid \psi = w_i)$, but $F_{Pr}(c' \mid \psi = w_j) \leq F_{Pr}(c \mid \psi = w_j)$ independent of

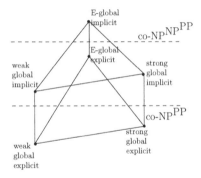

Fig. 4. Known complexity results

the ordering of the values of ψ. Note that the variable-and operator-nodes have binary values, making an ordering irrelevant[4], and the ordering on C and D is imposed by the conditional probability $\Pr(D\,|\,C)$. Thus, if we would be able to decide that there is an interpretation of the values of the variables of $\mathbf{B}_\phi$ such that $\mathbf{B}_\phi$ is globally monotone in distribution, we are able to decide $(\phi, \mathbf{X_E}, \mathbf{X_A}, \overline{\mathbf{X_M}})$. On the other hand, given that the network is globally monotone, we know that there cannot be an instantiation to $\mathbf{X_E}$ such that $(\phi, \mathbf{X_E}, \mathbf{X_A}, \overline{\mathbf{X_M}})$ is satisfied. Hence, GLOBAL E-MONOTONICITY is co-NP$^{\mathrm{NP}^{\mathrm{PP}}}$-hard.

It may not be obvious that the above construction can be made in polynomial time. Note that, however large X_E may become, both the conditional probabilities $\Pr(\psi\,|\,\mathbf{X_E})$ and $\Pr(C\,|\,V_\phi \wedge \psi)$ can be described using only a constant number of bits, since we explicitly allowed Γ to have implicit representations. Therefore, we need only time, polynomial in the size of the input (i.e., the NOT-EA-MAJSAT instance), to construct $\mathbf{B}_\phi$. □

6   Conclusion

In this paper, several variants of the MONOTONICITY problem in probabilistic networks were introduced. In Figure 4, the known complexity results for strong and weak global monotonicity variants, with explicit or implicit conditional probability distribution, and fixed or variable variable orderings are presented. The main result is the completeness proof of GLOBAL E-MONOTONICITY with implicit probability representation. It is established that this problem is complete for the class co-NP$^{\mathrm{NP}^{\mathrm{PP}}}$, a class for which few real-world problems are known to be complete. Unfortunately, a similar complexity result for the variant with explicit representation (or, with a representation where the variables are explicitly defined, while the probabilities are implicit) could not be established. This leaves us with a number of problems that are either in co-NP^{PP}, co-NP$^{\mathrm{NP}^{\mathrm{PP}}}$, or somewhere in between.

---

[4] if $\mathbf{B}_\phi$ is isotone for $x < \bar{x}$, it is antitone for $\bar{x} < x$ and vice versa.

Acknowledgements

The author wishes to thank Hans Bodlaender, Gerard Tel, and Linda van der Gaag, for fruitful discussions on this subject and useful comments on earlier drafts of this paper. Furthermore, the author is very grateful to three anonymous reviewers for their constructive critiques.

References

1. Garey, M.R., Johnson, D.S.: Computers and Intractability. A Guide to the Theory of NP-Completeness. W. H. Freeman and Co., San Francisco (1979)
2. Kappen, B., Wiegerinck, W., Akay, E., Nijman, M., Neijt, J., van Beek, A.: Promedas: A diagnostic decision support system. In: Proceedings of the 15th Belgian-Dutch Conference on Artificial Intelligence (BNAIC'03), pp. 455–456 (2003)
3. Littman, M.L., Majercik, S.M., Pitassi, T.: Stochastic boolean satisfiability. Journal of Automated Reasoning 27(3), 251–296 (2001)
4. Papadimitriou, C.H.: Computational Complexity. Addison-Wesley, Reading (1994)
5. Park, J.D., Darwiche, A.: Complexity results and approximation settings for MAP explanations. Journal of Artificial Intelligence Research 21, 101–133 (2004)
6. Pearl, J.: Probabilistic Reasoning in Intelligent Systems: Networks of Plausible Inference. Morgan Kaufmann, Palo Alto (1988)
7. Roth, D.: On the hardness of approximate reasoning. Artificial Intelligence 82(1-2), 273–302 (1996)
8. Shimony, S.E.: Finding MAPs for belief networks is NP-hard. Artificial Intelligence 68, 399–410 (1994)
9. Torán, J.: Complexity classes defined by counting quantifiers. Journal of the ACM 38, 752–773 (1991)
10. van der Gaag, L.C., Bodlaender, H.L., Feelders, A.: Monotonicity in Bayesian networks. In: Twentieth Conference on Uncertainty in Artificial Intelligence, pp. 569–576. AUAI Press (2004)
11. van der Gaag, L.C., Renooij, S., Witteman, C.L.M., Aleman, B.M.P., Taa, B.G.: Probabilities for a probabilistic network: a case study in oesophageal cancer. Artificial Intelligence in Medicine 25, 123–148 (2002)
12. Wagner, K.W.: The complexity of combinatorial problems with succinct input representation. Acta Informatica 23, 325–356 (1986)

Impossibility Results on Weakly Black-Box Hardness Amplification

Chi-Jen Lu[1], Shi-Chun Tsai[2], and Hsin-Lung Wu[2]

[1] Institute of Information Science, Academia Sinica, Taipei, Taiwan
cjlu@iis.sinica.edu.tw
[2] Department of Computer Science, National Chiao-Tung University, Hsinchu, Taiwan
{sctsai,hsinlung}@csie.nctu.edu.tw

Abstract. We study the task of hardness amplification which transforms a hard function into a harder one. It is known that in a high complexity class such as exponential time, one can convert worst-case hardness into average-case hardness. However, in a lower complexity class such as NP or sub-exponential time, the existence of such an amplification procedure remains unclear.

We consider a class of hardness amplifications called weakly black-box hardness amplification, in which the initial hard function is only used as a black box to construct the harder function. We show that if an amplification procedure in $\mathsf{TIME}(t)$ can amplify hardness beyond an $O(t)$ factor, then it must basically embed in itself a hard function computable in $\mathsf{TIME}(t)$. As a result, it is impossible to have such a hardness amplification with hardness measured against $\mathsf{TIME}(t)$. Furthermore, we show that, for any $k \in \mathbb{N}$, if an amplification procedure in $\Sigma_k \mathsf{P}$ can amplify hardness beyond a polynomial factor, then it must basically embed a hard function in $\Sigma_k \mathsf{P}$. This in turn implies the impossibility of having such hardness amplification with hardness measured against $\Sigma_k \mathsf{P}/\mathrm{poly}$.

1 Introduction

Randomness and hardness are two fundamental notions in complexity theory. They turn out to be closely related, and their connection has provided an important tool in our study of the BPP versus P problem. A major open problem in complexity theory, the BPP versus P problem asks whether or not all randomized polynomial-time algorithms can be derandomized into deterministic ones. A general framework for removing randomness from randomized algorithms is to construct the so-called psudorandom generator (PRG), which stretches a short random seed into a long random-looking string. Blum and Micali [4] and Yao [25] first showed that PRGs can be built from cryptographic one-way functions. In a seminal work, Nisan and Wigderson [17] made explicit the concept of trading hardness for randomness, and showed how to construct PRGs based on hard-to-compute Boolean functions. The construction of a PRG from an average-case hard function can be done efficiently [17]. To get a stronger result, one would like to start from a slightly hard function and transform it into a much harder one before using it to build a PRG. This is the task known as hardness amplification.

E. Csuhaj-Varjú and Z. Ésik (Eds.): FCT 2007, LNCS 4639, pp. 400–411, 2007.

Let us be more precise. We say that a Boolean function $f : \{0,1\}^n \to \{0,1\}$ is ε-hard against circuits of size s if any such a circuit attempting to compute f fails on at least ε fraction of inputs. We will focus more on the parameter ε and usually pay less attention to the parameter s. We call f worst-case hard, mildly hard, and average-case hard if ε is $1/2^n$, $1/\text{poly}(n)$, and $1/2 - 1/2^{\Omega(n)}$, respectively. The task of hardness amplification is to design a procedure that transforms any function f which is ε-hard against circuits of size s into a function $\bar{f}$ which is $\bar{\varepsilon}$-hard against circuits of size $\bar{s}$, with $\bar{\varepsilon} > \varepsilon$ and $\bar{s}$ close to s. One typically would like to have $\bar{f}$ and f in the same complexity class so as to obtain the relation between different hardness conditions within the same class. After a series of works [25,17,3,9], Impagliazzo and Wigderson [12] finally were able to convert a function in E which is worst-case hard into one which is average-case hard, with hardness against circuits of exponential size. As a result, they have BPP = P under the assumption that such a worst-case hard function exists. Since then, simpler proofs and better trade-offs have been obtained [20,11,21,22].

On the other hand, it is known that from a PRG one can obtain a worst-case hard function in E [17]. Thus, in a high complexity class such as E, the notions of pseudrandomness and various degrees of hardness are equivalent. Are they still equivalent in a lower complexity class such as NP? Using the technique in [10], one can transform a PRG into an average-case hard function using a procedure in NP, and by [17], one can also transform an average-case hard function into a PRG in polynomial time. This implies that the notions of average-case hardness and pseudorandomness remain equivalent for functions in NP. However, the relationship between worst-case and average-case hardness is not clear, as all the known transformations from a worst-case hard function to even a mildly hard one require exponential time (or linear space) [12,20,13,11,21,22]. In fact, it appears very difficult to bring down the complexity, and we would like to show that it is actually impossible. For this, we need to clarify what type of hardness amplification we are talking about, especially given the possibility that an average-case hard function may indeed exist.

Black-Box Constructions. One important type of hardness amplification is the *strongly black-box* hardness amplification. Such a hardness amplification from ε-hardness to $\bar{\varepsilon}$-hardness satisfies the following two conditions. First, the initial function f is given as a black box to construct the new function $\bar{f}$, in the sense that there is an oracle Turing machine $\text{AMP}^{(\cdot)}$ such that $\bar{f} = \text{AMP}^f$. Furthermore, the $\bar{\varepsilon}$-hardness of the new function $\bar{f}$ is also guaranteed in a black box way, in the sense that there is another oracle Turing machine $\text{DEC}^{(\cdot)}$ such that given any adversary A which computes $\bar{f}$ correctly on at least $(1 - \bar{\varepsilon})$ fraction of inputs, DEC using A as oracle can compute f correctly on at least $(1 - \varepsilon)$ fraction of inputs. We call AMP the encoding procedure and DEC the decoding procedure. In fact, almost all previous constructions of hardness amplification were done in such a strongly black-box way [25,3,9,12,20,13,18,8].

We consider a relaxation called *weakly black-box* hardness amplification, in which the black-box requirement on the hardness proof is dropped, while the black-box requirement on the encoding procedure is still kept. That is, the hardness of the new function $\bar{f}$ now can be proved in an arbitrary way, and no decoding procedure is required.

Previous Results. For the strongly black-box case, Viola [23] proved that no amplification procedures from worst-case hardness to mild hardness can be realized in PH. This was generalized by Lu et al. [14], who showed the impossibility of amplifying hardness from $(1 - \delta)/2$ to $(1 - \delta^k)/2$ in PH for any super-polynomial k. Furthermore, they showed that such a hardness amplification must be highly non-uniform in nature, in the sense that one must start from a function f which is hard against a very non-uniform complexity class even if one only wants to obtain a function $\bar{f}$ which is hard against a uniform complexity class [14].

Since the strongly black-box approach has its limitation, one may look for a weaker type of hardness amplification. Bogdanov and Trevisan [5] showed that even if the black-box constraint on the encoding procedure is dropped, one still cannot amplify from worst-case hardness to mild hardness for functions in NP unless PH collapses, when the decoding procedure is required to be computable in polynomial time and make only non-adaptive queries to the oracle.

The other possibility is to consider weakly black-box hardness amplification, in which the black-box constraint on the decoding procedure is dropped. Viola [24] proved that if a weakly black-box procedure amplifying from worst-case hardness to mild hardness can be realized in PH, then one can obtain from it a mildly hard function computable in PH. Although this can be seen as a negative result, it does not rule out the possibility of such a weakly black-box hardness amplification. In fact, it appears difficult to establish impossibility results in the weakly black-box model. This is because if an average-case hard function indeed exists, an amplification procedure may simply ignore the initial hard function and compute the average-case hard function from scratch. Then can one prove any meaningful impossibility result for a weakly black-box hardness amplification?

Our Results. We derive two negative results for weakly black-box hardness amplification. First, we prove that if a weakly black-box procedure realized in $\mathsf{TIME}(t)$ can amplify hardness by an $\omega(t)$ factor, from hardness $\varepsilon = \bar{\varepsilon}/\omega(t)$ against a complexity class $\mathcal{C} \subseteq \mathsf{SIZE}(2^{n/3})$ to hardness $\bar{\varepsilon}$ against any complexity class $\bar{\mathcal{C}}$, then one can obtain from it a function computable in $\mathsf{TIME}(t)$ with hardness about $\bar{\varepsilon}$ against $\bar{\mathcal{C}}$. Note that a function in $\mathsf{TIME}(t)$ cannot be hard against a class containing $\mathsf{TIME}(t)$. Therefore, we have an unconditional impossibility result: it is impossible to use a procedure realized in $\mathsf{TIME}(t)$ to transform a function which is $\bar{\varepsilon}/\omega(t)$-hard against a class $\mathcal{C} \subseteq \mathsf{SIZE}(2^{n/3})$ into a function which is $\bar{\varepsilon}$-hard against a class $\bar{\mathcal{C}} \supseteq \mathsf{TIME}(t)$. Note that with $t = 2^{o(n)}$, this gives an impossibility result for amplifying from worst-case hardness to mild hardness in sub-exponential time. We also extend this impossibility result to the case with $\mathcal{C}$ being any uniform complexity class equipped with an advice of length at most $2^{n/3}$. This says that such a weakly hardness amplification, just as in the strongly black-box case [14], must also be highly non-uniform in nature: it is impossible to have such a weakly hardness amplification if one starts from an initial function which is hard against a complexity class with only $2^{n/3}$ bits of non-uniformity (even of arbitrarily high uniform complexity).

Second, we prove that if a weakly black-box procedure realized in NP ($\Sigma_k\mathsf{P}$, respectively) can amplify hardness beyond a polynomial factor, from hardness $\varepsilon = \bar{\varepsilon}/n^{\omega(1)}$ against a complexity class $\mathcal{C} \subseteq \mathsf{SIZE}(2^{n/3})$ to hardness $\bar{\varepsilon}$ against a complexity class $\bar{\mathcal{C}}$ with $\bar{\mathcal{C}}/\mathrm{poly} = \bar{\mathcal{C}}$, then one can obtain from it a function computable in NP ($\Sigma_k\mathsf{P}$, respectively) with hardness about $\bar{\varepsilon}$ against $\bar{\mathcal{C}}$. This improves the result in [24], as the hard function obtained there seems to need at least the complexity of $\Sigma_2\mathsf{P}$ ($\Sigma_{k+1}\mathsf{P}$, respectively), one level higher than ours in PH. This in turn enables us to derive an unconditional impossibility result: it is impossible to use a procedure realized in NP ($\Sigma_k\mathsf{P}$, respectively) for such a hardness amplification, if one starts from an initial function which is hard against a complexity class with only $2^{n/3}$ bits of non-uniformity, but is required to produce a new function which is hard against a class containing $\mathsf{NP}/\mathrm{poly}$ ($\Sigma_k\mathsf{P}/\mathrm{poly}$, respectively).

Let us make some remarks on our results. We show that a low-complexity procedure basically cannot amplify hardness substantially in a weakly black-box way for high-complexity functions. One may question that if the goal is to establish an equivalence between two hardness conditions in a complexity class, then it suffices to have an amplification procedure in the same, instead of a lower, complexity class. However, we believe that it is interesting and worthwhile to know how low in complexity the amplification procedure can be, as it reflects how close the two hardness conditions really are.[1] For example, we can say that mild hardness and average-case hardness are close as one can be converted to the other in polynomial-time [12], while they may not be as close to pseudorandomness since the best transformation from pseudorandomness to them known so far requires the complexity of NP [10]. Note that the XOR lemma [25,17] and its derandomized version [12] show that a low complexity procedure is indeed capable of amplifying hardness for high-complexity (say, exponential-time) functions, as long as the hardness is only amplified within a certain factor. It is only when we want to go beyond the factor that a higher complexity amplification procedure is then required, as shown by our results.

2   Preliminaries

For any $n \in \mathbb{N}$, let $[n]$ denote the set $\{1, \ldots, n\}$ and let $\mathcal{U}_n$ denote the uniform distribution over $\{0,1\}^n$. For a finite set S, we also use S to denote the uniform distribution over S. We will sometimes view a Boolean function $f : \{0,1\}^n \to \{0,1\}$ as a 2^n-bit string (its truth table) and vice versa. For a string $x \in \{0,1\}^n$ and $i \in [n]$, we use x_i to denote the i'th bit of x. For two strings $u, v \in \{0,1\}^n$,

---

[1] Similar scenarios also appear elsewhere. For example, in the study of NP-complete problems, people at first only looked for polynomial-time reductions. It was found out later that most of the "natural" NP-complete problems are in fact NC^0-reducible to each other, so one may argue that these problems are very close to each other, or can even be seen as the same problem in some sense. Similar results also hold for several other standard complexity classes. See for example [1] and the references therein for more details.

let $\triangle(u, v)$ denote their relative Hamming distance $\frac{1}{n}|\{i \in [n] : u_i \neq v_i\}|$. All the logarithms in this paper will have base two.

We consider circuits with AND/OR/NOT gates, in which AND/OR gates are allowed to have unbounded fan-in. The size of a circuit is the number of its gates and the depth is the number of gates on the longest path from an input bit to the output gate. Let $\mathsf{AC}(d, s)$ ($\mathsf{SIZE}(s)$, respectively) denote the class of Boolean functions computed by circuits of depth d and size s (of arbitrary depth and size s, respectively). More information about complexity classes can be found in standard textbooks, such as [19]. As usual in complexity theory, when we talk about a function $f : \{0,1\}^n \to \{0,1\}^m$, we usually mean a sequence of functions $(f : \{0,1\}^n \to \{0,1\}^{m(n)})_{n \in \mathbb{N}}$, and when we make a statement about f, we usually mean that it holds for any sufficiently large $n \in \mathbb{N}$.

We define the hardness of a function against a complexity class as follows.

Definition 1. *We say that a function $f : \{0,1\}^n \to \{0,1\}$ is $(\varepsilon, \mathcal{C})$-hard, for a complexity class $\mathcal{C}$, if for any $C \in \mathcal{C}$, $\Pr_{x \in \mathcal{U}_n}[C(x) \neq f(x)] > \varepsilon$. We will call f ε-hard when the complexity class $\mathcal{C}$ is clear.*

The parameter ε in the definition above is allowed to be a function of n, so a better notation should be $\varepsilon(n)$, but for simplicity we drop the parameter n. In previous works, people usually consider hardness against circuits, i.e., with $\mathcal{C} = \mathsf{SIZE}(s)$ for some s. Since we will consider hardness against other complexity classes, we introduce this slightly more general definition.

Next, we define the notion of a weakly black-box hardness amplification.

Definition 2. *Let $\mathcal{C}$ and $\bar{\mathcal{C}}$ be complexity classes. We say that an oracle algorithm $\mathrm{AMP}^{(\cdot)} : \{0,1\}^{\bar{n}} \to \{0,1\}$ realizes a weakly black-box $(n, \varepsilon, \bar{\varepsilon}, \mathcal{C}, \bar{\mathcal{C}})$ hardness amplification, if given any $(\varepsilon, \mathcal{C})$-hard function $f : \{0,1\}^n \to \{0,1\}$, the function $\mathrm{AMP}^f : \{0,1\}^{\bar{n}} \to \{0,1\}$ is $(\bar{\varepsilon}, \bar{\mathcal{C}})$-hard. We say that such a hardness amplification is realized in some complexity class $\mathcal{A}$ if $\mathrm{AMP}^f \in \mathcal{A}^f$ for any f.*

Here, the reduction from the initial function f to the harder function is done in a black-box way, as the harder function AMP^f only uses f as an oracle.

We will need the following, known as the Borel-Cantelli Lemma (see e.g. [2]).

Lemma 1. *Let $E_1, E_2, \ldots$ be a sequence of probability events on the same probability space. Suppose that $\sum_{n=1}^{\infty} \Pr[E_n] < \infty$. Then $\Pr[\wedge_{k=1}^{\infty} \vee_{n \geq k} E_n] = 0$.*

3   Impossibility of Hardness Amplification in $\mathsf{TIME}(t)$

In this section, we show that if a weakly black-box hardness amplification realized in $\mathsf{TIME}(t)$ can amplify hardness beyond an $O(t)$ factor, then it must basically embed a hard function in it.

Theorem 1. *Suppose a weakly black-box $(n, \varepsilon, \bar{\varepsilon}, \mathcal{C}, \bar{\mathcal{C}})$ hardness amplification can be realized in $\mathsf{TIME}(t)$ with $2^{-n/2} \leq \varepsilon \leq o(\bar{\varepsilon}/t)$, $\mathcal{C} = \mathsf{SIZE}(2^{n/3})$, and $\bar{\mathcal{C}}$ being any complexity class. Then one can obtain from it an $(\bar{\varepsilon}/2, \bar{\mathcal{C}})$-hard function $\bar{A} : \{0,1\}^{\mathrm{poly}(n)} \to \{0,1\}$ computable in $\mathsf{TIME}(t)$.*

Using $t = 2^{o(n)}$, this implies that if such a hardness amplification from worst-case hardness to mild hardness can be realized in sub-exponential time (or sub-linear space), then it must basically embed a mildly hard function in it. Furthermore, since a function in $\mathsf{TIME}(t)$ cannot be hard against $\mathsf{TIME}(t)$, we have the following unconditional impossibility result on weakly black-box hardness amplification.

Corollary 1. *It is impossible to realize a weakly black-box* $(n, \varepsilon, \bar{\varepsilon}, \mathcal{C}, \bar{\mathcal{C}})$ *hardness amplification in* $\mathsf{TIME}(t)$, *with* $2^{-n/2} \leq \varepsilon \leq o(\bar{\varepsilon}/t)$, $\mathcal{C} = \mathsf{SIZE}(2^{n/3})$, *and* $\bar{\mathcal{C}} \supseteq \mathsf{TIME}(t)$.

Now we prove Theorem 1.

Proof. (of Theorem 1) Assume that such a weakly hardness amplification exists, with AMP realized in $\mathsf{TIME}(t)$. We will show that the function $\mathrm{AMP}^{\mathbf{0}}$ is $(\bar{\varepsilon}/2, \bar{\mathcal{C}})$-hard, where $\mathbf{0}$ is the constant zero function which always outputs zero for every input. The idea is to choose a certain kind of random function f and show that f is likely to be hard and AMP is unlikely to tell it apart from the function $\mathbf{0}$. A natural candidate is the following, which is obtained by adding random noise of certain rate to the function $\mathbf{0}$.[2]

Definition 3. *Let* $\mathcal{F}^{\delta}$ *denote the distribution of functions* $f : \{0,1\}^n \to \{0,1\}$ *such that for any* $x \in \{0,1\}^n$, $f(x) = 0$ *with probability* $1 - \delta$, *and* $f(x)$ *is given a random bit with probability* δ.

We next show that it gives us what we want, with $\delta = 4\varepsilon$.

Lemma 2. $\Pr_{f \in \mathcal{F}^{\delta}}[\mathrm{AMP}^f \text{ is } (\bar{\varepsilon}, \bar{\mathcal{C}})\text{-hard}] \geq 1 - 2^{-\Omega(n)}$.

Proof. Consider any $D \in \mathcal{C} = \mathsf{SIZE}(2^{n/3})$. Note that for any $x \in \{0,1\}^n$, we have $\Pr_f [D(x) \neq f(x)] \geq \delta/2 = 2\varepsilon$. This implies that $\mathrm{E}_f[\Pr_x [D(x) \neq f(x)]] = \mathrm{E}_x[\Pr_f [D(x) \neq f(x)]] \geq 2\varepsilon$, and thus $\Pr_f [\Pr_x [D(x) \neq f(x)] < \varepsilon] \leq 2^{-\Omega(\varepsilon 2^n)}$ by a Chernoff bound. Then $\Pr_f [f \text{ is not } (\varepsilon, \mathcal{C})\text{-hard}]$ equals

$$\Pr_f \left[\exists D \in \mathsf{SIZE}(2^{n/3}) : \Pr_x [D(x) \neq f(x)] < \varepsilon \right] \leq 2^{O(2^{n/3} \cdot n/3)} \cdot 2^{-\Omega(\varepsilon 2^n)} \leq 2^{-\Omega(n)},$$

by a union bound. So, $\Pr_f[\mathrm{AMP}^f \text{ is not } (\bar{\varepsilon}, \bar{\mathcal{C}})\text{-hard}] \leq \Pr_f[f \text{ is not } (\varepsilon, \mathcal{C})\text{-hard}] \leq 2^{-\Omega(n)}$. □

Lemma 3. $\Pr_{f \in \mathcal{F}^{\delta}}[\Delta(\mathrm{AMP}^f, \mathrm{AMP}^{\mathbf{0}}) \leq \bar{\varepsilon}/2] \geq 1 - o(1)$.

Proof. For any input $\bar{x}$, $\mathrm{AMP}^f(\bar{x}) \neq \mathrm{AMP}^{\mathbf{0}}(\bar{x})$ only when $\mathrm{AMP}^f(\bar{x})$ ever makes an oracle query x to f with $f(x) \neq 0$. AMP runs in time t and can make at most t queries to the oracle, so for every $\bar{x}$, $\Pr_f[\mathrm{AMP}^f(\bar{x}) \neq \mathrm{AMP}^{\mathbf{0}}(\bar{x})] \leq t \cdot \delta = o(\bar{\varepsilon})$. Then $\Pr_{f, \bar{x}}[\mathrm{AMP}^f(\bar{x}) \neq \mathrm{AMP}^{\mathbf{0}}(\bar{x})] = o(\bar{\varepsilon})$, and by Markov's inequality, $\Pr_f[\Pr_{\bar{x}}[\mathrm{AMP}^f(\bar{x}) \neq \mathrm{AMP}^{\mathbf{0}}(\bar{x})] \geq \bar{\varepsilon}/2] = o(1)$.

---

[2] A similar idea also appeared in [15] for the problem of amplifying hardness of one-way permutations.

From these two lemmas, there exists a function f such that AMP^f is $(\bar{\varepsilon}, \bar{\mathcal{C}})$-hard and $\triangle(\text{AMP}^f, \text{AMP}^0) \leq \bar{\varepsilon}/2$. This implies that the function AMP^0 is $(\bar{\varepsilon}/2, \bar{\mathcal{C}})$-hard. Since AMP^0 clearly belongs to $\text{TIME}(t)$, we have Theorem 1. $\qquad \square$

In fact we can have a similar impossibility result even if we replace the class $\text{SIZE}(2^{n/3})$ by any uniform complexity class $\mathcal{B}$ equipped with an advice of length $2^{n/3}$, denoted as $\mathcal{B}/2^{n/3}$. This means that such a weakly black-box hardness amplification can only work in a highly non-uniform setting.

Theorem 2. *It is impossible to realize a weakly black-box $(n, \varepsilon, \bar{\varepsilon}, \mathcal{C}, \bar{\mathcal{C}})$ hardness amplification in $\text{TIME}(t)$, with $2^{-n/2} \leq \varepsilon \leq o(\bar{\varepsilon}/(t \cdot n^2))$, $\mathcal{C} = \mathcal{B}/2^{n/3}$ for any uniform complexity class $\mathcal{B}$, and $\bar{\mathcal{C}} \supseteq \text{TIME}(t)$.*

Proof. The proof is largely based on that for Theorem 1, but here we need to treat things in a more careful way. As now we often need to talk about a sequence of functions, one on each input length, we change the notation slightly by adding a subscript n to a function of input length n. That is, now we write $f_n : \{0,1\}^n \to \{0,1\}$, and write f for the sequence of functions $(f_n)_{n \in \mathbb{N}}$. Similarly, we use $\mathcal{F}_n^\delta$ for the distribution of functions $f_n : \{0,1\}^n \to \{0,1\}$ in the proof of Theorem 1, and $\mathcal{F}^\delta$ for the sequence of distributions $(\mathcal{F}_n^\delta)_{n \in \mathbb{N}}$.

Lemma 4. *With measure one over $f \in \mathcal{F}^\delta$, AMP^f is $(\bar{\varepsilon}, \bar{\mathcal{C}})$-hard.*

Proof. Consider any Turing machine M in the uniform complexity class $\mathcal{B}$. Consider any input length n and let E_n denote the event that there exists an advice $\nu \in \{0,1\}^{2^{n/3}}$ such that $\Pr_{x \in \mathcal{U}_n}[M^\nu(x) \neq f_n(x)] < \varepsilon$. Then as in Lemma 2, one can show that $\Pr_{f_n \in \mathcal{F}_n^\delta}[E_n] \leq 2^{O(2^{n/3} \cdot n/3)} \cdot 2^{-\Omega(\varepsilon 2^n)} < 1/n^2$. Since $\sum_{n=1}^{\infty} 1/n^2 < \infty$, the event that E_n happens for infinitely many n has measure zero over $f \in \mathcal{F}^\delta$, by the Borel-Cantelli Lemma. Since there are only countable many Turing machines M's, the event that f is not $(\varepsilon, \mathcal{C})$-hard has measure zero over $f \in \mathcal{F}^\delta$. Finally, AMP^f is not $(\bar{\varepsilon}, \bar{\mathcal{C}})$-hard only when f is not $(\varepsilon, \mathcal{C})$-hard, so we have the lemma. $\qquad \square$

Lemma 5. *With measure one over $f_n \in \mathcal{F}_n^\delta$, $\triangle(\text{AMP}^{f_n}, \text{AMP}^{0_n}) \geq \bar{\varepsilon}/2$ for only finitely many n.*

Proof. Similarly to the proof of Lemma 3, but now with $\varepsilon \leq o(\bar{\varepsilon}/(t \cdot n^2))$, one can show that for any n, $\Pr_{f_n, \bar{x}}[\text{AMP}^{f_n}(\bar{x}) \neq \text{AMP}^{0_n}(\bar{x})] = o(\bar{\varepsilon}/n^2)$, and $\Pr_{f_n}[\triangle(\text{AMP}^{f_n}, \text{AMP}^{0_n}) \geq \bar{\varepsilon}/2] < 1/n^2$. Since $\sum_{n=1}^{\infty} 1/n^2 < \infty$, the lemma immediately follows from the Borel-Cantelli Lemma. $\qquad \square$

As in Theorem 1, the two lemmas above imply that the function AMP^0 is $(\bar{\varepsilon}/2, \bar{\mathcal{C}})$-hard. Since AMP^0 belongs to $\text{TIME}(t)$, it cannot be hard against any $\bar{\mathcal{C}} \supseteq \text{TIME}(t)$, and we have Theorem 2.

4   Impossibility Results in $\Sigma_k \text{P}$

In this section, we consider weakly black-box hardness amplification realized in $\Sigma_k \text{P}$. We will show that if it can amplify hardness beyond a certain factor, then it must basically embed a hard function in it.

Theorem 3. *Suppose a weakly black-box $(n, \varepsilon, \bar{\varepsilon}, \mathcal{C}, \bar{\mathcal{C}})$ hardness amplification can be realized in NP (Σ_kP, respectively), with $2^{-n/2} \leq \varepsilon \leq \bar{\varepsilon}^2/n^{\omega(1)}$, $\mathcal{C} = \text{SIZE}(2^{n/3})$, and $\bar{\mathcal{C}}$ satisfying $\bar{\mathcal{C}}/\text{poly} = \bar{\mathcal{C}}$. Then one can obtain from it an $(\bar{\varepsilon}/3, \bar{\mathcal{C}})$-hard function $\bar{A} : \{0,1\}^{\text{poly}(n)} \rightarrow \{0,1\}$ computable in NP (Σ_kP, respectively).*

Since a function in NP (Σ_kP, respectively) cannot be hard against NP/poly (Σ_kP/poly, respectively), we have the following unconditional impossibility result on weakly black-box hardness amplification.

Corollary 2. *It is impossible to realize a weakly black-box $(n, \varepsilon, \bar{\varepsilon}, \mathcal{C}, \bar{\mathcal{C}})$ hardness amplification in NP (Σ_kP, respectively), with $2^{-n/2} \leq \varepsilon \leq \bar{\varepsilon}^2/n^{\omega(1)}$, $\mathcal{C} = \text{SIZE}(2^{n/3})$, and any $\bar{\mathcal{C}} \supseteq$ NP (Σ_kP, respectively) satisfying $\bar{\mathcal{C}}/\text{poly} = \bar{\mathcal{C}}$.*

We will need the notion of random restriction [6,7]. A restriction ρ on N variables is an element of $\{0, 1, \star\}^N$, or seen as a function $\rho : [N] \rightarrow \{0, 1, \star\}$. A variable is fixed by ρ if it receives a value in $\{0,1\}$ while a variable remains free if it receives the symbol $\star$. For a string $y \in \{0,1\}^N$ and a restriction $\rho \in \{0, 1, \star\}^N$, let $y\!\restriction_\rho \in \{0,1\}^N$ be the restriction of y with respect to ρ: for $i \in [N]$, the i'th bit of $y\!\restriction_\rho$ is y_i if $\rho_i = \star$ and is ρ_i if $\rho_i \in \{0,1\}$.

Suppose there exists such a weakly black-box hardness amplification, with AMP realized in NP (Σ_kP, resp.). Then AMP can be computed by an AC($c, 2^{n^c}$) circuit, for some constant c, with the truth table of the oracle function given as part of the input (c.f. [6]). We will show how to derive a hard function from it.

The idea, which basically follows Viola's [24], is as follows. We know that a random function f is likely to be hard and a hard f gives a hard function AMP^f, but we do not know which f to choose. One attempt is to include f as part of the input in the new function, but the description of f is too long. The idea is that by choosing a suitable random restriction $\bar{\rho}$, the function $f\!\restriction_{\bar{\rho}}$ is still likely to be hard, and so is the function AMP$^{f\restriction_{\bar{\rho}}}$. On the other hand, a random restriction is likely to kill off the effect of a random function f on AMP$^{f\restriction_{\bar{\rho}}}$, so it becomes possible to replace the random function by a pseudo-random one $\bar{f}$, which has a short description. Therefore, if we have a random restriction which has a short description and satisfies the properties above, we can define the new function which includes $\bar{\rho}$ and $\bar{f}$ as part of the input and computes the function AMP$^{\bar{f}\restriction_{\bar{\rho}}}$ as its output. The existence of such a random restriction is guaranteed by the following lemma of Viola's [24]. For our purpose here, we state it in a slightly more general form.

Lemma 6. *[24] For any $n \in \mathbb{N}$, any constant c, and any $\varepsilon, \bar{\varepsilon} \in (0,1)$ such that $2^{-n} \leq \varepsilon \leq \bar{\varepsilon}^2/n^{\omega(1)}$, there is a distribution $\bar{\mathcal{R}}$ on restrictions $\bar{\rho} : \{0,1\}^n \rightarrow \{0, 1, \star\}$ such that the following three conditions all hold.*

- *Every $\bar{\rho} \in \bar{\mathcal{R}}$ can be described by $\text{poly}(n)$ bits, and given such a description and $x \in \{0,1\}^n$, one can compute $\bar{\rho}(x)$ in time $\text{poly}(n)$.*
- *$\Pr_{\bar{\rho} \in \bar{\mathcal{R}}}[|\{x : \bar{\rho}(x) = \star\}| < 3\varepsilon 2^n] = o(\bar{\varepsilon})$.*
- *$\Pr_{\bar{\rho} \in \bar{\mathcal{R}}; y, y' \in \mathcal{U}_{2^n}}[C(y\!\restriction_{\bar{\rho}}) \neq C(y'\!\restriction_{\bar{\rho}})] = o(\bar{\varepsilon}^2)$, for any $C : \{0,1\}^{2^n} \rightarrow \{0,1\} \in$ AC($c, 2^{n^c}$).*

Let $\mathcal{F}$ denote the set of all functions $f : \{0,1\}^n \to \{0,1\}$. It remains to show that the random restriction given in Lemma 6 can achieve what we discussed before. First, by using the second item of Lemma 6, we show that the function $\mathrm{AMP}^{f\restriction_{\bar{\rho}}}$ is hard with high probability over $\bar{\rho} \in \mathcal{R}$ and $f \in \mathcal{F}$.

Lemma 7. $\Pr_{\bar{\rho}\in\mathcal{R}, f\in\mathcal{F}}[\mathrm{AMP}^{f\restriction_{\bar{\rho}}} \text{ is not } (\bar{\varepsilon}, \bar{\mathcal{C}})\text{-hard}] = o(\bar{\varepsilon})$.

Proof. Call a restriction $\bar{\rho} \in \mathcal{R}$ *good* if $|\{x : \bar{\rho}(x) = \star\}| \geq 3\bar{\varepsilon}2^n$. Consider any good $\bar{\rho}$ and any $D \in \mathcal{C} = \mathsf{SIZE}(2^{n/3})$. Note that for any x such that $\bar{\rho}(x) = \star$, $\Pr_f[D(x) \neq f\restriction_{\bar{\rho}}(x)] = 1/2$. So we have $\mathrm{E}_f[\Pr_x[D(x) \neq f\restriction_{\bar{\rho}}(x)]] = \Pr_{x,f}[D(x) \neq f\restriction_{\bar{\rho}}(x)] \geq 3\varepsilon/2$, and $\Pr_f[\Pr_x[D(x) \neq f\restriction_{\bar{\rho}}(x)] < \varepsilon] \leq 2^{-\Omega(\varepsilon 2^n)}$, by a Chernoff bound. As a result, for any good $\bar{\rho}$, $\Pr_f[f\restriction_{\bar{\rho}} \text{ is not } (\varepsilon, \mathcal{C})\text{-hard}]$ is

$$\Pr_f\left[\exists D \in \mathsf{SIZE}(2^{n/3}) : \Pr_x[D(x) \neq f\restriction_{\bar{\rho}}(x)] < \varepsilon\right] \leq 2^{O(2^{n/3}\cdot n/3)} \cdot 2^{-\Omega(\varepsilon 2^n)} = o(\bar{\varepsilon}).$$

From Lemma 6, $\Pr_{\bar{\rho}\in\mathcal{R}}[\bar{\rho} \text{ is not good}] = o(\bar{\varepsilon})$, and by definition, $\mathrm{AMP}^{f\restriction_{\bar{\rho}}}$ is $(\bar{\varepsilon}, \bar{\mathcal{C}})$-hard whenever $f\restriction_{\bar{\rho}}$ is $(\varepsilon, \mathcal{C})$-hard. Therefore, $\Pr_{\bar{\rho}, f}[\mathrm{AMP}^{f\restriction_{\bar{\rho}}} \text{ is not } (\bar{\varepsilon}, \bar{\mathcal{C}})\text{-hard}]$ is at most $\Pr_{\bar{\rho}}[\bar{\rho} \text{ is not good}] + \Pr_{\bar{\rho}, f}[f\restriction_{\bar{\rho}} \text{ is not } (\varepsilon, \mathcal{C})\text{-hard} \mid \bar{\rho} \text{ is good}] = o(\bar{\varepsilon})$. □

From this lemma, we know that the function $\mathrm{AMP}^{f\restriction_{\bar{\rho}}}$ is hard for most $\bar{\rho} \in \mathcal{R}$ and $f \in \mathcal{F}$, but we do not know which $\bar{\rho}$ and f give a hard function. While $\bar{\rho}$ has a short description, f does not, so we cannot just include both $\bar{\rho}$ and f as part of the input. Viola's approach in [24] is to remove the dependence of f altogether, by considering the function A' which on input $(\bar{x}, \bar{\rho})$ outputs the majority value of $\mathrm{AMP}^{f\restriction_{\bar{\rho}}}(\bar{x})$ over $f \in \mathcal{F}$. The hardness of A' is guaranteed by the third item in Lemma 6, because for most $\bar{\rho}$ and for most f, the function $\mathrm{AMP}^{f\restriction_{\bar{\rho}}}$ is hard and $A'(\bar{\rho}, \cdot)$ is close to it. However, to compute such a majority value over $f \in \mathcal{F}$ costs one additional level in the polynomial hierarchy in [24], and with $\mathrm{AMP} \in \mathsf{NP}$ ($\Sigma_k\mathsf{P}$, respectively), Viola needs at least $\Sigma_2\mathsf{P}$ ($\Sigma_{k+1}\mathsf{P}$, respectively) to compute the function A'. Our idea is to replace the random function by a pseudorandom one.

Definition 4. *Let* $\mathrm{NIS} : \{0,1\}^r \to \{0,1\}^{2^n}$ *be Nisan's* $o(\bar{\varepsilon}^2)$-PRG *for* $\mathsf{AC}(c + 2, 2^{n^c+2})$, *with* $r = \mathrm{poly}(n)$ *[16]. Let* $\bar{\mathcal{F}}$ *be the class of functions* $\bar{f}_z : \{0,1\}^n \to \{0,1\}$, *with* $z \in \{0,1\}^r$, *defined as* $\bar{f}_z(x) = \mathrm{NIS}(z)_x$, *the* x'th *bit in* $\mathrm{NIS}(z)$.

There seems to be an obstacle in front of us. Unlike a random function, such a pseudo-random $\bar{f}$ is not hard at all. Then how do we guarantee the hardness of the function $\mathrm{AMP}^{\bar{f}\restriction_{\bar{\rho}}}$? We resolve this by showing that the function $\mathrm{AMP}^{\bar{f}\restriction_{\bar{\rho}}}$ is likely to be close to a hard function $\mathrm{AMP}^{f\restriction_{\bar{\rho}}}$. For this, we first show the following.

Lemma 8. *For any* $\bar{x} \in \{0,1\}^{\bar{n}}$, $\Pr_{\bar{\rho}\in\mathcal{R}, f\in\mathcal{F}, \bar{f}\in\bar{\mathcal{F}}}[\mathrm{AMP}^{f\restriction_{\bar{\rho}}}(\bar{x}) \neq \mathrm{AMP}^{\bar{f}\restriction_{\bar{\rho}}}(\bar{x})] = o(\bar{\varepsilon}^2)$.

Proof. Let $N = 2^n$. Fix any $\bar{x} \in \{0,1\}^{\bar{n}}$, and let $C : \{0,1\}^N \to \{0,1\}$ be the function which takes a function $g : \{0,1\}^n \to \{0,1\}$, seen as $g \in \{0,1\}^N$, as the input and outputs the value $\mathrm{AMP}^g(\bar{x})$. Clearly, $C \in \mathsf{AC}(c, 2^{n^c})$. Now for $\bar{\rho} \in \mathcal{R}$, let $\bar{C}_{\bar{\rho}} : \{0,1\}^N \times \{0,1\}^N \to \{0,1\}$ be the function such that $\bar{C}_{\bar{\rho}}(f, f') = 1$ if and only if $C(f \restriction_{\bar{\rho}}) \neq C(f' \restriction_{\bar{\rho}})$, which is computable by an $\mathsf{AC}(c + 2, 2^{n^c+2})$ circuit. Since NIS is an $o(\bar{\varepsilon}^2)$-PRG for such circuits,

$$\left| \Pr_{\bar{\rho},f,\bar{f}} \left[\bar{C}_{\bar{\rho}}(f, \bar{f}) = 1 \right] - \Pr_{\bar{\rho},f,f'} \left[\bar{C}_{\bar{\rho}}(f, f') = 1 \right] \right|$$

$$\leq \mathop{\mathrm{E}}_{\bar{\rho},f} \left[\left| \Pr_z \left[\bar{C}_{\bar{\rho}}(f, \mathrm{NIS}(z)) = 1 \right] - \Pr_{f'} \left[\bar{C}_{\bar{\rho}}(f, f') = 1 \right] \right| \right] = o(\bar{\varepsilon}^2).$$

By Lemma 6, $\Pr_{\bar{\rho},f,f'} \left[\bar{C}_{\bar{\rho}}(f, f') = 1 \right] = o(\bar{\varepsilon}^2)$, so we have $\Pr_{\bar{\rho},f,\bar{f}} \left[\bar{C}_{\bar{\rho}}(f, \bar{f}) = 1 \right]$ $\leq \Pr_{\bar{\rho},f,f'} \left[\bar{C}_{\bar{\rho}}(f, f') = 1 \right] + o(\bar{\varepsilon}^2) = o(\bar{\varepsilon}^2)$. □

From this, we can show that the function $\mathrm{AMP}^{\bar{f} \restriction \bar{\rho}}$ is hard for most $\bar{\rho} \in \bar{\mathcal{R}}$ and $\bar{f} \in \bar{\mathcal{F}}$.

Lemma 9. $\Pr_{\bar{\rho} \in \bar{\mathcal{R}}, \bar{f} \in \bar{\mathcal{F}}}[\mathrm{AMP}^{\bar{f} \restriction \bar{\rho}}$ *is not* $(\bar{\varepsilon}/2, \bar{\mathcal{C}})$-*hard*$] = o(\bar{\varepsilon})$.

Proof. From Lemma 7, we know that $\Pr_{\bar{\rho},f}[\mathrm{AMP}^{f \restriction \bar{\rho}}$ is not $(\bar{\varepsilon}, \bar{\mathcal{C}})$-hard$] = o(\bar{\varepsilon})$. From Lemma 8, we know that $\Pr_{\bar{\rho},f,\bar{f},\bar{x}}[\mathrm{AMP}^{f \restriction \bar{\rho}}(\bar{x}) \neq \mathrm{AMP}^{\bar{f} \restriction \bar{\rho}}(\bar{x})] = o(\bar{\varepsilon}^2)$, and by Markov's inequality, we have $\Pr_{\bar{\rho},f,\bar{f}}[\triangle(\mathrm{AMP}^{f \restriction \bar{\rho}}, \mathrm{AMP}^{\bar{f} \restriction \bar{\rho}}) > \bar{\varepsilon}/2] = o(\bar{\varepsilon})$. Note that $\mathrm{AMP}^{\bar{f} \restriction \bar{\rho}}$ is $(\bar{\varepsilon}/2, \bar{\mathcal{C}})$-hard whenever $\mathrm{AMP}^{f \restriction \bar{\rho}}$ is $(\bar{\varepsilon}, \bar{\mathcal{C}})$-hard and $\triangle(\mathrm{AMP}^{f \restriction \bar{\rho}}, \mathrm{AMP}^{\bar{f} \restriction \bar{\rho}}) \leq \bar{\varepsilon}/2$. Thus, $\Pr_{\bar{\rho},\bar{f}}[\mathrm{AMP}^{\bar{f} \restriction \bar{\rho}}$ is not $(\bar{\varepsilon}/2, \bar{\mathcal{C}})$-hard$]$ is at most

$$\Pr_{\bar{\rho},f} \left[\mathrm{AMP}^{f \restriction \bar{\rho}} \text{ is not } (\bar{\varepsilon}, \bar{\mathcal{C}})\text{-hard} \right] + \Pr_{\bar{\rho},f,\bar{f}} \left[\triangle(\mathrm{AMP}^{f \restriction \bar{\rho}}, \mathrm{AMP}^{\bar{f} \restriction \bar{\rho}}) > \bar{\varepsilon}/2 \right] = o(\bar{\varepsilon}). □$$

From the lemma above, we know that $\mathrm{AMP}^{\bar{f} \restriction \bar{\rho}}$ is hard for most $\bar{\rho}$ and $\bar{f}$. We do not know which $\bar{\rho}$ and $\bar{f}$ give a hard function, but since they have short description, we can include them as part of the input. Define the function $\bar{A}$: $\{0,1\}^{\bar{n}} \times \bar{\mathcal{R}} \times \bar{\mathcal{F}} \to \{0,1\}$ as $\bar{A}(\bar{x}, \bar{\rho}, \bar{f}) = \mathrm{AMP}^{\bar{f} \restriction \bar{\rho}}(\bar{x})$. Note that the input length of $\bar{A}$ is at most $\mathrm{poly}(n)$ as $\bar{\rho}$ and $\bar{f}$ can be described by $\mathrm{poly}(n)$ bits.

Lemma 10. *The function* $\bar{A}$ *is* $(\bar{\varepsilon}/3, \bar{\mathcal{C}})$-*hard*.

Proof. Consider any $\bar{D} : \{0,1\}^{\bar{n}} \times \bar{\mathcal{R}} \times \bar{\mathcal{F}} \to \{0,1\} \in \bar{\mathcal{C}}$. Note that for any $\bar{\rho} \in \bar{\mathcal{R}}$ and $\bar{f} \in \bar{\mathcal{F}}$ such that $\mathrm{AMP}^{\bar{f} \restriction \bar{\rho}}$ is $(\bar{\varepsilon}/2, \bar{\mathcal{C}})$-hard, $\Pr_{\bar{x}}[\bar{D}(\bar{x}, \bar{\rho}, \bar{f}) = \bar{A}(\bar{x}, \bar{\rho}, \bar{f})] = \Pr_{\bar{x}}[\bar{D}(\bar{x}, \bar{\rho}, \bar{f}) = \mathrm{AMP}^{\bar{f} \restriction \bar{\rho}}(\bar{x})] < 1 - \bar{\varepsilon}/2$, since the function $\bar{D}(\cdot, \bar{\rho}, \bar{f})$ belongs to $\bar{\mathcal{C}}/\mathrm{poly} = \bar{\mathcal{C}}$. Thus, for any $\bar{D} \in \bar{\mathcal{C}}$, $\Pr_{\bar{x},\bar{\rho},\bar{f}} \left[\bar{D}(\bar{x}, \bar{\rho}, \bar{f}) = \bar{A}(\bar{x}, \bar{\rho}, \bar{f}) \right]$ is less than

$$\Pr_{\bar{\rho},\bar{f}} \left[\mathrm{AMP}^{\bar{f} \restriction \bar{\rho}} \text{ is not } (\bar{\varepsilon}/2, \bar{\mathcal{C}})\text{-hard} \right] + 1 - \bar{\varepsilon}/2 \leq 1 - \bar{\varepsilon}/3,$$

which means that $\bar{A}$ is $(\bar{\varepsilon}/3, \bar{\mathcal{C}})$-hard. □

Now we are ready to prove Theorem 3.

Proof. (of Theorem 3) By Lemma 10, $\bar{A}$ is $(\bar{\varepsilon}/3, \bar{\mathcal{C}})$-hard. Note that given $\bar{f} \in \bar{\mathcal{F}}$, $\bar{\rho} \in \bar{\mathcal{R}}$, and any $x \in \{0,1\}^n$, one can compute $\bar{f}\!\restriction_{\bar{\rho}}(x)$ in time poly(n). Therefore, if AMP can be realized in NP (Σ_kP, resp.), the function $\bar{A}$ is computable in NP (Σ_kP, resp.), and we have the theorem.   □

Similar to Theorem 2, we have the following.

Theorem 4. *It is impossible to realize a weakly black-box* $(n, \varepsilon, \bar{\varepsilon}, \mathcal{C}, \bar{\mathcal{C}})$ *hardness amplification in* NP *(Σ_kP, resp.) with* $2^{-n/2} \leq \varepsilon \leq \bar{\varepsilon}^2/n^{\omega(1)}$, $\mathcal{C} = \mathcal{B}/2^{n/3}$ *for any uniform complexity class* $\mathcal{B}$, *and any* $\bar{\mathcal{C}} \supseteq$ NP$/$poly *(Σ_kP$/$poly, resp.) satisfying* $\bar{\mathcal{C}}/$poly $= \bar{\mathcal{C}}$.

Proof. Similarly to how we modify the proof of Theorem 1 to prove Theorem 2, we can also modify the proof of Theorem 3 to prove Theorem 4. Here we also change the notation slightly, by writing $\bar{f}_n \in \bar{\mathcal{F}}_n$ and $\bar{\rho}_n \in \bar{\mathcal{R}}_n$ for functions and restrictions on inputs of length n, respectively, and writing $\bar{f}, \bar{\rho}, \bar{\mathcal{F}}, \bar{\mathcal{R}}$ for the sequences $(f_n)_{n \in \mathbb{N}}, (\bar{\rho}_n)_{n \in \mathbb{N}}, (\bar{\mathcal{F}}_n)_{n \in \mathbb{N}}, (\bar{\mathcal{R}}_n)_{n \in \mathbb{N}}$, respectively. Again, we will apply the Borel-Cantelli Lemma, and it is easy to check that the proof for Theorem 3 can be modified to prove the following lemma.

Lemma 11. *With measure one over* $\bar{\rho} \in \bar{\mathcal{R}}$ *and* $\bar{f} \in \bar{\mathcal{F}}$, AMP$^{\bar{f}\restriction_{\bar{\rho}}}$ *is* $(\bar{\varepsilon}/2, \bar{\mathcal{C}})$-hard.

Unlike for Theorem 3, we now cannot show that the function $\bar{A}$ is hard. Nevertheless, from the lemma above, we know that for any large enough n, there exists $\bar{\rho}_n \in \bar{\mathcal{R}}_n$ and $\bar{f}_n \in \bar{\mathcal{F}}_n$ such that the function $\bar{A}(\bar{\rho}_n, \bar{f}_n, \cdot) = \text{AMP}^{\bar{f}_n \restriction_{\bar{\rho}_n}}(\cdot)$ is hard. We can see such $\bar{\rho}_n$'s and $\bar{f}_n$'s as advice strings, which are of length poly(n), and as a result we have a hard function which is computable in NP$/$poly (Σ_kP$/$poly, resp.). Since such a function cannot be hard against any $\bar{\mathcal{C}} \supseteq$ NP$/$poly (Σ_kP$/$poly, resp.), we have Theorem 4.   □

References

1. Agrawal, M., Allender, E., Rudich, S.: Reductions in circuit complexity: an isomorphism theorem and a gap theorem. Journal of Computer and System Sciences 57, 127–143 (1998)
2. Billingsley, P.: Probability and measure, 3rd edn. Wiley & Sons, Chichester (1995)
3. Babai, L., Fortnow, L., Nisan, N., Wigderson, A.: BPP has subexponential time simulations unless EXPTIME has publishable proofs. Computational Complexity 3(4), 307–318 (1993)
4. Blum, M., Micali, S.: How to generate cryptographically strong sequences of pseudo random bits. In: Proceedings of the 23rd Annual IEEE Symposium on Foundations of Computer Science, pp. 112–117. IEEE Computer Society Press, Los Alamitos (1982)
5. Bogdanov, A., Trevisan, L.: On worst-case to average-case reductions for NP problems. In: Proceedings of the 44th Annual Symposium on Foundations of Computer Science, Cambridge, Massachusetts, pp. 11–14 (2003)
6. Furst, M.L., Saxe, J.B., Sipser, M.: Parity, circuits, and the polynomial-time hierarchy. Mathematical Systems Theory 17(1), 13–27 (1984)

7. Håstad, J.: Computational limitations for small depth circuits. PhD thesis, MIT Press (1986)
8. Healy, A., Vadhan, S., Viola, E.: Using nondeterminism to amplify hardness. In: Proceedings of the 36th ACM Symposium on Theory of Computing, pp. 192–201. ACM Press, New York (2004)
9. Impagliazzo, R.: Hard-core distributions for somewhat hard problems. In: Proceedings of the 36th Annual IEEE Symposium on Foundations of Computer Science, pp. 538–545. IEEE Computer Society Press, Los Alamitos (1995)
10. Impagliazzo, R., Levin, L.: No better ways to generate hard NP instances than picking uniformly at random. In: Proceedings of the 31st Annual IEEE Symposium on Foundations of Computer Science, pp. 812–821. IEEE Computer Society Press, Los Alamitos (1990)
11. Impagliazzo, R., Shaltiel, R., Wigderson, A.: Extractors and pseudo-random generators with optimal seed length. In: Proceedings of the 32nd ACM Symposium on Theory of Computing, pp. 1–10. ACM Press, New York (2000)
12. Impagliazzo, R., Wigderson, A.: P=BPP if E requires exponential circuits: Derandomizing the XOR lemma. In: Proceedings of the 29th ACM Symposium on Theory of Computing, pp. 220–229. ACM Press, New York (1997)
13. Klivans, A., van Melkebeek, D.: Graph nonisomorphism has subexponential size proofs unless the polynomial-time hierarchy collapses. In: Proceedings of the 31st ACM Symposium on Theory of Computing, pp. 659–667. ACM Press, New York (1999)
14. Lu, C.-J., Tsai, S.-C., Wu, H.-L.: On the complexity of hardness amplification. In: Proceedings of the 20th Annual IEEE Conference on Computational Complexity, pp. 170–182. IEEE Computer Society Press, Los Alamitos (2005)
15. Lin, H., Trevisan, L., Wee, H.: On hardness amplification of one-way functions. In: Proceedings of the 2nd Theory of Cryptography Conference, pp. 34–49 (2005)
16. Nisan, N.: Pseudorandom bits for constant depth circuits. Combinatorica 11(1), 63–70 (1991)
17. Nisan, N., Wigderson, A.: Hardness vs Randomness. Journal of Computer and System Sciences 49(2), 149–167 (1994)
18. O'Donnell, R.: Hardness amplification within NP. In: Proceedings of the 34th ACM Symposium on Theory of Computing, pp. 751–760. ACM Press, New York (2002)
19. Papadimitriou, C.: Computational Complexity. Addison-Wesley, Reading (1994)
20. Sudan, M., Trevisan, L., Vadhan, S.: Pseudorandom generators without the XOR lemma. Journal of Computer and System Sciences 62(2), 236–266 (2001)
21. Shaltiel, R., Umans, C.: Simple extractors for all min-entropies and a new pseudo-random generator. In: Proceedings of the 42nd Annual IEEE Symposium on Foundations of Computer Science, pp. 648–657. IEEE Computer Society Press, Los Alamitos (2001)
22. Umans, C.: Pseudo-random generators for all hardnesses. Journal of Computer and System Sciences 67(2), 419–440 (2003)
23. Viola, E.: The complexity of constructing pseudorandom generators from hard functions. Computational Complexity 13(3-4), 147–188 (2004)
24. Viola, E.: On constructing parallel pseudorandom generators from one-way Functions. In: Proceedings of the 20th Annual IEEE Conference on Computational Complexity, pp. 183–197. IEEE Computer Society Press, Los Alamitos (2005)
25. Yao, A.C.-C.: Theory and applications of trapdoor functions. In: Proceedings of the 23rd Annual IEEE Symposium on Foundations of Computer Science, pp. 80–91. IEEE Computer Society Press, Los Alamitos (1982)

Maximal and Minimal Scattered Context Rewriting

Alexander Meduna[*] and Jiří Techet[**]

Department of Information Systems, Faculty of Information Technology,
Brno University of Technology, Božetěchova 2, Brno 61266, Czech Republic
techet@fit.vutbr.cz

Abstract. As their name suggest, during a *maximal derivation step*, a scattered context grammar G rewrites the maximal number of nonterminals while during a *minimal derivation step*, G rewrites the minimal number of nonterminals. This paper demonstrates that if the propagating scattered context grammars derive their sentences by making either of these two derivation steps, then they characterize the family of context sensitive languages.

1 Introduction

Since their introduction (see [3]), the propagating scattered context grammars have always represented an intensively investigated type of semi-parallel rewriting systems (see [1,2,3,5,7,8,9,13,14]). The language family generated by these grammars is included in the family of context-sensitive languages; on the other hand, the question of whether this inclusion is proper represents an open problem in the formal language theory. As a result, the theory has modified these grammars in several ways and demonstrated that these modified grammars characterize the family of context-sensitive languages (see [1,2,9]). The present paper introduces two new modifications of this kind.

As a matter of fact, the simple modifications discussed in this paper only modify the way the propagating scattered context grammars perform derivations while keeping their grammatical concept unchanged. More specifically, this modification requires that during every derivation step, a production containing maximal or minimal number of nonterminals on its left-hand side is chosen from the set of all applicable productions. The paper demonstrates that these grammars characterize the family of context-sensitive languages if they work in this modified way.

Consequently, if in the future the formal language theory demonstrates that any propagating scattered context grammar making maximal or minimal derivations can be transformed to an equivalent propagating scattered context grammar making ordinary derivations, it also demonstrates that these grammars generate the family of all context-sensitive languages and, thereby, solves

---

[*] Supported by GAČR grant 201/07/0005 and Research Plan MSM 021630528.
[**] Supported by GAČR grant 102/05/H050 and FRVŠ grant FR762/2007/G1.

E. Csuhaj-Varjú and Z. Ésik (Eds.): FCT 2007, LNCS 4639, pp. 412–423, 2007.

the long-standing open problem formulated above. From this point of view, the characterization achieved in this paper is of some interest to the formal language theory.

2   Preliminaries

We assume that the reader is familiar with the language theory (see [6,10,11,12]). V^* represents the free monoid generated by V. The unit of V^* is denoted by ε. Set $V^+ = V^* - \{\varepsilon\}$. For $w \in V^*$, $|w|$ and alph(w) denote the length of w and the set of symbols occurring in w, respectively. For $L \subseteq V^*$,

$$\text{alph}(L) = \{a \, : \, a \in \text{alph}(w), w \in L\} \; .$$

For $x_1, \ldots, x_n \in V^*$, $\Pi((x_1, \ldots, x_n), i) = x_i$. For a finite set of integers, I, max(I) and min(I) denote the maximal and the minimal element of I, respectively. Let $\Delta(t)$ be the set of all permutations of $\{1, \ldots, t\}$. For some $n, m \geq 0$, define

$$\text{permute}(n, m) = \{(i_1, \ldots, i_{n+m}) \in \Delta(n + m) \, : \, 1 \leq i_k < i_l \leq n \text{ implies } k < l\} \; .$$

For $x_1, \ldots, x_n \in V^*$, $(i_1, \ldots, i_n) \in \Delta(n)$, define

$$\text{reorder}((x_1, \ldots, x_n), (i_1, \ldots, i_n)) = (x_{i_1}, \ldots, x_{i_n}) \; .$$

A *state grammar* (see [4]) is a sixtuple, $G = (V, T, K, P, S, p_0)$, where K is a finite set of states, V is an alphabet, $T \subseteq V$, $S \in V - T$, $p_0 \in K$, P is a finite set of productions of the form $(A, p) \rightarrow (x, q)$, where $p, q \in K$, $A \in V - T$, $x \in V^+$. Let lhs$((A, p) \rightarrow (x, q))$ and rhs$((A, p) \rightarrow (x, q))$ denote (A, p) and (x, q), respectively. If $(A, p) \rightarrow (x, q) \in P$, $u = (rAs, p)$, and $v = (rxs, q)$, where $r, s \in V^*$, and for every $(B, p) \rightarrow (y, t) \in P$, $B \notin \text{alph}(r)$, then G makes a *derivation step* from u to v according to $(A, p) \rightarrow (x, q)$, symbolically written as $u \Rightarrow v \; [(A, p) \rightarrow (x, q)]$ in G or, simply, $u \Rightarrow v$. Let $\Rightarrow^+$ and $\Rightarrow^*$ denote the transitive closure of $\Rightarrow$ and the transitive-reflexive closure of $\Rightarrow$, respectively. The *language of* G is denoted by $L(G)$ and defined as

$$L(G) = \{x \in T^* \, : \, (S, p_0) \Rightarrow^* (x, q) \text{ for some } q \in K\} \; .$$

3   Definitions

A *scattered context grammar* (see [1,2,3,5,7,8,9,13,14]), a *SC* grammar for short, is a quadruple, $G = (V, T, P, S)$, where V is an alphabet, $T \subseteq V$, $S \in V - T$, and P is a finite set of productions such that each production has the form $(A_1, \ldots, A_n) \rightarrow (x_1, \ldots, x_n)$, for some $n \geq 1$, where $A_i \in V - T$, $x_i \in V^*$, for every $1 \leq i \leq n$. If every production $(A_1, \ldots, A_n) \rightarrow (x_1, \ldots, x_n) \in P$ satisfies $x_i \in V^+$ for all $1 \leq i \leq n$, G is a *propagating scattered context grammar*, a *PSC* grammar for short. If $(A_1, \ldots, A_n) \rightarrow (x_1, \ldots, x_n) \in P$, $u = u_1 A_1 u_2 \ldots u_n A_n u_{n+1}$, and $v = u_1 x_1 u_2 \ldots u_n x_n u_{n+1}$, where $u_i \in V^*$, for all $1 \le i \le n$, then G makes a *derivation step* from u to v according to $(A_1, \ldots, A_n) \to (x_1, \ldots, x_n)$, symbolically written as

$$u \Rightarrow v \, [(A_1, \ldots, A_n) \to (x_1, \ldots, x_n)]$$

in G or, simply, $u \Rightarrow v$. Set

$$\mathrm{len}((A_1, \ldots, A_n) \to (x_1, \ldots, x_n)) = |A_1 \ldots A_n| = n \ .$$

Let $\Rightarrow^+$ and $\Rightarrow^*$ denote the transitive closure of $\Rightarrow$ and the transitive-reflexive closure of $\Rightarrow$, respectively. The *language of G* is denoted by $L(G)$ and defined as $L(G) = \{x \in T^* : S \Rightarrow^* x\}$.

Let $G = (V, T, P, S)$ be a scattered context grammar. Define a *maximal derivation step*, $_{\max}\!\Rightarrow$, as $u \, _{\max}\!\Rightarrow v \, [p]$, $p \in P$ if and only if $u \Rightarrow v \, [p]$ and there is no $r \in P$, $\mathrm{len}(r) > \mathrm{len}(p)$, such that $u \Rightarrow w \, [r]$. Similarly, define a *minimal derivation step*, $_{\min}\!\Rightarrow$, as $u \, _{\min}\!\Rightarrow v \, [p]$, $p \in P$ if and only if $u \Rightarrow v \, [p]$ and there is no $r \in P$, $\mathrm{len}(r) < \mathrm{len}(p)$, such that $u \Rightarrow w \, [r]$. Define the transitive and transitive-reflexive closure of maximal and minimal derivation steps in the standard way. The *language of G which uses maximal* and *minimal derivations* is denoted by $L_{\max}(G)$ and $L_{\min}(G)$ and defined as

$$L_{\max}(G) = \{x \in T^* : S \, _{\max}\!\Rightarrow^* x\}$$

and

$$L_{\min}(G) = \{x \in T^* : S \, _{\min}\!\Rightarrow^* x\} \ ,$$

respectively. The corresponding language families are denoted by $\mathcal{L}(PSC, \max)$ and $\mathcal{L}(PSC, \min)$.

4   Results

In this section, we demonstrate that propagating scattered context grammars which use maximal and minimal derivations characterize the family of context-sensitive ($\mathcal{L}(CS)$) languages.

Theorem 1. $\mathcal{L}(CS) = \mathcal{L}(PSC, \max)$.

Proof. Let L be a context-sensitive language. As state grammars characterize the family of context-sensitive languages (see [4]), we suppose that L is described by a state grammar, $\bar{G} = (\bar{V}, T, K, \bar{P}, \bar{S}, p_0)$. Set

$$Y = \{\langle A, q \rangle \, : \, A \in \bar{V} - T, q \in K\} \ ,$$

and $Z = \{\bar{a} \, : \, a \in T\}$. Define the isomorphism, α, form $\bar{V}^*$ to $((\bar{V} - T) \cup Z)^*$ as $\alpha(A) = A$ for all $A \in \bar{V} - T$ and $\alpha(a) = \bar{a}$ for all $a \in T$. Set $V = \bar{V} \cup Y \cup Z \cup \{S, X\}$. Define the propagating scattered context grammar, G, as

$$G = (V, T, P, S) \ ,$$

where P is constructed as follows:

1. For every $x \in L(\bar{G})$, $|x| \leq 2$, add
   $(S) \to (x)$ to P;
2. For every

   $$(x, q) \in \{(x, q) : (\bar{S}, p_0) \Rightarrow^+_{\bar{G}} (x, q) \text{ for some } q \in K$$
   $$\text{and } 3 \leq |x| \leq \min(\{3, \max(\{|\Pi(\text{rhs}(p), 1)| : p \in \bar{P}\})\})\} ,$$

   where
   (a) $x \in T^*$, add
       $(S) \to (x)$ to P;
   (b) $x = x_1 A x_2$, $A \in \bar{V} - T$, $x_1, x_2 \in \bar{V}^*$, add
       $(S) \to (\alpha(x_1)\langle A, q\rangle\alpha(x_2))$ to P;
3. For every $(A, p) \to (x, q)$, $(B, p) \to (y, r) \in \bar{P}$, $C \in \bar{V}$, $\Gamma_{21} \in \text{permute}(2, 1)$,

   $$z = \text{reorder}((B, \langle A, p\rangle, \alpha(C)), \Gamma_{21}) ,$$

   add
   $z \to (X, X, X)$ to P;
4. For every $(A, p) \to (x, q) \in \bar{P}$, $B \in \bar{V} - T$, $C \in \bar{V}$, $\Gamma_{11} \in \text{permute}(1, 1)$,

   $$y = \text{reorder}(((\langle A, p\rangle, \alpha(C)), \Gamma_{11}) ,$$

   add
   (a) $(B, \langle A, p\rangle) \to (\langle B, q\rangle, \alpha(x))$,
   (b) $(\langle A, p\rangle, B) \to (\alpha(x), \langle B, q\rangle)$ to P;
   (c) If $x = vBw$, $v, w \in \bar{V}^*$, for every

       $$z = \text{reorder}((\alpha(v)\langle B, q\rangle\alpha(w), \alpha(C)), \Gamma_{11}) ,$$

       add
       $y \to z$ to P;
   (d) For every

       $$u = \text{reorder}((\alpha(x), \alpha(C)), \Gamma_{11}) ,$$

       add
       $y \to u$ to P;
5. For every $a \in T$, add
   $(\bar{a}) \to (a)$ to P.

Basic Idea. The state grammar, $\bar{G}$, is simulated by the propagating scattered context grammar, G, which performs maximal derivations. Productions introduced in step (1) of the construction are used to generate sentences $w \in L(\bar{G})$, $|w| \leq 2$, while the productions introduced in (2) start the simulation of a derivation of $\bar{G}$'s sentences, w, $|w| \geq 3$. Let $(A, p) \to (x, q)$ be a production of $\bar{G}$, which is applicable to a sentential form $(w_1 A w_2, p)$ generated by $\bar{G}$. The sentential form $(w_1 A w_2, p)$ in $\bar{G}$ corresponds to the sentential form $\alpha(w_1)\langle A, p\rangle\alpha(w_2)$ in G. To simulate an application of $(A, p) \to (x, q)$ in G, it is checked first whether the production is applied to the leftmost nonterminal of the sentential form for the given state p. If not, some production from (3) is applicable. This production is applied because it has the highest priority of all productions, and its application introduces the symbol X to the sentential form, which blocks the derivation. The successful derivation proceeds by a production from (4a), (4b), and (4c), which nondeterministically selects the following nonterminal to be rewritten and appends the new state to it. The production which finishes the derivation of a sentence in $\bar{G}$ is simulated by a production from (4d), which removes the compound nonterminal, $\langle \ldots \rangle$, from the sentential form. Finally, each symbol $\bar{a}, a \in T$ is rewritten to a.

Formal Proof

Claim 1. Every $x \in L(\bar{G})$, $|x| \leq 2$ is generated by G as follows:

$$S \;_{\max}\Rightarrow x \; [p_1] \; ,$$

where p_1 is one of the productions introduced in step (1) of the construction.   □

Claim 2. Every

$$(\bar{S}, p_0) \Rightarrow_{\bar{G}}^+ (x, q) \; ,$$

where $q \in K$, $x \in T^+$,

$$3 \leq |x| \leq \min(\{3, \max(\{|\Pi(\mathrm{rhs}(p), 1)| \; : \; p \in \bar{P}\})\})$$

is generated by G as follows:

$$S \;_{\max}\Rightarrow x \; [p_{2a}] \; ,$$

where p_{2a} is one of the productions introduced in step (2a) of the construction.   □

Claim 3. Every

$$(\bar{S}, p_0) \Rightarrow_{\bar{G}}^+ (x, q) \Rightarrow_{\bar{G}}^+ (u, r) \; ,$$

where $q, r \in K$, $u \in T^+$, $x = v_0 A w_0$, $A \in \bar{V} - T$, $v_0, w_0 \in \bar{V}^*$,

$$3 \leq |x| \leq \min(\{3, \max(\{|\Pi(\mathrm{rhs}(p), 1)| \; : \; p \in \bar{P}\})\}) \; ,$$

can only be generated by G as follows:

$$
\begin{aligned}
S \;_{\max}&\Rightarrow & \alpha(v_0)\langle A, q \rangle \alpha(w_0) & \quad [p_{2b}] \\
\;_{\max}&\Rightarrow^* & y & \quad [\Xi_4] \\
\;_{\max}&\Rightarrow & z & \quad [p_{4d}] \\
\;_{\max}&\Rightarrow^{|u|} & u & \quad [\Xi_5] \; ,
\end{aligned}
$$

where $y \in Z^* Y Z^*$, $z = \alpha(u)$; p_{2b} and p_{4d} is one of the productions introduced in steps (2b) and (4d), respectively, and Ξ_4 and Ξ_5 are sequences of productions introduced in steps (4a), (4b), (4c), and (5), respectively.

Proof. Observe that the productions introduced in steps (1) and (2) of the construction are the only productions containing S on their left-hand sides and no other productions contain S on their right-hand sides. To generate a sentence u, $|u| \geq 3$, the derivation has to start with

$$S \;_{max}\!\Rightarrow \alpha(v_0)\langle A, q\rangle\alpha(w_0) \; [p_{2b}] \;,$$

and productions from (1) and (2) are not used during the rest of the derivation.

Further observe that none of the productions introduced in step (3) of the construction can be applied during a successful generation as no productions rewrite the nonterminal, X, which is contained on the right-hand side of every production from step (3).

To generate a sentence over T, all symbols from $\bar{V} - T$ have to be removed from the sentential form. Only productions from step (4) can be used for their replacement as they contain symbols from $\bar{V}-T$ on their left-hand sides. Further, productions (4a), (4b), (4c) contain one symbol from Y both on their left and their right-hand sides, while productions from (4d) contain a symbol form Y only on their left-hand sides. Therefore, after the application of a production from (4d), none of the productions from step (4) is applicable. Because for every production p_4 and p_5 introduced in step (4) and (5), respectively, it holds that $len(p_4) > len(p_5)$, no production from step (5) is applied while some production from step (4) is applicable. As a result, the corresponding part of the derivation looks as follows:

$$\alpha(v_0)\langle A, q\rangle\alpha(w_0) \;_{max}\!\Rightarrow^* y \;[\varXi_4]$$
$$_{max}\!\Rightarrow z \;[p_{4d}] \;.$$

At this point, $z = \alpha(u)$ in a successful derivation. Productions from step (5) replace every $\bar{a} \in \mathrm{alph}(z)$ with a in $|u|$ steps so we obtain

$$z \;_{max}\!\Rightarrow^{|u|} u \;[\varXi_5] \;.$$

Putting together the previous observations, we obtain the formulation of Claim 3 so the claim holds.   □

Claim 4. In a successful derivation, every

$$\alpha(v_0)\langle B_0, q_0\rangle\alpha(w_0)$$
$$_{max}\!\Rightarrow \alpha(v_1)\langle B_1, q_1\rangle\alpha(w_1) \;[p_0]$$
$$\vdots$$
$$_{max}\!\Rightarrow \alpha(v_n)\langle B_n, q_n\rangle\alpha(w_n) \;[p_{n-1}]$$

is performed in G if and only if

$$(v_0 B_0 w_0, q_0)$$
$$\Rightarrow_{\bar{G}} (v_1 B_1 w_1, q_1) \;[(B_0, q_0) \rightarrow (x_1, q_1)]$$
$$\vdots$$
$$\Rightarrow_{\bar{G}} (v_n B_n w_n, q_n) \;[(B_{n-1}, q_{n-1}) \rightarrow (x_n, q_n)]$$

is performed in $\bar{G}$, where $v_i, w_i \in \bar{V}^*$, $B_i \in \bar{V} - T$, $q_i \in K$ for all $i \in \{0, \ldots, n\}$ for some $n \geq 0$, $x_1, \ldots, x_n \in \bar{V}^+$, and $p_0, \ldots, p_{n-1}$ are productions introduced in steps (4a), (4b), (4c) of the construction.

Proof

Only If: We show that

$$\alpha(v_0)\langle B_0, q_0\rangle\alpha(w_0) \ _{\max}\Rightarrow^m \alpha(v_m)\langle B_m, q_m\rangle\alpha(w_m)$$

implies

$$(v_0 B_0 w_0, q_0) \Rightarrow^m_{\bar{G}} (v_m B_m w_m, q_m)$$

by induction on m.

Basis: Let $m = 0$. Then,

$$\alpha(v_0)\langle B_0, q_0\rangle\alpha(w_0) \ _{\max}\Rightarrow^0 \alpha(v_0)\langle B_0, q_0\rangle\alpha(w_0)$$

and clearly

$$(v_0 B_0 w_0, q_0) \Rightarrow^0_{\bar{G}} (v_0 B_0 w_0, q_0) \ .$$

Induction Hypothesis: Suppose that the claim holds for all k-step derivations, where $k \leq m$, for some $m \geq 0$.

Induction Step: Let us consider a derivation

$$\alpha(v_0)\langle B_0, q_0\rangle\alpha(w_0) \ _{\max}\Rightarrow^{m+1} \alpha(v_{m+1})\langle B_{m+1}, q_{m+1}\rangle\alpha(w_{m+1}) \ .$$

Since $m + 1 \geq 1$, there is some

$$\alpha(v_m)\langle B_m, q_m\rangle\alpha(w_m) \in ((\bar{V} - T) \cup Z)^*Y((\bar{V} - T) \cup Z)^*$$

and a production p_m such that

$$\alpha(v_0)\langle B_0, q_0\rangle\alpha(w_0) \ _{\max}\Rightarrow^m \alpha(v_m)\langle B_m, q_m\rangle\alpha(w_m)$$
$$_{\max}\Rightarrow \alpha(v_{m+1})\langle B_{m+1}, q_{m+1}\rangle\alpha(w_{m+1}) \ [p_m] \ .$$

By the induction hypothesis, there is a derivation

$$(v_0 B_0 w_0, q_0) \Rightarrow^m_{\bar{G}} (v_m B_m w_m, q_m) \ .$$

The production p_m is one of the productions introduced in steps (4a) through (4c) and may be of the following three forms depending on the placement of B_{m+1}:

- $(B_{m+1}, \langle B_m, q_m\rangle) \rightarrow (\langle B_{m+1}, q_{m+1}\rangle, \alpha(x_{m+1}))$ for $v_m = v'_m B_{m+1} v''_m$,
- $(\langle B_m, q_m\rangle, B_{m+1}) \rightarrow (\alpha(x_{m+1}), \langle B_{m+1}, q_{m+1}\rangle)$ for $w_m = w'_m B_{m+1} w''_m$,
- $(\langle B_m, q_m\rangle, \alpha(A)) \rightarrow (\alpha(x'_{m+1})\langle B_{m+1}, q_{m+1}\rangle\alpha(x''_{m+1}), \alpha(A))$ or $(\alpha(A), \langle B_m, q_m\rangle) \rightarrow (\alpha(A), \alpha(x'_{m+1})\langle B_{m+1}, q_{m+1}\rangle\alpha(x''_{m+1}))$ for $x_{m+1} = x'_{m+1} B_{m+1} x''_{m+1}$,

where $A \in \bar{V}$, and $x_{m+1}, x'_{m+1}, x''_{m+1} \in \bar{V}^*$. Their construction is based on $\bar{P}$ so there is a production $(B_m, q_m) \to (x_{m+1}, q_{m+1}) \in \bar{P}$.

As we simulate G's derivation by $\bar{G}$, we have to demonstrate that for a given state, q_m, the leftmost nonterminal in a sentential form is rewritten in G. We prove it by contradiction. Suppose that there is a production, $p'_m \in P$, from step (4) which rewrites some $B'_m \in \bar{V} - T$ in a state q_m, and $B'_m \in \text{alph}(v_m)$. Then, there exists $(B'_m, q_m) \to (x'_{m+1}, q'_{m+1}) \in \bar{P}$, and, therefore, there also exist productions introduced in step (3) of the construction which are based on $(B_m, q_m) \to (x_{m+1}, q_{m+1})$ and $(B'_m, q_m) \to (x'_{m+1}, q'_{m+1})$. These productions have the following forms:

- $(B'_m, \langle B_m, q_m \rangle, \alpha(A)) \to (X, X, X)$,
- $(B'_m, \alpha(A), \langle B_m, q_m \rangle) \to (X, X, X)$,
- $(\alpha(A), B'_m, \langle B_m, q_m \rangle) \to (X, X, X)$,

where $A \in \bar{V}$. Because $|\alpha(v_m)\langle B_m, q_m \rangle \alpha(w_m)| \geq 3$, one of these productions is applicable. As productions introduced in step (3) have higher precedence than productions introduced in step (4), one of them is applied which introduces X to the sentential form. This symbol, however, can never be removed from the sentential form so the derivation is not successful.

As a result, the leftmost nonterminal for a state q_m is rewritten in G, so $(B_m, q_m) \to (x_{m+1}, q_{m+1})$ is used in $\bar{G}$ and we obtain

$$(v_m B_m w_m, q_m) \Rightarrow_{\bar{G}} (v_{m+1} B_{m+1} w_{m+1}, q_{m+1}) \; [(B_m, q_m) \to (x_{m+1}, q_{m+1})] \; .$$

If: We show that

$$(v_0 B_0 w_0, q_0) \Rightarrow_{\bar{G}}^m (v_m B_m w_m, q_m)$$

implies

$$\alpha(v_0)\langle B_0, q_0 \rangle \alpha(w_0) \;_{\max}\!\!\Rightarrow^m \alpha(v_m)\langle B_m, q_m \rangle \alpha(w_m)$$

by induction on m.

Basis: Let $m = 0$. Then

$$(v_0 B_0 w_0, q_0) \Rightarrow_{\bar{G}}^0 (v_0 B_0 w_0, q_0) \; .$$

Clearly,

$$\alpha(v_0)\langle B_0, q_0 \rangle \alpha(w_0) \;_{\max}\!\!\Rightarrow^0 \alpha(v_0)\langle B_0, q_0 \rangle \alpha(w_0) \; .$$

Induction Hypothesis: Suppose that the claim holds for all k-step derivations, where $k \leq m$, for some $m \geq 0$.

Induction Step: Consider a derivation

$$(v_0 B_0 w_0, q_0) \Rightarrow_{\bar{G}}^{m+1} (v_{m+1} B_{m+1} w_{m+1}, q_{m+1}) \; .$$

Since $m + 1 \geq 1$, there is some $(v_m B_m w_m, q_m)$, $v_m, w_m \in \bar{V}^*$, $B_m \in \bar{V} - T$, and a production $(B_m, q_m) \to (x_{m+1}, q_{m+1})$ such that

$$\begin{aligned} (v_0 B_0 w_0, q_0) &\Rightarrow_{\bar{G}}^m (v_m B_m w_m, q_m) \\ &\Rightarrow_{\bar{G}} (v_{m+1} B_{m+1} w_{m+1}, q_{m+1}) \; [(B_m, q_m) \to (x_{m+1}, q_{m+1})] \; . \end{aligned}$$

By the induction hypothesis, there is a derivation

$$\alpha(v_0)\langle B_0, q_0\rangle\alpha(w_0) \ _{max}\Rightarrow^m \alpha(v_m)\langle B_m, q_m\rangle\alpha(w_m) \ .$$

Because $(B_m, q_m) \rightarrow (x_{m+1}, q_{m+1})$ rewrites the leftmost rewritable symbol, B_m, for a given state, q_m, there is no production $(B'_m, q_m) \rightarrow (x'_{m+1}, q'_{m+1})$ satisfying $B'_m \in alph(v_m)$. As a result, none of the productions from step (3) is applicable.

For every $(B_m, q_m) \rightarrow (x_{m+1}, q_{m+1}) \in \bar{P}$ there are productions of the following three forms in G whose use depends on the placement of B_{m+1}:

1. $(B_{m+1}, \langle B_m, q_m\rangle) \rightarrow (\langle B_{m+1}, q_{m+1}\rangle, \alpha(x_{m+1}))$ for $v_m = v'_m B_{m+1} v''_m$,
2. $(\langle B_m, q_m\rangle, B_{m+1}) \rightarrow (\alpha(x_{m+1}), \langle B_{m+1}, q_{m+1}\rangle)$ for $w_m = w'_m B_{m+1} w''_m$,
3. $(\langle B_m, q_m\rangle, \alpha(A)) \rightarrow (\alpha(x'_{m+1})\langle B_{m+1}, q_{m+1}\rangle\alpha(x''_{m+1}), \alpha(A))$ or
   $(\alpha(A), \langle B_m, q_m\rangle) \rightarrow (\alpha(A), \alpha(x'_{m+1})\langle B_{m+1}, q_{m+1}\rangle\alpha(x''_{m+1}))$ for
   $x_{m+1} = x'_{m+1} B_{m+1} x''_{m+1}$,

where $A \in \bar{V}$, and $x_{m+1}, x'_{m+1}, x''_{m+1} \in \bar{V}^*$. As $|\alpha(v_m)\langle B_m, q_m\rangle\alpha(w_m)| \geq 3$, one of them is applicable in G so we obtain

$$\alpha(v_m)\langle B_m, q_m\rangle\alpha(w_m) \ _{max}\Rightarrow \alpha(v_{m+1})\langle B_{m+1}, q_{m+1}\rangle\alpha(w_{m+1}) \ . \qquad \square$$

By Claim 1 through 4 it follows that $\mathcal{L}(CS) \subseteq \mathcal{L}(PSC, max)$. As PSC grammars do not contain ε-productions, their derivations can be simulated by linear bounded automata. As a result, $\mathcal{L}(PSC, max) \subseteq \mathcal{L}(CS)$. Therefore, $\mathcal{L}(CS) = \mathcal{L}(PSC, max)$. $\qquad \square$

Theorem 2. $\mathcal{L}(CS) = \mathcal{L}(PSC, min)$.

Proof. Let L be a context-sensitive language described by a state grammar, $\bar{G} = (\bar{V}, T, K, \bar{P}, \bar{S}, p_0)$. Set

$$Y = \{\langle A, q\rangle : A \in \bar{V} - T, q \in K\} \ ,$$

and $Z = \{\bar{a} : a \in T\}$. Define the isomorphism, α, form $\bar{V}^*$ to $((\bar{V} - T) \cup Z)^*$ as $\alpha(A) = A$ for all $A \in \bar{V} - T$ and $\alpha(a) = \bar{a}$ for all $a \in T$. Set $V = \bar{V} \cup Y \cup Z \cup \{S, X\}$. Define the propagating scattered context grammar, G', as

$$G' = (V, T, P', S) \ ,$$

where P' is constructed as follows:

1. For every $x \in L(\bar{G})$, $|x| \leq 3$, add
   $(S) \rightarrow (x)$ to P';
2. For every

$$(x, q) \in \{(x, q) : (\bar{S}, p_0) \Rightarrow^+_{\bar{G}} (x, q) \text{ for some } q \in K$$
$$\text{and } 4 \leq |x| \leq min(\{4, max(\{|\Pi(rhs(p), 1)| : p \in \bar{P}\})\})\} \ ,$$

where

(a) $x \in T^*$, add

$(S) \rightarrow (x)$ to P';

(b) $x = x_1 A x_2$, $A \in \bar{V} - T$, $x_1, x_2 \in \bar{V}^*$, add

$(S) \rightarrow (\alpha(x_1)\langle A, q\rangle\alpha(x_2))$ to P';

3. For every $(A, p) \rightarrow (x, q)$, $(B, p) \rightarrow (y, r) \in \bar{P}$, add

$(B, \langle A, p\rangle) \rightarrow (X, X)$ to P';

4. For every $(A, p) \rightarrow (x, q) \in \bar{P}$, $B \in \bar{V} - T$, $D, E \in \bar{V}$, $\Gamma_{21} \in \mathrm{permute}(2, 1)$, $\Gamma_{12} \in \mathrm{permute}(1, 2)$,

$$u = \mathrm{reorder}((B, \langle A, p\rangle, \alpha(D)), \Gamma_{21}), \quad u' = \mathrm{reorder}((\langle B, q\rangle, \alpha(x), \alpha(D)), \Gamma_{21}),$$
$$r = \mathrm{reorder}((\langle A, p\rangle, B, \alpha(D)), \Gamma_{21}), \quad r' = \mathrm{reorder}((\alpha(x), \langle B, q\rangle, \alpha(D)), \Gamma_{21}),$$

$$y = \mathrm{reorder}((\langle A, p\rangle, \alpha(D), \alpha(E)), \Gamma_{12}) ,$$

add

(a) $u \rightarrow u'$,

(b) $r \rightarrow r'$ to P';

(c) If $x = vBw$, $v, w \in \bar{V}^*$, for every

$$z = \mathrm{reorder}((\alpha(v)\langle B, q\rangle\alpha(w), \alpha(D), \alpha(E)), \Gamma_{12}) ,$$

add

$y \rightarrow z$ to P';

(d) For every

$$u = \mathrm{reorder}((\alpha(x), \alpha(D), \alpha(E)), \Gamma_{12}) ,$$

add

$y \rightarrow u$ to P';

5. For every $a, b, c, d \in T$, add

(a) $(\bar{a}, \bar{b}, \bar{c}, \bar{d}) \rightarrow (a, \bar{b}, \bar{c}, \bar{d})$,

(b) $(\bar{a}, \bar{b}, \bar{c}, \bar{d}) \rightarrow (a, b, c, d)$ to P'.

Claim 5. Every

$$(\bar{S}, p_0) \Rightarrow_{\bar{G}}^+ (x, q) \Rightarrow_{\bar{G}}^+ (u, r) ,$$

where $q, r \in K$, $u \in T^+$, $x = v_0 A w_0$, $A \in \bar{V} - T$, $v_0, w_0 \in \bar{V}^*$,

$$4 \le |x| \le \min(\{4, \max(\{|\Pi(\mathrm{rhs}(p), 1)| : p \in \bar{P}\})\}) ,$$

can only be generated by G' as follows:

$$
\begin{array}{lll}
S \ _{\min}\!\!\Rightarrow & \alpha(v_0)\langle A, q\rangle\alpha(w_0) & [p_{2b}] \\
\ _{\min}\!\!\Rightarrow^* & y & [\varXi_4] \\
\ _{\min}\!\!\Rightarrow & z & [p_{4d}] \\
\ _{\min}\!\!\Rightarrow^{|u|-4} & v & [\varXi_5] \\
\ _{\min}\!\!\Rightarrow & u & [p_{5b}] ,
\end{array}
$$

where $y \in Z^* Y Z^*$, $z = \alpha(u)$, $v \in (T \cup Z)^+$; p_{2b}, p_{4d}, and p_{5b} is one of the productions introduced in steps (2b), (4d), and (5b), respectively, and $\varXi_4$ and $\varXi_5$ are sequences of productions introduced in steps (4a), (4b), (4c), and (5a), respectively.

Proof. The proof of the beginning of the derivation,

$$\begin{aligned} S \,_{\min}&\Rightarrow\ \alpha(v_0)\langle A, q\rangle\alpha(w_0)\ [p_{2b}]\\ _{\min}&\Rightarrow^*\ y\ \ \ \ \ \ \ \ \ \ \ \ \ \ \ \ [\varXi_4]\\ _{\min}&\Rightarrow\ z\ \ \ \ \ \ \ \ \ \ \ \ \ \ \ \ [p_{4d}]\ ,\end{aligned}$$

is analogous to the proof of Claim 3 (in terms of minimal derivations), and is left to the reader.

Recall that z satisfies $z = \alpha(u)$. Each of the productions introduced in step (5a) of the construction replaces one occurrence of $\bar{a}$ with a for some $a \in T$, and, finally, the application of a production from step (5b) replaces the remaining four nonterminals with their terminal variants. Therefore,

$$\begin{aligned} z \,_{\min}&\Rightarrow^{|u|-4}\ v\ [\varXi_5]\\ _{\min}&\Rightarrow\ \ \ \ \ u\ [p_{5b}]\ ,\end{aligned}$$

so the claim holds. $\qquad\qquad\qquad\qquad\qquad\qquad\qquad\qquad\qquad\qquad\qquad\qquad$ □

Notice that $\mathrm{len}(p_3) < \mathrm{len}(p_4) < \mathrm{len}(p_5)$ for every production p_3, p_4, and p_5 introduced in steps (3), (4), and (5), respectively, so the priorities (and the use) of the productions from the individual steps are the same as in case of grammars which use maximal derivations. As a result, formulations of Claim 1, 2, and 4 can be changed in terms of minimal derivations. As their proofs resemble the proofs of the claims mentioned above, these proofs are left to the reader. Therefore, $\mathcal{L}(CS) \subseteq \mathcal{L}(PSC, \min)$ and for the same reason as in Theorem 1, $\mathcal{L}(PSC, \min) \subseteq \mathcal{L}(CS)$, so $\mathcal{L}(CS) = \mathcal{L}(PSC, \min)$. $\qquad\qquad\qquad$ □

References

1. Fernau, H.: Scattered context grammars with regulation. Annals of Bucharest University, Mathematics-Informatics Series 45(1), 41–49 (1996)
2. Gonczarowski, J., Warmuth, M.K.: Scattered versus context-sensitive rewriting. Acta Informatica 27, 81–95 (1989)
3. Greibach, S., Hopcroft, J.: Scattered context grammars. Journal of Computer and System Sciences 3, 233–247 (1969)
4. Kasai, T.: An hierarchy between context-free and context-sensitive languages. Journal of Computer and System Sciences 4(5), 492–508 (1970)
5. Meduna, A.: A trivial method of characterizing the family of recursively enumerable languages by scattered context grammars. EATCS Bulletin 56, 104–106 (1995)
6. Meduna, A.: Automata and Languages: Theory and Applications. Springer, London (2000)
7. Meduna, A.: Generative power of three-nonterminal scattered context grammars. Theoretical Computer Science 246, 276–284 (2000)
8. Meduna, A., Techet, J.: Generation of sentences with their parses: the case of propagating scattered context grammars. Acta Cybernetica 17, 11–20 (2005)
9. Milgram, D., Rosenfeld, A.: A note on scattered context grammars. Information Processing Letters 1, 47–50 (1971)
10. Revesz, G.E.: Introduction to Formal Language Theory. McGraw-Hill, New York (1983)

11. Rozenberg, G., Salomaa, A.: Handbook of Formal Languages, vol. 1–3. Springer, Heidelberg (1997)
12. Salomaa, A.: Formal Languages. Academic Press, New York (1973)
13. Vaszil, G.: On the descriptional complexity of some rewriting mechanisms regulated by context conditions. Theoretical Computer Science 330, 361–373 (2005)
14. Virkkunen, V.: On scattered context grammars. Acta Universitatis Ouluensis Series A, Mathematica 6, 75–82 (1973)

Strictly Deterministic CD-Systems of Restarting Automata

H. Messerschmidt and F. Otto

Fachbereich Elektrotechnik/Informatik, Universität Kassel
34109 Kassel, Germany
{hardy,otto}@theory.informatik.uni-kassel.de

Abstract. A CD-system of restarting automata is called *strictly deterministic* if all its component systems are deterministic, and if there is a unique successor system for each component. Here we show that the strictly deterministic CD-systems of restarting automata are strictly more powerful than the corresponding deterministic types of restarting automata, but that they are strictly less powerful than the corresponding deterministic types of nonforgetting restarting automata. In fact, we present an infinite hierarchy of language classes based on the number of components of strictly deterministic CD-systems of restarting automata.

1 Introduction

The restarting automaton was introduced by Jančar et. al. as a formal tool to model the *analysis by reduction*, which is a technique used in linguistics to analyze sentences of natural languages [4]. This technique consists in a stepwise simplification of a given sentence in such a way that the correctness or incorrectness of the sentence is not affected. It is applied primarily in languages that have a free word order. Already several programs used in Czech and German (corpus) linguistics are based on the idea of restarting automata [9,12].

A (one-way) restarting automaton, RRWW-automaton for short, is a device M that consists of a finite-state control, a flexible tape containing a word delimited by sentinels, and a read/write window of a fixed size. This window is moved from left to right until the control decides (nondeterministically) that the content of the window should be rewritten by some *shorter* string. In fact, the new string may contain auxiliary symbols that do not belong to the input alphabet. After a rewrite, M can continue to move its window until it either halts and accepts, or halts and rejects, or restarts, that is, it places its window over the left end of the tape, and reenters the initial state. Thus, each computation of M can be described through a sequence of cycles.

Many restricted variants of restarting automata have been studied and put into correspondence to more classical classes of formal languages. For a recent survey see [10] or [11]. Also further extensions of the model have been considered. In particular, in [8] Messerschmidt and Stamer introduced the *nonforgetting restarting automaton*, which, when executing a restart operation, simply changes

E. Csuhaj-Varjú and Z. Ésik (Eds.): FCT 2007, LNCS 4639, pp. 424–434, 2007.

its internal state as with any other operation, instead of resetting it to the initial state. Further, in [7] the authors introduced cooperating distributed systems (CD-systens) of restarting automata and proved that CD-systems of restarting automata working in mode $= 1$ correspond to nonforgetting restarting automata.

Here we concentrate on CD-systems of restarting automata that are deterministic. It is known that deterministic restarting automata with auxiliary symbols accept exactly the Church-Rosser languages (see, e.g., [10,11]), while nonforgetting deterministic restarting automata are strictly more powerful [6]. However, for CD-systems of restarting automata the notion of determinism can be defined in various different ways. A CD-system $\mathcal{M} := ((M_i, \sigma_i)_{i \in I}, I_0)$ of restarting automata could be called *deterministic* if within each computation of the system $\mathcal{M}$, each configuration has at most a single successor configuration. This is a global view on determinism. On the other hand, we could follow the way determinism is used in CD-grammar systems (see, e.g., [2,3]) and call $\mathcal{M}$ already *deterministic* if all component automata M_i $(i \in I)$ are deterministic. This is a local view on determinism. Here we study a third option, called *strict determinism*, where we require not only that all component automata M_i $(i \in I)$ are deterministic, but also that the successor set σ_i is a singleton for each $i \in I$. This is again a global, but more restricted, view.

We will see that, in analogy to the situation for nondeterministic restarting automata, the globally deterministic CD-systems of restarting automata, when working in mode $= 1$, correspond exactly to nonforgetting deterministic restarting automata. Further, the expressive power of strictly deterministic CD-systems of restarting automata lies strictly in between that of deterministic restarting automata and that of nonforgetting deterministic restarting automata. In fact, based on the number of component systems of strictly deterministic CD-systems, we will obtain a proper infinite hierarchy of language classes.

This paper is structured as follows. In Section 2 we introduce nonforgetting restarting automata and CD-systems of restarting automata. Then, in Section 3, we define the various types of deterministic CD-systems of restarting automata formally and establish the announced relationship between nonforgetting deterministic restarting automata and globally deterministic CD-systems of restarting automata. In Section 4 we compare the expressive power of strictly deterministic CD-systems to that of globally deterministic CD-systems and to that of deterministic restarting automata. Also we present the announced hierarchy on the number of component systems. The paper concludes with a short discussion pointing out some open problems for future work.

2   Definitions

An RRWW-*automaton* is a one-tape machine that is described by an 8-tuple $M = (Q, \Sigma, \Gamma, \math162, \$, q_0, k, \delta)$, where Q is a finite set of states, Σ is a finite input alphabet, Γ is a finite tape alphabet containing Σ, the symbols $\math162, \$ \notin \Gamma$ serve as markers for the left and right border of the work space, respectively, $q_0 \in Q$ is the initial state, $k \geq 1$ is the size of the *read/write window*, and δ is the *transition relation* that associates a finite set of *transition steps* to each pair (q, u) consisting of a state $q \in Q$ and a possible contents u of the read/write window. There are four types of transition steps:

- *Move-right steps* of the form (q', MVR), where $q' \in Q$, which cause M to shift the read/write window one position to the right and to enter state q'.
- *Rewrite steps* of the form (q', v), where $q' \in Q$, and v is a string satisfying $|v| < |u|$. This step causes M to replace the content u of the read/write window by the string v, thereby shortening the tape, and to enter state q'. Further, the read/write window is placed immediately to the right of the string v. However, some additional restrictions apply in that the border markers ¢ and $ must not disappear from the tape nor that new occurrences of these markers are created.
- *Restart steps* of the form $\mathsf{Restart}$, which cause M to place the read/write window over the left end of the tape, so that the first symbol it sees is the left border marker ¢, and to reenter the initial state q_0.
- *Accept steps* of the form Accept, which cause M to halt and accept.

If $\delta(q, u) = \emptyset$ for some pair (q, u), then M necessarily halts, and we say that M *rejects* in this situation. There is one additional restriction that the transition relation must satisfy: ignoring move operations, *rewrite steps and restart steps alternate* in any computation of M, with a rewrite step coming first. However, it is more convenient to describe M by a finite set of so-called *meta-instructions* (see below).

A *configuration* of M is described by a string $\alpha q \beta$, where $q \in Q$, and either $\alpha = \varepsilon$ (the empty word) and $\beta \in \{¢\} \cdot \Gamma^* \cdot \{\$\}$ or $\alpha \in \{¢\} \cdot \Gamma^*$ and $\beta \in \Gamma^* \cdot \{\$\}$; here q represents the current state, $\alpha\beta$ is the current content of the tape, and it is understood that the head scans the first k symbols of β or all of β when $|\beta| \leq k$. A *restarting configuration* is of the form $q_0 ¢ w \$$, where $w \in \Gamma^*$; if $w \in \Sigma^*$, then $q_0 ¢ w \$$ is an *initial configuration*.

In general, an RRWW-automaton is nondeterministic, that is, to some configurations several different instructions may apply. If that is not the case, then the automaton is called *deterministic*.

A *rewriting meta-instruction* for M has the form $(E_1, u \to v, E_2)$, where E_1 and E_2 are regular expressions, and $u, v \in \Gamma^*$ are words satisfying $|u| > |v|$. To execute a cycle M chooses a meta-instruction of the form $(E_1, u \to v, E_2)$. On trying to execute this meta-instruction M will get stuck (and so reject) starting from the *restarting configuration* $q_0 ¢ w \$$, if w does not admit a factorization of the form $w = w_1 u w_2$ such that $¢ w_1 \in E_1$ and $w_2 \$ \in E_2$. On the other hand, if w does have factorizations of this form, then one such factorization is chosen nondeterministically, and $q_0 ¢ w \$$ is transformed into $q_0 ¢ w_1 v w_2 \$$. This computation is called a *cycle* of M. It is expressed as $w \vdash^c_M w_1 v w_2$. In order to describe the tails of accepting computations we use *accepting meta-instructions* of the form (E_1, Accept), which simply accepts the strings from the regular language E_1.

An input word $w \in \Sigma^*$ is *accepted* by M, if there is a computation which, starting with the initial configuration $q_0 \mathbb{c} w \$$, consists of a finite sequence of cycles that is followed by an application of an accepting meta-instruction. By $L(M)$ we denote the language consisting of all words accepted by M.

We are also interested in various restricted types of restarting automata. They are obtained by combining two types of restrictions:

(a) Restrictions on the movement of the read/write window (expressed by the first part of the class name): RR- denotes no restriction, R- means that each rewrite step is immediately followed by a restart.
(b) Restrictions on the rewrite-instructions (expressed by the second part of the class name): -WW denotes no restriction, -W means that no auxiliary symbols are available (that is, $\Gamma = \Sigma$), -ε means that no auxiliary symbols are available and that each rewrite step is simply a deletion (that is, if the rewrite operation $u \rightarrow v$ occurs in a meta-instruction of M, then v is obtained from u by deleting some symbols).

A *cooperating distributed system of* RRWW-*automata*, CD-RRWW-system for short, consists of a finite collection $\mathcal{M} := ((M_i, \sigma_i)_{i \in I}, I_0)$ of RRWW-automata $M_i = (Q_i, \Sigma, \Gamma_i, \mathbb{c}, \$, q_0^{(i)}, k, \delta_i)$ $(i \in I)$, *successor relations* $\sigma_i \subseteq I$ $(i \in I)$, and a subset $I_0 \subseteq I$ of *initial indices*. Here it is required that $Q_i \cap Q_j = \emptyset$ for all $i, j \in I$, $i \neq j$, that $I_0 \neq \emptyset$, that $\sigma_i \neq \emptyset$ for all $i \in I$, and that $i \notin \sigma_i$ for all $i \in I$. Further, let m be one of the following *modes of operation*, where $j \geq 1$:

$$= j : \text{execute exactly } j \text{ cycles;}$$
$$\text{t} : \text{continue until no more cycle can be executed.}$$

The computation of $\mathcal{M}$ in mode $= j$ on an input word x proceeds as follows. First an index $i_0 \in I_0$ is chosen nondeterministically. Then the RRWW-automaton M_{i_0} starts the computation with the initial configuration $q_0^{(i_0)} \mathbb{c} x \$$, and executes j cycles. Thereafter an index $i_1 \in \sigma_{i_0}$ is chosen nondeterministically, and M_{i_1} continues the computation by executing j cycles. This continues until, for some $l \geq 0$, the machine M_{i_l} accepts. Should at some stage the chosen machine M_{i_l} be unable to execute the required number of cycles, then the computation fails.

In mode t the chosen automaton M_{i_l} continues with the computation until it either accepts, in which case $\mathcal{M}$ accepts, or until it can neither execute another cycle nor an accepting tail, in which case an automaton $M_{i_{l+1}}$ with $i_{l+1} \in \sigma_{i_l}$ takes over. Should this machine not be able to execute a cycle or an accepting tail, then the computation of $\mathcal{M}$ fails.

By $L_{\text{m}}(\mathcal{M})$ we denote the language that the CD-RRWW-system $\mathcal{M}$ accepts in mode m. It consists of all words $x \in \Sigma^*$ that are accepted by $\mathcal{M}$ in mode m as described above. If X is any of the above types of restarting automata, then a CD-X-system is a CD-RRWW-system for which all component automata are of type X.

The following simple example shows that CD-R-systems have much more expressive power than R-automata. As R-automata restart immediately after executing a rewrite operation, rewriting meta-instructions for them are of the form

$(E, u \rightarrow v)$, where E is a regular language, and $u \rightarrow v$ is a rewrite step of M (see, e.g., [11]).

Example 1. Let $\mathcal{M} := ((M_1, \{2\}), (M_2, \{1\}), \{1\})$ be a CD-R-system, where M_1 is described by the following three meta-instructions:

$$(\math0 \cdot \{a, b\}^* \cdot c \cdot \# \cdot \{a, b\}^*, c \cdot \$ \rightarrow \$), \ c \in \{a, b\}, \text{ and } (\math0 \cdot \# \cdot \$, \mathsf{Accept}),$$

and M_2 is given through the meta-instructions:

$$(\math0 \cdot \{a, b\}^*, c \cdot \# \rightarrow \#), \ c \in \{a, b\}.$$

In mode $=1$ the machines M_1 and M_2 alternate, with M_1 beginning the computation. Starting with a word of the form $u\#v$, where $u, v \in \{a, b\}^+$, M_1 always deletes the last letter of the second factor, provided it coincides with the last letter of the first factor, and M_2 simply deletes the last letter of the first factor. It follows that $L_{=1}(\mathcal{M})$ coincides with the language $L_{\mathrm{copy}} := \{ w\#w \mid w \in \{a, b\}^* \}$. On the other hand, it is easily seen that L_{copy} is not accepted by any R-automaton.

The *nonforgetting restarting automaton* is a generalization of the restarting automaton that is obtained by combining restart transitions with a change of state just like the move-right and rewrite transitions. This allows a nonforgetting restarting automaton M to carry some information from one cycle to the next. We use the notation $(q_1, x) \vdash^c_M (q_2, y)$ to denote a cycle of M that transforms the restarting configuration $q_1\math0 x\$$ into the restarting configuration $q_2\math0 y\$$.

3   Various Notions of Determinism

A CD-system $\mathcal{M} := ((M_i, \sigma_i)_{i \in I}, I_0)$ of restarting automata is called *locally deterministic* if M_i is a deterministic restarting automaton for each $i \in I$. As the successor system is chosen nondeterministically from among all systems M_j with $j \in \sigma_i$, computations of a locally deterministic CD-system of restarting automata are in general not completely deterministic.

To avoid this remaining nondeterminism we strengthen the above definition. We call a CD-system $\mathcal{M} := ((M_i, \sigma_i)_{i \in I}, I_0)$ *strictly deterministic* if I_0 is a singleton, if M_i is a deterministic restarting automaton and if $|\sigma_i| = 1$ for each $i \in I$. Observe that the CD-R-system of Example 1 is strictly deterministic.

However, the restriction of having at most a single possible successor for each component system is a rather serious one, as we will see below. Thus, we define a third notion. A CD-system $\mathcal{M} := ((M_i, \sigma_i)_{i \in I}, I_0)$ is called *globally deterministic* if I_0 is a singleton, if M_i is a deterministic restarting automaton for each $i \in I$, and if, for each $i \in I$, each restart operation of M_i is combined with an index from the set σ_i. Thus, when M_i finishes a part of a computation according to the actual mode of operation by executing the restart operation $\delta_i(q, u) = (\mathsf{Restart}, j)$, where $j \in \sigma_i$, then the component M_j takes over. In this way it is guaranteed that all computations of a globally deterministic CD-system are deterministic. However, for a component system M_i there can still be several possible successor systems. This is reminiscent of the way in which nonforgetting restarting automata work.

We use the prefix det-global to denote globally deterministic CD-systems, and the prefix det-strict to denote strictly deterministic CD-systems. For each type of restarting automaton $X \in \{R, RR, RW, RRW, RWW, RRWW\}$, it is easily seen that the following inclusions hold:

$$\mathcal{L}(\text{det-X}) \subseteq \mathcal{L}_m(\text{det-strict-CD-X}) \subseteq \mathcal{L}_m(\text{det-global-CD-X}).$$

Concerning the globally deterministic CD-systems, we have the following results, which correspond to the results for nondeterministic CD-systems established in [7].

Theorem 1. *If M is a nonforgetting deterministic restarting automaton of type* X *for some* $X \in \{R, RR, RW, RRW, RWW, RRWW\}$, *then there exists a globally deterministic CD-system $\mathcal{M}$ of restarting automata of type* X *such that $L_{=1}(\mathcal{M}) = L(M)$ holds.*

For the converse we even have the following stronger result.

Theorem 2. *For each* $X \in \{R, RR, RW, RRW, RWW, RRWW\}$, *if $\mathcal{M}$ is a globally deterministic* CD-X-*system, and if j is a positive integer, then there exists a nonforgetting deterministic* X-*automaton M such that $L(M) = L_{=j}(\mathcal{M})$ holds.*

Thus, we see that globally deterministic CD-systems of restarting automata working in mode = 1 are just as powerful as deterministic nonforgetting restarting automata. It remains to study CD-systems that work in mode t.

Theorem 3. *Let* $X \in \{RR, RRW, RRWW\}$, *and let $\mathcal{M}$ be a globally deterministic* CD-X-*system. Then there exists a nonforgetting deterministic* X-*automaton M such that $L(M) = L_t(\mathcal{M})$ holds.*

It is not clear whether the latter result extends to CD-systems of R(W)(W)-automata. The problem with these types of restarting automata stems from the fact that within a cycle such an automaton will in general not see the complete tape content.

4   Strictly Deterministic CD-Systems

Here we study the expressive power of strictly deterministic CD-systems of restarting automata. As seen in Example 1 the copy language L_{copy} is accepted by a strictly deterministic CD-R-system with two components. This language is not growing context-sensitive [1,5]. As deterministic RRWW-automata only accept Church-Rosser languages, which are a proper subclass of the growing context-sensitive languages, this yields the following separation result.

Proposition 1. *For all* $X \in \{R, RR, RW, RRW, RWW, RRWW\}$,

$$\mathcal{L}(\text{det-X}) \subset \mathcal{L}_{=1}(\text{det-strict-CD-X}).$$

Let $L_{\text{copy}^m} := \{ w(\#w)^{m-1} \mid w \in \Sigma_0^+ \}$ be the m-fold copy language. Analogously to Example 1 it can be shown that this language is accepted by a strictly deterministic CD-R-system with m components that works in mode $= 1$. The next example deals with a generalization of these languages.

Example 2. Let $L_{\text{copy}^*} := \{ w(\#w)^n \mid w \in (\Sigma_0^2)^+, n \geq 1 \}$ be the iterated copy language, where $\Sigma_0 := \{a, b\}$. This language is accepted by a strictly deterministic CD-RWW-system $\mathcal{M} = ((M_1, \{2\}), (M_2, \{1\}), \{1\})$ with input alphabet $\Sigma := \Sigma_0 \cup \{\#\}$ and tape alphabet $\Gamma := \Sigma \cup \Gamma_0$, where $\Gamma_0 := \{A_{a,a}, A_{a,b}, A_{b,a}, A_{b,b}\}$. The RWW-automata M_1 and M_2 are given through the following meta-instructions, where $c, d, e, f \in \Sigma_0$:

$$
\begin{aligned}
M_1: \quad & (\mathcal{c} \cdot (\Sigma_0^2)^* \cdot cd \cdot \# \cdot (\Sigma_0^2)^*, cd \cdot \# \to A_{c,d} \cdot \#), \\
& (\mathcal{c} \cdot (\Sigma_0^2)^* \cdot cd \cdot \# \cdot (\Sigma_0^2)^*, cd \cdot \$ \to A_{c,d} \cdot \$), \\
& (\mathcal{c} \cdot (\Sigma_0^2)^* \cdot cd \cdot \# \cdot (\Sigma_0^2)^*, cdA_{e,f} \to A_{c,d}A_{e,f}), \\
& (\mathcal{c} \cdot \Gamma_0^* \cdot A_{c,d} \cdot \# \cdot (\Sigma_0^2)^*, cd \cdot \# \to A_{c,d} \cdot \#), \\
& (\mathcal{c} \cdot \Gamma_0^* \cdot A_{c,d} \cdot \# \cdot (\Sigma_0^2)^*, cd \cdot \$ \to A_{c,d} \cdot \$), \\
& (\mathcal{c} \cdot \Gamma_0^* \cdot A_{c,d} \cdot \# \cdot (\Sigma_0^2)^*, cdA_{e,f} \to A_{c,d}A_{e,f}), \\
& (\mathcal{c} \cdot \Gamma_0^+ \cdot \$, \text{Accept}), \\
M_2: \quad & (\mathcal{c} \cdot (\Sigma_0^2)^+, cd \cdot \# \to \#), \quad (\mathcal{c}, cd \cdot \# \to \varepsilon), \\
& (\mathcal{c} \cdot \Gamma_0^+, A_{c,d} \cdot \# \to \#), \quad (\mathcal{c}, A_{c,d} \cdot \# \to \varepsilon).
\end{aligned}
$$

In mode $= 1$, the two components M_1 and M_2 are used alternatingly, with M_1 starting the computation. Let $x := w_1 \# w_2 \# \ldots \# w_m$ be the given input, where $w_1, w_2, \ldots, w_m \in (\Sigma_0^2)^+$ and $m \geq 2$. First w_1 is compared to w_2 by processing these strings from right to left, two letters in each round. During this process w_1 is erased, while w_2 is encoded using the letters from Γ_0. Next the encoded version of w_2 is used to compare w_2 to w_3, again from right to left. This time the encoded version of w_2 is erased, while w_3 is encoded. This continues until all syllables w_i have been considered. It follows that $L_{=1}(\mathcal{M}) = L_{\text{copy}^*}$ holds.

For accepting the language L_{copy^*} without using auxiliary symbols we have a CD-system of restarting automata that is globally deterministic.

Lemma 1. *The language* L_{copy^*} *is accepted by a globally deterministic CD-R-system working in mode* $= 1$.

Proof. Let $\mathcal{M} = ((M_i, \sigma_i)_{i \in I}, I_0)$ be the CD-R-system that is specified by $I := \{0, 1, 2, 3, 4, 5, 6\}$, $I_0 := \{0\}$, $\sigma(0) := \{5\}$, $\sigma(1) := \{2, 6\}$, $\sigma(2) := \{1, 6\}$, $\sigma(3) := \{4, 5\}$, $\sigma(4) := \{3, 5\}$, $\sigma(5) := \{1\}$, $\sigma(6) := \{3\}$, and M_0 to M_6 are given through the following meta-instructions, where $c, d \in \Sigma_0$:

$$
\begin{aligned}
M_0: \quad & (\mathcal{c} \cdot ((\Sigma_0^2)^+ \cdot \Sigma_0 \cdot c \cdot \#)^+ \cdot (\Sigma_0^2)^+ \cdot \Sigma_0, c \cdot \$ \to \$, \text{Restart}(5)), \\
& (\mathcal{c} \cdot (u \cdot \#)^+ \cdot u \cdot \$, \text{Accept}) \text{ for all } u \in \Sigma_0^2,
\end{aligned}
$$

$M_1 : (\mathcal{c} \cdot ((\Sigma_0^2)^+ \cdot c \cdot \#)^+ \cdot (\Sigma_0^2)^+, c \cdot \$ \to \$, \text{Restart}(6)),$
$\quad\quad (\mathcal{c} \cdot ((\Sigma_0^2)^+ \cdot c \cdot \#)^+ \cdot (\Sigma_0^2)^+ \cdot c, d \cdot \# \to \#, \text{Restart}(2)),$

$M_2 : (\mathcal{c} \cdot ((\Sigma_0^2)^+ \cdot c \cdot \#)^+ \cdot (\Sigma_0^2)^+, c \cdot \$ \to \$, \text{Restart}(6)),$
$\quad\quad (\mathcal{c} \cdot ((\Sigma_0^2)^+ \cdot c \cdot \#)^+ \cdot (\Sigma_0^2)^+ \cdot c, d \cdot \# \to \#, \text{Restart}(1)),$

$M_3 : (\mathcal{c} \cdot ((\Sigma_0^2)^+ \cdot \Sigma_0 \cdot c \cdot \#)^+ \cdot (\Sigma_0^2)^+ \cdot \Sigma_0, c \cdot \$ \to \$, \text{Restart}(5)),$
$\quad\quad (\mathcal{c} \cdot ((\Sigma_0^2)^* \cdot \Sigma_0 \cdot c \cdot \#)^+ \cdot (\Sigma_0^2)^* \cdot \Sigma_0 \cdot c, d \cdot \# \to \#, \text{Restart}(4)),$
$\quad\quad (\mathcal{c} \cdot (u \cdot \#)^+ \cdot \$, \text{Accept})$ for all $u \in \Sigma_0^2,$

$M_4 : (\mathcal{c} \cdot ((\Sigma_0^2)^+ \cdot \Sigma_0 \cdot c \cdot \#)^+ \cdot (\Sigma_0^2)^+ \cdot \Sigma_0, c \cdot \$ \to \$, \text{Restart}(5)),$
$\quad\quad (\mathcal{c} \cdot ((\Sigma_0^2)^* \cdot \Sigma_0 \cdot c \cdot \#)^+ \cdot (\Sigma_0^2)^* \cdot \Sigma_0 \cdot c, d \cdot \# \to \#, \text{Restart}(3)),$
$\quad\quad (\mathcal{c} \cdot (u \cdot \#)^+ \cdot \$, \text{Accept})$ for all $u \in \Sigma_0^2,$

$M_5 : (\mathcal{c} \cdot \Sigma_0^+, c \cdot \# \to \#, \text{Restart}(1)),$

$M_6 : (\mathcal{c} \cdot \Sigma_0^+, c \cdot \# \to \#, \text{Restart}(3)).$

Clearly M_0 to M_6 are deterministic R-automata. Given an input of the form $w\#w\#\cdots\#w$, where $|w| = 2m > 2$, M_0 verifies that all syllables are of even length, and that they all end in the same letter, say c. This letter c is deleted from the last syllable, and M_5 is called, which simply deletes the last letter (that is, c) from the first syllable. Now M_1 is called, which in cooperation with M_2, removes the last letter from all the other syllables. Finally the tape content $w_1\#w_1\#\cdots\#w_1$ is reached, where $w = w_1 c$. In this situation M_1 (or M_2) notices that all syllables are of odd length, and that they all end with the same letter, say d, which it then removes from the last syllable. Now using M_6, M_3, and M_4 this letter is removed from all other syllables. This process continues until either an error is detected, in which case $\mathcal{M}$ rejects, or until a tape content of the form $u\#u\#\cdots\#u$ is reached for a word $u \in \Sigma_0^2$, in which case $\mathcal{M}$ accepts. Thus, we see that $\mathcal{M}$ accepts the language L_{copy^*} working in mode $= 1$.   □

Contrasting the positive results above we have the following result.

Theorem 4. *The language* L_{copy^*} *is not accepted by any strictly deterministic CD-RRW-system that is working in mode* $= 1$.

Proof. Let $\mathcal{M} = ((M_i, \sigma_i)_{i \in I}, I_0)$ be a strictly deterministic CD-RRW-system that accepts the language L_{copy^*} in mode $= 1$. We can assume that $I = \{0, 1, \ldots, m\}$, that $I_0 = \{0\}$, that $\sigma_i = \{i+1\}$ for all $i = 0, 1, \ldots, m-1$, and that $\sigma_m = \{s\}$ for some $s \in I$. Thus, each computation of $\mathcal{M}$ has the following structure:

$$w_0 \vdash_{\mathcal{M}}^{c^s} w_s \vdash_{M_s}^{c} w_{s+1} \vdash_{\mathcal{M}}^{c^{m-s-1}} w_m \vdash_{M_m}^{c} w_{m+1} \vdash_{M_s}^{c} w_{m+2} \vdash_{\mathcal{M}}^{c^{m-s-1}} \cdots,$$

that is, it is composed of a head $w_0 \vdash_{\mathcal{M}}^{c^s} w_s$ that consists of s cycles and of a sequence of meta-cycles of the form $w_s \vdash_{M_s}^{c} w_{s+1} \vdash_{\mathcal{M}}^{c^{m-s-1}} w_m \vdash_{M_m}^{c} w_{m+1}$ that consist of $m - s + 1$ cycles each.

Let $x := w\#w(\#w)^n$ be an input word with $w \in (\Sigma_0^2)^*$, where $|w|$ and the exponent n are sufficiently large. Then $x \in L_{\text{copy}^*}$, and hence, the computation of $\mathcal{M}$ that begins with the restarting configuration $q_0^{(0)} \mathcal{c} x \$$ is accepting. We will now analyze this computation. The factors w of x and their descendants in this computation will be denoted as *syllables*. To simplify the discussion we use indices to distinguish between different syllables.

$\mathcal{M}$ must compare each syllable w_i to all the other syllables. As $|w_i| = |w|$ is large, it can compare w_i to w_j for some $j \neq i$ only piecewise. However, during this process it needs to distinguish the parts that have already been compared from those parts that have not. This can only be achieved by rewriting w_i and w_j accordingly. Since no auxiliary symbols are available, this must be done by rewrite operations that either delete those parts of w_i and w_j that have already been compared, or that use the symbol $\#$ to mark the active positions within w_i and w_j. Hence, it takes at least $|w|/k$ many rewrite operations on w_i and the same number of rewrite operations on w_j to complete the comparison, where k is the maximal size of the read/write window of a component system of $\mathcal{M}$. As each rewrite operation is length-reducing, we see that after w_i and w_j have been compared completely, the remaining descendants of w_i and of w_j are of length at most $(1 - 1/k) \cdot |w|$, that is, information on w_i and on w_j has been lost during this process. It follows that $\mathcal{M}$ actually needs to compare w_i to all other syllables simultaneously.

Now assume that, for some j, no rewrite operation of the j-th meta-cycle is performed on the first two syllables of x. Then from that point on, no rewrite operation will be performed on the first syllable for many more meta-cycles. Indeed, as all component systems of $\mathcal{M}$ are deterministic, a change has to be propagated all the way from the third syllable back to the first syllable by a sequence of rewrites before another rewrite operation can affect w_1. This, however, means that at least $|w_2|/k$ many rewrite operations are applied to the second syllable, while no rewrite operation is applied to the first syllable. As observed above this destroys information on w_2 such that it is not possible anymore to verify whether or not w_1 and w_2 were identical.

It follows that at least one rewrite operation is applied to the first two syllables in each meta-cycle. As each rewrite operation is length-reducing, this implies that after at most $2 \cdot |w| + 1$ many meta-cycles the first two syllables have been completely erased. However, altogether these meta-cycles only execute $(2 \cdot |w| + 1) \cdot (m - s + 1)$ many rewrite operations, that is, only some of the syllables of x have been compared to w_1 and to w_2 during this process, provided that $n > (2 \cdot |w| + 1) \cdot (m - s + 1)$. It follows that $L_{=1}(\mathcal{M}) \neq L_{\text{copy}^*}$. □

Corollary 1. *For all types* $X \in \{R, RR, RW, RRW\}$,

$$\mathcal{L}_{=1}(\text{det-strict-CD-X}) \subset \mathcal{L}_{=1}(\text{det-global-CD-X}).$$

Using similar techniques as in the proof of Theorem 4 the following result can be shown.

Theorem 5. L_{copy^m} *is not accepted by any strictly deterministic* CD-RRW-*system with less than* m *components working in mode* $= 1$.

By $\mathcal{L}_{=1}(\text{det-strict-CD-X}(m))$ we denote the class of languages that are accepted by strictly deterministic CD-systems of restarting automata of type X that have

m components and that work in mode $= 1$. As the m-fold copy language L_{copy^m} is accepted by a strictly deterministic CD-R-system with m components that is working in mode $= 1$, the above result yields the following proper inclusion.

Corollary 2. *For all types* $X \in \{R, RR, RW, RRW\}$ *and all* $m \geq 1$,

$$\mathcal{L}_{=1}(\text{det-strict-CD-X}(m)) \subset \mathcal{L}_{=1}(\text{det-strict-CD-X}(m+1)).$$

Thus, for each type $X \in \{R, RR, RW, RRW\}$, we have an infinite hierarchy between the class of languages accepted by deterministic restarting automata of type X and the class of languages accepted by nonforgetting deterministic restarting automata of that type. However, strictly deterministic CD-systems working in mode t are more expressive.

Proposition 2. *The language* L_{copy^*} *is accepted by a strictly deterministic* CD-RW-*system working in mode* t.

Proof. L_{copy^*} is accepted by the CD-RW-system $\mathcal{M} = ((M_i, \sigma_i)_{i \in I}, I_0)$ that is specified by $I := \{0, 1, 2\}$, $I_0 := \{0\}$, $\sigma(0) := \{1\}$, $\sigma(1) := \{2\}$, $\sigma(2) := \{0\}$. Here M_0, M_1, and M_2 are given through the following meta-instructions, where $\Sigma_0 := \{a, b\}$ and $c, d, e \in \Sigma_0$:

$$
\begin{aligned}
M_0 \; &: (\mathcal{c} \cdot ((\Sigma_0^2)^+ \cdot cd \cdot \#)^+ \cdot (\Sigma_0^2)^+, cd \cdot \$ \to \# \$), \\
&\quad (\mathcal{c} \cdot cd \cdot (\# \cdot cd)^+ \cdot \$, \text{Accept}), \\
M_1 \; &: (\mathcal{c} \cdot (\Sigma_0^2)^+ \cdot (\#\# \cdot (\Sigma_0^2)^+)^*, cd \cdot \# \cdot e \to \#\# \cdot e), \\
M_2 \; &: (\mathcal{c} \cdot (\Sigma_0^2)^+ \cdot (\# \cdot (\Sigma_0^2)^+)^*, \#\# \to \#), \\
&\quad (\mathcal{c} \cdot (\Sigma_0^2)^+ \cdot (\# \cdot (\Sigma_0^2)^+)^+, \#\$ \to \$).
\end{aligned}
$$
$\square$

5 Concluding Remarks

We have seen that for restarting automata without auxiliary symbols the strictly deterministic CD-systems yield an infinite hierarchy that lies strictly in between the deterministic restarting automata and the nonforgetting deterministic restarting automata. However, the following related questions remain open:

1. Does this result extend to restarting automata with auxiliary symbols?
2. Are the locally deterministic CD-systems of restarting automata strictly more expressive than the globally deterministic CD-systems of restarting automata of the same type?
3. A nondeterministic CD-system of restarting automata is called *strict* if there is only a single initial system (that is, $|I_0| = 1$), and if the set of successors is a singleton for each component. It is easily seen that strict CD-systems are more expressive than restarting automata. However, is there a proper hierarchy of strict CD-systems, based on the number of component systems, that lies in between the (nondeterministic) restarting automata and the nonforgetting restarting automata?

References

1. Buntrock, G.: Wachsende kontext-sensitive Sprachen. Habilitationsschrift, Fakultät für Mathematik und Informatik, Universität Würzburg (1996)
2. Csuhaj-Varju, E., Dassow, J., Kelemen, J., Păun, G.: Grammar Systems. A Grammatical Approach to Distribution and Cooperation. Gordon and Breach, London (1994)
3. Dassow, J., Păun, G., Rozenberg, G.: Grammar systems. In: Rozenberg, G., Salomaa, A. (eds.) Handbook of Formal Languages, vol. 2, pp. 155–213. Springer, Heidelberg (1997)
4. Jančar, P., Mráz, F., Plátek, M., Vogel, J.: Restarting automata. In: Reichel, H. (ed.) FCT 1995. LNCS, vol. 965, pp. 283–292. Springer, Heidelberg (1995)
5. Lautemann, C.: One pushdown and a small tape. In: Wagner, K. (ed.) Dirk Siefkes zum 50. Geburtstag, Technische Universität Berlin and Universität Augsburg, pp. 42–47 (1988)
6. Messerschmidt, H., Otto, F.: On nonforgetting restarting automata that are deterministic and/or monotone. In: Grigoriev, D., Harrison, J., Hirsch, E.A. (eds.) CSR 2006. LNCS, vol. 3967, pp. 247–258. Springer, Heidelberg (2006)
7. Messerschmidt, H., Otto, F.: Cooperating distributed systems of restarting automata. Intern. J. Found. Comput. Sci. (to appear)
8. Messerschmidt, H., Stamer, H.: Restart-Automaten mit mehreren Restart-Zuständen. In: Bordihn, H. (ed.) Workshop Formale Methoden in der Linguistik und 14. Theorietag Automaten und Formale Sprachen, Proc., pp. 111–116. Institut für Informatik, Universität Potsdam (2004)
9. Oliva, K., Květoň, P., Ondruška, R.: The computational complexity of rule-based part-of-speech tagging. In: Matoušek, V., Mautner, P. (eds.) TSD 2003. LNCS (LNAI), vol. 2807, pp. 82–84, Springer, Heidelberg (2003)
10. Otto, F.: Restarting automata and their relations to the Chomsky hierarchy. In: Ésik, Z., Fülöp, Z. (eds.) DLT 2003. LNCS, vol. 2710, pp. 55–74. Springer, Heidelberg (2003)
11. Otto, F.: Restarting automata. In: Ésik, Z., Martin-Vide, C., Mitrana, V. (eds.) Recent Advances in Formal Languages and Applications. Studies in Computational Intelligence, vol. 25, pp. 269–303. Springer, Heidelberg (2006)
12. Plátek, M., Lopatková, M., Oliva, K.: Restarting automata: Motivations and applications. In: Holzer, M. (ed.) Workshop Petrinets und 13. Theorietag Automaten und Formale Sprachen, Institut für Informatik, Technische Universität München, Garching, pp. 90–96 (2003)

Product Rules in Semidefinite Programming

Rajat Mittal and Mario Szegedy

Rutgers University

Abstract. In recent years we witness the proliferation of semidefinite programming bounds in combinatorial optimization [1,5,8], quantum computing [9,2,3,6,4] and even in complexity theory [7]. Examples to such bounds include the semidefinite relaxation for the maximal cut problem [5], and the quantum value of multi-prover interactive games [3,4]. The first semidefinite programming bound, which gained fame, arose in the late seventies and was due to László Lovász [11], who used his theta number to compute the Shannon capacity of the five cycle graph. As in Lovász's upper bound proof for the Shannon capacity and in other situations the key observation is often the fact that the new parameter in question is multiplicative with respect to the product of the problem instances. In a recent result R. Cleve, W. Slofstra, F. Unger and S. Upadhyay show that the quantum value of XOR games multiply under parallel composition [4]. This result together with [3] strengthens the parallel repetition theorem of Ran Raz [12] for XOR games. Our goal is to classify those semidefinite programming instances for which the optimum is multiplicative under a naturally defined product operation. The product operation we define generalizes the ones used in [11] and [4]. We find conditions under which the product rule always holds and give examples for cases when the product rule does not hold.

1 Introduction

The Shannon capacity of a graph G is defined by $\lim_{n\to\infty} \text{stbl}(G^n)^{1/n}$, where $\text{stbl}(G)$ denotes the maximal independence set size of G. In his seminal paper of 1979, L. Lovász solved the open question that asked if the Shannon capacity of the five cycle, C_5 is $\sqrt{5}$ [11]. The proof was based on that $\text{stbl}(C_5^2) = 5$ and that the independence number of any graph G is upper bounded by a certain semidefinite programming bound, that he called $\vartheta(G)$. Lovász showed that $\vartheta(C_5) = \sqrt{5}$, and that ϑ is multiplicative: $\vartheta(G \times G') = \vartheta(G) \times \vartheta(G')$, for any two graphs, G and G'. These facts together with the super-multiplicativity of $\text{stbl}(G)$ are clearly sufficient to imply the conjecture.

In a recent result R. Cleve, W. Slofstra, F. Unger and S. Upadhyay show that the quantum value of XOR games multiply under parallel composition [4]. The quantum value of a XOR game arises as the solution of an associated semidefinite program [14] and upper bounds the classical value of the game. The result, when combined with the fact there is a relation between the classical and quantum values of a multi-prover game [3] gives a new proof for the parallel repetition

E. Csuhaj-Varjú and Z. Ésik (Eds.): FCT 2007, LNCS 4639, pp. 435–445, 2007.
© Springer-Verlag Berlin Heidelberg 2007

theorem of Ran Raz [12] at least for XOR games, which is stronger than the original theorem of Raz when the game value approaches 1.

These successful applications of semidefinite programming bounds together with other ones, such as bounding acceptance probabilities achievable with various computational devices for independent copies of a given computational problem (generally known as "direct sum theorems"), point to the great use of product theorems for semidefinite programming.

In spite of these successes we do not know of any work which systematically investigates the conditions under which such product theorems hold. This is what we attempt to do in this article. While we do not manage to classify all cases, we hope that our study will serve as a starting point for such investigations. We define a brand of semidefinite programming instances with significantly large subclasses that obey the product rule. In Theorems 1 and 2 we describe two cases when product theorems hold, while in Proposition 3 we give an example when it does not. We also raise several questions that intuit that product theorems always hold for "positive" instances, although that what should be the notion of positivity is not yet clear. Our goal is to provoke ideas, and set the scene for what one day might hopefully becomes a complete classification.

2   Affine Semidefinite Program Instances

We will investigate a brand of semidefinite programming instances, which is described by a triplet $\pi = (J, A, b)$, where

- J is a matrix of dimension $n \times n$;
- $A = (A^{(1)}, \ldots, A^{(m)})$ is a list of m matrices, each of dimension $n \times n$. We may view A as a three-dimensional matrix A_{kij} of dimensions $n \times n \times m$, where the last index corresponds to the upper index in the list;
- b is a vector of length m.

With π we associate a semidefinite programming instance with optimal value $\alpha(\pi)$:

$$\alpha(\pi) = \{\max J * X \mid AX = b \quad \text{and} \quad X \succeq 0\} \tag{1}$$

We define dimension of the instance as the dimension of A. Here variable matrix X has the same dimension $(n \times n)$ as J and also the elements of the list A. To avoid complications we assume that all matrices involved are symmetric. The operator that we denote by $*$ is the dot product $(tr(J^T X) = \sum_{ij} J_{ij} X_{ij})$ of matrices, so it results in a scalar. The set of m linear constraints are often of some simple form, e.g. in the case of Lovász's theta number all constraints are either of the form $X_{ij} = 0$ or $Tr(X) = 1$. In our framework the constraints can generally be of the form $\sum_{i,j} A_{kij} X_{ij} = b_k$, and the only restriction they have compared to the most general form of semidefinite programming instances is that all relations are strictly equations as opposed to inequalities *and* equations. These types of instances we call *affine*. In our notation the "scalar product" AX simply means the vector $(A^{(1)} * X, \ldots, A^{(m)} * X)$.

We will need the dual of π, which we denote by π^* (for the method to express the dual see for example [13]):

$$\{\min y.b \mid yA - J \succeq 0\} \qquad (2)$$

where y is a row vector of length m. Here yA is the matrix $\sum_{k=1}^{m} y_k A^{(k)}$. The well known duality theorem for semidefinite programming states that the value of the dual agrees with the value of the primal.

3   Product Instance

Definition 1. *Let $\pi_1 = (J_1, A_1, b_1)$ and $\pi_2 = (J_2, A_2, b_2)$ be two semidefinite instances with dimensions (n_1, n_1, m_1) and (n_2, n_2, m_2), respectively. We define the product instance as $\pi_1 \times \pi_2 = (J_1 \otimes J_2, A_1 \otimes A_2, b_1 \otimes b_2)$, where $A_1 \otimes A_2$ is by definition the list $(A_1^{(k)} \otimes A_2^{(l)})_{k,l}$ of length $m_1 m_2$ of $n_1 n_2 \times n_1 n_2$ matrices. The product instance has dimensions $(n_1 n_2, n_1 n_2, m_1 m_2)$.*

Although the above is a fairly natural definition, as it was pointed out in [10] in the special case of the Lovász's theta number, a slightly different definition gives the same optimal value, which is useful in some cases. The idea is that in lucky cases, when b_1 and/or b_2 have zeros, we may add new equations (extra to ones in Definition 1) to the primal system representing the product instance without changing its optimum value. The new instances that arise this way we call *weak product* and denote by "$\times_w$," even though there is a little ambiguity in the definition (it will only be clear from the context to an individual instance which equations we wish to add). Since if we add extra constraints to a maximization problem, the objective value does not increase, we have that

Proposition 1. $\alpha(\pi_1 \times_w \pi_2) \leq \alpha(\pi_1 \times \pi_2)$.

In Section 6 we give precise definitions for weak products and investigate their properties further. For the forthcoming sections we restrict ourselves to the product as defined in Definition 1.

4   Product Solution

Definition 2. *A subclass $\mathcal{C}$ of affine instances is said to obey the product rule if $\alpha(\pi_1 \times \pi_2) = \alpha(\pi_1)\alpha(\pi_2)$ for every $\pi_1, \pi_2 \in \mathcal{C}$.*

In section 5.4 we will give an example to an affine instance whose square does not obey the product rule. Therefore, for the product rule to hold we need to look for proper subclasses of all affine instances.

Let π_1 and π_2 be two affine instances with optimal solutions X_1 and X_2 for the primal and optimal solutions y_1 and y_2 for the dual. The first instinct for proving the product theorem would be to show that $X_1 \otimes X_2$ is a solution of the product instance with objective value $\alpha(\pi_1)\alpha(\pi_2)$, and $y_1 \otimes y_2$ is a solution of the dual of the product instance with the same value. The above two potential solutions for the product instance and its dual we call *product-solution* and *dual product-solution*. In other words, in order to show that the product rule holds for π_1 and π_2 it is sufficient to prove:

1. Feasibility of the product-solution: $(A_1 \otimes A_2)(X_1 \otimes X_2) = b_1 \otimes b_2$;
2. Feasibility of the dual product-solution: $y_1 \otimes y_2 (A_1 \otimes A_2) - J_1 \otimes J_2 \succeq 0$;
3. Objective value of the primal product-solution: $(J_1 \otimes J_2) * (X_1 \otimes X_2) = (J_1 * X_1)(J_2 * X_2)$;
4. Objective value of the dual product-solution: $(y_1 \otimes y_2).(b_1 \otimes b_2) = (y_1.b_1)(y_2.b_2)$.

We also need the positivity of $X_1 \otimes X_2$, but this is automatic from the positivity of X_1 and X_2. Which of 1–4 fail to hold in general? Basic linear algebra gives that conditions 1, 3 and 4 hold without any further assumption. Thus we already have that:

Proposition 2. *Let π_1 and π_2 be two affine instances. Then $\alpha(\pi_1 \times \pi_2) \geq \alpha(\pi_1)\alpha(\pi_2)$.*

In what follows, we will examine cases when Condition 2 also holds.

5   The Missing Condition

In the sequel we will present two different sufficient conditions for Condition 2 of the previous section and we also derive a necessary condition for it (which is also sufficient if we restrict our attention to an instance and its square), but the latter expression uses y_1 and y_2, like Condition 2 itself. It remains a task for the future to develop a necessary and sufficient condition whose criterion is formulated solely in terms of the problem instances π_1 and π_2.

5.1   Positivity of Matrix J

Our first simple condition is the positivity of J.

Theorem 1. *Assume that both J_1 and J_2 are positive semidefinite. Then $\alpha(\pi_1 \times \pi_2) = \alpha(\pi_1)\alpha(\pi_2)$.*

Proof. As we noted in Section 4 it is sufficient to show that Condition 2 of that section holds. By our assumptions on y_1 and y_2 we have that $y_1A_1 - J_1$ and $y_2A_2 - J_2$ are positive semi-definite. So $y_1A_1 + J_1$ and $y_2A_2 + J_2$ are also positive semi-definite, since they arise as sums of two positive matrices. For instance, $y_1A_1 + J_1 = (y_1A_1 - J_1) + 2J_1$. The above implies that

$$(y_1A_1 - J_1) \otimes (y_2A_2 + J_2) = y_1A_1 \otimes y_2A_2 - J_1 \otimes y_2A_2 + y_1A_1 \otimes J_2 - J_1 \otimes J_2 \succeq 0. \quad (3)$$

Also

$$(y_1A_1 + J_1) \otimes (y_2A_2 - J_2) = y_1A_1 \otimes y_2A_2 - y_1A_1 \otimes J_2 + J_1 \otimes y_2A_2 - J_1 \otimes J_2 \succeq 0 \quad (4)$$

Taking the average of the right hand sides of Equations (3) and (4) we obtain that

$$y_1 A_1 \otimes y_2 A_2 - J_1 \otimes J_2 \succeq 0, \tag{5}$$

which is the desired Condition 2. (Note: It is easy to see that $y_1 A_1 \otimes y_2 A_2 = y_1 \otimes y_2 (A_1 \otimes A_2)$.)

Lovász theta number ([11]) is an example that falls into this category. Consider the definition of Lovász theta number in [13]. Then J is the all $1's$ matrix, which is positive semidefinite. The matrix remains positive definite even if we consider the weighted version of the theta number [10], in which case J is of the form ww^T for some column wector w.

5.2   All $A^{(k)}$ Are Block Diagonal, and J Is Block Anti-diagonal

The argument in the previous section is applicable whenever $y_c A_c + J_c$ ($c \in \{1, 2\}$) are known to be positive semidefinite matrices. Let us state this explicitly:

Lemma 1. *Whenever $y_c A_c + J_c$ ($c \in \{1, 2\}$) are positive definite, where y_1 and y_2 are the optimal solutions of π_1^* and π_2^*, respectively, then the product theorem holds for π_1 and π_2.*

This is the avenue Cleve et. al. take in [4]. Following their lead, but slightly generalizing their argument we show:

Lemma 2. *For a semidefinite programming instance $\pi = (A, J, b)$ if the matrix J is block anti-diagonal and if y is a feasible solution of the dual such that yA is block diagonal then $yA + J \succeq 0$.*

Block diagonal and anti-diagonal matrices have the following structure:

Block anti-diagonality   Block diagonality

$$\begin{pmatrix} 0 & M \\ M^T & 0 \end{pmatrix} \qquad \begin{pmatrix} P & 0 \\ 0 & Q \end{pmatrix}$$

In our definition block diagonal and anti-diagonal matrices have two by two blocks. We require that if J is block anti-diagonal and yA is block-diagonal, then their rows and columns be divided to blocks in exactly the same way.

We will prove our claim by contradiction. Suppose yA and J are of the required form but $yA + J$ is not positive semidefinite. Then there exists a vector w, in block form $w = (w', w'')$ for which $w^T (yA + J)w$ is negative (we treat all vectors as column vectors). Define $v = (w', -w'')$. Now

$$v^T (yA - J)v =$$
$$(w', -w'')^T yA(w', -w'') - (w', -w'')^T J(w', -w'') =$$
$$(w', w'')^T yA(w', w'') + (w', w'')^T J(w', w'') =$$
$$w^T (yA + J)w < 0.$$

This implies that $yA - J$ is not positive semidefinite, which is a contradiction since by our assumption y is a solution of π^*. We can generalize the proof for case when J is of the form $J_1 + J_2$, where J_1 is of the form as before and J_2 is positive semidefinite (yA should still be block diagonal). Notice that the block diagonality of yA automatically holds if $A = (A^{(1)}, \ldots, A^{(m)})$, where each $A^{(k)}$ is block diagonal. We summarize the findings of this section in the following theorem:

Theorem 2. *Let $\pi_1 = (A_1, J_1, b_1)$ and $\pi_2 = (A_2, J_2, b_2)$ be affine instances such that for $c \in \{1, 2\}$:*

1. $A_c = (A_c^{(1)}, \ldots, A_c^{(m)})$, where each $A_c^{(k)}$ is block diagonal;
2. $J_c = J_c' + J_c''$ ($c \in \{1, 2\}$), where J_c' is block anti-diagonal and J_c'' is positive.

(All blocked matrices have the same block divisions.) Then for π_1 and π_2 the product theorem holds.

5.3 A Necessary Condition for the Feasibility of $y_1 \otimes y_2$

In this section we show that the condition in Lemma 1 is not only sufficient, but also necessary (or at least "half of it"), if we insist on the "first instinct" proof method.

Lemma 3. *For two instances π_1 and π_2, let y_1 and y_2 be optimal solutions of π_1^* and π_2^*, respectively. Then $y_1 \otimes y_2$ is a feasible solution of the dual of the product instance (i.e. Condition 2 of section 4 holds) only if at least one of $y_c A_c + J_c$ ($c \in \{1, 2\}$) are positive definite.*

Proof. Let us assume the contrary. Then we have vectors w_c ($c \in \{1, 2\}$) such that $w_c^T (y_c A_c + J_c) w_c < 0$ ($c \in \{1, 2\}$). Our assumptions imply that $w_c^T (y_c A_c - J_c) w_c \geq 0$ ($c \in \{1, 2\}$). Now it holds that

$$(w_1 \otimes w_2)^T ((y_1 A_1 - J_1) \otimes (y_2 A_2 + J_2))(w_1 \otimes w_2) < 0$$
$$\Rightarrow (w_1 \otimes w_2)^T (y_1 A_1 \otimes y_2 A_2 - J_1 \otimes J_2 - J_1 \otimes y_2 A_2 + y_1 A_1 \otimes J_2)(w_1 \otimes w_2) < 0$$
$$\Rightarrow (w_1 \otimes w_2)^T (y_1 A_1 \otimes y_2 A_2 - J_1 \otimes J_2)(w_1 \otimes w_2) +$$
$$(w_1 \otimes w_2)^T (y_1 A_1 \otimes J_2 - J_1 \otimes y_2 A_2)(w_1 \otimes w_2) < 0$$

By similar argument, considering now the inequality

$$(w_1 \otimes w_2)^T ((y_1 A_1 + J_1) \otimes (y_2 A_2 - J_2))(w_1 \otimes w_2) < 0,$$

we can show that

$$(w_1 \otimes w_2)^T (y_1 A_1 \otimes y_2 A_2 - J_1 \otimes J_2)(w_1 \otimes w_2) +$$
$$(w_1 \otimes w_2)^T (-y_1 A_1 \otimes J_2 + J_1 \otimes y_2 A_2)(w_1 \otimes w_2) < 0$$

By averaging the two inequalities we get that

$$(w_1 \otimes w_2)^T (y_1 A_1 \otimes y_2 A_2 - J_1 \otimes J_2)(w_1 \otimes w_2) < 0$$

This contradicts to the assumption of the lemma that $y_1 \otimes y_2$ is a feasible solution of $\pi_1 \times \pi_2$ (which in turn implies that $y_1 A_1 \otimes y_2 A_2 - J_1 \otimes J_2$ is positive definite).

One might suspect that the full converse of Lemma 1 holds, i.e. in the case of the feasibility of $y_1 \otimes y_2$ both $y_1 A_1 + J_1$ and $y_2 A_2 + J_2$ should be positive semi-definite, but in the next section we give a counter-example to this.

5.4   Maximum Eigenvalue of a Matrix

In this section we give an example when the product theorem does not hold. The example is the maximal eigenvalue function of a matrix, which, in contrast to the similar notion of spectral norm, is not multiplicative. Indeed, let M be a matrix with maximal eigenvalue 1 and minimal eigenvalue -2. Then, using the fact that under tensor product the spectra of matrices multiply, we get that $M \otimes M$ has maximal eigenvalue $4 \neq 1^2$ (the corresponding spectral norms would be 2 for M and 4 for $M \otimes M$).

Proposition 3. *The maximal eigenvalue of a matrix can be formulated as the optimal value of an affine semidefinite programming instance. This instance is not multiplicative.*

Proof. First notice that

$$\text{maxeigenvalue}(M) = \{\min \lambda \mid \lambda I - M \succeq 0\}. \tag{6}$$

This is a dual (minimization) instance. Observe that $m = 1$, $n' = n$, $A = (I)$, $J = M$ and $b = 1$. For the sake of completeness we describe the primal problem:

$$\text{maxeigenvalue}(M) = \{\max \sum_{1 \leq i,j \leq n} M_{ij} X_{ij} \mid \text{Tr} X = 1; \ X \succeq 0\}. \tag{7}$$

The product instance associated with two matrices, M_1 and M_2, has parameters $I = I_1 \otimes I_2$, $M = M_1 \otimes M_2$ and $b = 1$. Since I is an identity matrix of appropriate dimensions, the optimum value of this instance is exactly the maximal eigenvalue of $M_1 \otimes M_2$. On the other hand, as was stated in the beginning of the section, the maximal eigenvalue problem is not multiplicative.

It is educational to see where the condition of Proposition 3 fails. Recall that $J = M$, $A = (I)$ and $y = \lambda$ (the maximal eigenvalue of M). The point is that even when $\lambda I - M$ is positive, $\lambda I + M$ is not necessarily. On the other hand, if M is positive then $\lambda I - M \succeq 0 \Rightarrow \lambda I + M \succeq 0$, and indeed the maximum eigenvalue of positive matrices multiply under tensor product. As a perhaps far-fetched conjecture we ask:

Conjecture 1. For an affine instance $\pi = (A, J, b)$ define

$$\alpha^+(\pi) = \{\max |J * X| \mid AX = b \ \text{ and } \ X \succeq 0\}.$$

Is it true that α^+ is always multiplicative? Here α^+ represents a generalized "spectral norm."

We can extend the above example to show that in Lemma 3 we cannot exchange the "one of" to "both." Let M_1 be the matrix with eigenvalues -2 and 1 and let M_2 be the matrix with eigenvalues 0 and 1. Then $y_1 = 1$ and $y_2 = 1$, so $y_1 \otimes y_2 = 1$, which is a solution of

$$\{\min \lambda \mid \lambda I - M_1 \otimes M_2 \succeq 0\}, \tag{8}$$

even though $I + M_1$ is not positive semidefinite.

6   The Weak Product

A surprising observation about the theta number of Lovász, well described in [10], is that it is multiplicative with two different notions of products:

Definition 3 (Strong product "×" of graphs). $(u', u'') -- (v',v'') \, or \, (u', u'') = (v', v'') \, in \, G' \times G'' \, if \, and \, only \, if \, (u' -- v' \, or \, u' = v' \, in \, G') \, and \, (u'' -- v'' \, or \, u'' = v'' \, in \, G'').$

and

Definition 4 (Weak product "$\times_w$" of graphs). $G' \times_w G'' = \overline{\overline{G'} \times \overline{G''}}.$

Recall that $\vartheta(G)$ is defined by [13] (by J we denote the matrix with all 1 elements):

$$\vartheta(G) = \{\max J * X \mid I * X = 1; \ \forall (i, j) \in E(G): \ X_{i,j} = 0; X \succeq 0\}. \tag{9}$$

That is, every edge gives a new linear constraint, increasing m by one. In general, $E(G' \times_w G'') \supseteq E(G' \times G'')$, because $(u', u'') -- (v', v'')$ is an edge of $G' \times G''$ if and only if both of its projections are edges or identical coordinates, but $(u', u'') \neq (v', v'')$. On the other hand, $(u', u'') -- (v', v'')$ is an edge of $G' \times_w G''$ if and only if there exists at least one projection which is an edge.

It is easy to see that the constraint in Expression (9) for $\vartheta(G' \times G'')$ has a constraint for every constraint pair in the corresponding expression for G' and G', so the strong product is the one that corresponds to our usual product notion that appears in previous sections. In contrast, when we write down Expression (9) for $\vartheta(G' \times_w G'')$, we see a lot of extra constrains.

How do they arise? In general, assume that we know that the product solution $X_1 \otimes X_2$ is the optimal solution for $\pi_1 \times \pi_2$ (which is indeed the case under the conditions we considered in earlier sections). Assume furthermore that some coordinate i of b_1 is zero. Then $A_1^{(i)} * X_1 = 0$. Now we may take any $n_2 \times n_2$ matrix B, and it will hold that

$$(A_1^{(i)} \otimes B) * (X_1 \otimes X_2) = (A_1^{(i)} * X_1)(B * X_2) = 0.$$

Therefore adding matrices of the form $A_1^{(i)} \otimes B$ to $A_1 \otimes A_2$ and setting the the corresponding entry of the longer b vector of the product instance to zero will not influence the objective value. The same can be said about about exchanging the roles of π_1 and π_2.

We can easily see that the weak product in the case of the theta number arises this way. That what equations to the product system we wish to add this way is a matter of taste, and we believe it depends on the specific class of semidefinite programming instances under study. We summarize the finding of this section in the following proposition

Proposition 4. *Assume that for affine instances π_1 and π_2 the multiplicative rule holds. Then if define a system $\pi_1 \times_w \pi_2$ that we call "weak product" by conveniently adding arbitrary number of new constrains to the system that follow the construction rules described above (in particular, every added constraint should be associated with a zero entry of b_1 or b_2), the multiplicative rule will also hold for the weak product.*

The above lemma explains why the theta number of Lovász is multiplicative with respect to the weak product of graphs.

7   Some Open Problems

We formulate some further open problems all coming from the intuition that there must be a notion of "positive" affine instances for which the product theorem always holds.

Conjecture 2. Is it true that if for an instance π it holds that $\alpha(\pi^2) = \alpha(\pi)^2$, then for every $d > 2$ integer it holds that $\alpha(\pi^d) = \alpha(\pi)^d$.

The next question relates to monotonicity:

Conjecture 3. Let $\pi_1 = (A_1, J_1, b_1)$ and $\pi_2 = (A_2, J_2, b_2)$ be the affine instances for which the product theorem holds. Then it also holds for the instance pair $\pi_1' = (A_1, J_1 + J, b_1)$ and $\pi_2' = (A_2, J_2 + J', b_2)$, where J and J' are positive matrices.

The following question suggests that the more negative J is, the more special A has to be. In particular, if J is not positive then at least some A is excluded.

Conjecture 4. For every strictly non-positive J (i.e. J has a negative eigenvalue) there are A and b such that for the instance $\pi = (A, J, b)$ it holds that $\alpha(\pi^2) \neq \alpha(\pi)^2$.

On the other hand, we may conjecture that whether the product theorem holds or not is entirely independent of b:

Conjecture 5. Let $\pi_1 = (A_1, J_1, b_1)$ and $\pi_2 = (A_2, J_2, b_2)$ be the affine instances for which the product theorem holds. Then it also holds for the instance pair $\pi_1' = (A_1, J_1, b_1 + b)$ and $\pi_2' = (A_2, J_2, b_2 + b')$ for any b and b'.

Another question is: What are those instances π (if there are any) for which the product theorem always holds with any other instance?

8   Conclusions

We have started to systematically investigate product theorems for affine instances of semidefinite programming. Our theorems imply the important result of Cleve. et al. [4] about the multiplicativity of the quantum value for the XOR games and the multiplicativity of the theta number of Lovász [11]. Although their proof came both logically and chronologically first, the mere fact that the proposed theory has such immediate consequences, in our opinion serves as a worthwhile motivation for its development. Added to this that various direct sum results for different computational models would also be among the immediate consequences of the theory, we conclude that we have hit upon a basic research topic with immediate and multiple applications in computer science. The issue, therefore, at this point is not the number of potential applications, which seems abundant, but rather the relative scarcity of positive results. In the paper we have formulated conjectures that we hope will raise interest in researchers who intend to study this topic further.

References

1. Arora, S., Rao, S., Vazirani, U.: Expander Flows, Geometric Embeddings and Graph Partitioning. In: Proceedings of Symposium on the Theory of Computing (2004)
2. Barnum, H., Saks, M.E., Szegedy, M.: Quantum query complexity and semi-definite programming. In: IEEE Conference on Computational Complexity, pp. 179–193 (2003)
3. Cleve, R., Hoyer, P., Toner, B., Watrous, J.: Consequences and Limits of Nonlocal Strategies. In: IEEE Conference on Computational Complexity, pp. 236–249 (2004)
4. Cleve, R., Slofstra, W., Unger, F., Upadhyay, S.: Strong parallel repetition Theorem for Quantum XOR Proof Systems: quant-ph (August 2006)
5. Goemans, M.X., Williamson, D.P.: Approximation Algorithms for MAX-3-CUT and Other Problems Via Complex Semidefinite Programming. Journal of Computer and System Sciences (Special Issue for STOC 2001), 68, 442–470 (2004) (Preliminary version in Proceedings of 33rd STOC, Crete, 443–452 (2001))
6. Hoyer, P., Lee, T., Spalek, R.: Negative weights makes adversaries stronger, quant-ph/0611054
7. Laplante, S., Lee, T., Szegedy, M.: The Quantum Adversary Method and Classical Formula Size Lower Bounds. Computational Complexity 15(2), 163–196 (2006)
8. Karger, D.R., Motwani, R., Sudan, M.: Approximate Graph Coloring by Semidefinite Programming. J. ACM 45(2), 246–265 (1998)
9. Kitaev(unpublished proof). Quantum Coin Tossing. Slides at the archive of MSRI Berkeley
10. Knuth, D.E.: The Sandwich Theorem. Electron. J. Combin. (1994)

11. Lovász, L.: On the Shannon capacity of a graph. IEEE Transactions on Information Theory (January 1979)
12. Raz, R.: A Parallel Repetition Theorem. SIAM Journal of Computing 27(3), 763–803 (1998)
13. Szegedy, M.: A note on the theta number of Lovász and the generalized Delsarte bound. In: FOCS (1994)
14. Tsirelson, B.S.: Quantum analogues of the Bell inequalities: The case of two spatially separated domains. Journal of Soviet Mathematics 36, 557–570 (1987)

Expressive Power of LL(k) Boolean Grammars[*]

Alexander Okhotin

Academy of Finland and Department of Mathematics,
University of Turku, FIN-20014 Turku, Finland
alexander.okhotin@utu.fi

Abstract. The family of languages generated by Boolean grammars and usable with recursive descent parsing is studied. It is demonstrated that Boolean LL languages over a unary alphabet are regular, while Boolean LL subsets of $\Sigma^* a^*$ obey a certain periodicity property, which, in particular, makes the language $\{a^n b^{2^n} \mid n \geqslant 0\}$ nonrepresentable. It is also shown that $\{a^n b^n cs \mid n \geqslant 0, s \in \{a, b\}\}$ is not generated by any linear conjunctive LL grammar, while linear Boolean LL grammars cannot generate $\{a^n b^n c^* \mid n \geqslant 0\}$.

1 Introduction

Boolean grammars [5] are an extension of the context-free grammars, in which the rules may contain explicit Boolean operations. While context-free grammars can combine syntactical conditions using only disjunction (effectively specified by multiple rules for a single symbol), Boolean grammars additionally allow conjunction and negation. The extended expressive power of Boolean grammars and their intuitive clarity make them a much more powerful tool for specifying languages than the context-free grammars. Another important fact is that the main context-free parsing algorithms, such as the Cocke–Kasami–Younger, the recursive descent and the generalized LR, can be extended to Boolean grammars without an increase in computational complexity [5,6,7].

Though the practical properties of Boolean grammars seem to be as good as in the case of the more restricted context-free grammars, theoretical questions for Boolean grammars present a greater challenge. Already a formal definition of Boolean grammars involves certain theoretical problems [2,5]. A major gap in the knowledge on these grammars is the lack of methods of proving non-representability of languages [5]. Even though the family generated by Boolean grammars has been proved to be contained in $DTIME(n^3) \cap DSPACE(n)$, there is still no proof that any context-sensitive language lies outside of this class.

Results of the latter kind are hard to obtain for many interesting classes of automata and grammars. Consider the family of *trellis automata*, also known as one-way real-time cellular automata, which were studied since 1970s, and which have recently been proved to be equal in power to a subclass of Boolean grammars [4]. No methods of establishing nonrepresentability of languages in this class were

---

[*] Supported by the Academy of Finland under grant 118540.

E. Csuhaj-Varjú and Z. Ésik (Eds.): FCT 2007, LNCS 4639, pp. 446–457, 2007.
© Springer-Verlag Berlin Heidelberg 2007

known for two decades, until the first such result by Yu [10], who established a pumping lemma for a special case. Only a decade later the first context-free language not recognized by these automata was found by Terrier [9]. Another example is given by growing context-sensitive languages, for which a method of proving nonrepresentability was discovered by Jurdzinski and Loryś [1].

The purpose of this paper is to establish some limitations of the expressive power of the subcase of Boolean grammars to which the recursive descent parsing is applicable: the *LL(k) Boolean grammars* [7]. Already for this class, obtaining nonrepresentability proofs presents a challenge: consider that there exists an LL(1) linear conjunctive grammar for the language of computations of a Turing machine, which rules out a general pumping lemma. There also exists an LL(1) Boolean grammar for a P-complete language [8], which shows computational nontriviality of this class. This paper proposes several methods for proving nonrepresentability of languages by these grammars, which become the first results of such kind in the field of Boolean grammars.

Following a definition of Boolean grammars in Section 2, recursive descent parsers for Boolean grammars and their simple formal properties are described in Sections 3 and 4. In Section 5 it is proved that Boolean LL grammars over a unary alphabet generate only regular languages. Section 6 considers subsets of $\Sigma^* a^*$ representable by Boolean LL grammars and establishes a periodicity property of such languages, which, in particular, implies nonrepresentability of the language $\{a^n b^{2^n} \mid n \geqslant 0\}$. Stronger nonrepresentability results for two subclasses of Boolean LL grammars with linear concatenation are obtained in Sections 7 and 8. Based on these results, in Section 9, a detailed hierarchy of language families is obtained.

2   Boolean Grammars and Their Non-left-Recursive Subset

Definition 1 ([5]). *A Boolean grammar is a quadruple $G = (\Sigma, N, P, S)$, where Σ and N are disjoint finite nonempty sets of terminal and nonterminal symbols respectively; P is a finite set of rules of the form*

$$A \rightarrow \alpha_1 \& \ldots \& \alpha_m \& \neg\beta_1 \& \ldots \& \neg\beta_n, \tag{1}$$

where $m + n \geqslant 1$, $\alpha_i, \beta_i \in (\Sigma \cup N)^*$; $S \in N$ is the start symbol of the grammar.

Let us further assume that $m \geqslant 1$ and $n \geqslant 0$ in every rule (1). Note that if $m = 1$ and $n = 0$ in every such rule, then a context-free grammar is obtained. An intermediate family of *conjunctive grammars* [3] has $m \geqslant 1$ and $n = 0$ in every rule. *Linear* subclasses of Boolean, conjunctive and context-free grammars are defined by the additional requirement that $\alpha_i, \beta_i \in \Sigma^* \cup \Sigma^* N \Sigma^*$.

For each rule (1), the objects $A \rightarrow \alpha_i$ and $A \rightarrow \neg\beta_j$ (for all i, j) are called conjuncts, positive and negative respectively, and α_i and β_j are their bodies. The notation $A \rightarrow \pm\alpha_i$ and $A \rightarrow \pm\beta_j$ is used to refer to a positive or a negative conjunct with the specified body.

The intuitive semantics of a Boolean grammar is fairly clear: a rule (1) specifies that every string that satisfies each of the conditions α_i and none of the conditions β_i is therefore generated by A. However, constructing a mathematical definition of a Boolean grammar has proved to be a rather nontrivial task. Generally, a grammar is interpreted as a system of language equations in variables N, in which the equation for each $A \in N$ is

$$A = \bigcup_{A \to \alpha_1 \& \ldots \& \alpha_m \& \neg\beta_1 \& \ldots \& \neg\beta_n \in P} \left[\bigcap_{i=1}^{m} \alpha_i \cap \bigcap_{j=1}^{n} \overline{\beta_j} \right] \qquad (2)$$

The vector $(\ldots, L_G(A), \ldots)$ of languages generated by the nonterminals of the grammar is defined to be a solution of this system. Since this system, in general, may have no solutions or multiple solutions, this definition requires more precise conditions, which have been a subject of research [2,5].

Fortunately, for the subclass of Boolean grammars studied in this paper, the formal definition is much simplified. For a recursive descent parser to work correctly, a grammar needs to satisfy the following strong requirement.

Definition 2 ([7]). *Let $G = (\Sigma, N, P, S)$ be a Boolean grammar. The relation of context-free reachability in one step, $\rightsquigarrow$, is a binary relation on the set of strings with a marked substring $\{\alpha\langle\beta\rangle\gamma \mid \alpha, \beta, \gamma \in (\Sigma \cup N)^*\}$, defined as*

$$\alpha\langle\beta A\gamma\rangle\delta \rightsquigarrow \alpha\beta\eta\langle\sigma\rangle\theta\gamma\delta,$$

for all $\alpha, \beta, \gamma, \delta \in (\Sigma \cup N)^*$, $A \in N$ and for all conjuncts $A \to \pm\eta\sigma\theta$.

Definition 3 ([7]). *A Boolean grammar $G = (\Sigma, N, P, S)$ is said to be strongly non-left-recursive if and only if for all $A \in N$ and $\theta, \eta \in (\Sigma \cup N)^*$, such that $\varepsilon\langle A\rangle\varepsilon \rightsquigarrow^+ \theta\langle A\rangle\eta$, it holds that $\varepsilon \notin L_{G_+}(\theta)$, where G_+ is a conjunctive grammar obtained from G by removing all negative conjuncts from every rule.*

The height of a nonterminal A, denoted $h(A)$, is the greatest number of steps in a derivation $\varepsilon\langle A\rangle\varepsilon \rightsquigarrow^* \theta\langle B\rangle\eta$, where $\varepsilon \in L_{G_+}(\theta)$ and $B \in N$.

For every strongly non-left-recursive grammar, the corresponding system of equations (2) has a unique solution [7]. Then, for every $A \in N$, $L_G(A)$ is defined as the value of A in this solution. Let $L(G) = L_G(S)$.

Consider the following two simple examples of Boolean grammars (both are strongly non-left-recursive):

Example 1. The language $\{a^n b^n c^n \mid n \geqslant 0\}$ is linear conjunctive. The grammar, the corresponding system of equations and its unique solution are as follows:

$S \to A\&C$	$S = A \cap C$	$L_G(S) = \{a^n b^n c^n \mid n \geqslant 0\}$
$A \to aA \mid D$	$A = aA \cup D$	$L_G(A) = \{a^i b^j c^k \mid j = k\}$
$D \to bDc \mid \varepsilon$	$D = bDc \cup \varepsilon$	$L_G(D) = \{b^m c^m \mid m \geqslant 0\}$
$C \to aCc \mid B$	$C = aCc \cup B$	$L_G(C) = \{a^i b^j c^k \mid i = k\}$
$B \to bB \mid \varepsilon$	$B = bB \cup \varepsilon$	$L_G(B) = b^*$

It is based upon the representation of $\{a^n b^n c^n \mid n \geqslant 0\}$ as $L_G(A) \cap L_G(C)$.

Example 2. The following grammar [8] generates a *P*-complete language:

$$S \to E \& \neg AbS \& \neg CS \qquad C \to aCAb \mid b$$
$$A \to aA \mid \varepsilon \qquad\qquad E \to aE \mid bE \mid \varepsilon$$

Note that the entire family generated by Boolean grammars is contained in $DTIME(n^3) \subset P$ [5], and hence this language is among the hardest of its kind.

3   Boolean Recursive Descent Parser

A recursive descent parser for a Boolean grammar uses a parsing table similar to the well-known context-free LL table. Let $k \geqslant 1$. For a string w, define

$$First_k(w) = \begin{cases} w, & \text{if } |w| \leqslant k \\ \text{first } k \text{ symbols of } w, \text{ if } |w| > k \end{cases}$$

This definition can be extended to languages as $First_k(L) = \{First_k(w) | w \in L\}$. Define $\Sigma^{\leqslant k} = \{w \mid w \in \Sigma^*, |w| \leqslant k\}$.

Definition 4 ([7]). *A string $v \in \Sigma^*$ is said to follow $\sigma \in (\Sigma \cup N)^*$ if $\varepsilon\langle S\rangle\varepsilon \leadsto \theta\langle\sigma\rangle\eta$ for some $\theta, \eta \in (\Sigma \cup N)^*$, such that $v \in L_G(\eta)$.*

Definition 5 ([7]). *Let $G = (\Sigma, N, P, S)$ be a strongly non-left-recursive Boolean grammar, let $k > 0$. An LL(k) table for G is a function $T_k : N \times \Sigma^{\leqslant k} \to P \cup \{-\}$, such that for every rule $A \to \varphi$ and $u, v \in \Sigma^*$, for which $u \in L_G(\varphi)$ and v follows A, it holds that $T_k(A, First_k(uv)) = A \to \varphi$.*

A Boolean grammar is said to be LL(k) if such a table exists.

Both grammars in Examples 1–2 are LL(1). For the grammar in Example 1, consider that $\varepsilon\langle S\rangle\varepsilon \leadsto^+ a\langle B\rangle c$, and therefore $T(B, c) = B \to \varepsilon$ and $T(B, b) = B \to bB$, while the value of $T(B, a)$ can be anything in $\{B \to \varepsilon, B \to bB, -\}$.

A recursive descent parser, as in the context-free case, contains a procedure for each terminal and nonterminal symbol. There are two global variables used by all procedures: the input string $w = w_1 w_2 \ldots w_{|w|}$ and a positive integer p pointing at a position in this string. Each procedure $s()$, where $s \in \Sigma \cup N$, starts with some initial value of this pointer, $p = i$, and eventually either returns, setting the pointer to $p = j$ (where $i \leqslant j \leqslant |w|$), or *raises an exception*, in the sense of an exception handling model of, e.g., C++.

The text of procedures corresponding to every terminal $a \in \Sigma$ and to every nonterminal $A \in N$ is given in Figure 1 along with the main procedure [7]. Procedures $a()$ are as in the context-free case. Procedure $A()$ chooses a rule using the parsing table and then proceeds checking the conjuncts one by one. The code for the first conjunct $A \to \alpha_1$ stores the initial value of the pointer in the variable *start*, and remembers the end of the substring recognized by α_1 in the variable *end*. Every subsequent positive conjunct $A \to \alpha_i$ is tried starting from the same position *start*, and the variable *end* is used to check that it consumes exactly the same substring. The code for every negative conjunct

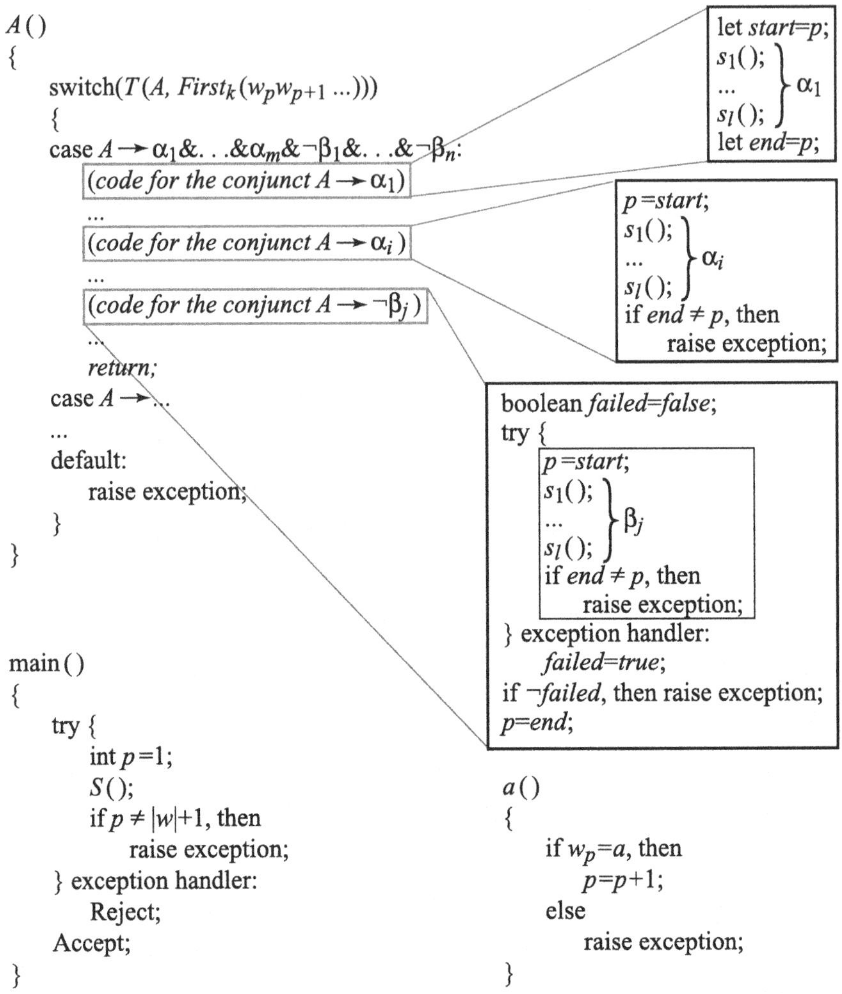

```
A()
{
    switch(T(A, First_k(w_p w_{p+1} ...)))
    {
    case A → α_1 &...& α_m & ¬β_1 &...& ¬β_n:
        (code for the conjunct A → α_1)
        ...
        (code for the conjunct A → α_i)
        ...
        (code for the conjunct A → ¬β_j)
        ...
        return;
    case A → ...
        ...
    default:
        raise exception;
    }
}


main()
{
    try {
        int p=1;
        S();
        if p ≠ |w|+1, then
            raise exception;
    } exception handler:
        Reject;
    Accept;
}
```

```
let start=p;
s_1();
...              } α_1
s_l();
let end=p;
```

```
p=start;
s_1();
...              } α_i
s_l();
if end ≠ p, then
    raise exception;
```

```
boolean failed=false;
try {
    p=start;
    s_1();
    ...          } β_j
    s_l();
    if end ≠ p, then
        raise exception;
} exception handler:
    failed=true;
if ¬failed, then raise exception;
p=end;
```

```
a()
{
    if w_p=a, then
        p=p+1;
    else
        raise exception;
}
```

Fig. 1. Boolean recursive descent parser: procedures $A()$, $a()$ and $main()$

tries to recognize a substring in the same way, but reports a successful parse if and only if the recognition is unsuccessful, thus implementing negation.

The correctness of Boolean recursive descent has been established as follows:

Lemma 1 ([7]). *Let $k \geqslant 1$. Let $G = (\Sigma, N, P, S)$ be an LL(k) Boolean grammar. Let $T : N \times \Sigma^{\leqslant k} \to P \cup \{-\}$ be an LL(k) table for G, let the set of procedures be constructed with respect to G and T. Then, for every $y, z, \tilde{z} \in \Sigma^*$ and $s_1, \ldots, s_\ell \in \Sigma \cup N$ ($\ell \geqslant 0$), such that z follows $s_1 \ldots s_\ell$ and $First_k(z) = First_k(\tilde{z})$, the code $s_1(); \ldots; s_\ell()$, executed on the input $y\tilde{z}$,*

- *returns, consuming y, if $y \in L_G(s_1 \ldots s_\ell)$;*
- *raises an exception, if $y \notin L_G(s_1 \ldots s_\ell)$.*

4 Simple Formal Properties

A few technical results need to be established for use in the subsequent arguments. There is nothing interesting in their proofs.

Definition 6. *An LL(k) Boolean grammar $G = (\Sigma, N, P, S)$ is said to be well-behaved, if, for every $A \in N$, (i) $L(A) \neq \varnothing$ and (ii) there exist $\theta, \eta \in (\Sigma \cup N)^*$, such that $\varepsilon\langle S\rangle\varepsilon \rightsquigarrow \theta\langle A\rangle\eta$. The grammar $G = (\Sigma, \{S\}, \{S \to aS\}, S)$ generating $\varnothing$ is also considered well-behaved.*

Lemma 2. *For every LL(k) Boolean grammar $G = (\Sigma, N, P, S)$ there exists an equivalent well-behaved LL(k) Boolean grammar.*

Lemma 3. *Let $G = (\Sigma, N, P, S)$ be a well-behaved LL(k) Boolean grammar. Then, for every $A \in N$, the grammar $G' = (\Sigma, N, P, A)$ is a well-behaved LL(k) Boolean grammar, and $L_{G'}(C) = L_G(C)$ for all $C \in N$.*

Lemma 4. *Let $G = (\Sigma, N, P, S)$ be a well-behaved LL(k) Boolean grammar, let $a \in \Sigma$. Then there exists a well-behaved LL(k) Boolean grammar for $L(G) \cap a^*$.*

Lemma 5. *For every well-behaved LL(k) Boolean grammar G there exists and can be effectively constructed a well-behaved LL(k) Boolean grammar G', such that $L(G') = L(G)$ and for each nonterminal A in G' there is either one or more rules of the form $A \to B_1\&\dots\&B_m\&\neg C_1\&\dots\&\neg C_n$, or a single rule of the form $A \to BC$, $A \to a$ or $A \to \varepsilon$.*

5 Boolean LL(k) Grammars over a Unary Alphabet

Context-free grammars over a one-letter alphabet are known to generate only regular languages, and linear conjunctive grammars have the same property. On the other hand, Boolean grammars can generate the nonregular language $\{a^{2^n} \mid n \geqslant 1\}$ [5]. We shall now demonstrate that Boolean LL(k) grammars over the unary alphabet generate only regular languages, and hence are weaker in power than Boolean grammars of the general form.

Theorem 1. *Every Boolean LL(k) language over a unary alphabet is regular.*

The following lemma is an essential component of the proof:

Lemma 6. *Let $G = (\Sigma, N, P, S)$ be a well-behaved LL(k) Boolean grammar, let $B \in N$, $a \in \Sigma$ and let some string in $a^k \Sigma^*$ follow B. Then there exists at most one number $i \geqslant 0$, such that $a^i \in L_G(B)$.*

Proof. Supposing the contrary, let $a^{i_1}, a^{i_2} \in L(B)$, where $0 \leqslant i_1 \leqslant i_2$, and let a^m, where $m \geqslant k$, be a string that follows B. Consider the string $a^{m+(j_2-j_1)}$, for which $m+(j_2-j_1) \geqslant k$ as well. By Lemma 1, $a^{i_1} \in L(B)$ implies that $B()$ returns on the input $w_1 = a^{i_1}a^{m+(j_2-j_1)}$, consuming a^{i_1}. On the other hand, $a^{i_2} \in L(B)$ implies that $B()$ should return on $w_2 = a^{i_2}a^m$, consuming a^{i_2}. Since $w_1 = w_2$, the computations of $B()$ on w_1 and w_2 are actually the same computation, and hence i_1 and i_2 must coincide, proving the claim.   □

Proof (Theorem 1). According to Lemmata 2 and 5, there is no loss of generality in the assumption that the given language is generated by a well-behaved LL(k) Boolean grammar $G = (\{a\}, N, P, S)$, such that for every $A \in N$ the set P contains either one or more rules of the form $A \to D_1 \& \ldots \& D_q \& \neg E_1 \& \ldots \& \neg E_r$, or a single rule of the form $A \to BC$, $A \to a$ or $A \to \varepsilon$.

Let us prove that in every rule $A \to BC$, if $L(C) \not\subseteq a^{\leqslant k-1}$, then $L(B)$ is a singleton. By assumption, there exists $a^j \in L(C)$, where $j \geqslant k$. Let a^ℓ be a string that follows A, then $a^{j+\ell} \in a^k a^*$ follows B. Since we know that $L(B) \subseteq a^*$ and $L(B) \neq \varnothing$, Lemma 6 states that $|L(B)| = 1$.

Now let us reconstruct the grammar to show that $L(G)$ is regular. For every rule $A \to BC$, such that $L(B)$ is a singleton, replace the rule with the rule $A \to a^i C$, where $L(B) = \{a^i\}$. If $L(B)$ is not a singleton, then $L(C) \subseteq a^{\leqslant k}$ by the claim above, and the rule can be equivalently replaced with $\{A \to a^i B \mid a^i \in L(C)\}$. The system of language equations corresponding to the transformed grammar has the same set of solutions as the original system, that is, it has a unique solution $(\ldots, L_G(A), \ldots)$. Since the system uses one-sided concatenation, all components of this solution are regular. □

Corollary 1. *Let $G = (\Sigma, N, P, S)$ be a well-behaved Boolean LL(k) grammar. Then, for every $a \in \Sigma$ and for every $A \in N$, the language $L_G(A) \cap a^*$ is regular.*

Proof. Consider the grammar $G' = (\Sigma, N, P, A)$. According to Lemma 3, G' is a well-behaved Boolean LL(k) grammar generating $L_G(A)$. Then, by Lemma 4, there exists a well-behaved Boolean LL(k) grammar G'', such that $L(G'') = L_G(A) \cap a^*$. This language is regular by Theorem 1. □

6 Nonrepresentability Results for Subsets of $\Sigma^* a^*$

Let us establish a method of proving nonrepresentability of some languages over non-unary alphabets. This method exploits long blocks of identical letters in a certain similarity to Yu's [10] nonrepresentability argument for trellis automata.

Theorem 2 (Periodicity theorem). *For every Boolean LL(k) language $L \subseteq \Sigma^*$ there exist constants $d, d', p \geqslant 0$, such that for all $w \in \Sigma^*$, $a \in \Sigma$, $n \geqslant d \cdot |w| + d'$ and $i \geqslant 0$, $wa^n \in L$ if and only if $wa^{n+ip} \in L$.*

Proof. By Lemmata 2 and 5, assume that L is generated by a well-behaved LL(k) Boolean grammar $G = (\Sigma, N, P, S)$, in which, for every $A \in N$, there is either one or more rules of the form $A \to D_1 \& \ldots \& D_q \& \neg E_1 \& \ldots \& \neg E_r$, or a unique rule of the form $A \to BC$, $A \to a$ or $A \to \varepsilon$.

According to Corollary 1, for every $A \in N$ and $a \in \Sigma$, the set $L_G(A) \cap a^*$ is regular. Let $d(A, a) \geqslant 0$ and $p(A, a) \geqslant 1$ be numbers, such that $L(A) \cap a^*$ is ultimately periodic starting from $d(A, a)$ and with the least period $p(A, a)$. Define $p = \text{lcm}_{A,a}\, p(A, a)$, $d_0 = \max_{A,a} d(A, a)$ and $d = d_0 \cdot |N|$.

The first claim is that if $A()$ returns on a string of the form wa^* without seeing the end of the string, then the number of a's in the tail cannot exceed $d \cdot (|w| + 1)$. To prove this inductively, a more elaborate formulation is needed:

Claim 2.1. *If $A()$ returns on $wa^n a^t$ (where $w \in \Sigma^*$, $n \geqslant 0$, $t \geqslant k$) consuming wa^n, and any string in $a^k a^*$ follows A, then*

$$n < d \cdot |w| + d_0 \cdot |\mathcal{X}| + 1 \tag{3}$$

where $\mathcal{X} \subseteq N$ is the set of all nonterminals X, such that in course of the computation of $A()$ the procedure $X()$ is ever called on $wa^n a^t$.

The proof is an inductive analysis of the computation of $A()$, which reveals that the parser must be matching the symbols of w to the first symbols of a^n. The essence of this claim is that if there are too many as, then the parser cannot keep count.

Claim 2.2. *Let $w \in \Sigma^+$, $0 \leqslant t < k$ and $n \geqslant d \cdot (|w| + 1) + d_0 + p + k - t + 1$, and suppose a^t follows A. Then, if $wa^n \in L_G(A)$, then $wa^{n+p} \in L_G(A)$ and $wa^{n-p} \in L_G(A)$.*

The proof of this claim analyzes the generation of wa^n, using Claim 2.1 to single out a nonterminal C generating a sufficiently long sequence of as. Then the periodicity of $L(C) \cap a^*$ is used to pump this sequence.

Finally, define $d' = d + d_0 + p + k - t + 1$, and the statement of the theorem follows from Claim 2.2.   □

Corollary 2. *If a language of the form $\{a^n b^{f(n)} \mid n \geqslant 1\}$, where $f : \mathbb{N} \to \mathbb{N}$ is an integer function, is Boolean LL, then the function $f(n)$ is bounded by $C \cdot n$ for some constant $C \geqslant 1$.*

Proof. By Theorem 2, there exist constants $d, d', p \geqslant 0$, such that for every string $a^n b^\ell \in L$ with $\ell \geqslant dn + d'$, it holds that $a^n b^{\ell+p} \in L$. Since, for every a^n, the language L contains only one string of the form $a^n b^*$, this condition should never hold, that is, for every $a^n b^\ell \in L$ we must have $\ell < dn + d'$. In other words, $f(n) < dn + d' \leqslant (d + d')n$, and setting $C = d + d'$ proves the theorem.   □

Example 3. The linear conjunctive language $\{a^n b^{2^n} \mid n \geqslant 0\}$ is not Boolean LL.

7   Linear Conjunctive LL Grammars

Consider the family of languages generated by linear conjunctive grammars satisfying the definition of an LL(k) Boolean grammar. Though these grammars are capable of specifying such a sophisticated language as the language of computations of a Turing machine, it turns out that some very simple languages are beyond their scope:

Theorem 3. *Let Σ be an alphabet, let $a, b \in \Sigma$ ($a \neq b$). Then, for every language $L \subseteq \Sigma^*$, $L \cdot \{a, b\}$ is linear conjunctive LL if and only if L is regular.*

Proof. The proof in one direction is trivial: if L is regular, then so is $L \cdot \{a, b\}$, and a finite automaton for the latter language can be transcribed as an LL(1) linear context-free grammar. Let us show that an LL(k) linear conjunctive grammar for $L \cdot \{a, b\}$ can be effectively transformed to a finite automaton for L. Let $G = (\Sigma, N, P, S)$ be an LL(k) linear conjunctive grammar for L, let $T : N \times \Sigma^{\leqslant k} \to P$ be a parsing table.

The first claim states that as long as a procedure $B()$ cannot see the end of the input in the beginning of the computation (that is, it is outside of the range of the lookahead), it must read the entire input. Otherwise it would have to decide in the beginning whether the last symbol is a or b, which cannot be done before seeing this last symbol.

Claim 3.1. *Let $w \in L$ and $s \in \{a, b\}$. If the successful computation of the parser on ws contains a call to $B()$ on a suffix yz, with $ws = xyz$ and $|yz| > k$, which returns, consuming y, then $z = \varepsilon$ and ε follows B.*

Let us now define a new grammar $G' = (\Sigma, N, P', S)$ as follows. Let m be the greatest length of the right-hand side of a rule in P. Every rule of the form $A \to u$ ($u \in \Sigma^*$) or $A \to u_1 B_1 \& \ldots \& u_n B_n$ in P is in P' as well. In addition, for every $A \in N$ and for every $w \in L_G(A)$, such that $|w| \leqslant k + m$, the set P' contains a rule $A \to w$. Let us prove that $L(G') = L(G)$.

Claim 3.2. *For every $A \in N$, $L_{G'}(A) \subseteq L_G(A)$.*

Claim 3.3. *Let $w \in L$ and $s \in \{a, b\}$ and consider the recursive descent parser for G. If its successful computation on ws contains a call to $A()$ on a suffix yz (with $ws = xyz$), which returns, consuming y, and z follows A, then $y \in L_{G'}(A)$.*

The proof of this claim can be summarized as follows: if y is sufficiently short, it is generated by a rule $A \to y$, and if y is long enough, then Claim 3.1 is applicable, and it implies that y is derived using a rule of the form $A \to u_1 B_1 \& \ldots \& u_n B_n$.

Claim 3.4. $L(G') = L(G)$.

We have thus shown that the language $L \cdot \{a, b\}$ is generated by a conjunctive grammar G' with one-sided concatenation, and therefore it is regular. Hence, L is regular as well, which completes the proof of the theorem.   □

Example 4. The context-free LL language $\{a^n b^n cs \mid n \geqslant 0, s \in \{a, b\}\}$, which is at the same time linear context-free, is not linear conjunctive LL.

Theorem 4. *The family of LL linear conjunctive languages is closed under intersection. It is not closed under union, concatenation and reversal.*

Proof. The closure under intersection is given by a direct application of conjunction. If *LinConjLL* were closed under union, then $\{a^n b^n ca \mid n \geqslant 0\} \cup \{a^n b^n cb \mid n \geqslant 0\}$ would be in *LinConjLL*, which would contradict Example 4. If *LinConjLL* were closed under concatenation, then $\{a^n b^n c \mid n \geqslant 0\} \cdot \{a, b\}$ would be in *LinConjLL*, which would again contradict Example 4. Suppose *LinConjLL* is closed under reversal. Then the language $(\{scb^n a^n \mid n \geqslant 0, s \in \{a, b\}\})^R$ would be in *LinConjLL* in violation of Example 4.   □

8   LL Linear Boolean Grammars

Linear Boolean grammars are known to have the same expressive power as linear conjunctive grammars [4,5]. In contrast, their LL subsets differ in power:

Example 5. The following LL(1) linear Boolean grammar generates the language $\{a^n b^n cs \mid n \geqslant 0, s \in \{a, b\}\}$:

$$
\begin{aligned}
S &\to X \& \neg T \\
T &\to X \& \neg Aca \& \neg Acb \\
A &\to aAb \mid \varepsilon \\
X &\to aX \mid bX \mid cX \mid \varepsilon
\end{aligned}
$$

Note that, by Lemma 4, there is no equivalent LL(k) linear conjunctive grammar.

This separate class of languages thus deserves a separate study. The following nonrepresentability result, proved by a counting argument, is useful in assessing their expressive power.

Lemma 7. *The language $\{a^n b^n c^\ell \mid n, \ell \geqslant 0\}$ is not LL linear Boolean.*

Proof. Suppose there exists a well-behaved LL(k) linear Boolean grammar $G = (\Sigma, N, P, S)$ for this language. Assume, without loss of generality, that every conjunct in this grammar is of the form $A \to \pm B$, $A \to \pm sB$, $A \to \pm Ct$, $A \to s$ or $A \to \varepsilon$, where $s, t \in \Sigma$ and $B, C \in N$.
    First infer the following property from Theorem 2:

Claim 7.1. *There exist numbers $d, p \geqslant 0$, such that for every nonterminal $B \in N$ and for all $n \geqslant 1$, $\ell \geqslant dn$ and $i \geqslant 0$,*

$$
b^n c^{\ell + ip} \in L(B) \quad \text{if and only if} \quad b^n c^\ell \in L(B).
$$

It is first proved that for any fixed numbers $m, \ell \geqslant 0$ and for a nonterminal A, the membership of strings of the form $a^m b^n c^\ell$ (with various $n \geqslant 0$) in $L(A)$ depends upon the membership of strings $b^n c^\ell$ in the languages $L(B)$, for different $B \in N$, and the dependence function is unique for all values of n.

Claim 7.2. *For every $A \in N$, for every $m \geqslant 0$, and for every $\ell \geqslant m \cdot |N| + h(A)$, there exists a Boolean function $f_{A,m,\ell}(\dots, x_{B,i}, \dots)$, where $B \in N$ and $0 \leqslant i \leqslant \ell$, such that for every $n \leqslant m$, $a^m b^n c^\ell \in L(A)$ if and only if $f_{A,m,\ell}(\dots, \sigma_{D,i}, \dots)$, where $\sigma_{D,i} = 1$ if and only if $b^n c^i \in L(D)$.*

Next, let us improve the statement of Claim 7.2, so that every function $f_A (\dots, x_{D,\ell}, \dots)$ depends upon a bounded number of Boolean variables, which does not increase with m and ℓ.

Claim 7.3. *For every $A \in N$, for every $m \geqslant 0$, and for every $\ell \geqslant m \cdot |N| + h(A)$, there exists a Boolean function $f_{A,m,\ell}(\dots, x_{B,i}, \dots)$, where $B \in N$ and $0 \leqslant i < d + p$, such that for every $n \leqslant m$, $a^m b^n c^\ell \in L(A)$ if and only if $f_{A,m,\ell}(\dots, \sigma_{D,i}, \dots)$, where $\sigma_{D,i} = 1$ if and only if $b^n c^i \in L(D)$.*

Now define $m_0 = 2^{2^{|N| \cdot (d+p)}}$ and $\ell_0 = |N| \cdot m_0 + \max_A h(A)$. For every $m \in \{0, \ldots, m_0\}$, denote $\widehat{f}_m = f_{S,m,\ell_0}$. Each function $\widehat{f}_m$ depends on $|N| \cdot (d+p)$ variables. There exist $2^{2^{|N| \cdot (d+p)}} = m_0$ distinct Boolean functions over this number of variables, and hence our set of $m_0 + 1$ such functions must contain a pair of duplicates:

Claim 7.4. *There exist two numbers m and $\widetilde{m}$, with $0 \leqslant m < \widetilde{m} \leqslant m_0$, such that $\widehat{f}_m \equiv \widehat{f}_{\widetilde{m}}$.*

Consider strings of the form $a^* b^m c^{\ell_0}$. For all $i \in \{\ell_0 - (d+p-1), \ldots, \ell_0\}$ and $D \in N$, define $\sigma_{D,i} = 1$ if $b^m c^i \in L_G(B)$ and $\sigma_{D,i} = 0$ otherwise. Take a true statement $a^m b^m c^{\ell_0} \in L(G)$. By Claim 7.3, $a^m b^m c^{\ell_0} \in L(G)$ if and only if $\widehat{f}_m(\ldots, \sigma_{D,i}, \ldots) = 1$. The latter, according to Claim 7.4, is equivalent to $\widehat{f}_{\widetilde{m}}(\ldots, \sigma_{D,i}, \ldots) = 1$, which, by Claim 7.3 again, holds if and only if $a^{\widetilde{m}} b^m c^{\ell_0} \in L(G)$, which is not true. A contradiction of the form *"true if and only if false"* has thus been obtained, which proves the lemma. □

Theorem 5. *The family of LL linear Boolean languages is closed under all Boolean operations. It is not closed under concatenation and reversal.*

Proof. Intersection and complementation can be specified directly, the closure under union follows by de Morgan's laws.

Suppose *LinBoolLL* is closed under concatenation. Then the language $\{a^n b^n \mid n \geqslant 0\} \cdot c^*$ is in *LinBoolLL*, which contradicts Lemma 7.

Similarly, if *LinBoolLL* were closed under reversal, then $(\{c^\ell b^n a^n \mid n, \ell \geqslant 0\})^R$ would be in *LinBoolLL*, again contradicting Lemma 7. □

9   Hierarchy

The results of this paper lead to a detailed comparison of different subfamilies of $LL(k)$ Boolean grammars with each other and with different subfamilies of Boolean grammars. The following theorem summarizes these results.

Theorem 6

1. *LinCFLL* $\subset$ *LinConjLL*, with $\{a^n b^n c^n \mid n \geqslant 0\} \in$ *LinConjLL* \ *LinCFLL*.
2. *LinCFLL* $\subset$ *CFLL*, with $\{a^n b^n c \mid n \geqslant 0\} \cdot \{a, b\} \in$ *CFLL* \ *LinCFLL*.
3. *LinConjLL* $\subset$ *ConjLL*, with $\{a^n b^n c \mid n \geqslant 0\} \cdot \{a, b\} \in$ *ConjLL* \ *LinConjLL*.
4. *CFLL* $\subset$ *ConjLL*, with $\{a^n b^n c^n \mid n \geqslant 0\} \in$ *ConjLL* \ *CFLL*.
5. *LinConjLL* $\subset$ *LinBoolLL*, with $\{a^n b^n c \mid n \geqslant 0\} \cdot \{a, b\} \in$ *LinBoolLL* \ *LinConjLL*.
6. *LinBoolLL* $\subset$ *BoolLL*, with $\{a^n b^n c^\ell \mid n, \ell \geqslant 0\} \in$ *BoolLL* \ *LinBoolLL*.
7. *LinCFLL* $\subset$ *LinCF*, with $\{a^n b^n c \mid n \geqslant 0\} \cdot \{a, b\} \in$ *LinCF* \ *LinCFLL*.
8. *LinBoolLL* $\subset$ *LinConj*, with $\{a^n b^n c^\ell \mid n, \ell \geqslant 0\} \in$ *LinConj* \ *LinConjLL*.
9. *CFLL* $\subset$ *CF*, with $\{a^n i b^{in} \mid n \geqslant 0, i \in \{1, 2\}\} \in$ *CF* \ *CFLL*.
10. *ConjLL* $\subset$ *Conj*, with $\{a^n b^{2^n} \mid n \geqslant 1\} \in$ *Conj* \ *ConjLL*.
11. *BoolLL* $\subset$ *Bool*, with $\{a^n b^{2^n} \mid n \geqslant 1\} \in$ *Bool* \ *BoolLL*.

The resulting inclusion diagram is given in Figure 2, in which arrows with a question mark denote inclusions not known to be proper, the rest being proper.

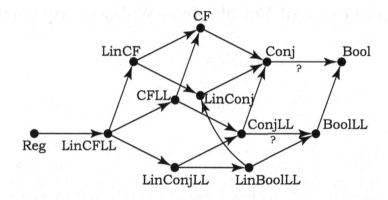

Fig. 2. Expressive power of subfamilies of Boolean grammars

Let us note some open problems. It remains to determine whether the families ConjLL and BoolLL are different. For some useful abstract languages generated by Boolean grammars, most notably for $\{wcw \,|\, w \in \{a,b\}^*\}$, it is important to know whether they are Boolean LL(k). Finally, while this paper establishes the first negative results for Boolean LL(k) grammars, it remains to invent a method of proving languages to be nonrepresentable by Boolean grammars of the general form. The lack of such a method limits our understanding of Boolean grammars.

References

1. Jurdzinski, T., Loryś, K.: Church-Rosser Languages vs. UCFL. In: Widmayer, P., Triguero, F., Morales, R., Hennessy, M., Eidenbenz, S., Conejo, R. (eds.) ICALP 2002. LNCS, vol. 2380, pp. 147–158. Springer, Heidelberg (2002)
2. Kountouriotis, V., Nomikos, Ch., Rondogiannis, P.: Well-founded semantics for Boolean grammars. In: Ibarra, O.H., Dang, Z. (eds.) DLT 2006. LNCS, vol. 4036, pp. 203–214. Springer, Heidelberg (2006)
3. Okhotin, A.: Conjunctive grammars. Journal of Automata, Languages and Combinatorics 6(4), 519–535 (2001)
4. Okhotin, A.: On the equivalence of linear conjunctive grammars to trellis automata. RAIRO Informatique Théorique et Applications 38(1), 69–88 (2004)
5. Okhotin, A.: Boolean grammars. Information and Computation 194(1), 19–48 (2004)
6. Okhotin, A.: Generalized LR parsing algorithm for Boolean grammars. International Journal of Foundations of Computer Science 17(3), 629–664 (2006)
7. Okhotin, A.: Recursive descent parsing for Boolean grammars. Acta Informatica (to appear)
8. Okhotin, A.: A simple P-complete problem and its representations by language equations. Machines, Computations and Universality (MCU 2007, Orléans, France, September 10–14, 2007) accepted
9. Terrier, V.: On real-time one-way cellular array. Theoretical Computer Science 141, 331–335 (1995)
10. Yu, S.: A property of real-time trellis automata. Discrete Applied Mathematics 15(1), 117–119 (1986)

Complexity of Pebble Tree-Walking Automata

Mathias Samuelides and Luc Segoufin

LIAFA, Paris 7
INRIA, Paris 11

Abstract. We consider tree-walking automata using k pebbles. The pebbles are either *strong* (can be lifted from anywhere) or *weak* (can be lifted only when the automaton is on it). For each k, we give the precise complexities of the problems of emptiness and inclusion of tree-walking automata using k pebbles.

1 Introduction

There are two natural ways to extend the classical finite state string automata on finite binary trees.

In the first one, which is the most studied one in the literature (see for instance [5]), the automata have parallel control and process the tree bottom-up. It forms a robust class of automata (it has minimization and determinization) and the class of languages accepted by them enjoys most of the nice properties of the string case. For instance it is closed under all Boolean operations and it corresponds to MSO definability. Tree languages accepted by bottom-up tree automata are called *regular*.

The second kind of tree automata has only one control. It is a sort of sequential automaton which moves from node to node in a tree, along its edges. They are called tree-walking automata and, in a sense, generalize the notion of two-way string automata by making use of all possible directions allowed in a tree [1,9]. However they are not determinizable [2] and have a rather weak expressive power [3]. For this reason pebble tree automata were introduced in [7] as a model with an interesting intermediate expressive power between tree-walking automata and bottom-up tree automata. A pebble tree automaton is a tree-walking automaton with a finite set $\{1, \ldots, k\}$ of pebbles which it can drop at and lift from a node. There is a stack discipline restriction though: pebble i can only be dropped at the current node if pebbles $i + 1, \ldots, k$ are already on the tree. Likewise, if pebbles $i, \ldots, k$ are on the tree only pebble i can be lifted. In the first model of pebble automata the pebbles were only allowed to be lifted by the automaton when its head is on it, but recently, in order to capture logics with transitive closure on trees, a stronger model of pebble automata was introduced in [8]. In the strong model pebbles are viewed as pointers and can be lifted from everywhere. Perhaps surprisingly, in [4] it was shown that the two models of pebble tree automata have the same expressive power. More precisely it was shown that for each k and each pebble tree automaton using k pebbles with the strong behavior, there exists a pebble tree automaton using k pebbles with the

E. Csuhaj-Varjú and Z. Ésik (Eds.): FCT 2007, LNCS 4639, pp. 458–469, 2007.

weak behavior accepting the same tree language. However the current translation yields an automaton whose size is a tower of $k-1$ exponents. It seems, but this has not been proved yet, that pebble tree automata using k strong pebbles are $(k-1)$-exponentially more succinct than pebble tree automata using k weak pebbles.

It is still conceivable that the class of pebble automata forms a robust class of tree languages. The main open issues are whether they are determinizable and whether they are closed under complement (the former would imply the later as mentioned in [13]). If they would be closed under complement, the family of tree languages accepted by pebble automata would also correspond to definability in unary transitive closure logic on trees [7]. The recent new interest in this family comes from their close relationship with some aspects of XML languages: They are a building block of *pebble transducers* which were used to capture XML transformations (cf. [12,10]).

In this paper we study the complexity of emptiness test for pebble automata and the complexity of testing whether one pebble automaton is included into another.

From each pebble automaton, weak or strong, it is easy to compute an equivalent MSO formula [7]. This shows that they define only regular tree languages and immediately yields a non-elementary test for emptiness and inclusion. This non-elementary complexity is unavoidable as shown in [12].

We are interested in the problem when k, the number of pebbles, is fixed and not part of the input. Emptiness and inclusion for tree-walking automata (the case $k=0$) are EXPTIME-complete. The upper-bound follows from the exponential time transformations of tree-walking automata and their complement into top-down tree automata given in [6,15]. The lower-bound is implicit in [14]. For $k>0$, we extend these results and show that both emptiness and inclusion are k-EXPTIME-complete. For each k, we prove the upper-bounds using the strong model and the lower-bounds using the weak and deterministic model. Therefore all variants considered here yield k-EXPTIME-complete problems.

The upper-bounds are proved by constructing, in time k-exponential, a bottom-up tree automaton for the language recognized by a tree-walking automata using k-pebbles and another one for the complement language. This is done by induction on the number of pebbles using an intermediate model which combines a tree-walking behavior with a bottom-up one. This induction is quite simple in the case where all pebbles have a weak behavior. In this case the subrun between the drop and the lift of the last pebble starts and ends at the same node and can therefore be replaced by a regular test. It is then possible to remove the last pebble by computing a product automaton. In the case of strong pebbles, subruns start and end at different nodes of the tree, and this complicates the construction.

The lower-bounds are proved by simulating a run of an alternating Turing machine using $(k-1)$-EXPSPACE by a deterministic pebble automaton using k pebbles with weak behaviors.

For each k, the complexities obtained for strong and weak pebble automata are the same. However we conjecture that pebble automaton using k strong pebbles is $(k-1)$-exponentially more succinct than pebble automaton using k weak pebbles. Therefore the strong model is more interesting as it achieves similar performances in terms of expressive power and of complexity but with a more succinct presentation.

When restricted to string models, our results show that both emptiness and inclusion for pebble automata using k pebbles are $(k-1)$-EXPSPACE-complete. Pebble string automata were already studied in [11] where it was shown that a pebble string automaton using k weak pebbles is k-exponentially more succinct than an one-way finite state automaton (the use of pebbles in [11] is actually even more restricted than the weak behavior mentioned above). The coding for proving our lower bounds is inspired from this result.

2   Definitions

The trees we consider are finite binary trees, with nodes labeled over a finite alphabet Σ. We insist that each internal node (non-leaf node) has exactly two children. A set of trees over a given alphabet is called a **tree language**.

Definition 2.1. *A* **bottom-up automaton** $\mathcal{B}$ *is a tuple* $(\Sigma, Q, q_0, F, \delta)$, *where* Q *is a finite set of states,* Σ *is a finite alphabet,* q_0 *is the initial state,* F *is the set of accepting states and* $\delta \subseteq (\Sigma \times Q \times Q) \times Q$ *is the transition relation.*

A **run** of a bottom-up automaton $\mathcal{B}$ on a tree t is a function ρ from the set of nodes of t to Q such that for every node x of label σ,

- If x is a leaf, then $((\sigma, q_0, q_0), \rho(x)) \in \delta$.
- If x has two children x_1 and x_2, then $((\sigma, \rho(x_1), \rho(x_2)), \rho(x)) \in \delta$.

A run of $\mathcal{B}$ on t is **accepting**, and the tree t is **accepted** by $\mathcal{B}$, if the state at the root is accepting. The family of tree languages defined by bottom-up automata is called the class of **regular tree languages**.

Pebble tree automata. Informally a pebble tree automaton walks through its input tree from node to node along its edges. Additionally it has a fixed set of pebbles, numbered from 1 to k that it can place in the tree. At each time, pebbles $i, \cdots, k$ are placed on some nodes of the tree, for some i. In one step the automaton can stay at the current node, move to its parent , to its left or to its right child, or it can lift pebble i or place pebble $i-1$ on the current node. Which of these transitions can be applied depends on the current state, the set of pebbles at the current node, the label and the type of the current node (root, left or right child and leaf or internal node).

We consider two kinds of pebble automata which differ in the way they can lift the pebble. In the weak model a pebble can be lifted only if it is on the current node. In the strong model this restriction does not apply.

Remark: In both models the placement of the pebbles follows a stack discipline: only the pebble with the number i can be lifted and only the pebble with number $i - 1$ can be placed. This restriction is essential as otherwise we would obtain k-head automata that recognize non-regular tree languages.

We turn to the formal definition of pebble automata. The set TYPES $= \{r, 0, 1\} \times \{l, i\}$ describes the possible **types** of a node. Here, r stands for the root, 0 for a left child, 1 for a right child, l for a leaf and i for an internal node. We indicate the possible kinds of moves of a pebble automaton by elements of the set MOVES $= \{\epsilon, \uparrow, \swarrow, \searrow, \text{lift}, \text{drop}\}$, where informally $\uparrow$ stands for 'move to parent', ϵ stands for 'stay', $\swarrow$ stands for 'move to the left child' and $\searrow$ stands for 'move to the right child'. Clearly drop refers to dropping a pebble and lift to lifting a pebble. Finally if S is a set then $\mathcal{P}(S)$ denotes the powerset of S.

Definition 2.2. *A* **pebble tree automaton** *using k pebbles is a tuple* $\mathcal{A} = (\Sigma, Q, I, F, \delta)$, *where Q is a finite set of states, $I, F \subseteq Q$ are respectively the set of* **initial** *and* **terminal** *states and δ is the transition relation of the form*

$$\delta \subseteq \big(Q \times \text{TYPES} \times \{1, \ldots, k+1\} \times \mathcal{P}(\{1, \cdots, k\}) \times \Sigma \big) \times (Q \times \text{MOVES})$$

A tuple $(q, \beta, i, S, \sigma, q', m) \in \delta$ means that if $\mathcal{A}$ is in state q with pebbles $i, \cdots, k$ on the tree, the current node contains exactly the pebbles from S, has type β and is labeled by σ then $\mathcal{A}$ can enter state q' and move according to m.

A pebble set of $\mathcal{A}$ is a set $P \subseteq \{1, \cdots, k\}$. For a tree t, a P-pebble assignment is a function which maps each $j \in P$ to a node in t.   For $0 \leq i \leq k$, an i-**configuration** c of $\mathcal{A}$ on t is a tuple (x, q, f), where x is a node, q a state and f an $\{i+1, \cdots, k\}$-pebble assignment. In this case x is called the current node, q the current state and f the current pebble assignment. We also write $(x, q, x_{i+1}, \cdots, x_k)$ if $f(j) = x_j$ for each $j \geq i + 1$. We write $c \xrightarrow{\mathcal{A}, t} c'$ when $\mathcal{A}$ can make a (single step) transition from the configuration c to c' according to its transition relation. The relation $\xrightarrow{\mathcal{A}, t}$ is defined in the obvious way following the intuition described above for δ. However in the weak model of pebble tree automata there is a restriction on the lift-operation: a lift transition can be applied to an i-configuration (x, q, f) only if $f(i + 1) = x$, i.e., if pebble $i + 1$ is on the current node. In the strong model this restriction does not hold. A run is a nonempty sequence $c_1, \cdots, c_l$ of configurations such that $c_j \xrightarrow{\mathcal{A}, t} c_{j+1}$ holds for each $j < l$. We write $c \xrightarrow{\mathcal{A}, t \, *} c'$ when there is a run of A from c to c'.

Instead of having a set of accepting states and acceptance at the root as usual, we assume that a walking automaton has a set of **terminal** states. Once a terminal state is reached, the automaton immediately stops walking the tree. When the automaton is used as an acceptor for trees, we further assume a partition of the terminal states into accepting and rejecting ones (note that the automaton always rejects if no terminal state is ever reached). As an acceptor for tree languages, this hypothesis does not make any difference as it is always possible to go back to the root of the tree once a terminal state is reached.

A pebble tree automaton is **deterministic** if δ is a partial function from $Q \times \text{TYPES} \times \{1, \cdots, k+1\} \times \mathcal{P}(\{1, \cdots, k\}) \times \Sigma$ to $Q \times \text{MOVES}$.

We use PA_k to denote the (strong) pebble automata using k pebbles, wPA_k the weak pebble automata using k pebbles, DPA_k and wDPA_k for the corresponding deterministic automata. By default we assume the strong case. A pebble automaton without pebbles is just a tree-walking automaton. We write TWA and DTWA for PA_0 and DPA_0.

Complexities. In this paper k-EXPTIME refers to the set of problems solvable by a Turing machine using a time which is a tower of k exponentials of a polynomial in the size of its input. In order to avoid a case analysis we sometime write 0-EXPTIME for PTIME. Similarly we define k-EXPSPACE with 0-EXPSPACE for PSPACE.

3   From Pebble Automaton to Bottom-Up Automaton

Given $\mathcal{A} \in \text{PA}_k$ we construct in this section a bottom-up tree automaton $\mathcal{B}$ recognizing the same language as $\mathcal{A}$ and a bottom-up tree automaton $\mathcal{C}$ recognizing the language of trees rejected by $\mathcal{A}$. The constructions are performed in k-EXPTIME. They are done by induction on k. During the induction we shall make use of the following intermediate model of automata which combines a tree-walking behavior with a bottom-up one.

Intuitively a wPABU_k is a wPA_k that can simulate a bottom-up automaton while placing its last pebble on the current node. More formally we have:

Definition 3.1. *A* wPABU_k *is a pair* $(\mathcal{A}, \mathcal{B})$ *where*

- *$\mathcal{B}$ is a (non deterministic) bottom-up automaton on* $\Sigma \times \mathcal{P}(\{1, \cdots, k\})$
- *$\mathcal{A}$ is a* wPA_k *such that the transitions that drop pebble 1 are of the form*

$$(Q_\mathcal{A} \times \text{TYPES} \times \{2\} \times \mathcal{P}(\{2, \cdots, k\}) \times \Sigma \times Q_\mathcal{B}) \times (Q_\mathcal{A} \times \{\text{drop}\})$$

where $Q_\mathcal{A}$ *and* $Q_\mathcal{B}$ *are the set of states of* $\mathcal{A}$ *and* $\mathcal{B}$.

A wPABU_k $(\mathcal{A}, \mathcal{B})$ behaves like a pebble tree automaton until it wants to drop pebble 1. When it drops pebble 1, it immediately simulates $\mathcal{B}$ on the current pebbled tree and resumes its walking behavior with a state which depends on the state reached by $\mathcal{B}$ at the root of the tree as specified by the transition above.

More formally, for $0 \le i \le k$, an i-configuration of $(\mathcal{A}, \mathcal{B})$ on a tree t is a tuple (x, q, f) where x is a node of t, q a state of $\mathcal{A}$ and f an $\{i+1, \cdots, k\}$-pebble assignment. Let t be a tree, x a node of t of type τ and of label a and let $c = (x, q, f)$ and $c' = (x', q', f')$ be i-configurations of $(\mathcal{A}, \mathcal{B})$. A single step transition $c \xrightarrow{(\mathcal{A}, \mathcal{B}), t} c'$ of $(\mathcal{A}, \mathcal{B})$ is defined as $c \xrightarrow{\mathcal{A}, t} c'$ if $\mathcal{A}$ does not drop pebble 1 while making the transition step from c to c'. Otherwise we must have $x' = x$, $f' = f \cup \{(1, x)\}$ and $((q, \tau, 2, f^{-1}(x), a, q_b), (q', \text{drop}))$ is a transition of $(\mathcal{A}, \mathcal{B})$ where q_b is the state accessed by a run of $\mathcal{B}$ at the root of the pebbled tree (t, f').

In order to handle the strong behaviors of pebbles we need to extend this definition. The idea is to use BU^* automata instead of bottom-up automata. Intuitively a BU^* automaton is a bottom-up automaton that can select a node. More formally this means:

Definition 3.2. *A* BU* *automaton is a tuple* $(Q, q_0, Q_f, Q', \delta)$ *such that* $Q' \subseteq Q$ *and* (Q, q_0, Q_f, δ) *is a bottom-up tree automaton such that for each tree t and each accepting run ρ on t there is an unique node $x \in t$ with $\rho(x) \in Q'$.*

A PABU$^*_k$ is like a wPABU$_k$ but the pebble tree automaton part is strong and the bottom-up part is a BU*. When dropping the last pebble, the pebble automaton simulates the bottom-up part and resumes its run at the node selected by the BU* automaton by lifting the last pebble. More formally this gives:

Definition 3.3. *A* PABU$^*_k$ *is a pair* $(\mathcal{A}, \mathcal{B})$ *where*

- $\mathcal{B}$ *is a (non deterministic)* BU* *on* $\Sigma \times \mathcal{P}(\{1, \cdots, k\})$
- $\mathcal{A}$ *is a* PA$_k$ *such that the transitions dropping and lifting the last pebble are replaced by transitions of the form*

$$(Q_{\mathcal{A}} \times \text{TYPES} \times \{2\} \times \mathcal{P}(\{2, \cdots, k\}) \times \Sigma \times Q_{f\mathcal{B}}) \times Q_{\mathcal{A}}$$

where $Q_{\mathcal{A}}$ is the set of states of $\mathcal{A}$ and $Q_{f\mathcal{B}}$ is the set of accepting states of $\mathcal{B}$.

More formally, for $1 \leq i \leq k$, an i-configuration of $(\mathcal{A}, \mathcal{B})$ is a tuple (x, q, f) where x is a node, q a state of $\mathcal{A}$ and f an $\{i+1, \cdots, k\}$-pebble assignment.

Let t be a tree, x be a node of t of type τ and of label a and let $c = (x, q, f)$ and $c' = (x', q', f')$ be configurations of $(\mathcal{A}, \mathcal{B})$. A single step transition $c \xrightarrow{(\mathcal{A},\mathcal{B}),t} c'$ of $(\mathcal{A}, \mathcal{B})$ is defined as $c \xrightarrow{\mathcal{A},t} c'$ if $\mathcal{A}$ does not drop pebble 1 while making the transition step from c to c'. Otherwise we must have $f' = f$, x' is the node selected by $\mathcal{B}$ on the pebbled tree $(t, f \cup \{(1, x)\})$ and $((q, \tau, 2, f^{-1}(x), a, q_b), q')$ is a transition of $(\mathcal{A}, \mathcal{B})$ where q_b is the state accessed by a run of $\mathcal{B}$ at the root of the pebbled tree $(t, f \cup \{(1, x)\})$.

A node x in a tree t is a **marked node** if x is the unique node of t having the type or pebble assignment of x. If m is the marking type or pebble assignment, we then say that t is marked by m, and that x is the m-node of t. For instance a tree is always marked by its root. At any moment during the run of a tree-walking automaton, a tree is marked by any of the pebbles which are currently dropped.

Given a PABU$^*_k$ $(\mathcal{A}, \mathcal{B})$ and a BU* $\mathcal{C}$, we say that $\mathcal{C}$ **simulates exactly** $(\mathcal{A}, \mathcal{B})$ on trees marked by m if: (i) each set of pairs of states of $\mathcal{A}$ is an accepting state of $\mathcal{C}$, (ii) for each tree t marked by node u, $\mathcal{C}$ reaches the root of t after selecting node v, in an accepting state $q_f = \{(q, q') \mid (q, u, \_) \xrightarrow{(\mathcal{A},\mathcal{B}),t \ *} (q', v, \_), q' \text{ is terminal}\}$. In other words the state reached by $\mathcal{C}$ at the root contains exactly the beginning and the ending states of all the terminating runs of $\mathcal{A}$ between nodes u and v. Note that this implies that $\mathcal{C}$ is unambiguous once the choice of v is made.

In a sense, our first lemma below extends the result of [6,15] for TWA, and translates a wPABU$_1$ and its complement into a bottom-up automaton. This is done with the same complexity bounds as for TWA: One exponential. The proof is omitted in this abstract. The idea is classical, the bottom-up tree automaton has to compute all possible loops of the tree-walking automaton while moving up the tree. The main new difficulty is to take care of the loops which involve the use of the pebble.

Lemma 3.4. *Let $(\mathcal{A}, \mathcal{B})$ be a wPABU$_1$. For any marking m, we can construct in time exponential in $|(\mathcal{A}, \mathcal{B})|$, $\mathcal{C} \in$ BU* such that $\mathcal{C}$ simulates exactly $(\mathcal{A}, \mathcal{B})$ on trees marked by m.*

We now extend the previous lemma to the strong pebble case. This is done by reducing the strong pebble case to the weak pebble one. Given two walking automata $\mathcal{A}$ and $\mathcal{B}$ we say that $\mathcal{B}$ *simulates exactly* $\mathcal{A}$ if (i) the set of states of $\mathcal{A}$ is included into the set of states of $\mathcal{B}$, (ii) for all tree t and all nodes u and v of t we have $(u, q, \_) \xrightarrow{\mathcal{A}, t} (v, q', \_)$ iff $(u, q, \_) \xrightarrow{\mathcal{B}, t} (v, q', \_)$ for all pair (q, q') of states of $\mathcal{A}$.

Lemma 3.5. *Given $(\mathcal{A}, \mathcal{B}) \in$ PABU$_1^*$, we can construct in polynomial time $(\mathcal{A}', \mathcal{B}')$ in wPABU$_1$ simulating exactly $(\mathcal{A}, \mathcal{B})$.*

Proof. The idea is as follows, $\mathcal{A}'$ will simulate $\mathcal{A}$ until the pebble is dropped. Then $\mathcal{A}'$ will move step by step the pebble in the tree until it reaches the position where the pebble is lifted. The difficulty is that $\mathcal{A}'$ cannot find out which node is selected by $\mathcal{B}$, until $\mathcal{A}'$ is on that node, and that as soon as $\mathcal{A}'$ moves the pebble, the simulation of $\mathcal{B}$ is no longer valid. To cope with this situation, $\mathcal{A}'$ will maintain extra information in its state and only simulate $\mathcal{B}$ partially.

Assume now that $\mathcal{A}$ drops the pebble on the node x_d, evaluates $\mathcal{B}$ and resumes its run from node x_l after lifting the pebble. Let $Q_\mathcal{B}$ be the set of states of $\mathcal{B}$.

We show how to simulate this behavior using a wPABU$_1$ $(\mathcal{A}', \mathcal{B}')$. On the tree t there is an unique path from x_d to x_l. The goal of $\mathcal{A}'$ is to transfer step by step the pebble on that path. To do this, at any time, assuming its pebble is on position x in the path, it will remember in its state (i) the state q_r reached by $\mathcal{B}$ at the root of the tree when it was simulated by $\mathcal{A}$, (ii) the state q_x reached by $\mathcal{B}$ on the current node x when it was simulated by $\mathcal{A}$, and (iii) the direction from x to the next node on the path from x_d to x_l.

This information is computed and maintained using $\mathcal{B}'$.

To do this $\mathcal{B}'$ will do the following. It first guesses a state $q_x \in Q_\mathcal{B}$ and a direction $\Delta \in \{$ DOWNLEFT, DOWNRIGHT, UPRIGHT, UPLEFT, INIT, HERE$\}$, which are expected to match those currently stored in the state of $\mathcal{A}'$ ($\mathcal{A}'$ will verify this in the next step), except for the first time where Δ is INIT. We then distinguish three cases:

- Case 1: Δ is INIT. Then $\mathcal{B}'$ simulates $\mathcal{B}$ and ends in a state containing: q_x the state reached by $\mathcal{B}$ at the position of the pebble, $\Delta' \in \{$ DOWNLEFT, DOWNRIGHT, UPRIGHT, UPLEFT, HERE$\}$ the direction from the pebble to the selected node and q_r the state reached by $\mathcal{B}$ at the root. Note that Δ' could be HERE if the selected position is the current position of the pebble.
- Case 2: Δ is UPRIGHT (the case UPLEFT is similar).
  In this case, $\mathcal{B}'$ simulates $\mathcal{B}$ unless it reaches x (marked by the pebble). When x is reached, if $\mathcal{B}$ already selected its node in the left subtree of x, then $\mathcal{B}'$ rejects. Otherwise, the current state is ignored and $\mathcal{B}'$ recomputes the current state assuming the state q_x at the left child of x. It remembers the new state q_x' and the direction Δ' from x to the selected node and resumes

its simulation of $\mathcal{B}$. At the root $\mathcal{B}'$ ends in a state containing q_x, Δ and the state reached by $\mathcal{B}$ at the root during the current simulation together with q'_x and Δ'.

- Case 3: Δ is DOWNRIGHT (the case DOWNLEFT is similar).
  In this case $\mathcal{B}'$ simulates $\mathcal{B}$ until it reaches x. When x is reached $\mathcal{B}'$ knows whether a node has been indeed selected in the right subtree of x. If this is not the case, or if the current state is not q_x, it rejects. If this is the case, $\mathcal{B}'$ remembers the state q'_x $\mathcal{B}$ has reached at the right child of x and also the direction Δ' from this right child to the node x_l. At the root $\mathcal{B}'$ accepts in a state containing q_x and Δ together with the state reached by $\mathcal{B}$ at the root during the current simulation together with q'_x and Δ'.

We can now define $\mathcal{A}'$. $\mathcal{A}'$ simulates $\mathcal{A}$ until the pebble is dropped at position x_d. At position x_d, when the pebble is first dropped it simulates $\mathcal{B}'$, verifies that $\mathcal{B}'$ indeed guessed the INIT case and stores the output of $\mathcal{B}'$ in its state. It then does the following until a HERE case is reached. It moves the pebble one step according to Δ, simulates $\mathcal{B}'$, verifies that the guessed values of $\mathcal{B}'$ are consistent with what it currently has in its state. If not it rejects, if yes it updates those values according to the output of $\mathcal{B}'$. When $\mathcal{B}'$ outputs HERE then $\mathcal{A}'$ lifts the pebble and resumes the simulation of $\mathcal{A}$.   □

We are now ready for the main induction loop.

Lemma 3.6. *Let $(\mathcal{A}, \mathcal{B})$ in PABU_k^*. Let m be any marking. We can construct in time k-exponential in $|(\mathcal{A}, \mathcal{B})|$, a BU^* $\mathcal{C}$ that simulates exactly $(\mathcal{A}, \mathcal{B})$ on trees marked by m.*

Proof. This is done by induction on n. The case $k = 1$ is given by combining Lemma 3.5 with Lemma 3.4. Assume now that the lemma is proved for k and we will prove it for $k + 1$. Let $\mathcal{A}'$ be the PA_1 defined from $\mathcal{A}$ as follows. The states of $\mathcal{A}'$ are all the states of $\mathcal{A}$ corresponding to configurations where all the pebbles $k, \cdots, 2$ are dropped. The transitions of $\mathcal{A}'$ are all the transitions of $\mathcal{A}$ restricted to the states of $\mathcal{A}'$. The terminal states of $\mathcal{A}'$ are exactly those that lift pebble 2. The marking is the position of pebble 2. Then $(\mathcal{A}', \mathcal{B})$ and $(\mathcal{A}, \mathcal{B})$ have exactly the same runs from a node u where pebble 2 is dropped to a node v where it is next lifted. From Lemma 3.5 we obtain $(\mathcal{A}'', \mathcal{B}')$ in wPABU_1 simulating exactly $(\mathcal{A}', \mathcal{B})$ assuming pebbles $k, \cdots, 2$ are already on the tree. Now by Lemma 3.4 we obtain in exponential time a BU^* automaton $\mathcal{C}'$ that simulates exactly $(\mathcal{A}'', \mathcal{B}')$ on trees marked by pebble 2, assuming pebbles $k, \cdots, 2$ are already on the tree. Let now $(\mathcal{A}''', \mathcal{C}')$ be the PABU_{k-1}^* defined from $\mathcal{A}$ as follows. The states of $\mathcal{A}'''$ are all the states of $\mathcal{A}$ corresponding to all configurations where pebble 1 is not dropped. The terminal states of $\mathcal{A}'''$ are the terminal states of $\mathcal{A}$. The transitions of $\mathcal{A}'''$ are all the transitions of $\mathcal{A}$ restricted to the states of $\mathcal{A}'''$ where all the transitions dropping pebble 2 are now replaced by a simulation of $\mathcal{C}'$. It is easy to verify that $(\mathcal{A}''', \mathcal{C}')$ simulates exactly $(\mathcal{A}, \mathcal{B})$, and we obtain the desired $\mathcal{C}$ by induction.   □

Theorem 3.7. *Let $\mathcal{A}$ in PA_k. We can construct in time k-exponential in $|\mathcal{A}|$ a bottom-up automaton C accepting the same language as $\mathcal{A}$ and a bottom-up automaton $\overline{C}$ accepting the complement of the language accepted by $\mathcal{A}$.*

Proof. We shall make use of this lemma which shows that we can always assume that the last pebble is weak. Its proof is a straightforward adaptation of the proof of Lemma 3.5.

Lemma 3.8. *Let $\mathcal{A}$ be in PA_k. We can construct in time polynomial in $|\mathcal{A}|$ an automaton $\mathcal{B} \in \mathrm{PA}_k$ accepting the same tree language as $\mathcal{A}$ and such that the pebble 1 of $\mathcal{B}$ is weak.*

Let $\mathcal{A}$ in PA_k. Let $\mathcal{A}'$ in PA_k recognizing the same language as $\mathcal{A}$ but with pebble 1 weak as given by Lemma 3.8. Let $\mathcal{A}''$ be the wPA_1 defined from $\mathcal{A}'$ as follows. The states of $\mathcal{A}''$ are all the states of $\mathcal{A}'$ corresponding to configurations where all the pebbles $n, \cdots, 2$ are dropped. The transitions of $\mathcal{A}''$ are all the transitions of $\mathcal{A}'$ restricted to states of $\mathcal{A}''$. The terminal states of $\mathcal{A}''$ are exactly those that lift pebble 2. Let $\mathcal{B}$ be the trivial bottom-up tree automaton with only one state that does nothing. Then $(\mathcal{A}'', \mathcal{B})$ is a wPABU_1 having the same runs as $\mathcal{A}'$ from a node u where pebble 2 is dropped to a node v where it is next lifted. Now by Lemma 3.4 we obtain in exponential time a BU^* automaton $\mathcal{C}'$ that simulates exactly $(\mathcal{A}'', \mathcal{B})$ on trees marked by pebble 2. Let now $(\mathcal{A}''', \mathcal{C}')$ be the PABU^*_{k-1} defined from $\mathcal{A}$ as follows. The states of $\mathcal{A}'''$ are all the states of $\mathcal{A}$ corresponding to all configurations where pebble 1 is not dropped. The terminal states of $\mathcal{A}'''$ are the terminal states of $\mathcal{A}$. The transitions of $\mathcal{A}'''$ are all the transitions of $\mathcal{A}$ restricted to the states of $\mathcal{A}'''$ where all the transitions dropping pebble 2 are now replaced with a simulation of $\mathcal{C}'$. It is easy to verify that $(\mathcal{A}''', \mathcal{C}')$ simulates exactly $\mathcal{A}$. Let $\mathcal{D}$ be the BU^* obtained by applying Lemma 3.6 on trees marked by the root. Note that $\mathcal{D}$ always marks the root and therefore can be seen as a bottom-up tree automaton. By construction of $\mathcal{D}$, the state reached by $\mathcal{D}$ at the root of any input tree contains exactly all the pair (q, q') so that if $\mathcal{A}$ starts at the root in state q then it comes back at the root in state q'. It is now immediate to define C and $\overline{C}$ from $\mathcal{D}$ by choosing appropriately the set of accepting states. $\qquad\square$

The **emptiness problem** for pebble tree automata is the problem of checking, given a pebble tree automaton, whether it accepts at least one tree. The **inclusion problem** for pebble tree automata is the problem of checking, given two tree automata $\mathcal{A}$ and $\mathcal{B}$ whether any tree accepted by $\mathcal{A}$ is also accepted by $\mathcal{B}$. As the emptiness problem for bottom-up tree automata is in PTIME we immediately derive from Theorem 3.7 an upper-bound for the emptiness and inclusion problems for PA, and therefore for wPA.

Theorem 3.9. *Let $k > 0$. The emptiness and the inclusion problems for PA_k are in k-EXPTIME.*

4 Lower Bounds

In this section we show that the complexities obtained previously are tight. We actually prove a stronger result as we show that the lower bounds already hold in the weak pebble model and with deterministic control.

For $k > 0$, let $\exp(k, n)$ be the function defined by $\exp(1, n) = 2^n$ and $\exp(k, n) = 2^{\exp(k-1,n)}$. A k-**number** of size n is defined recursively as follows. If $k = 1$ it is a tree formed by a root of label $\sharp$ followed by a unary tree forming a sequence of n bits, defining a number from 0 to $2^n - 1$ (this tree can be made binary by adding enough dummy extra nodes). For $k > 1$ it is a tree t having the following properties: (i) The root of t is labeled by $\sharp$, (ii) The path from the root of t to the rightmost leaf (excluding the root) contains exactly $\exp(k-1, n)$ nodes having label in $\{0, 1\}$. This path will be called: **rightmost branch**, (iii) The left child of each node x of the rightmost branch is a $(k-1)$-number that encodes the distance from $\sharp$ to x (the topmost branching node is assumed to have distance zero from $\sharp$).

We can easily see that the $exp(k-1, n)$ bits in the rightmost branch in t define a number from 0 to $\exp(k, n) - 1$.

In the rest of this section we blur the distinction between the root of a k-number and the integer it encodes. We will make use of the following terminology. If x is a k-number, the nodes of the rightmost branch of x are called the **bits** of x. For each such bit, the $(k-1)$-number branching off that node is called a **position** of x.

Let f be a function associating to a node x of a tree t a set $f(x)$ of nodes of t. Such a function f is said to be **determined** if there is a DTWA with a specific state q_S such that, when started at x in a tree t, it sequentially investigates all the nodes in $f(x)$, being in state q_S at a node y iff $y \in f(x)$. Typically determined functions are the set of positions of the k-number which is located immediately above or below x. In the following we will only use very simple determined functions f. In particular the size of the DTWA involved has a size which will not depend on the parameters k and n.

The main technical lemma is the following one which is inspired by the succinctness result of [11].

Lemma 4.1. *Let $n > 0$ and $k > 0$.*

1. *There exists a wDPA$_{(k-1)}$, of size polynomial in n, such that when started on a node x of a tree t, it returns to x in a state which is accepting iff the subtree rooted in x forms a k-number of size n.*
2. *For each determined function f, there exists a wDPA$_k$, of size polynomial in n, such that when started on a marked node x of a tree t, it returns to x in a state which is accepting iff there is a node $y \in f(x)$ such that x and y form the same k-number of size n.*
3. *There exists a wDPA$_k$, of size polynomial in n, such that when started on a node x of a tree t, it returns to x in a state which is accepting iff the k-number of size n rooted at the left child of x is the successor of the k-number of size n rooted at the left child of the right child of x.*

Proof. Fix $n > 0$. All items are proved simultaneously by induction on k. For point 2 let $\mathcal{A}_f$ be the DTWA for f.

If $k = 1$, point 1 is clear. The automaton uses n states to check that the tree has the correct depth and then comes back to the initial place. For point 2, the automaton successively drops the pebble on each node y of the set of nodes in $f(x)$ using $\mathcal{A}_f$, simulates the automaton obtained for point 1 to check that the subtrees of x and y are indeed 1-number. Once this is done it processes the subtrees of x and y bit per bit, going back and forth between x and y (recall that x is marked by hypothesis and that y is marked by the pebble), checking for equality. The current position being processed is stored in the state and this requires $O(n)$ states. For point 3 the pebble is dropped on the appropriate child of x and we proceed as for point 2, simulating addition with 1 instead of checking equality.

Assume now that $k > 1$. Consider point 1. By induction it is easy to verify with a wDPA$_{(k-2)}$ that the subtree rooted in x has the right shape: it starts with a $\sharp$ and is a sequence of $(k - 1)$-number. It remains to check that this sequence codes all $(k - 1)$-number in the order from 0 to $\exp(k - 1, n) - 1$. For each node y of the rightmost branch of x the automaton drops pebble k on y and simulates the wDPA$_{(k-1)}$ obtained from point 3 by induction in order to verify that the positions of y and of the right child of y are successive $(k - 1)$-number. Once this is done for each bit y of x the automaton goes back to x by going up in the tree until a $\sharp$ is found.

Consider now point 2. The automaton first checks by induction that x is the root of a k-number. Then for each node $y \in f(x)$ it does the following. It first checks whether y is the root of k-number and if this is the case it drops pebble k successively on each position z of y. Let g be the function which associates to z the set of positions of x. It is easy to see that g is determined by the deterministic automaton which from z goes back to x, which is marked, and then successively investigates all position of x by going down to the right child. By induction on point 2 the automaton can find with the remaining $k - 1$ pebbles, among the positions of x, the one with the same $(k - 1)$-number as z and checks that the associated bits are equal. If the bits are different the automaton comes back to z, lifts pebble k and moves back to y by going up until it finds a $\sharp$ and then proceeds with the next node of $f(x)$. If the bits match the automaton comes back to z, lifts pebble k and proceeds with the next bit of y. Once all bits of y are processed successfully, it goes back to x and accepts.

Consider finally point 3. This is done as is point 2 above with the following differences. The set of nodes $f(x)$ is a singleton and the node x does not have to be marked anymore as it can be recover from the position of pebble k. Moreover, instead of checking equality of two $(k - 1)$-number the automaton simulates addition with 1. □

Using this lemma the coding of alternating Turing machines is rather straightforward.

Theorem 4.2. *Let $k \geq 1$. The emptiness problem (and hence the inclusion problem) for wDPA$_k$ (and hence for PA$_k$) is k-EXPTIME-hard.*

5   Discussion

It is not too hard to see that when restricted to strings, the techniques developed in this paper imply:

Theorem 5.1. *For $k \geq 1$ (the case $k = 0$ is equivalent to the case $k = 1$).*

1. *The emptiness and inclusion problems for wDPA_k over strings are $(k-1)$-EXPSPACE-hard.*
2. *The emptiness and inclusion problems for PA_k over strings are in $(k-1)$-EXPSPACE.*

Over trees our result and the one of [4] show that both the weak model and the strong model of pebble have the same expressive power and the same complexities. We believe that pebble tree automata using k strong pebbles are $(k-1)$-exponentially more succinct than pebble tree automata using k weak pebbles. It would be interesting to settle this issue.

References

1. Aho, A.V., Ullman, J.D.: Translations on a Context-Free Grammar. Information and Control 19(5), 439–475 (1971)
2. Bojańczyk, M., Colcombet, T.: Tree-Walking Automata Cannot Be Determinized. Theor. Comput. Sci. 350(2-3), 164–173 (2006)
3. Bojańczyk, M., Colcombet, T.: Tree-walking automata do not recognize all regular languages. In: STOC (2005)
4. Bojańczyk, M., Samuelides, M., Schwentick, T., Segoufin, L.: Expressive power of pebble automata. In: ICALP (2006)
5. Comon, H., et al.: Tree Automata Techniques and Applications. Available at http://www.grappa.univ-lille3.fr/tata
6. Cosmadakis, S.S., Gaifman, H., Kanellakis, P.C., Vardi, M.Y.: Decidable Optimization Problems for Database Logic Programs. In: STOC (1988)
7. Engelfriet, J., Hoogeboom, H.J.: Tree-walking pebble automata. In: Karhumäki, J., et al. (ed.) Jewels are forever, pp. 72–83. Springer, Heidelberg (1999)
8. Engelfriet, J., Hoogeboom, H.J.: Nested Pebbles and Transitive Closure. In: STACS (2006)
9. Engelfriet, J., Hoogeboom, H.J., Van Best, J.-P.: Trips on Trees. Acta Cybern. 14(1), 51–64 (1999)
10. Engelfriet, J., Maneth, S.: A comparison of pebble tree transducers with macro tree transducers. Acta Inf. 39(9), 613–698 (2003)
11. Globerman, N., Harel, D.: Complexity Results for Two-Way and Multi-Pebble Automata and their Logics. Theor. Comput. Sci. 169(2), 161–184 (1996)
12. Milo, T., Suciu, D., Vianu, V.: Typechecking for XML transformers. J. Comput. Syst. Sci. 66(1), 66–97 (2003)
13. Muscholl, A., Samuelides, M., Segoufin, L.: Complementing deterministic tree-walking automata. IPL 99(1), 33–39 (2006)
14. Neven, F.: Extensions of Attribute Grammars for Structured Documents Queries. In: DBPL (1999)
15. Vardi, M.Y.: A note on the reduction of two-way automata to one-way automata. IPL 30, 261–264 (1989)

Some Complexity Results for Prefix Gröbner Bases in Free Monoid Rings

Andrea Sattler-Klein

Technische Universität Kaiserslautern, Fachbereich Informatik
Postfach 3049, 67653 Kaiserslautern, Germany
sattler@informatik.uni-kl.de

Abstract. We establish the following complexity results for prefix Gröbner bases in free monoid rings: 1. $|\mathcal{R}| \cdot size(p)$ reduction steps are sufficient to normalize a given polynomial p w.r.t. a given right-normalized system $\mathcal{R}$ of prefix rules compatible with some total admissible well-founded ordering $>$. 2. $O(|\mathcal{R}|^2 \cdot size(\mathcal{R}))$ basic steps are sufficient to transform a given terminating system $\mathcal{R}$ of prefix rules into an equivalent right-normalized system. 3. $O(|\mathcal{R}|^3 \cdot size(\mathcal{R}))$ basic steps are sufficient to decide whether or not a given terminating system $\mathcal{R}$ of prefix rules is a prefix Gröbner basis. The latter result answers an open question posed by Zeckzer in [10].

1 Introduction

The importance of the theory of Gröbner bases for ideals in commutative polynomial rings over fields as introduced by Buchberger in 1965 has led to various generalizations. An important one is the theory of prefix Gröbner bases introduced by Madlener and Reinert in [2] (see also [5]) for handling right-ideals in monoid and group rings. Their work generalizes the theory introduced by Mora for Gröbner bases in non-commutative polynomial rings [3] (see also [4]) and has recently been further generalized to modules over monoid rings in [1].

Based on the ideas of Madlener and Reinert, Zeckzer has developed the system MRC, a system for computing prefix Gröbner bases in monoid and group rings (see [6],[7],[10]). While in general the procedure for computing prefix Gröbner bases may not terminate, its termination is guaranteed for free monoid rings. Therefore, the class of prefix Gröbner bases in free monoid rings are of particular interest.

In the following we will restrict our attention on prefix Gröbner basis in free monoid rings and study the complexity of some related problems and algorithms. When doing this we will abstract from the underlying field operations.

A fundamental algorithm needed when dealing with prefix Gröbner bases is one for computing normal forms. It is a well known fact that the number of reduction steps needed for computing a normal form of a given polynomial p w.r.t. a given prefix Gröbner basis $\mathcal{R}$ can be exponential in the size of the input, i.e. in the size $size(p) + size(\mathcal{R})$ (see e.g. [10]). Thus, one question that arises iswhether there exists an interesting subclass of prefix Gröbner bases that allows a more efficient normal form algorithm.

E. Csuhaj-Varjú and Z. Ésik (Eds.): FCT 2007, LNCS 4639, pp. 470–481, 2007.

Here, we will investigate the class of right-normalized prefix Gröbner bases with regard to this question. It will turn out that polynomially many reduction steps are sufficient for computing normal forms in this case. More precisely, we will establish the following upper bound on the number of reduction steps needed to normalize a given polynomial p w.r.t. a given right-normalized terminating system $\mathcal{R}$ of prefix rules: $|\mathcal{R}| \cdot size(p)$. Thus, for a right-normalized terminating system $\mathcal{R}$ of prefix rules, the number of reduction steps needed by the normal form algorithm does not depend on the sizes of the rules in $\mathcal{R}$, but only on the number of rules in $\mathcal{R}$.

The next question that then arises is how a terminating system $\mathcal{R}$ of prefix rules, can be efficiently transformed into a corresponding right-normalized system. We will answer this question by presenting an algorithm that solves this problem in polynomially many basic steps. These results about right-normalized systems will be presented in Section 3.

In Section 4 we will turn our attention to the problem to decide for a given finite and terminating system $\mathcal{R}$ of prefix rules in a free monoid ring whether or not $\mathcal{R}$ is a prefix Gröbner basis. The standard way to solve this decision problem is to test if all S-polynomials of $\mathcal{R}$ can be reduced to 0. Since the computation of a normal form for a given polynomial p w.r.t. a given finite and terminating system $\mathcal{R}$ of prefix rules may require exponentially many basic steps in general, the time complexity for this standard decision algorithm is not bounded above by a polynomial function.

Based on the presented results concerning right-normalized systems we will develop a more efficient decision algorithm for the problem: The new algorithm decides in $O(|\mathcal{R}|^3 \cdot size(\mathcal{R}))$ basic steps whether or not a given finite and terminating system $\mathcal{R}$ of prefix rules is a prefix Gröbner basis. This result gives an answer to one of the open problems listed by Zeckzer in [10]: Herein Zeckzer asks whether or not the described decision problem can be solved in polynomial time.

Due to lack of space we will omit all proofs in the following. We refer to the full version of the paper for the proofs (see [9]).

2   Preliminaries

In the following we introduce the basic definitions and foundations that are needed when considering prefix Gröbner bases in free monoid rings from a rewriter's point of view. For further reading concerning prefix Gröbner bases we refer to [2], [5] and [10].

Let Σ be a finite alphabet and let K be a computable field. Then Σ^* denotes the set of all *strings* (*words*) over Σ including the empty string ε, i.e., Σ^* is the free monoid generated by Σ. For $u, v \in \Sigma^*$ and $\Gamma \subseteq \Sigma^*$, $u\Gamma v$ denotes the set $\{ uwv \mid w \in \Gamma \}$. Moreover, for a set $\Gamma \subseteq \Sigma^*$ and a number $n \in \mathbb{N}_0$, Γ^n denotes the set $\{ u_1 u_2 ... u_n \mid u_1, u_2, ..., u_n \in \Gamma \}$. An ordering $>$ on Σ^* is called *admissible* if $u > v$ implies $xuy > xvy$ for all $u, v, x, y \in \Sigma^*$, and it is called *wellfounded* if there is no infinite descending chain $u_1 > u_2 > u_3 > ...$.

For a finite set $\Gamma \subseteq \Sigma^*$ and a total ordering on Σ^*, $max_> \Gamma$ denotes the largest string of Γ w.r.t. $>$.

The *free monoid ring* $K[\Sigma^*]$ is the ring of all formal sums (called *polynomials*) $\sum_{i=1}^n \alpha_i * w_i$ ($n \in \mathbb{N}_0$) with *coefficients* $\alpha_i \in K - \{0\}$ and *terms* $w_i \in \Sigma^*$ such that for all $i, j \in \{1, ..., n\}$ with $i \neq j$, $w_i \neq w_j$ holds. The products $\alpha_i * w_i$ ($\alpha_i \in K - \{0\}$, $w_i \in \Sigma^*$) are called *monomials* and the set of all terms occurring in a polynomial p is denoted by $T(p)$. Instead of $1 * w_i$ we will also sometimes simply write w_i. For a polynomial $p = \sum_{i=1}^n \alpha_i * w_i$, a string $x \in \Sigma^*$ and $\beta \in K$, $\beta \cdot p \circ x$ denotes the polynomial $\sum_{i=1}^n (\beta \cdot \alpha_i) * w_i x$. Moreover, for a finite set $\Gamma \subseteq \Sigma^*$, $\sum \Gamma$ denotes the polynomial $\sum_{w \in \Gamma} 1 * w$.

A pair $(\alpha * t, r)$ with $\alpha \in K - \{0\}$, $t \in \Sigma^*$ and $r \in K[\Sigma^*]$ is called a *rule*. Given a total wellfounded admissible ordering $>$ on Σ^* we associate with each non-zero polynomial $p \in K[\Sigma^*]$ a rule $(l, r) \in K\Sigma^* \times K[\Sigma^*]$ with $l = \alpha * t$ ($\alpha \in K - \{0\}$, $t \in \Sigma^*$), namely the one that satisfies the following two properties: 1. $l - r = p$, 2. (l, r) is *compatible* with $>$, i.e., $t > s$ for all $s \in T(r)$. Accordingly, we associate with a set $F \subseteq K[\Sigma^*]$ of polynomials the set of corresponding rules that are compatible with $>$. For a rule $(l, r) \in K\Sigma^* \times K[\Sigma^*]$ we also write $l \to r$. If the coefficient of the left-hand side of a rule (l, r) associated with a polynomial p is 1 then (l, r) as well as p are called *monic*. A set of rules $\mathcal{R} \subseteq K\Sigma^* \times K[\Sigma^*]$ is called *monic* if each rule of $\mathcal{R}$ is monic.

If (l, r) is a rule of $\mathcal{R} \subseteq K\Sigma^* \times K[\Sigma^*]$ then the term of the monomial l is called the *head term* of the rule (l, r) and of the polynomial $l - r$, respectively. The head term of the polynomial $l - r$ is denoted by $HT(l - r)$. The set of all head terms of $\mathcal{R}$ is denoted by $HT(\mathcal{R})$ and the set of all right-hand sides of $\mathcal{R}$ is denoted by $RHS(\mathcal{R})$. Moreover, for a set T of terms $PSUF(T)$ denotes the set of proper suffixes of the terms in T.

A set of rules $\mathcal{R} \subseteq K\Sigma^* \times K[\Sigma^*]$ induces a reduction relation $\to_{\mathcal{R}}$ on $K[\Sigma^*]$ which is defined in the following way: For $p, q \in K[\Sigma^*]$, $p \to_{\mathcal{R}} q$ if and only if there exists a rule $(\alpha * t, r) \in \mathcal{R}$ (with $\alpha \in K$ and $t \in \Sigma^*$), a monomial $\beta * s$ in p (with $\beta \in K, s \in \Sigma^*$) and a string $x \in \Sigma^*$ such that 1. $tx = s$ and 2. $q = p - \beta * s + (\beta \cdot \alpha^{-1}) \cdot r \circ x$. We also write $p \xrightarrow[\beta * s]{} _{\mathcal{R}} q$ in this case to indicate the monomial that is substituted by the reduction step and say that the rule $\alpha * t \to r$ *(prefix) reduces* p to q in one step. If $\alpha * t \to r$ (with $\alpha \in K, t \in \Sigma^*$ and $r \in K[\Sigma^*]$) is a rule, $\beta \in K$ and $x \in \Sigma^*$ then $(\beta \cdot \alpha) * tx \to_{\mathcal{R}} \beta \cdot r \circ x$ is called an *instance* of the rule $\alpha * t \to r$. A polynomial $p \in K[\Sigma^*]$ is called *(prefix) reducible* w.r.t. a set of rules $\mathcal{R} \subseteq K\Sigma^* \times K[\Sigma^*]$ if there exists a polynomial $q \in K[\Sigma^*]$ with $p \to_{\mathcal{R}} q$. Otherwise, p is called $\mathcal{R}$-*irreducible*.

As usually, $\to_{\mathcal{R}}^*$ denotes the reflexive and transitive closure of $\to_{\mathcal{R}}$, i.e., $p \to_{\mathcal{R}}^* q$ means that p can be reduced to q in n reduction steps for some $n \in \mathbb{N}_0$. We also write $p \to_{\mathcal{R}}^n q$ if p reduces to q in n steps and we denote by $D_{\to_{\mathcal{R}}}(p, q)$ the minimum of the set $\{n \in \mathbb{N}_0 \mid p \to_{\mathcal{R}}^n q\}$ in this case. If $p \to_{\mathcal{R}}^* q$ holds, then q is called a *descendant* of p. An irreducible descendant of p is called a *normal form* of p. If p has a unique normal form w.r.t. $\mathcal{R}$ then this normal form is denoted by $NF_{\mathcal{R}}(p)$. Moreover, $\leftrightarrow_{\mathcal{R}}^*$ denotes the reflexive, symmetric and transitive closure of $\rightarrow_{\mathcal{R}}$. Two sets of rules $\mathcal{R}, \mathcal{S} \subseteq K\Sigma^* \times K[\Sigma^*]$ are called *equivalent* if $\leftrightarrow_{\mathcal{R}}^* = \leftrightarrow_{\mathcal{S}}^*$.

If $(\alpha * t, r_1)$ and $(\beta * s, r_2)$ $(\alpha, \beta \in K$ and $t, s \in \Sigma^*)$ are two rules of $\mathcal{R} \subseteq K\Sigma^* \times K[\Sigma^*]$ such that $t = sx$ for some $x \in \Sigma^*$ then $(r_1, (\alpha \cdot \beta^{-1}) \cdot r_2 \circ x)$ is a *critical pair* (of $\mathcal{R}$) and the corresponding polynomial $r_1 - (\alpha \cdot \beta^{-1}) \cdot r_2 \circ x$ is called a *(prefix) S-polynomial* (of $\mathcal{R}$). The set of all S-polynomials of $\mathcal{R}$ is denoted by $SPOL(\mathcal{R})$. A set of rules $\mathcal{R} \subseteq K\Sigma^* \times K[\Sigma^*]$ is called *confluent* if for all $p, q, r \in K[\Sigma^*]$ the following holds: If q and r are descendants of p then they are *joinable* in $\mathcal{R}$, i.e., they have a common descendant w.r.t. $\mathcal{R}$. Moreover, $\mathcal{R}$ is called *noetherian* (or *terminating*) if no infinite chain of the form $p_0 \rightarrow_{\mathcal{R}} p_1 \rightarrow_{\mathcal{R}} p_2 \rightarrow_{\mathcal{R}} \ldots$ exists. If $\mathcal{R}$ is compatible with a total wellfounded admissible ordering then it is noetherian. If in addition, each critical pair of $\mathcal{R}$ is joinable in $\mathcal{R}$, or in other words, each S-polynomial of $\mathcal{R}$ is $\mathcal{R}$-reducible to 0, then $\mathcal{R}$ is confluent. $\mathcal{R} \subseteq K\Sigma^* \times K[\Sigma^*]$ is called *left-normalized* if for all $(l, r) \in \mathcal{R}$, l is irreducible w.r.t. $\mathcal{R} - \{(l, r)\}$. Moreover, $\mathcal{R}$ is called *right-normalized* if for all $(l, r) \in \mathcal{R}$, r is irreducible w.r.t. $\mathcal{R}$ and it is called *interreduced* if it is left- and right-normalized.

Let $F \subseteq K[\Sigma^*]$ be a set of non-zero polynomials, let $>$ be a total wellfounded admissible ordering on Σ^* and let $\mathcal{R} \subseteq K\Sigma^* \times K[\Sigma^*]$ be the associated set of rules. Then a set of rules $\mathcal{S} \subseteq K\Sigma^* \times K[\Sigma^*]$ is called a *prefix Gröbner basis* for F (or for $\mathcal{R}$) w.r.t. $>$ if the following holds: 1. $\leftrightarrow_{\mathcal{S}}^* = \leftrightarrow_{\mathcal{R}}^*$, 2. $\mathcal{S}$ is compatible with $>$, 3. $\mathcal{S}$ is confluent. If $\mathcal{S}$ is a prefix Gröbner basis for a set $F \subseteq K[\Sigma^*]$, then a polynomial p is an element of the right-ideal generated by F if and only if its uniquely determined $\mathcal{S}$-normal form is equal to 0. For a set $F \subseteq K[\Sigma^*]$ $(\mathcal{R} \subseteq K\Sigma^* \times K[\Sigma^*])$ of non-zero polynomials (of rules) and a given total wellfounded admissible ordering $>$ on Σ^* there exists a uniquely determined finite, monic set $\mathcal{R}^* \subseteq K\Sigma^* \times K[\Sigma^*]$ that is an interreduced prefix Gröbner basis for F $(\mathcal{R})$ w.r.t. $>$. Since in a left-normalized set $\mathcal{R} \subseteq K\Sigma^* \times K[\Sigma^*]$ there are no critical pairs, any left-normalized set $\mathcal{R} \subseteq K\Sigma^* \times K[\Sigma^*]$ compatible with some total wellfounded admissible ordering $>$ is a prefix Gröbner basis. On the other hand, the set $\mathcal{R}$ associated with $F \subseteq K[\Sigma^*]$ and $>$ can be effectively transformed in a finite prefix Gröbner basis for F by normalizing the left-hand sides.

Obviously, if in a set $\mathcal{R} \subseteq K\Sigma^* \times K[\Sigma^*]$ of rules, each rule $(\alpha * t, r)$ (with $\alpha \in K, t \in \Sigma^*$) is replaced by $(1 * t, \alpha^{-1} \cdot r)$ then the resulting system is a monic system that is equivalent to $\mathcal{R}$. Therefore, we will assume in the following that the rules of a set $\mathcal{R} \subseteq K\Sigma^* \times K[\Sigma^*]$ are always monic ones.

Since for our complexity analysis we will not take into account the field operations that have to be performed we define the size of a set of rules independently of the coefficients occurring: The *size* of the empty word is defined by $size(\epsilon) := 1$, while the *size* of a nonempty word w is its length. Moreover, for a non-zero polynomial $p \in K[\Sigma^*]$, the size is defined by $size(p) := \sum_{t \in T(p)} size(t)$ and for $p = 0$, $size(p) := 1$. Further, for a set $\mathcal{R} \subseteq K\Sigma^* \times K[\Sigma^*]$ of rules, $size(\mathcal{R})$ is defined as $\sum_{(l,r) \in \mathcal{R}} (size(l) + size(r))$.

3   Normalform Computation

It is a well known fact that for a given prefix Gröbner basis $\mathcal{R} \subseteq K\Sigma^* \times K[\Sigma^*]$ and a given polynomial $p \in K[\Sigma^*]$, the number of reduction steps needed to compute a normal form of p w.r.t. $\mathcal{R}$ can be exponential in the size of the input, i.e. in the size $size(\mathcal{R}) + size(p)$. In particular, this phenomenon can occur even in case $\mathcal{R}$ is compatible w.r.t. the length ordering on Σ^*. This shows the following example from [8] which we will use in this paper to illustrate some further phenomena.

Example 1. Let K be an arbitrary computable field, let $\Sigma = \{g, f, x, y\}$ and let $> \subseteq \Sigma^* \times \Sigma^*$ be the length ordering on Σ^*. Moreover, for $n \in \mathbb{N}_0$, let $\mathcal{R}_n \subseteq K\Sigma^* \times K[\Sigma^*]$ be defined as follows:

$$\mathcal{R}_n = \{g^2 f \rightarrow x + y\} \cup \{g^{2i+2} f \rightarrow g^{2i} fx + g^{2i} fy \mid 1 \le i \le n\}.$$

Then for all $n \in \mathbb{N}_0$, $\mathcal{R}_n$ is compatible with $>$ and left-normalized. Hence, it is a prefix Gröbner basis. Moreover, for all $n \in \mathbb{N}_0$ the following holds:

1. $size(\mathcal{R}_n) = 3n^2 + 10n + 5$
2. $NF_{\mathcal{R}_n}(g^{2n+2} f) = \sum \{x, y\}^{n+1}$
3. $D_{\rightarrow_{\mathcal{R}_n}}(g^{2n+2} f, \sum\{x, y\}^{n+1}) = 2^{n+1} - 1$

Thus, the question that arises is whether there exist interesting subclasses of terminating set of rules of $K\Sigma^* \times K[\Sigma^*]$ where the computation of normal forms can be done more efficiently.

The systems $\mathcal{R}_n$ $(n \in \mathbb{N}_0)$ given in Example 1 are left-normalized, but not right-normalized. Thus, it is a natural question to ask whether or not this fact is an essential one.

In the following we will answer this question by showing that for each right-normalized set $\mathcal{R} \subseteq K\Sigma^* \times K[\Sigma^*]$ that is compatible with some total admissible wellfounded ordering $>$ polynomially many reduction steps are in fact sufficient for computing normal forms. To this end, we first consider the reduction strategy $\hookrightarrow_{\mathcal{R}}$ which prefers large terms (w.r.t. $>$).

Definition 1. *Let $>$ be a total admissible wellfounded ordering on Σ^* and let $\mathcal{R} \subseteq K\Sigma^* \times K[\Sigma^*]$ be a set of rules compatible with $>$. Then the relation $\hookrightarrow_{\mathcal{R}} \subseteq K[\Sigma^*] \times K[\Sigma^*]$ is defined as follows: If $p, p' \in K[\Sigma^*]$, then $p \hookrightarrow_{\mathcal{R}} p'$ iff $p \xrightarrow{}_{\alpha*t} \mathcal{R} \, p'$ where $t = max_>\{s \in T(p) \mid s \text{ is } \rightarrow_{\mathcal{R}}\text{-reducible}\}$ and $\alpha \in K$.*

In [8] it has been proved that for each left-normalized system $\mathcal{R}$ that is compatible with some total admissible wellfounded ordering $>$, the reduction strategy $\hookrightarrow_{\mathcal{R}}$ is optimal with regard to the lengths of the normalizing reduction sequences in that it is possible to construct to each normalizing $\rightarrow_{\mathcal{R}}$-sequence of length k $(k \in \mathbb{N}_0)$ a corresponding normalizing $\hookrightarrow_{\mathcal{R}}$-sequence of length $\le k$. But, as illustrated above, the lengths of these sequences cannot be bounded by a polynomial function in general.

On the other hand, it has turned out that the reduction strategy $\hookrightarrow_{\mathcal{R}}$ is not optimal in general: For a non-left-normalized set of rules $\mathcal{R}$, the reduction strategy $\hookrightarrow_{\mathcal{R}}$ can be very inefficient even if $\mathcal{R}$ is a prefix Gröbner basis (see [8]).

However, in the following we will prove that for a given right-normalized set $\mathcal{R} \subseteq K\Sigma^* \times K[\Sigma^*]$ compatible with a given total admissible wellfounded ordering $>$ and a given polynomial $p \in K[\Sigma^*]$, the length of each $\hookrightarrow_{\mathcal{R}}$-sequence starting with p is bounded above by $|\mathcal{R}| \cdot size(p)$. To this end we first consider the following property of the relation $\hookrightarrow_{\mathcal{R}}$ (see [8]).

Lemma 1. *Let $>$ be a total admissible wellfounded ordering on Σ^* and let $\mathcal{R} \subseteq K\Sigma^* \times K[\Sigma^*]$ be a set of rules compatible with $>$. Moreover, let $p_0, p_1, \hat{p_1}, p_2 \in K[\Sigma^*]$ be polynomials and let $\alpha_0 * t_0, \alpha_1 * t_1$ be monomials.*

If $\quad p_0 \underset{\alpha_0 * t_0}{\overset{\hookrightarrow}{\mathcal{R}}} p_1 \quad \hookrightarrow^*_{\mathcal{R}} \quad \hat{p_1} \quad \underset{\alpha_1 * t_1}{\overset{\hookrightarrow}{\mathcal{R}}} p_2 \quad then \quad t_0 > t_1 \,.$

From this fact we can easily derive the following bound for the lengths of the $\hookrightarrow_{\mathcal{R}}$-sequences.

Corollary 1. *Let $>$ be a total admissible wellfounded ordering on Σ^* and let $\mathcal{R} \subseteq K\Sigma^* \times K[\Sigma^*]$ be a set of rules compatible with $>$. Moreover, let $p, p' \in K[\Sigma^*]$ be polynomials with $p \hookrightarrow^n_{\mathcal{R}} p'$ for some $n \in \mathbb{N}_0$. Then the following holds:*

$$n \leq |\{t \in T(q) \mid q \in K[\Sigma^*] \text{ with } p \to^*_{\mathcal{R}} q \text{ and } t \text{ is } \to_{\mathcal{R}}\text{-reducible}\}|$$

Thus, the question that arises is how the set of reducible terms of the descendants of a given polynomial p with respect to a given right-normalized terminating set $\mathcal{R}$ can be characterized.

In order to derive an appropriate characterization we first consider the set of all terms of a descendant q of p with respect to $\mathcal{R}$. As the next lemma shows the structure of the terms of q is rather simple.

Lemma 2. *Let $\mathcal{R} \subseteq K\Sigma^* \times K[\Sigma^*]$ be a right-normalized terminating set of rules and let $p, q \in K[\Sigma^*]$ be polynomials with $p \to^*_{\mathcal{R}} q$. Then the following holds:*

$$T(q) \subseteq T(p) \cup T(RHS(\mathcal{R})) \cup (T(RHS(\mathcal{R})) \circ PSUF(T(p)))$$

By using this result and the fact that the terms of $T(RHS(\mathcal{R}))$ are $\to_{\mathcal{R}}$-irreducible since $\mathcal{R}$ is right-normalized we can derive the following characterization for the reducible terms of a descendant q of p provided ε is not a head term of $\mathcal{R}$.

Corollary 2. *Let $\mathcal{R} \subseteq K\Sigma^* \times K[\Sigma^*]$ be a right-normalized terminating set of rules and let $p, q \in K[\Sigma^*]$ be polynomials with $p \to^*_{\mathcal{R}} q$. If $\varepsilon \notin HT(\mathcal{R})$ then:*

$$\{t \in T(q) \mid t \text{ is } \to_{\mathcal{R}}\text{-reducible}\} \subseteq HT(\mathcal{R}) \cup (HT(\mathcal{R}) \circ PSUF(T(p)))$$

This corollary together with Corollary 1 gives a polynomial upper bound for the lengths of the $\hookrightarrow_{\mathcal{R}}$-sequences w.r.t. a right-normalized terminating set $\mathcal{R}$ for the case when $\varepsilon \notin HT(\mathcal{R})$ holds:

If $p, p' \in K[\Sigma^*]$ and $n \in \mathbb{N}_0$ such that $p \hookrightarrow_{\mathcal{R}}^n p'$, then

$$n \leq |\{ t \in T(q) \mid q \in K[\Sigma^*] \text{ with } p \to_{\mathcal{R}}^* q \text{ and } t \text{ is } \to_{\mathcal{R}}\text{-reducible} \}|$$
$$\leq |HT(\mathcal{R}) \cup (HT(\mathcal{R}) \circ PSUF(T(p)))|$$
$$\leq |\mathcal{R}| + |\mathcal{R}| \cdot (size(p) - 1) = |\mathcal{R}| \cdot size(p)$$

On the other hand, if $\varepsilon \in HT(\mathcal{R})$, then each term can be reduced to 0 in one $\hookrightarrow_{\mathcal{R}}$-step. Thus, the derived polynomial bound is true in this case too.

By a more differentiated analysis of the situations that may arise when normalizing a polynomial p with respect to a right-normalized terminating set $\mathcal{R}$ using the reduction strategy $\hookrightarrow_{\mathcal{R}}$ we can derive a little bit better upper bound for n.

Theorem 1. *Let $>$ be a total admissible wellfounded ordering on Σ^* and let $\mathcal{R} \subseteq K\Sigma^* \times K[\Sigma^*]$ be a non-empty right-normalized set of rules compatible with $>$. Moreover, let $p, p' \in K[\Sigma^*]$ be polynomials with $p \neq 0$ such that $p \hookrightarrow_{\mathcal{R}}^n p'$ for some $n \in \mathbb{N}_0$. Then the following holds:*

$$n \leq |\mathcal{R}| \cdot size(p) - |\mathcal{R}| + 1$$

We want to emphasize two aspects of this result: First of all, this result holds for an arbitrary right-normalized set $\mathcal{R} \subseteq K\Sigma^* \times K[\Sigma^*]$ compatible with some total admissible wellfounded ordering and does not require that $\mathcal{R}$ is a prefix Gröbner basis. Secondly, the bound shows that the sizes and forms of the rules of $\mathcal{R}$ do not have an essential influence on the lengths of the reduction sequences that can be performed: It is only the cardinality of $\mathcal{R}$ that plays an important role in this context.

Of course, at a first sight one might wonder why Theorem 1 which is based on a rather technical proof (see [9]) is presented here, although the corresponding bound only is asymptotically equivalent to the other bound presented before. However, the reason for doing this is the fact that it can be proved that the second one of the derived bounds is a very sharp one.

Lemma 3. *Let K be an arbitrary computable field. For all $k, i \geq 1$ there exist an alphabet Σ, an interreduced prefix Gröbner basis $\mathcal{R} \subseteq K\Sigma^* \times K[\Sigma^*]$ and a polynomial $p \in K[\Sigma^*]$ such that the following holds:*

- $|\mathcal{R}| = k + 1$
- $size(p) = i$
- $|\mathcal{R}| \cdot size(p) - |\mathcal{R}| + 1$ *reduction steps are needed to reduce p to normal form*

Proof
Let K be an arbitrary computable field and let $k, i \geq 1$. Let $\Sigma = \{a, x_1, x_2, ..., x_k, \bar{x}_1, \bar{x}_2, ..., \bar{x}_k\}$ and let $> \subseteq \Sigma^* \times \Sigma^*$ be the length-lexicographical ordering induced by $a > x_1 > x_2 > ... > x_k > \bar{x}_1 > \bar{x}_2 > ... > \bar{x}_k$. Moreover, let $\mathcal{R} \subseteq K\Sigma^* \times K[\Sigma^*]$ be defined as follows

$$\mathcal{R} = \{ a \to \varepsilon + \Sigma_{j=1}^k x_j \} \cup \{ x_j a \to \bar{x}_j a \mid j \in \{1, ..., k\} \}$$

*and let $p = a^i$. Then, as it is shown in the full version of the paper (see [9]), $\mathcal{R}$
and p have the desired properties.*

Note that the systems $\mathcal{R}$ constructed in the proof of the last lemma are not only
right-normalized, but also left-normalized, and hence they are prefix Gröbner
bases. Thus, the proof shows that even for interreduced prefix Gröbner bases
the bound given in Theorem 1 is sharp.

According to Theorem 1 for an arbitrary but fixed, right-normalized ter-
minating set $\mathcal{R} \subseteq K\Sigma^* \times K[\Sigma^*]$ compatible with some appropriate ordering,
the number of reduction steps needed for computing normal forms w.r.t. $\hookrightarrow_{\mathcal{R}}$
is linearly bounded. This shows that right-normalized sets of rules should be
preferred in practise when dealing with prefix Gröbner bases, since they allow
an efficient normal form algorithm. But, what is the time complexity of the
problem to normalize the right-hand sides of a terminating set $\mathcal{R} \subseteq K\Sigma^* \times K[\Sigma^*]$?

Of course, the number of reduction steps that will be performed when normalizing the right-hand sides of a terminating set $\mathcal{R}$ can depend essentially on
the reduction strategy used and on the order in which the rules are treated. To
see this, we consider Example 1 again: For each $n \in \mathbb{N}_0$, the system $\mathcal{R}_n$ is a prefix Gröbner basis and thus, each polynomial $p \in K[\Sigma^*]$ has exactly one normal
form w.r.t. $\to_{\mathcal{R}_n}$. Thus, normalizing the right-hand sides of the rules of $\mathcal{R}_n$ yields
a uniquely determined right-normalized set $\mathcal{T}_n$ independently of the strategies
used. The set $\mathcal{T}_n$ has the following form: $\mathcal{T}_n = \{ g^{2i+2}f \to \sum\{x,y\}^{i+1} \mid 0 \le
i \le n\}$. Hence, for $n \in \mathbb{N}_0$, the size of $\mathcal{T}_n$ grows exponentially in n as well as in
$size(\mathcal{R}_n)$.

However, for the number of reduction steps needed to generate the set $\mathcal{T}_n$ the
order in which the right-hand sides of $\mathcal{R}_n$ are reduced is essential: If the rules of
$\mathcal{R}_n$ are treated in decreasing order with respect to the lengths of their left-hand
sides, then $2 \cdot \sum_{i=1}^{n}(2^i - 1) = 2 \cdot (2^{n+1} - 2 - n) = 2^{n+2} - 2n - 4$ reduction
steps are needed for computing $\mathcal{T}_n$. In contrast to this, linearly many reduction
steps, namely $\sum_{i=1}^{n} 2 = 2n$, will only be performed when the right-hand sides
of the rules of $\mathcal{R}_n$ are normalized in increasing order w.r.t. the lengths of the
corresponding left-hand sides.

Can this observation be generalized, i.e. is it always a good strategy to treat
the rules in such an order that the left-hand sides increase w.r.t. $>$ when normalizing the right-hand sides of a set $\mathcal{R} \subseteq K\Sigma^* \times K[\Sigma^*]$?

Of course, this might depend on the reduction strategy used. However, if we proceed in the way described then it suffices to normalize each right-hand side w.r.t. a
right-normalized subset of the current set. Hence then, as Theorem 1 shows, the
reduction strategy $\hookrightarrow_{\mathcal{R}}$ leads to polynomially bounded reduction sequences.

These considerations suggest to use the following algorithm for transforming
a given set of rules $\mathcal{R} \subseteq K\Sigma^* \times K[\Sigma^*]$ compatible with some given total admissible wellfounded ordering $>$ into a corresponding equivalent right-normalized
set $\mathcal{T}$.

Algorithm: NORMALIZE_RHS

INPUT: A total admissible wellfounded ordering $>$ on Σ^* and
            a set $\mathcal{R} = \{\, l_i \rightarrow r_i \mid 1 \leq i \leq n \,\} \subseteq K\Sigma^* \times K[\Sigma^*]$ $(n \in \mathbb{N})$
            that is compatible with $>$.

OUTPUT: A set $\mathcal{T} = \{\, l_i \rightarrow r_i' \mid 1 \leq i \leq n \,\}$ where for all $i \in \{\, 1, .., n \,\}$,
            r_i' is a $\mathcal{R}$-normal form of r_i.

begin
Sort the rules of $\mathcal{R}$ w.r.t. $>$ such that $l_1 \leq l_2 \leq \ldots \leq l_n$;
For $i := 2$ **to** n **do**
$\quad r_i' := \text{NORMALIZE}_{\hookrightarrow}(r_i, \{\, l_1 \rightarrow r_1', \ldots, l_{i-1} \rightarrow r_{i-1}' \,\})$;
$\mathcal{T} := \{\, l_i \rightarrow r_i' \mid 1 \leq i \leq n \,\}$
end

Here, the subalgorithm NORMALIZE$_{\hookrightarrow}$ computes for a given polynomial $p \in K[\Sigma^*]$ and a given terminating set of rules $\mathcal{S} \subseteq K\Sigma^* \times K[\Sigma^*]$ a normal form of p w.r.t. $\mathcal{S}$ using the reduction strategy $\hookrightarrow_{\mathcal{S}}$.

And indeed, analysis of the algorithm shows that the number of basic steps performed by the algorithm NORMALIZE_RHS on a given input is polynomially bounded.

Theorem 2. *Algorithm* NORMALIZE_RHS *computes to a given input* $(>, \mathcal{R})$ *in* $O(|\mathcal{R}|^2 \cdot size(\mathcal{R}))$ *basic steps an equivalent right-normalized set* $\mathcal{T}$ *that is compatible with* $>$.

4   Prefix Gröbner Basis Check

A given terminating set $\mathcal{R} \subseteq K\Sigma^* \times K[\Sigma^*]$ is a prefix Gröbner Basis if and only if $\mathcal{R}$ is confluent, i.e. if and only if all S-polynomials can be reduced to 0 w.r.t. $\rightarrow_{\mathcal{R}}^*$. The corresponding standard decision algorithm can be found for instance in [5] and [10]. What is the time complexity of the algorithm?

Due to the fact that the number of reduction steps needed to compute a normal form of a polynomial p w.r.t. a set $\mathcal{R}$ of prefix rules can be exponential in the size of the input, it is not difficult to see that the number of reduction steps needed by the standard decision procedure is not bounded above by a polynomial function (cf. [10]). To illustrate this we consider Example 1 again: If we add to the alphabet Σ the new symbol h and extend for $n \in \mathbb{N}_0$, the set $\mathcal{R}_n$ by the two rules $h^{2n+4} \rightarrow g^{2n+2}f$ and $h^{2n+4} \rightarrow 0$, then the resulting set $\mathcal{S}_n$ contains two S-polynomials, namely $g^{2n+2}f$ and, for reasons of symmetry, $-g^{2n+2}f$. As shown in Example 1 for all $n \in \mathbb{N}_0$ we have:

$$D_{\rightarrow_{\mathcal{R}_n}}(g^{2n+2}f, \sum\{x, y\}^{n+1}) = 2^{n+1} - 1 .$$

Since the two new rules are obviously not applicable during a reduction sequence starting with $g^{2n+2}f$ we get:

$$D_{\rightarrow_{\mathcal{S}_n}}(g^{2n+2}f, \sum\{x, y\}^{n+1}) = D_{\rightarrow_{\mathcal{R}_n}}(g^{2n+2}f, \sum\{x, y\}^{n+1}) = 2^{n+1} - 1 .$$

Thus, the standard decision algorithm will perform $\geq 2 \cdot (2^{n+1} - 1)$ reduction steps when applied on input $\mathcal{S}_n$. Of course, for reasons of symmetry it would suffice to reduce one of the S-polynomials only. But even then, the number of reduction steps performed by the algorithm will be exponential in the size of the input.

In the following we will show how the results achieved in the previous section can be used to derive a more efficient algorithm solving the described decision problem. The main idea of the algorithm is to right-normalize the set $\mathcal{R}$ in a first step since for a right-normalized prefix Gröbner basis, normal forms can be computed using polynomially many basic steps (see Theorem 1).

As the next lemma shows the resulting right-normalized system $\mathcal{T}$ is confluent if and only if $\mathcal{R}$ is so.

Lemma 4. *Let $n \in \mathbb{N}$ and let $\mathcal{R} = \{\, l_i \rightarrow r_i \mid 1 \leq i \leq n \,\} \subseteq K\Sigma^* \times K[\Sigma^*]$ be a terminating set of rules. Moreover, let $\bar{r_1}$ be an $\mathcal{R}$-normal form of r_1 and let $\mathcal{T} = \{\, l_1 \rightarrow \bar{r_1} \,\} \cup \{\, l_i \rightarrow r_i \mid 2 \leq i \leq n \,\}$. Then the following holds:*

1. $\leftrightarrow_{\mathcal{R}}^* = \leftrightarrow_{\mathcal{T}}^*$
2. $\mathcal{R}$ is confluent   iff $\mathcal{T}$ is confluent

Thus, $\mathcal{R}$ is a prefix Gröbner basis if and only if all S-polynomials of $\mathcal{T}$ can be reduced to 0 w.r.t. $\rightarrow_{\mathcal{T}}^*$. These observations lead to the following decision algorithm.

Algorithm: IS_PGB

INPUT: A total admissible wellfounded ordering $>$ on Σ^*
          and a non-empty set $\mathcal{R} \subseteq K\Sigma^* \times K[\Sigma^*]$ compatible with $>$.
OUTPUT: *answer* = *yes* if $\mathcal{R}$ is a prefix Gröbner basis, *answer* = *no* otherwise.

begin
$\mathcal{T} := \text{NORMALIZE\_RHS}(>, \mathcal{R})$;
answer := *yes*;
$C := SPOL(\mathcal{T})$;
while (*answer* = *yes*) **and** ($C \neq \emptyset$) **do**
        begin
        Choose an element p of C;
        $C := C - \{\, p \,\}$;
        $\bar{p} := \text{NORMALIZE}_\rightarrow(p, \mathcal{T})$;
        if $\bar{p} \neq 0$ **then** *answer* := *no*;
        end;
end

What is the time complexity of this algorithm IS_PGB? Using the complexity results of the previous section the following result can be established.

Theorem 3. *Algorithm IS_PGB decides on input $(>, \mathcal{R})$ in $O(|\mathcal{R}|^3 \cdot size(\mathcal{R}))$ basic steps whether or not the given set $\mathcal{R}$ compatible with $>$ is a prefix Gröbner basis.*

5   Concluding Remarks

Transforming a terminating set $\mathcal{R} \subseteq K\Sigma^* \times K[\Sigma^*]$ into an equivalent left- and right-normalized system $\mathcal{T}$ by interreduction may require exponentially many reduction steps (see [8]). Moreover, normalizing the right-hand sides of $\mathcal{R}$ may result in a right-normalized system whose size is exponential in the size of the original system. However, as proved in Theorem 2, polynomially many basic steps are sufficient to transform $\mathcal{R}$ into an equivalent right-normalized system.

In [10] Zeckzer has proved that $O(m \cdot l^2 \cdot |\mathcal{R}|^2)$ head reduction steps are sufficient to compute a head normal form of a polynomial p with respect to $\mathcal{R}$ (i.e. a descendant whose head term is irreducible w.r.t. $\mathcal{R}$) in case $\mathcal{R}$ is an interreduced prefix Gröbner basis. Here m is the length of the maximal term of $\mathcal{R} \cup \{p\}$ w.r.t $>$ and l is the maximal number of terms in a polynomial of the set $\{ l - r \mid (l,r) \in \mathcal{R} \} \cup \{p\}$.

Our results show that this bound can be improved essentially in that $|\mathcal{R}| \cdot size(p)$ reduction steps are sufficient to compute a head normal form of p and even a normal form of p and that this even holds in case $\mathcal{R}$ is not left-normalized, but only right-normalized.

An essential property of the new bound is that it does not depend on the sizes of the rules in $\mathcal{R}$, but only on the cardinality of $\mathcal{R}$. Thus, if $\mathcal{T}$ is obtained from a terminating set $\mathcal{R} \subseteq K\Sigma^* \times K[\Sigma^*]$ by normalizing the right-hand sides of $\mathcal{R}$, then the number of basic steps needed for computing a normal form of a polynomial p w.r.t. $\mathcal{T}$ is bounded by $|\mathcal{R}| \cdot size(p)$ although the size of $\mathcal{T}$ may be exponential in the size of $\mathcal{R}$.

These new complexity results for right-normalized systems are the basis for the algorithm IS_PGB which decides in polynomially many basic steps whether or not a given set $\mathcal{R} \subseteq K\Sigma^* \times K[\Sigma^*]$ is a prefix Gröbner basis. This shows that the corresponding standard algorithm used in practise is very inefficient.

References

1. Ackermann, P., Kreuzer, M.: Gröbner Basis Cryptosystems. Journ. AAECC 17, 173–194 (2006)
2. Madlener, K., Reinert, B.: On Gröbner Bases in Monoid and Group Rings. In: Proc. ISSAC'93, pp. 54–263. ACM Press, New York (1993)
3. Mora, T.: Gröbner Bases for Non-Commutative Polynomial Rings. In: Calmet, J. (ed.) Algebraic Algorithms and Error-Correcting Codes. LNCS, vol. 229, pp. 353–362. Springer, Heidelberg (1986)
4. Mora, T.: An Introduction to Commutative and Noncommutative Gröbner Bases. Theoretical Computer Science 134, 131–173 (1994)
5. Reinert, B.: On Gröbner Bases in Monoid and Group Rings. PhD thesis, Universität Kaiserslautern (1995)
6. Reinert, B., Zeckzer, D.: MRC - A System for Computing Gröbner Bases in Monoid and Group Rings. In: Presented at the 6th Rhine Workshop on Computer Algebra, Sankt Augustin (1998)
7. Reinert, B., Zeckzer, D.: MRC - Data Structures and Algorithms for Computing in Monoid and Group Rings. Journ. AAECC 10(1), 41–78 (1999)

8. Sattler-Klein, A.: An Exponential Lower Bound for Prefix Gröbner Bases in Free Monoid Rings. In: Thomas, W., Weil, P. (eds.) STACS 2007. LNCS, vol. 4393, pp. 308–319. Springer, Heidelberg (2007)
9. Sattler-Klein, A.: Some Complexity Results for Prefix Gröbner Bases in Free Monoid Rings. Internal Report, Universität Kaiserslautern (to appear)
10. Zeckzer, D.: Implementation, Applications, and Complexity of Prefix Gröbner Bases in Monoid and Group Rings. PhD thesis, Universität Kaiserslautern (2000)

Fast Asymptotic FPTAS for Packing Fragmentable Items with Costs

Hadas Shachnai and Omer Yehezkely

Department of Computer Science, The Technion, Haifa 32000, Israel
{hadas,omery}@cs.technion.ac.il

Abstract. Motivated from recent applications in community TV networks and VLSI circuit design, we study variants of the classic bin packing problem, in which a set of items needs to be packed in a minimum number of unit-sized bins, allowing items to be *fragmented*. This can potentially reduce the total number of bins used, however, item fragmentation does not come for free. In *bin packing with size preserving fragmentation (BP-SPF)*, there is a bound on the total number of fragmented items. In *bin packing with size increasing fragmentation (BP-SIF)*, fragmenting an item increases the input size (due to a header/footer of fixed size that is added to each fragment). Both BP-SPF and BP-SIF do not belong to the class of problems that admit a *polynomial time approximation scheme (PTAS)*.

In this paper, we develop fast *asymptotic fully polynomial time approximation schemes (AFPTAS)* for both problems. The running times of our schemes are *linear* in the input size. As special cases, our schemes yield AFPTASs for classical bin packing and for variable-sized bin packing, whose running times improve the best known running times for these problems.

1   Introduction

In the classical *bin packing* problem, n items $(a_1, \ldots, a_n)$ of sizes $s(a_1), \ldots, s(a_n)$ $\in (0, 1]$ need to be packed in a minimal number of unit-sized bins. Bin packing is well known to be NP-hard. We consider variants of bin packing in which items may be fragmented (into two or more pieces). This can potentially reduce the number of bins used; however, item fragmentation does not come for free. We study the following two variants.

Size preserving fragmentation (BP-SPF): an item a_i can be split into two fragments: a_{i_1}, a_{i_2}, such that $s(a_i) = s(a_{i_1}) + s(a_{i_2})$. The resulting fragments can also split in the same way. Each split has a unit cost and the total cost cannot exceed a given *budget* $C \geq 0$. In the special case where $C = 0$ we get an instance of classical bin packing.

Size increasing fragmentation (BP-SIF): a header (or a footer) of fixed size, $\Delta > 0$, is attached to each (whole or fragmented) item; thus, the capacity required for packing an item of size $s(a_i)$ is $s(a_i) + \Delta$. Upon fragmenting an

E. Csuhaj-Varjú and Z. Ésik (Eds.): FCT 2007, LNCS 4639, pp. 482–493, 2007.

item, each fragment gets a header; that is, if a_i is replaced by two items such that $s(a_i) = s(a_{i_1}) + s(a_{i_2})$, then packing a_{i_j} requires capacity $s(a_{i_j}) + \Delta$.

The above two variants of the bin packing problem capture many practical scenarios, including message transmission in community TV networks, VLSI circuit design and preemptive scheduling on parallel machines with setup times/setup costs. For more details on these applications see [25].

Both BP-SIF and BP-SPF are known to be NP-hard (see in [16] and [25]), therefore, we focus on finding efficient approximate solutions.

1.1 Related Work

It is well known (see, e.g., [19]) that bin packing does not belong to the class of NP-hard problems that admit a PTAS. In fact, bin packing cannot be approximated within factor $\frac{3}{2} - \varepsilon$, for any $\varepsilon > 0$, unless P=NP [9]. However, there exists an *asymptotic* PTAS *(APTAS)* which uses, for any instance I, $(1 + \varepsilon) OPT(I) + k$ bins for some fixed k, where $OPT(I)$ is the number of bins used in any optimal solution. Fernandez de la Vega and Lueker [6] presented an APTAS with $k = 1$. Alternatively, a *dual* PTAS, which uses $OPT(I)$ bins of size $(1 + \varepsilon)$ was given by Hochbaum and Shmoys [12]. Such a dual PTAS can also be derived from the work of Epstein and Sgall [5] on multiprocessor scheduling, since bin packing is dual to the *minimum makespan* problem.

Karmarkar and Karp [14] presented an AFPTAS for bin packing which uses $(1+\varepsilon)OPT(I)+O(1/\varepsilon^2)$ bins. The scheme of [14] is based on rounding the (fractional) solution of a *linear programming (LP)* relaxation of bin packing. To solve this linear program in polynomial time, despite the fact that it has a vast number of variables, the authors use a variant of the ellipsoid method due to Grötschel, Lovász and Schrijver (GLS) [11]. The resulting running time is $O(\varepsilon^{-8}n \log n)$. An AFPTAS with substantially improved running time of $O(n \log \varepsilon^{-1} + \varepsilon^{-6} \log^6 \varepsilon^{-1})$ was proposed by Plotkin et al. [23].

In *variable-sized bin packing*, we have a set of items whose sizes are in $(0, 1]$, and a set of bin sizes in $(0, 1]$ (including the size 1) available for packing the items. We need to pack the items in a set of bins of the given sizes, such that the total bin capacity used is minimized. The variable-sized bin packing problem was first investigated by Friesen and Langston [8], who gave several approximation algorithms, the best of which has asymptotic worst case ratio of $4/3$. Murgolo [21] presented an AFPTAS, which solves a covering linear program using the techniques of [14], namely, the dual program is solved approximately using the modified GLS algorithm. Comprehensive surveys on the bin packing problem and its variants appear, e.g., in [2,27] (see also the recent work of Epstein and Levin [4] on dynamic approximation schemes for the *online* bin packing problem, and the references therein.)

Mandal et al. introduced in [16] the BP-SIF problem and showed that it is NP-hard. Menakerman and Rom [18] and Naaman and Rom [22] were the first to develop algorithms for bin packing with item fragmentation, however, the problems studied in [18] and [22] are different from our problems. For a version of BP-SPF in which the number of bins is given, and the objective is to minimize the total cost incurred by fragmentation, the paper [18] studies the performance of simple algorithms such as First-Fit, Next-Fit and First-Fit-Decreasing, and shows that for any instance which can be packed in N bins using f^* splits of items, each of these algorithms might end up using $f^* + N - 1$ splits.

The paper [25] presents dual PTASs and APTASs for BP-SPF and BP-SIF. The dual PTASs pack all the items in $OPT(I)$ bins of size $(1 + \varepsilon)$, and the APTASs use at most $(1 + \varepsilon)OPT(I) + 1$ bins. All of these schemes have running times that are polynomial in n and exponential in $1/\varepsilon$. The paper also shows that each of the problems admits a *dual* AFPTAS. The proposed schemes pack the items in $OPT(I) + O(1/\varepsilon^2)$ bins of size $(1 + \varepsilon)$. The schemes are based on solving mixed packing and covering LP with negative entries, using the ellipsoid method; this results in the running time of $O(\varepsilon^{-12}n \log n)$, which renders the schemes highly impractical. The question whether the two variants of bin packing admitted *(non-dual)* AFPTASs remained open. In this paper, we resolve this question by presenting for the two problems approximation schemes which pack the items in $(1 + \varepsilon)OPT(I) + O(\varepsilon^{-1} \log \varepsilon^{-1})$ unit-sized bins. The proposed schemes improve significantly the running times of the schemes in [25], to times that are *linear* in the input size. Since these schemes are *combinatorial*, they are also easier to implement.

There has been some related work in the area of preemptive scheduling on parallel machines. The paper [24] presents a tight bound on the number of preemptions required for a schedule of minimum makespan, and a PTAS for minimizing the makespan of a schedule with job-wise or total bound on the number of preemptions. For the special case of preemptive scheduling of jobs on identical parallel machines, so as to minimize the *makespan*, the paper uses the property of *primitive* optimal schedules for developing an approximation scheme; however, the scheme is based on dynamic programming applied to a *discretized* instance. Such discretization cannot be applied to our problems, since the resulting packing may exceed the bin capacities.

1.2 Our Results

In this paper we develop *fast* AFPTASs for BP-SPF and BP-SIF. Our schemes pack the items in $(1 + \varepsilon)OPT(I) + O(\varepsilon^{-1} \log \varepsilon^{-1})$ bins, in time that is *linear* in n, the number of items, and polynomial in $1/\varepsilon$ (see Theorem 1). As special cases, our scheme for BP-SPF yields AFPTASs for classical bin packing and for variable-sized bin packing, whose running times improve the best known running times for these problems (see Theorems 2 and 3). We note that the improvement for variable-sized bin packing is substantial, since the best known running time for this problem is dominated by the running time of the modified GLS algorithm, which is $O(\varepsilon^{-8}n \log n)$ [21].

Techniques: A major difficulty in packing with item fragmentation is in defining the *fragment sizes*. One possible approach is to discretize these sizes (as in [25]); however, this leads to a *dual* AFPTAS. Another approach is to *guess* the bin configurations and to solve a linear program to find the fragment sizes, but then the scheme cannot be *fully* polynomial, as the number of configurations that need to be considered is exponential in $1/\varepsilon$. To get around this difficulty, we initially transform the input I to one that can be packed with *no fragmentation*; later, we transform the resulting packing to a valid packing of I. We use an interesting structural property of the two variants of bin packing, namely, the existence of optimal *primitive* packings (see Section 2.1), to establish a relation between packing with item fragmentation and variable-sized bin packing. In particular, we model any feasible packing satisfying this property as an undirected graph; each connected component is represented by an *oversized* bin, whose size is equal to the number of bins in this component. The items are then packed in bins of variable sizes, with no fragmentation. The scheme completes by generating from each oversized bin a set of ordinary (i.e., unit-sized) bins, allowing some of the items to fragment, while maintaining the constraints on the number of bins/fragmented items. We expect that our non-standard approach will find more uses for other problems that involve splitting of input elements, such as preemptive scheduling of jobs on parallel machines. Recently, Fishkin et al. [7] studied this problem where preemptions incur delays, due to *migrations* of jobs from one machine to another, and showed the existence of optimal *primitive* schedules for certain subclasses of instances. More generally, our schemes may be useful for other scheduling problems in which there exist optimal schedules satisfying McNaughton's rule [17].

Due to space constraints, some of the proofs are omitted. The detailed results appear in [26].

2 An AFPTAS for BP-SPF

2.1 Preliminaries

At the heart of our schemes lies an interesting structural property of optimal solutions for our problems. Define the *bin packing graph* of a given packing as an undirected graph, where each bin i is represented by a vertex v_i; there is an edge (v_i, v_j) if bin i and bin j share fragments of the same item. Note that a fragment-free packing induces a graph with no edges. A *primitive packing* is a feasible packing in which (i) each bin has at most two fragments of items, (ii) each item can be fragmented only once, and (iii) the respective bin packing graph is a collection of paths. Note that the last condition implies that in any connected component of the bin packing graph there are two bins including only a single fragment. The following was shown in [25].

Lemma 1. *Any instance of BP-SPF has an optimal primitive packing.*

Consider a primitive solution for a BP-SPF instance. Suppose that some connected component consists of $c \geq 1/\varepsilon$ bins, then we can partition this component to εc components, each containing $1/\varepsilon$ bins. To avoid violating the budget of C on the number of fragmented items, whenever we need to fragment an item, we can add a new bin in which the item is packed with no fragmentation. Overall, we may add at most εc new bins. Thus, we have shown.

Lemma 2. *Any primitive solution for BP-SPF which uses N bins can be replaced by a solution in which each connected component is of size at most $\lceil 1/\varepsilon \rceil$, and the number of bins used is at most $N(1 + \varepsilon)$.*

Recall that in BP-SPF we are given a set of n items $I = (a_1, a_2, ..., a_n)$, where a_i has the size $s(a_i) \in (0, 1]$. The number of fragmented items is bounded by C. The goal is to pack all items using minimal number of bins and at most C splits. The following is an outline of our scheme for BP-SPF. (i) Preprocess the input to obtain a *fixed* number of item sizes. (ii) Guess $OPT(I) = d$, the number of bins used by an optimal packing of I. (iii) Solve a linear programming relaxation of the resulting instance of packing the items (with no fragmentations) into *ordinary* and *oversized* bins. (iv) Round the (fractional) solution for the LP and pack the large items according to the integral solution. (v) Pack the remaining large items and the small items, one at a time (see below).

2.2 Preprocessing the Input

Initially, we partition the items into two groups by their sizes: the *large* items have size at least ε; all other items are *small*. We then transform the instance I to an instance I' in which the number of distinct items sizes is fixed. This can be done by using the shifting technique (see, e.g., in [27]). Generally, the items are sorted in non-decreasing order by sizes, then, the ordered list is partitioned into at most $1/\varepsilon^2$ subsets, each containing $H = \lceil n\varepsilon^2 \rceil$ items. The size of each item is rounded up to the size of the largest item in its subset. This yields an instance in which the number of distinct item sizes is $m = n/H \leq 1/\varepsilon^2$. Denote by n_j the number of items in size group j, $1 \leq j \leq m$.

We proceed to define for the large items the *configuration matrix*, A, for bins in which there are no fragmented items. In particular, for the shifted large items, a *bin configuration* is a vector of size $m \leq 1/\varepsilon^2$, in which the j-th entry gives h_j, the number of items of size group j packed in the bin. The configuration matrix A consists of the set of all possible bin configurations, where each configuration is a column in A; therefore, the number of columns in A is $q \leq (1/\varepsilon)^{1/\varepsilon^2}$. Next, we define a matrix B including as columns the configurations of *oversized* bins. Each of these bins has a size in the range $[2, C + 1]$. An oversized bin represents a connected component in some optimal primitive packing of the input. Each configuration of an oversized bin gives the number of items of each size group in this bin; thus, the number of columns in B is $s = O((C/\varepsilon)^{1/\varepsilon^2})$.

2.3 Solving the Linear Program

In the following we describe the linear program that we formulate for finding a (fractional) packing of the items with no fragmentation. We number the oversized bin configurations by $1, \ldots, s$. Note that the capacities of the oversized bins used will determine the number of fragmented items in the solution output by the scheme. Let f_k be the minimum capacity of an oversized bin having the k-th configuration, then the number of fragmented items in the connected component corresponding to this bin is $f_k - 1$. Denote by $\mathbf{x}, \mathbf{y}$ the variable vectors giving the number of ordinary and oversized bins having certain configuration, respectively. We want to minimize the total bin capacity used for packing the input, such that the number of fragmented items does not exceed the budget C.

Having guessed correctly the value of d, we need to find a feasible solution for the following program.

$$(LP) \qquad \sum_{i=1}^{q} A_{ji}x_i + \sum_{k=1}^{s} B_{jk}y_k \geq n_j \quad \text{for } j = 1, \ldots, m$$

$$\sum_{i=1}^{q} x_i + \sum_{k=1}^{s} f_k y_k \leq d$$

$$\sum_{k=1}^{s} (f_k - 1)y_k \leq C$$

$$x_i \geq 0 \qquad \text{for } i = 1, \ldots, q$$

$$y_k \geq 0 \qquad \text{for } k = 1, \ldots, s$$

The first set of constraints guarantees that we pack all the items in size group j, for all $1 \leq j \leq m$; the two last constraints guarantee that the total number of bins used is at most d, and that the number of fragmented items does not exceed the budget C.

A technique developed by Young [28], for obtaining fast approximately feasible solutions for mixed linear programs, yields a solution which may violate the packing constraints in the above program at most by factor of ε, namely, $\sum_i x_i + \sum_k f_k y_k \leq d(1+\varepsilon)$, and $\sum_k (f_k-1)y_k \leq C(1+\varepsilon)$ (see also in [13]).[1] Generally, the technique is based on repeated calls to an oracle, which solves in each iteration (one or more) instances of the knapsack problem. Thus, the heart of the scheme is in efficient implementation of the oracle. In solving the LP for our problem, the oracle needs to solve a set of $C+1$ instances of the *multiple choice knapsack (MCK)* problem. In the c-th instance, $1 \leq c \leq C+1$, we are given a bin of capacity c and m items; the i-th item, a_i, has the size $\varepsilon \leq s(a_i) \leq 1$ and the profit $0 < p(a_i) \leq 1$. We need to find a feasible packing of the items in the bin, by selecting any number of copies of each item, such that the overall profit is maximized. The oracle should solve this problem within $1 + \varepsilon$ from the optimal, for each bin size $1 \leq c \leq C+1$. We call this set of instances of the MCK problem *all_MCK*.

By Lemma 2, we may assume that the maximum bin size in our all_MCK problem is $\lceil 1/\varepsilon \rceil + 1$. Since each item a_i, has a size $s(a_i) \geq \varepsilon$, we can pack at most $(1/\varepsilon)^2$ copies of each item. Therefore, for each bin size $1 \leq c \leq \lceil 1/\varepsilon \rceil$, we can solve the knapsack problem with $n = m/\varepsilon^2$ items. Using a fast FPTAS of Kellerer and Pferscy [15], this can be done in $O(\varepsilon^{-3} m \log \varepsilon^{-1})$ steps.

Alternatively, we can solve each MCK instance exactly, by formulating the problem, with a bin of size c, as the following *integer program (IP)*.

---

[1] We show below that such a solution can be fixed to maintain the budget constraint in our BP-SPF instance.

$$\{maximize \sum_{i=1}^{m} p(a_i)z_i \ subject \ to : \ \sum_{i=1}^{m} s(a_i)z_i \leq c, \ z_i \geq 0 \ 1 \leq i \leq m\}$$

where z_i is the number of copies selected from the i-th item. As shown in [3], such IP in fixed dimension can be optimally solved in $O(mM)$ steps, where M is the longest binary representation of any input element.[2] We repeat this for $c = 1, 2, \ldots, C + 1$. Thus, we have shown

Lemma 3. *For any $\varepsilon > 0$, the above all_MCK problem can be solved within factor of $(1+\varepsilon)$ from the optimal in $O(\frac{m}{\varepsilon} \cdot \min\{\varepsilon^{-2} \log \varepsilon^{-1}, M\})$ steps, where M is the longest binary representation of any input element.*

Finally, given a feasible $(1+\varepsilon)$-approximate solution for LP, we apply a technique of Beling and Megiddo [1] for transforming a given solution for a system of linear equations to a *basic* solution.

2.4   Packing the Items

Given the (fractional) solution for LP, we obtain integral vectors $\mathbf{x}'$ and $\mathbf{y}'$, by rounding down the entries in the vectors $\mathbf{x}$ and $\mathbf{y}$, respectively; that is, $\mathbf{x}' = \lfloor \mathbf{x} \rfloor$, and $\mathbf{y}' = \lfloor \mathbf{y} \rfloor$. Note that since we have a basic solution, at most $m + 2$ variables can be assigned non-zero values. We start by packing the items according to the rounded solution (defined by $\mathbf{x}'$ and $\mathbf{y}'$), using the respective bin configurations. Next, for each non-integral variable y_k, we add a bin of size $\lfloor (y_k - y_k') \cdot f_k \rfloor$. Denote this set of new bins by R. We proceed by packing the remaining large items into R, using the First Fit algorithm. Note that the number of (unit-sized) bins used so far does not exceed the number of bins in the solution defined by $\mathbf{x}$ and $\mathbf{y}$. If any large item remains unpacked, we add unit-sized bins in which these items are packed using First Fit. We then add the small items in arbitrary order, using Next Fit.[3]

Our scheme completes by dividing each of the oversized bins into a set of ordinary bins. In the following we show that we can pack all items efficiently.

Lemma 4. *Using the rounded solution for the LP, we can pack all items in the instance in $O(n + m \log m)$ steps.*

3   Analysis of the Scheme

We first show that, although (i) we obtain only *approximately* feasible solution for LP, and (ii) rounding the solution for LP may result in adding oversized bins, the packing output by our scheme maintains the budget constraint.

Lemma 5. *The total number of fragmented items is at most C.*

---

[2] An instance solved by the oracle consists of a set of rationals that give $s(a_i), p(a_i)$ for $1 \leq i \leq m$. The size of the binary representation of each rational is the sum of the sizes of its numerator and denominator.

[3] Detailed descriptions of First Fit and Next Fit can be found, e.g., in [2].

Proof. Recall that the technique in [28] yields an *approximately feasible* solution, in which the packing constraints may be violated by factor of ε. Thus, given an approximately feasible solution for LP, either the number of bins or the number of fragmented items may be increased by factor of ε. We note that any extra fragmentation can be replaced by an extra bin, namely, we can pack the fragmented item in a new bin with no fragmentation. This will result in at most $\varepsilon \cdot OPT(I)$ new bins. Therefore we may assume that in our solution for the LP at most C items are fragmented.

We now show that while packing the items by the (rounded) integral solution, we do not exceed the bound of C. In particular, we need to show that the number of items fragmented due to the usage of extra bins while packing the large items, is at most C. Consider the variable y_k, and let $0 < g_k = y_k - \lfloor y_k \rfloor < 1$. Recall that f_k is the capacity of the oversized bin having the k-th configuration, $1 \le f_k \le C + 1$. We define a new oversized bin whose capacity is $\lfloor f_k g_k \rfloor$. The number of fragmented items due to this new bin is $\lfloor f_k g_k \rfloor - 1 \le (f_k - 1) \cdot g_k$. Since the right hand side in the last inequality is the number of items fragmented due to the fractional part of y_k (in the solution for LP), we have not increased the total number of fragmented items. This holds for any $1 \le k \le s$. ∎

The proof of the next lemma is given in [26].

Lemma 6. *The input is packed in at most $(1 + \varepsilon)^2(1 + 2\varepsilon)OPT(I) + 4(m + 2)$ bins.*

We now analyze the running time of the scheme.

Lemma 7. *The above scheme can be implemented in linear time.*

Proof. The running time of the scheme is determined by the following steps. (*i*) Preprocess the input: we first partition the items to 'large' and 'small'; then, we apply shifting to the large items to obtain $m = O(1/\varepsilon^2)$ distinct items sizes. This can be implemented in $O(n \log m)$ steps (see, e.g., in [23]). (*ii*) Applying the technique of Young [28] for solving the LP, we get that the number of calls to the oracle is $O(m \log m\varepsilon^{-2})$. By Lemma 3, overall, this requires $O(m \log m\varepsilon^{-2} \cdot \frac{m}{\varepsilon} \cdot \min\{\varepsilon^{-2} \log \varepsilon^{-1}, M\})$ steps for solving the LP. Applying a technique of Beling and Megiddo [1], which transforms the solution for LP into a basic one, requires $O(m^{2.62} \log m \cdot \varepsilon^{-2})$ steps. Since we need to 'guess' d, the optimal number of bins, we solve the LP $O(\log n)$ times. (*iii*) Finally, we pack the items and divide each oversized bin into a set of ordinary bins. By Lemma 4, this can be done in $O(n + m \log m)$ steps. Hence, we get that the total running time of the scheme is

$$O(n \log m + \log n \cdot m^2 \log m \cdot \varepsilon^{-2} \cdot (m^{0.62} + \varepsilon^{-1} \cdot \min\{\varepsilon^{-2} \log \varepsilon^{-1}, M\})), \quad (1)$$

and since m, the number of distinct item sizes, is fixed, we get the statement of the lemma. ∎

3.1 Combining Shifting with Geometric Grouping

In the following we show that the linear grouping technique used for shifting the item sizes can be replaced by *geometric grouping* [14], while maintaining the bound on the total number of fragmented items. Recall that, in geometric grouping, the interval $(0, 1]$ is partitioned to sub-intervals of geometrically decreasing sizes, such that the smallest interval is $(0, \delta]$, for some $0 < \delta \leq \varepsilon$. This results in a set of sub-intervals $(0, \delta], (\delta, 2\delta], \ldots, (1/4, 1/2], (1/2, 1]$. Let $k = size(I) \cdot \varepsilon / \log_2 \varepsilon^{-1}$, where $size(I)$ is the total size of the items in the instance I. We partition the set of items whose sizes are in $(1/2, 1]$ to groups of k items. We then proceed to the sub-interval $(1/4, 1/2]$ and use groups of $2k$ items, and continue sequentially to the other sub-intervals; in each sub-interval, the number of items per group increases by factor of 2. Now, we apply shifting to the items in each sub-interval as follows. The sizes of the items in each group are rounded up to the smallest size of an item in the next size group in this sub-interval. This results in $H = O(\varepsilon^{-1} \log \varepsilon^{-1})$ size groups.

Lemma 8. *Applying shifting combined with geometric grouping, the items can be packed in the bins using at most C fragmentations.*

To obtain the overall running time of the scheme, we take in (1) $m = O(\varepsilon^{-1} \log \varepsilon^{-1})$, and use the next technical lemma.

Lemma 9. *For any $x, y, z \geq 1$ $O(n \log \varepsilon^{-1} + \log^x n \cdot \varepsilon^{-y} \log^z \varepsilon^{-1})$ can be bounded by $O(n \log \varepsilon^{-1} + \varepsilon^{-y} \log^{x+z} \varepsilon^{-1})$.*

Theorem 1. *There is an AFPTAS for BP-SPF which packs the items in $(1 + \varepsilon)OPT(I) + O(\varepsilon^{-1} \log \varepsilon^{-1})$ bins, and whose running time is $O(n \log \varepsilon^{-1} + \varepsilon^{-5} \log^4 \varepsilon^{-1} \cdot \min\{\varepsilon^{-2} \log \varepsilon^{-1}, M\})$, where M is the longest binary representation of any input element.*

In [26], we show that our scheme for BP-SPF can be modified to apply for BP-SIF.

4 Application to Bin Packing and Variable-Sized Bin Packing

As special cases, our scheme for BP-SPF can be applied to classical bin packing and variable-sized bin packing, improving the best known running times for these problems.

Theorem 2. *There is an AFPTAS for bin packing which packs the items in at most $(1 + \varepsilon)OPT(I) + O(\varepsilon^{-1} \log \varepsilon^{-1})$ bins in time $O(n \log \varepsilon^{-1} + \varepsilon^{-4} \log^3 \varepsilon^{-1} \cdot \min\{\varepsilon^{-2}, m^{0.62}M\})$, where M is the longest binary representation of any input element.*

Proof. We can solve bin packing as a special case of BP-SPF in which $C = 0$. We use the scheme as given in Section 2. Initially, we omit from the input items of size smaller than ε, and apply the scheme for the remaining items. The running time of the scheme is determined by the following steps. (i) Preprocessing the input. We apply as before geometric grouping, which can be done in $O(n \log m)$ steps. (ii) Solving approximately the linear program for bin packing. In applying the technique of Young [28], we can model the LP as a fractional covering problem with a simplex as a block $\{(x_C)|\sum x_C = r\}$ with guessed objective value r. Thus, using a result of [13] (see also [10]) the number of iterations, and the number of non-zero variables in the solution can be bounded by $O(m(\log m + \varepsilon^{-2}))$. In each call, the oracle needs to find a $(1 + \varepsilon)$-approximate solution for the following instance of MCK. Given is a set of m items, such that each item a_i has a size $\varepsilon \leq s(a_i) \leq 1$ and profit $0 < p(a_i) \leq 1$. We may select from each item unbounded number of copies. We want to pack a set of item copies in a unit-sized bin such that the total profit is maximized. Since the total number of items in any feasible packing is at most $1/\varepsilon$, we can solve the problem as an instance of the knapsack problem, with $n = m/\varepsilon$ items. Using the scheme of [15], this can be done in $O((\frac{m}{\varepsilon} + \varepsilon^{-3})\log \varepsilon^{-1})$ steps. Alternatively, we can solve the IP for MCK in $O(mM)$ steps, where M is the longest binary representation of any input element. Thus, we solve the LP in $O(m(\log m + \varepsilon^{-2}) \cdot \min\{(\frac{m}{\varepsilon} + \varepsilon^{-3})\log \varepsilon^{-1}, mM\}))$ steps. Now, we use the technique of [1] to convert the solution for the LP to a basic solution. This is done in $O(m^{2.62}(\log m + \varepsilon^{-2}))$ steps. Since we need to 'guess' d, the optimal number of bins, we solve the LP $O(\log n)$ times. (iii) The items are packed in $O(n + m \log m)$ steps. Summarizing, we get that the running time of the scheme is $O(n \log m + \log n \cdot m(\log m + \varepsilon^{-2}) \cdot \min\{\varepsilon^{-3}\log \varepsilon^{-1}, m^{1.62}M\}))$. Since $m = O(\varepsilon^{-1}\log \varepsilon^{-1})$, we get the running time $O(n \log \varepsilon^{-1} + \log n \cdot \varepsilon^{-4}\log^2 \varepsilon^{-1} \cdot \min\{\varepsilon^{-2}, m^{0.62}M\})$. Applying Lemma 9, we get the statement of the theorem. Finally, the small items are added in linear time, using Next-Fit. ∎

We now show how our scheme for BP can be modified to solve *variable-sized* bin packing. Given an instance of variable-sized bin packing, with the set of bin sizes $B_1, \ldots, B_N$, such that $0 < B_j \leq 1$, and there exists j such that $B_j = 1$, we omit bins of sizes smaller than ε. also, we partition the items by their sizes: the *large* items have size at least ε^2; all other items are *small*. Initially, we find a packing of the large items. In the preprocessing step of the scheme, we use geometric grouping to reduce the number of distinct item sizes to $O(\varepsilon^{-2}\log \varepsilon^{-1})$. Let $\mathbf{x}$ denote the number of bins of each possible configuration (see in Section 2.2), and $\mathbf{f}$ is the vector of bin sizes. Then, having guessed correctly d, the total capacity of an optimal solution, we need to solve the linear program $\{LP' : B\mathbf{x} \geq \mathbf{n}, \mathbf{f} \cdot \mathbf{x} \leq d\}$, where $\mathbf{n}$ is the vector giving the number of items of each size. We can apply our scheme for BP with LP' replacing LP. The oracle needs to solve in each iteration an all_MCK instance, for all possible bin sizes in $[\varepsilon, 1]$. To implement the oracle efficiently, we use the next lemma (proof omitted).

Lemma 10. *Given a feasible packing of a variable-sized bin packing instance in bins of possible sizes $B_j \in [\varepsilon, 1]$, $1 \leq j \leq N$, using the total bin capacity D, there exists a packing that uses $O(\log \varepsilon^{-1}/\varepsilon)$ distinct bin sizes, such that the total bin capacity used is at most $D(1 + \varepsilon)$.*

Since each large item has the size at least ε^2, in solving MCK for each bin size, we can use the fast scheme of [15] for the knapsack problem with $n = m/\varepsilon^2$ items. As before, we can solve the problem exactly in $O(mM)$ steps. The oracle needs to solve this problem repeatedly, for $O(\varepsilon^{-1}\log \varepsilon^{-1})$ bin sizes. Then the solution for LP' can be converted to a basic one in $O(m^{2.62}(\log m + \varepsilon^{-2}))$ steps [1]. Since we need to guess the value of d, the total capacity used by an optimal solution, we solve LP' $O(\log n)$ times. Finally, we note that preprocessing the input, i.e., reducing the number of distinct item sizes and the number of distinct bin sizes, requires $O((N+n)\log m)$ steps. Also, the items are packed in $O(n+m)$ steps. To get the total running time of the scheme, we summarize the above steps, taking $m = O(\varepsilon^{-2}\log\varepsilon^{-1})$, and apply Lemma 9.[4]

Theorem 3. *There is an AFPTAS for variable-sized bin packing which packs the items using a total bin capacity of at most $(1+\varepsilon)OPT(I) + O(\varepsilon^{-2}\log\varepsilon^{-1})$, where $OPT(I)$ is the optimal capacity, and whose running time is $O((N + n)\log\varepsilon^{-1} + \varepsilon^{-8}\log^3\varepsilon^{-1} \cdot \min\{\varepsilon^{-2}\log\varepsilon^{-1}, \varepsilon^{-0.24}M\})$, where M is the longest binary representation of any input element.*

References

1. Beling, P., Megiddo, N.: Using fast matrix multiplication to find basic solutions. Theoretical Computer Science 205, 307–316 (1993)
2. Coffman Jr, E.G., Garey, M.R., Johnson, D.S.: Approximation algorithms for bin packing: a survey. In: Hochbaum, D.S. (ed.) Approximation Algorithms for NP-Hard Problems, pp. 46–93. PWS Publishing, Boston, MA (1997)
3. Eisenbrand, F.: Fast integer programming in fixed dimension. In: Proc. of ESA (2003)
4. Epstein, L., Levin, A.: A Robust APTAS for the Classical Bin Packing Problem. In: Proc. of ICALP (2006)
5. Epstein, L., Sgall, J.: Approximation schemes for scheduling on uniformly related and identical parallel machines. In: Proc. of ESA (1999)
6. de la Vega, W.F., Lueker, G.S.: Bin packing can be solved within $1 + \varepsilon$ in linear time. Combinatorica 1, 349–355 (1981)
7. Fishkin, A.V., Jansen, K., Sevastianov, S., Sitters, R.: Preemptive Scheduling of Independent Jobs on Identical Parallel Machines Subject to Migration Delay. In: Proc. of ESA (2005)
8. Friesen, D.K., Langston, M.A.: Variable sized bin packing. SIAM J. on Computing 15, 222–230 (1986)
9. Garey, M.R., Johnson, D.S.: Computers and Intractability: A guide to the theory of NP-completeness. W. H. Freeman and Company, San Francisco (1979)
10. Grigoriadis, M.D., Khachiyan, L.G., Porkolab, L., Villavicencio, J.: Approximate Max-Min Resource Sharing for Structured Concave Optimization. SIAM J. on Optimization 11(4), 1081–1091 (2000)
11. Grötschel, M., Lovász, L., Schrijver, A.: The ellipsoid method and its consequences in combinatorial optimization. Combinatorica 1, 169–197 (1981)
12. Hochbaum, D.S., Shmoys, D.B.: Using dual approximation algorithms for scheduling problems: Practical and theoretical results. J. of the ACM 34(1), 144–162 (1987)
13. Jansen, K., Porkolab, L.: On Preemptive Resource Constrained Scheduling: Polynomial-time Approximation Schemes. In: Proc. of IPCO, pp. 329–349 (2002)
14. Karmarkar, N., Karp, R.M.: An efficient approximation scheme for the one dimensional bin packing problem. In: Proc. of FOCS (1982)

---

[4] We give the detailed scheme in the full version of the paper.

15. Kellerer, H., Pferschy, U.: A New Fully Polynomial Approximation Scheme for the Knapsack Problem. J. of Combinatorial Optimization 3, 59–71 (1999)
16. Mandal, C.A., Chakrabarti, P.P, Ghose, S.: Complexity of fragmentable object bin packing and an application. Computers and Mathematics with Applications 35(11), 91–97 (1998)
17. McNaughton, R.: Scheduling with deadlines and loss functions. Manage. Sci. 6, 1–12 (1959)
18. Menakerman, N., Rom, R.: Bin Packing Problems with Item Fragmentations. In: Proc. of WADS (2001)
19. Motwani, R.: Lecture notes on approximation algorithms. Technical report, Dept. of Computer Science, Stanford Univ., CA (1992)
20. Multimedia Cable Network System Ltd.: Data-Over-Cable Service Interface Specification (2000), http://www.cablelabs.com
21. Murgolo, F.D.: An Efficient Approximation Scheme for Variable-Sized Bin Packing. SIAM J. Comput. 16(1), 149–161 (1987)
22. Naaman, N., Rom, R.: Packet Scheduling with Fragmentation. In: Proc. of INFO-COM'02, pp. 824–831 (2002)
23. Plotkin, S.A., Shmoys, D.B., Tardos, É.: Fast Approximation Algorithms for Fractional Packing and Covering Problems. In: Proc. of FOCS (1995)
24. Shachnai, H., Tamir, T., Woeginger, G.J.: Minimizing Makespan and Preemption Costs on a System of Uniform Machines. Algorithmica 42, 309–334 (2005)
25. Shachnai, H., Tamir, T., Yehezkely, O.: Approximation Schemes for Packing with Item Fragmentation. Theory of Computing Systems (to appear)
26. Shachnai, H., Yehezkely, O.: Fast Asymptotic FPTAS for Packing Fragmentable Items with Costs. full version
    http://www.cs.technion.ac.il/~hadas/PUB/frag_afptas.pdf
27. Vazirani, V.V.: Bin Packing. In: Approximation Algorithms, pp. 74–78. Springer, Heidelberg (2001)
28. Young, N.E.: Sequential and Parallel Algorithms for Mixed Packing and Covering. In: Proc. of FOCS, pp. 538–546 (2001)

An $O(1.787^n)$-Time Algorithm for Detecting a Singleton Attractor in a Boolean Network Consisting of AND/OR Nodes

Takeyuki Tamura[*] and Tatsuya Akutsu[*]

Bioinformatics Center, Institute for Chemical Research, Kyoto University,
Gokasho, Uji, Kyoto, Japan, 611-0011
{tamura,takutsu}@kuicr.kyoto-u.ac.jp
http://sunflower.kuicr.kyoto-u.ac.jp/member.html.en

Abstract. The Boolean network (BN) is a mathematical model of genetic networks. It is known that detecting a singleton attractor, which is also called a fixed point, is NP-hard even for AND/OR BNs (i.e., BNs consisting of AND/OR nodes), where singleton attractors correspond to steady states. Though a naive algorithm can detect a singleton attractor for an AND/OR BN in $O(n2^n)$ time, no $O((2 - \epsilon)^n)$ ($\epsilon > 0$) time algorithm was known even for an AND/OR BN with non-restricted indegree, where n is the number of nodes in a BN. In this paper, we present an $O(1.787^n)$ time algorithm for detecting a singleton attractor of a given AND/OR BN, along with related results.

1 Introduction

Computational analysis of biological networks is becoming an important topic in various areas such as bioinformatics, computational biology and systems biology. For that purpose, various kinds of mathematical models of biological networks have been proposed. Among them, the *Boolean network* (BN, in short), which is a model of genetic networks, has received much attention [3,4,6,10,11]. It is a very simple model: each node (e.g., gene) takes either 0 (inactive) or 1 (active) and the states of nodes change synchronously according to regulation rules given as Boolean functions [8,18].

Attractors in a BN have also received much attention since an attractor corresponds to a stable state, and stable states play an important role in biological systems. In particular, extensive studies have been done for analyzing the number and length of attractors [5,11,16]. Most of existing studies on attractors focus on average case features of random BNs with low indegree (connectivity). However, not much attention has been paid on analysis of attractors in a specific BN. In particular, to our knowledge, only several studies have been done on algorithms for detecting attractors in a given BN.

---

[*] Research is supported by "Education and Research Organization for Genome Information Science" with support from MEXT (Ministry of Education, Culture, Sports, Science and Technology).

E. Csuhaj-Varjú and Z. Ésik (Eds.): FCT 2007, LNCS 4639, pp. 494–505, 2007.

Akutsu et al. proved that detecting a singleton attractor (i.e., an attractor with period 1) is NP-hard by a polynomial time reduction from SAT (the satisfiability problem of Boolean formulas in conjunctive normal form) [2]. Milano and Roli independently proposed a similar reduction [14]. Akutsu further improved these hardness results and showed that detection of a singleton attractor remains NP-hard even for BNs with maximum indegree two [1]. Zhang et al. developed algorithms with guaranteed average case time complexity [20]. For example, it is shown that in the average case, one of the algorithms identifies all singleton attractors in $O(1.19^n)$ time for a random BN with maximum indegree two. However, these algorithms may take $O(2^n)$ or more time in the worst case even if there exist only a small number of singleton attractors. Recently, Leone et al. applied SAT algorithms to identify singleton attractors in a BN [13]. However, they did not focus on the time complexity issue. Akutsu studied the time complexity of that approach and showed that detection of a singleton attractor for a BN with maximum indegree k can be reduced to $(k + 1)$-SAT [1].

As mentioned above, there is a close relationship between the attractor detection problem and the SAT problem. SAT is a well-known NP-complete problem and extensive studies have been done for developing $O(c^n)$ time algorithms with smaller c for k-SAT, where n is the number of variables and each clause in k-SAT consists of at most k literals. To our knowledge, the fastest algorithms for 3-SAT and 4-SAT developed by Iwama and Tamaki run in $O(1.324^n)$ time and in $O(1.474^n)$ time, respectively [9]. However, no $O((2 - \epsilon)^n)$ ($\epsilon > 0$) time algorithms are known for general SAT. On the other hand, Hirsh developed an $\tilde{O}(1.239^m)$ time algorithm for SAT with m-clauses [7], which was further improved to $\tilde{O}(1.234^m)$ time by Yamamoto [19], where $\tilde{O}(f(m))$ means $O(f(m)poly(m, n))$. However, these algorithms cannot be directly applied to our problem although we utilize the algorithm in [19] as a subroutine.

In this paper, we present an $O(1.787^n)$ time algorithm for detecting a singleton attractor of a given AND/OR BN, in which a Boolean function assigned to each node is restricted to be a conjunction or disjunction of literals as shown in Fig.1 (a). Since we consider BNs with non-restricted indegree, it is reasonable to assume some restriction on Boolean functions (otherwise, it would take $\Theta(2^n)$ bits to represent a Boolean function). It seems that many Boolean functions corresponding to gene regulation rules are expressed as conjunctions or disjunctions of literals (if these can be regarded as Boolean functions). Furthermore, as mentioned in subsection 5.1, every Boolean function can be represented as a combination of "AND", "OR" and "NOT". Therefore, an AND/OR BN is considered to be a simpler good model although the larger computation time may be needed for general BNs if such combinations are used.

The organization of this paper is as follows. In Section 2, we briefly review a Boolean network and the attractor detection problem along with basic facts. We then present the main algorithm and its analysis in Section 3, where an improved analysis is performed in Section 4. In Section 5, we discuss an extension for a general BN and give an algorithm for a special case. Finally, we conclude with future work.

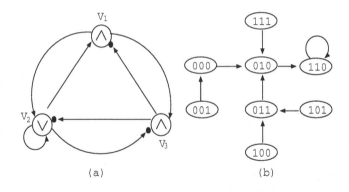

Fig. 1. (a) An example of AND/OR BN where $v_1(t+1) = v_2(t) \wedge \overline{v_3(t)}$, $v_2(t+1) = v_1(t) \vee v_2(t) \vee \overline{v_3(t)}$ and $v_3(t+1) = v_1(t) \wedge \overline{v_2(t)}$ are satisfied. "∧", "∨" and "•" mean "AND", "OR" and "NOT" respectively. (b) The state transition of $[v_1, v_2, v_3]$. (c) The corresponding truth table. $[1, 1, 0]$ is a singleton attractor since it has a self-loop.

2   Preliminaries

A BN $N(V, F)$ consists of a set of n nodes V and a set of n Boolean functions F, where $V = \{v_1, v_2, \ldots, v_n\}$ and $F = \{f_1, f_2, \ldots, f_n\}$. In general, V and F correspond to a set of genes and a set of gene regulatory rules respectively. Let $v_i(t)$ represent the state of v_i at time t. The overall expression level of all the genes in the network at time t is given by the vector $v(t) = [v_1(t), v_2(t), \ldots, v_n(t)]$. This vector is referred as the *Gene Activity Profile* (GAP) of the network at time t, where $v_i(t) = 1$ means that the i-th gene is expressed and $v_i(t) = 0$ means that the i-th gene is not expressed. Since $v(t)$ ranges from $[0, 0, \ldots, 0]$ (all entries are 0) to $[1, 1, \ldots, 1]$ (all entries are 1), there are 2^n possible states. The regulatory rules among the genes are given as $v_i(t+1) = f_i(v(t))$ for $i = 1, 2, \ldots, n$. When the state of gene v_i at time $t+1$ depends on the states of k_i genes at time t, the *indegree* of gene v_i is k_i and denoted by $id(v_i)$. The number of genes that are directly influenced by gene v_i is called the *outdegree* of gene v_i and denoted by $od(v_i)$. The states of all genes are updated simultaneously according to the corresponding Boolean functions. A consecutive sequence of GAPs $v(t), v(t+1), \ldots, v(t+p)$ is called an attractor with period p if $v(t) = v(t+p)$. When $p = 1$, an attractor is called a *singleton attractor*. When $p > 1$, it is called a *cyclic attractor*.

For example, a BN where $v_1(t+1) = v_2(t) \wedge \overline{v_3(t)}$, $v_2(t+1) = v_1(t) \vee v_2(t) \vee \overline{v_3(t)}$ and $v_3(t+1) = v_1(t) \wedge \overline{v_2(t)}$ is given in Fig. 1 (a). Note that "•" means "NOT". The state transition of $[v_1, v_2, v_3]$ is as shown in Fig. 1 (b). The corresponding truth table is shown in Fig. 1 (c). $[1, 1, 0]$ is a singleton attractor since $v(t+1) = [1, 1, 0]$ when $v(t) = [1, 1, 0]$.

In this paper, we treat Boolean functions which can be represented by either $(v_{i_1}{}^{a_1} \wedge v_{i_2}{}^{a_2} \wedge \cdots \wedge v_{i_{k_i}}{}^{a_{k_i}})^b$ or $(v_{i_1}{}^{a_1} \vee v_{i_2}{}^{a_2} \vee \cdots \vee v_{i_{k_i}}{}^{a_{k_i}})^b$ where $v^a = (v + a)$ mod 2. Note that v_{i_j}, a_j and b are assigned only either 0 or 1. If every Boolean function of a BN satisfies the above condition, we call it *AND/OR Boolean network*. The number of nodes in AND/OR BN is obtained by counting "AND" and "OR". For example, in Fig. 1 (a), the AND/OR BN has 3 nodes. If no confusion arises, we treat a AND/OR BN as a directed graph as shown in Fig. 1 (a) and denote $N(V, E)$ where V is a set of nodes and E is a set of directed edges.

If a BN is acyclic and does not have self-loops, there is a polynomial time algorithm to detect an attractor [2,20]. In such a case, the number of attractors is only one and it is a singleton attractor. On the other hand, if a BN is acyclic and has self-loops, detection of an attractor is NP-hard [2]. In this paper, we allow that a BN has self-loops.

In our main algorithm for detecting a singleton attractor, there are steps, which we call *consistency checks*, to determine whether or not 0-1 assignments for nodes contradict 0-1 assignments for their parent nodes. For example, in Fig. 1 (a), if $v_1(t) = 1$ and $v_2(t) = 0$ are assigned, the consistency check detects a contradiction since $v_1(t + 1) = 0 \neq v_1(t)$. The following lemma shows that consistency checks can be done in ignorable time since our main algorithm takes an exponential time of n and $O(n^k a^n) \ll O((a + \epsilon)^n)$ holds for any $a > 1$ and $\epsilon > 0$, where k is a small positive integer.

Lemma 1. *A consistency check for a GAP or a partial GAP can be done in $O(n^2)$ time.*

Proof. Since a proof for a GAP can be applied to a partial GAP, it suffices to prove for a GAP. Suppose that a GAP is assigned at time t. Let $v_{i_1}, v_{i_2}, \ldots, v_{i_{k_i}}$ be the parent nodes of v_i. A consistency check can be done by determining whether or not the 0-1 assignment contradicts $v_i(t) = f_i(v_{i_1}(t), v_{i_2}(t), \ldots, v_{i_{k_i}}(t))$ for all i $(1 \le i \le n)$. Since indegrees are at most n, it takes $O(n)$ time for every node. Note that multi-edges are not allowed from the definition. Then, the lemma holds. □

In this paper, we treat only singleton attractors. Since $v(t) = v(t+1)$ must hold for a singleton attractor, it suffices to consider only time step t. Thus, we omit t from here on.

3 $O(1.792^n)$ Time Algorithm

In this section, we present an $O(1.792^n)$ time algorithm which detects a singleton attractor of a given AND/OR BN. The $O(1.787^n)$ time algorithm, which is to be shown in the next section, can be obtained by improving the analysis of this algorithm. Although the detection of a singleton attractor for a BN with maximum indegree k can be reduced to $(k + 1)$-SAT [1], it cannot be directly applied to our problem since no $O((2 - \epsilon)^n)$ $(\epsilon > 0)$ time algorithms are known for SAT with general k.

Let (V, E) denote the structure of a given BN. An edge $(u, v) \in E$ is called a *non-assigned edge* if no assignment has been done on any of u and v. It should be noted there exist at most 3 consistent assignments (among 4 possible assignments) on (u, v) even if there exist self-loops since either a conjunction of literals or a disjunction of literals is assigned to v. We show below the algorithm, which is to be later explained using an example.

1. Let all the nodes be non-assigned.
2. While there exists a non-assigned edge (u, v), examine all possible 3 assignments on (u, v) recursively.
3. Let U be the set of nodes whose values were already determined. Let $W = V - U$.
4. If $|U| > \alpha n$, examine all possible assignments on W and then perform consistency check.
   Otherwise, compute an appropriate assignment on W by using Yamamoto's algorithm and then perform consistency check.

It is to be noted that the subgraph induced by W is a set of isolated nodes (with self-loops). Therefore, each node v in W is classified into the following types:

type I: the value of v is directly determined from assignment on U,
type II: the value of v is not directly determined from assignment on U,

where type I nodes consists of the following:

- The value of v is determined from the values of the input nodes to v,
- v is an input of AND node u and 1 is assigned to u,
- v is an input of OR node u and 0 is assigned to u.

Based on this fact, we can use $\tilde{O}(1.234^m)$ time SAT algorithm for m-clauses to compute an appropriate assignment on W in the following way, where $\tilde{O}(f(m))$ means $O(f(m)poly(m))$. Suppose that $v_{i_1}, \cdots, v_{i_p}$ in W are type II input nodes to node $u \in U$. We assume w.l.o.g. that u is an AND node to which 0 is already assigned (we can treat analogously the case where u is an OR node). Furthermore, we can assume w.l.o.g. that u is defined as $u = l_{i_1} \wedge l_{i_2} \wedge \cdots \wedge l_{i_p}$ where l_{i_j} is either v_{i_j} or $\overline{v_{i_j}}$. Then, the constraint of $l_{i_1} \wedge l_{i_2} \wedge \cdots \wedge l_{i_p} = 0$ can be rewritten as a SAT clause $\overline{l_{i_1}} \vee \overline{l_{i_2}} \vee \cdots \vee \overline{l_{i_p}}$. Therefore, we can use the SAT algorithm to find an assignment on W that leads to a singleton attractor.

From the above, it is straight-forward to see the correctness of the algorithm. Thus, we analyze the time complexity. Let $K = |U|$ just after STEP 2.

Lemma 2. *Recursive execution of STEP 2 generates $O(1.733^K)$ assignments.*

Proof. Since 3 assignments are examined per two nodes, the number of possible assignments generated at STEP 2 is bounded by $f(K)$ where $f(K)$ is defined by

$$f(2) = 3,$$
$$f(k) = 3 \cdot f(k-2).$$

Then, $f(K)$ is $O(3^{K/2})$, which is at most $O(1.733^K)$.   □

Lemma 3. *If the former part of STEP 4 is executed, the total number of examined assignments is $O(2^{n-K} \cdot 1.733^K)$.*

Theorem 1. *Detection of a singleton attractor can be done in $O(1.792^n)$ time for AND/OR BNs.*

Proof. If the former part of STEP 4 is executed, the computation time is $O(2^{n-K} \cdot 1.733^K)$. Otherwise, it is $O(1.234^K \cdot 1.733^K)$ since SAT problems with $O(K)$ clauses should be solved in STEP 4. By solving $1.234^K = 2^{n-K}$, we obtain $K = 0.767n$. Therefore, we set $\alpha = 0.767$ and have the following;

$$1.234^{0.767n} \cdot 1.733^{0.767n} < 1.792^n,$$
$$2^{n-0.767n} \cdot 1.733^{0.767n} < 1.792^n.$$ □

Example 1. In an example shown in Fig. 2, suppose (b, d) and (f, h) are selected at STEP 2. Then, four clauses may be constructed by (a, g), (a, c), (c, g, i), and (g, i). If $d =$"$\wedge$" and $d = 0$, $(\bar{a} \vee \bar{c})$ is constructed and both a and c become type II nodes. Note that b is not included in the clause since the value of b has already been assigned at STEP 2. On the other hand, if $d = 1$, 1 is assigned to a and c and then both a and c become type I nodes. If the former part of STEP 4 is executed, all possible assignments for a, c, e, g, i, which has 2^5 cases, are examined. Otherwise the SAT problem is solved by Yamamoto's algorithm [19].

4 Improved Analysis

In this section, we present an $O(1.787^n)$ time algorithm which detects a singleton attractor of a given AND/OR BN. This algorithm can be obtained by improving the analysis of the $O(1.792^n)$ time algorithm of the previous section. We estimate the number of SAT clauses produced by STEP 2 of the previous algorithm more accurately.

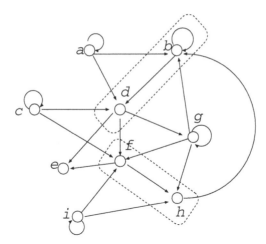

Fig. 2.

For example, in Fig. 3 (a), v_4 is a parent of v_8. Since v_4 is "$\vee$" and v_8 is "$\wedge$", the possible assignments for $[v_4, v_8]$ are $[0, 0]$, $[1, 0]$, $[1, 1]$. Note that $[0, 1]$ does not satisfy the condition of a singleton attractor since v_8 must be 0 when v_4 is 0. When $v_4 = 0$, the assignment for $[v_1, v_2, v_3]$ is determined as $[0, 0, 1]$ as shown in Fig. 3 (b). Similarly, when $v_8 = 1$, the assignment for $[v_4, v_5, v_6, v_7]$ must be $[1, 0, 1, 1]$ as shown in Fig. 3 (c). However, when $v_4 = 1$, the assignment for $[v_1, v_2, v_3]$ is not determined uniquely but $(v_1 \vee v_2 \vee \overline{v_3}) = 1$ must hold as shown in Fig. 3 (d) in order to satisfy the condition of a singleton attractor. In such a case, we say that v_4 *adds* a SAT clause of $(v_1 \vee v_2 \vee \overline{v_3})$. Similarly, when $v_8 = 0$, a SAT clause of $(v_5 \vee \overline{v_6} \vee \overline{v_7})$ is added as shown in Fig. 3 (e). Note that v_4 is not included in the SAT clause since the value of v_4 has already been assigned in STEP 2. Thus, the numbers of SAT clauses which are added in the cases of $[v_4, v_8] = [0, 0]$, $[1, 0]$, $[1, 1]$ are 1, 2 and 1 respectively as shown in Fig. 3 (f).

By applying the above discussion to any two neighboring nodes, the numbers of added SAT clauses can be bounded for each case. For example, suppose that both of v_4 and v_8 are "$\wedge$". In such a case, the numbers of SAT clauses which are added by $[v_4, v_8] = [0, 0]$, $[1, 0]$, $[1, 1]$ are 2, 1 and 0 respectively. By examining all cases, it is seen that the computation time in the case where the latter part of STEP 4 is executed is bounded by

$$\sum_{k=0}^{\frac{K}{2}} 1.234^{(\frac{K}{2}+k)} \cdot 2^{(\frac{K}{2}-k)} \cdot {}_{\frac{K}{2}}C_k$$

where k is the number of cases (i.e., node pairs) in which 2 clauses are added.

Theorem 2. *Detection of a singleton attractor can be done in $O(1.787^n)$ time for AND/OR BNs.*

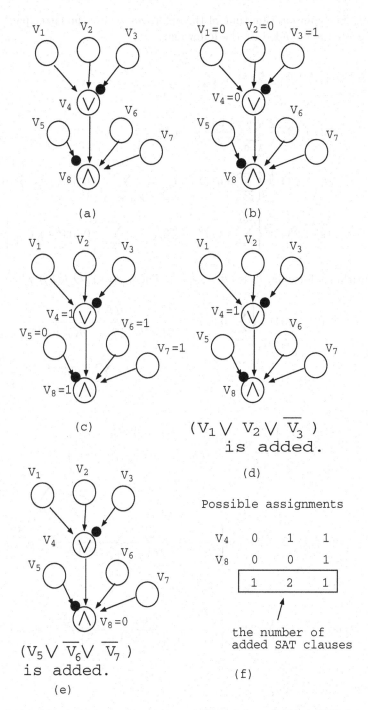

Fig. 3. Relationships between assignments and the number of added SAT clauses

Proof. The computation time of the algorithm where the latter part of STEP 4 is executed is bounded by the following;

$$\sum_{k=0}^{\frac{K}{2}} 1.234^{(\frac{K}{2}+k)} \cdot 2^{(\frac{K}{2}-k)} \cdot {}_{\frac{K}{2}}C_k$$

$$= (2 \cdot 1.234)^{\frac{K}{2}} \sum_{k=0}^{\frac{K}{2}} \left(\frac{1.234}{2}\right)^k \cdot {}_{\frac{K}{2}}C_k$$

$$= (2 \cdot 1.234)^{\frac{K}{2}} \left\{ \sum_{k=0}^{\beta K} 0.617^k \cdot {}_{\frac{K}{2}}C_k + \sum_{k=\beta K+1}^{\frac{K}{2}} 0.617^k \cdot {}_{\frac{K}{2}}C_k \right\} \quad (0 \le \beta < \frac{1}{4})$$

$$= (2 \cdot 1.234)^{\frac{K}{2}} \left\{ \sum_{k=0}^{\beta K} 0.617^k \cdot {}_{\frac{K}{2}}C_{\beta K} + \sum_{k=\beta K+1}^{\frac{K}{2}} 0.617^k \cdot {}_{\frac{K}{2}}C_{\frac{K}{4}} \right\} = g(k)$$

By Stirling's formula, $n!$ is $\tilde{O}(n^n \cdot e^{-n})$. Then, $g(k)$ is $\tilde{O}(h(k))$ where

$$h(k) = (2 \cdot 1.234)^{\frac{K}{2}} \left\{ \sum_{k=0}^{\beta K} 0.617^k \cdot \frac{(\frac{K}{2})^{\frac{K}{2}}}{(\beta K)^{\beta K} \cdot (\frac{K}{2} - \beta K)^{(\frac{K}{2} - \beta K)}} \right.$$

$$\left. + \sum_{k=\beta K+1}^{\frac{K}{2}} 0.617^k \cdot \frac{(\frac{K}{2})^{\frac{K}{2}}}{(\frac{K}{4})^{\frac{K}{4}} \cdot (\frac{K}{4})^{\frac{K}{4}}} \right\}$$

$$= (2 \cdot 1.234)^{\frac{K}{2}} \left\{ \sum_{k=0}^{\beta K} 0.617^k \cdot \frac{(\frac{1}{2})^{\frac{K}{2}}}{(\beta^{2\beta})^{\frac{K}{2}} \cdot \{(\frac{1}{2} - \beta)^{(1-2\beta)}\}^{\frac{K}{2}}} \right.$$

$$\left. + \sum_{k=\beta K+1}^{\frac{K}{2}} 0.617^k \cdot 2^{\frac{K}{2}} \right\}$$

$$< poly(K) \cdot (2 \cdot 1.234)^{\frac{K}{2}} \left\{ \frac{(\frac{1}{2})^{\frac{K}{2}}}{(\beta^{2\beta})^{\frac{K}{2}} \cdot \{(\frac{1}{2} - \beta)^{(1-2\beta)}\}^{\frac{K}{2}}} + 0.617^{\beta K} \cdot 2^{\frac{K}{2}} \right\}$$

$$= poly(K) \cdot (2 \cdot 1.234)^{\frac{K}{2}} \left\{ \left\{ \frac{0.5}{(\beta^{2\beta}) \cdot \{(\frac{1}{2} - \beta)^{(1-2\beta)}\}} \right\}^{\frac{K}{2}} + (0.617^{2\beta} \cdot 2)^{\frac{K}{2}} \right\}$$

$$\tag{1}$$

When $\beta = 0.12823$,

$$\frac{0.5}{(\beta^{2\beta}) \cdot \{(\frac{1}{2} - \beta)^{(1-2\beta)}\}} < 1.768$$

and

$$0.617^{2\beta} \cdot 2 < 1.768$$

hold. Thus, (1) is bounded by

$$< O((1.234 \cdot 2 \cdot 1.768)^{\frac{K}{2}}) < O(2.089^K).$$

If the former part of STEP 4 is executed, the computation time is $O(2^{n-K} \cdot 1.733^K)$. Otherwise, it is $O(2.089^K)$. By solving $2^{n-K} \cdot 1.733^K = 2.089^K$, we have $K = 0.7877$. Therefore, we set $\alpha = 0.7877$ and have the following;

$$2^{n-0.7877n} \cdot 1.733^{0.7877n} < 1.7866^n.$$

$$2.089^{0.7877n} < 1.7866^n. \qquad \square$$

5   Extension and Special Case

5.1   Extension to a General BN

We have considered AND/OR BNs so far. That restriction is reasonable because there are 2^{2^n} types of Boolean functions for a general BN, and thus it would take more than $O(2^n)$ time to identify a Boolean function if such complicated Boolean functions are allowed. However, it may be required to handle non-AND/OR Boolean functions. Thus, we briefly consider how to cope with BNs with general Boolean functions.

For the singleton attractor detection problem, every BN can be transformed into an AND/OR BN although additional nodes are needed. If the number of additional nodes is less than $0.193n$, the computation time of our algorithm is bounded by $O(1.787^{1.193n})$ for a general BN. It is still $O((2-\epsilon)^n)$ ($\epsilon > 0$). Since canalizing functions and nested canalizing functions are known to be good models for regulatory rules of eukaryotic genes [12,17], the number of such additional nodes are considered to be not large for real biological networks.

5.2   A Linear Time Algorithm for a Tree-Like BN

In this subsection, we present a linear time algorithm to detect a singleton attractor of a given AND/OR BN with a rooted tree-architecture *including self-loops*. Although we did not use this algorithm for those of Sections 3 and 4, it may be useful as a subroutine for designing faster algorithms.

Let $T(v_i)$ ($1 \leq i \leq n$) be the subtree whose root is v_i. Let v_{i_j} ($1 \leq j \leq id(v_i)$) be a parent node of v_i where $v_{i_j} = v_i$ can hold for some j (if v_i has a self-loop).

Theorem 3. *Detection of a singleton attractor can be done in $O(n)$ time for an AND/OR BN with a rooted tree-architecture including self-loops.*

Proof. Suppose that directions of edges are root-leaf order. Assign both 0 and 1 to the root node. The detection can be done by examining nodes in root-leaf order in $O(n)$ time. Then, we can assume that directions of edges are leaf-root order. The algorithm is by dynamic programming. If there are no "NOT"s, the following procedure is applied.

- When v_i is "AND",
  - $D(v_i, 1) = 1$ iff $D(v_{i_j}, 1) = 1$ for all v_{i_j}.
  - $D(v_i, 1) = 0$ iff $D(v_{i_j}, 0) = 1$ for some v_{i_j}.
- When v_i is "OR",
  - $D(v_i, 1) = 0$ iff $D(v_{i_j}, 0) = 1$ for all v_{i_j}.
  - $D(v_i, 1) = 1$ iff $D(v_{i_j}, 1) = 1$ for some v_{i_j},

where

- $D(v_i, 1) = 1$ if there exists a singleton attractor for $T(v_i)$ such that $v_i = 1$.
- $D(v_i, 1) = 0$ otherwise.

and

- $D(v_i, 0) = 1$ if there exists a singleton attractor for $T(v_i)$ such that $v_i = 0$.
- $D(v_i, 0) = 0$ otherwise.

Note that the discussion can be applied to the case where there are "NOT"s by changing 0/1 appropriately. It is clear that the computation time is $O(n)$.   □

6   Conclusion and Future Works

We performed a worst case analysis of the problem of detecting a singleton attractor of a given AND/OR BN. Even for AND/OR BNs, $O((2 - \epsilon)^n)$ $(\epsilon > 0)$ time exact algorithms are known only for cases where the maximum indegree is limited. In this paper, we proposed an $O(1.787^n)$ worst-case-time algorithm for a given AND/OR BN. For the singleton attractor detection problem, every BN can be transformed into an AND/OR BN although additional nodes are needed. If the number of additional nodes is less than $0.193n$, the computation time of our algorithm is still $O((2 - \epsilon)^n)$ $(\epsilon > 0)$ for general BNs. The number of such additional nodes are considered to be not large for real biological networks since canalizing functions and nested canalizing functions are known to be good models for regulatory rules of eukaryotic genes [12,17]. An experimental comparison of proposed algorithms is one of our future works.

This paper focused on the Boolean network as a of biological network model. However, the proposed techniques may be useful for designing algorithms which find steady states in other models and for theoretical analysis of such algorithms. For example, Mochizuki performed theoretical analysis on the number of steady states in some continuous biological networks based on nonlinear differential equations [15]. However, the central part of the analysis is done in a combinatorial manner and is very similar to that for Boolean networks. Therefore, it may be possible to develop fast algorithms for finding steady states in such continuous network models. Application and extension of the proposed techniques to other types of biological networks are important future works.

References

1. Akutsu, T.: On finding attractors in Boolean Networks using SAT algorithms. (manuscript)
2. Akutsu, T., Kuhara, S., Maruyama, O., Miyano, S.: A system for identifying genetic networks from gene expression patterns produced by gene disruptions and overexpressions. Genome Informatics 9, 151–160 (1998)
3. Akutsu, T., Miyano, S., Kuhara, S.: Inferring qualitative relations in genetic networks and metabolic pathways. Bioinformatics 16, 727–734 (2000)
4. Albert, R., Barabasi, A.-L.: Dynamics of complex systems: Scaling laws for the period of Boolean networks. Physical Review Letters 84, 5660–5663 (2000)
5. Drossel, B., Mihaljev, T., Greil, F.: Number and length of attractors in a critical Kauffman model with connectivity one. Physical Review Letters 94, 088701 (2005)
6. Glass, L., Kauffman, S.A.: The logical analysis of continuous, nonlinear biochemical control networks. Journal of Theoretical Biology 39, 103–129 (1973)
7. Hirsch, E.A.: New worst-case upper bounds for SAT. Journal of Automated Reasoning 24, 397–420 (2000)
8. Huang, S.: Gene expression profiling, genetic networks, and cellular states: an integrating concept for tumorigenesis and drug discovery. Journal of Molecular Medicine 77(6), 469–480 (1999)
9. Iwama, K., Tamaki, S.: Improved upper bounds for 3-SAT. In: Proc. 15th ACM-SIAM Symposium on Discrete Algorithms, p. 328 (2004)
10. Kauffman, S.: Metabolic stability and epigenesis in randomly connected genetic nets. Journal of Theoretical Biology 22, 437–467 (1968)
11. Kauffman, S.: The Origin of Order: Self-organization and selection in evolution. Oxford Univ. Press, New York (1993)
12. Kauffman, S., Peterson, C., Samuelsson, B., Troein, C.: Random Boolean network models and the yeast transcriptional network. Proceedings of the National Academy of Sciences 100(25), 14796–14799 (2003)
13. Leone, M., Pagnani, A., Parisi, G., Zagordi, O.: Finite size corrections to random Boolean networks, cond-mat/0611088 (2006)
14. Milano, M., Roli, A.: Solving the satisfiability problem through Boolean networks. In: Lamma, E., Mello, P. (eds.) AI*IA 99: Advances in Artificial Intelligence. LNCS (LNAI), vol. 1792, pp. 72–93. Springer, Heidelberg (2000)
15. Mochizuki, A.: An analytical study of the number of steady states in gene regulatory networks. J. Theoret. Biol. 236, 291–310 (2005)
16. Samuelsson, B., Troein, C.: Superpolynomial growth in the number of attractors in Kauffman networks. Physical Review Letters 90, 098701 (2003)
17. Shmulevich, I., Kauffman, S.: Activities and sensitivities in Boolean network models. Physical Review Letters 93(4), 048701 (2004)
18. Somogyi, R., Sniegoski, C.A.: Modeling the complexity of genetic networks: Understanding multigenic and pleitropic regulation. Complexity 1(6), 45–63 (1996)
19. Yamamoto, M.: An improved $\tilde{O}(1.234^m)$-time deterministic algorithm for SAT. In: Proc. International Symposium on Algorithms and Computation, pp. 644–653 (2005)
20. Zhang, S., Hayashida, M., Akutsu, T., Ching, W., Ng, M.K.: Algorithms for finding small attractors in Boolean networks, EURASIP Journal on Bioinformatics and Systems Biology (in press)

Author Index

Lecture Notes in Computer Science

For information about Vols. 1–4546

please contact your bookseller or Springer

Vol. 4592: Z. Kedad, N. Lammari, E. Métais, F. Meziane, Y. Rezgui (Eds.), Natural Language Processing and Information Systems. XIV, 442 pages. 2007.

Vol. 4591: J. Davies, J. Gibbons (Eds.), Integrated Formal Methods. IX, 660 pages. 2007.

Vol. 4590: W. Damm, H. Hermanns (Eds.), Computer Aided Verification. XV, 562 pages. 2007.

Vol. 4589: J. Münch, P. Abrahamsson (Eds.), Product-Focused Software Process Improvement. XII, 414 pages. 2007.

Vol. 4588: T. Harju, J. Karhumäki, A. Lepistö (Eds.), Developments in Language Theory. XI, 423 pages. 2007.

Vol. 4587: R. Cooper, J. Kennedy (Eds.), Data Management. XIII, 259 pages. 2007.

Vol. 4586: J. Pieprzyk, H. Ghodosi, E. Dawson (Eds.), Information Security and Privacy. XIV, 476 pages. 2007.

Vol. 4585: M. Kryszkiewicz, J.F. Peters, H. Rybinski, A. Skowron (Eds.), Rough Sets and Intelligent Systems Paradigms. XIX, 836 pages. 2007. (Sublibrary LNAI).

Vol. 4584: N. Karssemeijer, B. Lelieveldt (Eds.), Information Processing in Medical Imaging. XX, 777 pages. 2007.

Vol. 4583: S.R. Della Rocca (Ed.), Typed Lambda Calculi and Applications. X, 397 pages. 2007.

Vol. 4582: J. Lopez, P. Samarati, J.L. Ferrer (Eds.), Public Key Infrastructure. XI, 375 pages. 2007.

Vol. 4581: A. Petrenko, M. Veanes, J. Tretmans, W. Grieskamp (Eds.), Testing of Software and Communicating Systems. XII, 379 pages. 2007.

Vol. 4580: B. Ma, K. Zhang (Eds.), Combinatorial Pattern Matching. XII, 366 pages. 2007.

Vol. 4579: B. M. Hämmerli, R. Sommer (Eds.), Detection of Intrusions and Malware, and Vulnerability Assessment. X, 251 pages. 2007.

Vol. 4578: F. Masulli, S. Mitra, G. Pasi (Eds.), Applications of Fuzzy Sets Theory. XVIII, 693 pages. 2007. (Sublibrary LNAI).

Vol. 4577: N. Sebe, Y. Liu, Y.-t. Zhuang, T.S. Huang (Eds.), Multimedia Content Analysis and Mining. XIII, 513 pages. 2007.

Vol. 4576: D. Leivant, R. de Queiroz (Eds.), Logic, Language, Information and Computation. X, 363 pages. 2007.

Vol. 4575: T. Takagi, T. Okamoto, E. Okamoto, T. Okamoto (Eds.), Pairing-Based Cryptography – Pairing 2007. XI, 408 pages. 2007.

Vol. 4574: J. Derrick, J. Vain (Eds.), Formal Techniques for Networked and Distributed Systems – FORTE 2007. XI, 375 pages. 2007.

Vol. 4573: M. Kauers, M. Kerber, R. Miner, W. Windsteiger (Eds.), Towards Mechanized Mathematical Assistants. XIII, 407 pages. 2007. (Sublibrary LNAI).

Vol. 4572: F. Stajano, C. Meadows, S. Capkun, T. Moore (Eds.), Security and Privacy in Ad-hoc and Sensor Networks. X, 247 pages. 2007.

Vol. 4571: P. Perner (Ed.), Machine Learning and Data Mining in Pattern Recognition. XIV, 913 pages. 2007. (Sublibrary LNAI).

Vol. 4570: H.G. Okuno, M. Ali (Eds.), New Trends in Applied Artificial Intelligence. XXI, 1194 pages. 2007. (Sublibrary LNAI).

Vol. 4569: A. Butz, B. Fisher, A. Krüger, P. Olivier, S. Owada (Eds.), Smart Graphics. IX, 237 pages. 2007.

Vol. 4568: T. Ishida, S. R. Fussell, P. T. J. M. Vossen (Eds.), Intercultural Collaboration. XIII, 395 pages. 2007.

Vol. 4566: M.J. Dainoff (Ed.), Ergonomics and Health Aspects of Work with Computers. XVIII, 390 pages. 2007.

Vol. 4565: D.D. Schmorrow, L.M. Reeves (Eds.), Foundations of Augmented Cognition. XIX, 450 pages. 2007. (Sublibrary LNAI).

Vol. 4564: D. Schuler (Ed.), Online Communities and Social Computing. XVII, 520 pages. 2007.

Vol. 4563: R. Shumaker (Ed.), Virtual Reality. XXII, 762 pages. 2007.

Vol. 4562: D. Harris (Ed.), Engineering Psychology and Cognitive Ergonomics. XXIII, 879 pages. 2007. (Sublibrary LNAI).

Vol. 4561: V.G. Duffy (Ed.), Digital Human Modeling. XXIII, 1068 pages. 2007.

Vol. 4560: N. Aykin (Ed.), Usability and Internationalization, Part II. XVIII, 576 pages. 2007.

Vol. 4559: N. Aykin (Ed.), Usability and Internationalization, Part I. XVIII, 661 pages. 2007.

Vol. 4558: M.J. Smith, G. Salvendy (Eds.), Human Interface and the Management of Information, Part II. XXIII, 1162 pages. 2007.

Vol. 4557: M.J. Smith, G. Salvendy (Eds.), Human Interface and the Management of Information, Part I. XXII, 1030 pages. 2007.

Vol. 4556: C. Stephanidis (Ed.), Universal Access in Human-Computer Interaction, Part III. XXII, 1020 pages. 2007.

Vol. 4555: C. Stephanidis (Ed.), Universal Access in Human-Computer Interaction, Part II. XXII, 1066 pages. 2007.

Vol. 4554: C. Stephanidis (Ed.), Universal Acess in Human Computer Interaction, Part I. XXII, 1054 pages. 2007.

Vol. 4553: J.A. Jacko (Ed.), Human-Computer Interaction, Part IV. XXIV, 1225 pages. 2007.

Vol. 4552: J.A. Jacko (Ed.), Human-Computer Interaction, Part III. XXI, 1038 pages. 2007.

Vol. 4551: J.A. Jacko (Ed.), Human-Computer Interaction, Part II. XXIII, 1253 pages. 2007.

Vol. 4550: J.A. Jacko (Ed.), Human-Computer Interaction, Part I. XXIII, 1240 pages. 2007.

Vol. 4549: J. Aspnes, C. Scheideler, A. Arora, S. Madden (Eds.), Distributed Computing in Sensor Systems. XIII, 417 pages. 2007.

Vol. 4548: N. Olivetti (Ed.), Automated Reasoning with Analytic Tableaux and Related Methods. X, 245 pages. 2007. (Sublibrary LNAI).

Vol. 4547: C. Carlet, B. Sunar (Eds.), Arithmetic of Finite Fields. XI, 355 pages. 2007.